QUESTAR III
TEXTBOOK PROGRAM

$10.00 LATE FEE AFTER JULY 1ST

THIS BOOK NEEDS TO BE
RETURNED BY THE END OF JUNE

AMSCO'S
ALGEBRA 2
and
TRIGONOMETRY

Ann Xavier Gantert

AMSCO SCHOOL PUBLICATIONS, INC.,
a division of Perfection Learning®

Dedication

To Jessica Alexander and Uriel Avalos in gratitude for their invaluable work in preparing this text for publication.

Ann Xavier Gantert

The author has been associated with mathematics education in New York State as a teacher and an author throughout the many changes of the past fifty years. She has worked as a consultant to the Mathematics Bureau of the Department of Education in the development and writing of Sequential Mathematics and has been a coauthor of Amsco's *Integrated Mathematics* series, which accompanied that course of study.

Reviewers:

Richard Auclair
Mathematics Teacher
La Salle School
Albany, NY

Domenic D'Orazio
Mathematics Teacher
Midwood High School
Brooklyn, NY

Steven J. Balasiano
Assistant Principal,
Supervision Mathematics
Canarsie High School
Brooklyn, NY

George Drakatos
Mathematics Teacher
Baldwin Senior High School
Baldwin, NY

Debbie Calvino
Mathematics Supervisor,
  Grades 7–12
Valley Central High School
Montgomery, NY

Ronald Hattar
Mathematics Chairperson
Eastchester High School
Eastchester, NY

Raymond Scacalossi Jr.
Mathematics Coordinator
Manhasset High School
Manhasset, NY

Text Designer: Nesbitt Graphics, Inc.
Compositor: ICC Macmillan
Cover Design by Meghan J. Shupe
Cover Art by Radius Images (RM)

Please visit our Web sites at:
www.amscopub.com and *www.perfectionlearning.com*
When ordering this book, please specify:

1479906 *or* ALGEBRA 2 AND TRIGONOMETRY, *Hardbound*

ISBN 978-1-56765-702-9

PREFACE

Algebra 2 and Trigonometry is a new text for a course in intermediate algebra and trigonometry that continues the approach that has made Amsco a leader in presenting mathematics in a modern, integrated manner. Over the last decade, this approach has undergone numerous changes and refinements to keep pace with ever-changing technology.

This textbook is the final book in the three-part series in which Amsco parallels the integrated approach to the teaching of high school mathematics promoted by the National Council of Teachers of Mathematics in its *Principles and Standards for School Mathematics* and mandated by the New York State Board of Regents in the *Mathematics Core Curriculum*. The text presents a range of materials and explanations that are guidelines for achieving a high level of excellence in their understanding of mathematics.

In this book:

✔ **The real numbers** are reviewed and the understanding of operations with irrational numbers, particularly radicals, is expanded.

✔ **The graphing calculator** continues to be used as a routine tool in the study of mathematics. Its use enables the student to solve problems that require computation that more realistically reflects the real world. The use of the calculator replaces the need for tables in the study of trigonometry and logarithms.

✔ **Coordinate geometry** continues to be an integral part of the visualization of algebraic and trigonometric relationships.

✔ **Functions** represent a unifying concept throughout. The algebraic functions introduced in *Integrated Algebra 1* are reviewed, and exponential, logarithmic, and trigonometric functions are presented.

✔ **Algebraic skills** from *Integrated Algebra 1* are maintained, strengthened, and expanded as both a holistic approach to mathematics and as a bridge to advanced studies.

✔ **Statistics** includes the use of the graphing calculator to reexamine range, quartiles, and interquartile range, to introduce measures of dispersion such as variance and standard deviation, and to determine the curve that best represents a set of bivariate data.

✔ **Integration** of geometry, algebra, trigonometry, statistics, and other branches of mathematics begun in *Integrated Algebra 1* and *Geometry* is continued and further expanded.

✔ **Exercises** are divided into three categories. *Writing About Mathematics* encourages the student to reflect on and justify mathematical conjectures, to discover counterexamples, and to express mathematical ideas in his or her own words. *Developing Skills* provides routine practice exercises that enable the student and teacher to evaluate the student's ability to both manipulate mathematical symbols and understand mathematical relationships. *Applying Skills* provides exercises in which the new ideas of each section, together with previously learned skills, are used to solve problems that reflect real-life situations.

✔ **Problem solving**, a primary goal of all learning standards, is emphasized throughout the text. Students are challenged to apply what has been learned to the solution of both routine and non-routine problems.

✔ **Enrichment** is stressed both in the text and in the Teacher's Manual where many suggestion are given for teaching strategies and alternative assessment. The Manual provides opportunities for extended tasks and hands-on activities. Reproducible *Enrichment Activities* that challenge students to explore topics in greater depth are provided in each chapter of the Manual.

In this text, the real number system is expanded to include the complex numbers, and algebraic, exponential, logarithmic, and trigonometric functions are investigated. The student is helped to understand the many branches of mathematics, to appreciate the common threads that link these branches, and to recognize their interdependence.

The intent of the author is to make this book of greatest service to the average student through detailed explanations and multiple examples. Each section provides careful step-by-step procedures for solving routine exercises as well as the non-routine applications of the material. Sufficient enrichment material is included to challenge students of all abilities.

Specifically:

✔ Concepts are carefully developed using appropriate language and mathematical symbolism. General principles are stated clearly and concisely.

✔ Numerous examples serve as models for students with detailed explanations of the mathematical concepts that underlie the solution. Alternative approaches are suggested where appropriate.

✔ Varied and carefully graded exercises are given in abundance to develop skills and to encourage the application of those skills. Additional enrichment materials challenge the most capable students.

This text is offered so that teachers may effectively continue to help students to comprehend, master, and enjoy mathematics as they progress in their education.

CONTENTS

Chapter 4

RELATIONS AND FUNCTIONS — 119

Chapter 5

QUADRATIC FUNCTIONS AND COMPLEX NUMBERS — 186

Chapter 6

SEQUENCES AND SERIES   247

Chapter 7

EXPONENTIAL FUNCTIONS   286

Chapter 8

LOGARITHMIC FUNCTIONS   319

Chapter 12

TRIGONOMETRIC IDENTITIES

Chapter 13

TRIGONOMETRIC EQUATIONS

Chapter 14

TRIGONOMETRIC APPLICATIONS

THE INTEGERS

In golf tournaments, a player's standing after each hole is often recorded on the leaderboard as the number of strokes above or below a standard for that hole called a *par*. A player's standing is a positive number if the number of strokes used was greater than par and a negative number if the number of strokes used was less than par. For example, if par for the first hole is 4 strokes and a player uses only 3, the player's standing after playing the first hole is -1.

Rosie Barbi is playing in an amateur tournament. Her standing is recorded as 2 below par (-2) after sixteen holes. She shoots 2 below par on the seventeenth hole and 1 above par on the eighteenth. What is Rosie's standing after eighteen holes? Nancy Taylor, who is her closest opponent, has a standing of 1 below par (-1) after sixteen holes, shoots 1 below par on the seventeenth hole and 1 below par on the eighteenth. What is Nancy's standing after eighteen holes?

In this chapter, we will review the set of integers and the way in which the integers are used in algebraic expressions, equations, and inequalities.

1-1 WHOLE NUMBERS, INTEGERS, AND THE NUMBER LINE

The first numbers that we learned as children and probably the first numbers used by humankind are the natural numbers. Most of us began our journey of discovery of the mathematical world by *counting*, the process that lists, in order, the names of the **natural numbers** or the **counting numbers**. When we combine the natural numbers with the number 0, we form the set of **whole numbers**:

$$\{0, 1, 2, 3, 4, 5, 6, \ldots\}$$

These numbers can be displayed as points on the number line:

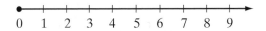

The number line shows us the order of the whole numbers; 5 is to the right of 2 on the number line because $5 > 2$, and 3 is to the left of 8 on the number line because $3 < 8$. The number 0 is the smallest whole number. There is no largest whole number.

The temperature on a winter day may be two degrees above zero or two degrees below zero. The altitude of the highest point in North America is 20,320 feet above sea level and of the lowest point is 282 feet below sea level. We represent numbers less than zero by extending the number line to the left of zero, that is, to numbers that are less than zero, and by assigning to every whole number a an opposite, $-a$, such that $a + (-a) = 0$.

DEFINITION _____

The **opposite** or **additive inverse** of a is $-a$, the number such that
$$a + (-a) = 0.$$

The set of **integers** is the union of the set of whole numbers and their opposites. The set of non-zero whole numbers is the positive integers and the opposites of the positive integers are the negative integers.

Let a, b, and c represent elements of the set of integers. Under the operation of addition, the following properties are true:

1. Addition is closed: $a + b$ is an integer

2. Addition is commutative: $a + b = b + a$

3. Addition is associative: $(a + b) + c = a + (b + c)$

4. Addition has an identity element, 0: $a + 0 = a$

5. Every integer has an inverse: $a + (-a) = 0$

We say that the integers form a **commutative group** under addition because the five properties listed above are true for the set of integers.

Subtraction

DEFINITION

$a - b = c$ if and only if $b + c = a$.

Solve the equation $b + c = a$ for c:

$$b + c = a$$
$$-b + b + c = a + (-b)$$
$$c = a + (-b)$$

Therefore, $a - b = a + (-b)$.

Absolute Value

A number, a, and its opposite, $-a$, are the same distance from zero on the number line. When that distance is written as a positive number, it is called the **absolute value** of a.

- If $a > 0$, then $|a| = a - 0 = a$
- If $a < 0$, then $|a| = 0 - a = -a$

Note: When $a < 0$, a is a negative number and its opposite, $-a$, is a positive number.

For instance, $5 > 0$. Therefore, $|5| = 5 - 0 = 5$.

$$-5 < 0. \text{ Therefore, } |{-5}| = 0 - (-5) = 5.$$

We can also say that $|a| = |{-a}| = a$ or $-a$, whichever is positive.

EXAMPLE I

Show that the opposite of $-b$ is b.

Solution The opposite of b, $-b$, is the number such that $b + (-b) = 0$.

Since addition is commutative, $b + (-b) = (-b) + b = 0$.

The opposite of $-b$ is the number such that $(-b) + b = 0$. Therefore, the opposite of $-b$ is b.

Exercises

Writing About Mathematics

1. Tina is three years old and knows how to count. Explain how you would show Tina that $3 + 2 = 5$.

2. Greg said that $|a - b| = |b - a|$. Do you agree with Greg? Explain why or why not.

Developing Skills

In 3–14, find the value of each given expression.

3. $|6|$

4. $|-12|$

5. $|8 - 3|$

6. $|3 - 8|$

7. $|5 + (-12)|$

8. $|-12 + (-(-5))|$

9. $|4 - 6 + (-2)|$

10. $|8 + (10 - 18)|$

11. $|3| - |3|$

12. $|8| - |-2| - |2|$

13. $-(|-2| + |3|)$

14. $|4 - 3| + |-1|$

In 15–18, use the definition of subtraction to write each subtraction as a sum.

15. $8 - 5 = 3$

16. $7 - (-2) = 9$

17. $-2 - 5 = -7$

18. $-8 - (-5) = -3$

19. Two distinct points on the number line represent the numbers a and b. If $|5 - a| = |5 - b| = 6$, what are the values of a and b?

Applying Skills

In 20–22, Mrs. Menendez uses computer software to record her checking account balance. Each time that she makes an entry, the amount that she enters is added to her balance.

20. If she writes a check for $20, how should she enter this amount?

21. Mrs. Menendez had a balance of $52 in her checking account and wrote a check for $75.

   a. How should she enter the $75?

   b. How should her new balance be recorded?

22. After writing the $75 check, Mrs. Menendez realized that she would be overdrawn when the check was paid by the bank so she transferred $100 from her savings account to her checking account. How should the $100 be entered in her computer program?

1-2 WRITING AND SOLVING NUMBER SENTENCES

Equations

A sentence that involves numerical quantities can often be written in the symbols of algebra as an equation. For example, let x represent any number. Then the sentence "Three less than twice a number is 15" can be written as:

$$2x - 3 = 15$$

When we translate from one language to another, word order often must be changed in accordance with the rules of the language into which we are translating. Here we must change the word order for "three less than twice a number" to match the correct order of operations.

The **domain** is the set of numbers that can replace the variable in an algebraic expression. A number from the domain that makes an equation true is a **solution** or **root** of the equation. We can find the solution of an equation by writing a sequence of **equivalent equations**, or equations that have the same solution set, until we arrive at an equation whose solution set is evident. We find equivalent equations by changing both sides of the given equation in the same way. To do this, we use the following properties of equality:

Properties of Equality

- **Addition Property of Equality:** If equals are added to equals, the sums are equal.
- **Subtraction Property of Equality:** If equals are subtracted from equals, the differences are equal.
- **Multiplication Property of Equality:** If equals are multiplied by equals, the products are equal.
- **Division Property of Equality:** If equals are divided by non-zero equals, the quotients are equal.

On the left side of the equation $2x - 3 = 15$, the variable is multiplied by 2 and then 3 is subtracted from the product. We will simplify the left side of the equation by "undoing" these operations in reverse order, that is, we will first add 3 and then divide by 2. We can check that the number we found is a root of the given equation by showing that when it replaces x, it gives us a correct statement of equality.

$$2x - 3 = 15$$
$$2x - 3 + 3 = 15 + 3$$
$$2x = 18$$
$$x = 9$$

Check
$$2x - 3 = 15$$
$$2(9) - 3 \stackrel{?}{=} 15$$
$$15 = 15 ✔$$

Often the definition of a mathematical term or a formula is needed to write an equation as the following example demonstrates:

EXAMPLE 1

Let $\angle A$ be an angle such that the complement of $\angle A$ is 6 more than twice the measure of $\angle A$. Find the measure of $\angle A$ and its complement.

Solution   To write an equation to find $\angle A$, we must know that two angles are complements if the sum of their measures is 90°.

Let $x =$ the measure of $\angle A$.

Then $2x + 6 =$ the measure of the complement of $\angle A$.

The sum of the measures of an angle and of its complement is 90.

$$
\begin{aligned}
x + 2x + 6 &= 90 \\
3x + 6 &= 90 \\
3x &= 84 \\
x &= 28 \\
2x + 6 &= 2(28) + 6 = 62
\end{aligned}
$$

Therefore, the measure of $\angle A$ is 28 and the measure of its complement is 62.

Check   The sum of the measures of $\angle A$ and its complement is $28 + 62$ or 90. ✔

Answer $m\angle A = 28$; the measure of the complement of $\angle A$ is 62.

EXAMPLE 2

Find the solution of the following equation: $|6x - 3| = 15$.

Solution   Since $|15| = |-15| = 15$, the algebraic expression $6x - 3$ can be equal to 15 or to -15.

$$
\begin{array}{ccc}
6x - 3 = 15 & \text{or} & 6x - 3 = -15 \\
6x - 3 + 3 = 15 + 3 & & 6x - 3 + 3 = -15 + 3 \\
6x = 18 & & 6x = -12 \\
x = 3 & & x = -2
\end{array}
$$

$$
\begin{array}{cc}
\text{Check: } x = 3 & \text{Check: } x = -2 \\
|6x - 3| = 15 & |6x - 3| = 15 \\
|6(3) - 3| \overset{?}{=} 15 & |6(-2) - 3| \overset{?}{=} 15 \\
|15| = 15 \ ✔ & |-15| = 15 \ ✔
\end{array}
$$

Answer   The solution set is $\{3, -2\}$.

Inequalities

A number sentence can often be an inequality. To find the solution set of an inequality, we use methods similar to those that we use to solve equations. We need the following two properties of inequality:

Properties of Inequality

- **Addition and Subtraction Property of Inequality:** If equals are added to or subtracted from unequals, the sums or differences are unequal in the same order.

- **Multiplication and Division Property of Inequality:** If unequals are multiplied or divided by positive equals, the products or quotients are unequal in the same order. If unequals are multiplied or divided by negative equals, the products or quotients are unequal in the opposite order.

EXAMPLE 3

Find all positive integers that are solutions of the inequality $4n + 7 < 27$.

Solution We solve this inequality by using a procedure similar to that used for solving an equation.

$$4n + 7 < 27$$
$$4n + 7 + (-7) < 27 + (-7)$$
$$4n < 20$$
$$n < 5$$

Since n is a positive integer, the solution set is $\{1, 2, 3, 4\}$. *Answer*

EXAMPLE 4 ██

Polly has $210 in her checking account. After writing a check for tickets to a concert, she has less than $140 in her account but she is not overdrawn. If each ticket cost $35, how many tickets could she have bought?

Solution   Let x = the number of tickets.

The cost of x tickets, $35x$, will be subtracted from $210, the amount in her checking account. Since she is not overdrawn after writing the check, her balance is at least 0 and less than $140.

$$0 \leq 210 - 35x < 140$$

$$\underline{-210 \quad -210 \qquad\qquad -210} \quad \text{Add } -210 \text{ to each member of the inequality.}$$

$$-210 \leq \quad -35x < -70$$

$$\frac{-210}{-35} \geq \quad \frac{-35x}{-35} > \frac{-70}{-35} \quad \text{Divide each member of the inequality by } -35.$$

$$6 \geq \qquad x > 2 \quad \text{Note that dividing by a negative number reverses the order of the inequality.}$$

Polly bought more than 2 tickets but at most 6.

Answer   Polly bought 3, 4, 5, or 6 tickets.   ▢

Exercises

Writing About Mathematics

1. Explain why the solution set of the equation $12 - |x| = 15$ is the empty set.

2. Are $-4x > 12$ and $x > -3$ equivalent inequalities? Justify your answer.

Developing Skills

In 3–17, solve each equation or inequality. Each solution is an integer.

3. $5x + 4 = 39$ **4.** $7x + 18 = 39$ **5.** $3b + 18 = 12$

6. $12 - 3y = 18$ **7.** $9a - 7 = 29$ **8.** $13 - x = 15$

9. $|2x + 4| = 22$ **10.** $|3 - y| = 8$ **11.** $|4a - 12| = 16$

12. $|2x + 3| - 8 = 15$ **13.** $7a + 3 > 17$ **14.** $9 - 2b \leq 1$

15. $3 < 4x - 1 < 11$ **16.** $0 < x - 3 < 4$ **17.** $5 \geq 4b + 9 \geq 17$

Applying Skills

In 18–23, write and solve an equation or an inequality to solve the problem.

18. Peter had 156 cents in coins. After he bought 3 packs of gum he had no more than 9 cents left. What is the minimum cost of a pack of gum?

19. In an algebra class, 3 students are working on a special project and the remaining students are working in groups of five. If there are 18 students in class, how many groups of five are there?

20. Andy paid a reservation fee of $8 plus $12 a night to board her cat while she was on vacation. If Andy paid $80 to board her cat, how many nights was Andy on vacation?

21. At a parking garage, parking costs $5 for the first hour and $3 for each additional hour or part of an hour. Mr. Kanesha paid $44 for parking on Monday. For how many hours did Mr. Kanesha park his car?

22. Kim wants to buy an azalea plant for $19 and some delphinium plants for $5 each. She wants to spend less than $49 for the plants. At most how many delphinium plants can she buy?

23. To prepare for a tennis match and have enough time for schoolwork, Priscilla can practice no more than 14 hours.. If she practices the same length of time on Monday through Friday, and then spends 4 hours on Saturday, what is the most time she can practice on Wednesday?

1-3 ADDING POLYNOMIALS

A **monomial** is a constant, a variable, or the product of constants and variables. Each algebraic expression, 3, a, ab, $-2a^2$, is a monomial.

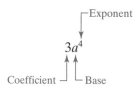

A **polynomial** is the sum of monomials. Each monomial is a **term** of the polynomial. The expressions $3a^2 + 7a - 2$ is a polynomial over the set of integers since all of the numerical coefficients are integers. For any integral value of a, $3a^2 + 7a - 2$ has an integral value. For example, if $a = -2$, then:

$$3a^2 + 7a - 2 = 3(-2)^2 + 7(-2) - 2$$
$$= 3(4) + 7(-2) - 2$$
$$= 12 - 14 - 2$$
$$= -4$$

The same properties that are true for integers are true for polynomials: we can use the commutative, associative, and distributive properties when working with polynomials. For example:

$$(3a^2 + 5a) + (6 - 7a) = (3a^2 + 5a) + (-7a + 6) \quad \text{Commutative Property}$$
$$= 3a^2 + (5a - 7a) + 6 \quad \text{Associative Property}$$
$$= 3a^2 + (5 - 7)a + 6 \quad \text{Distributive Property}$$
$$= 3a^2 - 2a + 6$$

Note: When the two polynomials are added, the two terms that have the same power of the same variable factor are combined into a single term.

Two terms that have the same variable and exponent or are both numbers are called **similar terms** or **like terms**. The sum of similar terms is a monomial.

$$3a^2 + 5a^2 \qquad\qquad -7ab + 3ab \qquad\qquad x^3 + 4x^3$$
$$= (3 + 5)a^2 \qquad\qquad = (-7 + 3)ab \qquad\qquad = (1 + 4)x^3$$
$$= 8a^2 \qquad\qquad\quad = -4ab \qquad\qquad\quad = 5x^3$$

Two monomials that are not similar terms cannot be combined. For example, $4x^3$ and $3x^2$ are not similar terms and the sum $4x^3 + 3x^2$ is not a monomial. A polynomial in simplest form that has two terms is a **binomial**. A polynomial in simplest form that has three terms is a **trinomial**.

Solving Equations and Inequalities

An equation or inequality often has a variable term on both sides. To solve such an equation or inequality, we must first write an equivalent equation or inequality with the variable on only one side.

For example, to solve the inequality $5x - 7 > 3x + 9$, we will first write an equivalent inequality that does not have a variable in the right side. Add the opposite of $3x$, $-3x$, to both sides. The terms $3x$ and $-3x$ are similar terms whose sum is 0.

$$5x - 7 > 3x + 9$$
$$-3x + 5x - 7 > -3x + 3x + 9 \qquad \text{Add } -3x, \text{ the opposite of } 3x, \text{ to both sides.}$$
$$2x - 7 > 9 \qquad\qquad\qquad -3x + 3x = (-3 + 3)x = 0x = 0$$
$$2x - 7 + 7 > 9 + 7 \qquad\qquad \text{Add 7, the opposite of } -7, \text{ to both sides.}$$
$$2x > 16 \qquad\qquad\qquad\quad \text{Divide both sides by 2. Dividing by a}$$
$$x > 8 \qquad\qquad\qquad\qquad \text{positive does not reverse the inequality.}$$

If x is an integer, then the solution set is $\{9, 10, 11, 12, 13, \ldots\}$.

EXAMPLE I

a. Find the sum of $x^3 - 5x + 9$ and $x - 3x^3$.

b. Find the value of each of the given polynomials and the value of their sum when $x = -4$.

Solution **a.** The commutative and associative properties allow us to change the order and the grouping of the terms.

$$(x^3 - 5x + 9) + (x - 3x^3) = (x^3 - 3x^3) + (-5x + x) + 9$$
$$= (1 - 3)x^3 + (-5 + 1)x + 9$$
$$= -2x^3 - 4x + 9 \text{ } Answer$$

b. $x^3 - 5x + 9$ $x - 3x^3$ $-2x^3 - 4x + 9$

$= (-4)^3 - 5(-4) + 9$ $= (-4) - 3(-4)^3$ $= -2(-4)^3 - 4(-4) + 9$

$= -64 + 20 + 9$ $= -4 + 192$ $= 128 + 16 + 9$

$= -35 \text{ } Answer$ $= 188 \text{ } Answer$ $= 153 \text{ } Answer$

EXAMPLE 2

Subtract $(3b^4 + b + 3)$ from $(b^4 - 5b + 3)$ and write the difference as a polynomial in simplest form.

Solution Subtract $(3b^4 + b + 3)$ from $(b^4 - 5b + 3)$ by adding the opposite of $(3b^4 + b + 3)$ to $(b^4 - 5b + 3)$.

$$(b^4 - 5b + 3) - (3b^4 + b + 3) = (b^4 - 5b + 3) + (-3b^4 - b - 3)$$
$$= (b^4 - 3b^4) + (-5b - b) + (3 - 3)$$
$$= -2b^4 - 6b \text{ } Answer$$

EXAMPLE 3

Pam is three times as old as Jody. In five years, Pam will be twice as old as Jody. How old are Pam and Jody now?

Solution   Let $x =$ Jody's age now

$3x =$ Pam's age now

$x + 5 =$ Jody's age in 5 years

$3x + 5 =$ Pam's age in 5 years

Pam's age in 5 years will be twice Jody's age in 5 years.

$$3x + 5 \quad = \quad 2(x + 5)$$
$$3x + 5 \quad = \quad 2x + 10$$
$$-2x + 3x + 5 \quad = \quad -2x + 2x + 10$$
$$x + 5 \quad = \quad 0 \quad + 10$$
$$x + 5 - 5 = \quad 10 - 5$$
$$x \quad = \quad 5$$
$$3x \quad = \quad 15$$

Answer   Jody is 5 and Pam is 15.

Exercises

Writing About Mathematics

1. Danielle said that there is no integer that makes the inequality $|2x + 1| < x$ true. Do you agree with Danielle? Explain your answer.

2. A binomial is a polynomial with two terms and a trinomial is a polynomial with three terms. Jess said that the sum of a trinomial and binomial is always a trinomial. Do you agree with Jess? Justify your answer.

Developing Skills

In 3–12, write the sum or difference of the given polynomials in simplest form.

3. $(3y - 5) + (2y - 8)$

4. $(x^2 + 3x - 2) + (4x^2 - 2x + 3)$

5. $(4x^2 - 3x - 7) + (3x^2 - 2x + 3)$

6. $(-x^2 + 5x + 8) + (x^2 - 2x - 8)$

7. $(a^2b^2 - ab + 5) + (a^2b^2 + ab - 3)$

8. $(7b^2 - 2b + 3) - (3b^2 + 8b + 3)$

9. $(3 + 2b + b^2) - (9 + 5b + b^2)$

10. $(4x^2 - 3x - 5) - (3x^2 - 10x + 3)$

11. $(y^2 - y - 7) + (3 - 2y + 3y^2)$

12. $(2a^4 - 5a^2 - 1) + (a^3 + a)$

In 13–22, solve each equation or inequality. Each solution is an integer.

13. $7x + 5 = 4x + 23$

14. $y + 12 = 5y - 4$

15. $7 - 2a = 3a + 32$

16. $12 + 6b = 2b$

17. $2x + 3 < x + 15$

18. $5y - 1 \geq 2y + 5$

19. $9y + 2 \leq 7y$

20. $14c > 80 - 6c$

21. $(b - 1) - (3b - 4) = b$

22. $-3 - 2x \geq 12 + x$

Applying Skills

23. An online music store is having a sale. Any song costs 75 cents and any ringtone costs 50 cents. Emma can buy 6 songs and 2 audiobooks for the same price as 5 ringtones and 3 audiobooks. What is the cost of an audiobook?

24. The length of a rectangle is 5 feet more than twice the width.

a. If x represents the width of the rectangle, represent the perimeter of the rectangle in terms of x.

b. If the perimeter of the rectangle is 2 feet more than eight times the width of the rectangle, find the dimensions of the rectangle.

25. On his trip to work each day, Brady pays the same toll, using either all quarters or all dimes. If the number of dimes needed for the toll is 3 more than the number of quarters, what is the toll?

1-4 SOLVING ABSOLUTE VALUE EQUATIONS AND INEQUALITIES

Absolute Value Equations

We know that if a is a positive number, then $|a| = a$ and that $|-a| = a$. For example, if $|x| = 3$, then $x = 3$ or $x = -3$ because $|3| = 3$ and $|-3| = 3$. We can use these facts to solve an **absolute value equation**, that is, an equation containing the absolute value of a variable.

For instance, solve $|2x - 3| = 17$. We know that $|17| = 17$ and $|-17| = 17$. Therefore $2x - 3$ can equal 17 or it can equal -17.

$$
\begin{array}{ccc}
2x - 3 = 17 & \text{or} & 2x - 3 = -17 \\
2x - 3 + 3 = 17 + 3 & & 2x - 3 + 3 = -17 + 3 \\
2x = 20 & & 2x = -14 \\
x = 10 & & x = -7
\end{array}
$$

The solution set of $|2x - 3| = 17$ is $\{-7, 10\}$.

In order to solve an absolute value equation, we must first isolate the absolute value expression. For instance, to solve $|4a + 2| + 7 = 21$, we must first add -7 to each side of the equation to isolate the absolute value expression.

$$
\begin{array}{c}
|4a + 2| + 7 = 21 \\
|4a + 2| + 7 - 7 = 21 - 7 \\
|4a + 2| = 14
\end{array}
$$

Now we can consider the two possible cases: $4a + 2 = 14$ or $4a + 2 = -14$.

$$
\begin{array}{ccc}
4a + 2 = 14 & \text{or} & 4a + 2 = -14 \\
4a + 2 - 2 = 14 - 2 & & 4a + 2 - 2 = -14 - 2 \\
4a = 12 & & 4a = -16 \\
a = 3 & & a = -4
\end{array}
$$

The solution set of $|4a + 2| + 7 = 21$ is $\{-4, 3\}$.

Note that the solution sets of the equations $|x + 3| = -5$ and $|x + 3| + 5 = 2$ are the empty set because absolute value is always positive or zero.

EXAMPLE 1

Find the solution of the following equation: $|4x - 2| = 10$.

Solution Since $|10| = |-10| = 10$, the algebraic expression $4x - 2$ can be equal to 10 or to -10.

$$4x - 2 = 10 \qquad \text{or} \qquad 4x - 2 = -10$$
$$4x - 2 + 2 = 10 + 2 \qquad\qquad 4x - 2 + 2 = -10 + 2$$
$$4x = 12 \qquad\qquad\qquad 4x = -8$$
$$x = 3 \qquad\qquad\qquad x = -2$$

Check: $x = 3$ $\qquad\qquad$ Check: $x = -2$
$$|4x - 2| = 10 \qquad\qquad |4x - 2| = 10$$
$$|4(3) - 2| \overset{?}{=} 10 \qquad\qquad |4(-2) - 2| \overset{?}{=} 10$$
$$|10| = 10 ✔ \qquad\qquad |-10| = 10 ✔$$

Answer The solution set is $\{3, -2\}$.

Absolute Value Inequalities

For any two given algebraic expressions, a and b, three relationships are possible: $a = b$, $a < b$, or $a > b$. We can use this fact to solve an **absolute value inequality** (an inequality containing the absolute value of a variable). For example, we know that for the algebraic expressions $|x - 4|$ and 3, there are three possibilities:

CASE 1 $|x - 4| = 3$

$$x - 4 = -3 \quad \text{or} \quad x - 4 = 3$$
$$x = 1 \quad \text{or} \quad x = 7$$

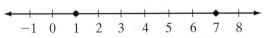

Note that the solution set of this inequality consists of the values of x that are 3 units from 4 in either direction.

CASE 2 $|x - 4| < 3$

The solution set of this inequality consists of the values of x that are less than 3 units from 4 in either direction, that is, $x - 4$ is less than 3 and greater than -3.

$$x - 4 > -3 \quad \text{and} \quad x - 4 < 3$$
$$x > 1 \quad \text{and} \quad x < 7$$

If x is an integer, the solution set is $\{2, 3, 4, 5, 6\}$. Note that these are the integers between the solutions of $|x - 4| = 3$.

CASE 3 $|x - 4| > 3$

The solution set of this inequality consists of the values of x that are more than 3 units from 4 in either direction, that is $x - 4$ is greater than 3 or less than -3.

$$x - 4 < -3 \quad \text{or} \quad x - 4 > 3$$
$$x < 1 \quad \text{or} \quad x > 7$$

If x is an integer, the solution set is $\{\ldots, -3, -2, -1, 0, 8, 9, 10, 11, \ldots\}$. Note that these are the integers that are less than the smaller solution of $|x - 4| = 3$ and greater than the larger solution of $|x - 4| = 3$.

We know that $|a| = a$ if $a \geq 0$ and $|a| = -a$ if $a < 0$. We can use these relationships to solve inequalities of the form $|x| < k$ and $|x| > k$.

Solve $|x| < k$ for positive k

If $x \geq 0$, $|x| = x$.

Therefore, $x < k$ and $0 \leq x < k$.

If $x < 0$, $|x| = -x$.

Therefore, $-x < k$ or $x > -k$.

This can be written $-k < x < 0$.

The solution set of $|x| < k$ is

$$-k < x < k.$$

Solve $|x| > k$ for positive k

If $x \geq 0$, $|x| = x$.

Therefore, $x > k$.

If $x < 0$, $|x| = -x$.

Therefore, $-x > k$ or $x < -k$.

The solution set of $|x| > k$ is

$$x < -k \text{ or } x > k.$$

▶ **If $|x| < k$ for any positive number k, then $-k < x < k$.**

▶ **If $|x| > k$ for any positive number k, then $x > k$ or $x < -k$.**

EXAMPLE 2

Solve for b and list the solution set if b is an integer: $|6 - 3b| - 5 > 4$

Solution (1) Write an equivalent inequality
with only the absolute value on
one side of the inequality:

$$|6 - 3b| - 5 > 4$$
$$|6 - 3b| > 9$$

(2) Use the relationship derived in
this section:

$$6 - 3b > 9 \quad \text{or} \quad 6 - 3b < -9$$

If $|x| > k$ for any positive number k,
then $x > k$ or $x < -k$.

(3) Solve each inequality for b:

$6 - 3b > 9$	$6 - 3b < -9$
$-3b > 3$	$-3b < -15$
$\frac{-3b}{-3} < \frac{3}{-3}$	$\frac{-3b}{-3} > \frac{-15}{-3}$
$b < -1$	$b > 5$

Answer $\{\ldots, -5, -4, -3, -2, 6, 7, 8, 9, \ldots\}$

Exercises

Writing About Mathematics

1. Explain why the solution of $|-3b| = 9$ is the same as the solution of $|3b| = 9$.

2. Explain why the solution set of $|2x + 4| + 7 < 3$ is the empty set.

Developing Skills

In 3–14, write the solution set of each equation.

3. $|x - 5| = 12$ **4.** $|x + 8| = 6$ **5.** $|2a - 5| = 7$

6. $|5b - 10| = 25$ **7.** $|3x - 12| = 9$ **8.** $|4y + 2| = 14$

9. $|35 - 5x| = 10$ **10.** $|-5a| + 7 = 22$ **11.** $|8 + 2b| - 3 = 9$

12. $|2x - 5| + 2 = 13$ **13.** $|4x - 12| + 8 = 0$ **14.** $|7 - x| + 2 = 12$

In 15–26, solve each inequality and write the solution set if the variable is an element of the set of integers.

15. $|x| > 9$ **16.** $|y + 2| > 7$ **17.** $|b + 6| \leq 5$

18. $|x - 3| < 4$ **19.** $|y + 6| > 13$ **20.** $|2b - 7| \geq 9$

21. $|6 - 3x| < 15$ **22.** $|8 + 4b| \geq 0$ **23.** $|5 - b| + 4 < 9$

24. $|11 - 2b| - 6 > 11$ **25.** $|6 - 3b| + 4 < 3$ **26.** $|7 - x| + 2 \leq 12$

Applying Skills

27. A carpenter is making a part for a desk. The part is to be 256 millimeters wide plus or minus 3 millimeters. This means that the absolute value of the difference between the dimension of the part and 256 can be no more than 3 millimeters. To the nearest millimeter, what are the acceptable dimensions of the part?

28. A theater owner knows that to make a profit as well as to comply with fire regulations, the number of tickets that he sells can differ from 225 by no more than 75. How many tickets can the theater owner sell in order to make a profit and comply with fire regulations?

29. A cereal bar is listed as containing 200 calories. A laboratory tested a sample of the bars and found that the actual calorie content varied by as much as 28 calories. Write and solve an absolute value inequality for the calorie content of the bars.

1-5 MULTIPLYING POLYNOMIALS

We know that the product of any number of equal factors can be written as a power of that factor. For example:

$$a \times a \times a \times a = a^4$$

In the expression a^4, a is the **base**, 4 is the **exponent**, and a^4 is the **power**. The exponent tells us how many times the base, a, is to be used as a factor.

To multiply powers with like bases, keep the same base and add the exponents. For example:

$$x^3 \times x^2 = (x \times x \times x) \times (x \times x) = x^5$$
$$3^4 \times 3^5 = (3 \times 3 \times 3 \times 3) \times (3 \times 3 \times 3 \times 3 \times 3) = 3^9$$

In general:

$$x^a \cdot x^b = x^{a+b}$$

Note that we are not performing the multiplication but simply counting how many times the base is used as a factor.

Multiplying a Monomial by a Monomial

The product of two monomials is a monomial. We use the associative and commutative properties of multiplication to write the product.

$$3a^2b(2abc) = 3(2)(a^2)(a)(b)(b)(c)$$
$$= 6a^3b^2c$$

Note: When multiplying $(a^2)(a)$ and $(b)(b)$, the exponent of a and of b is 1.

The square of a monomial is the product of each factor of the monomial used twice as a factor.

$$(3ab^2)^2 \qquad\qquad (-2x^3y)^2 \qquad\qquad -2(x^3y)^2$$
$$= (3ab^2)(3ab^2) \qquad = (-2x^3y)(-2x^3y) \qquad = -2(x^3y)(x^3y)$$
$$= 9a^2b^4 \qquad\qquad = 4x^6y^2 \qquad\qquad = -2x^6y^2$$

Multiplying a Polynomial by a Monomial

To multiply a monomial times a polynomial, we use the distributive property of multiplication over addition, $a(b + c) = ab + ac$:

$$-4(y - 7) \qquad\qquad 5x(x^2 - 3x + 2)$$
$$= -4y - 4(-7) \qquad\quad = 5x(x^2) + 5x(-3x) + 5x(2)$$
$$= -4y + 28 \qquad\qquad = 5x^3 - 15x^2 + 10x$$

Note: The product of a monomial times a polynomial has the same number of terms as the polynomial.

Multiplying a Polynomial by a Binomial

To multiply a binomial by a polynomial we again use the distributive property of multiplication over addition. First, recall that the distributive property $a(b + c) = ab + ac$ can be written as:

$$(b + c)a = ba + ca$$

Now let us use this form of the distributive property to find the product of two binomials, for example:

$$(b + c) \quad (a) \quad = b \ (a) \quad + c \ (a)$$
$$(x + 2)(x + 5) = x(x + 5) + 2(x + 5)$$
$$= x^2 + 5x + 2x + 10$$
$$= x^2 + 7x + 10$$

Multiplying two binomials (polynomials with two terms) requires four multiplications. We multiply each term of the first binomial times each term of the second binomial. The word **FOIL** helps us to remember the steps needed.

$$(x + 4)(x - 3) = x(x - 3) + 4(x - 3)$$
$$\qquad\qquad Ⓕ \quad Ⓞ \quad Ⓘ \quad Ⓛ$$
$$= x^2 - 3x + 4x - 12$$
$$= x^2 + x - 12$$

Product of the Ⓕirst terms
Product of the Ⓞutside terms
Product of the Ⓘnside terms
Product of the Ⓛast terms

Multiplying a Polynomial by a Polynomial

To multiply any two polynomials, multiply each term of the first polynomial by each term of the second. For example:

$$(a^2 + a - 3)(2a^2 + 3a - 1)$$
$$= a^2(2a^2 + 3a - 1) + a(2a^2 + 3a - 1) - 3(2a^2 + 3a - 1)$$
$$= (2a^4 + 3a^3 - a^2) + (2a^3 + 3a^2 - a) - (6a^2 - 9a + 3)$$
$$= 2a^4 + (3a^3 + 2a^3) + (-a^2 + 3a^2 - 6a^2) + (-a - 9a) + 3$$
$$= 2a^4 + 5a^3 - 4a^2 - 10a + 3$$

Note: Since each of the polynomials to be multiplied has 3 terms, there are 3×3 or 9 products. After combining similar terms, the polynomial in simplest form has five terms.

EXAMPLE 1

Write each of the following as a polynomial in simplest form.

a. $ab(a^2 + 2ab + b^2)$ **b.** $(3x - 2)(2x + 5)$

c. $(y + 2)(y - 2)$ **d.** $(2a + 1)(a^2 - 2a - 2)$

Solution   **a.** $ab(a^2 + 2ab + b^2) = a^3b + 2a^2b^2 + ab^3$ *Answer*

b. $(3x - 2)(2x + 5) = 3x(2x + 5) - 2(2x + 5)$
$$= 6x^2 + 15x - 4x - 10$$
$$= 6x^2 + 11x - 10 \text{ } Answer$$

c. $(y + 2)(y - 2) = y(y - 2) + 2(y - 2)$
$$= y^2 - 2y + 2y - 4$$
$$= y^2 - 4 \text{ } Answer$$

d. $(2a + 1)(a^2 - 2a - 2) = 2a(a^2 - 2a - 2) + 1(a^2 - 2a - 2)$
$$= 2a^3 - 4a^2 - 4a + a^2 - 2a - 2$$
$$= 2a^3 - 3a^2 - 6a - 2 \text{ } Answer$$

EXAMPLE 2

Write in simplest form: $(2b)^2 + 5b[2 - 3(b - 1)]$

Solution (1) Simplify the innermost parentheses first:

$$(2b)^2 + 5b[2 - 3(b - 1)]$$
$$= (2b)^2 + 5b[2 - 3b + 3]$$
$$= (2b)^2 + 5b[5 - 3b]$$

(2) Multiply the terms in the brackets: $= (2b)^2 + 25b - 15b^2$

(3) Simplify powers: $= 4b^2 + 25b - 15b^2$

(4) Add similar terms: $= 25b - 11b^2$ *Answer*

EXAMPLE 3

Solve and check: $y(y + 2) - 3(y + 4) = y(y + 1)$

Solution (1) Simplify each side of the equation:

$$y(y + 2) - 3(y + 4) = y(y + 1)$$
$$y^2 + 2y - 3y - 12 = y^2 + y$$
$$y^2 - y - 12 = y^2 + y$$

(2) Add $-y^2$ to both sides of the equation:

$$-y^2 + y^2 - y - 12 = -y^2 + y^2 + y$$
$$-y - 12 = y$$

(3) Add y to both sides of the equation:

$$y - y - 12 = y + y$$
$$-12 = 2y$$

(4) Divide both sides of the equation by 2:

$$-6 = y$$

(5) *Check:*

$$y(y + 2) - 3(y + 4) = y(y + 1)$$
$$-6(-6 + 2) - 3(-6 + 4) \overset{?}{=} -6(-6 + 1)$$
$$-6(-4) - 3(-2) \overset{?}{=} -6(-5)$$
$$24 + 6 \overset{?}{=} 30$$
$$30 = 30 \checkmark$$

Answer $y = -6$

Exercises

Writing About Mathematics

1. Melissa said that $(a + 3)^2 = a^2 + 9$. Do you agree with Melissa? Justify your answer.

2. If a trinomial is multiplied by a binomial, how many times must you multiply a monomial by a monomial? Justify your answer.

Developing Skills

In 3–23, perform the indicated operations and write the result in simplest form.

3. $2a^5b^2(7a^3b^2)$

4. $6c^2d(-2cd^3)$

5. $(6xy^2)^2$

6. $(\ 3c^4)^2$

7. $-(3c^4)^7$

8. $3b(5b - 4)$

9. $2x^2y(y - 2y^2)$

10. $(x + 3)(2x - 1)$

11. $(a - 5)(a + 4)$

12. $(3x + 1)(x - 2)$

13. $(a + 3)(a - 3)$

14. $(5b + 2)(5b - 2)$

15. $(a + 3)^2$

16. $(3b - 2)^2$

17. $(y - 1)(y^2 - 2y + 1)$

18. $(2x + 3)(x^2 + x - 5)$

19. $3a + 4(2a - 3)$

20. $b^2 + b(3b + 5)$

21. $4y(2y - 3) - 5(2 - y)$

22. $a^3(a^2 + 3) - (a^5 + 3a^3)$

23. $(z - 2)^3$

In 24–29, solve for the variable and check. Each solution is an integer.

24. $(2x + 1) + (4 - 3x) = 10$

25. $(3a + 7) - (a - 1) = 14$

26. $2(b - 3) + 3(b + 4) = b + 14$

27. $(x + 3)^2 = (x - 5)^2$

28. $4x(x + 2) - x(3 + 4x) = 2x + 18$

29. $y(y + 2) - y(y - 2) = 20 - y$

Applying Skills

30. The length of a rectangle is 4 more than twice the width, x. Express the area of the rectangle in terms of x.

31. The length of the longer leg, a, of a right triangle is 1 centimeter less than the length of the hypotenuse and the length of the shorter leg, b, is 8 centimeters less than the length of the hypotenuse.

a. Express a and b in terms of c, the length of the hypotenuse.

b. Express $a^2 + b^2$ as a polynomial in terms of c.

c. Use the Pythagorean Theorem to write a polynomial equal to c^2.

1-6 FACTORING POLYNOMIALS

The **factors** of a monomial are the numbers and variables whose product is the monomial. Each of the numbers or variables whose product is the monomial is a factor of the monomial as well as 1 and any combination of these factors. For example, the factors of $3a^2b$ are 1, 3, a, and b, as well as $3a$, $3b$, a^2, ab, $3a^2$, $3ab$, a^2b, and $3a^2b$.

Common Monomial Factor

A polynomial can be written as a monomial times a polynomial if there is at least one number or variable that is a factor of each term of the polynomial. For instance:

$$4a^4 - 10a^2 = 2a^2(2a^2 - 5)$$

Note: $2a^2$ is the *greatest* **common monomial factor** of the terms of the polynomial because 2 is the greatest common factor of 4 and 10 and a^2 is the *smallest* power of a that occurs in each term of the polynomial.

EXAMPLE 1

Factor:

Answers

a. $12x^2y^3 - 15xy^2 + 9y$ $= 3y(4x^2y^2 - 5xy + 3)$

b. $a^2b^3 + ab^2c$ $= ab^2(ab + c)$

c. $2x^2 - 8x + 10$ $= 2(x^2 - 4x + 5)$

Common Binomial Factor

We know that:

$$5ab \qquad + 3b \qquad = \qquad b(5a + 3)$$

If we replace b by $(x + 2)$ we can write:

$$5a(x + 2) + 3(x + 2) = (x + 2)(5a + 3)$$

Just as b is the common factor of $5ab + 3b$, $(x + 2)$ is the common factor of $5a(x + 2) + 3(x + 2)$. We call $(x + 2)$ the **common binomial factor**.

EXAMPLE 2

Find the factors of: $a^3 + a^2 - 2a - 2$

Solution Find the common factor of the first two terms and the common factor of the last two terms. Use the sign of the first term of each pair as the sign of the common factor.

$$a^3 + a^2 - 2a - 2 = a^2(a + 1) - 2(a + 1)$$
$$= (a + 1)(a^2 - 2) \text{ Answer}$$

Note: In the polynomial given in Example 2, the product of the first and last terms is equal to the product of the two middle terms: $a^3 \cdot (-2) = a^2 \cdot (-2a)$. This relationship will always be true if a polynomial of four terms can be factored into the product of two binomials.

Binomial Factors

We can find the binomial factors of a trinomial, if they exist, by reversing the process of finding the product of two binomials. For example:

$$(x + 3)(x - 2) = x(x - 2) + 3(x - 2)$$
$$= x^2 - 2x + 3x - 6$$
$$= x^2 + x - 6$$

Note that when the polynomial is written as the sum of four terms, the product of the first and last terms, $(x^2 \cdot -6)$, is equal to the product of the two middle terms, $(-2x \cdot 3x)$. We can apply these observations to factoring a trinomial into two binomials.

EXAMPLE 3

Factor $x^2 + 7x + 12$.

Solution **METHOD I**

(1) Write the trinomial as the sum of four terms by writing $7x$ as the sum of two terms whose product is equal to the product of the first and last terms:

$x^2 \cdot 12 = 12x^2$

$x \cdot 12x = 12x^2$ but $x + 12x \neq 7x$ ✗

$2x \cdot 6x = 12x^2$ but $2x + 6x \neq 7x$ ✗

$3x \cdot 4x = 12x^2$ and $3x + 4x = 7x$ ✔

(2) Rewrite the polynomial as the sum of four terms:

$x^2 + \quad 7x \quad + 12$
$= x^2 + 3x + 4x + 12$

(3) Factor out the common monomial from the first two terms and from the last two terms:

$= x(x + 3) + 4(x + 3)$

(4) Factor out the common binomial factor:

$= (x + 3)(x + 4)$

METHOD 2

This trinomial can also be factored by recalling how the product of two binomials is found.

(1) The first term of the trinomial is the product of the first terms of the binomial factors:

$$x^2 + 7x + 12 = (x \qquad)(x \qquad)$$

(2) The last term of the trinomial is the product of the last terms of the binomial factors. Write all possible pairs of factors for which this is true.

$$(x + 1)(x + 12) \qquad (x - 1)(x - 12)$$
$$(x + 2)(x + 6) \qquad (x - 2)(x - 6)$$
$$(x + 3)(x + 4) \qquad (x - 3)(x - 4)$$

(3) For each possible pair of factors, find the product of the outside terms plus the product of the inside terms.

$$12x + 1x = 13x \ ✗ \qquad -12x + (-1x) = -13x \ ✗$$
$$6x + 2x = 8x \ ✗ \qquad -6x + (-2x) = -8x \ ✗$$
$$3x + 4x = 7x \ ✔ \qquad -3x + (-4x) = -7x \ ✗$$

(4) The factors of the trinomial are the two binomials such that the product of the outside terms plus the product of the inside terms equals $+7x$.

$$x^2 + 7x + 12 = (x + 3)(x + 4)$$

Answer $(x + 3)(x + 4)$

EXAMPLE 4

Factor: $3x^2 - x - 4$

Solution (1) Find the product of the first and last terms: $3x^2(-4) = -12x^2$

(2) Find the factors of this product whose sum is the middle term: $-4x + 3x = -x$

(3) Write the trinomial with four terms, using this pair of terms in place of $-x$:

$$3x^2 - x - 4$$
$$= 3x^2 - 4x + 3x - 4$$

(4) Factor the common factor from the first two terms and from the last two terms:

$$= x(3x - 4) + 1(3x - 4)$$

(5) Factor the common binomial factor: $= (3x - 4)(x + 1)$

Answer $(3x - 4)(x + 1)$

Special Products and Factors

We know that to multiply a binomial by a binomial, we perform four multiplications. If all four terms are unlike terms, then the polynomial is in simplest form. Often, two of the four terms are similar terms that can be combined so that the product is a trinomial. For example:

$$(a^2 + 3)(a - 2) \qquad\qquad (x + 3)(x - 5)$$
$$= a^2(a - 2) + 3(a - 2) \qquad = x(x - 5) + 3(x - 5)$$
$$= a^3 - 2a^2 + 3a - 6 \qquad = x^2 - 5x + 3x - 15$$
$$\qquad\qquad\qquad\qquad = x^2 - 2x - 15$$

When the middle terms are additive inverses whose sum is 0, then the product of the two binomials is a binomial.

$$(a + 3)(a - 3) = a(a - 3) + 3(a - 3)$$
$$= a^2 - 3a + 3a - 9$$
$$= a^2 + 0a - 9$$
$$= a^2 - 9$$

Therefore, the product of the sum and difference of the same two numbers is the difference of their squares. In general, the factors of the difference of two perfect squares are:

$$a^2 - b^2 = (a + b)(a - b)$$

EXAMPLE 5

Factor:

	Think	*Write*
a. $4x^2 - 25$	$(2x)^2 - (5)^2$	$(2x + 5)(2x - 5)$
b. $16 - 9y^2$	$(4)^2 - (3y)^2$	$(4 + 3y)(4 - 3y)$
c. $36a^4 - b^4$	$(6a^2)^2 - (b^2)^2$	$(6a^2 + b^2)(6a^2 - b^2)$

When factoring a polynomial, it is important to make sure that each factor is a **prime polynomial** or has no factors other than 1 and itself. Once you have done this, the polynomial is said to be **completely factored**. For instance:

$$3ab^2 - 6ab + 3a \qquad\qquad x^4 - 16$$
$$= 3a(b^2 - 2b + 1) \qquad = (x^2 + 4)(x^2 - 4)$$
$$= 3a(b - 1)(b - 1) \qquad = (x^2 + 4)(x + 2)(x - 2)$$

EXAMPLE 6

Factor: $5a^3 - a^2 - 5a + 1$

Solution (1) The product of the first and last terms is equal to the product of the two middle terms. Therefore, the polynomial is a product of two binomials:

$$5a^3 \cdot 1 \overset{?}{=} -a^2 \cdot -5a$$
$$5a^3 = 5a^3 \ ✔$$

(2) Find a common factor of the first two terms and then of the last two terms. Then, factor out the common binomial factor:

$$5a^3 - a^2 - 5a + 1$$
$$= a^2(5a - 1) - 1(5a - 1)$$
$$= (5a - 1)(a^2 - 1)$$

(3) The binomial factor $(a^2 - 1)$ is the difference of two squares, which can be factored into the sum and difference of the equal factors of the squares:

$$= (5a - 1)(a^2 - 1)$$
$$= (5a - 1)(a + 1)(a - 1)$$

Answer $(5a - 1)(a + 1)(a - 1)$

Exercises

Writing About Mathematics

1. Joel said that the factors of $x^2 + bx + c$ are $(x + d)(x + e)$ if $de = c$ and $d + e = b$. Do you agree with Joel? Justify your answer.

2. Marietta factored $x^2 + 5x - 4$ as $(x + 4)(x + 1)$ because $4(1) = 4$ and $4 + 1 = 5$. Do you agree with Marietta? Explain why or why not.

Developing Skills

In 3–8, write each polynomial as the product of its greatest common monomial factor and a polynomial.

3. $8x^2 + 12x$

4. $6a^4 - 3a^3 + 9a^2$

5. $5ab^2 - 15ab + 20a^2b$

6. $x^3y^3 - 2x^3y^2 + x^2y^2$

7. $4a - 12ab + 16a^2$

8. $21a^2 - 14a + 7$

In 9–26, write each expression as the product of two binomials.

9. $y(y + 1) - 1(y + 1)$

10. $3b(b - 2) - 4(b - 2)$

11. $2x(y + 4) + 3(y + 4)$

12. $a^3 - 3a^2 + 3a - 9$

13. $2x^3 - 3x^2 - 4x + 6$

14. $y^3 + y^2 - 5y - 5$

15. $x^2 + 7x + x + 7$

16. $x^2 + 5x + 6$

17. $x^2 - 5x + 6$

18. $x^2 + 5x - 6$

19. $x^2 - x - 6$

20. $x^2 + 9x + 20$

21. $3x^2 - 5x - 12$

22. $2y^2 + 5y - 3$

23. $5b^2 + 6b + 1$

24. $6x^2 - 13x + 2$

25. $4y^2 + 4y + 1$

26. $9x^2 - 12x + 4$

In 27–39, factor each polynomial completely.

27. $a^3 + 3a^2 - a - 3$ **28.** $5x^2 - 15x + 10$ **29.** $b^3 - 4b$

30. $4ax^2 + 4ax - 24a$ **31.** $12c^2 - 3$ **32.** $x^4 - 81$

33. $x^4 - 16$ **34.** $2x^3 + 13x^2 + 15x$ **35.** $4x^3 - 10x^2 + 6x$

36. $z^4 - 12z^2 + 27$ **37.** $(c + 2)^2 - 1$ **38.** $4 - (y - 1)^2$

39. $x^2y - 16y$ **40.** $3(x - 1)^2 - 12$ **41.** $9 - 9(x + 2)^2$

Applying Skills

In 42–45, each polynomial represents the area of a rectangle. Write two binomials that could represent the length and width of the rectangle.

42. $4x^2 - 7x - 2$ **43.** $16x^2 - 25$ **44.** $9x^2 - 6x + 1$ **45.** $3x^2 + 5x - 2$

1-7 QUADRATIC EQUATIONS WITH INTEGRAL ROOTS

An equation such as $3x + 4 = 16$ is a linear equation in one variable, that is, an equation in which the variable occurs to the first power only. An equation such as $x^2 - 3x + 2 = 0$ is a **quadratic equation** or a **polynomial equation of degree two** because the highest power of the variable is two. A quadratic equation is in **standard form** when it is written as a polynomial equal to 0. In general, if $a \neq 0$, the standard form of a quadratic equation is

$$ax^2 + bx + c = 0$$

To write the quadratic equation $3 + 2x(x - 1) = 5$ in standard form, first simplify the left member and then add -5 to each side of the equation.

$$3 + x(x - 1) = 5$$
$$3 + x^2 - x = 5$$
$$3 + x^2 - x + (-5) = 5 + (-5)$$
$$x^2 - x - 2 = 0$$

Solving a Quadratic Equation

We know that $ab = 0$ if and only if $a = 0$ or $b = 0$. We can use this fact to solve a quadratic equation in standard form when the roots are integers. First, write the non-zero member of the equation as the product of factors, each of which contains the first power of the variable, and then set each factor equal to 0 to find the roots.

EXAMPLE I

Solve the equation $3 + x(x - 1) = 5$.

Solution

$3 + x(x - 1) = 5$	Write the equation in standard form.
$x^2 - x - 2 = 0$	
$(x - 2)(x + 1) = 0$	Factor the left side.

$$x - 2 = 0 \quad | \quad x + 1 = 0 \qquad \text{Set each factor equal to 0 and solve for } x.$$
$$x = 2 \quad | \quad x = -1$$

Check: $x = 2$	*Check:* $x = -1$
$3 + x(x - 1) = 5$	$3 + x(x - 1) = 5$
$3 + 2(2 - 1) \overset{?}{=} 5$	$3 + (-1)(-1 - 1) \overset{?}{=} 5$
$3 + 2(1) \overset{?}{=} 5$	$3 - 1(-2) \overset{?}{=} 5$
$3 + 2 \overset{?}{=} 5$	$3 + 2 \overset{?}{=} 5$
$5 = 5 ✔$	$5 = 5 ✔$

Answer $x = 2$ or -1

EXAMPLE 2

Solve for x: $2x^2 + 4x = 30$

Solution (1) Write the equation in standard form:

$$2x^2 + 4x = 30$$
$$2x^2 + 4x - 30 = 0$$

(2) Factor the left member:

$$2(x^2 + 2x - 15) = 0$$
$$2(x + 5)(x - 3) = 0$$

(3) Set each factor that contains the variable equal to zero and solve for x:

$$x + 5 = 0 \quad | \quad x - 3 = 0$$
$$x = -5 \quad | \quad x = 3$$

Answer $x = -5$ or 3

EXAMPLE 3

The length of a rectangle is 2 feet shorter than twice the width. The area of the rectangle is 84 square feet. Find the dimensions of the rectangle.

Solution   Let w = the width of the rectangle.

$2w - 2$ = the length of the rectangle.

$$\text{Area} = \text{length} \times \text{width}$$
$$84 = (2w - 2)(w)$$
$$84 = 2w^2 - 2w$$
$$0 = 2w^2 - 2w - 84$$
$$0 = 2(w^2 - w - 42)$$
$$0 = 2(w - 7)(w + 6)$$

$$0 = w - 7 \quad \Big| \quad 0 = w + 6$$
$$7 = w \quad \Big| \quad -6 = w$$

The width must be a positive number. Therefore, only 7 feet is a possible width for the rectangle. When $w = 7$, $2w - 2 = 2(7) - 2 = 12$.

The area of the rectangle is $7(12) = 84$ square feet.

Answer   The dimensions of the rectangle are 7 feet by 12 feet.

Exercises

Writing About Mathematics

1. Ross said that if $(x - a)(x - b) = 0$ means that $(x - a) = 0$ or $(x - b) = 0$, then $(x - a)(x - b) = 2$ means that $(x - a) = 2$ or $(x - b) = 2$. Do you agree with Ross? Explain why or why not.

2. If $(x - a)(x - b)(x - c) = 0$, is it true that $(x - a) = 0$, or $(x - b) = 0$ or $(x - c) = 0$? Justify your answer.

Developing Skills

In 3–17, solve and check each of the equations.

3. $x^2 - 4x + 3 = 0$

4. $x^2 - 7x + 10 = 0$

5. $x^2 - 5x - 6 = 0$

6. $x^2 + 6x + 5 = 0$

7. $x^2 + 10x - 24 = 0$

8. $x^2 - 9x = 10$

9. $4 - x(x - 3) = 0$

10. $x(x + 7) - 2 = 28$

11. $2x^2 - x = 12 + x$

12. $3x^2 - 5x = 36 - 2x$

13. $7 = x(8 - x)$

14. $9 = x(6 - x)$

15. $2x(x + 1) = 12$

16. $x(x - 2) + 2 = 1$

17. $3x(x - 10) + 80 = 5$

Applying Skills

18. Brad is 3 years older than Francis. The product of their ages is 154. Determine their ages.

19. The width of a rectangle is 12 feet less than the length. The area of the rectangle is 540 square feet. Find the dimensions of the rectangle.

20. The length of a rectangle is 6 feet less than three times the width. The area of the rectangle is 144 square feet. Find the dimensions of the rectangle.

21. The length of the shorter leg, a, of a right triangle is 6 centimeters less than the length of the hypotenuse, c, and the length of the longer leg, b, is 3 centimeters less than the length of the hypotenuse. Find the length of the sides of the right triangle.

22. The height h, in feet, of a golf ball shot upward from a ground level sprint gun is described by the formula $h = -16t^2 + 48t$ where t is the time in seconds. When will the ball hit the ground again?

1-8 QUADRATIC INEQUALITIES

DEFINITION _____

A **quadratic inequality** is an inequality that contains a polynomial of degree two.

When we solve a linear inequality, we use the same procedure that we use to solve a linear equation. Can we solve a *quadratic* inequality by using the same procedure that we use to solve a quadratic equation? How is the solution of the inequality $x^2 - 3x - 4 > 0$ similar to the solution of $x^2 - 3x - 4 = 0$?

To solve the equation, we factor the trinomial and write two equations in which each factor is equal to 0. To solve the inequality, can we factor the trinomial and write two inequalities in which each factor is greater than 0?

$$x^2 - 3x - 4 = 0 \qquad\qquad x^2 - 3x - 4 > 0$$
$$(x - 4)(x + 1) = 0 \qquad\qquad (x - 4)(x + 1) > 0$$
$$x - 4 = 0 \qquad\qquad x - 4 \overset{?}{>} 0$$
$$x + 1 = 0 \qquad\qquad x + 1 \overset{?}{>} 0$$

If the product of two factors is greater than 0, that is, positive, then it is true that each factor may be greater than 0 because the product of two positive numbers is positive. However, it is also true that each factor may be less than 0 because the product of two negative numbers is also positive. Therefore, when we solve a quadratic inequality, we must consider two possibilities:

$$x^2 - 3x - 4 > 0$$
$$(x - 4)(x + 1) > 0$$

$x - 4 > 0$ and $x + 1 > 0$	$x - 4 < 0$ and $x + 1 < 0$
$x > 4$ $\qquad x > -1$	$x < 4$ $\qquad x < -1$

If x is greater than 4 and greater than -1, then x is greater than 4. | If x is less than 4 and less than -1, then x is less than -1.

The solution set is $\{x : x > 4 \text{ or } x < -1\}$.

On the number line, the solutions of the equality $x^2 - 3x - 4 = 0$ are -1 and 4. These two numbers separate the number line into three intervals.

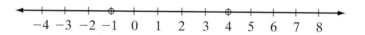

$$\begin{array}{ccccccccccccc} & -4 & -3 & -2 & -1 & 0 & 1 & 2 & 3 & 4 & 5 & 6 & 7 & 8 \end{array}$$

Choose a representative number from each interval:

Let $x = -3$:	**Let $x = 1$:**	**Let $x = 6$:**
$x^2 - 3x - 4 > 0$	$x^2 - 3x - 4 > 0$	$x^2 - 3x - 4 > 0$
$(-3)^2 - 3(-3) - 4 \overset{?}{>} 0$	$(1)^2 - 3(1) - 4 \overset{?}{>} 0$	$(6)^2 - 3(6) - 4 \overset{?}{>} 0$
$9 + 9 - 4 \overset{?}{>} 0$	$1 - 3 - 4 \overset{?}{>} 0$	$36 - 18 - 4 \overset{?}{>} 0$
$14 > 0$ ✔	$-6 \not> 0$ ✗	$14 > 0$ ✔

We find that an element from the interval $x < -1$ or an element from the interval $x > 4$ make the inequality true but an element from the interval $-1 < x < 4$ makes the inequality false.

A quadratic inequality in which the product of two linear factors is less than zero is also solved by considering two cases. A product is negative if the two factors have opposite signs. Therefore, we must consider the case in which the first factor is positive and the second factor is negative and the case in which the first factor is negative and the second factor is positive.

Procedure 1

To solve a quadratic inequality:

CASE 1 *The quadratic inequality is of the form* $(x - a)(x - b) > 0$

1. Let each factor be greater than 0 and solve the resulting inequalities.

2. Let each factor be less than 0 and solve the resulting inequalities.

3. Combine the solutions of the inequalities from steps 1 and 2 to find the solution set of the given inequality.

CASE 2 *The quadratic inequality is of the form* $(x - a)(x - b) < 0$

1. Let the first factor be greater than 0 and let the second factor be less than 0. Solve the resulting inequalities.

2. Let the first factor be less than 0 and let the second factor be greater than 0. Solve the resulting inequalities.

3. Combine the solutions of the inequalities from steps 1 and 2 to find the solution set of the given inequality.

A quadratic inequality can also be solved by finding the solutions to the corresponding equality. The solution to the inequality can be found by testing an element from each interval into which the number line is separated by the roots of the equality.

Procedure 2

To solve a quadratic inequality:

1. Find the roots of the corresponding equality.

2. The roots of the equality separate the number line into two or more intervals.

3. Test a number from each interval. An interval is part of the solution if the test number makes the inequality true.

EXAMPLE 1

List the solution set of $x^2 - 2x - 15 < 0$ if x is an element of the set of integers.

Solution Factor the trinomial. One of the factors is negative and other is positive.

$$x^2 - 2x - 15 < 0$$
$$(x - 5)(x + 3) < 0$$

$x - 5 > 0$ and $x + 3 < 0$	$x - 5 < 0$ and $x + 3 > 0$
$x > 5 \qquad\qquad x < -3$	$x < 5 \qquad\qquad x > -3$
There are no values of x that are both greater than 5 and less than -3.	The solution set is $\{x : -3 < x < 5\}$.

Check The numbers -3 and 5 separate the number line into three intervals. Choose a representative number from each interval.

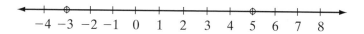

Let $x = -4$:

$$x^2 - 2x - 15 < 0$$

$$(-4)^2 - 2(-4) - 15 \overset{?}{<} 0$$

$$16 + 8 - 15 \overset{?}{<} 0$$

$$9 \not< 0 \text{ ✗}$$

Let $x = 1$:

$$x^2 - 2x - 15 < 0$$

$$(1)^2 - 2(1) - 15 \overset{?}{<} 0$$

$$1 - 2 - 15 \overset{?}{<} 0$$

$$-16 < 0 \text{ ✔}$$

Let $x = 7$:

$$x^2 - 2x - 15 < 0$$

$$(7)^2 - 2(7) - 15 \overset{?}{<} 0$$

$$49 - 14 - 15 \overset{?}{<} 0$$

$$20 \not< 0 \text{ ✗}$$

When we choose a representative number from each of these intervals, we find that an element from the interval $x < -3$ and an element from the interval $x > 5$ make the inequality false but an element from the interval $-3 < x < 5$ makes the inequality true.

Answer $\{-2, -1, 0, 1, 2, 3, 4\}$

A graphing calculator can be used to verify the quadratic inequality of the examples. For instance, enter the inequality $x^2 - 2x - 15 < 0$ into Y_1. Use the 2nd TEST menu to enter the inequality symbols $<, >, \leq,$ and $\geq$. Using the TRACE button, we can then verify that the integers from -2 to 4 make the inequality true (the "Y = 1" in the bottom of the graph indicates that the inequality is true) while integers less than -2 or greater than 4 make the inequality false (the "Y = 0" in the bottom of the graph indicates that the inequality is false).

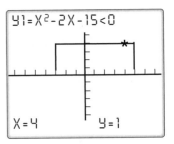

EXAMPLE 2

List the solution set of $x^2 + 6x + 8 \geq 0$ if x is an element of the set of integers.

Solution Factor the corresponding quadratic equality, $x^2 + 6x + 8 = 0$:

$$x^2 + 6x + 8 = 0$$
$$(x + 2)(x + 4) = 0$$

$$
\begin{array}{c|c}
x + 2 = 0 & x + 4 = 0 \\
x = -2 & x = -4
\end{array}
$$

The roots -2 and -4 separate the number line into three intervals: $x < -4$, $-4 < x < -2$, and $-2 < x$. Test a number from each interval to find the solution of the inequality:

Let $x = -5$:   **Let $x = -3$:**   **Let $x = 0$:**

$x^2 + 6x + 8 \overset{?}{\geq} 0$ $x^2 + 6x + 8 \overset{?}{\geq} 0$ $x^2 + 6x + 8 \overset{?}{\geq} 0$

$(-5)^2 + 6(-5) + 8 \overset{?}{\geq} 0$ $(-3)^2 + 6(-3) + 8 \overset{?}{\geq} 0$ $(0)^2 + 6(0) + 8 \overset{?}{\geq} 0$

$3 \geq 0 \checkmark$ $-1 \not\geq 0 \; \text{✗}$ $8 \geq 0 \checkmark$

The inequality is true in the intervals $x < -4$ and $x > -2$. However, since the inequality is less than or equal to, the roots also make the inequality true. Therefore, the solution set is $\{x : x \leq -4 \text{ or } x \geq -2\}$.

Answer $\{\ldots, -6, -5, -4, -2, -1, 0, \ldots\}$

Exercises

Writing About Mathematics

1. Rita said that when the product of three linear factors is greater than zero, all of the factors must be greater than zero or all of the factors must be less than zero. Do you agree with Rita? Explain why or why not.

2. Shelley said that if $(x - 7)(x - 5) < 0$, then $(x - 7)$ must be the negative factor and $(x - 5)$ must be the positive factor.

  a. Do you agree with Shelly? Explain why or why not.

  b. When the product of two factors is negative, is it always possible to tell which is the positive factor and which is the negative factor? Justify your answer.

Developing Skills

In 3–17, write the solution set of each inequality if x is an element of the set of integers.

3. $x^2 + 5x + 6 < 0$

4. $x^2 + 5x - 6 > 0$

5. $x^2 - 3x + 2 \leq 0$

6. $x^2 - 7x + 10 > 0$

7. $x^2 - x - 6 < 0$

8. $x^2 - 8x - 20 \geq 0$

9. $x^2 + x - 12 < 0$

10. $x^2 - 6x + 5 > 0$

11. $x^2 - 2x \geq 0$

12. $x^2 - x < 6$

13. $x^2 - 4x + 4 > 0$

14. $x^2 - 4x + 4 \geq 0$

15. $x^2 + x - 2 < 0$

16. $2x^2 - 2x - 24 \leq 0$

17. $2x^2 - 2x - 24 > 0$

Applying Skills

18. A rectangular floor can be covered completely with tiles that each measure one square foot. The length of the floor is 1 foot longer than the width and the area is less than 56 square feet. What are the possible dimensions of the floor?

19. A carton is completely filled with boxes that are 1 foot cubes. The length of the carton is 2 feet greater than the width and the height of the carton is 3 feet. If the carton holds at most 72 cubes, what are the possible dimensions of the carton?

CHAPTER SUMMARY

The set of **natural numbers** is the set $\{1, 2, 3, 4, 5, 6, \ldots\}$. The set of **whole numbers** is the union of the set of natural numbers and the number 0. The set of **integers** is the union of set of whole numbers and their opposites.

The **absolute value** of a is symbolized by $|a|$. If $a > 0$, then $|a| = a - 0 = a$. If $a < 0$, then $|a| = 0 - a = -a$.

The **domain** is the set of numbers that can replace the variable in an algebraic expression. A number from the domain that makes an equation or inequality true is a **solution** or **root** of the equation or inequality.

We use the following properties of equality to solve an equation:

- **Addition Property of Equality:** If equals are added to equals, the sums are equal.
- **Subtraction Property of Equality:** If equals are subtracted from equals, the differences are equal.
- **Multiplication Property of Equality:** If equals are multiplied by equals, the products are equal.
- **Division Property of Equality:** If equals are divided by equals, the quotients are equal.

We use the following properties of inequality to solve inequalities:

- **Addition and Subtraction Property of Inequality:** If equals are added to or subtracted from unequals, the sums or differences are unequal in the same order.
- **Multiplication and Division Property of Inequality:** If unequals are multiplied or divided by positive equals, the products or quotients are unequal in the same order. If unequals are multiplied or divided by negative equals, the products or quotients are unequal in the opposite order.

A **monomial** is a constant, a variable, or the product of constants and variables. The **factors** of a monomial are the numbers and variables whose product is the monomial. A **polynomial** is the sum of monomials. Each monomial is a **term** of the polynomial.

An absolute value equation or inequality can be solved by using the following relationships:

- If $|x| = k$ for any positive number k, then $x = -k$ or $x = k$.
- If $|x| < k$ for any positive number k, then $-k < x < k$.
- If $|x| > k$ for any positive number k, then $x > k$ or $x < -k$.

If $a \neq 0$, the standard form of a quadratic equation is $ax^2 + bx + c = 0$. A quadratic equation that has integral roots can be solved by factoring the polynomial of the standard form of the equations and setting each factor that contains the variable equal to zero.

An inequality of the form $(x - a)(x - b) > 0$ can be solved by letting each factor be positive and by letting each factor be negative. An inequality of the form $(x - a)(x - b) < 0$ can be solved by letting one factor be positive and the other be negative.

VOCABULARY

1-1 Natural numbers • Counting numbers • Whole numbers • Opposite • Additive inverse • Integers • Commutative group • Absolute value

1-2 Domain • Solution • Root • Equivalent equations

1-3 Monomial • Polynomial • Term • Similar terms • Like terms • Binomial • Trinomial

1-4 Absolute value equation • Absolute value inequality

1-5 Base • Exponent • Power • FOIL

1-6 Factor • Common monomial factor • Common binomial factor • Prime polynomials • Completely factored

1-7 Quadratic equation • Polynomial equation of degree two • Standard form

1-8 Quadratic Inequality

REVIEW EXERCISES

In 1–12, write each expression is simplest form.

1. $5x - 7x$

2. $4(2a + 3) - 9a$

3. $2d - (5d - 7)$

4. $5(b + 9) - 3b(10 - b)$

5. $x(x + 3) - 4(5 - x)$

6. $8 - 2(a^2 + a + 4)$

7. $7d(2d + c) + 3c(4d - c)$

8. $(2x - 1)(3x + 1) - 5x^2$

9. $c^2 - (c + 2)(c - 2)$

10. $(2x + 1)^2 - (2x + 1)^2$

11. $(-2x)^2 - 2x^2$

12. $4y^2 + 2y(3y - 2) - (3y)^2$

In 13–24, factor each polynomial completely.

13. $2x^2 + 8x + 6$

14. $3a^2 - 30a + 75$

15. $5x^3 - 15x^2 - 20x$

16. $10ab^2 - 40a$

17. $c^4 - 16$

18. $3y^3 - 12y^2 + 6y - 24$

19. $x^3 + 5x^2 - x - 5$

20. $x^4 - 2x^2 - 1$

21. $2x^2 - 18x + 36$

22. $x^3 - 3x^2 + 2x$

23. $5a^4 - 5b^4$

24. $5x^2 + 22x - 15$

In 25–40, solve each equation or inequality for x. For each inequality, the solution set is a subset of the set of integers.

25. $8x + 27 = 5x$

26. $3(x - 7) = 5 + x$

27. $2x - 9 < 5x - 21$

28. $-3 \leq 2x - 1 < 7$

29. $|2x + 5| = 9$

30. $7 - |x + 1| = 0$

31. $|3 - 6y| + 2 > 11$

32. $4 - |x + 3| < 2$

33. $x^2 - 9x + 20 = 0$

34. $x(12 - x) = 35$

35. $x^2 + 7x + 6 < 0$

36. $x^2 - 2x - 35 > 0$

37. $x^2 \leq 5x$

38. $x(x + 3) > 0$

39. $4x^2 - 16x + 12 \leq 0$

40. $2x^2 + 2x - 4 \geq 0$

41. Explain why the equation $|3x - 5| + 4 = 0$ has no solution in the set of integers.

42. The length of a rectangle is 4 centimeters less than three times the width. The perimeter of the rectangle is 88 centimeters. What are the dimensions of the rectangle?

43. The length of a rectangle is 6 feet more than three times the width. The area of the rectangle is 240 square feet. What are the dimensions of the rectangle?

44. The length of the longer leg of a right triangle is 4 inches more than twice the length of the shorter leg. The length of the hypotenuse is 6 inches more than twice the length of the shorter leg. What are the lengths of the legs of the right triangle?

45. The equation $h = -16t^2 + 80t$ gives the height, h, in feet after t seconds when a ball has been thrown upward at a velocity of 80 feet per second.

   a. Find the height of the ball after 3 seconds.

   b. After how many seconds will the ball be at a height of 64 feet?

Exploration

A whole number that is the sum of all of its factors except itself is called a **perfect number**. Euclid said that if $(2^k - 1)$ is a prime, then $N = 2^{k-1}(2^k - 1)$ is a perfect number. A perfect number of this form is called a **Euclidean perfect number**.

1. Use the formula for a Euclidean perfect number to find the first four perfect numbers.

2. Show that a Euclidean perfect number is always even.

3. Show that a Euclidean perfect number must have 6 or 8 as the units digit. (*Hint:* What are the possible units digits of $(2^k - 1)$? Of 2^{k-1}?)

THE RATIONAL NUMBERS

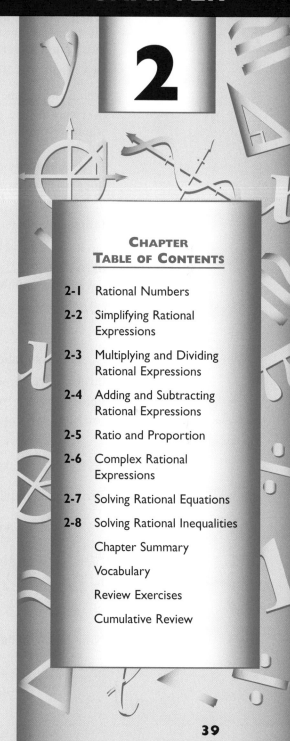

When a divided by b is not an integer, the quotient is a fraction. The Babylonians, who used a number system based on 60, expressed the quotients:

$$20 \div 8 \text{ as } 2 + \frac{30}{60} \text{ instead of } 2\frac{1}{2}$$

$$21 \div 8 \text{ as } 2 + \frac{37}{60} + \frac{30}{3,600} \text{ instead of } 2\frac{5}{8}$$

Note that this is similar to saying that 20 hours divided by 8 is 2 hours, 30 minutes and that 21 hours divided by 5 is 2 hours, 37 minutes, 30 seconds. This notation was also used by Leonardo of Pisa (1175–1250), also known as Fibonacci.

The base-ten number system used throughout the world today comes from both Hindu and Arabic mathematicians. One of the earliest applications of the base-ten system to fractions was given by Simon Stevin (1548–1620), who introduced to 16th-century Europe a method of writing decimal fractions. The decimal that we write as 3.147 was written by Stevin as 3 ⓪ 1 ① 4 ② 7 ③ or as 3 1 4 7. John Napier (1550–1617) later brought the decimal point into common usage.

2-1 RATIONAL NUMBERS

When persons travel to another country, one of the first things that they learn is the monetary system. In the United States, the dollar is the basic unit, but most purchases require the use of a fractional part of a dollar. We know that a penny is $0.01 or $\frac{1}{100}$ of a dollar, that a nickel is $0.05 or $\frac{5}{100} = \frac{1}{20}$ of a dollar, and a dime is $0.10 or $\frac{10}{100} = \frac{1}{10}$ of a dollar. Fractions are common in our everyday life as a part of a dollar when we make a purchase, as a part of a pound when we purchase a cut of meat, or as a part of a cup of flour when we are baking.

In our study of mathematics, we have worked with numbers that are not integers. For example, 15 minutes is $\frac{15}{60}$ or $\frac{1}{4}$ of an hour, 8 inches is $\frac{8}{12}$ or $\frac{2}{3}$ of a foot, and 8 ounces is $\frac{8}{16}$ or $\frac{1}{2}$ of a pound. These fractions are numbers in the set of rational numbers.

DEFINITION

A **rational number** is a number of the form $\frac{a}{b}$ where a and b are integers and $b \neq 0$.

For every rational number $\frac{a}{b}$ that is not equal to zero, there is a **multiplicative inverse** or **reciprocal** $\frac{b}{a}$ such that $\frac{a}{b} \cdot \frac{b}{a} = 1$. Note that $\frac{a}{b} \cdot \frac{b}{a} = \frac{ab}{ab}$. If the non-zero numerator of a fraction is equal to the denominator, then the fraction is equal to 1.

EXAMPLE I

Write the multiplicative inverse of each of the following rational numbers:

Answers

a. $\frac{3}{4}$ $\frac{4}{3}$

b. $\frac{-5}{8}$ $\frac{8}{-5} = -\frac{8}{5}$

c. 5 $\frac{1}{5}$

Note that in **b**, the reciprocal of a negative number is a negative number. ▪

Decimal Values of Rational Numbers

 The rational number $\frac{a}{b}$ is equivalent to $a \div b$. When a fraction or a division such as $25 \div 100$ is entered into a calculator, the decimal value is displayed. To express the quotient as a fraction, select Frac from the **MATH** menu. This can be done in two ways.

ENTER: 25 ÷ 100 **ENTER**   ENTER: 25 ÷ 100 **MATH** **1** **ENTER**
MATH **1** **ENTER**

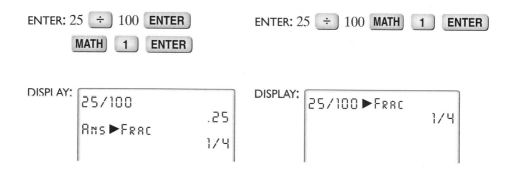

When a calculator is used to evaluate a fraction such as $\frac{8}{12}$ or $8 \div 12$, the decimal value is shown as .6666666667. The calculator has rounded the value to ten decimal places, the nearest ten-billionth. The true value of $\frac{8}{12}$, or $\frac{2}{3}$, is an infinitely repeating decimal that can be written as $0.\overline{6}$. The line over the 6 means that the digit 6 repeats infinitely. Other examples of infinitely repeating decimals are:

$$0.\overline{2} = 0.22222222\ldots = \tfrac{2}{9}$$
$$0.\overline{142857} = 0.142857142857\ldots = \tfrac{1}{7}$$
$$0.\overline{12} = 0.1212121212\ldots = \tfrac{4}{33}$$
$$0.1\overline{6} = 0.1666666666\ldots = \tfrac{1}{6}$$

Every rational number is either a finite decimal or an infinitely repeating decimal. Because a finite decimal such as 0.25 can be thought of as having an infinitely repeating 0 and can be written as $0.25\overline{0}$, the following statement is true:

▶ **A number is a rational number if and only if it can be written as an infinitely repeating decimal.**

EXAMPLE 2

Find the common fractional equivalent of $0.\overline{18}$.

Solution Let $x = 0.\overline{18} = 0.18181818\ldots$

How to Proceed

(1) Multiply the value of x by 100 to write a number in which the decimal point follows the first pair of repeating digits:

$$100x = 18.181818\ldots$$

(2) Subtract the value of x from both sides of this equation:

$$\begin{array}{r} 100x = 18.181818\ldots \\ -x = -0.181818\ldots \\ \hline 99x = 18 \end{array}$$

(3) Solve the resulting equation for x and simplify the fraction:

$$x = \frac{18}{99} = \frac{2}{11}$$

Check The solution can be checked on a calculator.

ENTER: 2 ÷ 11 **ENTER**

DISPLAY:

```
2/11
          .1818181818
```

Answer $\frac{2}{11}$

EXAMPLE 3

Express $0.12\overline{48}$ as a common fraction.

Solution: Let $x = 0.12484848\ldots$

How to Proceed

(1) Multiply the value of x by the power of 10 that makes the decimal point follow the first set of repeating digits. Since we want to move the decimal point 4 places, multiply by $10^4 = 10,000$:

$$10,000x = 1,248.4848\ldots$$

(2) Multiply the value of x by the power of 10 that makes the decimal point follow the digits that do not repeat. Since we want to move the decimal point 2 places, multiply by $10^2 = 100$:

$$100x = 12.4848\ldots$$

(3) Subtract the equation in step 2 from the equation in step 1:

$$\begin{array}{r} 10,000x = 1,248.4848\ldots \\ -100x = -12.4848\ldots \\ \hline 9,900x = 1,236 \end{array}$$

(4) Solve for x and reduce the fraction to lowest terms:

$$x = \frac{1,236}{9,900} = \frac{103}{825}$$

Answer $\frac{103}{825}$

Procedure

To convert an infinitely repeating decimal to a common fraction:

1. Write the equation: x = decimal value.

2. Multiply both sides of the equation in step 1 by 10^m, where m is the number of places to the right of the decimal point following the first set of repeating digits.

3. Multiply both sides of the equation in step 1 by 10^n, where n is the number of places to the right of the decimal point following the *non*-repeating digits. (If there are no non-repeating digits, then let $n = 0$.)

4. Subtract the equation in step 3 from the equation in step 2.

5. Solve the resulting equation for x, and simplify the fraction completely.

Exercises

Writing About Mathematics

1. a. Why is a coin that is worth 25 cents called a quarter?

b. Why is the number of minutes in a quarter of an hour different from the number of cents in a quarter of a dollar?

2. Explain the difference between the additive inverse and the multiplicative inverse.

Developing Skills

In 3–7, write the reciprocal (multiplicative inverse) of each given number.

3. $\frac{3}{8}$ **4.** $\frac{7}{12}$ **5.** $\frac{-2}{7}$ **6.** 8 **7.** 1

In 8–12, write each rational number as a repeating decimal.

8. $\frac{1}{6}$ **9.** $\frac{2}{9}$ **10.** $\frac{5}{7}$ **11.** $\frac{2}{15}$ **12.** $\frac{7}{8}$

In 13–22, write each decimal as a common fraction.

13. 0.125 **14.** $0.\overline{6}$ **15.** $0.\overline{2}$ **16.** $0.\overline{36}$ **17.** $0.\overline{108}$

18. $0.15\overline{6}$ **19.** $0.8\overline{3}$ **20.** $0.5\overline{7}$ **21.** $0.1\overline{36}$ **22.** $0.15\overline{90}$

2-2 SIMPLIFYING RATIONAL EXPRESSIONS

A rational number is the quotient of two integers. A **rational expression** is the quotient of two polynomials. Each of the following fractions is a rational expression:

$$\frac{3}{4} \qquad \frac{ab}{7} \qquad \frac{x+5}{2x} \qquad \frac{a^2-1}{4ab} \qquad \frac{y-2}{y^2-5y+6}$$

Division by 0 is not defined. Therefore, each of these rational expressions has no meaning when the denominator is zero. For instance:

- $\frac{x+5}{2x}$ has no meaning when $x = 0$.

- $\frac{a^2-1}{4ab}$ has no meaning when $a = 0$ or when $b = 0$.

- $\frac{y-2}{y^2-5y+6} = \frac{y-2}{(y-2)(y-3)}$ has no meaning when $y = 2$ or when $y = 3$.

EXAMPLE I

For what value or values of a is the fraction $\frac{2a-5}{3a^2-4a+1}$ undefined?

Solution A fraction is undefined or has no meaning when a factor of the denominator is equal to 0.

How to Proceed

(1) Factor the denominator: $\qquad 3a^2 - 4a + 1 = (3a-1)(a-1)$

(2) Set each factor equal to 0: $\qquad\qquad 3a - 1 = 0 \quad | \quad a - 1 = 0$

(3) Solve each equation for a: $\qquad\qquad\quad 3a = 1 \quad | \quad a = 1$

$$a = \tfrac{1}{3} \quad |$$

Answer The fraction is undefined when $a = \frac{1}{3}$ and when $a = 1$.

▶ If $\frac{a}{b}$ and $\frac{c}{d}$ are rational numbers with $b \neq 0$ and $d \neq 0$, then

$$\frac{a}{b} \cdot \frac{c}{d} = \frac{ac}{bd} \qquad \text{and} \qquad \frac{ac}{bd} = \frac{a}{b} \cdot \frac{c}{d} = \frac{a}{d} \cdot \frac{c}{b}$$

For example: $\frac{3}{4} \times \frac{1}{3} = \frac{3}{12}$ and $\frac{3}{12} = \frac{3 \times 1}{3 \times 4} = \frac{3}{3} \times \frac{1}{4} = 1 \times \frac{1}{4} = \frac{1}{4}$.

We can write a rational expression in simplest form by finding common factors in the numerator and denominators, as shown above.

EXAMPLE 2

Simplify: *Answers*

a. $\dfrac{2x}{6}$ $\qquad = \dfrac{2}{2} \cdot \dfrac{x}{3} = 1 \cdot \dfrac{x}{3} = \dfrac{x}{3}$

b. $\dfrac{3ab}{a(a-1)}$ $\qquad = \dfrac{a}{a} \cdot \dfrac{3b}{a-1} = 1 \cdot \dfrac{3b}{a-1} = \dfrac{3b}{a-1}\ (a \neq 0, 1)$

c. $\dfrac{y-2}{y^2 - 5y + 6}$ $\quad = \dfrac{y-2}{(y-2)(y-3)} = \dfrac{y-2}{y-2} \cdot \dfrac{1}{y-3} = 1 \cdot \dfrac{1}{y-3} = \dfrac{1}{y-3}\ (y \neq 2, 3)$

Note: We must eliminate any value of the variable or variables for which the denominator of the given rational expression is zero.

The rational expressions $\dfrac{x}{3}, \dfrac{3b}{a-1}$, and $\dfrac{1}{y-3}$ in the example shown above are in **simplest form** because there is no factor of the numerator that is also a factor of the denominator except 1 and -1. We say that the fractions have been reduced to **lowest terms**.

When the numerator or denominator of a rational expression is a monomial, each number or variable is a factor of the monomial. When the numerator or denominator of a rational expression is a polynomial with more than one term, we must factor the polynomial. Once both the numerator and denominator of the fraction are factored, we can reduce the fraction by identifying factors in the numerator that are also factors in the denominator.

In the example given above, we wrote:

$$\frac{y-2}{y^2 - 5y + 6} = \frac{y-2}{(y-2)(y-3)} = \frac{y-2}{y-2} \cdot \frac{1}{y-3} = 1 \cdot \frac{1}{y-3} = \frac{1}{y-3}\ (y \neq 2, 3)$$

We can simplify this process by *canceling* the common factor in the numerator and denominator.

$$\frac{\overset{1}{\cancel{(y-2)}}}{\underset{1}{\cancel{(y-2)}}(y-3)} = \frac{1}{y-3}\ (y \neq 2, 3)$$

Note that canceling $(y-2)$ in the numerator and denominator of the fraction given above is the equivalent of dividing $(y-2)$ by $(y-2)$. When any number or algebraic expression that is not equal to 0 is divided by itself, the quotient is 1.

Procedure

To reduce a fraction to lowest terms:

METHOD 1

1. Factor completely both the numerator and the denominator.

2. Determine the greatest common factor of the numerator and the denominator.

3. Express the given fraction as the product of two fractions, one of which has as its numerator and its denominator the greatest common factor determined in step 2.

4. Write the fraction whose numerator and denominator are the greatest common factor as 1 and use the multiplication property of 1.

METHOD 2

1. Factor both the numerator and the denominator.

2. Divide both the numerator and the denominator by their greatest common factor by canceling the common factor.

EXAMPLE 3

Write $\dfrac{3x - 12}{3x}$ in lowest terms.

Solution

METHOD 1

$$\frac{3x - 12}{3x} = \frac{3(x - 4)}{3x}$$

$$= \frac{3}{3} \cdot \frac{x - 4}{x}$$

$$= 1 \cdot \frac{x - 4}{x}$$

$$= \frac{x - 4}{x}$$

METHOD 2

$$\frac{3x - 12}{3x} = \frac{\overset{1}{3}(x - 4)}{\underset{1}{3x}}$$

$$= \frac{x - 4}{x}$$

Answer $\dfrac{x - 4}{x}$ $(x \neq 0)$

Factors That Are Opposites

The binomials $(a - 2)$ and $(2 - a)$ are opposites or additive inverses. If we change the order of the terms in the binomial $(2 - a)$, we can write:

$$(2 - a) = (-a + 2) = -1(a - 2)$$

We can use this factored form of $(2 - a)$ to reduce the rational expression $\frac{2 - a}{a^2 - 4}$ to lowest terms.

$$\frac{2 - a}{a^2 - 4} = \frac{-1(a - 2)}{(a + 2)(a - 2)} = \frac{-1\overset{1}{\cancel{(a - 2)}}}{(a + 2)\cancel{(a - 2)}} = \frac{-1}{(a + 2)} = -\frac{1}{a + 2} \ (a \neq 2, -2)$$

EXAMPLE 4

Simplify the expression: $\frac{8 - 2x}{x^2 - 8x + 16}$

Solution

METHOD 1

$$\frac{8 - 2x}{x^2 - 8x + 16} = \frac{2(4 - x)}{(x - 4)(x - 4)}$$

$$= \frac{2(-1)(x - 4)}{(x - 4)(x - 4)}$$

$$= \frac{(x - 4)}{(x - 4)} \cdot \frac{-2}{(x - 4)}$$

$$= \frac{-2}{x - 4}$$

METHOD 2

$$\frac{8 - 2x}{x^2 - 8x + 16} = \frac{2(4 - x)}{(x - 4)(x - 4)}$$

$$= \frac{2(-1)(x - 4)}{(x - 4)(x - 4)}$$

$$= \frac{-2\overset{1}{\cancel{(x - 4)}}}{(x - 4)\underset{1}{\cancel{(x - 4)}}}$$

$$= \frac{-2}{x - 4}$$

Answer $\frac{-2}{x - 4} \ (x \neq 4)$

Exercises

Writing About Mathematics

1. Abby said that $\frac{3x}{3x + 4}$ can be reduced to lowest terms by canceling $3x$ so that the result is $\frac{1}{4}$. Do you agree with Abby? Explain why or why not.

2. Does $\frac{2a - 3}{2a - 3} = 1$ for all real values of a? Justify your answer.

Developing Skills

In 3–10, list the values of the variables for which the rational expression is undefined.

3. $\frac{5a^2}{3a}$

4. $\frac{-2d}{6c}$

5. $\frac{a + 2}{ab}$

6. $\frac{x - 5}{x + 5}$

7. $\frac{2a}{2a - 7}$

8. $\frac{b + 3}{b^2 + b - 6}$

9. $\frac{4c}{2c^2 - 2c}$

10. $\frac{5}{x^3 - 5x^2 - 6x}$

In 11–30, write each rational expression in simplest form and list the values of the variables for which the fraction is undefined.

11. $\frac{6}{10}$

12. $\frac{5a^2b}{10a}$

13. $\frac{12xy^2}{3x^2y}$

14. $\frac{14b^4}{21b^3}$

15. $\frac{9cd^2}{12c^4d^2}$

16. $\frac{8a + 16}{12a}$

17. $\frac{9y^2 + 3y}{6y^2}$

18. $\frac{8ab - 4b^2}{6ab}$

19. $\frac{10d}{15d - 20d^2}$

20. $\frac{8c^2}{8c^2 + 16c}$

21. $\frac{3xy}{9xy + 6x^2y^3}$

22. $\frac{2a + 10}{3a + 15}$

23. $\frac{4a^2 - 16}{4a + 8}$

24. $\frac{x^2 - 7x + 12}{x^2 + 2x - 15}$

25. $\frac{5y^2 - 20}{y^2 + 4y + 4}$

26. $\frac{-7 + 7a}{21a^2 - 21}$

27. $\frac{a^3 - a^2 - a + 1}{a^2 - 2a + 1}$

28. $\frac{3 - (b + 1)}{4 - b^2}$

29. $\frac{4 - 2(x - 1)}{x^2 - 6x + 9}$

30. $\frac{5(1 - b) + 15}{b^2 - 16}$

2-3 MULTIPLYING AND DIVIDING RATIONAL EXPRESSIONS

Multiplying Rational Expressions

We know that $\frac{2}{3} \times \frac{4}{5} = \frac{8}{15}$ and that $\frac{3}{4} \times \frac{3}{2} = \frac{9}{8}$. In general, the product of two rational numbers $\frac{a}{b}$ and $\frac{c}{d}$ is $\frac{a}{b} \cdot \frac{c}{d} = \frac{ac}{bd}$ for $b \neq 0$ and $d \neq 0$.

This same rule holds for the product of two rational expressions:

▶ **The product of two rational expressions is a fraction whose numerator is the product of the given numerators and whose denominator is the product of the given denominators.**

For example:

$$\frac{3(a + 5)}{4a} \cdot \frac{12}{a} = \frac{36(a + 5)}{4a^2} \quad (a \neq 0)$$

This product can be reduced to lowest terms by dividing numerator and denominator by the common factor, 4.

$$\frac{3(a + 5)}{4a} \cdot \frac{12}{a} = \frac{36(a + 5)}{4a^2} = \frac{\overset{9}{\cancel{36}}(a + 5)}{\underset{1}{\cancel{4}}a^2} = \frac{9(a + 5)}{a^2}$$

We could have canceled the factor 4 before we multiplied, as shown below.

$$\frac{3(a + 5)}{4a} \times \frac{12}{a} = \frac{3(a + 5)}{\underset{1}{\cancel{4}}a} \times \frac{\overset{3}{\cancel{12}}}{a} = \frac{9(a + 5)}{a^2}$$

Note that a is not a common factor of the numerator and denominator because it is *one term* of the factor $(a + 5)$, not a factor of the numerator.

> ### Procedure
>
> **To multiply fractions:**
>
> **METHOD 1**
>
> 1. Multiply the numerators of the given fractions and the denominators of the given fractions.
>
> 2. Reduce the resulting fraction, if possible, to lowest terms.
>
> **METHOD 2**
>
> 1. Factor any polynomial that is not a monomial.
>
> 2. Cancel any factors that are common to a numerator and a denominator.
>
> 3. Multiply the resulting numerators and the resulting denominators to write the product in lowest terms.

EXAMPLE 1

    a. Find the product of $\dfrac{12b^2}{5b + 15} \cdot \dfrac{b^2 - 9}{3b^2 + 9b}$ in simplest form.

    b. For what values of the variable are the given fractions and the product undefined?

Solution **a. METHOD 1**

      How to Proceed

(1) Multiply the numerators of the fractions and the denominators of the fractions:

$$\frac{12b^2}{5b + 15} \cdot \frac{b^2 - 9}{3b^2 + 9b} = \frac{12b^2(b^2 - 9)}{(5b + 15)(3b^2 + 9b)}$$

(2) Factor the numerator and the denominator. Note that the factors of $(5b + 15)$ are $5(b + 3)$ and the factors of $(3b^2 + 9b)$ are $3b(b + 3)$. Reduce the resulting fraction to lowest terms:

$$= \frac{12b^2(b + 3)(b - 3)}{15b(b + 3)(b + 3)}$$

$$= \frac{3b(b + 3)}{3b(b + 3)} \cdot \frac{4b(b - 3)}{5(b + 3)}$$

$$= 1 \cdot \frac{4b(b - 3)}{5(b + 3)}$$

$$= \frac{4b(b - 3)}{5(b + 3)} \quad Answer$$

METHOD 2

How to Proceed

(1) Factor each binomial term completely:

$$\frac{12b^2}{5b + 15} \cdot \frac{b^2 - 9}{3b^2 + 9b} = \frac{12b^2}{5(b + 3)} \cdot \frac{(b + 3)(b - 3)}{3b(b + 3)}$$

(2) Cancel any factors that are common to a numerator and a denominator:

$$= \frac{\overset{4}{\cancel{12b^2}}}{5(b + 3)} \cdot \frac{\overset{1}{\cancel{(b + 3)}}(b - 3)}{\underset{1\ 1}{3\cancel{b}(b + 3)}\underset{1}{}}$$

(3) Multiply the remaining factors:

$$= \frac{4b(b - 3)}{5(b + 3)} \quad \textit{Answer}$$

b. The given fractions and their product are undefined when $b = 0$ and when $b = -3$. *Answer*

Dividing Rational Expressions

We can divide two rational numbers or two rational expressions by changing the division to a related multiplication. Let $\frac{a}{b}$ and $\frac{c}{d}$ be two rational numbers or rational expressions with $b \neq 0, c \neq 0, d \neq 0$. Since $\frac{c}{d}$ is a non-zero rational number or rational expression, there exists a multiplicative inverse $\frac{d}{c}$ such that $\frac{c}{d} \cdot \frac{d}{c} = 1$.

$$\frac{a}{b} \div \frac{c}{d} = \frac{\frac{a}{b}}{\frac{c}{d}} = \frac{\frac{a}{b} \cdot \frac{d}{c}}{\frac{c}{d} \cdot \frac{d}{c}} = \frac{\frac{a}{b} \cdot \frac{d}{c}}{1} = \frac{a}{b} \cdot \frac{d}{c}$$

We have just derived the following procedure.

> **Procedure**
>
> **To divide two rational numbers or rational expressions, multiply the dividend by the reciprocal of the divisor.**

For example:

$$\frac{x + 5}{2x^2} \div \frac{5}{x + 1} = \frac{x + 5}{2x^2} \cdot \frac{x + 1}{5}$$

$$= \frac{(x + 5)(x + 1)}{10x^2} \quad (x \neq -1, 0)$$

This product is in simplest form because the numerator and denominator have no common factors.

Recall that a can be written as $\frac{a}{1}$ and therefore, if $a \neq 0$, the reciprocal of a is $\frac{1}{a}$. For example:

$$\frac{2b - 2}{5} \div (b - 1) = \frac{2(b - 1)}{5} \cdot \frac{1}{b - 1} = \frac{2\overset{1}{\cancel{(b - 1)}}}{5} \cdot \frac{1}{\underset{1}{\cancel{b - 1}}} = \frac{2}{5} \ (b \neq 1)$$

EXAMPLE 2

Divide and simplify: $\dfrac{a^2 - 3a - 10}{5a} \div \dfrac{a^2 - 5a}{15}$

Solution

How to Proceed

(1) Use the reciprocal of the divisor to write the division as a multiplication:

$$\dfrac{a^2 - 3a - 10}{5a} \div \dfrac{a^2 - 5a}{15} = \dfrac{a^2 - 3a - 10}{5a} \cdot \dfrac{15}{a^2 - 5a}$$

(2) Factor each polynomial:

$$= \dfrac{(a - 5)(a + 2)}{5a} \cdot \dfrac{15}{a(a - 5)}$$

(3) Cancel any factors that are common to the numerator and denominator:

$$= \dfrac{\overset{1}{\cancel{(a - 5)}}(a + 2)}{5a} \cdot \dfrac{\overset{3}{\cancel{15}}}{a\overset{}{\cancel{(a - 5)}}}$$

(4) Multiply the remaining factors:

$$= \dfrac{3(a + 2)}{a^2}$$

Answer $\dfrac{3(a + 2)}{a^2}$ $(a \neq 0, 5)$

EXAMPLE 3

Perform the indicated operations and write the answer in simplest form:

$$\dfrac{3a}{a + 3} \div \dfrac{5a}{2a + 6} \cdot \dfrac{3}{a}$$

Solution Recall that multiplications and divisions are performed in the order in which they occur from left to right.

$$\dfrac{3a}{a + 3} \div \dfrac{5a}{2a + 6} \cdot \dfrac{3}{a} = \dfrac{3a}{a + 3} \cdot \dfrac{2(a + 3)}{5a} \cdot \dfrac{a}{3}$$

$$= \dfrac{3a}{a + 3} \cdot \dfrac{2(a + 3)}{5a} \cdot \dfrac{a}{3}$$

$$= \dfrac{6}{5} \cdot \dfrac{a}{3}$$

$$= \dfrac{2a}{5} \ (a \neq -3, 0) \ \textit{Answer}$$

Exercises

Writing About Mathematics

1. Joshua wanted to write this division in simplest form: $\frac{3}{x-2} \div \frac{4(x-2)}{7}$. He began by cancel-ing $(x-2)$ in the numerator and denominator and wrote following:

$$\frac{3}{\cancel{x-2}_{1}} \div \frac{4(\cancel{x-2}^{1})}{7} = \frac{3}{1} \div \frac{4}{7} = \frac{3}{1} \times \frac{7}{4} = \frac{21}{4}$$

Is Joshua's answer correct? Justify your answer.

2. Gabriel wrote $\frac{12x}{5x+10} \div \frac{4}{5} = \frac{12x \div 4}{(5x+10) \div 5} = \frac{3x}{x+2}$. Is Gabriel's solution correct? Justify your answer.

Developing Skills

In 3–12, multiply and express each product in simplest form. In each case, list any values of the variables for which the fractions are not defined.

3. $\frac{2}{3} \times \frac{3}{4}$

4. $\frac{5}{7a} \cdot \frac{3a}{20}$

5. $\frac{4y}{5x} \cdot \frac{x}{8y}$

6. $\frac{3a}{5} \cdot \frac{10}{9a}$

7. $\frac{b+1}{4} \cdot \frac{12}{5b+5}$

8. $\frac{a^2-100}{3a} \cdot \frac{a^2}{2a-20}$

9. $\frac{7y+21}{7y} \cdot \frac{3}{y^2-9}$

10. $\frac{a^2-5a+4}{3a+6} \cdot \frac{2a+4}{a^2-16}$

11. $\frac{2a+4}{6a} \cdot \frac{3a^2}{a^2+2a}$

12. $\frac{6-2x}{x^2-9} \cdot \frac{15+5x}{4x}$

In 13–24, divide and express each quotient in simplest form. In each case, list any values of the variables for which the fractions are not defined.

13. $\frac{3}{4} \div \frac{9}{20}$

14. $\frac{12}{a} \div \frac{6}{4a}$

15. $\frac{6b}{5c} \div \frac{3b}{10c}$

16. $\frac{a^2}{8a} \div \frac{3a}{4}$

17. $\frac{x-2}{3x} \div \frac{4x-8}{9}$

18. $\frac{6y^2-3y}{3y} \div \frac{4y^2-1}{2}$

19. $\frac{c^2-6c+9}{5c-15} \div \frac{c-3}{5}$

20. $\frac{w^2-w}{5w} \div \frac{w^2-1}{5}$

21. $\frac{4b+12}{b} \div (b+3)$

22. $\frac{a^2+8a+15}{4a} \div (a+3)$

23. $(2x+7) \div \frac{1}{2x^2+5x-7}$

24. $(a^2-1) \div \frac{2a+2}{a}$

In 25–30, perform the indicated operations and write the result in simplest form. In each case, list any values of the variables for which the fractions are not defined.

25. $\frac{3}{5} \times \frac{5}{9} \div \frac{4}{3}$

26. $\frac{3x}{x-1} \cdot \frac{x^2-1}{x} \div \frac{x+1}{3}$

27. $\frac{2a}{a+2} \cdot \frac{a^2-4}{4a^2} \div \frac{a-2}{a}$

28. $(x^2 - 2x + 1) \div \frac{x-1}{3} \cdot \frac{x+4}{3x}$

29. $(3b)^2 \div \frac{3b}{b+2} \cdot \frac{2b+4}{b}$

30. $\frac{x^2-3x+2}{4x} \cdot \frac{12x^2}{x^2-2x} \div \frac{x-1}{x}$

2-4 ADDING AND SUBTRACTING RATIONAL EXPRESSIONS

We know that $\frac{3}{7} = 3\left(\frac{1}{7}\right)$ and that $\frac{2}{7} = 2\left(\frac{1}{7}\right)$. Therefore:

$$\frac{3}{7} + \frac{2}{7} = 3\left(\frac{1}{7}\right) + 2\left(\frac{1}{7}\right)$$
$$= (3 + 2)\left(\frac{1}{7}\right)$$
$$= 5\left(\frac{1}{7}\right)$$
$$= \frac{5}{7}$$

We can also write:

$$\frac{x+2}{x} + \frac{2x-5}{x} = (x+2)\left(\frac{1}{x}\right) + (2x-5)\left(\frac{1}{x}\right)$$
$$= (x + 2 + 2x - 5)\left(\frac{1}{x}\right)$$
$$= \frac{3x-3}{x} \ (x \neq 0)$$

In general, the sum of any two fractions with a common denominator is a fraction whose numerator is the sum of the numerators and whose denominator is the common denominator.

$$\frac{3a}{a+1} + \frac{a}{a+1} = \frac{4a}{a+1} \ (a \neq -1)$$

In order to add two fractions that do *not* have a common denominator, we need to change one or both of the fractions to equivalent fractions that *do* have a common denominator. That common denominator can be the product of the given denominators.

For example, to add $\frac{1}{5} + \frac{1}{7}$, we change each fraction to an equivalent fraction whose denominator is 35 by multiplying each fraction by a fraction equal to 1.

$$\frac{1}{5} + \frac{1}{7} = \frac{1}{5} \times \frac{7}{7} + \frac{1}{7} \times \frac{5}{5}$$
$$= \frac{7}{35} + \frac{5}{35}$$
$$= \frac{12}{35}$$

To add the fractions $\frac{5}{2x}$ and $\frac{3}{y}$, we need to find a denominator that is a multiple of both $2x$ and of y. One possibility is their product, $2xy$. Multiply each fraction by a fraction equal to 1 so that the denominator of each fraction will be $2xy$:

$$\frac{5}{2x} = \frac{5}{2x} \cdot \frac{y}{y} = \frac{5y}{2xy} \quad \text{and} \quad \frac{3}{y} = \frac{3}{y} \cdot \frac{2x}{2x} = \frac{6x}{2xy}$$

$$\frac{5}{2x} + \frac{3}{y} = \frac{5y}{2xy} + \frac{6x}{2xy} = \frac{5y + 6x}{2xy} \quad (x \neq 0, y \neq 0)$$

The **least common denominator (LCD)** is often smaller than the product of the two denominators. It is the **least common multiple (LCM)** of the denominators, that is, the product of all factors of one or both of the denominators.

For example, to add $\frac{1}{2a + 2} + \frac{1}{a^2 - 1}$, first find the factors of each denominator. The least common denominator is the product of all of the factors of the first denominator times all factors of the second that are not factors of the first. Then multiply each fraction by a fraction equal to 1 so that the denominator of each fraction will be equal to the LCD.

Factors of $2a + 2$: $2 \cdot (a + 1)$
Factors of $a^2 - 1$: $(a + 1) \cdot (a - 1)$
LCD: $2 \cdot (a + 1) \cdot (a - 1)$

$$\frac{1}{2a + 2} = \frac{1}{2(a + 1)} \cdot \frac{a - 1}{a - 1} \quad \text{and} \quad \frac{1}{a^2 - 1} = \frac{1}{(a + 1)(a - 1)} \cdot \frac{2}{2}$$

$$= \frac{a - 1}{2(a + 1)(a - 1)} \qquad\qquad = \frac{2}{2(a + 1)(a - 1)}$$

$$\frac{1}{2a + 2} + \frac{1}{a^2 - 1} = \frac{a - 1 + 2}{2(a + 1)(a - 1)} = \frac{a + 1}{2(a + 1)(a - 1)}$$

Since this sum has a common factor in the numerator and denominator, it can be reduced to lowest terms.

$$\frac{a + 1}{2(a + 1)(a - 1)} = \frac{\overset{1}{\cancel{a + 1}}}{2\underset{1}{\cancel{(a + 1)}}(a - 1)} = \frac{1}{2(a - 1)} \quad (a \neq -1, a \neq 1)$$

Any polynomial can be written as a rational expression with a denominator of 1. To add a polynomial to a rational expression, write the polynomial as an equivalent rational expression.

For example, to write the sum $b + 3 + \frac{1}{2b}$ as a single fraction, multiply $(b + 3)$ by 1 in the form $\frac{2b}{2b}$.

$$(b + 3) + \frac{1}{2b} = \frac{b + 3}{1}\left(\frac{2b}{2b}\right) + \frac{1}{2b}$$

$$= \frac{2b^2 + 6b}{2b} + \frac{1}{2b}$$

$$= \frac{2b^2 + 6b + 1}{2b} \quad (b \neq 0)$$

EXAMPLE 1

Write the difference $\dfrac{x}{x^2 - 4x + 3} - \dfrac{x}{x^2 + 2x - 3}$ as a single fraction in lowest terms.

Solution   *How to Proceed*

(1) Find the LCD of the fractions:

$$x^2 - 4x + 3 = (x - 3) \cdot (x - 1)$$
$$x^2 + 2x - 3 = \qquad (x - 1) \cdot (x + 3)$$
$$\text{LCD} = (x - 3) \cdot (x - 1) \cdot (x + 3)$$

(2) Write each fraction as an equivalent fraction with a denominator equal to the LCD:

$$\dfrac{x}{x^2 - 4x + 3} = \dfrac{x}{(x - 3)(x - 1)} \cdot \dfrac{x + 3}{x + 3}$$
$$= \dfrac{x^2 + 3x}{(x - 3)(x - 1)(x + 3)}$$

$$\dfrac{x - 3}{x^2 + 2x - 3} = \dfrac{x - 3}{(x + 3)(x - 1)} \cdot \dfrac{x - 3}{x - 3}$$
$$= \dfrac{x^2 - 6x + 9}{(x - 3)(x - 1)(x + 3)}$$

(3) Subtract:

$$\dfrac{x}{x^2 - 4x + 3} - \dfrac{x - 3}{x^2 + 2x - 3}$$
$$= \dfrac{x^2 + 3x - (x^2 - 6x + 9)}{(x - 3)(x - 1)(x + 3)}$$

(4) Simplify:

$$= \dfrac{9x - 9}{(x - 3)(x - 1)(x + 3)}$$

(5) Reduce to lowest terms:

$$= \dfrac{9}{(x - 3)(x + 3)}$$

Answer $\dfrac{9}{(x - 3)(x + 3)}$ $(x \neq -3, 1, 3)$

EXAMPLE 2

Simplify: $\left(x - \dfrac{1}{x}\right)\left(1 + \dfrac{1}{x - 1}\right)$

Solution **STEP 1.** Rewrite each expression in parentheses as a single fraction.

$$x - \dfrac{1}{x} = x\left(\dfrac{x}{x}\right) - \dfrac{1}{x} \qquad \text{and} \qquad 1 + \dfrac{1}{x - 1} = 1\left(\dfrac{x - 1}{x - 1}\right) + \dfrac{1}{x - 1}$$
$$= \dfrac{x^2}{x} - \dfrac{1}{x} \qquad\qquad\qquad\qquad\qquad = \dfrac{x - 1}{x - 1} + \dfrac{1}{x - 1}$$
$$= \dfrac{x^2 - 1}{x} \qquad\qquad\qquad\qquad\qquad\quad = \dfrac{x - 1 + 1}{x - 1}$$
$$\qquad\qquad\qquad\qquad\qquad\qquad\qquad\qquad = \dfrac{x}{x - 1}$$

STEP 2. Multiply.

$$\left(\dfrac{x^2 - 1}{x}\right)\left(\dfrac{x}{x - 1}\right) = \left(\dfrac{(x + 1)(x - 1)}{x}\right)\left(\dfrac{x}{x - 1}\right)$$
$$= \left(\dfrac{(x + 1)\overset{1}{\cancel{(x - 1)}}}{\cancel{x}}\right)\left(\dfrac{\overset{1}{\cancel{x}}}{\underset{1}{\cancel{x - 1}}}\right)$$
$$= \dfrac{x + 1}{1}$$
$$= x + 1$$

Alternative   **STEP 1.** Multiply using the distributive property.
Solution

$$\left(x - \tfrac{1}{x}\right)\left(1 + \tfrac{1}{x-1}\right) = x\left(1 + \tfrac{1}{x-1}\right) - \tfrac{1}{x}\left(1 + \tfrac{1}{x-1}\right)$$

$$= x + \frac{x}{x-1} - \frac{1}{x} - \frac{1}{x(x-1)}$$

STEP 2. Add the fractions. The least common denominator is $x(x-1)$.

$$x + \frac{x}{x-1} - \frac{1}{x} - \frac{1}{x(x-1)} = x\left(\frac{x(x-1)}{x(x-1)}\right) + \frac{x}{(x-1)}\left(\frac{x}{x}\right) - \frac{1}{x}\left(\frac{x-1}{x-1}\right) - \frac{1}{x(x-1)}$$

$$= \frac{x^3 - x^2}{x(x-1)} + \frac{x^2}{x(x-1)} - \frac{x-1}{x(x-1)} - \frac{1}{x(x-1)}$$

$$= \frac{x^3 - x^2 + x^2 - x + 1 - 1}{x(x-1)}$$

$$= \frac{x^3 - x}{x(x-1)}$$

STEP 3. Simplify.

$$\frac{x^3 - x}{x(x-1)} = \frac{\overset{1}{\cancel{x}}(\cancel{x-1})(x+1)}{\underset{1}{\cancel{x}}(\underset{1}{\cancel{x-1}})} = x + 1$$

Answer $x + 1$ $(x \neq 0, 1)$

Exercises

Writing About Mathematics

1. Ashley said that $\frac{(a+2)(a-1)}{(a+3)(a-1)} = \frac{a+2}{a+3}$ for all values of a except $a = -3$. Do you agree with Ashley? Explain why or why not.

2. Matthew said that $\frac{a}{b} + \frac{c}{d} = \frac{ad + bc}{bd}$ when $b \neq 0, d \neq 0$. Do you agree with Matthew? Justify your answer.

Developing Skills

In 3–20, perform the indicated additions or subtractions and write the result in simplest form. In each case, list any values of the variables for which the fractions are not defined.

3. $\frac{x}{3} + \frac{2x}{3}$

4. $\frac{2x^2 + 1}{5x} - \frac{7x^2 - 1}{5x}$

5. $\frac{x}{7} + \frac{x}{3}$

6. $\frac{a-1}{5} - \frac{a+1}{4}$

7. $\frac{y+2}{2} + \frac{2y-3}{3}$

8. $\frac{a+5}{5a} - \frac{a-8}{8a}$

9. $\frac{2x+3}{6x} - \frac{x-2}{4x}$

10. $\frac{a+1}{3a} + \frac{3}{2}$

11. $3 + \frac{2}{x}$

12. $5 - \frac{1}{2y}$

13. $a - \frac{3}{2a}$

14. $\frac{1}{x} + 1$

15. $\frac{3}{x+2} + \frac{x-2}{x}$

16. $\frac{b}{b-1} - \frac{1}{2-2b}$

17. $\frac{1}{2-x} + \frac{2}{x-2}$

18. $\frac{1}{a^2 - a - 6} - \frac{1}{2a^2 - 7a + 3}$

19. $\frac{2}{a^2 - 4} - \frac{1}{a^2 + 2a}$

20. $\frac{1}{x} + \frac{1}{x-2} - \frac{2}{x^2 - 2x}$

Applying Skills

In 21–24, the length and width of a rectangle are expressed in terms of a variable.

a. Express each perimeter in terms of the variable.

b. Express each area in terms of the variable.

21. $l = 2x$ and $w = \frac{1}{x}$

22. $l = 3x + 3$ and $w = \frac{1}{3}$

23. $l = \frac{x}{x-1}$ and $w = \frac{3}{x-1}$

24. $l = \frac{x}{x+1}$ and $w = \frac{x}{x+2}$

2-5 RATIO AND PROPORTION

We often want to compare two quantities that use the same unit. For example, in a given class of 25 students, there are 11 students who are boys. We can say that $\frac{11}{25}$ of the students are boys or that the ratio of students who are boys to all students in the class is 11 : 25.

DEFINITION

A **ratio** is the comparison of two numbers by division. The ratio of a to b can be written as $\frac{a}{b}$ or as $a : b$ when $b \neq 0$.

A ratio, like a fraction, can be simplified by dividing each term by the same non-zero number. A ratio is in **simplest form** when the terms of the ratio are integers that have no common factor other than 1.

For example, to write the ratio of 3 inches to 1 foot, we must first write each measure in terms of the same unit and then divide each term of the ratio by a common factor.

$$\frac{3 \text{ inches}}{1 \text{ foot}} = \frac{3 \text{ inches}}{1 \text{ foot}} \times \frac{1 \text{ foot}}{12 \text{ inches}} = \frac{3}{12} = \frac{1}{4}$$

In lowest terms, the ratio of 3 inches to 1 foot is 1 : 4.

An equivalent ratio can also be written by multiplying each term of the ratio by the same non-zero number. For example, $4 : 7 = 4(2) : 7(2) = 8 : 14$.

In general, for $x \neq 0$:

$$a : b = ax : bx$$

EXAMPLE 1

The length of a rectangle is 1 yard and the width is 2 feet. What is the ratio of length to width of this rectangle?

Solution   The ratio must be in terms of the same measure.

$$\frac{1 \text{ yd}}{2 \text{ ft}} \times \frac{3 \text{ ft}}{1 \text{ yd}} = \frac{3}{2}$$

Answer   The ratio of length to width is $3 : 2$.

EXAMPLE 2

The ratio of the length of one of the congruent sides of an isosceles triangle to the length of the base is $5 : 2$. If the perimeter of the triangle is 42.0 centimeters, what is the length of each side?

Solution   Let AB and BC be the lengths of the congruent sides of isosceles $\triangle ABC$ and AC be the length of the base.

$AB : AC = 5 : 2 = 5x : 2x$

Therefore, $AB = 5x$,

$BC = 5x$,

and $AC = 2x$.

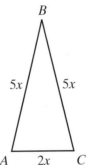

$AB + BC + AC = $ Perimeter	$AB = BC = 5(3.5)$	$AC = 2(3.5)$
$5x + 5x + 2x = 42$	$= 17.5$ cm	$= 7.0$ cm
$12x = 42$		
$x = 3.5$ cm		

Check $AB + BC + AC = 17.5 + 17.5 + 7.0 = 42.0$ cm ✔

Answer   The sides measure 17.5, 17.5, and 7.0 centimeters.

Proportion

An equation that states that two ratios are equal is called a **proportion**. For example, $3 : 12 = 1 : 4$ is a proportion. This proportion can also be written as $\frac{3}{12} = \frac{1}{4}$.

In general, if $b \neq 0$ and $d \neq 0$, then $a : b = c : d$ or $\frac{a}{b} = \frac{c}{d}$ are proportions. The first and last terms, a and d, are called the **extremes** of the proportion and the second and third terms, b and c, are the **means** of the proportion.

If we multiply both sides of the proportion by the product of the second and last terms, we can prove a useful relationship among the terms of a proportion.

$$\frac{a}{b} = \frac{c}{d}$$

$$bd\left(\frac{a}{b}\right) = bd\left(\frac{c}{d}\right)$$

$$\overset{1}{b}d\left(\frac{a}{\underset{1}{b}}\right) = b\overset{1}{d}\left(\frac{c}{\underset{1}{d}}\right)$$

$$da = bc$$

▶ **In any proportion, the product of the means is equal to the product of the extremes.**

EXAMPLE 3

In the junior class, there are 24 more girls than boys. The ratio of girls to boys is $5 : 4$. How many girls and how many boys are there in the junior class?

Solution   *How to Proceed*

(1) Use the fact that the number of girls is 24 more than the number of boys to represent the number of girls and of boys in terms of x:

Let x = the number of boys,

$x + 24$ = the number of girls.

(2) Write a proportion. Set the ratio of the number of boys to the number of girls, in terms of x, equal to the given ratio:

$$\frac{x + 24}{x} = \frac{5}{4}$$

(3) Use the fact that the product of the means is equal to the product of the extremes. Solve the equation:

$$5x = 4(x + 24)$$
$$5x = 4x + 96$$
$$x = 96$$
$$x + 24 = 96 + 24$$
$$= 120$$

Alternative	(1) Use the given ratio to represent the number of boys and the number of girls in terms of x:	Let $5x =$ the number of girls
Solution		$4x =$ the number of boys
	(2) Use the fact that the number of girls is 24 more than the number of boys to write an equation. Solve the equation for x:	$5x = 24 + 4x$
		$x = 24$
	(3) Use the value of x to find the number of girls and the number of boys:	$5x = 5(24) = 120$ girls
		$4x = 4(24) = 96$ boys

Answer There are 120 girls and 96 boys in the junior class.

Exercises

Writing About Mathematics

1. If $\frac{a}{b} = \frac{c}{d}$, then is $\frac{a}{c} = \frac{b}{d}$? Justify your answer.

2. If $\frac{a}{b} = \frac{c}{d}$, then is $\frac{a + b}{b} = \frac{c + d}{d}$? Justify your answer.

Developing Skills

In 3–10, write each ratio in simplest form.

3. $12 : 8$ **4.** $21 : 14$ **5.** $3 : 18$ **6.** $15 : 75$

7. $6a : 9a, a \neq 0$ **8.** $\frac{18}{27}$ **9.** $\frac{24}{72}$ **10.** $\frac{10x}{35x}, x \neq 0$

In 11–19, solve each proportion for the variable.

11. $\frac{x}{8} = \frac{6}{24}$ **12.** $\frac{x - 1}{15} = \frac{2}{5}$ **13.** $\frac{a - 2}{2a} = \frac{5}{14}$

14. $\frac{y + 3}{y + 8} = \frac{6}{15}$ **15.** $\frac{3x + 3}{16} = \frac{2x + 1}{10}$ **16.** $\frac{2}{x - 1} = \frac{x + 2}{2}$

17. $\frac{4x - 8}{3} = \frac{8}{x - 3}$ **18.** $\frac{x - 2}{x} = \frac{3}{x + 2}$ **19.** $\frac{x}{5} = \frac{x + 4}{x + 13}$

Applying Skills

20. The ratio of the length to the width of a rectangle is $5 : 4$. The perimeter of the rectangle is 72 inches. What are the dimensions of the rectangle?

21. The ratio of the length to the width of a rectangle is $7 : 3$. The area of the rectangle is 336 square centimeters. What are the dimensions of the rectangle?

22. The basketball team has played 21 games. The ratio of wins to losses is $5 : 2$. How many games has the team won?

23. In the chess club, the ratio of boys to girls is 6 : 5. There are 3 more boys than girls in the club. How many members are in the club?

24. Every year, Javier makes a total contribution of $125 to two local charities. The two donations are in the ratio of 3 : 2. What contribution does Javier make to each charity?

25. A cookie recipe uses flour and sugar in the ratio of 9 : 4. If Nicholas uses 1 cup of sugar, how much flour should he use?

26. The directions on a bottle of cleaning solution suggest that the solution be diluted with water. The ratio of solution to water is 1 : 7. How many cups of solution and how many cups of water should Christopher use to make 2 gallons (32 cups) of the mixture?

2-6 COMPLEX RATIONAL EXPRESSIONS

A **complex fraction** is a fraction whose numerator, denominator, or both contain fractions. Some examples of complex fractions are:

$$\frac{\frac{1}{3}}{2} \qquad \frac{5}{\frac{2}{7}} \qquad \frac{2\frac{3}{4}}{\frac{1}{8}}$$

A complex fraction can be simplified by multiplying by a fraction equal to 1; that is, by a fraction whose non-zero numerator and denominator are equal. The numerator and denominator of this fraction should be a common multiple of the denominators of the fractional terms. For example:

$$\frac{\frac{1}{3}}{2} = \frac{\frac{1}{3}}{2} \times \frac{3}{3} = \frac{1}{6} \qquad \frac{5}{\frac{2}{7}} = \frac{5}{\frac{2}{7}} \times \frac{7}{7} = \frac{35}{2} \qquad \frac{2\frac{3}{4}}{\frac{1}{8}} = \frac{2 + \frac{3}{4}}{\frac{1}{8}} \times \frac{8}{8} = \frac{16 + 6}{1} = 22$$

A complex fraction can also be simplified by dividing the numerator by the denominator.

$$\frac{\frac{1}{3}}{2} = \frac{1}{3} \div \frac{2}{1} = \frac{1}{3} \times \frac{1}{2} = \frac{1}{6} \qquad \frac{5}{\frac{2}{7}} = \frac{5}{1} \div \frac{2}{7} = 5 \times \frac{7}{2} = \frac{35}{2}$$

$$\frac{2\frac{3}{4}}{\frac{1}{8}} = \frac{\frac{8}{4} + \frac{3}{4}}{\frac{1}{8}} = \frac{\frac{11}{4}}{\frac{1}{8}} = \frac{11}{4} \div \frac{1}{8} = \frac{11}{4} \times \frac{8}{1} = \frac{88}{4} = 22$$

A **complex rational expression** has a rational expression in the numerator, the denominator, or both. For example, the following are complex rational expressions.

$$\frac{\frac{1}{a}}{a} \qquad \frac{x + 3}{\frac{1}{x}} \qquad \frac{1 + \frac{1}{b} - \frac{2}{b^2}}{\frac{1}{b}}$$

Like complex fractions, complex rational expressions can be simplified by multiplying numerator and denominator by a common multiple of the denominators in the fractional terms.

$$\frac{\frac{1}{a}}{a} = \frac{1}{a} \cdot \frac{a}{a}$$

$$= \frac{1}{a^2}$$

$$(a \neq 0)$$

$$\frac{x + 3}{\frac{1}{x}} = \frac{x + 3}{\frac{1}{x}} \cdot \frac{x}{x}$$

$$= \frac{x^2 + 3x}{1}$$

$$= x^2 + 3x$$

$$(x \neq 0)$$

$$\frac{1 + \frac{1}{b} - \frac{2}{b^2}}{1 - \frac{1}{b}} = \frac{1 + \frac{1}{b} - \frac{2}{b^2}}{1 - \frac{1}{b}} \cdot \frac{b^2}{b^2}$$

$$= \frac{b^2 + b - 2}{b^2 - b}$$

$$= \frac{(b + 2)(b - 1)}{b(b - 1)}$$

$$= \frac{(b + 2)\overset{1}{\cancel{(b - 1)}}}{b\underset{1}{\cancel{(b - 1)}}}$$

$$= \frac{b + 2}{b} \quad (b \neq 0, 1)$$

Alternatively, a complex rational expression can also be simplified by dividing the numerator by the denominator. Choose the method that is easier for each given expression.

Procedure

To simplify a complex fraction:
METHOD I
Multiply by $\frac{m}{m}$, where m is the least common multiple of the denominators of the fractional terms.
METHOD 2
Multiply the numerator by the reciprocal of the denominator of the fraction.

EXAMPLE I

Express $\dfrac{\frac{2}{a}}{\frac{3}{4a^2}}$ in simplest form.

Solution **METHOD I**

Multiply numerator and denominator of the fraction by the least common multiple of the denominators of the fractional terms. The least common multiple of a and $4a^2$ is $4a^2$.

$$\frac{\frac{2}{a}}{\frac{3}{4a^2}} = \frac{\frac{2}{a}}{\frac{3}{4a^2}} \cdot \frac{4a^2}{4a^2} = \frac{8a}{3}$$

METHOD 2

Write the fraction as the numerator divided by the denominator. Change the division to multiplication by the reciprocal of the divisor.

$$\frac{\frac{2}{a}}{\frac{3}{4a^2}} = \frac{2}{a} \div \frac{3}{4a^2} = \frac{2}{a} \cdot \frac{4a^2}{3} = \frac{2}{\overset{}{a}} \cdot \frac{4\overset{a}{a^2}}{3} = \frac{8a}{3}$$

Answer $\frac{8a}{3}$ $(a \neq 0)$

EXAMPLE 2

Simplify: $\dfrac{\frac{b}{10} + \frac{1}{2}}{\frac{b^2}{25} - 1}$

Solution

METHOD 1

The least common multiple of $10, 2,$ and 25 is 50.

$$\frac{\frac{b}{10} + \frac{1}{2}}{\frac{b^2}{25} - 1} = \frac{\frac{b}{10} + \frac{1}{2}}{\frac{b^2}{25} - 1} \cdot \frac{50}{50}$$

$$= \frac{5b + 25}{2b^2 - 50}$$

$$= \frac{5(b + 5)}{2(b + 5)(b - 5)}$$

$$= \frac{5}{2(b - 5)}$$

METHOD 2

$$\frac{\frac{b}{10} + \frac{1}{2}}{\frac{b^2}{25} - 1} = \frac{\frac{b}{10} + \frac{5}{10}}{\frac{b^2}{25} - \frac{25}{25}}$$

$$= \frac{\frac{b + 5}{10}}{\frac{b^2 - 25}{25}}$$

$$= \frac{b + 5}{10} \div \frac{b^2 - 25}{25}$$

$$= \frac{\overset{1}{b + 5}}{\underset{2}{10}} \cdot \frac{\overset{5}{25}}{\underset{1}{(b + 5)(b - 5)}}$$

$$= \frac{5}{2(b - 5)}$$

Answer $\dfrac{5}{2(b - 5)}$ $(b \neq -5, 5)$

Exercises

Writing About Mathematics

1. For what values of a is $\left(1 - \frac{1}{a}\right) \div \left(1 - \frac{1}{a^2}\right) = \dfrac{1 - \frac{1}{a}}{1 - \frac{1}{a^2}}$ undefined? Explain your answer.

2. Bebe said that since each of the denominators in the complex fraction $\dfrac{\frac{d}{4} + \frac{3}{5}}{2 - \frac{d^2}{2}}$ is a non-zero constant, the fraction is defined for all values of d. Do you agree with Bebe? Explain why or why not.

Developing Skills

In 3–20, simplify each complex rational expression. In each case, list any values of the variables for which the fractions are not defined.

3. $\dfrac{\frac{3}{3}}{\frac{3}{4}}$

4. $\dfrac{\frac{2}{5}}{4}$

5. $\dfrac{\frac{7}{8}}{1\frac{3}{4}}$

6. $\dfrac{\frac{2}{x}}{\frac{1}{2x}}$

7. $\dfrac{\frac{1}{6} + \frac{1}{4}}{\frac{1}{3} + \frac{1}{2}}$

8. $\dfrac{1 + \frac{1}{a}}{a + 1}$

9. $\dfrac{2 - \frac{2}{d}}{\frac{1}{d} - 1}$

10. $\dfrac{b - \frac{1}{b}}{\frac{1}{b} - 1}$

11. $\dfrac{3 - \frac{3}{b}}{b - 1}$

12. $\dfrac{y + \frac{1}{2}}{2y + 1}$

13. $\dfrac{4y - \frac{1}{y}}{y - \frac{1}{2}}$

14. $\dfrac{\frac{1}{2x} + \frac{1}{3x}}{\frac{1}{x^2}}$

15. $\dfrac{a - \frac{49}{a}}{a - 9 + \frac{14}{a}}$

16. $\dfrac{3 - \frac{9}{x}}{x - 8 + \frac{15}{x}}$

17. $\dfrac{1 + \frac{3}{b} + \frac{2}{b^2}}{1 - \frac{1}{b^2}}$

18. $\dfrac{1 + \frac{1}{y} - \frac{6}{y^2}}{1 + \frac{11}{y} + \frac{24}{y^2}}$

19. $\dfrac{\frac{1}{a} - 1}{1 - \frac{1}{a}}$

20. $\dfrac{5 - \frac{45}{a^2}}{\frac{3}{a} - 1}$

In 21–24, simplify each expression. In each case, list any values of the variables for which the fractions are not defined.

21. $\dfrac{3}{2x} - \dfrac{1 + \frac{1}{x}}{x + 1}$

22. $\dfrac{\frac{a}{4} + \frac{7a}{8}}{\frac{6a^2}{5} - \frac{3a^2}{10}} + \dfrac{3}{a}$

23. $\left(\dfrac{3}{a} + \dfrac{5}{a^2} \right) \div \left(\dfrac{10}{a} + 6 \right) + \dfrac{3}{4}$

24. $\left(6 + \dfrac{12}{b} \right) \div \left(3b - \dfrac{12}{b} \right) + \dfrac{b}{2 - b}$

2-7 SOLVING RATIONAL EQUATIONS

An equation in which there is a fraction in one or more terms can be solved in different ways. However, in each case, the end result must be an equivalent equation in which the variable is equal to a constant.

For example, solve: $\frac{x}{4} - 2 = \frac{x}{2} + \frac{1}{2}$. This is an equation with rational coefficients and could be written as $\frac{1}{4}x - 2 = \frac{1}{2}x + \frac{1}{2}$. There are three possible methods of solving this equation.

METHOD 1

Work with the fractions as given.
Combine terms containing the variable on one side of the equation and constants on the other. Find common denominators to add or subtract fractions.

$$\frac{x}{4} - 2 = \frac{x}{2} + \frac{1}{2}$$
$$\frac{x}{4} - \frac{x}{2} = \frac{1}{2} + 2$$
$$\frac{x}{4} - \frac{2x}{4} = \frac{1}{2} + \frac{4}{2}$$
$$-\frac{x}{4} = \frac{5}{2}$$
$$x = \frac{5}{2}(-4) = -\frac{20}{2} = -10$$

METHOD 2

Rewrite the equation without fractions.
Multiply both sides of the equation by the least common denominator. In this case, the LCD is 4.

$$\frac{x}{4} - 2 = \frac{x}{2} + \frac{1}{2}$$
$$4\left(\frac{x}{4} - 2\right) = 4\left(\frac{x}{2} + \frac{1}{2}\right)$$
$$x - 8 = 2x + 2$$
$$-x = 10$$
$$x = -10$$

METHOD 3

Rewrite each side of the equation as a ratio. Use the fact that the product of the means is equal to the product of the extremes.

$$\frac{x}{4} - 2 = \frac{x}{2} + \frac{1}{2}$$
$$\frac{x}{4} - \frac{8}{4} = \frac{x}{2} + \frac{1}{2}$$
$$\frac{x - 8}{4} = \frac{x + 1}{2}$$
$$4(x + 1) = 2(x - 8)$$
$$4x + 4 = 2x - 16$$
$$2x = -20$$
$$x = -10$$

These same procedures can also be used for a **rational equation**, an equation in which the variable appears in one or more denominators.

EXAMPLE 1

Solve the equation $\frac{1}{x} + \frac{1}{3} = \frac{3}{2x} - 1$.

Solution   Use Method 2.

How to Proceed

(1) Write the equation:

$$\frac{1}{x} + \frac{1}{3} = \frac{3}{2x} - 1$$

(2) Multiply both sides of the equation by the least common denominator, $6x$:

$$6x\left(\frac{1}{x} + \frac{1}{3}\right) = 6x\left(\frac{3}{2x} - 1\right)$$

(3) Simplify:

$$\frac{6x}{x} + \frac{6x}{3} = \frac{18x}{2x} - 6x$$

$$6 + 2x = 9 - 6x$$

(4) Solve for x:

$$8x = 3$$

$$x = \frac{3}{8}$$

Check

$$\frac{1}{x} + \frac{1}{3} = \frac{3}{2x} - 1$$

$$\frac{1}{\frac{3}{8}} + \frac{1}{3} \stackrel{?}{=} \frac{3}{2\left(\frac{3}{8}\right)} - 1$$

$$\left(1 \times \frac{8}{3}\right) + \frac{1}{3} \stackrel{?}{=} \left(3 \times \frac{4}{3}\right) - 1$$

$$\frac{8}{3} + \frac{1}{3} \stackrel{?}{=} 4 - 1$$

$$\frac{9}{3} \stackrel{?}{=} 3$$

$$3 = 3 \checkmark$$

Answer $x = \frac{3}{8}$

When solving a rational equation, it is important to check that the fractional terms are defined for each root, or solution, of the equation. Any solution for which the equation is undefined is called an **extraneous root**.

EXAMPLE 2

Solve for a: $\frac{a - 2}{a} = \frac{a + 2}{2a}$

Solution   Since this equation is a proportion, we can use that the product of the means is equal to the product of the extremes. Check the roots.

$$\frac{a - 2}{a} = \frac{a + 2}{2a}$$

$$2a(a - 2) = a(a + 2)$$

$$2a^2 - 4a = a^2 + 2a$$

$$a^2 - 6a = 0$$

$$a(a - 6) = 0$$

$$a = 0 \quad | \quad a - 6 = 0$$

$$a = 6$$

Check: $a = 0$

$$\frac{a - 2}{a} = \frac{a + 2}{2a}$$

$$\frac{0 - 2}{0} \stackrel{?}{=} \frac{0 + 2}{2(0)}$$

Each side of the equation is undefined for $a = 0$. Therefore, $a = 0$ is not a root.

Check: $a = 6$

$$\frac{a - 2}{a} = \frac{a + 2}{2a}$$

$$\frac{6 - 2}{6} \stackrel{?}{=} \frac{6 + 2}{2(6)}$$

$$\frac{4}{6} \stackrel{?}{=} \frac{8}{12}$$

$$\frac{2}{3} = \frac{2}{3} \checkmark$$

Answer $a = 6$

EXAMPLE 3

Solve: $\dfrac{x}{x+2} = \dfrac{3}{x} + \dfrac{4}{x(x+2)}$

Solution Use Method 2 to rewrite the equation without fractions. The least common denominator is $x(x+2)$.

$$\frac{x}{x+2} = \frac{3}{x} + \frac{4}{x(x+2)}$$

$$x(x+2)\left(\frac{x}{x+2}\right) = x(x+2)\left(\frac{3}{x}\right) + x(x+2)\left(\frac{4}{x(x+2)}\right)$$

$$\overset{1}{x\cancel{(x+2)}}\left(\frac{x}{\cancel{x+2}}\right) = \overset{1}{\cancel{x}}(x+2)\left(\frac{3}{\cancel{x}}\right) + \overset{1}{\cancel{x}}\overset{1}{\cancel{(x+2)}}\left(\frac{4}{\underset{1}{\cancel{x}}\underset{1}{\cancel{(x+2)}}}\right)$$

$$x^2 = 3x + 6 + 4$$

$$x^2 - 3x - 10 = 0$$

$$(x-5)(x+2) = 0$$

$$x - 5 = 0 \quad | \quad x + 2 = 0$$

$$x = 5 \quad | \quad x = -2$$

Check the roots:

<table>
<tr><td>

Check: $x = 5$

$$\frac{x}{x+2} = \frac{3}{x} + \frac{4}{x(x+2)}$$

$$\frac{5}{5+2} \overset{?}{=} \frac{3}{5} + \frac{4}{5(5+2)}$$

$$\frac{5}{7} \times \frac{5}{5} \overset{?}{=} \frac{3}{5} \times \frac{7}{7} + \frac{4}{35}$$

$$\frac{25}{35} \overset{?}{=} \frac{21}{35} + \frac{4}{35}$$

$$\frac{25}{35} = \frac{25}{35} ✔$$

Since $x = 5$ leads to a true statement, 5 is a root of the equation.

</td><td>

Check: $x = -2$

$$\frac{x}{x+2} = \frac{3}{x} + \frac{4}{x(x+2)}$$

$$\frac{-2}{-2+2} \overset{?}{=} \frac{3}{-2} + \frac{4}{-2(-2+2)}$$

$$\frac{-2}{0} \overset{?}{=} -\frac{3}{2} + \frac{4}{-2(0)}$$

Since $x = -2$ leads to a statement that is undefined, -2 is not a root of the equation.

</td></tr>
</table>

Answer 5 is the only root of $\dfrac{x}{x+2} = \dfrac{3}{x} + \dfrac{4}{x(x+2)}$.

It is important to check each root obtained by multiplying both members of the given equation by an expression that contains the variable. The derived equation may not be equivalent to the given equation.

EXAMPLE 4

Solve and check: $0.2y + 4 = 5 - 0.05y$

Solution The coefficients 0.2 and 0.05 are decimal fractions ("2 *tenths*" and "5 *hundredths*"). The denominator of 0.2 is 10 and the denominator of 0.05 is 100. Therefore, the least common denominator is 100.

$$0.2y + 4 = 5 - 0.05y$$
$$100(0.2y + 4) = 100(5 - 0.05y)$$
$$20y + 400 = 500 - 5y$$
$$25y = 100$$
$$y = 4$$

Check
$$0.2y + 4 = 5 - 0.05y$$
$$0.2(4) + 4 \overset{?}{=} 5 - 0.05(4)$$
$$0.8 + 4 \overset{?}{=} 5 - 0.2$$
$$4.8 = 4.8 ✔$$

Answer $y = 4$

EXAMPLE 5

On his way home from college, Daniel traveled 15 miles on local roads and 90 miles on the highway. On the highway, he traveled 30 miles per hour faster than on local roads. If the trip took 2 hours, what was Daniel's rate of speed on each part of the trip?

Solution Use $\frac{\text{distance}}{\text{rate}} = $ time to represent the time for each part of the trip.

Let $x = $ rate on local roads. Then $\frac{15}{x} = $ time on local roads.

Let $x + 30 = $ rate on the highway. Then $\frac{90}{x + 30} = $ time on the highway.

The total time is 2 hours.

$$\frac{15}{x} + \frac{90}{x + 30} = 2$$
$$x(x + 30)\left(\frac{15}{x}\right) + x(x + 30)\left(\frac{90}{x + 30}\right) = 2x(x + 30)$$
$$15(x + 30) + 90x = 2x^2 + 60x$$
$$15x + 450 + 90x = 2x^2 + 60x$$
$$0 = 2x^2 - 45x - 450$$

Recall that a trinomial can be factored by writing the middle term as the sum of two terms whose product is the product of the first and last term.

$$2x^2(-450) = -900x^2$$

Find the factors of this product whose sum is the middle term $-45x$.

$$15x + (-60x) = -45x$$

Write the trinomial with four terms, using this pair of terms in place of $-45x$, and factor the trinomial.

$$0 = 2x^2 - 45x - 450$$
$$0 = 2x^2 + 15x - 60x - 450$$
$$0 = x(2x + 15) - 30(2x + 15)$$
$$0 = (2x + 15)(x - 30)$$

$$2x + 15 = 0 \quad | \quad x - 30 = 0$$
$$2x = -15 \quad | \quad x = 30$$
$$x = -\frac{15}{2} \quad |$$

Reject the negative root since a rate cannot be a negative number. Use $x = 30$.

On local roads, Daniel's rate is 30 miles per hour and his time is $\frac{15}{30} - \frac{1}{2}$ hour.

On the highway, Daniel's rate is $30 + 30$ or 60 miles per hour. His time is $\frac{90}{60} = \frac{3}{2}$ hours. His total time is $\frac{1}{2} + \frac{3}{2} = \frac{4}{2} = 2$ hours.

Answer Daniel drove 30 mph on local roads and 60 mph on the highway.

Exercises

Writing About Mathematics

1. Samantha said that the equation $\frac{a-2}{a} = \frac{a+2}{2a}$ in Example 2 could be solved by multiplying both sides of the equation by $2a$. Would Samantha's solution be the same as the solution obtained in Example 2? Explain why or why not.

2. Brianna said that $\frac{3}{x-2} = \frac{5}{x+2}$ is a rational equation but $\frac{x-2}{3} = \frac{x+2}{5}$ is not. Do you agree with Brianna? Explain why or why not.

Developing Skills

In 3–20, solve each equation and check.

3. $\frac{1}{4}a + 8 = \frac{1}{2}a$

4. $\frac{3}{4}x = 14 - x$

5. $\frac{x+2}{5} = \frac{x-2}{3}$

6. $\frac{x}{5} - \frac{x}{10} = 7$

7. $\frac{2x}{3} + 1 = \frac{3x}{4}$

8. $x - 0.05x = 19$

9. $0.4x + 8 = 0.5x$

10. $0.2a = 0.05a + 3$

11. $1.2b - 3 = 7 - 0.05b$

12. $\frac{x}{x+5} = \frac{2}{3}$

13. $\frac{7}{2x-3} = \frac{4}{x}$

14. $\frac{3}{a} + \frac{1}{2} = \frac{11}{5a}$

15. $18 - \frac{4}{b} = 10$

16. $\frac{x}{2} = \frac{3}{2x+1}$

17. $1 = \frac{5}{x+3} + \frac{5}{(x+2)(x+3)}$

18. $\frac{a-1}{4} = \frac{8}{a+3}$

19. $\frac{4}{y+2} = 1 - \frac{8}{y(y+2)}$

20. $\frac{4}{3b-2} - \frac{7}{3b+2} = \frac{1}{9b^2-4}$

Applying Skills

21. Last week, the ratio of the number of hours that Joseph worked to the number of hours that Nicole worked was $2 : 3$. This week Joseph worked 4 hours more than last week and Nicole worked twice as many hours as last week. This week the ratio of the hours Joseph worked to the number of hours Nicole worked is $1 : 2$. How many hours did each person work each week?

22. Anthony rode his bicycle to his friend's house, a distance of 1 mile. Then his friend's mother drove them to school, a distance of 12 miles. His friend's mother drove at a rate that is 25 miles per hour faster than Anthony rides his bike. If it took Anthony $\frac{3}{5}$ of an hour to get to school, at what average rate does he ride his bicycle? (Use $\frac{\text{distance}}{\text{rate}} = $ time for each part of the trip to school.)

23. Amanda drove 40 miles. Then she increased her rate of speed by 10 miles per hour and drove another 40 miles to reach her destination. If the trip took $1\frac{4}{5}$ hours, at what rate did Amanda drive?

24. Last week, Emily paid $8.25 for x pounds of apples. This week she paid $9.50 for $(x + 1)$ pounds of apples. The price per pound was the same each week. How many pounds of apples did Emily buy each week and what was the price per pound? (Use $\frac{\text{total cost}}{\text{number of pounds}} = $ cost per pound for each week.)

2-8 SOLVING RATIONAL INEQUALITIES

Inequalities are usually solved with the same procedures that are used to solve equations. For example, we can solve this equation and this inequality by using the same steps.

$$\frac{1}{4}x + \frac{1}{8} = \frac{1}{3}x + \frac{3}{8}$$
$$24\left(\frac{1}{4}x\right) + 24\left(\frac{1}{8}\right) = 24\left(\frac{1}{3}x\right) + 24\left(\frac{3}{8}\right)$$
$$6x + 3 = 8x + 9$$
$$-2x = 6$$
$$x = -3$$

$$\frac{1}{4}x + \frac{1}{8} < \frac{1}{3}x + \frac{3}{8}$$
$$24\left(\frac{1}{4}x\right) + 24\left(\frac{1}{8}\right) < 24\left(\frac{1}{3}x\right) + 24\left(\frac{3}{8}\right)$$
$$6x + 3 < 8x + 9$$
$$-2x < 6$$
$$x > -3$$

All steps leading to the solution of this equation and this inequality are the same, but special care must be used when multiplying or dividing an inequality. Note that when we multiplied the inequality by 24, a positive number, the order of the inequality remained unchanged. In the last step, when we divided the inequality by -2, a negative number, the order of the inequality was reversed.

When we solve an inequality that has a variable expression in the denominator by multiplying both sides of the inequality by the variable expression, we must consider two possibilities: the variable represents a positive number or the variable represents a negative number. (The expression cannot equal zero as that would make the fraction undefined.)

For example, to solve $2 - \frac{x}{x + 1} > 3 - \frac{9}{x + 1}$, we multiply both sides of the inequality by $x + 1$. When $x + 1$ is positive, the order of the inequality remains unchanged. When $x + 1$ is negative, the order of the inequality is reversed.

If $x + 1 > 0$, then $x > -1$.	If $x + 1 < 0$, then $x < -1$.
$2 - \frac{x}{x + 1} > 3 - \frac{9}{x + 1}$	$2 - \frac{x}{x + 1} > 3 - \frac{9}{x + 1}$
$\frac{9}{x + 1} - \frac{x}{x + 1} > 3 - 2$	$\frac{9}{x + 1} - \frac{x}{x + 1} > 3 - 2$
$(x + 1)\frac{9 - x}{x + 1} > (x + 1)(1)$	$(x + 1)\frac{9 - x}{x + 1} < (x + 1)(1)$
$9 - x > x + 1$	$9 - x < x + 1$
$\frac{-2x}{-2} < \frac{-8}{-2}$	$\frac{-2x}{-2} > \frac{-8}{-2}$
$x < 4$	$x > 4$

Therefore, $x > -1$ and $x < 4$ or $-1 < x < 4$.

There are no values of x such that $x < -1$ and $x > 4$.

The solution set of the equation $2 - \frac{x}{x + 1} > 3 - \frac{9}{x + 1}$ is $\{x : -1 < x < 4\}$.

An alternative method of solving this inequality is to use the corresponding equation.

STEP 1. Solve the corresponding equation.

$$2 - \frac{x}{x + 1} = 3 - \frac{9}{x + 1}$$

$$(x + 1)(2) - (x + 1)\frac{x}{x + 1} = (x + 1)(3) - (x + 1)\frac{9}{x + 1}$$

$$2x + 2 - x = 3x + 3 - 9$$

$$-2x = -8$$

$$x = 4$$

STEP 2. Find any values of x for which the equation is undefined.

Terms $\frac{x}{x + 1}$ and $\frac{9}{x + 1}$ are undefined when $x = -1$.

STEP 3. To find the solutions of the corresponding inequality, divide the number line into three intervals using the solution to the equality, 4, and the value of x for which the equation is undefined, -1.

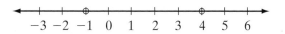

Choose a value from each section and substitute that value into the inequality.

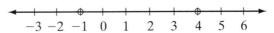

Let $x = -2$:	**Let $x = 0$:**	**Let $x = 5$:**
$2 - \frac{x}{x+1} > 3 - \frac{9}{x+1}$	$2 - \frac{x}{x+1} > 3 - \frac{9}{x+1}$	$2 - \frac{x}{x+1} > 3 - \frac{9}{x+1}$
$2 - \frac{-2}{-2+1} \overset{?}{>} 3 - \frac{9}{-2+1}$	$2 - \frac{0}{0+1} \overset{?}{>} 3 - \frac{9}{0+1}$	$2 - \frac{5}{5+1} \overset{?}{>} 3 - \frac{9}{5+1}$
$2 - (2) \overset{?}{>} 3 - (-9)$	$2 - (0) \overset{?}{>} 3 - (9)$	$2 - \frac{5}{6} \overset{?}{>} 3 - \frac{9}{6}$
$0 \not> 12$ ✗	$2 > -6$ ✔	$\frac{7}{6} \not> \frac{9}{6}$ ✗

The inequality is true for values in the interval $-1 < x < 4$ and false for all other values.

EXAMPLE I

Solve for a: $2 - \frac{3}{a} > \frac{5}{a}$

Solution **METHOD 1**

Multiply both sides of the equation by a. The sense of the inequality will remain unchanged when $a > 0$ and will be reversed when $a < 0$.

Let $a > 0$:	**Let $a < 0$:**
$a\left(2 - \frac{3}{a}\right) > a\left(\frac{5}{a}\right)$	$a\left(2 - \frac{3}{a}\right) < a\left(\frac{5}{a}\right)$
$2a - 3 > 5$	$2a - 3 < 5$
$2a > 8$	$2a < 8$
$a > 4$	$a < 4$
$a > 0$ and $a > 4 \rightarrow a > 4$	$a < 0$ and $a < 4 \rightarrow a < 0$

The solution set is $\{a : a < 0 \text{ or } a > 4\}$.

METHOD 2

Solve the corresponding equation for a.

$$2 - \frac{3}{a} = \frac{5}{a}$$
$$a\left(2 - \frac{3}{a}\right) = a\left(\frac{5}{a}\right)$$
$$2a - 3 = 5$$
$$2a = 8$$
$$a = 4$$

Partition the number line using the solution to the equation, $a = 4$, and the value of a for which the equation is undefined, $a = 0$. Check in the inequality a representative value of a from each interval of the graph.

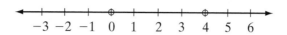

Let $a = -1$:

$$2 - \frac{3}{a} > \frac{5}{a}$$

$$2 - \frac{3}{-1} \overset{?}{>} \frac{5}{-1}$$

$$2 + 3 \overset{?}{>} -5$$

$$5 > -5 \checkmark$$

Let $a = 1$:

$$2 - \frac{3}{a} > \frac{5}{a}$$

$$2 - \frac{3}{1} \overset{?}{>} \frac{5}{1}$$

$$2 - 3 \overset{?}{>} 5$$

$$-1 \not> 5 \; \text{✗}$$

Let $a = 5$:

$$2 - \frac{3}{a} > \frac{5}{a}$$

$$2 - \frac{3}{5} \overset{?}{>} \frac{5}{5}$$

$$\frac{10}{5} - \frac{3}{5} \overset{?}{>} 1$$

$$\frac{7}{5} > 1 \checkmark$$

Answer $\{a : a < 0 \text{ or } a > 4\}$

Exercises

Writing About Mathematics

1. When the equation $2 - \frac{3}{b} = \frac{5}{b + 2}$ is solved for b, the solutions are -1 and 3. Explain why the number line must be separated into five segments by the numbers $-2, -1, 0,$ and 3 in order to check the solution set of the inequality $2 - \frac{3}{b} > \frac{5}{b + 2}$.

2. What is the solution set of $\frac{|x|}{x} < 0$? Justify your answer.

Developing Skills

In 3–14, solve and check each inequality.

3. $\frac{a}{4} > \frac{a}{2} + 6$

4. $\frac{y - 3}{5} < \frac{y + 2}{10}$

5. $\frac{3b - 4}{8} < \frac{4b - 3}{4}$

6. $\frac{2 - d}{7} > \frac{d - 2}{5}$

7. $\frac{a + 1}{4} - 2 > 11 - \frac{a}{6}$

8. $\frac{7}{2x} - \frac{2}{x} > \frac{3}{2}$

9. $5 - \frac{7}{y} < 2 + \frac{5}{y}$

10. $3 - \frac{2}{a + 1} < 5$

11. $\frac{4x}{x - 4} + 2 < \frac{2}{x - 4}$

12. $\frac{2}{x} + \frac{3}{x} < 10$

13. $\frac{x}{x + 5} - \frac{1}{x + 5} > 4$

14. $3 - \frac{9}{a + 1} > 5 - \frac{1}{a + 1}$

CHAPTER SUMMARY

A **rational number** is an element of the set of numbers of the form $\frac{a}{b}$ when a and b are integers and $b \neq 0$. For every rational number $\frac{a}{b}$ that is not equal to zero there is a **multiplicative inverse** or **reciprocal** $\frac{b}{a}$ such that $\frac{a}{b} \cdot \frac{b}{a} = 1$.

A number is a rational number if and only if it can be written as an infinitely repeating decimal.

A **rational expression** is the quotient of two polynomials. A rational expression has no meaning when the denominator is zero.

A rational expression is in **simplest form** or **reduced to lowest terms** when there is no factor of the numerator that is also a factor of the denominator except 1 and -1.

If $\frac{a}{b}$ and $\frac{c}{d}$ are two rational numbers with $b \neq 0$ and $d \neq 0$:

$$\frac{a}{b} \cdot \frac{c}{d} = \frac{ac}{bd} \qquad \frac{a}{b} \div \frac{c}{d} = \frac{a}{b} \cdot \frac{d}{c} = \frac{ad}{bc} \ (c \neq 0)$$

$$\frac{a}{b} + \frac{c}{d} = \frac{a}{b} \cdot \frac{d}{d} + \frac{c}{d} \cdot \frac{b}{b} = \frac{ad}{bd} + \frac{cb}{bd} = \frac{ad + cb}{bd}$$

A **ratio** is the comparison of two numbers by division. The ratio of a to b can be written as $\frac{a}{b}$ or as $a : b$ when $b \neq 0$.

An equation that states that two ratios are equal is called a **proportion**. In the proportion $a : b = c : d$ or $\frac{a}{b} = \frac{c}{d}$, the first and last terms, a and d, are called the **extremes**, and the second and third terms, b and c, are the **means**.

In any proportion, the product of the means is equal to the product of the extremes.

A **complex rational expression** has a rational expression in the numerator, the denominator or both. Complex rational expressions can be simplified by multiplying numerator and denominator by a common multiple of the denominators in the numerator and denominator.

A **rational equation**, an equation in which the variable appears in one or more denominators, can be simplified by multiplying both members by a common multiple of the denominators.

When a rational inequality is multiplied by the least common multiple of the denominators, two cases must be considered: when the least common multiple is positive and when the least common multiple is negative.

VOCABULARY

2-1 Rational number • Multiplicative inverse • Reciprocal

2-2 Rational expression • Simplest form • Lowest terms

2-4 Least common denominator (LCD) • Least common multiple (LCM)

2-5 Ratio • Ratio in simplest form • Proportion • Extremes • Means

2-6 Complex fraction • Complex rational expression

2-7 Rational equation • Extraneous root

REVIEW EXERCISES

1. What is the multiplicative inverse of $\frac{7}{2}$?

2. For what values of b is the rational expression $\frac{2(b+1)}{b(b^2-1)}$ undefined?

3. Write $\frac{5}{12}$ as an infinitely repeating decimal.

In 4–15, perform the indicated operations, express the answer in simplest form, and list the values of the variables for which the fractions are undefined.

4. $\frac{3}{5ab^2} \cdot \frac{10a^2b}{9}$

5. $\frac{2}{5a} + \frac{1}{4a}$

6. $\frac{3x+12}{4x} \cdot \frac{2}{x^2-16}$

7. $\frac{a^2+2a}{5a} \div \frac{a^2+7a+10}{a^2}$

8. $\frac{2}{a+1} + \frac{3}{a^2-1}$

9. $\frac{y}{y-3} - \frac{18}{y^2-9}$

10. $\frac{d^2+3d-18}{4d} \div \frac{2d+12}{8}$

11. $\frac{3}{5b} \cdot \frac{5b+10}{6} - \frac{1}{b}$

12. $\frac{1}{a+2} + \frac{a}{a^2+a} \div \frac{a}{a+1}$

13. $\left(\frac{1}{a-4} + \frac{3}{4-a}\right) \cdot \frac{16-a^2}{2}$

14. $\left(1 + \frac{1}{x+1}\right) \cdot \left(1 + \frac{3}{x^2-4}\right)$

15. $\left(x - \frac{1}{x}\right) \div (1 - x)$

In 16–19, simplify each complex rational expression and list the values of the variables (if any) for which the fractions are undefined.

16. $\dfrac{1+\frac{1}{2}}{2+\frac{1}{4}}$

17. $\dfrac{\frac{a+b}{b}}{1+\frac{b}{a}}$

18. $\dfrac{\frac{x}{12}-\frac{1}{2}}{\frac{x^2}{12}-3}$

19. $\dfrac{a-1-\frac{20}{a}}{1-\frac{16}{a^2}}$

In 20–27, solve and check each equation or inequality.

20. $\frac{a}{7} + \frac{3}{14} = \frac{5a}{2}$

21. $\frac{x}{5} + 9 = 12 - \frac{x}{4}$

22. $\frac{3}{y} + \frac{1}{2} = \frac{6}{y}$

23. $\frac{2x}{x-2} = \frac{3x}{x+1}$

24. $\frac{x}{x-3} + \frac{6}{x(x-3)} = \frac{5}{x-3}$

25. $\frac{2d+1}{3} = \frac{2d+2}{d} - 2$

26. $\frac{1}{x} + 3 > \frac{7}{x}$

27. $\frac{1}{2x+1} - 2 \le 8$

28. The ratio of boys to girls in the school chorus was 4 : 5. After three more boys joined the chorus, the ratio of boys to girls is 9 : 10. How many boys and how many girls are there now in the chorus?

29. Last week Stephanie spent $10.50 for cans of soda. This week, at the same cost per can, she bought three fewer cans and spent $8.40. How many cans of soda did she buy each week and what was the cost per can?

30. On a recent trip, Sarah traveled 80 miles at a constant rate of speed. Then she encountered road work and had to reduce her speed by 15 miles per hour for the last 30 miles of the trip. The trip took 2 hours. What was Sarah's rate of speed for each part of the trip?

31. The areas of two rectangles are 208 square feet and 182 square feet. The length of the larger is 2 feet more than the length of the smaller. If the rectangles have equal widths, find the dimensions of each rectangle. $\left(\text{Use } \frac{\text{area}}{\text{length}} = \text{width.}\right)$

Exploration

The early Egyptians wrote fractions as the sum of **unit fractions** (the reciprocals of the counting numbers) with no reciprocal repeated. For example:

$$\frac{3}{4} = \frac{2}{4} + \frac{1}{4} = \frac{1}{2} + \frac{1}{4} \quad \text{and} \quad \frac{2}{3} = \frac{1}{3} + \frac{1}{3} = \frac{1}{3} + \frac{1}{4} + \frac{1}{12}$$

Of course the Egyptians used hieroglyphs instead of the numerals familiar to us today. Whole numbers were represented as follows.

To write unit fractions, the symbol ⌒ was drawn over the whole number hieroglyph. For example:

1. Show that for $n \neq 0, \frac{1}{n} = \frac{1}{n + 1} + \frac{1}{n(n + 1)}$.

2. Write $\frac{3}{4}$ and $\frac{2}{3}$ using Egyptian fractions.

3. Write each of the fractions as the sum of unit fractions. (*Hint:* Use the results of Exercise 1.)

    a. $\frac{2}{5}$ **b.** $\frac{7}{10}$ **c.** $\frac{7}{12}$ **d.** $\frac{23}{24}$ **e.** $\frac{11}{18}$

Part I

Answer all questions in this part. Each correct answer will receive 2 credits. No partial credit will be allowed.

1. Which of the following is *not* true in the set of integers?
(1) Addition is commutative.
(2) Every integer has an additive inverse.
(3) Multiplication is distributive over addition.
(4) Every non-zero integer has a multiplicative inverse.

2. $2 - |3 - 5|$ is equal to
(1) 0   (2) 2   (3) 6   (4) 4

3. The sum of $3a^2 - 5a$ and $a^2 + 7a$ is
(1) $3 + 2a$ (2) $3a^2 + 2a$ (3) $4a^2 + 2a$ (4) $6a^2$

4. In simplest form, $2x(x + 5) - (7x + 1)$ is equal to
(1) $2x^2 + 3x + 1$ (3) $2x^2 + 12x + 1$
(2) $2x^2 + 3x - 1$ (4) $2x^2 - 7x + 4$

5. The factors of $2x^2 - x - 6$ are
(1) $(2x - 3)(x + 2)$ (3) $(2x + 3)(x - 2)$
(2) $(2x + 2)(x - 3)$ (4) $(2x - 3)(x - 2)$

6. The roots of the equation $x^2 - 7x + 10 = 0$ are
(1) 2 and 5   (3) -7 and 10
(2) -2 and -5 (4) -5 and 2

7. For $a \neq 0, 2, -2$, the fraction $\dfrac{\frac{2}{a} - 1}{a - \frac{4}{a}}$ is equal to

(1) $\dfrac{1}{a + 2}$ (3) $\dfrac{1}{a - 2}$
(2) $-\dfrac{1}{a + 2}$ (4) $-\dfrac{1}{a - 2}$

8. In simplest form, the quotient $\dfrac{b}{2b - 1} \div \dfrac{3b}{8b - 4}$ equals
(1) $\dfrac{3}{4}$ (3) $\dfrac{3b^2}{4(2b - 1)^2}$
(2) $\dfrac{4}{3}$ (4) $\dfrac{7a}{4(2b - 1)}$

9. For what values of a is the fraction $\dfrac{2a + 4}{a^2 - 2a - 35}$ undefined?
(1) $-2, -5, 7$ (2) $2, 5, -7$ (3) $-5, 7$ (4) $5, -7$

10. The solution set of the equation $|2x - 1| = 7$ is
(1) $\{-3, 4\}$ (2) $\varnothing$ (3) $\{-3\}$ (4) $\{4\}$

Part II

Answer all questions in this part. Each correct answer will receive 2 credits. Clearly indicate the necessary steps, including appropriate formula substitutions, diagrams, graphs, charts, etc. For all questions in this part, a correct numerical answer with no work shown will receive only 1 credit.

11. Solve and check: $|7 - 2x| \leq 3$

12. Perform the indicated operations and write the answer in simplest form:

$$\frac{3}{a + 5} - \frac{a - 3}{5} \div \frac{a^2 - 9}{15}$$

Part III

Answer all questions in this part. Each correct answer will receive 4 credits. Clearly indicate the necessary steps, including appropriate formula substitutions, diagrams, graphs, charts, etc. For all questions in this part, a correct numerical answer with no work shown will receive only 1 credit.

13. Write the following product in simplest form and list the values of x for which it is undefined: $\frac{5x + 5}{x^2 - 1} \cdot \frac{x^2 - x}{15}$

14. The length of a rectangle is 12 meters longer than half the width. The area of the rectangle is 90 square meters. Find the dimensions of the rectangle.

Part IV

Answer all questions in this part. Each correct answer will receive 6 credits. Clearly indicate the necessary steps, including appropriate formula substitutions, diagrams, graphs, charts, etc. For all questions in this part, a correct numerical answer with no work shown will receive only 1 credit.

15. Find the solution set and check: $2x^2 - 5x > 7$

16. Diego had traveled 30 miles at a uniform rate of speed when he encountered construction and had to reduce his speed to one-third of his original rate. He continued at this slower rate for 10 miles. If the total time for these two parts of the trip was one hour, how fast did he travel at each rate?

REAL NUMBERS AND RADICALS

Traditionally, we use units such as inches or meters to measure length, but we could technically measure with any sized unit. Two line segments are called *commeasurable* if there exists some unit with which both segments have integral measures. Therefore, the ratio of the lengths of the line segments is the ratio of integers, a rational number. Two line segments are *incommensurable* if there is no unit, no matter how small, with which both segments can have integral measures. Therefore, the ratio of the lengths is not a rational number, that is, the ratio is an irrational number.

The oldest known proof of the existence of incommensurable line segments is found in the tenth book of Euclid's *Elements* although the proof was known long before Euclid's time. Euclid established that the diagonal of a square and a side of a square are incommensurable. The ratio of their lengths is an irrational number. In arithmetic terms, what Euclid proved was that $\sqrt{2}$ was irrational. The concept of a real number, then, had its beginnings as the ratio of the lengths of line segment.

3-1 THE REAL NUMBERS AND ABSOLUTE VALUE

In Chapter 2 we learned that a number is rational if and only if it can be written as an infinitely repeating decimal. However, we know that there are infinite decimal numbers that do not repeat. For example, the following numbers have patterns that continue infinitely but do not repeat.

$$0.20200200020000200000\ldots \qquad 0.123456789101112131415\ldots$$

These numbers are called **irrational numbers**.

DEFINITION ――
An **irrational number** is an infinite decimal number that does not repeat.

Other examples of irrational numbers are numbers that are the square root of an integer that is not a perfect square. For example, $\sqrt{2}$, $\sqrt{3}$, $\sqrt{5}$, and $\sqrt{7}$ are all irrational numbers. The ratio of the circumference of a circle to its diameter, π, is also an irrational number.

DEFINITION ――
The union of the set of rational numbers and the set of irrational numbers is the set of **real numbers**.

Every point on the number line is associated with one and only one real number and every real number is associated with one and only one point on the number line. We say that there is a one-to-one correspondence between the real numbers and the points on the number line.

EXAMPLE 1

Which of the following numbers is irrational?

$$-7, \ \tfrac{5}{4}, \ 2.1\overline{54}, \ 0.12131415\ldots$$

Solution $-7 = \frac{-7}{1}$ and $\frac{5}{4}$ are each the ratio of integers and are therefore rational.

$2.1\overline{54} = 2.1545454\ldots$ is an infinite repeating decimal and is therefore rational.

$0.12131415\ldots$ is an infinite decimal that does not repeat and is therefore irrational.

Answer $0.12131415\ldots$

Graphing Inequalities on the Number Line

In the set of integers, the solution set of an inequality can often be listed.

For example, when the domain is the set of integers, the solution set of the inequality shown at the right is $\{-1, 0, 1, 2, 3, 4, 5\}$.

$$5 < x + 7 \qquad\ \leq 12$$
$$5 - 7 < x + 7 - 7 \leq 12 - 7$$
$$-2 < x \qquad\qquad \leq 5$$

If the domain is the set of real numbers, the elements of the solution set cannot be listed but can be shown on the number line.

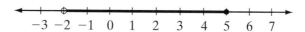

Note 1: The open circle at -2 indicates that this is a lower boundary of the set but not an element of the set. The darkened circle at 5 indicates that 5 is the upper boundary of the set and an element of the set.

Note 2: The set can be symbolized using **interval notation** as $(-2, 5]$. The curved parenthesis, (, at the left indicates that the lower boundary is *not* an element of the solution set and the bracket,], at the right indicates that the upper boundary is included in the set. The set includes all real numbers from -2 to 5, including 5 but not including -2.

Absolute Value Inequalities

We know that if $|x| = 3$, $x = 3$, or $x = -3$. These two points on the real number line separate the line into three segments.

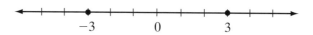

- The points between -3 and 3 are closer to 0 than are -3 and 3 and therefore have absolute values that are less than 3.

- The points to the left of -3 and to the right of 3 are farther from 0 than are -3 and 3 and therefore have absolute values that are greater than 3.

In Chapter 1, we learned that for any *positive* value of k:

- if $|x| < k$, then $-k < x < k$.
- if $|x| > k$, then $x > k$ or $x < -k$.

What about when k is negative? We can use the definition of absolute value to solve inequalities of the form $|x| < k$ and $|x| > k$.

Solve $|x| < k$ for negative k.	Solve $|x| > k$ for negative k.
Since $|x|$ is always non-negative and a non-negative number is always greater than a negative number, there is no value of x for which $|x| < k$ is true. The solution set is the empty set.	Since $|x|$ is always non-negative and a non-negative number is always greater than a negative number, $|x| > k$ is always true. The solution set is the set of real numbers.

SUMMARY

Inequality	Value of k	Solution Set
$\lvert x\rvert < k$	$k > 0$	$\{x : -k < x < k\}$
$\lvert x\rvert < k$	$k < 0$	$\varnothing$
$\lvert x\rvert > k$	$k > 0$	$\{x : x < -k \text{ or } x > k\}$
$\lvert x\rvert > k$	$k < 0$	$\{$All real numbers$\}$

EXAMPLE 2

Solve for x and graph the solution set: $\lvert 3a - 1\rvert < 5$.

Solution For a positive value of k, the solution of $\lvert x\rvert < k$ is $-k < x < k$.
Let $x = 3a - 1$ and $k = 5$.

How to Proceed

(1) Write the solution in terms of a:
$$-5 < 3a - 1 \quad < 5$$

(2) Add 1 to each member of the inequality:
$$-5 + 1 < 3a - 1 + 1 < 5 + 1$$
$$-4 < 3a \quad < 6$$

(3) Divide each member by 3:
$$-\tfrac{4}{3} < \tfrac{3a}{3} \quad < \tfrac{6}{3}$$
$$-\tfrac{4}{3} < a \quad < 2$$

(4) Graph the solution set:

Check We can check the solution on the graphing calculator.

Graph $Y_1 = \lvert 3X - 1\rvert$ and $Y_2 = 5$. Absolute value is entered in the calculator by using the abs (function (MATH ▶ 1). By pressing TRACE -4 ÷ 3 ENTER and TRACE 2 ENTER, we can see that the two functions intersect at points for which $x = -\tfrac{4}{3}$ and $x = 2$. From the graph, we can verify that $\left(-\tfrac{4}{3}, 2\right)$ is the solution to the inequality since the graph of Y_1 lies below the graph of Y_2 in this interval.

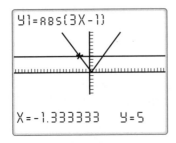

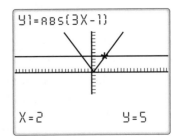

Writing About Mathematics

1. Tony said that $\dfrac{3}{1 - \frac{1}{5}}$ is irrational because it is not the ratio of integers and is therefore not a

rational number. Do you agree with Tony? Explain why or why not.

2. Maria said that since the solution of the inequality $|2x - 5| < 3$ can be found by using
$-3 < 2x - 5 < 3$, then the solution of the inequality $|2x - 5| > 3$ can be found by using
$-3 > 2x - 5 > 3$. Do you agree with Maria? Explain why or why not.

Developing Skills

In 3–14, determine whether each of the numbers is rational or irrational.

3. $\frac{0}{4}$

4. $2.17273747\ldots$

5. 3π

6. $\sqrt{17}$

7. $1\frac{3}{4}$

8. $\frac{\sqrt{2}}{2}$

9. $\sqrt{3} + 5$

10. $\frac{\sqrt{5}}{\sqrt{5}}$

11. $\frac{\sqrt{16}}{2}$

12. $0 + \pi$

13. $\frac{\pi}{\pi}$

14. $2\frac{1}{3} + \sqrt{3}$

In 15–26, find and graph the solution set of each inequality.

15. $|x| < 7$

16. $|a - 5| \geq 3$

17. $|2y + 5| > 9$

18. $|2 - 4b| \leq 6$

19. $|-5 - a| > 4$

20. $9 - |3x + 3| > 0$

21. $\left|5x - \frac{1}{2}\right| - \frac{3}{2} > 0$

22. $2|x + 2| \leq -3$

23. $2|x + 2| > -3$

24. $\left|\frac{5}{2}x + 2\right| \leq 0$

25. $\left|x + \frac{1}{2}\right| + 1 > \frac{1}{2}$

26. $3|2x - 2| + 2 \geq -5$

Applying Skills

27. The temperature on Mars roughly satisfies the inequality $|t - 75| \leq 145$ where t is the temperature in Fahrenheit. What is the range of temperatures on Mars?

28. Elevation of land in the United States is given by the inequality $|h - 10{,}019| \leq 10{,}301$ where h is the height in feet. What is the range of elevations in the United States?

3-2 ROOTS AND RADICALS

Square Root

We know that $4 = 2 \cdot 2$. Therefore, 4 is the square of 2, and 2 is a square root of 4. We also know that $4 = -2 \cdot (-2)$. Therefore, 4 is the square of -2, and -2 is a square root of 4. We write this as $4 = 2^2$ and $4 = (-2)^2$. Because there are two numbers whose square is 4, 4 has two square roots, written $\sqrt{4} = 2$ and $-\sqrt{4} = -2$. We can write these last two equalities as $\pm\sqrt{4} = \pm 2$.

DEFINITION _____

A **square root** of k is one of the two equal factors whose product is k.

Every positive real number has two square roots: a positive real number and its opposite, a negative real number. For example:

$$x^2 = 9 \qquad\qquad y^2 = 5$$
$$x = \pm\sqrt{9} = \pm 3 \qquad\qquad y = \pm\sqrt{5}$$
$$x = 3 \text{ or } x = -3 \qquad\qquad y = \sqrt{5} \text{ or } y = -\sqrt{5}$$

The **principal square root** of a positive real number is its positive square root. Thus, the principal square root of 9 is 3 and the principal square root of 5 is $\sqrt{5}$. In general, when referring to the square root of a number, we mean the principal square root.

When $\sqrt{5}$ is entered, a calculator will return the approximate decimal equivalent of $\sqrt{5}$ to as many decimal places as allowed by the calculator.

ENTER: [2nd] [√] 5 [ENTER]

DISPLAY:

```
√(5
        2.236067977
```

If we enter the decimal 2.236067977 into the calculator and square it, the calculator will return 4.999999998, an approximate value correct to nine decimal places. If we square 2.236067977 using pencil and paper we obtain a number with eighteen decimal places with 9 in the last decimal place. The exact decimal value of $\sqrt{5}$ is infinite and non-repeating.

Cube Root

DEFINITION _____

The **cube root** of k is one of the three equal factors whose product is k.

The cube root of k is written as $\sqrt[3]{k}$. For example, since $5 \cdot 5 \cdot 5 = 125$, 5 is one of the three equal factors whose product is 125 and $\sqrt[3]{125} = 5$. There is only one real number that is the cube root of 125. This real number is called the **principal cube root**. In a later chapter, we will enlarge the set of real numbers to form a set called the *complex numbers*. In the set of complex numbers, every real number has three cube roots, only one of which is a real number.

Note that $(-5) \cdot (-5) \cdot (-5) = -125$. Therefore, $-5 = \sqrt[3]{-125}$ and -5 is the principal cube root of -125.

The *n*th Root of a Number

DEFINITION _____

The **nth root** of k is one of the n equal factors whose product is k.

The nth root of k is written as $\sqrt[n]{k}$. In this notation, k is the **radicand**, n is the **index**, and $\sqrt[n]{k}$ is the **radical**. If k is positive, the **principal nth root** of k is the positive nth root.

If k is negative and n is odd, the **principal nth root** of k is negative. If k is negative and n is even, there is no **principal nth root** of k in the set of real numbers.

For example, in the expression $\sqrt[4]{16}$:

$$\text{Radical} \longrightarrow \overset{\displaystyle \ulcorner \text{Index}}{\underset{\llcorner \text{Radicand}}{\sqrt[4]{16}}}$$

Since $2 \cdot 2 \cdot 2 \cdot 2 = 16$, $\sqrt[4]{16} = 2$ and 2 is the principal fourth root of 16. Note that $(-2) \cdot (-2) \cdot (-2) \cdot (-2) = 16$, so $-\sqrt[4]{16} = -2$. The real number 16 has two fourth roots that are real numbers: 2 (the principal fourth root) and -2.

If the real number k is not the product of n equal factors that are rational numbers, then the principal nth root of k^n, $\sqrt[n]{k}$, is irrational. For example, since 13 is a prime and has only itself and 1 as factors, the principal nth root of 13, $\sqrt[n]{13}$, is an irrational number.

Since the product of an even number of positive factors is positive and the product of an even number of negative factors is positive, $\sqrt[n]{k}$ is a real number when n is even if and only if k is positive. For example $\sqrt[4]{81} = 3$ but $\sqrt[4]{-81}$ has no real roots.

To evaluate a radical on the calculator, we need to access the MATH menu. Entry 4, $\sqrt[3]{\ }($, is used to evaluate cube root. Entry 5, $\sqrt[x]{\ }$, is used to evaluate a radical with any index. When using entry 5, the index of the root must be entered first.

Evaluate $\sqrt[3]{12}$

ENTER: [MATH] [4] 12 [ENTER]

DISPLAY:
```
³√(12
         2.289428485
```

Evaluate $\sqrt[5]{35}$

ENTER: 5 [MATH] [5] 35 [ENTER]

DISPLAY:
```
5ˣ√35
         2.036168005
```

When a variable appears in the radicand, the radical can be simplified if the exponent of the variable is divisible by the index. For instance, since $8x^6 = (2x^2)(2x^2)(2x^2)$, $\sqrt[3]{8x^6} = 2x^2$.

EXAMPLE 1

If $a^2 = 169$, find all values of a.

Solution   Since $13 \cdot 13 = 169$ and $-13 \cdot (-13) = 169$, $a = \pm\sqrt{169} = \pm13$. *Answer*

EXAMPLE 2

Evaluate each of the following in the set of real numbers:

a. $\sqrt{49}$ **b.** $-\sqrt{121b^4}$ **c.** $\sqrt[3]{-\tfrac{1}{8}}$ **d.** $\sqrt[4]{-81}$

Solution   **a.** $7 \cdot 7 = 49$. Therefore, 7 is the principal square root and $\sqrt{49} = 7$.

b. $(-11b^2)(-11b^2) = 121b^4$. Therefore, $11b^2$ is the principal square root of $\sqrt{121b^4}$ and $-\sqrt{121b^4} = -11b^2$.

c. $\left(-\tfrac{1}{2}\right)\left(-\tfrac{1}{2}\right)\left(-\tfrac{1}{2}\right) = -\tfrac{1}{8}$. Therefore, $-\tfrac{1}{2}$ is the principal cube root of $-\tfrac{1}{8}$ and $\sqrt[3]{-\tfrac{1}{8}} = -\tfrac{1}{2}$.

d. A negative real number does not have four equal factors that are real numbers. $\sqrt[4]{-81}$ does not represent a real number.

Answers   **a.** 7   **b.** $-11b^2$ **c.** $-\tfrac{1}{2}$ **d.** not a real number

EXAMPLE 3

Find the length of the longer leg of a right triangle if the measure of the shorter leg is 9 centimeters and the measure of the hypotenuse is 41 centimeters.

Solution Use the Pythagorean Theorem, $a^2 + b^2 = c^2$. Let a be the length of the longer leg, b be the length of the shorter leg, and c be the length of the hypotenuse.

$$a^2 + b^2 = c^2$$
$$a^2 + 9^2 = 41^2$$
$$a^2 + 81 = 1{,}681$$
$$a^2 = 1{,}600$$
$$a = \pm 40$$

Reject the negative root.

Answer The length of the longer leg is 40 centimeters.

Exercises

Writing About Mathematics

1. a. Kevin said that if the index of a radical is even and the radicand is positive, then the radical has two real roots. Do you agree with Kevin? Explain why or why not.

   b. Kevin said that if the index of a radical is odd, then the radical has one real root and that the root is positive if the radicand is positive and negative if the radicand is negative. Do you agree with Kevin? Explain why or why not.

2. a. Sarah said that in the set of real numbers, $\sqrt{a}$ is one of the two equal factors whose product is a. Therefore, $\sqrt{a} \cdot \sqrt{a} = a$ for some values of a. Do you agree with Sarah? Explain why or why not.

   b. If you agree with Sarah, for which values of a is the statement true? Explain.

Developing Skills

In 3–10, tell whether each represents a number that is rational, irrational, or neither.

3. $\sqrt{25}$ **4.** $\sqrt{8}$ **5.** $\sqrt{-8}$ **6.** $\sqrt[3]{-8}$

7. $\sqrt{0}$ **8.** $\sqrt[4]{16}$ **9.** $\sqrt[5]{-243}$ **10.** $\sqrt{0.25}$

In 11–38, evaluate each expression in the set of real numbers.

11. $\sqrt{16}$ **12.** $\pm\sqrt{16}$ **13.** $-\sqrt{16}$ **14.** $\sqrt{625}$

15. $\sqrt{169}$ **16.** $-\sqrt{0.04}$ **17.** $\pm\sqrt{0.64}$ **18.** $\sqrt{1.44}$

19. $\sqrt[3]{27}$ **20.** $\sqrt[4]{16}$ **21.** $\sqrt[3]{-125}$ **22.** $-\sqrt[3]{125}$

23. $-\sqrt[3]{-125}$ **24.** $\sqrt[5]{-1}$ **25.** $\sqrt{\frac{4}{25}}$ **26.** $-\sqrt{\frac{49}{36}}$

27. $\sqrt[3]{\frac{8}{27}}$ **28.** $\sqrt[5]{-\frac{1}{32}}$ **29.** $\sqrt[3]{0.001}$ **30.** $\sqrt[4]{0.0256}$

31. $\sqrt{x^6}, x \geq 0$ **32.** $\sqrt{100c^4}, c \geq 0$ **33.** $\sqrt{0.25x^2}, x \geq 0$ **34.** $\sqrt[3]{\frac{1,000}{a^3}}, a \neq 0$

35. $-\sqrt[4]{\frac{b^8}{1,296}}, b \geq 0$ **36.** $\sqrt[5]{-0.00001y^5}, y \geq 0$

37. $\sqrt[k]{1}, k$ an integer greater than 2   **38.** $\sqrt[k]{x^{2k}}, k$ an integer greater than 2

In 39–42, find the set of real numbers for which the given radical is a real number.

39. $\sqrt{x-2}$ **40.** $\sqrt{9-3x}$ **41.** $\sqrt{4x+12}$ **42.** $\sqrt[4]{x+5}$

In 43–46, solve each equation for the variable.

43. $x^2 = 81$ **44.** $a^2 = 196$ **45.** $b^2 = 100$ **46.** $y^2 - 169 = 0$

Applying Skills

47. The area of a square is 14 square centimeters. What is the length of a side of the square?

48. Find the length of the hypotenuse of a right triangle if the length of the longer leg is 20 feet and the length of the shorter leg is 12 feet.

49. Find the length of the shorter leg of a right triangle if the length of the longer leg is 36 inches and the length of the hypotenuse is 39 inches.

50. What is the length of a side of a square if the length of a diagonal is $\sqrt{72}$ inches?

51. The length l, width w, and height h, of a rectangular carton are $l = 12$ feet, $w = 4$ feet, and $h = 3$ feet. The length of a diagonal, d, the distance from one corner of the box to the opposite corner, is given by the formula $d^2 = l^2 + w^2 + h^2$. What is the length of the diagonal of the carton?

3-3 SIMPLIFYING RADICALS

We know that $\sqrt{4 \cdot 9} = \sqrt{36} = 6$ and that $\sqrt{4} \cdot \sqrt{9} = 2 \cdot 3 = 6$. Therefore,

$$\sqrt{4 \cdot 9} = \sqrt{4} \cdot \sqrt{9}.$$

We can state this relationship in general terms. For all non-negative real numbers a and b:

$$\sqrt{a \cdot b} = \sqrt{a} \cdot \sqrt{b}$$

This relationship can be used when $\sqrt{a}$ and $\sqrt{b}$ are rational numbers and when they are irrational numbers. When $\sqrt{a \cdot b}$ is an irrational number, we can often write it in simpler form. For instance:

$$\sqrt{450} \qquad\qquad \sqrt{450} \qquad\qquad \sqrt{450}$$
$$= \sqrt{9} \cdot \sqrt{50} \qquad = \sqrt{25} \cdot \sqrt{18} \qquad = \sqrt{225} \cdot \sqrt{2}$$
$$= 3\sqrt{50} \qquad\qquad = 5\sqrt{18} \qquad\qquad = 15\sqrt{2}$$

Each of these irrational numbers is equal to $\sqrt{450}$. We say that $15\sqrt{2}$ is the **simplest form** of $\sqrt{450}$ because the radicand has no factor that is a perfect square. (Alternatively, a radical is in simplest form when the radicand is a *prime number* or the product of prime numbers with each prime factor occurring once.) Note that the radicand in each of the other forms of $\sqrt{450}$ has a factor that is a perfect square and can be simplified.

$$\sqrt{450} = 3\sqrt{50} \qquad\qquad \sqrt{450} = 5\sqrt{18}$$
$$= 3\sqrt{25}\sqrt{2} \qquad\qquad = 5\sqrt{9}\sqrt{2}$$
$$= 3 \cdot 5\sqrt{2} \qquad\qquad = 5 \cdot 3\sqrt{2}$$
$$= 15\sqrt{2} \qquad\qquad\quad = 15\sqrt{2}$$

We know that $x^3 \cdot x^3 = x^6$. Then for $x \geq 0$, $\sqrt{x^6} = x^3$. In general, $x^n \cdot x^n = x^{2n}$. Therefore, for $x \geq 0$:

$$\sqrt{x^{2n}} = x^n$$

In particular, when $n = 1$, we have that $\sqrt{x^2} = x$.
We can also write x^5 as $x^4 \cdot x$. Therefore,

$$\sqrt{x^5} = \sqrt{x^4 \cdot x} \qquad \sqrt{12y^7} = \sqrt{4y^6} \cdot \sqrt{3y} \qquad \sqrt{50a^3b^5} = \sqrt{25a^2b^4} \cdot \sqrt{2ab}$$
$$= \sqrt{x^4} \cdot \sqrt{x} \qquad\quad = 2y^3\sqrt{3y} \qquad\qquad\quad = 5ab^2\sqrt{2ab}$$
$$= x^2\sqrt{x} \qquad\qquad\quad (y \geq 0) \qquad\qquad\qquad (a \geq 0 \text{ and } b \geq 0)$$
$$(x \geq 0)$$

Fractional Radicands

When the radicand of an irrational number is a fraction, the radical is in simplest form when it is written with an integer under the radical sign. For example, $\sqrt{\frac{2}{3}}$ does not have a perfect square factor in either the numerator or the denominator of the radicand. The radical is not in simplest form. To simplify this radical, write $\frac{2}{3}$ as an equivalent fraction with a denominator that is a perfect square.

$$\sqrt{\tfrac{2}{3}} = \sqrt{\tfrac{2}{3} \times \tfrac{3}{3}} = \sqrt{\tfrac{6}{9}} = \sqrt{\tfrac{1}{9} \times \tfrac{6}{1}} = \sqrt{\tfrac{1}{9}} \times \sqrt{6} = \tfrac{1}{3}\sqrt{6} \text{ or } \tfrac{\sqrt{6}}{3}$$

Note that since $\frac{a}{c} = \left(\frac{1}{c}\right)\left(\frac{a}{1}\right) = \left(\frac{1}{c}\right)(a)$ and the square root of a product is equal to the product of the square roots of the factors:

$$\sqrt{\tfrac{a}{c}} = \sqrt{\tfrac{1}{c} \cdot \tfrac{a}{1}} = \sqrt{\tfrac{1}{c}} \cdot \sqrt{\tfrac{a}{1}} = \tfrac{1}{\sqrt{c}}\sqrt{a} = \tfrac{\sqrt{a}}{\sqrt{c}}$$

Therefore, for any non-negative a and positive c:

$$\sqrt{\tfrac{a}{c}} = \tfrac{\sqrt{a}}{\sqrt{c}}$$

This leads to an alternative method of simplifying a fractional radicand:

$$\sqrt{\tfrac{5}{7}} = \tfrac{\sqrt{5}}{\sqrt{7}} \times \tfrac{\sqrt{7}}{\sqrt{7}} = \tfrac{\sqrt{35}}{\sqrt{49}} = \tfrac{\sqrt{35}}{7}$$

or

$$\sqrt{\tfrac{a}{c}} = \tfrac{\sqrt{a}}{\sqrt{c}} \times \tfrac{\sqrt{c}}{\sqrt{c}} = \tfrac{\sqrt{ac}}{\sqrt{c^2}} = \tfrac{\sqrt{ac}}{c}$$

EXAMPLE 1

Write each expression in simplest form.

a. $\sqrt{\tfrac{8}{9}}$ **b.** $\sqrt{\tfrac{4}{5}}$ **c.** $\sqrt{\tfrac{3}{8}}$

Solution   **a.** Here the denominator is a perfect square and 8 has a perfect square factor, 4:

$$\sqrt{\tfrac{8}{9}} = \sqrt{\tfrac{4}{9} \times 2} = \sqrt{\tfrac{4}{9}} \times \sqrt{2} = \tfrac{2}{3}\sqrt{2} \;\; \textit{Answer}$$

b. Here the numerator is a perfect square but the denominator is not. We must multiply by $\tfrac{5}{5}$ to write the radicand as a fraction with a denominator that is a perfect square:

$$\sqrt{\tfrac{4}{5}} = \sqrt{\tfrac{4}{5} \times \tfrac{5}{5}} = \sqrt{\tfrac{20}{25}} = \sqrt{\tfrac{4}{25} \times 5} = \sqrt{\tfrac{4}{25}} \times \sqrt{5} = \tfrac{2}{5}\sqrt{5} \;\; \textit{Answer}$$

c. Here neither the numerator nor the denominator is a perfect square. We must first write the radicand as a fraction whose denominator is a perfect square. Since 16 is the smallest perfect square that is a multiple of 8, we can multiply by $\tfrac{2}{2}$ to obtain a radicand with a denominator that is a perfect square, 16:

$$\sqrt{\tfrac{3}{8}} = \sqrt{\tfrac{3}{8} \times \tfrac{2}{2}} = \sqrt{\tfrac{6}{16}} = \tfrac{\sqrt{6}}{\sqrt{16}} = \tfrac{\sqrt{6}}{4} \;\; \textit{Answer}$$

Note: Part **b** could also be solved by an alternative method. Rewrite the radical as the quotient of two radicals. Multiply by $\tfrac{\sqrt{5}}{\sqrt{5}}$. Then simplify:

$$\sqrt{\tfrac{4}{5}} = \tfrac{\sqrt{4}}{\sqrt{5}} \times \tfrac{\sqrt{5}}{\sqrt{5}}$$
$$= \tfrac{\sqrt{4}\sqrt{5}}{\sqrt{25}}$$
$$= \tfrac{2\sqrt{5}}{5} \;\; \textit{Answer}$$

To write square root radicals with a variable in the denominator, we want the variable in the denominator to be a perfect square, that is, to have an exponent that is even. For example, simplify $\sqrt{\frac{9a}{8b^3}}$ if $a > 0$ and $b > 0$. To do this we must write the radicand of the denominator as a perfect square. The smallest perfect square that is a multiple of 8 is 8×2 or 16 and the smallest perfect square that is a multiple of b^3 is $b^3 \cdot b$ or b^4. In order to have a denominator of $\sqrt{16b^4}$ we must multiply the given radical by $\sqrt{\frac{2b}{2b}}$.

$$\sqrt{\frac{9a}{8b^3}} = \sqrt{\frac{9a}{8b^3} \cdot \frac{2b}{2b}} = \sqrt{\frac{18ab}{16b^4}} = \sqrt{\frac{9}{16b^4}} \cdot \sqrt{2ab} = \frac{3}{4b^2}\sqrt{2ab}$$

Alternatively, we could rewrite the radical as the quotient of radicals, multiply by $\frac{\sqrt{8b^3}}{\sqrt{8b^3}}$, and then simplify:

$$\sqrt{\frac{9a}{8b^3}} = \frac{\sqrt{9a}}{\sqrt{8b^3}} \cdot \frac{\sqrt{8b^3}}{\sqrt{8b^3}} = \frac{\sqrt{72ab^3}}{\sqrt{(8b^3)^2}} = \frac{6b\sqrt{2ab}}{8b^3} = \frac{3\sqrt{2ab}}{4b^2}$$

Roots with Index *n*

The rules that we have derived for square roots are true for roots with any index.

$$\sqrt[n]{ab} = \sqrt[n]{a} \cdot \sqrt[n]{b} \quad \textbf{and} \quad \sqrt[n]{\frac{a}{b}} = \frac{\sqrt[n]{a}}{\sqrt[n]{b}}$$

In order to write $\sqrt[n]{k}$ in simplest form, we must find the largest factor of k of the form a^n. For instance:

$$\sqrt[3]{40} = \sqrt[3]{8 \times 5} = \sqrt[3]{8} \times \sqrt[3]{5} = 2\sqrt[3]{5}$$

Recall that if the index of a root is n, then if a is a rational number, $\sqrt[n]{a^k}$ is a rational number when k is divisible by n. For example:

$$\sqrt[3]{a^6} = a^2 \qquad \sqrt[4]{b^{12}} = b^3 \qquad \sqrt[5]{q^5} = q^1 = q \qquad \sqrt[n]{x^{an}} = x^a$$

In order to write $\sqrt[3]{x^8}$ in simplest form, find the largest multiple of 3 that is less than 8. That multiple is 6. Write x^8 as $x^6 \cdot x^2$.

$$\sqrt[3]{x^8} = \sqrt[3]{x^6} \cdot \sqrt[3]{x^2} = x^2 \cdot \sqrt[3]{x^2}$$

In order to simplify $\sqrt[n]{\frac{p}{q}}$, we must write $\frac{p}{q}$ as an equivalent fraction whose denominator is of the form r^n. For instance:

$$\sqrt[4]{\frac{1}{2}} = \sqrt[4]{\frac{1}{2} \times \frac{2^3}{2^3}} = \sqrt[4]{\frac{2^3}{2^4}} = \frac{\sqrt[4]{2^3}}{\sqrt[4]{2^4}} = \frac{\sqrt[4]{8}}{2} \text{ or } \frac{1}{2}\sqrt[4]{8}$$

$$\sqrt[4]{\frac{3a}{2b^3c^2}} = \sqrt[4]{\frac{3a}{2b^3c^2} \cdot \frac{8bc^2}{8bc^2}} = \sqrt[4]{\frac{24abc^2}{16b^4c^4}}$$

$$= \sqrt[4]{\frac{1}{16b^4c^4}}\sqrt[4]{24abc^2}$$

$$= \frac{1}{2bc}\sqrt[4]{24abc^2}$$

EXAMPLE I

Write each of the following in simplest radical form:

a. $\sqrt{48}$ **b.** $\sqrt{\frac{9}{20b}}$ **c.** $\sqrt[3]{54y^8}$

Solution **a.** The factors of 48 are $2 \times 24, 3 \times 16, 4 \times 12,$ and 6×8. The pair of factors with the largest perfect square is 3×16:

$$\sqrt{48} = \sqrt{16 \times 3} = \sqrt{16} \times \sqrt{3} = 4\sqrt{3} \ \textit{Answer}$$

b. The smallest multiple of $20b$ that is a perfect square is $100b^2$. Multiply the radicand by $\frac{5b}{5b}$:

$$\sqrt{\frac{9}{20b}} = \sqrt{\frac{9}{20b} \cdot \frac{5b}{5b}} = \sqrt{\frac{45b}{100b^2}}$$
$$= \sqrt{\frac{9}{100b^2} \cdot 5b}$$
$$= \sqrt{\frac{9}{100b^2}} \cdot \sqrt{5b}$$
$$= \frac{3}{10b}\sqrt{5b} \ \textit{Answer}$$

c. The factors of 54 are $2 \times 27, 3 \times 18,$ and 6×9. The factor 27 is the cube of 3. The largest multiple of the index that is less than the exponent of y is 6:

$$\sqrt[3]{54y^8} = \sqrt[3]{27y^6 \cdot 2y^2} = \sqrt[3]{27y^6} \cdot \sqrt[3]{2y^2} = 3y^2 \cdot \sqrt[3]{2y^2} \ \textit{Answer}$$

EXAMPLE 2

a. Write $2\sqrt{50b^3}$ in simplest radical form.

b. For what values b does $2\sqrt{50b^3}$ represent a real number?

Solution **a.** $2\sqrt{50b^3} = 2\sqrt{25b^2} \cdot \sqrt{2b}$

$\qquad\qquad = 2(5b)\sqrt{2b}$

$\qquad\qquad = 10b\sqrt{2b} \ \textit{Answer}$

b. $2\sqrt{50b^3}$ is a real number when $50b^3$ is positive or 0, that is, for $b \geq 0$. *Answer*

EXAMPLE 3

Write $\sqrt{0.2}$ in simplest radical form.

Solution **METHOD I**
Write 0.2 as the product of a perfect square and a prime:

$$\sqrt{0.2} = \sqrt{0.20}$$
$$= \sqrt{0.04 \times 5}$$
$$= \sqrt{0.04} \times \sqrt{5}$$
$$= 0.2\sqrt{5}$$

METHOD 2

Write 0.2 as a common fraction, $\frac{2}{10}$. Write $\frac{2}{10}$ as an equivalent fraction with a denominator that is a perfect square, 100:

$$\sqrt{0.2} = \sqrt{\tfrac{2}{10}}$$
$$= \sqrt{\tfrac{2}{10} \times \tfrac{10}{10}}$$
$$= \sqrt{\tfrac{20}{100}}$$
$$= \sqrt{\tfrac{4}{100}} \times \sqrt{5}$$
$$= \tfrac{2}{10}\sqrt{5}$$

Answer $0.2\sqrt{5}$ or $\frac{2}{10}\sqrt{5}$

Note: For a decimal fraction, the denominator will be a perfect square when there are two decimal places (hundredths), four decimal places (ten-thousandths) or any even number of decimal places (a denominator of 10^{2n} for n a counting number). For example, $\sqrt{0.01} = 0.1$, $\sqrt{0.0001} = 0.01$, $\sqrt{0.000001} = 0.001$.

Exercises

Writing About Mathematics

1. Explain the difference between $-\sqrt{36}$ and $\sqrt{-36}$.

2. If a is a negative number, is $-\sqrt[3]{-8a^3}$ a positive number, a negative number, or not a real number? Explain your answer.

Developing Skills

In 3–38, write each radical in simplest radical form. Variables in the radicand of an even index are non-negative. Variables occurring in the denominator of a fraction are non-zero.

3. $\sqrt{12}$ **4.** $\sqrt{50}$ **5.** $\sqrt{32}$ **6.** $\sqrt{8b^3}$

7. $\sqrt{98c^4}$ **8.** $3\sqrt{20y^5}$ **9.** $5\sqrt{200xy^2}$ **10.** $4\sqrt{363x^5y^7}$

11. $\frac{1}{2}\sqrt{72ab^5}$ **12.** $\sqrt[3]{16}$ **13.** $\sqrt[3]{24}$ **14.** $\sqrt[3]{40a^4}$

15. $\sqrt[3]{375x^5y^6}$ **16.** $\sqrt[4]{48a^9b^3}$ **17.** $\sqrt{\frac{4a^4}{25}}$ **18.** $\sqrt{\frac{3b^3}{49}}$

19. $\sqrt{\frac{6}{81y^6}}$ **20.** $\sqrt{\frac{a^5}{2}}$ **21.** $\sqrt{\frac{a^3}{5b}}$ **22.** $\sqrt{\frac{1}{6xy}}$

23. $\sqrt{\frac{3x}{4y}}$ **24.** $\sqrt{\frac{3}{5b^5}}$ **25.** $\sqrt{\frac{5a}{18}}$ **26.** $\sqrt{\frac{15}{8b^3}}$

27. $\sqrt{\dfrac{4a^3}{27b^3}}$ **28.** $\sqrt{\dfrac{9}{50xy^7}}$ **29.** $\sqrt{\dfrac{8x}{5y^3}}$ **30.** $\sqrt{0.08}$

31. $\sqrt{0.5}$ **32.** $\sqrt{2.42ab^5}$ **33.** $\sqrt{0.001x^3}$ **34.** $\sqrt{128x^2}$

35. $\sqrt{300c}$ **36.** $\sqrt[3]{\dfrac{a^3}{3}}$ **37.** $\sqrt[4]{\dfrac{2a^8}{b^2c^3}}$ **38.** $\sqrt[4]{32x^5y^4}$

Applying Skills

39. The lengths of the legs of a right triangle are 8 centimeters and 12 centimeters. Express the length of the hypotenuse in simplest radical form.

40. The length of one leg of an isosceles right triangle is 6 inches. Express the length of the hypotenuse in simplest radical form.

41. The length of the hypotenuse of a right triangle is 24 meters and the length of one leg is 12 meters. Express the length of the other leg in simplest radical form.

42. The dimensions of a rectangle are 15 feet by 10 feet. Express the length of the diagonal in simplest radical form.

43. The area of a square is 150 square feet. Express the length of a side of the square in simplest radical form.

44. The area of a circular pool is $5x^2y^4\pi$ square meters. Express the radius of the pool in simplest radical form.

45. The area of a triangle is $\sqrt[5]{243x^5y^{10}}$ square units. If the length of a side of the triangle is $\sqrt[5]{x^5}$, express the length of the altitude to that side in simplest radical form.

46. Tom has a trunk that is 18 inches wide, 32 inches long, and 16 inches high. Which of the following objects could Tom store in the trunk? (There may be more than one answer.)

    (1) A walking stick 42 inches long

    (2) A fishing rod 37 inches long

    (3) A yardstick

    (4) A baseball bat 34 inches long

3-4 ADDING AND SUBTRACTING RADICALS

Recall that the sum or difference of fractions or of similar algebraic terms is found by using the distributive property.

$$\frac{2}{5} + \frac{4}{5} = \frac{1}{5} \times 2 + \frac{1}{5} \times 4 \qquad\qquad 2x + 4x = x(2 + 4)$$
$$= \frac{1}{5}(2 + 4) \qquad\qquad\qquad\qquad = x(6)$$
$$= \frac{1}{5}(6) \qquad\qquad\qquad\qquad\qquad = 6x$$
$$= \frac{6}{5}$$

We do not write all of the steps shown above when adding fractions or algebraic terms. The principles that justify each of these steps assure us that the results are correct.

We can apply the same principles to the addition or subtraction of radical expressions. To express the sum or difference of two radicals as a single radical, the radicals must have the same index and the same radicand, that is, they must be **like radicals**. We can use the same procedure that we use to add or subtract like terms.

$$2\sqrt{5} + 4\sqrt{5} \qquad\qquad 4\sqrt{3b} - 3\sqrt{3b} \qquad\qquad \sqrt[3]{2} + 7\sqrt[3]{2}$$
$$= \sqrt{5}(2 + 4) \qquad\qquad = \sqrt{3b}(4 - 3) \qquad\qquad = \sqrt[3]{2}(1 + 7)$$
$$= \sqrt{5}(6) \qquad\qquad = \sqrt{3b}(1) \qquad\qquad = \sqrt[3]{2}(8)$$
$$= 6\sqrt{5} \qquad\qquad = \sqrt{3b} \qquad\qquad = 8\sqrt[3]{2}$$

Two radicals that do not have the same radicand or do not have the same index are **unlike radicals**. The sum or difference of two unlike radicals cannot be expressed as a single radical if they cannot be written as equivalent like radicals. Each of the following, $\sqrt{2} + \sqrt{3}$, $\sqrt[3]{2} + \sqrt{2}$, and $\sqrt[4]{2} + \sqrt{3}$, are the sums of unlike radicals and are in simplest radical form.

Simplifying Unlike Radicals

The fractions $\frac{1}{3}$ and $\frac{1}{4}$ do not have common denominators but we can add these two fractions by writing them as equivalent fractions with a common denominator. The radicals $\sqrt{8}$ and $\sqrt{50}$ have the same index but do not have the same radicand. We can add these radicals if they can be written in simplest form with a common radicand.

$$\sqrt{8} = \sqrt{4}\sqrt{2} = 2\sqrt{2} \qquad \text{and} \qquad \sqrt{50} = \sqrt{25}\sqrt{2} = 5\sqrt{2}$$
$$\sqrt{8} + \sqrt{50} = 2\sqrt{2} + 5\sqrt{2} = 7\sqrt{2}$$
$$\sqrt{12b^3} = \sqrt{4b^2}\sqrt{3b} = 2b\sqrt{3b} \qquad \text{and} \qquad \sqrt{27b^3} = \sqrt{9b^2}\sqrt{3b} = 3b\sqrt{3b}$$
$$\sqrt{12b^3} + \sqrt{27b^3} = 2b\sqrt{3b} + 3b\sqrt{3b} = 5b\sqrt{3b}$$

Note that it is not always possible to express two unlike radicals as like radicals. For example, the sum $\sqrt{2} + \sqrt{3}$ cannot be written with the same radicand and therefore the sum cannot be written as one radical.

EXAMPLE I

Write $\sqrt{12} + \sqrt{20} - \sqrt{45} + \sqrt{\frac{1}{3}}$ in simplest form.

Solution Simplify each radical and combine like radicals.

$$\sqrt{12} + \sqrt{20} - \sqrt{45} + \sqrt{\tfrac{1}{3}} = \sqrt{4}\sqrt{3} + \sqrt{4}\sqrt{5} - \sqrt{9}\sqrt{5} + \sqrt{\tfrac{3}{9}}$$
$$= 2\sqrt{3} + 2\sqrt{5} - 3\sqrt{5} + \tfrac{1}{3}\sqrt{3}$$
$$= 2\sqrt{3} + \tfrac{1}{3}\sqrt{3} + 2\sqrt{5} - 3\sqrt{5}$$
$$= \tfrac{7}{3}\sqrt{3} - \sqrt{5} \;\; Answer$$

EXAMPLE 2

Simplify: $3x\sqrt{\frac{1}{3x}} + \sqrt{300x}$

Solution Simplify each radical.

$$3x\sqrt{\tfrac{1}{3x}} = 3x\sqrt{\tfrac{1}{3x}\cdot\tfrac{3x}{3x}} = 3x\sqrt{\tfrac{3x}{9x^2}} = 3x\tfrac{\sqrt{3x}}{3x} = 3\overset{1}{x}\tfrac{\sqrt{3x}}{\underset{1}{3x}} = \sqrt{3x}$$
$$\sqrt{300x} = \sqrt{100}\sqrt{3x} = 10\sqrt{3x}$$

Add the simplified radicals.

$$3x\sqrt{\tfrac{1}{3x}} + \sqrt{300x} = \sqrt{3x} + 10\sqrt{3x} = 11\sqrt{3x} \;\; (x > 0) \;\; Answer$$

EXAMPLE 3

Solve and check: $4x - \sqrt{8} = \sqrt{72}$

Solution

$$4x - \sqrt{8} = \sqrt{72}$$
$$4x - \sqrt{8} + \sqrt{8} = \sqrt{72} + \sqrt{8}$$
$$4x = \sqrt{36}(\sqrt{2}) + \sqrt{4}(\sqrt{2})$$
$$4x = 6\sqrt{2} + 2\sqrt{2}$$
$$4x = 8\sqrt{2}$$
$$\tfrac{4x}{4} = \tfrac{8\sqrt{2}}{4}$$
$$x = 2\sqrt{2}$$

Check

$$4x - \sqrt{8} = \sqrt{72}$$
$$4(2\sqrt{2}) - \sqrt{4}(\sqrt{2}) \overset{?}{=} \sqrt{36}(\sqrt{2})$$
$$8\sqrt{2} - 2\sqrt{2} \overset{?}{=} 6\sqrt{2}$$
$$6\sqrt{2} = 6\sqrt{2} \;\checkmark$$

Answer $x = 2\sqrt{2}$

Exercises

Writing About Mathematics

1. Danielle said that $3x\sqrt{\frac{1}{3x}}$ could be simplified by writing $3x\sqrt{\frac{1}{3x}}$ as $\sqrt{\frac{9x^2}{3x}} = \sqrt{3x}$. Do you agree with Danielle? Justify your answer.

2. Does $\sqrt{16} + \sqrt{48} = \sqrt{64}$? Justify your answer.

Developing Skills

In 3–38 write each expression in simplest form. Variables in the radicand with an even index are non-negative. Variables occurring in the denominator of a fraction are non-zero.

3. $\sqrt{2} + 5\sqrt{2}$

4. $6\sqrt{5} - 4\sqrt{5}$

5. $8\sqrt{3} + \sqrt{3}$

6. $5\sqrt{7} - \sqrt{7}$

7. $\sqrt{50} + \sqrt{2}$

8. $3\sqrt{5y} - \sqrt{20y}$

9. $\sqrt{250a^2} + \sqrt{10a^2}$

10. $8\sqrt{11b^4} - \sqrt{99b^4}$

11. $\sqrt{24xy^2} + \sqrt{54xy^2}$

12. $\sqrt{200a^7} - \sqrt{50a^7}$

13. $\sqrt{98c^5} - \sqrt{18c^5}$

14. $x\sqrt{32x} + \sqrt{128x^3}$

15. $4b\sqrt{24b^3} + \sqrt{54b^5}$

16. $3x^3\sqrt{80} + 2\sqrt{125x^6}$

17. $\sqrt{5} + \sqrt{\frac{1}{5}}$

18. $\sqrt{24} + 2\sqrt{\frac{3}{2}}$

19. $14\sqrt{\frac{1}{7}} + \sqrt{28}$

20. $\sqrt{\frac{1}{2x}} + \sqrt{\frac{1}{2x}}$

21. $a\sqrt{45} + \sqrt{20a^2} - 5\sqrt{2a}$

22. $x\sqrt{600} - 2\sqrt{24x^2} + 4x\sqrt{96}$

23. $2\sqrt{3y} - 5y^2 + 4\sqrt{3y} + \sqrt{36y^4}$

24. $\sqrt{162a^4b^3} + 3 - ab\sqrt{18a^2b} - 1$

25. $\sqrt{12} - \sqrt{24} + \sqrt{48} + \sqrt{27}$

26. $5\sqrt{\frac{1}{5}} - \sqrt{\frac{1}{10}} + \sqrt{20}$

27. $\sqrt{\frac{1}{6}} + \sqrt{\frac{8}{3}} - \sqrt{\frac{2}{3}}$

28. $\sqrt[3]{2} + \sqrt[3]{16}$

29. $\sqrt[3]{54} + \sqrt[3]{128}$

30. $\sqrt[4]{48} - \sqrt[4]{3}$

31. $\sqrt{9x} + \sqrt{25x}$

32. $\sqrt{100y} - \sqrt{25y}$

33. $\sqrt{8a} - \sqrt{2a}$

34. $\sqrt{18b^2} + \sqrt{800b^2}$

35. $\sqrt{63a^2} - \sqrt{45a^2}$

36. $\sqrt{4ab^2} - \sqrt{ab^2}$

37. $\sqrt{50x^3} + \sqrt{200x^3}$

38. $\sqrt{49x^3} - 2x\sqrt{4x}$

In 39–42, solve and check each equation.

39. $5x - \sqrt{3} = \sqrt{48}$

40. $12y + \sqrt{32} = \sqrt{200}$

41. $4a + \sqrt{6} = a + \sqrt{96}$

42. $y + \sqrt{20} = \sqrt{45} - 2y$

Applying Skills

In 43–47, express each answer in simplest radical form.

43. The lengths of the sides of a triangle are $\sqrt{75}$ inches, $\sqrt{27}$ inches, and $\sqrt{108}$ inches. What is the perimeter of the triangle?

44. The length of each of the two congruent sides of an isosceles triangle is $\sqrt{500}$ feet and the length of the third side is $\sqrt{45}$ feet. What is the perimeter of the triangle?

45. The lengths of the legs of a right triangle are $\sqrt{18}$ centimeters and $\sqrt{32}$ centimeters.

a. Find the length of the hypotenuse.

b. Find the perimeter of the triangle.

46. The length of each leg of an isosceles right triangle is $\sqrt{98}$ inches.

a. Find the length of the hypotenuse.

b. What is the perimeter of the triangle?

47. The dimensions of a rectangle are $\sqrt{250}$ meters and $\sqrt{1,440}$ meters.

a. Express the perimeter of the rectangle in simplest radical form.

b. Express the length of the diagonal of the rectangle in simplest radical form.

3-5 MULTIPLYING RADICALS

Recall that if a and b are non-negative numbers, $\sqrt{ab} = \sqrt{a} \cdot \sqrt{b}$. Therefore, by the symmetric property of equality, we can say that $\sqrt{a} \cdot \sqrt{b} = \sqrt{ab}$. Recall also that for any positive number a, $\sqrt{a} \cdot \sqrt{a} = \sqrt{a^2} = a$. We can use these rules to multiply radicals.

For example:

$$\sqrt{4} \cdot \sqrt{25} = \sqrt{100} = 10$$
$$\sqrt{2} \cdot \sqrt{2} = \sqrt{4} = 2$$
$$\sqrt{8} \cdot \sqrt{2} = (\sqrt{4} \cdot \sqrt{2}) \cdot \sqrt{2} = 2(\sqrt{2} \cdot \sqrt{2}) = 2(2) = 4$$
$$\sqrt{6a^3} \cdot \sqrt{18a} = \sqrt{108a^4} = \sqrt{36a^4} \cdot \sqrt{3} = 6a^2\sqrt{3} \quad (a \geq 0)$$

Note: $\sqrt{-2} \times \sqrt{-8} \neq \sqrt{16}$ because $\sqrt{-2}$ and $\sqrt{-8}$ are not real numbers.

The rule for the multiplication of radicals is true for any two radicals with the same index if the radicals are real numbers. That is, since $\sqrt[n]{ab} = \sqrt[n]{a} \cdot \sqrt[n]{b}$ is true for non-negative real numbers, the following is also true by the symmetric property of equality:

$$\sqrt[n]{a} \cdot \sqrt[n]{b} = \sqrt[n]{ab} \ (a, b \geq 0)$$

Note: This rule does not apply when the index n is even and a or b are negative. For instance, $\sqrt[4]{-2} \times \sqrt[4]{-8} \neq \sqrt[4]{16}$ because $\sqrt[4]{-2}$ and $\sqrt[4]{-8}$ are not real numbers.

EXAMPLE I

Simplify:

Answers

a. $\sqrt[3]{8} \cdot \sqrt[3]{27}$

$= \sqrt[3]{216} = 6$

or

$= 2 \cdot 3 = 6$

b. $\sqrt[4]{48x^2} \cdot \sqrt[4]{\frac{x^2}{3}}$

$= \sqrt[4]{48x^2 \cdot \frac{x^2}{3}} = \sqrt[4]{16x^4} = 2x \ (x \geq 0)$

Multiplying Sums That Contain Radicals

The distributive property for multiplication over addition or subtraction is true for all real numbers. Therefore, we can apply it to irrational numbers that contain radicals.

For example:

$$\sqrt{3}(2 + \sqrt{3}) = \sqrt{3}(2) + \sqrt{3}(\sqrt{3}) = 2\sqrt{3} + 3$$
$$(2 + \sqrt{5})(1 + \sqrt{5}) = 2(1 + \sqrt{5}) + \sqrt{5}(1 + \sqrt{5})$$
$$= 2 + 2\sqrt{5} + \sqrt{5} + 5 = 7 + 3\sqrt{5}$$
$$(3 + \sqrt{2})(3 - \sqrt{2}) = 3(3 - \sqrt{2}) + \sqrt{2}(3 - \sqrt{2})$$
$$= 9 - 3\sqrt{2} + 3\sqrt{2} - 2 = 7$$

In each of these examples, we are multiplying irrational numbers. We know that $\sqrt{3}$, $\sqrt{5}$ and $\sqrt{2}$ are all irrational numbers. The sum of a rational number and an irrational number is always an irrational number. Therefore, $2 + \sqrt{3}$, $2 + \sqrt{5}, 1 + \sqrt{5}, 3 + \sqrt{2}$, and $3 - \sqrt{2}$ are all irrational numbers. The product of irrational numbers may be a rational or an irrational number.

EXAMPLE 2

Express each of the following products in simplest form:

a. $\sqrt{5}(\sqrt{10})$ **b.** $(3 + \sqrt{6a})(1 + \sqrt{2a})$ **c.** $(\sqrt{5} + \sqrt{7})(\sqrt{5} - \sqrt{7})$

Solution **a.** $\sqrt{5}(\sqrt{10}) = \sqrt{50} = \sqrt{25} \cdot \sqrt{2} = 5\sqrt{2}$ *Answer*

b. $(3 + \sqrt{6a})(1 + \sqrt{2a}) = 3(1 + \sqrt{2a}) + \sqrt{6a}(1 + \sqrt{2a})$

$$= 3 + 3\sqrt{2a} + \sqrt{6a} + \sqrt{12a^2}$$
$$= 3 + 3\sqrt{2a} + \sqrt{6a} + \sqrt{4a^2}(\sqrt{3})$$
$$= 3 + 3\sqrt{2a} + \sqrt{6a} + 2a\sqrt{3} \text{ Answer}$$

c. $(\sqrt{5} + \sqrt{7})(\sqrt{5} - \sqrt{7}) = \sqrt{5}(\sqrt{5} - \sqrt{7}) + \sqrt{7}(\sqrt{5} - \sqrt{7})$

$$= 5 - \sqrt{35} + \sqrt{35} - 7$$
$$= -2 \text{ Answer}$$

EXAMPLE 3

The length of a side, s, of an equilateral triangle is $\sqrt{5}$ inches and the length of the altitude, h, is $\sqrt{\frac{15}{4}}$ inches. Find the area of the triangle.

Solution

$$\text{Area of a triangle} = \tfrac{1}{2}bh$$
$$= \tfrac{1}{2}(\sqrt{5})\left(\sqrt{\tfrac{15}{4}}\right)$$
$$= \tfrac{1}{2}\left(\sqrt{\tfrac{75}{4}}\right)$$
$$= \tfrac{1}{2}\left(\sqrt{\tfrac{25}{4}} \cdot \sqrt{3}\right)$$
$$= \tfrac{1}{2}\left(\tfrac{5}{2}\sqrt{3}\right)$$
$$= \tfrac{5}{4}\sqrt{3} \text{ sq inches } \textit{Answer}$$

Exercises

Writing About Mathematics

1. Brandon said that if a is a positive real number, then $3a$, $4a$, and $5a$ are the lengths of the sides of a right triangle. Therefore, $3\sqrt{2}$, $4\sqrt{2}$, and $5\sqrt{2}$ are the lengths of the sides of a right triangle. Do you agree with Brandon? Justify your answer.

2. Jennifer said that if a is a positive real number, then $\sqrt[4]{a^2} = \sqrt{a}$. Do you agree with Jennifer? Justify your answer.

Developing Skills

In 3–41, express each product in simplest form. Variables in the radicand with an even index are non-negative.

3. $\sqrt{2} \cdot \sqrt{8}$

4. $\sqrt{5} \cdot \sqrt{45}$

5. $\sqrt{3} \cdot \sqrt{27}$

6. $\sqrt{8} \cdot \sqrt{12}$

7. $-\sqrt{10} \cdot \sqrt{18}$

8. $3\sqrt{2} \cdot \sqrt{10}$

9. $\sqrt{\frac{1}{3}} \cdot \sqrt{24}$

10. $\sqrt{21} \cdot \sqrt{\frac{4}{3}}$

11. $8\sqrt{6} \cdot \sqrt{\frac{5}{12}}$

12. $\left(\sqrt{12}\right)^2$

13. $\left(3\sqrt{3}\right)^2$

14. $\left(-2\sqrt{5}\right)^2$

15. $\sqrt{x^3} \cdot \sqrt{4x}$

16. $2\sqrt{ab} \cdot 2\sqrt{ab^2}$

17. $\sqrt{5y} \cdot \sqrt{4y^3}$

18. $\sqrt{x^5 y^3} \cdot \sqrt{3xy}$

19. $7\sqrt{a} \cdot 5\sqrt{\frac{a}{9}}$

20. $\sqrt{\frac{x}{2}} \cdot \sqrt{\frac{x^2}{2}}$

21. $\sqrt{\frac{a}{3}} \cdot \sqrt{\frac{a^2}{4}}$

22. $\sqrt[3]{2} \cdot \sqrt[3]{4}$

23. $\sqrt[3]{15a^2} \cdot \sqrt[3]{9a^4}$

24. $\sqrt[4]{27} \cdot \sqrt[4]{3}$

25. $\sqrt{2}\left(2 + \sqrt{2}\right)$

26. $\sqrt{5}\left(1 - \sqrt{10}\right)$

27. $\sqrt{8}\left(6 + \sqrt{2}\right)$

28. $\sqrt{5a}\left(\sqrt{5a} - 3\right)$

29. $\sqrt{12xy^3}\left(\sqrt{3xy} + 3\right)$

30. $\left(1 + \sqrt{5}\right)\left(3 - \sqrt{5}\right)$

31. $\left(9 + \sqrt{2b}\right)\left(1 + \sqrt{2b}\right)$

32. $\left(7 + \sqrt{5y}\right)\left(3 - \sqrt{5y}\right)$

33. $\left(7 + \sqrt{5b}\right)\left(7 - \sqrt{5b}\right)$

34. $\left(x - \sqrt[4]{3y}\right)\left(2x - \sqrt[4]{3y}\right)$

35. $\left(\sqrt{6} + 6\right)\left(\sqrt{6} - 7\right)$

36. $\left(\sqrt{6} + 6c\right)\left(\sqrt{6} - 6c\right)$

37. $\left(a + \sqrt{b}\right)\left(a - \sqrt{b}\right)$

38. $\left(1 - \sqrt{3}\right)^2$

39. $\left(3 + \sqrt{5ab^3}\right)^2$

40. $\left(1 - \sqrt{7}\right)\left(1 + \sqrt{7}\right)\left(1 + \sqrt{7}\right)$

41. $\left(2 - \sqrt{5}\right)\left(2 + \sqrt{5}\right)^2$

Applying Skills

42. The length of a side of a square is $48\sqrt{2}$ meters. Express the area of the square in simplest form.

43. The dimensions of a rectangle are $12\sqrt{2}$ feet by $\sqrt{50}$ feet. Express the area of the rectangle in simplest form.

44. The dimensions of a rectangular solid are $\sqrt{5}$ inches by $\left(2 + \sqrt{3}\right)$ inches by $\left(2 - \sqrt{3}\right)$ inches. Express the volume of the solid in simplest form.

45. The lengths of the legs of a right triangle in feet are $\left(3 + \sqrt{3}\right)$ and $\left(3 - \sqrt{3}\right)$.

   a. Find the length of the hypotenuse of the triangle.

   b. Express, in simplest form, the perimeter of the triangle.

   c. What is the area of the triangle?

46. The radius of the surface of a circular pool is $\left(2 + \sqrt{xy^5}\right)$ meters. Express the area of the pool in simplest form.

3-6 DIVIDING RADICALS

We can use what we know about the relationship between division and multiplication and the rule for the multiplication of radicals to write a rule for the division of radicals. We know that for any non-negative a and positive b, $\sqrt{a} \div \sqrt{b} = \frac{\sqrt{a}}{\sqrt{b}}$. Therefore, if a and b are positive real numbers, then:

$$\sqrt{a} \div \sqrt{b} = \sqrt{a} \cdot \sqrt{\tfrac{1}{b}} = \sqrt{\tfrac{a}{b}} \quad \text{or} \quad \frac{\sqrt{a}}{\sqrt{b}} = \sqrt{\tfrac{a}{b}}$$

Recall that a similar rule is true for roots with any index, and so we can write:

$$\sqrt[n]{a} \div \sqrt[n]{b} = \frac{\sqrt[n]{a}}{\sqrt[n]{b}} = \sqrt[n]{\tfrac{a}{b}}$$

Note: $\sqrt[3]{\frac{-8}{27}} = \frac{\sqrt[3]{-8}}{\sqrt[3]{27}} = \frac{-2}{3} = -\frac{2}{3}$ is a true statement because $\sqrt[3]{\frac{-8}{27}}$, $\sqrt[3]{-8}$, and $\sqrt[3]{27}$ are all real numbers. However, we cannot write $\sqrt{\frac{-4}{9}} = \frac{\sqrt{-4}}{\sqrt{9}}$ because $\sqrt{\frac{-4}{9}}$ and $\sqrt{-4}$ are not real numbers.

EXAMPLE 1

Express $\sqrt{27a} \div \sqrt{3a^3}$ in simplest radical form when a is a positive real number.

Solution
$$\sqrt{27a} \div \sqrt{3a^3} = \frac{\sqrt{27a}}{\sqrt{3a^3}} = \sqrt{\frac{27a}{3a^3}} = \sqrt{\frac{9}{a^2} \cdot \frac{\overset{1}{\cancel{3a}}}{\underset{1}{\cancel{3a}}}} = \sqrt{\frac{9}{a^2}} = \frac{3}{a} \quad \textit{Answer}$$

Alternative Solution
$$\sqrt{27a} \div \sqrt{3a^3} = \frac{\sqrt{27a}}{\sqrt{3a^3}} = \frac{\sqrt{9}(\sqrt{3a})}{\sqrt{a^2}(\sqrt{3a})} = \frac{3\overset{1}{\cancel{\sqrt{3a}}}}{a\underset{1}{\cancel{\sqrt{3a}}}} = \frac{3}{a} \quad \textit{Answer}$$

EXAMPLE 2

Write $\frac{\sqrt{15}}{\sqrt{45}}$ in simplest radical form.

Solution
$$\frac{\sqrt{15}}{\sqrt{45}} = \sqrt{\frac{15}{45}} = \sqrt{\frac{1}{3}} = \sqrt{\frac{1}{3} \times \frac{3}{3}} = \sqrt{\frac{3}{9}} = \frac{\sqrt{3}}{3} \quad \textit{Answer}$$

Alternative Solution
$$\frac{\sqrt{15}}{\sqrt{45}} = \frac{\sqrt{3} \times \sqrt{5}}{\sqrt{9} \times \sqrt{5}} = \frac{\sqrt{3}}{3} \times 1 = \frac{\sqrt{3}}{3} \quad \textit{Answer}$$

EXAMPLE 3

Simplify: $\dfrac{12\sqrt{10a} + 8\sqrt{15b}}{4\sqrt{5ab}}$

Solution Use the distributive property of division over addition.

$$\dfrac{12\sqrt{10a} + 8\sqrt{15b}}{4\sqrt{5ab}} = \dfrac{12\sqrt{10a}}{4\sqrt{5ab}} + \dfrac{8\sqrt{15b}}{4\sqrt{5ab}}$$

$$= 3\sqrt{\dfrac{2}{b}} + 2\sqrt{\dfrac{3}{a}}$$

$$= 3\sqrt{\dfrac{2}{b} \cdot \dfrac{b}{b}} + 2\sqrt{\dfrac{3}{4} \cdot \dfrac{a}{a}}$$

$$= 3\sqrt{\dfrac{2b}{b^2}} + 2\sqrt{\dfrac{3a}{a^2}}$$

$$= \dfrac{3}{b}\sqrt{2b} + \dfrac{2}{a}\sqrt{3a} \quad (a > 0, b > 0) \quad \textit{Answer}$$

Exercises

Writing About Mathematics

1. Jonathan said that $\dfrac{\sqrt{10}}{2} = \sqrt{5}$. Do you agree with Jonathan? Justify your answer.

2. Show that the quotient of two irrational numbers can be either rational or irrational.

Developing Skills

In 3–29 write each quotient in simplest form. Variables in the radicand with an even index are non-negative. Variables occurring in the denominator of a fraction are non-zero.

3. $\sqrt{24} \div \sqrt{6}$

4. $\sqrt{75} \div \sqrt{3}$

5. $\sqrt{72} \div \sqrt{8}$

6. $\sqrt{50a^3} \div \sqrt{5a}$

7. $\sqrt{24x^2} \div \sqrt{3x^3}$

8. $\dfrac{\sqrt{150}}{\sqrt{3}}$

9. $\dfrac{\sqrt{54}}{\sqrt{2}}$

10. $\dfrac{\sqrt{300}}{\sqrt{25}}$

11. $\dfrac{\sqrt{35a^3}}{\sqrt{10a}}$

12. $\dfrac{\sqrt{80x^2y}}{\sqrt{30xy^2}}$

13. $\dfrac{\sqrt{27b}}{\sqrt{6b^2}}$

14. $\dfrac{3}{\sqrt{3x}}$

15. $\dfrac{7}{\sqrt{7y}}$

16. $\dfrac{\sqrt{12a^2}}{\sqrt{4a}}$

17. $\dfrac{\sqrt{18c^3}}{\sqrt{9c}}$

18. $\dfrac{4\sqrt{2} + 8\sqrt{12}}{2\sqrt{2}}$

19. $\dfrac{3\sqrt{10} - 9\sqrt{50}}{3\sqrt{5}}$

20. $\dfrac{\sqrt{72} + \sqrt{54}}{\sqrt{18}}$

21. $\dfrac{\sqrt{20} - \sqrt{5}}{\sqrt{5}}$

22. $\dfrac{\sqrt{48} + \sqrt{3}}{\sqrt{3}}$

23. $\dfrac{\sqrt{10} + \sqrt{15}}{\sqrt{10}}$

24. $\dfrac{5 + 6\sqrt{5}}{\sqrt{5}}$

25. $\dfrac{\sqrt[3]{27x^3} + \sqrt[3]{36x^5}}{\sqrt[3]{3x^3}}$

26. $\dfrac{\sqrt[4]{ab^4}}{\sqrt[4]{a^2b^4}}$

27. $\dfrac{\sqrt[4]{c^6}}{\sqrt[5]{c^5}}$

28. $\dfrac{\sqrt[3]{24w^2}}{\sqrt[3]{3w^4}}$

29. $\dfrac{\sqrt[4]{64x^4} + \sqrt[4]{40x^6}}{\sqrt[4]{x^6}}$

Applying Skills

30. The area of a rectangle is $25\sqrt{35}$ square feet and the width is $10\sqrt{5}$ feet. Find the length of the rectangle in simplest radical form.

31. The area of a right triangle is $6\sqrt{2}$ square centimeters and the length of one leg is $\sqrt{12}$ centimeters.

    a. What is the length of the other leg?

    b. What is the length of the hypotenuse?

3-7 RATIONALIZING A DENOMINATOR

To **rationalize the denominator** of a fraction means to write the fraction as an equivalent fraction with a denominator that is a rational number.

A fraction is in simplest form when the denominator is a positive integer, that is, a rational number. For example, we found the simplest form of a fraction such as $\frac{\sqrt{5}}{\sqrt{2}}$ by multiplying the numerator and denominator by $\sqrt{2}$ so that the denominator would be an integer.

$$\sqrt{\tfrac{5}{2}} = \sqrt{\tfrac{5}{2} \times \tfrac{2}{2}} = \sqrt{\tfrac{10}{4}} = \tfrac{\sqrt{10}}{\sqrt{4}} = \tfrac{\sqrt{10}}{2} = \tfrac{1}{2}\sqrt{10}$$

How do we write the fraction $\frac{3}{2 + \sqrt{6}}$ in simplest form, that is, with a rational denominator? To do so, we must multiply numerator and denominator of the fraction by some numerical expression that will make the denominator a rational number.

Recall that $(a + b)(a - b) = a^2 - b^2$. For example:

$$(1 + \sqrt{2})(1 - \sqrt{2}) \qquad (5 - \sqrt{3})(5 + \sqrt{3}) \qquad (2 + \sqrt{6})(2 - \sqrt{6})$$
$$= 1^2 - (\sqrt{2})^2 \qquad\quad = 5^2 - (\sqrt{3})^2 \qquad\quad = 2^2 - (\sqrt{6})^2$$
$$= 1 - 2 \qquad\qquad\quad\; = 25 - 3 \qquad\qquad = 4 - 6$$
$$= -1 \qquad\qquad\qquad\; = 22 \qquad\qquad\quad = -2$$

The binomials $(a + b)$ and $(a - b)$ are called **conjugates**. We can use this last example to simplify $\frac{3}{2 + \sqrt{6}}$. If we multiply the fraction by 1 in the form $\frac{2 - \sqrt{6}}{2 - \sqrt{6}}$, we will have a fraction whose denominator is an integer.

$$\frac{3}{2 + \sqrt{6}} = \frac{3}{2 + \sqrt{6}} \times \frac{2 - \sqrt{6}}{2 - \sqrt{6}} = \frac{6 - 3\sqrt{6}}{2^2 - (\sqrt{6})^2} = \frac{6 - 3\sqrt{6}}{4 - 6} = \frac{6 - 3\sqrt{6}}{-2} = \frac{-6 + 3\sqrt{6}}{2}$$

If we use a calculator to find a rational approximation for the given fraction and for the simplest form of the given fraction, these approximations will be equal.

ENTER: 3 [÷] [(] 2 + [2nd] [√]
6 [)] [)] [ENTER]

ENTER: [(] −6 [+] 3 [2nd] [√]
6 [)] [)] [÷] 2 [ENTER]

DISPLAY:
```
3/(2+√(6))
        .6742346142
```

DISPLAY:
```
(-6+3√(6))/2
        .6742346142
```

Recall that when the given numerical expression is written as a fraction, the line of the fraction is a grouping symbol. When entering these fractions into the calculator, the numerator or denominator that is written as a sum or a difference must be enclosed by parentheses.

To write the fraction $\frac{4}{2a + \sqrt{8b}}$ with a rational denominator, multiply the fraction by $\frac{2a - \sqrt{8b}}{2a - \sqrt{8b}}$ and simplify the result.

$$\frac{4}{2a + \sqrt{8b}} \cdot \frac{2a - \sqrt{8b}}{2a - \sqrt{8b}} = \frac{4(2a - \sqrt{8b})}{4a^2 - 8b}$$
$$= \frac{4(2a - 2\sqrt{2b})}{4(a^2 - 2b)}$$
$$= \frac{8(a - \sqrt{2b})}{4(a^2 - 2b)}$$
$$= \frac{2(a - \sqrt{2b})}{a^2 - 2b}$$

The fraction $\frac{2(a - \sqrt{2b})}{a^2 - 2b}$ has a rational denominator if a and b are rational numbers.

EXAMPLE 1

Write $\frac{7}{2\sqrt{7}}$ as an equivalent fraction with a rational denominator.

Solution When $\sqrt{7}$ is multiplied by itself, the product is 7, a rational number. Multiply the fraction by 1 in the form $\frac{\sqrt{7}}{\sqrt{7}}$.

$$\frac{7}{2\sqrt{7}} = \frac{7}{2\sqrt{7}} \times \frac{\sqrt{7}}{\sqrt{7}} = \frac{7\sqrt{7}}{2(7)} = \frac{\overset{1}{7}\sqrt{7}}{2(7)} = \frac{\sqrt{7}}{2} \text{ Answer}$$

EXAMPLE 2

Rationalize the denominator of the fraction $\frac{3 + \sqrt{2}}{3 - \sqrt{2}}$.

Solution The conjugate of $3 - \sqrt{2}$ is $3 + \sqrt{2}$. Multiply the fraction by 1 in the form $\frac{3 + \sqrt{2}}{3 + \sqrt{2}}$.

$$\frac{3 + \sqrt{2}}{3 - \sqrt{2}} = \frac{3 + \sqrt{2}}{3 - \sqrt{2}} \times \frac{3 + \sqrt{2}}{3 + \sqrt{2}} = \frac{9 + 3\sqrt{2} + 3\sqrt{2} + 2}{9 - 2} = \frac{11 + 6\sqrt{2}}{7} \quad \text{Answer}$$

Note: The numerator is $(3 + \sqrt{2})^2$ or $3^2 + 2(3\sqrt{2}) + (\sqrt{2})^2$.

EXAMPLE 3

Find the sum: $\frac{\sqrt{2}}{3 - \sqrt{6}} + \frac{6}{\sqrt{8}}$

Solution Rationalize the denominator of each fraction.

$$\frac{\sqrt{2}}{3 - \sqrt{6}} = \frac{\sqrt{2}}{3 - \sqrt{6}} \times \frac{3 + \sqrt{6}}{3 + \sqrt{6}} = \frac{3\sqrt{2} + \sqrt{12}}{3^2 - (\sqrt{6})^2} = \frac{3\sqrt{2} + \sqrt{4}(\sqrt{3})}{9 - 6} = \frac{3\sqrt{2} + 2\sqrt{3}}{3}$$

$$\frac{6}{\sqrt{8}} = \frac{6}{\sqrt{8}} \times \frac{\sqrt{2}}{\sqrt{2}} = \frac{6\sqrt{2}}{\sqrt{16}} = \frac{6\sqrt{2}}{4} = \frac{3\sqrt{2}}{2}$$

Add the fractions. In order to add fractions, we need a common denominator. The common denominator of $\frac{3\sqrt{2} + 2\sqrt{3}}{3}$ and $\frac{3\sqrt{2}}{2}$ is 6.

$$\frac{\sqrt{2}}{3 - \sqrt{6}} + \frac{6}{\sqrt{8}} = \frac{3\sqrt{2} + 2\sqrt{3}}{3} \times \frac{2}{2} + \frac{3\sqrt{2}}{2} \times \frac{3}{3}$$

$$= \frac{6\sqrt{2} + 4\sqrt{3}}{6} + \frac{9\sqrt{2}}{6}$$

$$= \frac{6\sqrt{2} + 4\sqrt{3} + 9\sqrt{2}}{6}$$

$$= \frac{15\sqrt{2} + 4\sqrt{3}}{6} \quad \text{Answer}$$

EXAMPLE 4

Solve for x and check the solution: $x\sqrt{2} = 3 - x$

Solution	*How to Proceed*	
	(1) Write an equivalent equation with only terms in x on the left side of the equation and only constant terms on the right side. Add x to each side of the equation:	$x\sqrt{2} = 3 - x$ $x\sqrt{2} + x = 3 - x + x$ $x\sqrt{2} + x = 3$
	(2) Factor the left side of the equation:	$x(\sqrt{2} + 1) = 3$
	(3) Divide both sides of the equation by the coefficient of x:	$\dfrac{x(\sqrt{2} + 1)}{\sqrt{2} + 1} = \dfrac{3}{\sqrt{2} + 1}$
	(4) Rationalize the denominator:	$x = \dfrac{3}{\sqrt{2} + 1} \times \dfrac{\sqrt{2} - 1}{\sqrt{2} - 1}$ $x = \dfrac{3(\sqrt{2} - 1)}{2 - 1}$ $x = \dfrac{3(\sqrt{2} - 1)}{1}$ $x = 3\sqrt{2} - 3$
	(5) *Check:*	

$$x\sqrt{2} = 3 - x$$
$$(3\sqrt{2} - 3)(\sqrt{2}) \overset{?}{=} 3 - (3\sqrt{2} - 3)$$
$$3(2) - 3\sqrt{2} \overset{?}{=} 3 - 3\sqrt{2} + 3$$
$$6 - 3\sqrt{2} = 6 - 3\sqrt{2} \checkmark$$

Exercises

Writing About Mathematics

1. Justin simplified $\dfrac{7}{2\sqrt{7}}$ by first writing 7 as $\sqrt{49}$ and then dividing numerator and denominator by $\sqrt{7}$.

a. Show that Justin's solution is correct.

b. Can $\dfrac{7}{2\sqrt{5}}$ be simplified by using the same procedure? Explain why or why not.

2. To rationalize the denominator of $\dfrac{4}{2 + \sqrt{8}}$, Brittany multiplied by $\dfrac{2 - \sqrt{8}}{2 - \sqrt{8}}$ and Justin multiplied by $\dfrac{1 - \sqrt{2}}{1 - \sqrt{2}}$. Explain why both are correct.

Developing Skills

In 3–38, rationalize the denominator and write each fraction in simplest form. All variables represent positive numbers.

3. $\dfrac{1}{\sqrt{3}}$ **4.** $\dfrac{5}{\sqrt{10}}$ **5.** $\dfrac{4}{\sqrt{2}}$ **6.** $\dfrac{4}{2\sqrt{3}}$

7. $\dfrac{15}{5\sqrt{3}}$ **8.** $\dfrac{4}{8\sqrt{6}}$ **9.** $\dfrac{12}{\sqrt{27}}$ **10.** $\dfrac{6}{\sqrt{12}}$

11. $\dfrac{8}{\sqrt{24}}$ **12.** $\dfrac{2\sqrt{2}}{4\sqrt{3}}$ **13.** $\dfrac{5\sqrt{5}}{15\sqrt{2}}$ **14.** $\dfrac{\sqrt{24}}{2\sqrt{6}}$

15. $\dfrac{1}{3 + \sqrt{5}}$ **16.** $\dfrac{1}{5 - \sqrt{2}}$ **17.** $\dfrac{1}{1 + \sqrt{3}}$ **18.** $\dfrac{4}{3 - \sqrt{3}}$

19. $\dfrac{3}{1 + \sqrt{5}}$ **20.** $\dfrac{4}{4 + \sqrt{7}}$ **21.** $\dfrac{3}{\sqrt{5} - 2}$ **22.** $\dfrac{9}{\sqrt{7} + 2}$

23. $\dfrac{\sqrt{2}}{2 - \sqrt{2}}$ **24.** $\dfrac{6}{3 + \sqrt{3}}$ **25.** $\dfrac{\sqrt{5x}}{\sqrt{5x} - 2}$ **26.** $\dfrac{\sqrt{20y}}{y\sqrt{5} + 1}$

27. $\dfrac{2 + \sqrt{2}}{3 - \sqrt{2}}$ **28.** $\dfrac{6 - \sqrt{7}}{4 - \sqrt{7}}$ **29.** $\dfrac{\sqrt{10} - 1}{\sqrt{10} + 1}$ **30.** $\dfrac{7 + \sqrt{5}}{7 - \sqrt{5}}$

31. $\dfrac{a + 2}{b - \sqrt{2}}$ **32.** $\dfrac{2\sqrt{x} - 5\sqrt{y}}{\sqrt{x} + \sqrt{y}}$ **33.** $\dfrac{4}{\sqrt{z} + 8}$ **34.** $\dfrac{\sqrt{a}}{\sqrt{a} - 2}$

35. $\dfrac{2}{\sqrt{x}} + \dfrac{2}{\sqrt{y}}$ **36.** $\dfrac{1}{\sqrt{x} + 6} - \dfrac{2}{\sqrt{6}}$ **37.** $\dfrac{2\sqrt{a}}{\sqrt{b}} + \dfrac{2\sqrt{b}}{\sqrt{a}}$ **38.** $\dfrac{3}{x + \sqrt{2}} + \dfrac{5}{\sqrt{x}}$

In 39–42: **a.** Write each fraction in simplest radical form. **b.** Use a calculator to find a rational approximation for the given fraction. **c.** Use a calculator to find a rational approximation for the fraction in simplest form.

39. $\dfrac{4}{\sqrt{6}}$ **40.** $\dfrac{2 + \sqrt{3}}{\sqrt{3}}$

41. $\dfrac{4}{\sqrt{3} - 1}$ **42.** $\dfrac{3 + \sqrt{7}}{3 - \sqrt{7}}$

In 43–46, solve and check each equation.

43. $2a + \sqrt{50} = \sqrt{98}$ **44.** $5x - \sqrt{12} = \sqrt{108} - 3x$

45. $y\sqrt{3} + 1 = 3 - y$ **46.** $7 - b\sqrt{8} = b\sqrt{5} + 4$

Applying Skills

47. The area of a rectangle is 24 square inches. The length of the rectangle is $\sqrt{5} + 1$ inches. Express the width of the rectangle in simplest form.

48. The perimeter of an isosceles triangle is $\sqrt{50}$ feet. The lengths of the sides are in the ratio $3 : 3 : 4$. Find the length of each side of the triangle.

3-8 SOLVING RADICAL EQUATIONS

An equation that contains at least one radical term with a variable in the radicand is called a **radical equation**. For example, $\sqrt{2x - 3} = 5$ is a radical equation. Since the radical is a square root, we can solve this equation by squaring both sides of the equation.

Solution:

$$\sqrt{2x - 3} = 5$$
$$(\sqrt{2x - 3})^2 = 5^2$$
$$2x - 3 = 25$$
$$2x = 28$$
$$x = 14$$

Check:

$$\sqrt{2x - 3} = 5$$
$$\sqrt{2(14) - 3} \stackrel{?}{=} 5$$
$$\sqrt{28 - 3} \stackrel{?}{=} 5$$
$$\sqrt{25} \stackrel{?}{=} 5$$
$$5 = 5 ✔$$

Note that squaring both sides of a radical equation does not always result in an equivalent equation. For example, if the given equation had been $\sqrt{2x - 3} = -5$, the equation obtained by squaring both sides would have been $2x - 3 = 25$, the same as the equation obtained by squaring both sides of $\sqrt{2x - 3} = 5$. The solution of $2x - 3 = 25$ is not a root of $\sqrt{2x - 3} = -5$. The radical $\sqrt{2x - 3}$ represents the principal root of $2x - 3$, which is a positive number. Therefore, the equation $\sqrt{2x - 3} = -5$ has no real root. There is no real number such that $\sqrt{2x - 3} = -5$.

Often an equation must be rewritten as an equivalent equation with the radical alone on one side of the equation before squaring both sides of the equation. For example, to solve the equation $x - \sqrt{5 - x} = 3$, we must first add $-x$ to both sides of the equation to isolate the radical.

$$x - \sqrt{5 - x} = 3$$
$$-\sqrt{5 - x} = 3 - x$$
$$(-\sqrt{5 - x})^2 = (3 - x)^2$$
$$5 - x = 9 - 6x + x^2$$
$$0 = 4 - 5x + x^2$$
$$0 = (4 - x)(1 - x)$$
$$0 = 4 - x \mid 0 = 1 - x$$
$$x = 4 \qquad x = 1$$

Check: x = 4

$$x - \sqrt{5 - x} = 3$$
$$4 - \sqrt{5 - 4} \stackrel{?}{=} 3$$
$$4 - \sqrt{1} \stackrel{?}{=} 3$$
$$4 - 1 \stackrel{?}{=} 3$$
$$3 = 3 ✔$$

Check: x = 1

$$x - \sqrt{5 - x} = 3$$
$$1 - \sqrt{5 - 1} \stackrel{?}{=} 3$$
$$1 - \sqrt{4} \stackrel{?}{=} 3$$
$$1 - 2 \stackrel{?}{=} 3$$
$$-1 \neq 3 ✗$$

The root of this equation is $x = 4$.

Note that after squaring both sides of the equation, we have a quadratic equation that has two real roots. One of these is the root of the given equation, $x - \sqrt{5 - x} = 3$ which is equivalent to $-\sqrt{5 - x} = 3 - x$. The other is the root of the equation $-\sqrt{5 - x} = x - 3$. When we square both sides of either of these equations, we obtain the same equation, $5 - x = 9 - 6x + x^2$.

EXAMPLE 1

Solve and check: $4 + \sqrt{1 - 3x} = 12$

Solution

How to Proceed

(1) Isolate the radical by adding -4 to both sides of the equation:

(2) Square both sides of the equation:

$$4 + \sqrt{1 - 3x} = 12$$
$$\sqrt{1 - 3x} = 8$$
$$(\sqrt{1 - 3x})^2 = 8^2$$
$$1 - 3x = 64$$

(3) Add -1 to both sides of the equation:

$$-3x = 63$$

(4) Divide both sides of the equation by -3:

$$x = -21$$

(5) *Check:*

$$4 + \sqrt{1 - 3x} = 12$$
$$4 + \sqrt{1 - 3(-21)} \overset{?}{=} 12$$
$$4 + \sqrt{1 + 63} \overset{?}{=} 12$$
$$4 + \sqrt{64} \overset{?}{=} 12$$
$$4 + 8 \overset{?}{=} 12$$
$$12 = 12 \checkmark$$

Answer $x = -21$

EXAMPLE 2

Find the value of a such that $\sqrt[3]{4 - 2a} = -2$.

Solution

How to Proceed

(1) The radical is already isolated. Since the radical is a cube root, cube both sides of the equation:

$$\sqrt[3]{4 - 2a} = -2$$
$$(\sqrt[3]{4 - 2a})^3 = (-2)^3$$
$$4 - 2a = -8$$

(2) Solve the resulting equation for a:

$$-2a = -12$$
$$a = 6$$

(3) *Check:*

$$\sqrt[3]{4 - 2a} = -2$$
$$\sqrt[3]{4 - 2(6)} \overset{?}{=} -2$$
$$\sqrt[3]{4 - 12} \overset{?}{=} -2$$
$$\sqrt[3]{-8} \overset{?}{=} -2$$
$$-2 = -2 \checkmark$$

Answer $a = 6$

EXAMPLE 3

What is the solution set of $x = 1 + \sqrt{15 - 7x}$?

Solution

How to Proceed

(1) Write the equation with the radical alone on one side of the equation by adding -1 to both sides of the equation:

$$x = 1 + \sqrt{15 - 7x}$$
$$x - 1 = \sqrt{15 - 7x}$$

(2) Square both sides of the equation:

$$(x - 1)^2 = \left(\sqrt{15 - 7x}\right)^2$$
$$x^2 - 2x + 1 = 15 - 7x$$

(3) Write the quadratic equation in standard form:

$$x^2 + 5x - 14 = 0$$

(4) Factor the polynomial.

$$(x + 7)(x - 2) = 0$$

(5) Set each factor equal to 0 and solve for x:

$$x + 7 = 0 \quad | \quad x - 2 = 0$$
$$x = -7 \quad | \quad x = 2$$

(6) *Check:*

Let $x = -7$:	**Let $x = 2$:**
$x = 1 + \sqrt{15 - 7x}$	$x = 1 + \sqrt{15 - 7x}$
$-7 \stackrel{?}{=} 1 + \sqrt{15 - 7(-7)}$	$2 \stackrel{?}{=} 1 + \sqrt{15 - 7(2)}$
$-7 \stackrel{?}{=} 1 + \sqrt{15 + 49}$	$2 \stackrel{?}{=} 1 + \sqrt{15 - 14}$
$-7 \stackrel{?}{=} 1 + \sqrt{64}$	$2 \stackrel{?}{=} 1 + \sqrt{1}$
$-7 \stackrel{?}{=} 1 + 8$	$2 \stackrel{?}{=} 1 + 1$
$-7 \neq 9$ ✗	$2 = 2$ ✔

-7 is not a solution of $x = 1 + \sqrt{15 - 7x}$ and 2 is a solution of $x = 1 + \sqrt{15 - 7x}$.

Answer The solution set of $x = 1 + \sqrt{15 - 7x}$ is $\{2\}$.

EXAMPLE 4

Solve the equation $\sqrt{x+5} = 1 + \sqrt{x}$.

How to Proceed

(1) Square both sides of the equation. The right side will be a trinomial that has a radical in one of its terms:

$$(\sqrt{x+5})^2 = (1 + \sqrt{x})^2$$
$$x + 5 = 1 + 2\sqrt{x} + x$$

(2) Simplify the equation obtained in step 1 and isolate the radical:

$$4 = 2\sqrt{x}$$

(3) Square both sides of the equation obtained in step 2:

$$4^2 = (2\sqrt{x})^2$$
$$16 = 4x$$

(4) Solve for the variable:

$$x = 4$$

(5) *Check:*

$$\sqrt{x+5} = 1 + \sqrt{x}$$
$$\sqrt{4+5} \stackrel{?}{=} 1 + \sqrt{4}$$
$$3 = 3 ✔$$

Answer $x = 4$

Exercises

Writing About Mathematics

1. Explain why $\sqrt{x+3} < 0$ has no solution in the set of real numbers while $\sqrt{x+3} \geq 0$ is true for all real numbers greater than or equal to -3.

2. In Example 3, are $x - 1 = \sqrt{15 - 7x}$ and $(x - 1)^2 = (\sqrt{15 - 7x})^2$ equivalent equations? Explain why or why not.

Developing Skills

In 3–38, solve each equation for the variable, check, and write the solution set.

3. $\sqrt{a} = 5$

4. $\sqrt{x} = 7$

5. $4\sqrt{y} = 12$

6. $\sqrt{4y} = 12$

7. $2\sqrt{b} = 8$

8. $\sqrt{2b} = 8$

9. $1 + \sqrt{x} = 3$

10. $\sqrt{1 + x} = 3$

11. $5 + \sqrt{a} = 7$

12. $\sqrt{5 + a} = 7$

13. $3 - \sqrt{y} = 1$

14. $\sqrt{3 - y} = 1$

15. $3 - \sqrt{2x + 5} = 0$

16. $8 + \sqrt{2x - 1} = 15$

17. $\sqrt{5x + 2} = \sqrt{9x - 14}$

18. $\sqrt{20 - 2x} = \sqrt{5x - 8}$

19. $x = \sqrt{4x + 5}$

20. $y = \sqrt{y + 2}$

21. $x + 3 = \sqrt{1 - 3x}$

22. $a - 2 = \sqrt{2a - 1}$

23. $b - 3 = \sqrt{3b - 11}$

24. $3 - b = \sqrt{3b - 11}$

25. $x = 1 + \sqrt{x + 11}$

26. $x + \sqrt{x + 1} = 5$

27. $x + 4\sqrt{x} = 5$

28. $3 + \sqrt{4x - 3} = 2x$

29. $2 + \sqrt{3x - 2} = 3x$

30. $\sqrt[3]{x} = 2$

31. $5 + \sqrt[3]{a + 2} = 3$

32. $2 + \sqrt[3]{3b - 2} = 6$

33. $1 - \sqrt[3]{x + 7} = 4$

34. $-10 + \sqrt[4]{n - 2} = -8$

35. $\sqrt{x + 5} = 1 + \sqrt{x}$

36. $1 + \sqrt{2x} = \sqrt{3x + 1}$

37. $\sqrt{x + 4} - \sqrt{x - 1} = 1$

38. $\sqrt{5x} - \sqrt{x + 4} = 2$

Applying Skills

39. The lengths of the legs of an isosceles right triangle are x and $\sqrt{6x + 16}$. What are the lengths of the legs of the triangle?

40. The width of a rectangle is x and the length is $\sqrt{x - 1}$. If the width is twice the length, what are the dimensions of the rectangle?

41. In $\triangle ABC$, $AB = \sqrt{7x + 5}$, $BC = \sqrt{5x + 15}$, and $AC = \sqrt{2x}$.

    a. If $AB = BC$, find the length of each side of the triangle.

    b. Express the perimeter of the triangle in simplest radical form.

CHAPTER SUMMARY

An **irrational number** is an infinite decimal number that does not repeat. The union of the set of rational numbers and the set of irrational numbers is the set of **real numbers**.

If the domain is the set of real numbers, the elements of the solution set of an inequality usually cannot be listed but can be shown on the number line.

If $a \geq 0$ and $|x| = a$, then $x = a$ or $x = -a$. These two points on the real number line separate the line into three segments. The segment between a and $-a$ is the graph of the solution set of $|x| < a$ and the other two segments are the graph of the solution set of $|x| > a$.

A **square root** of a positive number is one of the two equal factors whose product is that number. The **principal square root** of a positive real number is its positive square root. The **cube root** of k is one of the three equal factors whose product is k. The **nth root** of k is one of the n equal factors whose product is k. The nth root of k is written as $\sqrt[n]{k}$ where k is the **radicand**, n is the **index**, and $\sqrt[n]{k}$ is the **radical**.

An irrational number of the form $\sqrt[n]{a}$ is in simplest form when a is an integer that has no integral factor of the form b^n.

To express the sum or difference of two radicals as a single radical, the radicals must have the same index and the same radicand, that is, they must be **like radicals**. The distributive property is used to add like radicals.

For any non-negative real numbers a and b:

$$\sqrt{a \cdot b} = \sqrt{a} \cdot \sqrt{b}$$

$$\sqrt{a} \div \sqrt{b} = \frac{\sqrt{a}}{\sqrt{b}} = \sqrt{\frac{a}{b}} \quad (b \neq 0)$$

$$\sqrt{a} \cdot \sqrt{a} = \sqrt{a^2} = a$$

For any non-negative real number a and natural number n:

$$\sqrt{a^{2n}} = a^n$$

If $\sqrt[n]{a}$ and $\sqrt[n]{b}$ are real numbers, then:

$$\sqrt[n]{a \cdot b} = \sqrt[n]{a} \cdot \sqrt[n]{b}$$

$$\sqrt[n]{a} \div \sqrt[n]{b} = \frac{\sqrt[n]{a}}{\sqrt[n]{b}} = \sqrt[n]{\frac{a}{b}} \quad (b \neq 0)$$

To **rationalize the denominator** of a fraction means to write the fraction as an equivalent fraction with a denominator that is a rational number.

The binomials $a + b$ and $a - b$ are conjugate binomials. When the denominator of a fraction is of the form $a \pm \sqrt{b}$ or $\sqrt{a} \pm \sqrt{b}$, we can rationalize the fraction by multiplying by a fraction whose numerator and denominator are the conjugates of the given denominator.

An equation that contains at least one radical term with a variable in the radicand is called a **radical equation**. To solve an equation that contains a square root radical, isolate the radical and square both sides of the equation. The derived equation may or may not be equivalent to the given equation and a check is necessary.

VOCABULARY

3-1 Irrational numbers • Real numbers • Interval notation

3-2 Square root • Principal square root • Cube root • nth root • Radicand • Index • Radical • Principal nth root

3-3 Simplest form

3-4 Like radicals • Unlike radicals

3-7 Rationalize the denominator • Conjugates

3-8 Radical equation

REVIEW EXERCISES

In 1–8, find and graph the solution set of each inequality.

1. $|1 - 4x| < 2$

2. $|2x + 3| \leq 8$

3. $\left| x - \frac{1}{2} \right| \leq 11$

4. $|-4x + 5| > 13$

5. $|4 - 5x| \geq 7$

6. $\left|\frac{x+6}{7}\right| > 1$

7. $\left|\frac{5x-3}{4}\right| \leq 2$

8. $|x - 1| + 2 < 3$

In 9–41, write each expression in simplest radical form. Variables representing indexes are positive integers. Variables in the radicand of an even index are non-negative. Variables occurring in the denominator of a fraction are non-zero.

9. $\sqrt{128}$

10. $\sqrt{\frac{3}{4}}$

11. $\sqrt{75} + \sqrt{300}$

12. $\sqrt{45} - \sqrt{20}$

13. $3\sqrt{12} + 5\sqrt{27}$

14. $\frac{1}{2}\sqrt{72} + \frac{1}{3}\sqrt{18}$

15. $\sqrt{24} + \sqrt{20} + \sqrt{54}$

16. $\sqrt{8} \times \sqrt{18}$

17. $3\sqrt{5} \times \sqrt{125}$

18. $\sqrt{2}(3 + \sqrt{8})$

19. $\sqrt{40}(\sqrt{10} - \sqrt{20})$

20. $6\sqrt{2}(1 + \sqrt{50})$

21. $(1 + \sqrt{2})(1 - \sqrt{2})$

22. $(5 - \sqrt{3})(5 + \sqrt{3})$

23. $(2 + \sqrt{3})(1 + \sqrt{3})$

24. $\frac{\sqrt{24}}{\sqrt{12}}$

25. $\frac{10 + \sqrt{5}}{\sqrt{5}}$

26. $\frac{14}{\sqrt{8} + 1}$

27. $\frac{\sqrt{3}}{4 - \sqrt{3}}$

28. $\sqrt{9a} + \sqrt{25a}$

29. $\sqrt{8b^3} + \sqrt{50b^3}$

30. $\sqrt[4]{12x^5}$

31. $\sqrt{\frac{a}{b}}$

32. $16\sqrt{x^5} - \sqrt{16x^5}$

33. $\sqrt[4]{256x^4} - \sqrt[4]{36x^4}$

34. $\sqrt{x^3 y^4} \cdot \sqrt{xy}$

35. $\sqrt[3]{18a^5} \cdot \sqrt[3]{3a^4}$

36. $\frac{\sqrt{64a^5}}{\sqrt{4a}}$

37. $\frac{\sqrt[3]{27b^7}}{\sqrt[3]{3b}}$

38. $\frac{1}{3 - \sqrt{x}}$

39. $\frac{4 - \sqrt{a}}{4 + \sqrt{a}}$

40. $\frac{\sqrt[n]{a^{n+4}}}{\sqrt[n]{a^4}}$

41. $\sqrt[n]{x^{n+2}} \cdot \sqrt[n]{x^{n-2}}$, where n is an integer > 2

In 42–49, solve each equation and check.

42. $\sqrt{2x - 1} = 5$

43. $\sqrt{3x - 1} = \sqrt{2x + 4}$

44. $3 + \sqrt{x + 7} = 8$

45. $8 + \sqrt{x - 5} = 11$

46. $y = \sqrt{7y - 12}$

47. $b = 3 + \sqrt{29 - 5b}$

48. $x + 1 + \sqrt{6 - 2x} = 0$

49. $\sqrt{a^2 - 2} = \sqrt{a + 4}$

50. The lengths of the sides of a triangle are 8 feet, $\sqrt{75}$ feet, and $(4 + \sqrt{12})$ feet. Express the perimeter of the triangle is simplest radical form.

51. The length of each leg of an isosceles right triangle is $2\sqrt{2}$ meters.

a. Find the length of the hypotenuse of the triangle in simplest radical form.

b. Find the perimeter of the triangle in simplest radical form.

52. Show that x, $\sqrt{2x + 1}$, and $x + 1$ can be the lengths of the sides of a right triangle for all $x > 0$.

53. The area of a rectangle is 12 square feet and the length is $2 + \sqrt{3}$ feet.

    a. Find the width of the rectangle is simplest radical form.

    b. Find the perimeter of the rectangle in simplest radical form.

Exploration

As mentioned in the chapter opener, Euclid proved that $\sqrt{2}$ is irrational. Following is a proof by contradiction based on Euclid's reasoning.

Assume that $\sqrt{2}$ is a rational number. Then we can write $\sqrt{2} = \frac{a}{b}$ with a and b integers and $b \neq 0$. Also assume that $\frac{a}{b}$ is a fraction in simplest form. Therefore:

$$2 = \frac{a^2}{b^2} \text{ or } a^2 = 2b^2$$

If $a^2 = 2b^2$, then the square of a is an even number. Since $a^2 = a \cdot a$ and a^2 is even, then a must also be even. An even number can be written as two times some other integer, so we can write $a = 2k$. Substituting into the original equation $2 = \frac{a^2}{b^2}$:

$$2 = \frac{(2k)^2}{b^2}$$
$$2 = \frac{4k^2}{b^2}$$
$$2b^2 = 4k^2$$
$$b^2 = 2k^2$$

This means that b is also even. Therefore, $\frac{a}{b}$ is *not* in simplest form because a and b have a common factor, 2. This contradicts the assumption that $\frac{a}{b}$ is in simplest form, so $\sqrt{2}$ must be irrational.

There are many other proofs that $\sqrt{2}$ is irrational. Research one of these proofs and discuss your findings with the class.

CUMULATIVE REVIEW   CHAPTERS 1–3

Part I

Answer all questions in this part. Each correct answer will receive 2 credits. No partial credit will be allowed.

1. Which of the following is not a rational number?

    (1) $\frac{1}{2}$ (2) $0.\overline{25}$ (3) $\sqrt{0.04}$ (4) $\sqrt{0.4}$

2. When $3a^2 - 5a$ is subtracted from $a^2 + 4$, the difference is

(1) $-3a^2 + 5a + 4$ (3) $-2a^2 + 5a + 4$

(2) $-2a^2 - 5a + 4$ (4) $2a^2 - 5a - 4$

3. When factored completely, $12a^2 - 3a$ is equal to

(1) $3a(4a)$ (3) $3(4a^2 - a)$

(2) $a(12a - 3)$ (4) $3a(4a - 1)$

4. The factors of $a^2 + 5a - 6$ are

(1) $(a + 3)(a - 2)$ (3) $(a + 6)(a - 1)$

(2) $(a - 3)(a + 2)$ (4) $(a - 6)(a + 1)$

5. If $4(x + 6) = x - 12$, then x is equal to

(1) -12 (2) -6 (3) 6 (4) 12

6. The solution set of $|y - 3| = 7$ is

(1) $\{10\}$ (2) $\{4, -10\}$ (3) $\{-4, 10\}$ (4) $\{-4\}$

7. The solution set of $3x^2 - x = 2$ is

(1) $\{1, 2\}$ (2) $\left\{\frac{2}{3}, 1\right\}$ (3) $\left\{-\frac{2}{3}, -1\right\}$ (4) $\left\{-\frac{2}{3}, 1\right\}$

8. In simplest form, $\dfrac{1 + \frac{1}{2}}{1 - \frac{1}{4}}$ is equal to

(1) $\frac{3}{8}$ (2) $\frac{1}{2}$ (3) $\frac{2}{3}$ (4) 2

9. If $\frac{3}{x} - 5 > \frac{8}{x}$, then

(1) $-1 < x < 1$ (3) $0 < x < 1$

(2) $x > -1$ or $x < -1$ (4) $-1 < x < 0$

10. In simplest form, $\sqrt{24} + \sqrt{150}$ is equal to

(1) $7\sqrt{6}$ (2) $7\sqrt{12}$ (3) $10\sqrt{6}$ (4) $\sqrt{174}$

Part II

Answer all questions in this part. Each correct answer will receive 2 credits. Clearly indicate the necessary steps, including appropriate formula substitutions, diagrams, graphs, charts, etc. For all questions in this part, a correct numerical answer with no work shown will receive only 1 credit.

11. Divide and write the quotient in simplest form.

$$\frac{a^2 - 16}{a^2 - a - 12} \div \frac{a^2 + 4a}{2a}$$

12. Solve and check: $2x^2 - 9x = 5$.

Part III

Answer all questions in this part. Each correct answer will receive 4 credits. Clearly indicate the necessary steps, including appropriate formula substitutions, diagrams, graphs, charts, etc. For all questions in this part, a correct numerical answer with no work shown will receive only 1 credit.

13. Find the solution set of the inequality $|3b + 6| < 7$ and graph the solution on the number line.

14. On a test, the number of questions that Tyler answered correctly was 4 more than twice the number that he answered incorrectly. If the ratio of correct answers to incorrect answers is $8 : 3$, how many questions did Tyler answer correctly?

Part IV

Answer all questions in this part. Each correct answer will receive 6 credits. Clearly indicate the necessary steps, including appropriate formula substitutions, diagrams, graphs, charts, etc. For all questions in this part, a correct numerical answer with no work shown will receive only 1 credit.

15. On her way to work, Rachel travels two miles on local roads and 12 miles on the highway. Her speed on the highway is twice her speed on local roads. Her driving time for the trip to work is 16 minutes. What is Rachel's rate of speed for each part of the trip?

16. The length of Bill's garden is 2 yards more than four times the width. The area of the garden is 30 square yards. What are the dimensions of the garden?

RELATIONS AND FUNCTIONS

Long-distance truck drivers keep very careful watch on the length of time and the number of miles that they drive each day. They know that this relationship is given by the formula $rt = d$. On some trips, they are required to drive for a fixed length of time each day, and that makes $\frac{d}{r}$ a constant. When this is the case, we say that the distance traveled and the rate of travel are *directly proportional*. When the rate of travel increases, the distance increases and when the rate of travel decreases, the distance decreases. On other trips, truck drivers are required to drive a fixed distance each day, and that makes rt equal a constant. When this the case, we say that the time and the rate of travel are *inversely proportional*. When the rate of travel is increased, the time is decreased and when the rate of travel is decreased, the time is increased. Each of these cases defines a function.

In this chapter, we will study some fundamental properties of functions, and in the chapters that follow, we will apply these properties to some important classes of functions.

4-1 RELATIONS AND FUNCTIONS

Every time that we write an algebraic expression such as $2x + 3$, we are defining a set of ordered pairs, $(x, 2x + 3)$. The expression $2x + 3$ means that for every input x, there is an output that is 3 more than twice x. In **set-builder notation**, we write this set of ordered pairs as $\{(x, y) : y = 2x + 3\}$. This is read "the set of ordered pairs (x, y) such that y is equal to 3 more than twice x." The colon $(:)$ represents the phrase "such that" and the description to the right of the colon defines the elements of the set. Another set of ordered pairs related to the algebraic expression $2x + 3$ is $\{(x, y) : y \geq 2x + 3\}$.

DEFINITION _____

A **relation** is a set of ordered pairs.

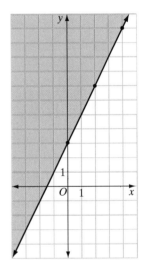

The inequality $y \geq 2x + 3$ defines a relation. For every real number x, there are many possible values of y. The set of ordered pairs that make the inequality true is the union of the pairs that make the equation $y = 2x + 3$ true and the pairs that make the inequality $y > 2x + 3$ true. This is an infinite set. We can list typical pairs from the set but cannot list all pairs.

To show the solution set of the inequality $y \geq 2x + 3$ as points on the coordinate plane, we choose some values of x and y such as $(0, 3)$, $(2, 7)$, and $(4, 11)$ that make the equality true and draw the line through these points. All of the pairs on the line and above the line make the inequality $y \geq 2x + 3$ true.

The equation $y = 2x + 3$ determines a set of ordered pairs. In this relation, for every value of x, there is exactly one value of y. This relation is called a *function*.

DEFINITION _____

A **function** is a relation in which no two ordered pairs have the same first element.

The ordered pairs of the function $y = 2x + 3$ depend on the set from which the first element, x, can be chosen. The set of values of x is called the **domain of the function**. The set of values for y is called the **range of the function**.

DEFINITION _____

A **function from set A to set B** is the set of ordered pairs (x, y) such that every $x \in A$ is paired with exactly one $y \in B$.

The set A is the domain of the function. However, the range of the function is *not* necessarily equal to the set B. The range is the set of values of y, a *subset* of B. The symbol $\in$ means "is an element of," so $x \in A$ means "x is an element of set A."

For example, let $\{(x, y) : y = 2x + 3\}$ be a function from the set of integers to the set of integers. Every integer is a value of x but not every integer is a value of y. When x is an integer, $2x + 3$ is an odd integer. Therefore, when the domain of the function $\{(x, y) : y = 2x + 3\}$ is the set of integers, the range is the set of odd integers.

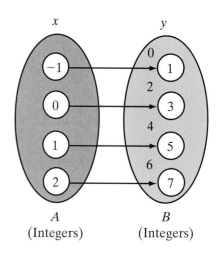

x y

A
(Integers)

B
(Integers)

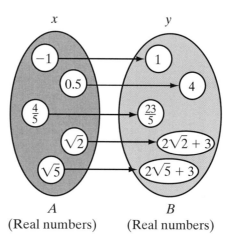

x y

A
(Real numbers)

B
(Real numbers)

On the other hand, let $\{(x, y) : y = 2x + 3\}$ be a function from set A to set B. Let set A be the set of real numbers and set B be the set of real numbers. In this case, every real number can be written in the form $2x + 3$ for some real number x. Therefore, the range is the set of real numbers. We say that the function $\{(x, y) : y = 2x + 3\}$ is *onto* when the range is equal to set B.

DEFINITION _____

A function from set A to set B is **onto** if the range of the function is equal to set B.

The following functions have been studied in earlier courses. Unless otherwise stated, each function is from the set of real numbers to the set of real numbers. In general, we will consider set A and set B to be the set of real numbers.

Constant Function

$y = a$, with a a constant.

Example: $y = 3$

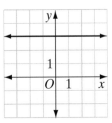

The range is {3}.
The function is *not* onto.

Linear Function

$y = mx + b$, with m and b constants.

Example: $y = 2x + 3$

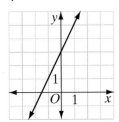

The range is the set of real numbers.
The function is onto.

Quadratic Function

$y = ax^2 + bx + c$ with a, b, and c constants and $a \neq 0$.

Example: $y = x^2 - 4x + 3$

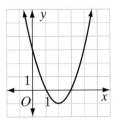

The range is the set of real numbers ≥ -1.
The function is *not* onto.

Absolute Value Function

$y = a + |bx + c|$ with a, b, and c constants.

Example: $y = -2 + |x + 1|$

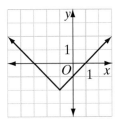

The range is the set of real numbers ≥ -2.
The function is *not* onto.

Square Root Function

$$y = \sqrt{ax + b}, \text{ with } ax + b \geq 0.$$

- The domain is the set of real numbers greater than or equal to $\frac{-b}{a}$.
- The range is the set of non-negative real numbers.

Example: $y = \sqrt{x}$

This is a function from the set of non-negative real numbers to the set of non-negative real numbers.

The range is the set of non-negative real numbers.

The function is onto.

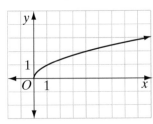

Vertical Line Test

A relation is a function if no two pairs of the relation have the same first element. On a graph, two points lie on the same vertical line if and only if the coordinates of the points have the same first element. Therefore, if a vertical line intersects the graph of a relation in two or more points, the relation is not a function. This is referred to as the **vertical line test**.

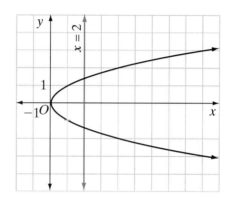

The graph of the relation $\{(x, y) : y = \pm\sqrt{x} \text{ and } x \geq 0\}$ is shown at the right. The graph shows that the line whose equation is $x = 2$ intersects the graph in two points. Every line whose equation is $x = a$ for $a > 0$ also intersects the graph of the relation in two points. Therefore, the relation is not a function.

EXAMPLE 1

For each given set, determine:

a. Is the set a relation?   **b.** Is the set a function?

(1) $\{(x, y) : y > x\}$ if $x \in \{\text{real numbers}\}$

(2) $\{x : x > 4\}$ if $x \in \{\text{real numbers}\}$

(3) $\{(x, y) : y = |x|\}$ if $x \in \{\text{real numbers}\}$

Solution (1) **a.** $\{(x, y) : y > x\}$ is a relation because it is a set of ordered pairs.

b. $\{(x, y) : y > x\}$ is not a function because many ordered pairs have the same first elements. For example, $(1, 2)$ and $(1, 3)$ are both ordered pairs of this relation.

(2) **a.** $\{x : x > 4\}$ is not a relation because it is not a set of ordered pairs.

b. $\{x : x > 4\}$ is not a function because it is not a relation.

(3) **a.** $\{(x, y) : y = |x|\}$ is a relation because it is a set of ordered pairs.

b. $\{(x, y) : y = |x|\}$ is a function because for every real number x there is exactly one real number that is the absolute value. Therefore, no two pairs have the same first element.

In part (3) of Example 1, the graph of $y = |x|$ passes the vertical line test, which we can see using a graphing calculator.

ENTER: [Y=] [MATH] [▶] [1] [X,T,Θ,n] [GRAPH]

DISPLAY: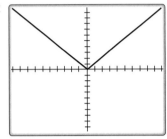

EXAMPLE 2

The graph of a relation is shown at the right.

a. Determine whether or not the graph represents a function.

b. Find the domain of the graph.

c. Find the range of the graph.

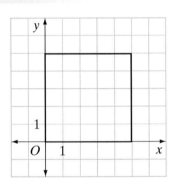

Solution **a.** The relation is not a function. Points $(1, 0)$ and $(1, 5)$ as well as many other pairs have the same first element. *Answer*

b. The domain is $\{x : 0 \le x \le 5\}$. *Answer*

c. The range is $\{y : 0 \le y \le 5\}$. *Answer*

EXAMPLE 3

An artist makes hand-painted greeting cards that she sells online for three dollars each. In addition, she charges three dollars for postage for each package of cards that she ships, regardless of size. The total amount, y, that a customer pays for a package of cards is a function of the number of cards purchased, x.

a. Write the relation described above as a function in set-builder notation.

b. What is the domain of the function?

c. Write four ordered pairs of the function.

d. What is the range of the function?

e. Is the function onto?

Solution **a.** The price per card is $3.00, so x cards cost $3x$. The cost of postage, $3.00, is a constant. Therefore, the total cost, y, for x cards is $3x + 3$. In set-builder notation, this is written as:

$$\{(x, y) : y = 3x + 3\}$$

b. The first element, x, which represents the number of cards shipped, must be a positive integer. Therefore, the domain is the set of positive integers.

c. Many ordered pairs are possible. For example:

x	$3x + 3$	y	(x, y)
1	$3(1) + 3$	6	$(1, 6)$
2	$3(2) + 3$	9	$(2, 9)$
3	$3(3) + 3$	12	$(3, 12)$
4	$3(4) + 3$	15	$(4, 15)$

d. Since $3x + 3$ is a multiple of 3, the range is the set of positive integers greater than 3 that are multiples of 3.

e. The domain is the set of positive integers. This is a function from the set of positive integers to the set of positive integers. But the range is the set of positive multiples of 3. The function is not onto because the range is not equal to the set of positive integers.

EXAMPLE 4

Find the largest domain of each function:

a. $y = \dfrac{1}{x + 5}$ **b.** $y = \dfrac{x + 1}{x^2 - 2x - 3}$ **c.** $y = \dfrac{x}{5x^2 + 1}$

Solution The functions are well defined wherever the denominators are *not* equal to 0. Therefore, the domain of each function consists of the set real numbers such that the denominator is not equal to 0.

a. $x + 5 = 0$ or $x = -5$. Therefore, the domain is $\{x : x \neq -5\}$.

b.
$$x^2 - 2x - 3 = 0$$
$$(x - 3)(x + 1) = 0$$

$$x - 3 = 0 \quad | \quad x + 1 = 0$$
$$x = 3 \quad | \quad x = -1$$

Therefore, the domain is $\{x : x \neq -1, 3\}$.

c. The denominator $5x^2 + 1$ cannot be equal to zero. Therefore, the domain is the set of real numbers.

Answers **a.** Domain $= \{x : x \neq -5\}$ **b.** Domain $= \{x : x \neq -1, 3\}$
c. Domain $= \{\text{Real numbers}\}$

Exercises

Writing About Mathematics

1. Explain why $\{(x, y) : x = y^2\}$ is not a function but $\{(x, y) : \sqrt{x} = y\}$ is a function.

2. Can $y = \sqrt{x}$ define a function from the set of positive integers to the set of positive integers? Explain why or why not.

Developing Skills

In 3–5: **a.** Explain why each set of ordered pairs is or is not a function. **b.** List the elements of the domain. **c.** List the elements of the range.

3. $\{(1, 1), (2, 4), (3, 9), (4, 16)\}$

4. $\{(1, -1), (0, 0), (1, 1)\}$

5. $\{(-2, 5), (-1, 5), (0, 5), (1, 5), (2, 5)\}$

In 6–11: **a.** Determine whether or not each graph represents a function. **b.** Find the domain for each graph. **c.** Find the range for each graph.

6.
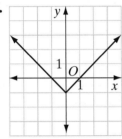

7.
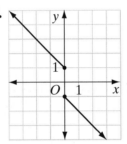

8.
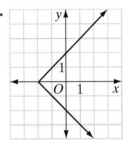

9.
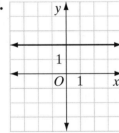

10.
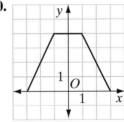

11.
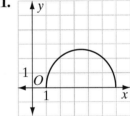

In 12–23, each set is a function from set A to set B. **a.** What is the largest subset of the real numbers that can be set A, the domain of the given function? **b.** If set A = set B, is the function onto? Justify your answer.

12. $\{(x, y) : y = -183\}$

13. $\{(x, y) : y = 5 - x\}$

14. $\{(x, y) : y = x^2\}$

15. $\{(x, y) : y = -x^2 + 3x - 2\}$

16. $\left\{(x, y) : y = \sqrt{2x}\right\}$

17. $\{(x, y) : y = |4 - x|\}$

18. $\left\{(x, y) : y = \frac{1}{x}\right\}$

19. $\left\{(x, y) : y = \sqrt{3 - x}\right\}$

20. $\left\{(x, y) : y = \frac{1}{\sqrt{x + 1}}\right\}$

21. $\left\{(x, y) : y = \frac{1}{x^2 + 1}\right\}$

22. $\left\{(x, y) : y = \frac{x + 1}{x - 1}\right\}$

23. $\left\{(x, y) : y = \frac{x - 5}{|x - 3|}\right\}$

Applying Skills

24. The perimeter of a rectangle is 12 meters.

  a. If x is the length of the rectangle and y is the area, describe, in set-builder notation, the area as a function of length.

  b. Use integral values of x from 0 to 10 and find the corresponding values of y. Sketch the graph of the function.

  c. What is the largest subset of the real numbers that can be the domain of the function?

25. A candy store sells candy by the piece for 10 cents each. The amount that a customer pays for candy, y, is a function of the number of pieces purchased, x.

  a. Describe, in set-builder notation, the cost in cents of each purchase as a function of the number of pieces of candy purchased.

  b. Yesterday, no customer purchased more than 8 pieces of candy. List the ordered pairs that describe possible purchases yesterday.

  c. What is the domain for yesterday's purchases?

  d. What is the range for yesterday's purchases?

4-2 FUNCTION NOTATION

The rule that defines a function may be expressed in many ways. The letter f, or some other lowercase letter, is often used to designate a function. The symbol $f(x)$ represents the value of the function when evaluated for x. For example, if f is the function that pairs a number with three more than twice that number, f can also be written in any one of the following ways:

1. $f = \{(x, y) : y = 2x + 3\}$	The function f is the set of ordered pairs $(x, 2x + 3)$.
2. $f(x) = 2x + 3$	The function value paired with x is $2x + 3$.
3. $y = 2x + 3$	The second element of the pair (x, y) is $2x + 3$.
4. $\{(x, 2x + 3)\}$	The set of ordered pairs whose second element is 3 more than twice the first.
5. $f: x \rightarrow 2x + 3$	Under the function f, x maps to $2x + 3$.
6. $x \xrightarrow{f} 2x + 3$	The image of x for the function f is $2x + 3$.

The expressions in 2, 3, and 4 are abbreviated forms of that in 1. Note that y and $f(x)$ are both symbols for the second element of an ordered pair of the function whose first element is x. That second element is often called the function value. The equation $f(x) = 2x + 3$ is the most common form of function notation. For the function $f = \{(x, y) : y = 2x + 3\}$, when $x = 5$, $y = 13$ or $f(5) = 13$. The use of parentheses in the symbol $f(5)$ does *not* indicate multiplication. The symbol f(5) represents the function *evaluated* at 5, that is, f(5) is the second element of the ordered pair of the function f when the first element is 5. In the symbol f(x), x can be replaced by any number in the domain of f or by any algebraic expression that represents a number in the domain of f.

Therefore, if $f(x) = 2x + 3$:

1. $f(-4) = 2(-4) + 3 = -5$

2. $f(a + 1) = 2(a + 1) + 3 = 2a + 5$

3. $f(2x) = 2(2x) + 3 = 4x + 3$

4. $f(x^2) = 2x^2 + 3$

We have used f to designate a function throughout this section. Other letters, such as g, h, p, or q can also be used. For example, $g(x) = 2x$ defines a function. Note that if $f(x) = 2x + 3$ and $g(x) = 2x$, then $f(x)$ and $g(x) + 3$ define the same set of ordered pairs and $f(x) = g(x) + 3$.

EXAMPLE 1

Let f be the set of ordered pairs such that the second element of each pair is 1 more than twice the first.

a. Write f(x) in terms of x.

b. Write four different ways of symbolizing the function.

c. Find f(7).

Solution   **a.** $f(x) = 1 + 2x$

b. $\{(x, y) : y = 1 + 2x\}$ $y = 1 + 2x$

$x \overset{f}{\rightarrow} 1 + 2x$ $f: x \rightarrow 1 + 2x$

c. $f(7) = 1 + 2(7) = 15$

Exercises

Writing About Mathematics

1. Let $f(x) = x^2$ and $g(x + 2) = (x + 2)^2$. Are f and g the same function? Explain why or why not.

2. Let $f(x) = x^2$ and $g(x + 2) = x^2 + 2$. Are f and g the same function? Explain why or why not.

Developing Skills

In 3–8, for each function: **a.** Write an expression for f(x). **b.** Find f(5).

3. $y = x - 2$

4. $x \xrightarrow{f} x^2$

5. f: $x \rightarrow |3x - 7|$

6. $\{(x, 5x)\}$

7. $y = \sqrt{x - 1}$

8. $x \xrightarrow{f} \frac{2}{x}$

In 9–14, $y = f(x)$. Find f(−3) for each function.

9. $f(x) = 7 - x$

10. $y = 1 + x^2$

11. $f(x) = -4x$

12. $f(x) = 4$

13. $y = \sqrt{1 - x}$

14. $f(x) = |x + 2|$

15. The graph of the function f is shown at the right.

Find:

a. f(2)

b. f(−5)

c. f(−2)

d. $f\left(3\frac{1}{2}\right)$

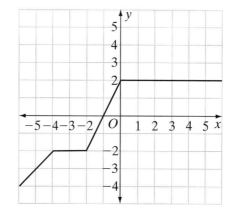

Applying Skills

16. The sales tax t on a purchase is a function of the amount a of the purchase. The sales tax rate in the city of Eastchester is 8%.

   a. Write a rule in function notation that can be used to determine the sales tax on a purchase in Eastchester.

   b. What is a reasonable domain for this function?

   c. Find the sales tax when the purchase is $5.00.

   d. Find the sales tax when the purchase is $16.50.

17. A muffin shop's weekly profit is a function of the number of muffins m that it sells. The equation approximating the weekly profit is $f(m) = 0.60m - 900$.

   a. Draw the graph showing the relationship between the number of muffins sold and the profit.

   b. What is the profit if 2,000 muffins are sold?

   c. How many muffins must be sold for the shop to make a profit of $900?

4-3 LINEAR FUNCTIONS AND DIRECT VARIATION

The graph of the **linear function**
$f = \{(x, y) : y = 2x - 1\}$ is shown to the right.

- The graph has a y-intercept of -1, that is, it intersects the y-axis at $(0, -1)$.

- The graph has a slope of 2, that is, the ratio of the change in y to the change in x is $2 : 1$.

- The domain and range of this function are both the set of real numbers.

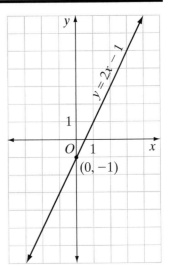

Since it is a function, no two pairs have the same first element. We know that this is true because no vertical line can intersect the graph of f in more than one point. Note also that no horizontal line can intersect the graph of f in more than one point. This is called the **horizontal line test**. All points with the same second element lie on the same horizontal line, that is, if $f(x_1) = f(x_2)$, then $x_1 = x_2$. Therefore, for the function f, no two pairs have the same second element. We say that the function f is a *one-to-one function*. Every element of the domain is paired with exactly one element of the range and every element of the range is paired with exactly one element of the domain. One-to-one can also be written as 1-1.

DEFINITION

A function f is said to be **one-to-one** if no two ordered pairs have the same second element.

For any linear function from the set of real numbers to the set of real numbers, the domain and range are the set of real numbers. Since the range is equal to the set of real numbers, a linear function is onto.

▶ **Every non-constant linear function is one-to-one and onto.**

EXAMPLE 1

Determine if each of the following functions is one-to-one.

a. $y = 4 - 3x$ **b.** $y = x^2$ **c.** $y = |x + 3|$

Solution A function is one-to-one if and only if no two pairs of the function have the same second element.

a. $y = 4 - 3x$ is a linear function whose graph is a slanted line. A slanted line can be intersected by a horizontal line in exactly one point. Therefore, $y = 4 - 3x$ is a one-to-one function.

b. $y = x^2$ is not one-to-one because $(1, 1)$ and $(-1, 1)$ are two ordered pairs of the function.

c. $y = |x + 3|$ is not one-to-one because $(1, 4)$ and $(-7, 4)$ are two ordered
pairs of the function.

EXAMPLE 2

Sketch the graph of $y = -2x + 3$ and show that no two points have the same
second coordinate.

Solution   The graph of $y = -2x + 3$ is a line with a
y-intercept of 3 and a slope of -2. Two
points have the same second coordinate if
and only if they lie on the same horizontal
line. A horizontal line intersects a slanted
line in exactly one point. In the diagram,
the graph of $y = 5$ intersects the graph of
$y = -2x + 3$ in exactly one point. The function passes the horizontal line test. No two
points on the graph of $y = -2x + 3$ have
the same second element and the function
is one-to-one.

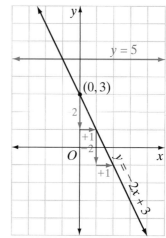

Transformations of Linear Functions

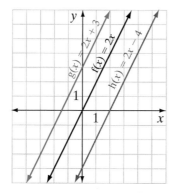

Compare the graphs of $f(x) = 2x$,
$g(x) = 2x + 3$, and $h(x) = 2x - 4$.

1. $g(x) = f(x) + 3$. The graph of g is
   the graph of f shifted 3 units up.

2. $h(x) = f(x) - 4$. The graph of g is
   the graph of f shifted 4 units
   down.

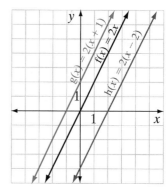

Compare the graphs of $f(x) = 2x$,
$g(x) = 2(x + 1)$, and $h(x) = 2(x - 2)$.

1. $g(x) = f(x + 1)$. The graph of g is
   the graph of f shifted 1 unit to
   the left in the x direction.

2. $h(x) = f(x - 2)$. The graph of g is
   the graph of f shifted 2 units to
   the right in the x direction.

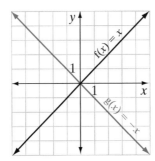

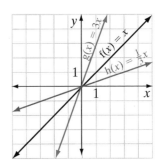

Compare the graphs of f(x) = x and g(x) = −x.

- g(x) = −f(x). The graph of g is the graph of f reflected in the x-axis.

Compare the graphs of f(x) = x, g(x) = 3x, and h(x) = $\frac{1}{3}$x.

1. g(x) = 3f(x). The graph of g is the graph of f stretched vertically by a factor of 3.

2. h(x) = $\frac{1}{3}$f(x). The graph of h is the graph of f compressed vertically by a factor of $\frac{1}{3}$.

Recall that in general, if f(x) is a linear function, then:

▶ **The graph of f(x) + a is the graph of f(x) moved |a| units up when a is positive and |a| units down when a is negative.**

▶ **The graph of f(x + a) is the graph of f(x) moved |a| units to the left when a is positive and |a| units to the right when a is negative.**

▶ **The graph of −f(x) is the graph of f(x) reflected in the x-axis.**

▶ **When a > 1, the graph of af(x) is the graph of f(x) stretched vertically by a factor of a.**

▶ **When 0 < a < 1, the graph of af(x) is the graph of f(x) compressed vertically by a factor of a.**

Direct Variation

If a car moves at a uniform rate of speed, the ratio of the distance that the car travels to the time that it travels is a constant. For example, when the car moves at 45 miles per hour,

$$\frac{d}{t} = 45$$

where d is distance in miles and t is time in hours.

The equation $\frac{d}{t} = 45$ can also be written as the function {(t, d) : d = 45t}. This function is a linear function whose domain and range are the set of real numbers. The function is one-to-one and onto. Although the function d = 45t can have a domain and range that are the set of real numbers, when t represents the time traveled, the domain must be the set of positive real numbers. If the car travels no more than 10 hours a day, the domain would be $0 \le t \le 10$ and the range would be $0 \le d \le 450$.

When the ratio of two variables is a constant, we say that the variables are **directly proportional** or that the variables **vary directly**. Two variables that are directly proportional increase or decrease by the same factor. Every **direct variation** of two variables is a linear function that is one-to-one.

EXAMPLE 3

Jacob can type 55 words per minute.

a. Write a function that shows the relationship between the number of words typed, w, and the number of minutes spent typing, m.

b. Is the function one-to-one?

c. If Jacob types for no more than 2 hours at a time, what are the domain and range of the function?

Solution   **a.** "Words per minute" can be written as $\frac{\text{words}}{\text{minutes}} = \frac{w}{m} = 55$. This is a direct variation function that can be written as $\{(m, w) : w = 55m\}$.

b. The function is a linear function and every linear function is one-to-one.

c. If Jacob types no more than 2 hours, that is, 120 minutes, at a time, the domain is $0 \le m \le 120$ and the range is $0 \le w \le 6{,}600$.

Exercises

Writing About Mathematics

1. Megan said that if $a > 1$ and $g(x) = \frac{1}{a}f(x)$, then the graph of $f(x)$ is the graph of $g(x)$ stretched vertically by the factor a. Do you agree with Megan? Explain why or why not.

2. Kyle said that if r is directly proportional to s, then there is some non-zero constant, c, such that $r = cs$ and that $\{(s, r)\}$ is a one-to-one function. Do you agree with Kyle? Explain why or why not.

Developing Skills

In 3–6, each set represents a function. **a.** What is the domain of each function? **b.** What is the range of each function? **c.** Is the function one-to-one?

3. $\{(1, 4), (2, 7), (3, 10), (4, 13)\}$

4. $\{(0, 8), (2, 6), (4, 4), (6, 2)\}$

5. $\{(2, 7), (3, 7), (4, 7), (5, 7), (6, 7)\}$

6. $\{(0, 3), (-1, 5), (-2, 7), (-3, 9), (-4, 11)\}$

In 7–12, determine if each graph represents a one-to-one function.

7.

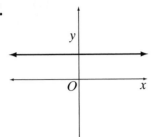

8.

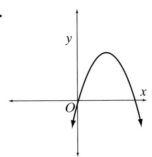

9.

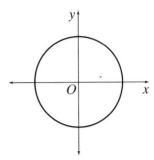

10.

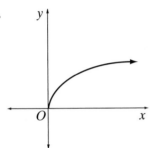

11.

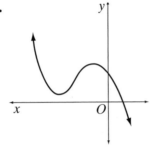

12.

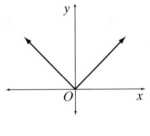

In 13–20: **a.** Graph each function. **b.** Is the function a direct variation? **c.** Is the function one-to-one?

13. $y = 3x$

14. $y = 6 - x$

15. $y = -x$

16. $y = \frac{8}{x}$

17. $y = \frac{1}{2}x$

18. $y = \sqrt{x}$

19. $\frac{y}{x} = 2$

20. $\frac{x}{y} = 2$

21. Is every linear function a direct variation?

22. Is the direct variation of two variables always a linear function?

Applying Skills

In 23–28, write an equation of the direct variation described.

23. The cost of tickets, c, is directly proportional to the number of tickets purchased, n. One ticket costs six dollars.

24. The distance in miles, d, that Mr. Spencer travels is directly proportional to the length of time in hours, t, that he travels at 35 miles per hour.

25. The length of a line segment in inches, i, is directly proportional to the length of the segment in feet, f.

26. The weight of a package of meat in grams, g, is directly proportional to the weight of the package in kilograms, k.

27. Water is flowing into a swimming pool at the rate of 25 gallons per minute. The number of gallons of water in the pool, g, is directly proportional to the number of minutes, m, that the pool has been filling from when it was empty.

28. At 9:00 A.M., Christina began to add water to a swimming pool at the rate of 25 gallons per minute. When she began, the pool contained 80 gallons of water. Christina stopped adding water to the pool at 4:00 P.M. Let *g* be the number of gallons of water in the pool and *t* be the number of minutes that have past since 9:00 A.M.

a. Write an equation for *g* as a function of *t*.

b. What is the domain of the function?

c. What is the range of the function?

d. Is the function one-to-one?

e. Is the function an example of direct variation? Explain why or why not.

Hands-On Activity 1

Refer to the graph of f shown to the right.

1. Sketch the graph of $g(x) = f(-x)$.

2. Sketch the graph of $h(x) = -f(x)$.

3. Describe the relationship between the graph of $f(x)$ and the graph of $g(x)$.

4. Describe the relationship between the graph of $f(x)$ and the graph of $h(x)$.

5. Sketch the graph of $p(x) = 3f(x)$.

6. Describe the relationship between the graph of $f(x)$ and the graph of $p(x)$.

7. Describe the relationship between the graph of $f(x)$ and the graph of $-p(x)$.

8. Sketch the graph of $f(x) = x + 1$ and repeat steps 1 through 7.

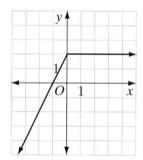

Hands-On Activity 2

Let f be the function whose graph is shown to the right.

1. Sketch the graph of $g(x) = f(x + 2)$.

2. Sketch the graph of $h(x) = f(x - 4)$.

3. Describe the relationship between the graph of f and the graph of $p(x) = f(x + a)$ when *a* is positive.

4. Describe the relationship between the graph of f and the graph of $p(x) = f(x + a)$ when *a* is negative.

5. Sketch the graph of $f(x) + 2$.

6. Sketch the graph of $f(x) - 4$.

7. Describe the relationship between the graph of f and the graph of $f(x) + a$ when *a* is positive.

8. Describe the relationship between the graph of f and the graph of $f(x) + a$ when *a* is negative.

9. Sketch the graph of $2f(x)$.

10. Describe the relationship between $f(x)$ and $af(x)$ when a is a constant greater than 0.

11. Sketch the graph of $-f(x)$.

12. Describe the relationship between $f(x)$ and $-f(x)$.

13. Sketch the graph of $g(x) = f(-x)$.

14. Describe the relationship between $f(x)$ and $g(x)$.

4-4 ABSOLUTE VALUE FUNCTIONS

Let $f(x) = |x|$ be an **absolute value function** from the set of real numbers to the set of real numbers. When the domain is the set of real numbers, the range of f is the set of non-negative real numbers. Therefore, the absolute value function is not onto. For every positive number a, there are two ordered pairs (a, a) and $(-a, a)$ that are pairs of the function. The function is not one-to-one. We say that the function is **many-to-one** because many (in this case two) different first elements, a and $-a$, have the same second element a.

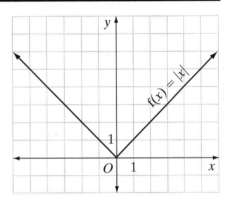

We use single letters such as f and g to name a function. For special functions three letters are often used to designate the function. For example, on the graphing calculator, the absolute value function is named abs. The absolute value function, abs(, is located in the MATH menu under the NUM heading. For instance, to evaluate $|2(1) - 5|$:

ENTER: MATH   ▶   1   2   (1   )   −   5   )   ENTER

DISPLAY:

```
ABS(2(1)-5)
            3
```

Note that the opening parenthesis is supplied by the calculator. We enter the closing parenthesis after we have entered the expression enclosed between the absolute value bars.

EXAMPLE I

Graph the function $y = |x + 3|$ on a graphing calculator.

Solution To display the graph of $y = |x + 3|$, first enter the function value into Y_1. Then graph the function in the ZStandard window.

ENTER: Y= MATH ► 1 3   ENTER: ZOOM 6

X,T,Θ,*n* + 3)

DISPLAY: DISPLAY:
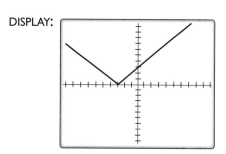

```
Plot1 Plot2 Plot3
\Y₁◼abs(X+3)
\Y₂=
\Y₃=
\Y₄=
\Y₅=
\Y₆=
\Y₇=
```

Note that the minimum value of y is 0. The x value when $y = 0$ is $x + 3 = 0$ or $x = -3$.

The graph of an absolute value function can be used to find the roots of an absolute value equation or inequality.

EXAMPLE 2

Use a graph to find the solution set of the following:

a. $|2x + 1| = 3$ **b.** $|2x + 1| < 3$ **c.** $|2x + 1| > 3$.

Solution Draw the graph of $y = |2x + 1|$.

On the same set of axes, draw the graph of $y = 3$.

a. The points of intersection of the graphs are $(-2, 3)$ and $(1, 3)$, so $|2x + 1| = 3$ for $x = -2$ or $x = 1$.

b. Between $x = -2$ and $x = 1$, the graph of $y = |2x + 1|$ lies below the graph of $y = 3$, so $|2x + 1| < 3$ for $-2 < x < 1$.

c. For $x < -2$ and $x > 1$, the graph of $y = |2x + 1|$ lies above the graph of $y = 3$, so $|2x + 1| > 3$ for $x < -2$ or $x > 1$.

Answers **a.** $\{-2, 1\}$ **b.** $\{x : -2 < x < 1\}$ **c.** $\{x : x < -2 \text{ or } x > 1\}$

EXAMPLE 3

Find the coordinates of the ordered pair of the function $y = |4x - 2|$ for which y has a minimum value.

Solution   Since $y = |4x - 2|$ is always non-negative, its minimum y-value is 0.

When $|4x - 2| = 0, 4x - 2 = 0$. Solve this linear equation for x:

$$4x - 2 = 0$$
$$4x = 2$$
$$x = \frac{2}{4} = \frac{1}{2}$$

Calculator Solution   Enter the function $y = |4x - 2|$ into Y_1 and then press GRAPH. Press 2nd CALC to view the CALCULATE menu and choose 2: zero. Use the arrows to enter a left bound to the left of the minimum, a right bound to the right of the minimum, and a guess near the minimum. The calculator will display the coordinates of the minimum, which is where the graph intersects the y-axis.

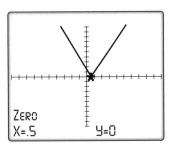

Answer   The minimum occurs at $\left(\frac{1}{2}, 0\right)$.

Exercises

Writing About Mathematics

1. If the domain of the function $f(x) = |3 - x|$ is the set of real numbers less than 3, is the function one-to-one? Explain why or why not.

2. Eric said that if $f(x) = |2 - x|$, then $y = 2 - x$ when $x \leq 2$ and $y = x - 2$ when $x > 2$. Do you agree with Eric? Explain why or why not.

Developing Skills

In 3–6, find the coordinates of the ordered pair with the smallest value of y for each function.

3. $y = |x|$ **4.** $y = |2x + 8|$ **5.** $f(x) = \left|\frac{x}{2} - 7\right|$ **6.** $g(x) = |5 - x|$

In 7–10, the domain of each function is the set of real numbers. **a.** Sketch the graph of each function. **b.** What is the range of each function?

7. $y = |x - 1|$ **8.** $y = |3x + 9|$ **9.** $f(x) = |4x| + 1$ **10.** $g(x) = \left|3 - \frac{x}{2}\right| - 3$

Applying Skills

11. $A(2, 7)$ is a fixed point in the coordinate plane. Let $B(x, 7)$ be any point on the same horizontal line. If $AB = h(x)$, express $h(x)$ in terms of x.

12. Along the New York State Thruway there are distance markers that give the number of miles from the beginning of the thruway. Brianna enters the thruway at distance marker 150.

   a. If $m(x)$ = the number of miles that Brianna has traveled on the thruway, write an equation for $m(x)$ when she is at distance marker x.

   b. Write an equation for $h(x)$, the number of hours Brianna required to travel to distance marker x at 65 miles per hour

13. **a.** Sketch the graph of $y = |x|$.

   b. Sketch the graph of $y = |x| + 2$

   c. Sketch the graph of $y = |x| - 2$.

   d. Describe the graph of $y = |x| + a$ in terms of the graph of $y = |x|$.

14. **a.** Sketch the graph of $y = |x|$.

   b. Sketch the graph of $y = |x + 2|$.

   c. Sketch the graph of $y = |x - 2|$.

   d. Describe the graph of $y = |x + a|$ in terms of the graph of $y = |x|$.

15. **a.** Sketch the graph of $y = |x|$.

   b. Sketch the graph of $y = -|x|$.

   c. Describe the graph of $y = -|x|$ in terms of the graph of $y = |x|$.

16. **a.** Sketch the graph of $y = |x|$.

   b. Sketch the graph of $y = 2|x|$.

   c. Sketch the graph of $y = \frac{1}{2}|x|$.

   d. Describe the graph of $y = a|x|$ in terms of the graph of $y = |x|$.

17. **a.** Draw the graphs of $y = |x + 3|$ and $y = 5$.

   b. From the graph drawn in **a**, determine the solution set of $|x + 3| = 5$.

   c. From the graph drawn in **a**, determine the solution set of $|x + 3| > 5$.

   d. From the graph drawn in **a**, determine the solution set of $|x + 3| < 5$.

18. **a.** Draw the graphs of $y = -|x - 4|$ and $y = -2$.

   b. From the graph drawn in **a**, determine the solution set of $-|x - 4| = -2$.

   c. From the graph drawn in **a**, determine the solution set of $-|x - 4| > -2$.

   d. From the graph drawn in **a**, determine the solution set of $-|x - 4| < -2$.

4-5 POLYNOMIAL FUNCTIONS

Recall that a monomial is a constant, a variable, or the product of constants and variables. For example, $5x$ is a monomial in x. A polynomial is a monomial or the sum of monomials. The expression $x^3 - 5x^2 - 4x + 20$ is a polynomial in x.

A function f is a *polynomial function* when f(x) is a polynomial in x. For example, f(x) = $x^3 - 5x^2 - 4x + 20$ is a polynomial function.

DEFINITION _____

A **polynomial function f of degree n** is a function of the form
$$f(x) = a_n x^n + a_{n-1} x^{n-1} + \cdots + a_0, a_n \neq 0.$$

The simplest polynomial function is the constant function, $y = $ a constant. For example, $y = 5$ is a constant function whose graph is a horizontal line.

The linear function is a polynomial function of degree one. When $m \neq 0$, the graph of the linear function $y = mx + b$ is a slanted line. The y-intercept of the line is b and the slope of the line is m.

Quadratic Functions

The **quadratic function** is a polynomial function of degree two. When $a \neq 0$, the graph of the quadratic function $y = ax^2 + bx + c$ is a curve called a **parabola**.

When a is positive, the parabola opens upward and has a minimum value of y. When $a = 1, b = 2$, and $c = -3$, the quadratic function is $y = x^2 + 2x - 3$.

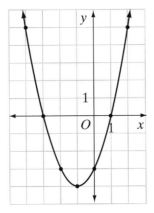

x	$x^2 + 2x - 3$	y
-4	$(-4)^2 + 2(-4) - 3$	5
-3	$(-3)^2 + 2(-3) - 3$	0
-2	$(-2)^2 + 2(-2) - 3$	-3
-1	$(-1)^2 + 2(-1) - 3$	-4
0	$(0)^2 + 2(0) - 3$	-3
1	$(1)^2 + 2(1) - 3$	0
2	$(2)^2 + 2(2) - 3$	5

The **turning point**, also called the **vertex**, for this parabola is $(-1, -4)$ and the **axis of symmetry of the parabola** is the line $x = -1$, the vertical line through the turning point.

When a is negative, the parabola opens downward and has a maximum value of y. When $a = -1, b = -2$, and $c = 3$, the quadratic function is $y = -x^2 - 2x + 3$.

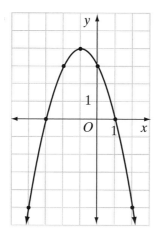

x	$-x^2 - 2x + 3$	y
-4	$-(-4)^2 - 2(-4) + 3$	-5
-3	$-(-3)^2 - 2(-3) + 3$	0
-2	$-(-2)^2 - 2(-2) + 3$	3
-1	$-(-1)^2 - 2(-1) + 3$	4
0	$-(0)^2 - 2(0) + 3$	3
1	$-(1)^2 - 2(1) + 3$	0
2	$-(2)^2 - 2(2) + 3$	-5

The turning point for this parabola is $(-1, 4)$ and the axis of symmetry is the line $x = -1$, the vertical line through the turning point.

Recall from previous courses that for the parabola that is the graph of $f(x) = ax^2 + bx + c$:

1. The minimum or maximum value of the parabola occurs at the axis of symmetry.

2. The formula for the axis of symmetry of the parabola is:

$$x = \frac{-b}{2a}$$

Transformations of Quadratic Functions

The quadratic function $g(x) = x^2 + 6x + 10$ can be written as $g(x) = (x + 3)^2 + 1$.

If: $f(x) = x^2$

$g(x) = (x + 3)^2 + 1$

Then: $g(x) = f(x + 3) + 1$

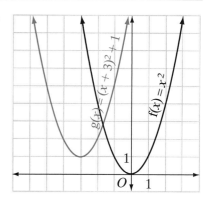

The function $f(x + 3) + 1$ moves every point of $f(x)$ 3 units to the left and 1 unit up. In particular, since the vertex of f is $(0, 0)$, the vertex of g is $(-3, 1)$. Note that for every point on f, there is a corresponding point on g that is 3 units to the left and 1 unit up as shown in the graph on the right.

How does the graph of $g(x) = 2f(x)$ and the graph of $h(x) = \frac{1}{2}f(x)$ compare with the graph of $f(x)$? The graphs of $f(x) = x^2$, $g(x) = 2x^2$, and $h(x) = \frac{1}{2}x^2$ are shown to the right. Compare points on $f(x)$, $g(x)$, and $h(x)$ that have the same x-coordinate. The graph of g is the graph of f stretched vertically by a factor of 2. In other words, the y-value of a point on the graph of g is twice the y value of the point on f and the y value of a point on h is one-half the y value of the point on f.

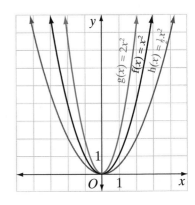

In general, if $f(x)$ is a quadratic function, then:

▶ **The vertex of a quadratic function of the form $f(x) = a(x - h)^2 + k$ is (h, k).**

▶ **The function $g(x) = f(x - h) + k$ moves every point of $f(x)$ $|h|$ units to the left if h is positive and to the right if h is negative and $|k|$ units up if k is positive and down if k is negative.**

▶ **When $f(x)$ is multiplied by a when $a > 1$, the graph is stretched in the vertical direction and when $f(x)$ is multiplied by a when $0 < a < 1$, $f(x)$ is compressed in the vertical direction.**

EXAMPLE 1

Graph $y = \frac{1}{3}(x - 4)^2 + 3$.

Solution   The graph of $y = \frac{1}{3}(x - 4)^2 + 3$ is the graph of $y = x^2$ compressed vertically by a factor of $\frac{1}{3}$, shifted to the right by 4 units and up by 3 units. The vertex of the parabola is $(4, 3)$ and it opens upward as shown in the graph.

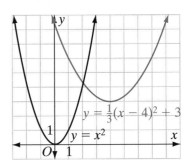

Roots of a Polynomial Function

The **roots** or **zeros** of a polynomial function are the values of x for which $y = 0$. These are the x-coordinates of the points where the graph of the function intersects the x-axis. For the function $y = x^2 + 2x - 3$ as well as for the function $y = -x^2 - 2x + 3$, the roots or zeros of the function are -3 and 1.

These roots can also be found by letting $y = 0$ and solving the resulting quadratic functions for x.

$$y = x^2 + 2x - 3$$
$$0 = x^2 + 2x - 3$$
$$0 = (x + 3)(x - 1)$$

$$x + 3 = 0 \quad | \quad x - 1 = 0$$
$$x = -3 \quad | \quad x = 1$$

$$y = -x^2 - 2x + 3$$
$$0 = -(x^2 + 2x - 3)$$
$$0 = -(x + 3)(x - 1)$$

$$x + 3 = 0 \quad | \quad x - 1 = 0$$
$$x = -3 \quad | \quad x = 1$$

Higher Degree Polynomial Functions

A polynomial function of degree two, a quadratic function, has one turning point. A polynomial function of degree three has at most two turning points. The graph at the right is the graph of the polynomial function $y = x^3 + x^2 - x - 1$. This graph has a turning point at $x = -1$ and at a point between $x = 0$ and $x = 1$.

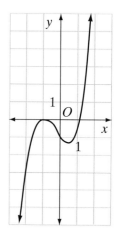

A polynomial function of degree three can have one or three real roots or zeros. It appears that the polynomial function $y = x^3 + x^2 - x - 1$ has only two real roots. However, the root -1, at which the graph is tangent to the x-axis, is called a **double root**, a root that occurs twice. We can see that this is true by solving the polynomial function algebraically. To find the roots of a polynomial function, let y equal zero.

$$y = x^3 + x^2 - x - 1$$
$$0 = x^3 + x^2 - x - 1$$
$$0 = x^2(x + 1) - 1(x + 1)$$
$$0 = (x + 1)(x^2 - 1)$$
$$0 = (x + 1)(x + 1)(x - 1)$$

$$x + 1 = 0 \quad | \quad x + 1 = 0 \quad | \quad x - 1 = 0$$
$$x = -1 \quad | \quad x = -1 \quad | \quad x = 1$$

The polynomial function of degree three has three real roots, but two of these roots are equal. Note that if 1 is a root of the polynomial function, $(x - 1)$ is a factor of the polynomial, and if -1 is a double root of the polynomial function, $(x + 1)(x + 1)$ or $(x + 1)^2$ is a factor of the polynomial. In general:

▶ **If a is a root of a polynomial function, then $(x - a)$ is a factor of the polynomial.**

Note also that the polynomial function $y = x^3 + x^2 - x - 1$ has three roots. Any polynomial function of degree three has 1 or 3 real roots. If it has two real roots, one of them is a double root. These real roots may be irrational numbers. In general:

▶ **A polynomial function of degree n has at most n distinct real roots.**

Examining the Roots of Polynomial from the Graph

We found the roots of the polynomial $y = x^3 + x^2 - x - 1$ by factoring, but not all polynomials can be factored over the set of integers. However, we can use a calculator to sketch the graph of the function and examine the roots using the zero function.

For example, to examine the roots of $f(x) = x^4 - 4x^3 - x^2 + 16x - 12$, enter the function into Y_1 and view the graph in the ZStandard window. The graph appears to cross the x-axis at -2, 1, 2, and 3. We can verify that these are the roots of the function.

Press **2nd**   **CALC** to view the CALCULATE menu and choose 2: zero.

Use the arrows to enter a left bound to the left of one of the zero values, a right bound to the right of the zero value, and a guess near the zero value. The calculator will display the coordinates of the point at which the graph intersects the x-axis. Repeat to find the other roots.

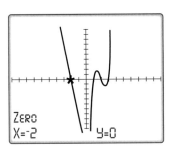

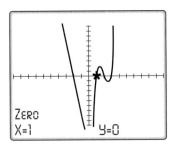

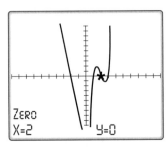

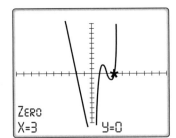

We can verify that these are the roots of the function by evaluating the function at each value:

$$f(-2) = (-2)^4 - 4(-2)^3 - (-2)^2 + 16(-2) - 12$$
$$= 16 - 4(-8) - 4 - 32 - 12$$
$$= 0$$

$$f(2) = (2)^4 - 4(2)^3 - (2)^2 + 16(2) - 12$$
$$= 16 - 4(8) - 4 + 32 - 12$$
$$= 0$$

$$f(1) = (1)^4 - 4(1)^3 - (1)^2 + 16(1) - 12$$
$$= 1 - 4 - 1 + 16 - 12$$
$$= 0$$

$$f(3) = (3)^4 - 4(3)^3 - (3)^2 + 16(3) - 12$$
$$= 81 - 4(27) - 9 + 48 - 12$$
$$= 0$$

 Note: The calculator approach gives exact values only when the root is an integer or a rational number that can be expressed as a finite decimal. The calculator returns approximate values for rational roots with infinite decimal values and for irrational roots.

EXAMPLE 2

From the graph, determine the roots of the polynomial function p(x).

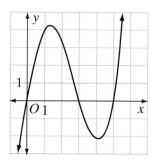

Solution The graph intersects the x-axis at 0, 3, and 5.

Therefore, the roots or zeros of the function are 0, 3, and 5.

EXAMPLE 3

Find the real roots of $f(x) = x^4 - 2x^3 - 2x^2 - 2x - 3$ graphically.

Solution Enter the function into Y_1 and graph the function into the standard viewing window. The graph appears to cross the x-axis at 3 and -1. We can verify that these are the roots of the function. Press **2nd** **CALC** **2** to use the zero function. Use the arrows to enter a left bound to the left of one of the zero values, a right bound to the right of the zero value, and a guess near the zero value. Examine one zero value at a time. Repeat to find the other roots.

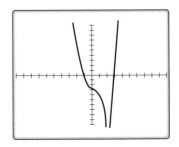

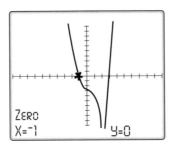 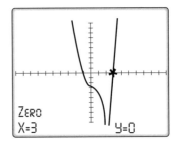

Answer The real roots of the function are -1 and 3.

EXAMPLE 4

Approximate the real roots of $f(x) = \frac{1}{3}(x^3 + x^2 - 11x - 3)$ to the nearest tenth.

Solution The graph appears to cross the x-axis at -3.7, -0.3, and 3. We can verify that these are valid approximations of the roots of the function by using the graphing calculator. Enter the function into Y_1 and graph the function into the standard viewing window. Press **2nd** **CALC** **2** to use the zero

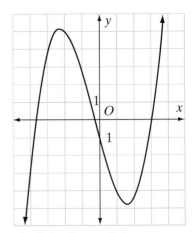

function. Use the arrows to enter a left bound to the left of one of the zero values, a right bound to the right of the zero value, and a guess near the zero value. Examine one zero value at a time. Repeat to find the other roots.

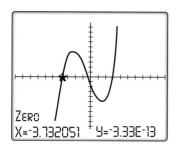 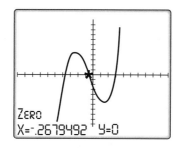 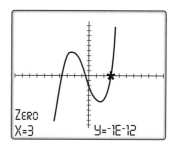

Answer The real roots of the function are approximately equal to -3.73, -0.27, and 3.

Exercises

Writing About Mathematics

1. Tiffany said that the polynomial function $f(x) = x^4 + x^2 + 1$ cannot have real roots. Do you agree with Tiffany? Explain why or why not.

2. The graph of $y = x^2 - 4x + 4$ is tangent to the x-axis at $x = 2$ and does not intersect the x-axis at any other point. How many roots does this function have? Explain your answer.

Developing Skills

In 3–11, each graph shows a polynomial functions from the set of real numbers to the set of real numbers.

a. Find the real roots or zeros of the function from the graph, if they exist.

b. What is the range of the function?

c. Is the function onto?

d. Is the function one-to-one?

3.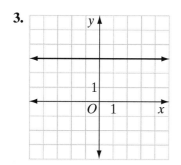

4.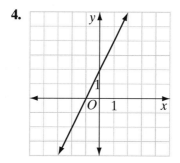

5.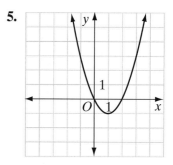

6.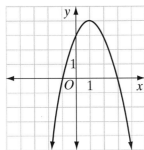

7.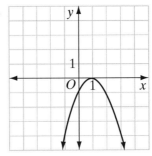

8.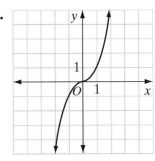

9.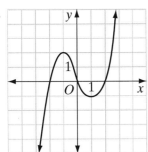

10.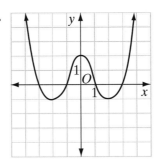

11.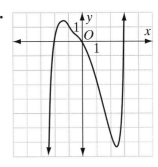

In 12–17, use a graph to find the solution set of each inequality.

12. $x^2 + 2x - 3 < 0$

13. $x^2 + 4x + 3 \leq 0$

14. $x^2 - x - 2 > 0$

15. $x^2 - 2x + 1 < 0$

16. $-x^2 + 6x - 5 < 0$

17. $-x^2 - 3x + 4 \geq 0$

Applying Skills

18. The perimeter of a rectangle is 24 feet.

  a. If x is the measure of one side of the rectangle, represent the measure of an adjacent side in terms of x.

  b. If y is the area of the rectangle, express the area in terms of x.

  c. Draw the graph of the function written in **b**.

  d. What are the dimensions of the rectangle with the largest area?

19. The sum of the lengths of the legs of a right triangle is 20 feet.

  a. If x is the measure of one of the legs, represent the measure of the other leg in terms of x.

  b. If y is the area of the triangle, express the area in terms of x.

  c. Draw the graph of the function written in **b**.

  d. What are the dimensions of the triangle with the largest area?

20. A polynomial function of degree three, p(x), intersects the x-axis at $(-4, 0), (-2, 0)$, and $(3, 0)$ and intersects the y-axis at $(0, -24)$. Find p(x).

21. a. Sketch the graph of $y = x^2$.

   b. Sketch the graph of $y = x^2 + 2$

   c. Sketch the graph of $y = x^2 - 3$.

   d. Describe the graph of $y = x^2 + a$ in terms of the graph of $y = x^2$.

   e. What transformation maps $y = x^2$ to $y = x^2 + a$?

22. a. Sketch the graph of $y = x^2$.

   b. Sketch the graph of $y = (x + 2)^2$.

   c. Sketch the graph of $y = (x - 3)^2$.

   d. Describe the graph of $y = (x + a)^2$ in terms of the graph of $y = x^2$.

   e. What transformation maps $y = x^2$ to $y = (x + a)^2$?

23. a. Sketch the graph of $y = x^2$.

   b. Sketch the graph of $y = -x^2$.

   c. Describe the graph of $y = -x^2$ in terms of the graph of $y = x^2$.

   d. What transformation maps $y = x^2$ to $y = -x^2$?

24. a. Sketch the graph of $y = x^2$.

   b. Sketch the graph of $y = 3x^2$

   c. Sketch the graph of $y = \frac{1}{3}x^2$.

   d. Describe the graph of $y = ax^2$ in terms of the graph of $y = x^2$ when $a > 1$.

   e. Describe the graph of $y = ax^2$ in terms of the graph of $y = x^2$ when $0 < a < 1$.

25. For the parabola whose equation is $y = ax^2 + bx + c$, the equation of the axis of symmetry is $x = \frac{-b}{2a}$. The turning point of the parabola lies on the axis of symmetry. Therefore its x-coordinate is $\frac{-b}{2a}$. Substitute this value of x in the equation of the parabola to find the y-coordinates of the turning point. Write the coordinates of the turning point in terms of a, b, and c.

4-6 THE ALGEBRA OF FUNCTIONS

Function Arithmetic

Fran and Greg work together to produce decorative vases. Fran uses a potter's wheel to form the vases and Greg glazes and fires them. Last week, they completed five jobs. The table on the following page shows the number of hours spent on each job.

	Hours Worked		
Job number	**Fran**	**Greg**	**Total**
1	3	7	10
2	2	5	7
3	4	8	12
4	$\frac{1}{2}$	3	$3\frac{1}{2}$
5	$1\frac{1}{2}$	6	$7\frac{1}{2}$

Let x equal the job number, $f(x)$ = the number of hours that Fran worked on job x, $g(x)$ = the number of hours that Greg worked on job x, and $t(x)$ = the total number of hours worked on job x.

Therefore:

$f(1) = 3$	$g(1) = 7$	$t(1) = 10$
$f(2) = 2$	$g(2) = 5$	$t(2) = 7$
$f(3) = 4$	$g(3) = 8$	$t(3) = 12$
$f(4) = \frac{1}{2}$	$g(4) = 3$	$t(4) = 3\frac{1}{2}$
$f(5) = 1\frac{1}{2}$	$g(5) = 6$	$t(5) = 7\frac{1}{2}$

For each function, the domain is the set $\{1, 2, 3, 4, 5\}$ and $t(x) = f(x) + g(x)$ or $t(x) = (f + g)(x)$. We can perform arithmetic operations with functions. In general, if $f(x)$ and $g(x)$ are functions, then:

$$(f + g)(x) = f(x) + g(x)$$

The domain of $(f + g)$ is the set of elements common to the domain of f and of g, that is, the domain of $(f + g)$ is the intersection of the domains of f and g.

For example, if $f(x) = 2x^2$ and $g(x) = 5x$, then:

$$(f + g)(x) = f(x) + g(x) = 2x^2 + 5x$$

If the domains of f and g are both the set of real numbers, then the domain of $(f + g)$ is also the set of real numbers.

The difference, product, and quotient of two functions, as well as the product of a constant times a function, can be defined in a similar way.

If $f(x)$ and $g(x)$ are functions, then:

1. The difference of f and g is expressed as:

$$(f - g)(x) = f(x) - g(x)$$

Example: Let $f(x) = 2x^2$ and $g(x) = 5x$.
Then $(f - g)(x) = f(x) - g(x) = 2x^2 - 5x$.

2. The product of f and g is expressed as:

$$(fg)(x) = f(x) \cdot g(x)$$

Example: Let $f(x) = 2x^2$ and $g(x) = 5x$.
Then $(fg)(x) = f(x) \cdot g(x) = 2x^2(5x) = 10x^3$.

3. The quotient of f and g is expressed as:

$$\left(\frac{f}{g}\right)(x) = \frac{f(x)}{g(x)}$$

Example: Let $f(x) = 2x^2$ and $g(x) = 5x$. Then $\left(\frac{f}{g}\right)(x) = \frac{f(x)}{g(x)} = \frac{2x^2}{5x} = \frac{2x}{5}$.

4. The product of f and a constant, a, is expressed as:

$$af(x)$$

Example: Let $f(x) = 2x^2$ and $a - 3$. Then $3f(x) = 3(2x^2) = 6x^2$.

Note: If $g(x) = a$, a constant, then $(fg)(x) = f(x)g(x) = f(x) \cdot (a) = af(x)$. That is, if one of the functions is a constant, then function multiplication is the product of a function and a constant.

    If the domain of f and of g is the set of real numbers, then the domain of $(f + g)$, $(f - g)$, (fg), and $af(x)$ is the set of real numbers and the domain of $\left(\frac{f}{g}\right)$ is the set of numbers for which $g(x) \neq 0$.
    In the examples given above, the domain of f and of g is the set of real numbers. The domain of $f + g$, of $f - g$, and of fg is the set of real numbers. The domain of $\frac{f}{g}$ is the set of real numbers for which $g(x) \neq 0$, that is, the set of non-zero real numbers.

EXAMPLE 1

Let $f(x) = 3x$ and $g(x) = \sqrt{x}$.

a. Write an algebraic expression for h(x) if $h(x) = 2f(x) \div g(x)$.

b. What is the largest possible domain for f, 2f, g, and h that is a subset of the set of real numbers?

c. What are the values of $f(2)$, $g(2)$, and $h(2)$?

Solution **a.** $h(x) = \frac{2f(x)}{g(x)} = \frac{2(3x)}{\sqrt{x}} = \frac{6x}{\sqrt{x}} \cdot \frac{\sqrt{x}}{\sqrt{x}} = \frac{6x\sqrt{x}}{x} = 6\sqrt{x}$

b. The domain of f and of 2f is the set of real numbers.

The domain of g is the set of non-negative real numbers.

The domain of h is the set of positive real numbers, that is, the set of all numbers in the domains of both f and g for which $g(x) \neq 0$.

c. $f(2) = 3(2) = 6$ $g(2) = \sqrt{2}$ $h(2) = 6\sqrt{2}$

Transformations and Function Arithmetic

Let $h(x) = 3$ be a constant function. The second element of each ordered pair is the constant, 3. The domain of h is the set of real numbers and the range is {3}. The graph of $h(x) = 3$ is a horizontal line 3 units above the x-axis.

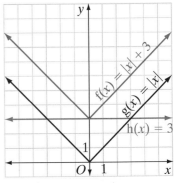

Let $g(x) = |x|$. The domain of g is the set of real numbers and the range is the set of non-negative real numbers.

Let $f(x) = g(x) + h(x) = |x| + 3$. The domain of $f(x)$ is the set of real numbers. For each value of x, $f(x)$ is 3 more than $g(x)$. The graph of $f(x)$ is the graph of $g(x)$ translated 3 units in the vertical direction.

In general, for any function g:

▶ **If $h(x) = a$, a constant, and $f(x) = g(x) + h(x) = g(x) + a$, then the graph of $f(x)$ is the graph of $g(x)$ translated a units in the vertical direction.**

Let h and g be defined as before.

Let $f(x) = g(x) \cdot h(x) = gh(x) = 3|x|$. The domain of $f(x)$ is the set of real numbers. For each value of x, $f(x)$ is 3 times $g(x)$. The graph of $f(x)$ is the graph of $g(x)$ stretched by the factor 3 in the vertical direction.

In general, for any function g:

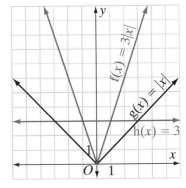

▶ **If $h(x) = a$, and a is a positive constant, then $f(x) = g(x) \cdot h(x) = ag(x)$. The graph of $f(x)$ is the graph of $g(x)$ stretched by the factor a in the vertical direction when $a > 1$ and compressed by a factor of a when $0 < a < 1$.**

Let $h(x) = -1$ be a constant function. The second element of each pair is the constant, -1. The domain of h is the set of real numbers and the range is {-1}.

Let $g(x) = |x|$. The domain of g is the set of real numbers and the range is the set of non-negative real numbers.

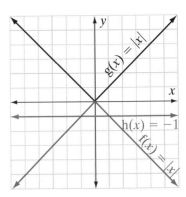

Let $f(x) = g(x) \cdot h(x) = gh(x) = -1|x| = -|x|$. The domain of $f(x)$ is the set of real numbers. For each value of x, $f(x)$ is -1 times $g(x)$, that is, the opposite of $g(x)$. The graph of $f(x)$ is the graph of $g(x)$ reflected in the x-axis.

In general, for any function g:

▶ **If $h(x) = -1$, then $f(x) = g(x) \cdot h(x) = -g(x)$. The graph of $f(x)$ is the graph of $g(x)$ reflected in the x-axis.**

Horizontal translations will be covered in the next section when we discuss function composition.

EXAMPLE 2

The graphs of $f(x)$, $g(x)$, and $h(x)$ are shown at the right. The function $g(x)$ is the function $f(x)$ compressed vertically by the factor $\frac{1}{2}$, and the function $h(x)$ is the function $g(x)$ moved 3 units up.

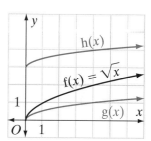

If $f(x) = \sqrt{x}$, what is the equation of $h(x)$?

Solution The vertex of the parabola that is the graph of $f(x)$ is the point $(0, 0)$. When $f(x)$ is compressed by the factor $\frac{1}{2}$,

$$g(x) = \tfrac{1}{2}f(x) = \tfrac{1}{2}\sqrt{x}$$

Answer When $g(x)$ is moved 3 units up, $h(x) = g(x) + 3 = \tfrac{1}{2}f(x) + 3 = \tfrac{1}{2}\sqrt{x} + 3$. ▪

Note: In Example 2, the order that you perform the transformations is important. If we had translated vertically first, the resulting function would have been $\frac{1}{2}(f(x) + 3)$ or $\frac{1}{2}\sqrt{x} + \frac{3}{2}$.

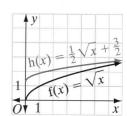

Exercises

Writing About Mathematics

1. Eric said that if $f(x) = |2 - x|$ and $g(x) = |x - 2|$, then $(f + g)(x) = 0$. Do you agree with Eric? Explain why or why not.

2. Give an example of a function g for which $2g(x) \neq g(2x)$. Give an example of a function f for which $2f(x) = f(2x)$.

Developing Skills

3. Let f = {(0, 5), (1, 4), (2, 3), (3, 2), (4, 1), (5, 0)}
and g = {(1, 1), (2, 4), (3, 9), (4, 16), (5, 25) (6, 36)}.

a. What is the domain of f? **b.** What is the domain of g?

c. What is the domain of (g − f)? **d.** List the ordered pairs of (g − f) in set notation.

e. What is the domain of $\frac{g}{f}$? **f.** List the ordered pairs of $\frac{g}{f}$ in set notation.

In 4–7, the graph of y = f(x) is shown. **a.** Find f(0) for each function. **b.** Find x when f(x) = 0. For each function, there may be no values, one value, or more than one value. **c.** Sketch the graph of y = f(x) + 2. **d.** Sketch the graph of y = 2f(x). **e.** Sketch the graph of y = −f(x).

4.

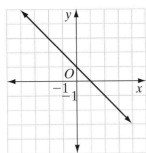

5.

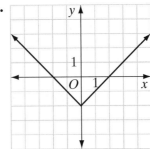

6.

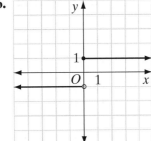

7.

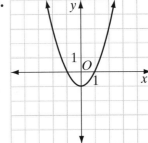

In 8–13, the domain of f(x) = 4 − 2x and of g(x) = x^2 is the set of real numbers and the domain of h(x) = $\frac{1}{x}$ is the set of non-zero real numbers.

a. Write each function value in terms of x.

b. Find the domain of each function.

8. (g + f)(x) **9.** (g − h)(x) **10.** (gh)(x)

11. $\left(\frac{g}{f}\right)(x)$ **12.** (g + 3f)(x) **13.** (−gf)(x)

Applying Skills

14. When Brian does odd jobs, he charges ten dollars an hour plus two dollars for transportation. When he works for Mr. Atkins, he receives a 15% tip based on his wages for the number of hours that he works but not on the cost of transportation.

a. Find $c(x)$, the amount Brian charges for x hours of work.

b. Find $t(x)$, Brian's tip when he works for x hours.

c. Find $e(x)$, Brian's earnings when he works for Mr. Atkins for x hours, if $e(x) = c(x) + t(x)$.

d. Find $e(3)$, Brian's earnings when he works for Mr. Atkins for 3 hours.

15. Mrs. Cucci makes candy that she sells for $8.50 a pound. When she ships the candy to out-of-town customers, she adds a flat shipping charge of $2.00 plus an additional $0.50 per pound.

a. Find $c(x)$, the cost of x pounds of candy.

b. Find $s(x)$, the cost of shipping x pounds of candy.

c. Find $t(x)$, the total bill for x pounds of candy that has been shipped, if $t(x) = c(x) + s(x)$.

d. Find $t(5)$, the total bill for five pounds of candy that is shipped.

16. Create your own function $f(x)$ and show that $f(x) + f(x) = 2f(x)$. Explain why this result is true in general.

4-7 COMPOSITION OF FUNCTIONS

In order to evaluate a function such as $y = \sqrt{x + 3}$, it is necessary to perform two operations: first we must add 3 to x, and then we must take the square root of the sum. The set of ordered pairs $\{(x, y) : y = \sqrt{x + 3}\}$ is the **composition** of two functions. If we let $f(x) = x + 3$ and $g(x) = \sqrt{x}$, we can form the function $h(x) = \sqrt{x + 3}$. We evaluate the function h as shown below:

x	$\xrightarrow{f}$	$x + 3$	$\xrightarrow{g}$	$\sqrt{x + 3}$	$(x, h(x))$
-2	$\xrightarrow{f}$	1	$\xrightarrow{g}$	$\sqrt{1} = 1$	$(-2, 1)$
0	$\xrightarrow{f}$	3	$\xrightarrow{g}$	$\sqrt{3}$	$(0, \sqrt{3})$
1	$\xrightarrow{f}$	4	$\xrightarrow{g}$	$\sqrt{4} = 2$	$(1, 2)$
$\frac{3}{4}$	$\xrightarrow{f}$	$\frac{15}{4}$	$\xrightarrow{g}$	$\sqrt{\frac{15}{4}} = \frac{\sqrt{15}}{2}$	$\left(\frac{3}{4}, \frac{\sqrt{15}}{2}\right)$
5	$\xrightarrow{f}$	8	$\xrightarrow{g}$	$\sqrt{8} = 2\sqrt{2}$	$(5, 2\sqrt{2})$

The function h is a **composite function**. It is the set of ordered pairs that result from mapping x to $f(x)$ and then mapping $f(x)$ to $g(f(x))$. We say that $h(x) = g(f(x))$ or that $h(x) = g \circ f(x)$. The symbol $\circ$ is read "of." The function h is the composition of g following f. The function f is applied first, followed by g. We evaluate the composition of functions from right to left.

In the introductory example, g is defined only for non-negative real numbers. However, f is a function from the set of real numbers to the set of real numbers. In order for the composition $g \circ f$ to be well defined, we must restrict the domain of f so that its range is a subset of the domain of g. Since the largest possible domain of g is the set of non-negative real numbers, the range of f must be the non-negative real numbers, that is, $f(x) \geq 0$ or $x + 3 \geq 0$ and $x \geq -3$. Therefore, when we restrict f to $\{x : x \geq -3\}$, the composition $g \circ f$ is well defined. The domain of h is $\{x : x \geq -3\}$ and the range of h is $\{y : y \geq 0\}$.

▶ **The domain of the composition $g \circ f$ is the set of all x in the domain of f where $f(x)$ is in the domain of g.**

EXAMPLE 1

Given: f = {(0, 1), (1, 3), (2, 5), (3, 7)} and g = {(1, 0), (3, 3), (5, 1), (7, 2)}

Find: **a.** $f \circ g$ **b.** $g \circ f$

Solution **a.** To find $(f \circ g)(x)$, work from right to left. First find $g(x)$ and then find $f(g(x))$.

x	$\xrightarrow{g}$	$g(x)$	$\xrightarrow{f}$	$(f \circ g)(x)$
1	$\xrightarrow{g}$	0	$\xrightarrow{f}$	1
3	$\xrightarrow{g}$	3	$\xrightarrow{f}$	7
5	$\xrightarrow{g}$	1	$\xrightarrow{f}$	3
7	$\xrightarrow{g}$	2	$\xrightarrow{f}$	5

b. To find $(g \circ f)(x)$, work from right to left. First find $f(x)$ and then find $g(f(x))$.

x	$\xrightarrow{f}$	$f(x)$	$\xrightarrow{g}$	$(g \circ f)(x)$
0	$\xrightarrow{f}$	1	$\xrightarrow{g}$	0
1	$\xrightarrow{f}$	3	$\xrightarrow{g}$	3
2	$\xrightarrow{f}$	5	$\xrightarrow{g}$	1
3	$\xrightarrow{f}$	7	$\xrightarrow{g}$	2

Answers **a.** $f \circ g$ = {(1, 1), (3, 7), (5, 3), (7, 5)}

b. $g \circ f$ = {(0, 0), (1, 3), (2, 1), (3, 2)}

EXAMPLE 2

Let $h(x) = (p \circ q)(x)$, $p(x) = |x|$, and $q(x) = x - 4$.

a. Find $h(-3)$.   **b.** What is the domain of h?   **c.** What is the range of h?

Solution **a.** $h(-3) = p(q(-3)) = p(-3 - 4) = p(-7) = |-7| = 7$

b. Since the range of q is the same as the domain of p, all elements of the domain of q can be used as the domain of $h = (p \circ q)$. The domain of q is the set of real numbers, so the domain of h is the set of real numbers.

c. All elements of the range of $h = (p \circ q)$ are elements of the range of p. The range of p is the set of non-negative real numbers.

Answers **a.** $h(-3) = 7$

b. The domain of h is the set of real numbers.

c. The range of h is the set of non-negative real numbers.

EXAMPLE 3

Show that the composition of functions is *not* commutative, that is, it is not necessarily true that $f(g(x)) = g(f(x))$.

Solution If the composition of functions is commutative, then $f(g(x)) = g(f(x))$ for all $f(x)$ and all $g(x)$. To show the composition of functions is *not* commutative, we need to show one counterexample, that is, one pair of functions $f(x)$ and $g(x)$ such that $f(g(x)) \neq g(f(x))$.

Let $f(x) = 2x$ and $g(x) = x + 3$.

$$\begin{aligned} f(g(x)) &= f(x + 3) \\ &= 2(x + 3) \\ &= 2x + 6 \end{aligned} \qquad \begin{aligned} g(f(x)) &= g(2x) \\ &= 2x + 3 \end{aligned}$$

Let $x = 5$.

Then $f(g(5)) = 2(5) + 6 = 16$.

Let $x = 5$.

Then $g(f(5)) = 2(5) + 3 = 13$.

Therefore, $f(g(x)) \neq g(f(x))$ and the composition of functions is not commutative.

∎

EXAMPLE 4

Let $p(x) = x + 5$ and $q(x) = x^2$.

a. Write an algebraic expression for $r(x) = (p \circ q)(x)$.
b. Write an algebraic expression for $f(x) = (q \circ p)(x)$.
c. Evaluate $r(x)$ for $\{x : -2 \leq x \leq 3\}$ if $x \in \{\text{Integers}\}$.
d. Evaluate $f(x)$ for $\{x : -2 \leq x \leq 3\}$ if $x \in \{\text{Integers}\}$.
e. For these two functions, is $(p \circ q)(x) = (q \circ p)(x)$?

Solution **a.** $r(x) = (p \circ q)(x) = p(x^2) = x^2 + 5$
b. $f(x) = (q \circ p)(x) = q(x + 5) = (x + 5)^2 = x^2 + 10x + 25$

c.

x	$r(x)$	$(x, r(x))$
−2	$(-2)^2 + 5 = 9$	$(-2, 9)$
−1	$(-1)^2 + 5 = 6$	$(-1, 6)$
0	$(0)^2 + 5 = 5$	$(0, 5)$
1	$(1)^2 + 5 = 6$	$(1, 6)$
2	$(2)^2 + 5 = 9$	$(2, 9)$
3	$(3)^2 + 5 = 14$	$(3, 14)$

d.

x	$f(x)$	$(x, f(x))$
−2	$(-2)^2 + 10(-2) + 25 = 9$	$(-2, 9)$
−1	$(-1)^2 + 10(-1) + 25 = 16$	$(-1, 16)$
0	$(0)^2 + 10(0) + 25 = 25$	$(0, 25)$
1	$(1)^2 + 10(1) + 25 = 36$	$(1, 36)$
2	$(2)^2 + 10(2) + 25 = 49$	$(2, 49)$
3	$(3)^2 + 10(3) + 25 = 64$	$(3, 64)$

e. No. From the tables, since $(p \circ q)(3) = 14$ and $(q \circ p)(3) = 64$, $(p \circ q)(x) \neq (q \circ p)(x)$.

Answers **a.** $r(x) = x^2 + 5$ **b.** $f(x) = x^2 + 10x + 25$

c. $\{(-2, 9), (-1, 6), (0, 5), (1, 6), (2, 9), (3, 14)\}$

d. $\{(-2, 9), (-1, 16), (0, 25), (1, 36), (2, 49), (3, 64)\}$

e. No

Transformations and Function Composition

Let $p(x) = x + 5$ and $q(x) = x - 5$. The domains of p and q are the set of real numbers and the ranges are also the set of real numbers.

Let $g(x) = x^2$. The domain of g is the set of real numbers and the range is the set of non-negative real numbers.

Let $f(x) = (g \circ q)(x) = g(x + 5) = (x + 5)^2$.

Let $h(x) = (g \circ q)(x) = g(x - 5) = (x - 5)^2$. The domains of f and h are the set of real numbers and the ranges are the set of non-negative real numbers. If we sketch the graph of $g(x) = x^2$ and the graph of $f(x) = (x + 5)^2$, note that $f(x)$ is the graph of $g(x)$ moved 5 units to the left. Similarly, the graph of $h(x) = (x - 5)^2$ is the graph of $g(x)$ moved 5 units to the right.

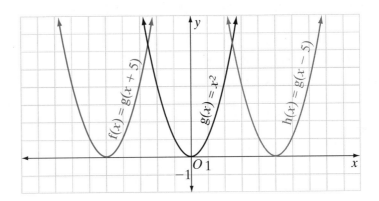

In general, for any function g:

▶ **If $h(x) = g(x + a)$ for a constant a, the graph of $h(x)$ is the graph of $g(x)$ moved $|a|$ units to the left when a is positive and $|a|$ units to the right when a is negative.**

Compare the functions and their graphs on the following page.

$f(x) = \sqrt{x}$ $g(x + 1) = \sqrt{x + 1}$ $h(x) = \sqrt{x} + 1$ $p(x) = \sqrt{x + 1}$

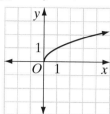   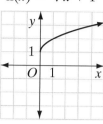   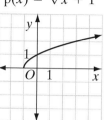

Domain: $x \geq 0$ Domain: $x \geq -1$ Domain: $x \geq 0$ Domain: $x \geq -1$

Note that:

- $g(x)$ is the same function as $f(x)$, $x \to \sqrt{x}$ is the same as $x + 1 \to \sqrt{x + 1}$.
- $h(x)$ is $f(x)$ moved 1 unit up.
- $p(x)$ is $f(x)$ moved 1 unit to the left.

Exercises

Writing About Mathematics

1. Marcie said that if $f(x) = x^2$, then $f(a + 1) = (a + 1)^2$. Do you agree with Marcie? Explain why or why not.

2. Explain the difference between $fg(x)$ and $f(g(x))$.

Developing Skills

In 3–10, evaluate each composition for the given values if $f(x) = 3x$ and $g(x) = x - 2$.

3. $f(g(4))$ **4.** $g(f(4))$ **5.** $f \circ g(-2)$ **6.** $g \circ f(-2)$

7. $f(f(5))$ **8.** $g(g(5))$ **9.** $f\left(g\left(\frac{2}{3}\right)\right)$ **10.** $g\left(f\left(\frac{2}{3}\right)\right)$

In 11–18: **a.** Find $h(x)$ when $h(x) = g(f(x))$. **b.** What is the domain of $h(x)$? **c.** What is the range of $h(x)$? **d.** Graph $h(x)$.

11. $f(x) = 2x + 1, g(x) = 4x$ **12.** $f(x) = 3x, g(x) = x - 1$

13. $f(x) = x^2, g(x) = 4 + x$ **14.** $f(x) = 4 + x, g(x) = x^2$

15. $f(x) = x^2, g(x) = \sqrt{x}$ **16.** $f(x) = |2 + x|, g(x) = -x$

17. $f(x) = 5 - x, g(x) = |x|$ **18.** $f(x) = 3x - 1, g(x) = \frac{1}{3}(x + 1)$

In 19–22, let $f(x) = |x|$. Find $f(g(x))$ and $g(f(x))$ for each given function.

19. $g(x) = x + 3$ **20.** $g(x) = 2x$ **21.** $g(x) = 2x + 3$ **22.** $g(x) = 5 - x$

23. For which of the functions given in 19–22 does $f(g(x)) = g(f(x))$?

24. If $p(x) = 2$ and $q(x) = x + 2$, find $p(q(5))$ and $q(p(5))$.

25. If $h(x) = 2(x + 1)$ and $h(x) = f(g(x))$, what are possible expressions for $f(x)$ and for $g(x)$?

Applying Skills

26. Let c(x) represent the cost of an item, x, plus sales tax and d(x) represent the cost of an item less a discount of $10.

a. Write c(x) using an 8% sales tax.

b. Write d(x).

c. Find c ∘ d(x) and d ∘ c(x). Does c ∘ d(x) = d ∘ c(x)? If not, explain the difference between the two functions. When does it makes sense to use each function?

d. Which function can be used to find the amount that must be paid for an item with a $10 discount and 8% tax?

27. The relationship between temperature and the rate at which crickets chirp can be approximated by the function n(x) = 4x − 160 where n is the number of chirps per minute and x is the temperature in degrees Fahrenheit. On a given summer day, the temperature outside between the hours of 6 A.M. and 12 P.M. can be modeled by the function f(x) = 0.55x^2 + 1.66x + 50 where x is the number of hours elapsed from 6 P.M.

a. Find the composite function n ∘ f.

b. What is the rate of chirping at 11 A.M.?

4-8 INVERSE FUNCTIONS

Identity Function

A function of the form $y = mx + b$ is a linear function. When we let $m = 1$ and $b = 0$, the function is $y = 1x + 0$ or $y = x$. This is a function in which the second element, y, of each ordered pair is equal to the first element.

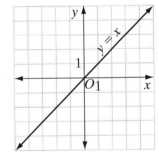

▶ The function $y = x$ is called the *identity function*, I, such that I(x) = x.

The graph of the identity function is shown to the right.

DEFINITION _____
The **identity function (I)** is the function that maps every element of the domain to the same element of the range.

The domain of the identity can be the set of real numbers or any subset of the set of real numbers. The domain and range of the identity function are the same set.

Recall that for the set of real numbers, the identity for addition is 0 because for each real number a, $a + 0 = 0 + a = a$. We can say that adding the identity to any real number a leaves a unchanged.

Similarly, for the set of real numbers, the identity for multiplication is 1 because for each real number a, $a \cdot 1 = 1 \cdot a = a$. We can say that multiplying the identity for multiplication by any real number a leaves a unchanged.

Likewise, the identity function is $I(x)$ because for any function f:

$$(f \circ I) = (I \circ f) = f$$

The composition of the identity function with any function f leaves f unchanged. For example, let $f(x) = 2x + 3$ and $I(x)$ be the identity function, that is, $I(x) = x$.

$(I \circ f)(x)$					$(f \circ I)(x)$				
$x \xrightarrow{f}$		$f(x)$	$\xrightarrow{I}$	$f(x)$	$x \xrightarrow{I}$	x	$\xrightarrow{f}$		$f(x)$
$0 \xrightarrow{f}$		$2(0) + 3 = 3$	$\xrightarrow{I}$	3	$0 \xrightarrow{I}$	0	$\xrightarrow{f}$		$2(0) + 3 = 3$
$-1 \xrightarrow{f}$		$2(-1) + 3 = 1$	$\xrightarrow{I}$	1	$-1 \xrightarrow{I}$	-1	$\xrightarrow{f}$		$2(-1) + 3 = 1$
$2 \xrightarrow{f}$		$2(2) + 3 = 7$	$\xrightarrow{I}$	7	$2 \xrightarrow{I}$	2	$\xrightarrow{f}$		$2(2) + 3 = 7$
$5 \xrightarrow{f}$		$2(5) + 3 = 13$	$\xrightarrow{I}$	13	$5 \xrightarrow{I}$	5	$\xrightarrow{f}$		$2(5) + 3 = 13$
$x \xrightarrow{f}$		$2x + 3$	$\xrightarrow{I}$	$2x + 3$	$x \xrightarrow{I}$	x	$\xrightarrow{f}$		$2x + 3$

$$(I \circ f)(x) = 2x + 3 \text{ and } (f \circ I)(x) = 2x + 3,$$
$$\text{so } (I \circ f)(x) = (f \circ I)(x).$$

Let $f(x)$ be any function whose domain and range are subsets of real numbers. Let $y = x$ or $I(x) = x$ be the identity function. Then:

$$x \xrightarrow{f} f(x) \xrightarrow{I} f(x) \quad \text{and} \quad x \xrightarrow{I} x \xrightarrow{f} f(x)$$

Therefore:

$$(I \circ f)(x) = (f \circ I)(x) = f(x)$$

Inverse Functions

In the set of real numbers, two numbers are additive inverses if their sum is the additive identity, 0, and two numbers are multiplicative inverses if their product is the multiplicative identity, 1. We can define inverse functions in a similar way.

The function inverse of f is g if for all x in the domains of $(f \circ g)$ and of $(g \circ f)$, the composition $(f \circ g)(x) = (g \circ f)(x) = I(x) = x$. If g is the function inverse of f, then f is the function inverse of g.

DEFINITION

The functions f and g are **inverse functions** if f and g are functions and $(f \circ g) = (g \circ f) = I$.

For instance, let $f(x) = 2x + 1$ and $g(x) = \frac{x - 1}{2}$.

(f ∘ g)(x)

x	$\overset{g}{\to}$	$g(x)$	$\overset{f}{\to}$	$f(g(x)) = I(x)$	$(x, I(x))$
-2	$\overset{g}{\to}$	$\frac{-2 - 1}{2} = -\frac{3}{2}$	$\overset{f}{\to}$	$2\left(-\frac{3}{2}\right) + 1 = -3$	$(-3, -3)$
3	$\overset{g}{\to}$	$\frac{3 - 1}{2} = 1$	$\overset{f}{\to}$	$2(1) + 1 = 3$	$(3, 3)$
5	$\overset{g}{\to}$	$\frac{5 - 1}{2} = 2$	$\overset{f}{\to}$	$2(2) + 1 = 5$	$(5, 5)$
x	$\overset{g}{\to}$	$\frac{x - 1}{2}$	$\overset{f}{\to}$	$2\left(\frac{x - 1}{2}\right) + 1 = x$	(x, x)

$$(f \circ g)(x) = x$$

(g ∘ f)(x)

x	$\overset{f}{\to}$	$f(x)$	$\overset{g}{\to}$	$g(f(x)) = I(x)$	$(x, I(x))$
-2	$\overset{f}{\to}$	$2(-2) + 1 = -3$	$\overset{g}{\to}$	$\frac{-3 - 1}{2} = -2$	$(-2. -2)$
3	$\overset{f}{\to}$	$2(3) + 1 = 7$	$\overset{g}{\to}$	$\frac{7 - 1}{2} = 3$	$(3, 3)$
5	$\overset{f}{\to}$	$2(5) + 1 = 11$	$\overset{g}{\to}$	$\frac{11 - 1}{2} = 5$	$(5, 5)$
x	$\overset{f}{\to}$	$2x + 1$	$\overset{g}{\to}$	$\frac{(2x + 1) - 1}{2} = x$	(x, x)

$$(g \circ f)(x) = x$$

Since $(f \circ g)(x) = x$ and $(g \circ f)(x) = x$, f and g are inverse functions. The symbol f^{-1} is often used for the inverse of the function f. We showed that when $f(x) = 2x + 1$ and $g(x) = \frac{x - 1}{2}$, f and g are inverse functions. We can write:

$$g(x) = f^{-1}(x) = \frac{x - 1}{2} \quad \text{or} \quad f(x) = g^{-1}(x) = 2x + 1$$

The graph of f^{-1} is the reflection of the graph of f over the line $y = x$. For every point (a, b) on the graph of f, there is a point (b, a) on the graph of f^{-1}. For example, $(1, 3)$ is a point on f and $(3, 1)$ is a point on f^{-1}.

If $f = \{(x, y) : y = 2x + 1\}$, then $f^{-1} = \{(x, y) : x = 2y + 1\}$. Since it is customary to write y in terms of x, solve $x = 2y + 1$ for y.

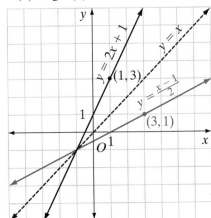

The inverse of a function can be found by interchanging x and y, and then solving for y:

$$y = 2x + 1$$
$$x = 2y + 1$$
$$x - 1 = 2y$$
$$\frac{x - 1}{2} = y$$

In set-builder notation, $f^{-1} = \left\{(x,y) : y = \frac{x-1}{2}\right\}$.

Not every function has an inverse. For instance:

Let $f = \{(-3, 9), (-2, 4), (-1, 1), (0, 0), (1, 1), (2, 4), (3, 9)\}$.

If we interchange the first and second elements of each pair, we obtain the following set:

$$g = \{(9, -3), (4, -2), (1, -1), (0, 0), (1, 1), (4, 2), (9, 3)\}$$

Although g is a relation, it is not a function. Therefore, g cannot be the inverse function of f.

▶ **A function has an inverse function if and only if for every first element there is exactly one second element and for every second element there is exactly one first element. (Alternatively, a function has an inverse if and only it is one-to-one.)**

Since the existence of the inverse of a function depends on whether or not the function is one-to-one, we can use the horizontal line test to determine when a function has an inverse.

Recall that a function that is not one-to-one is said to be many-to-one.

EXAMPLE I

Let $h(x) = 2 - x$. Find $h^{-1}(x)$ and show that h and h^{-1} are inverse functions.

Solution Let $h = \{(x, y) : y = 2 - x\}$. Then $h^{-1} = \{(x, y) : x = 2 - y\}$.

Solve $x = 2 - y$ for y in terms of x.

$$x = 2 - y$$
$$x - 2 = -y$$
$$-1(x - 2) = -1(-y)$$
$$-x + 2 = y$$
$$h^{-1}(x) = -x + 2$$

Show that $h(h^{-1}(x)) = h^{-1}(h(x)) = I(x)$.

$$h(h^{-1}(x)) = h(-x + 2) \qquad h^{-1}(h(x)) = h^{-1}(2 - x)$$
$$= 2 - (-x + 2) \qquad\qquad = -(2 - x) + 2$$
$$= 2 + x - 2 \qquad\qquad\quad = -2 + x + 2$$
$$= x ✔ \qquad\qquad\qquad\quad = x ✔$$

EXAMPLE 2

If $g(x) = \frac{1}{3}x + 1$, find $g^{-1}(x)$.

Solution   *How to Proceed*

(1) For function g, write y in terms of x: $y = \frac{1}{3}x + 1$

(2) To find the inverse of a function, $x = \frac{1}{3}y + 1$
     interchange x and y. Write the function
     g^{-1} by interchanging x and y:

(3) Solve for y in terms of x: $x = \frac{1}{3}y + 1$

$$3x = y + 3$$

$$3x - 3 = y$$

Answer $g^{-1}(x) = 3x - 3$

Inverse of an Absolute Value Function

If $y = |x|$ is the absolute value function, then to form
an inverse, we interchange x and y. But the relation
$x = |y|$ is not a function because for every non-zero
value of x, there are two pairs with the same first ele-
ment. For example, if y is 3, then x is $|3|$ or 3 and one
pair of the relation is $(3, 3)$. If y is -3, then x is $|-3|$ or
3 and another pair of the relation is $(3, -3)$. The relation has two pairs, $(3, 3)$ and $(3, -3)$, with the same
first element and is not a function. This is shown on
the graph at the right.

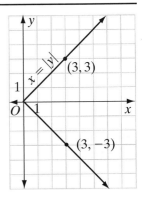

▶ **When the domain of an absolute value function is the set of real numbers,
the function is not one-to-one and has no inverse function.**

EXAMPLE 3

The composite function $h = f \circ g$ maps x to $|4 - x|$.

a. What is the function g?

b. What is the function f?

c. Find $h(1)$ and $h(7)$.

d. Is the function h one-to-one?

e. Does h have an inverse function?

Solution **a.** To evaluate this function, we first find $4 - x$. Therefore, $g(x) = 4 - x$.

b. After finding $4 - x$, we find the absolute value. Therefore, $f(x) = |x|$.

c.
$$\begin{aligned} h(1) &= |4 - x| & h(7) &= |4 - x| \\ &= |4 - 1| & &= |4 - 7| \\ &= |3| & &= |-3| \\ &= 3 & &= 3 \end{aligned}$$

d. The function is not one-to-one because at least two pairs, $(1, 3)$ and $(7, 3)$, have the same second element.

e. The function h does not have an inverse because it is not one-to-one. For every ordered pair (x, y) of a function, (y, x) is an ordered pair of the inverse. Two pairs of h are $(1, 3)$ and $(7, 3)$. Two pairs of an inverse would be $(3, 1)$ and $(3, 7)$. A function cannot have two pairs with the same first element. Therefore, h has no inverse function. Any function that is not one-to-one cannot have an inverse function. ◾

Inverse of a Quadratic Function

The domain of a polynomial function of degree two (or a quadratic function) is the set of real numbers. The range is a subset of the real numbers. A polynomial function of degree two is *not* one-to-one and is *not* onto.

    A function that is not one-to-one does not have an inverse function under composition. However, it is sometimes possible to select a subset of the domain for which the function is one-to-one and does have an inverse for that subset of the original domain. For example, we can restrict the domain of $y = x^2 + 2x - 3$ to those values of x that are greater than or equal to the x coordinate of the turning point.

    For $x \geq -1$, there is exactly one y value for each x value. Some of the pairs of the function $y = x^2 + 2x - 3$ when $x \geq -1$ are $(-1, -4)$, $(0, -3)$, $(1, 0)$, and $(2, 5)$. Then $(-4, -1)$, $(-3, 0)$, $(0, 1)$, and $(5, 2)$ would be pairs of the inverse function, $x = y^2 + 2y - 3$, when $y \geq -1$.

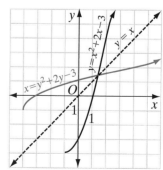

    Note that the inverse is the reflection of the function over the line $y = x$.

Exercises

Writing About Mathematics

1. Taylor said that if (a, b) is a pair of a one-to-one function f, then (b, a) must be a pair of the inverse function f^{-1}. Do you agree with Taylor? Explain why or why not.

2. Christopher said that $f(x) = |x - 2|$ and $g(x) = |x + 2|$ are inverse functions after he showed that $f(g(2)) = 2$, $f(g(5)) = 5$, and $f(g(7)) = 7$. Do you agree that f and g are inverse functions? Explain why or why not.

Developing Skills

In 3–10, find each of the function values when $f(x) = 4x$.

3. $I(3)$

4. $I(5)$

5. $I(-2)$

6. $I(f(2))$

7. $f(I(3))$

8. $f(f^{-1}(-6))$

9. $f^{-1}(f(-6))$

10. $f(f^{-1}(\sqrt{2}))$

In 11–16, determine if the function has an inverse. If so, list the pairs of the inverse function. If not, explain why there is no inverse function.

11. $\{(0, 8), (1, 7), (2, 6), (3, 5), (4, 4)\}$

12. $\{(1, 4), (2, 7), (1, 10), (4, 13)\}$

13. $\{(0, 8), (2, 6), (4, 4), (6, 2)$

14. $\{(2, 7), (3, 7), (4, 7), (5, 7), (6, 7)\}$

15. $\{(-1, 3), (-1, 5), (-2, 7), (-3, 9), (-4, 11)\}$

16. $\{(x, y) : y = x^2 + 2 \text{ for } 0 \leq x \leq 5\}$

In 17–20: **a.** Find the inverse of each given function. **b.** Describe the domain and range of each given function and its inverse in terms of the largest possible subset of the real numbers.

17. $f(x) = 4x - 3$

18. $g(x) = x - 5$

19. $f(x) = \frac{x + 5}{3}$

20. $f(x) = \sqrt{x}$

21. If $f = \{(x, y) : y = 5x\}$ is a direct variation function, find f^{-1}.

22. If $g = \{(x, y) : y = 7 - x\}$, find g^{-1} if it exists. Is it possible for a function to be its own inverse?

23. Does $y = x^2$ have an inverse function if the domain is the set of real numbers? Justify your answer.

In 24–26, sketch the inverse of the given function.

24.
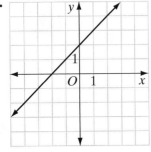

25.
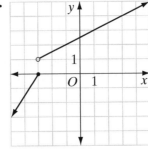

26.
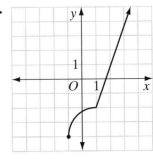

Applying Skills

27. On a particular day, the function that converts American dollars, x, to Indian rupees, $f(x)$, is $f(x) = 0.2532x$. Find the inverse function that converts rupees to dollars. Verify that the functions are inverses.

28. When the function $g(x) = x^2 + 8x + 18$ is restricted to the interval $x \geq -4$, the inverse is $g^{-1}(x) = \sqrt{x - 2} - 4$.

a. Graph g for values of $x \geq -4$. Graph g^{-1} on the same set of axes.

b. What is the domain of g? What is the range of g?

c. What is the domain of g^{-1}? What is the range of g^{-1}?

d. Describe the relationship between the domain and range of g and its inverse.

4-9 CIRCLES

A **circle** is the set of points at a fixed distance from a fixed point. The fixed distance is the **radius** of the circle and the fixed point is the **center** of the circle.

The length of the line segment through the center of the circle with endpoints on the circle is a **diameter** of the circle. The definition of a circle can be used to write an equation of a circle in the coordinate plane.

Center-Radius Form of the Equation of a Circle

Recall the formula for the distance between two points in the coordinate plane. Let $B(x_1, y_1)$ and $A(x_2, y_2)$ be any two points in the coordinate plane. $C(x_2, y_1)$ is the point on the same vertical line as A and on the same horizontal line as B. The length of the line segment joining two points on the same horizontal line is the absolute value of the difference of their x-coordinates. Therefore, $BC = |x_1 - x_2|$.

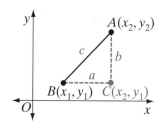

The length of the line segment joining two points on the same vertical line is the absolute value of the difference of their y-coordinates. Therefore, $AC = |y_1 - y_2|$.

Triangle ABC is a right triangle whose hypotenuse is $\overline{AB}$. Use the Pythagorean Theorem with $a = BC$, $b = AC$, and $c = AB$ to express AB in terms of the coordinates of its endpoints.

$$c^2 = a^2 + b^2$$
$$(AB)^2 = (BC)^2 + (AC)^2$$
$$(AB)^2 = (|x_1 - x_2|)^2 + (|y_1 - y_2|)^2$$
$$AB = \sqrt{(x_1 - x_2)^2 + (y_1 - y_2)^2}$$

If d is the distance from A to B, then:

$$d = \sqrt{(x_1 - x_2)^2 + (y_1 - y_2)^2}$$

To write the equation for the circle of radius r whose center is at $C(h, k)$, let $P(x, y)$ be any point on the circle. Using the distance formula, CP is equal to

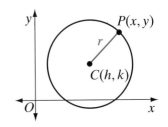

$$CP = \sqrt{(x - h)^2 + (y - k)^2}$$

However, CP is also equal to the radius, r.

$$\sqrt{(x - h)^2 + (y - k)^2} = r$$

$$\left(\sqrt{(x - h)^2 + (y - k)^2}\right)^2 = r^2$$

$$(x - h)^2 + (y - k)^2 = r^2$$

This is the **center-radius form of the equation of a circle.**

For any circle with radius r and center at the origin, $(0, 0)$, we can write the equation of the circle by letting $h = 0$ and $k = 0$.

$$(x - h)^2 + (y - k)^2 = r^2$$
$$(x - 0)^2 + (y - 0)^2 = r^2$$

$$x^2 + y^2 = r^2$$

The equation of the circle with radius 3 and center at the origin is

$$(x - 0)^2 + (y - 0)^2 = (3)^2$$

or

$$x^2 + y^2 = 9$$

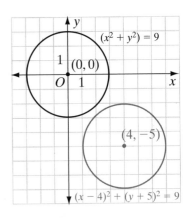

The graph intersects the x-axis at $(-3, 0)$ and $(3, 0)$ and the y-axis at $(0, -3)$ and $(0, 3)$. The equation of the circle with radius 3 and center at $(4, -5)$ is

$$(x - 4)^2 + (y - (-5))^2 = 3^2$$

or

$$(x - 4)^2 + (y + 5)^2 = 9$$

Note that when the graph of the circle with radius 3 and center at the origin is moved 4 units to the right and 5 units down, it is the graph of the equation $(x - 4)^2 + (y + 5)^2 = 9$.

A circle is drawn on the coordinate plane. We can write an equation of the circle if we know the coordinates of the center and the coordinates of one point on the circle. In the diagram, the center of the circle is at $C(-2, 3)$ and one point on the circle is $P(1, 4)$. To find the equation of the circle, we can use the distance formula to find CP, the radius of the circle.

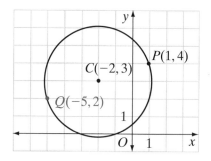

$$CP = \sqrt{(-2 - 1)^2 + (3 - 4)^2}$$
$$= \sqrt{(-3)^2 + (-1)^2}$$
$$= \sqrt{9 + 1}$$
$$= \sqrt{10}$$

Therefore, the equation of the circle is

$$(x - (-2))^2 + (y - 3)^2 = (\sqrt{10})^2$$
$$(x + 2)^2 + (y - 3)^2 = 10$$

Note that $Q(-5, 2)$ is also a point on the circle. If we substitute the coordinates of Q for x and y in the equation, we will have a true statement.

$$(x + 2)^2 + (y - 3)^2 = 10$$
$$(-5 + 2)^2 + (2 - 3)^2 \overset{?}{=} 10$$
$$(-3)^2 + (-1)^2 \overset{?}{=} 10$$
$$9 + 1 \overset{?}{=} 10$$
$$10 = 10 \checkmark$$

Standard Form of the Equation of a Circle

The **standard form of the equation of a circle** is:

$$x^2 + y^2 + Dx + Ey + F = 0$$

We can write the center-radius form in standard form by expanding the squares of the binomials.

$$(x - h)^2 + (y - k)^2 = r^2$$
$$x^2 - 2hx + h^2 + y^2 - 2ky + k^2 = r^2$$

This equation in standard form is $x^2 + y^2 - 2hx - 2ky + h^2 + k^2 - r^2 = 0$.

Note that in the standard form of the equation of a circle, the coefficients of x^2 and y^2 are both equal and usually equal to 1. If we compare the two equations, $D = -2h$, $E = -2k$, and $F = h^2 + k^2 - r^2$.

EXAMPLE I

a. Find the center-radius form of the equation $x^2 + y^2 - 4x + 6y - 3 = 0$.

b. Determine the center and radius of the circle.

Solution **a.** $D = -4 = -2h$. Therefore, $h = 2$ and $h^2 = 4$.

$E = 6 = -2k$. Therefore, $k = -3$ and $k^2 = 9$.

$F = -3 = h^2 + k^2 - r^2$. Therefore, $-3 = 4 + 9 - r^2$ or $r^2 = 16$.

The equation of the circle in center-radius form is $(x - 2)^2 + (y + 3)^2 = 16$.

b. This is a circle with center at $(2, -3)$ and radius equal to $\sqrt{16} = 4$.

Answers **a.** $(x - 2)^2 + (y + 3)^2 = 16$ **b.** Center $= (2, -3)$, radius $= 4$

EXAMPLE 2

The endpoints of a diameter of a circle are $P(6, 1)$ and $Q(-4, -5)$. Write the equation of the circle: **a.** in center-radius form **b.** in standard form.

Solution **a.** The center of a circle is the midpoint of a diameter. Use the midpoint formula to find the coordinates of the center, M.

$$M(x_m, y_m) = \left(\frac{x_1 + x_2}{2}, \frac{y_1 + y_2}{2}\right)$$
$$= \left(\frac{6 + (-4)}{2}, \frac{1 + (-5)}{2}\right)$$
$$= \left(\frac{2}{2}, \frac{-4}{2}\right)$$
$$= (1, -2)$$

Use the distance formula to find MP, the radius, r, of the circle.

Let $(x_1, y_1) = (6, 1)$, the coordinates of P and $(x_2, y_2) = (1, -2)$, the coordinates of M.

$$r = MP = \sqrt{(6 - 1)^2 + (1 - (-2))^2}$$
$$= \sqrt{5^2 + 3^2}$$
$$= \sqrt{25 + 9}$$
$$= \sqrt{34}$$

Substitute the coordinates of the center and the radius of the circle in the center-radius form of the equation.

$$(x - h)^2 + (y - k)^2 = r^2$$
$$(x - 1)^2 + (y - (-2))^2 = (\sqrt{34})^2$$
$$(x - 1)^2 + (y + 2)^2 = 34$$

b. Expand the square of the binomials of the center-radius form of the equation.

$$(x - 1)^2 + (y + 2)^2 = 34$$
$$x^2 - 2x + 1 + y^2 + 4y + 4 = 34$$
$$x^2 + y^2 - 2x + 4y + 1 + 4 - 34 = 0$$
$$x^2 + y^2 - 2x + 4y - 29 = 0$$

Answers **a.** $(x - 1)^2 + (y + 2)^2 = 34$ **b.** $x^2 + y^2 - 2x + 4y - 29 = 0$

EXAMPLE 3

In standard form, the equation of a circle is $x^2 + y^2 + 3x - 4y - 14 = 0$. Write the equation in center-radius form and find the coordinates of the center and radius of the circle.

Solution   *How to Proceed*

(1) Complete the square of $x^2 + 3x$. We "complete the square" by writing the expression as a trinomial that is a perfect square. Since $(x + a)^2 = x^2 + 2ax + a^2$, the constant term needed to complete the square is the square of one-half the coefficient of x:

$$x^2 + 3x + \left(\tfrac{3}{2}\right)^2 = x^2 + 3x + \tfrac{9}{4}$$
$$= \left(x + \tfrac{3}{2}\right)^2$$

(2) Complete the square of $y^2 - 4y$. The constant term needed to complete the square is the square of one-half the coefficient of y:

$$y^2 - 4y + \left(\tfrac{-4}{2}\right)^2 = y^2 - 4y + 4$$
$$= (y - 2)^2$$

(3) Add 14 to both sides of the given equation:

$$x^2 + y^2 + 3x - 4y - 14 = 0$$
$$x^2 + y^2 + 3x - 4y = 14$$

(4) To both sides of the equation, add the constants needed to complete the squares:

$$x^2 + 3x + \tfrac{9}{4} + y^2 - 4y + 4 = 14 + \tfrac{9}{4} + 4$$

(5) Write the equation in center-radius form. Determine the center and radius of the circle from the center-radius form:

$$\left(x + \tfrac{3}{2}\right)^2 + (y - 2)^2 = \tfrac{56}{4} + \tfrac{9}{4} + \tfrac{16}{4}$$
$$\left(x + \tfrac{3}{2}\right)^2 + (y - 2)^2 = \tfrac{81}{4}$$
$$\left(x + \tfrac{3}{2}\right)^2 + (y - 2)^2 = \left(\tfrac{9}{2}\right)^2$$

Answer The equation of the circle in center-radius form is $\left(x + \tfrac{3}{2}\right)^2 + (y - 2)^2 = \left(\tfrac{9}{2}\right)^2$.

The center is at $\left(-\tfrac{3}{2}, 2\right)$ and the radius is $\tfrac{9}{2}$.

Exercises

Writing About Mathematics

1. Is the set of points on a circle a function? Explain why or why not.

2. Explain why $(x - h)^2 + (y - k)^2 = -4$ is not the equation of a circle.

Developing Skills

In 3–10, the coordinates of point P on the circle with center at C are given. Write an equation of each circle: **a.** in center-radius form **b.** in standard form.

3. $P(0, 2), C(0, 0)$ **4.** $P(0, -3), C(0, 0)$ **5.** $P(4, 0), C(0, 0)$ **6.** $P(3, 2), C(4, 2)$

7. $P(-1, 5), C(-1, 1)$ **8.** $P(0, -3), C(6, 5)$ **9.** $P(1, 1), C(6, 13)$ **10.** $P(4, 2), C(0, 1)$

In 11–19, write the equation of each circle.

11.

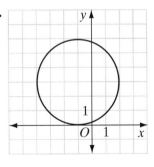

12.

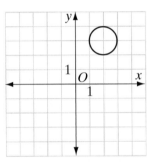

13.

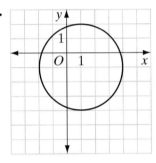

14.

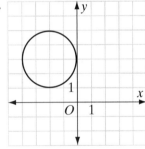

15.

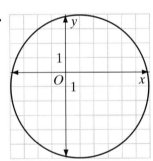

16.

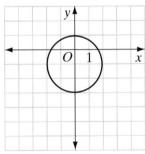

17.

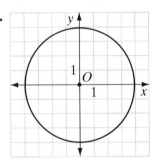

18.

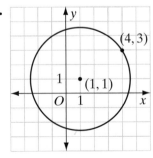

19.

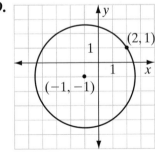

In 20–27: **a.** Write each equation in center-radius form. **b.** Find the coordinates of the center. **c.** Find the radius of the circle.

20. $x^2 + y^2 - 25 = 0$

21. $x^2 + y^2 - 2x - 2y - 7 = 0$

22. $x^2 + y^2 + 2x - 4y + 1 = 0$

23. $x^2 + y^2 - 6x + 2y - 6 = 0$

24. $x^2 + y^2 + 6x - 6y + 6 = 0$

25. $x^2 + y^2 - 8y = 0$

26. $x^2 + y^2 + 10x - 5y - 32 = 0$

27. $x^2 + y^2 + x - 3y - 2 = 0$

Applying Skills

28. An architect is planning the entryway into a courtyard as an arch in the shape of a semi-circle with a radius of 8 feet. The equation of the arch can be written as $(x - 8)^2 + y^2 = 64$ when the domain is the set of non-negative real numbers less than or equal to 16 and the range is the set of positive real numbers less than or equal to 8.

    a. Draw the arch on graph paper.

    b. Can a box in the shape of a cube whose edges measure 6 feet be moved through the arch?

    c. Can a box that is a rectangular solid that measures 8 feet by 8 feet by 6 feet be moved through the arch?

29. What is the measure of a side of a square that can be drawn with its vertices on a circle of radius 10?

30. What are dimensions of a rectangle whose length is twice the width if the vertices are on a circle of radius 10?

31. Airplane passengers have been surprised to look down over farmland and see designs, called *crop circles*, cut into cornfields. Suppose a farmer wishes to make a design consisting of three **concentric circles**, that is, circles with the same center but different radii. Write the equations of three concentric circles centered at a point in a cornfield with coordinates $(2, 2)$.

Hands-On-Activity

Three points determine a circle. If we know the coordinates of three points, we should be able to determine the equation of the circle through these three points. The three points are the vertices of a triangle. Recall that the perpendicular bisectors of the sides of a triangle meet in a point that is equidistant from the vertices of the triangle. This point is the center of the circle through the three points.

    To find the circle through the points $P(5, 3)$, $Q(-3, 3)$, and $R(3, -3)$, follow these steps. Think of P, Q, and R as the vertices of a triangle.

    1. Find the midpoint of $\overline{PQ}$.

    2. Find the slope of $\overline{PQ}$ and the slope of the line perpendicular to $\overline{PQ}$.

    3. Find the equation of the perpendicular bisector of $\overline{PQ}$.

    4. Repeat steps 1 through 3 for $\overline{QR}$.

    5. Find the point of intersection of the perpendicular bisectors of $\overline{PQ}$ and $\overline{QR}$. Call this point C, the center of the circle.

    6. Find $CP = CQ = CR$. The distance from C to each of these points is the radius of the circle.

    7. Use the coordinates of C and the radius of the circle to write the equation of the circle. Show that P, Q, and R are points on the circle by showing that their coordinates make the equation of the circle true.

4-10 INVERSE VARIATION

Forty people live in an apartment complex. They often want to divide themselves into groups to gather information that will be of interest to all of the tenants. In the past they have worked in 1 group of 40, 2 groups of 20, 4 groups of 10, 5 groups of 8, 8 groups of 5, and 10 groups of 4. Notice that as the number of groups increases, the number of people in each group decreases and when the number of groups decreases, the number of people in each group increases. We say that the number of people in each group and the number of groups are *inversely proportional* or that the two quantities *vary inversely*.

DEFINITION _____

Two numbers, x and y, **vary inversely** or are **inversely proportional** when $xy = c$, a non-zero constant.

Let x_1 and y_1 be two numbers such that $x_1 y_1 = c$, a constant. Let x_2 and y_2 be two other numbers such that $x_2 y_2 = c$, the same constant. Then:

$$x_1 y_1 = x_2 y_2 \qquad \text{or} \qquad \frac{x_1}{x_2} = \frac{y_2}{y_1}$$

Note that when this **inverse variation** relationship is written as a proportion, the order of the terms in x is opposite to the order of the terms in y.

If x and y vary inversely and $xy = 20$, then the domain and range of this relation are the set of non-zero real numbers. The graph of this relation is shown below.

x	y
20	1
10	2
5	4
4	5
2	10
1	20

x	y
−20	−1
−10	−2
−5	−4
−4	−5
−2	−10
−1	−20

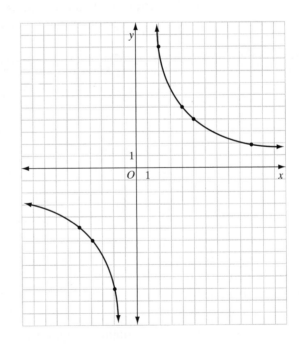

Note:

1. No vertical line intersects the graph in more than one point, so the relation is a function.

2. No horizontal line intersects the graph in more than one point, so the function is one-to-one.

3. This is a function from the set of non-zero real numbers to the set of non-zero real numbers. The range is the set of non-zero real numbers. The function is onto.

4. If we interchange x and y, the function remains unchanged. The function is its own inverse.

5. The graph is called a **hyperbola**.

EXAMPLE 1

The areas of $\triangle ABC$ and $\triangle DEF$ are equal. The length of the base of $\triangle ABC$ is four times the length of the base of $\triangle DEF$. How does the length of the altitude to the base of $\triangle ABC$ compare with the length of the altitude to the base of $\triangle DEF$?

Solution   Let b_1 be the length of the base and h_1 be the height of $\triangle ABC$.

Let b_2 be the length of the base and h_2 be the height of $\triangle DEF$.

(1) Area of $\triangle ABC$ = area of $\triangle DEF$. $\frac{1}{2}b_1h_1 = \frac{1}{2}b_2h_2$

The lengths of the bases and the $b_1h_1 = b_2h_2$
heights of these triangles are
inversely proportional:

(2) The length of the base of $\triangle ABC$ $b_1 = 4b_2$
is four times the length of the base
of $\triangle DEF$:

(3) Substitute $4b_2$ for b_1 in the equation $4b_2h_1 = b_2h_2$
$b_1h_1 = b_2h_2$:

(4) Divide both sides of the equation $h_1 = \frac{1}{4}h_2$
by $4b_2$:

Answer   The height of $\triangle ABC$ is one-fourth the height of $\triangle DEF$.

Note: When two quantities are inversely proportional, as one quantity changes by a factor of a, the other quantity changes by a factor of $\frac{1}{a}$.

EXAMPLE 2

Mrs. Vroman spends $3.00 to buy cookies for her family at local bakery. When the price of a cookie increased by 10 cents, the number of cookies she bought decreased by 1. What was the original price of a cookie and how many did she buy? What was the increased price of a cookie and how many did she buy?

Solution For a fixed total cost, the number of items purchased varies inversely as the cost of one item.

Let $x_1 = p =$ the original price of one cookie in cents

$y_1 = n =$ the number of cookies purchased at the original price

$x_2 = p + 10 =$ the increased price of one cookie in cents

$y_2 = n - 1 =$ the number of cookies purchased at the increased price

(1) Use the product form for inverse variation: $\quad x_1 y_1 = x_2 y_2$

(2) Substitute values in terms of x and y: $\quad pn = (p + 10)(n - 1)$

(3) Multiply the binomials: $\quad pn = pn - p + 10n - 10$

(4) Solve for p in terms of n: $\quad 0 = -p + 10n - 10$

$$p = 10n - 10$$

(5) Since she spends $3 or 300 cents for cookies, $pn = 300$. Write this equation in terms of n by substituting $10n - 10$ for p:

$$pn = 300$$
$$(10n - 10)(n) = 300$$
$$10n^2 - 10n = 300$$
$$10n^2 - 10n - 300 = 0$$

(6) Solve for n:

$$n^2 - n - 30 = 0$$
$$(n - 6)(n + 5) = 0$$
$$n - 6 = 0 \mid n + 5 = 0$$
$$n = 6 \mid \quad n = -5 \text{ (reject this root)}$$

(7) Solve for p:

$$p = 10n - 10$$
$$p = 10(6) - 10$$
$$p = 50$$

Answer Originally a cookie cost 50 cents and Mrs. Vroman bought 6 cookies. The price increased to 60 cents and she bought 5 cookies.

Exercises

Writing About Mathematics

1. Is the function $f = \{(x, y) : xy = 20\}$ a polynomial function? Explain why or why not.

2. Explain the difference between direct variation and inverse variation.

Developing Skills

In 3–5, each graph is a set of points whose x-coordinates and y-coordinates vary inversely. Write an equation for each function.

3.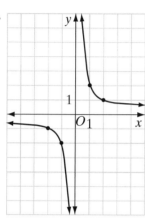

4.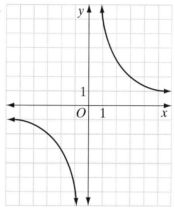

5.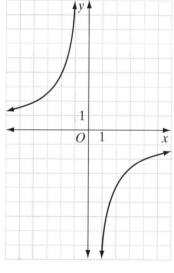

In 6–12, tell whether the variables vary directly, inversely, or neither.

6. On his way to work, Randy travels 10 miles at r miles per hour for h hours.

7. A driver travels for 4 hours between stops covering d miles at a rate of r miles per hour.

8. Each day, Jaymee works for 6 hours typing p pages of a report at a rate of m minutes per page.

9. Each day, Sophia works for h hours typing p pages of a report at a rate of 15 minutes per page.

10. Each day, Brandon works for h hours typing 40 pages of a report at a rate of m minutes per page.

11. Each day, I am awake a hours and I sleep s hours.

12. A bank pays 4% interest on all savings accounts. A depositor receives I dollars in interest when the balance in the savings account is P dollars.

Applying Skills

13. Rectangles *ABCD* and *EFGH* have the same area. The length of *ABCD* is equal to twice the length of the *EFGH*. How does the width of *ABCD* compare to the width of *EFGH*?

14. Aaron can ride his bicycle to school at an average rate that is three times that of the rate at which he can walk to school. How does the time that it takes Aaron to ride to school compare with the time that it takes him to walk to school?

15. When on vacation, the Ross family always travels the same number of miles each day.

    a. Does the time that they travel each day vary inversely as the rate at which they travel?

    b. On the first day the Ross family travels for 3 hours at an average rate of 60 miles per hour and on the second day they travel for 4 hours. What was their average rate of speed on the second day?

16. Megan traveled 165 miles to visit friends. On the return trip she was delayed by construction and had to reduce her average speed by 22 miles per hour. The return trip took 2 hours longer. What was the time and average speed for each part of the trip?

17. Ian often buys in large quantities. A few months ago he bought several cans of frozen orange juice for $24. The next time Ian purchased frozen orange juice, the price had increased by $0.10 per can and he bought 1 less can for the same total price. What was the price per can and the numbers of cans purchased each time?

CHAPTER SUMMARY

A **relation** is a set of ordered pairs.

A **function** is a relation such that no two ordered pairs have the same first element. A **function from set *A* to set *B*** is the set of ordered pairs (x, y) such that every $x \in A$ is paired with exactly one $y \in B$. Set *A* is the **domain** of the function. The **range** of the function is a subset of set *B* whose elements are the second elements of the pairs of the function. A function is **onto** if the range is set *B*.

Function Notation

1. f = $\{(x, y) : y = 2x + 3\}$	The function f is the set of ordered pairs $(x, 2x + 3)$.
2. f$(x) = 2x + 3$	The function value paired with x is $2x + 3$.
3. $y = 2x + 3$	The second element of the pair (x, y) is $2x + 3$.
4. $\{(x, 2x + 3)\}$	The set of ordered pairs whose second element is 3 more than twice the first.
5. f: $x \rightarrow 2x + 3$	Under the function f, x maps to $2x + 3$.
6. $x \xrightarrow{f} 2x + 3$	The image of x for the function f is $2x + 3$.

In general, if f(x) and g(x) are functions,

$$(f + g)(x) = f(x) + g(x)$$

$$(f - g)(x) = f(x) - g(x)$$

$$(fg)(x) = f(x)g(x)$$

$$\left(\frac{f}{g}\right)(x) = \frac{f(x)}{g(x)}$$

The domain of $(f + g), (f - g)$, and (fg) is the set of elements common to the domain of f and of g, and the domain of $\frac{f}{g}$ is the set of elements common to the domains of f and of g for which $g(x) \neq 0$.

A **composite function** $h(x) = g(f(x))$ is the set of ordered pairs that are the result of mapping x to $f(x)$ and then mapping $f(x)$ to $g(f(x))$. The function $h(x) = g(f(x))$ can also be written as $g \circ f(x)$. The function h is the **composition** of g following f.

The **identity function, I(x)**, is that function that maps every element of the domain to the same element of the range.

A function f is said to be **one-to-one** if no two ordered pairs have the same second element.

Every one-to-one function $f(x)$ has an inverse function $f^{-1}(x)$ if $f(f^{-1}(x)) = f^{-1}(f(x)) = I(x)$. The range of f is the domain of f^{-1} and the range of f^{-1} is the domain of f.

When the ratio of two variables is a constant, we say that the variables are **directly proportional** or that the variables **vary directly**.

When the domain of an absolute value function, $y = |x|$, is the set of real numbers, the function is not one-to-one and has no inverse function.

A **polynomial function of degree** n is a function of the form

$$f(x) = a_n x^n + a_{n-1} x^{n-1} + \cdots + a_0, a_n \neq 0$$

The **roots** or **zeros** of a polynomial function are the values of x for which $y = 0$. These are the x-coordinates of the points at which the graph of the function intersects the x-axis.

The **center-radius form** of the equation of a circle is $(x - h)^2 + (y - k)^2 = r^2$ where (h, k) is the center of the circle and r is the radius. The **standard form** of the equation of a circle is:

$$x^2 + y^2 + Dx + Ey + F = 0$$

Two numbers x and y **vary inversely** or are **inversely proportional** when $xy = a$ and a is a non-zero constant. The domain and range of $xy = a$ or $y = \frac{a}{x}$ are the set of non-zero real numbers.

For every real number a, the graph of $f(x + a)$ is the graph of $f(x)$ moved $|a|$ units to the left when a is positive and $|a|$ units to the right when a is negative.

For every real number a, the graph of $f(x) + a$ is the graph of $f(x)$ moved $|a|$ units up when a is positive and $|a|$ units down when a is negative.

The graph of $-f(x)$ is the graph of $f(x)$ reflected in the x-axis.

For every positive real number a, the graph of $af(x)$ is the graph of $f(x)$ stretched by a factor of a in the y direction if $a > 1$ or compressed by a factor of a if $0 < a < 1$.

VOCABULARY

4-1 Set-builder notation • Relation • Function • Domain of a function • Range of a function • Function from set A to set B • ∈ • Onto • Vertical line test

4-3 Linear Function • Horizontal line test • One-to-one • Directly proportional • Vary directly • Direct variation

4-4 Absolute value function • Many-to-one

4-5 Polynomial function f of degree n • Quadratic function • Parabola • Turning point • Vertex • Axis of symmetry of the parabola • Roots • Zeros • Double root

4-7 Composition • Composite function

4-8 Identity function (I) • Inverse functions • f^{-1}

4-9 Circle • Radius • Center • Diameter • Center-radius form of the equation of a circle • Standard form of the equation of a circle • Concentric circles

4-10 Vary inversely • Inversely proportional • Inverse variation • Hyperbola

REVIEW EXERCISES

In 1–5: Determine if the relation is or is not a function. Justify your answer.

1. $\{(0, 0), (1, 1), (1, -1), (4, 2), (4, -2), (3, 4), (5, 7)\}$

2. $\{(x, y) : y = \sqrt{x}\}$ **3.** $\{(x, y) : x^2 + y^2 = 9\}$

4. $\{(x, y) : y = 2x + 3\}$ **5.** $\{(x, y) : y = -\frac{4}{x}\}$

In 6–11, each equation defines a function from the set of real numbers to the set of real numbers. **a.** Is the function one-to-one? **b.** Is the function onto? **c.** Does the function have an inverse function? If so, write an equation of the inverse.

6. $y = 4x + 3$ **7.** $y = x^2 - 2x + 5$

8. $y = |x + 2|$ **9.** $y = x^3$

10. $y = -1$ **11.** $y = \sqrt{x - 4}$

In 12–17, from each graph, determine any rational roots of the function.

12.

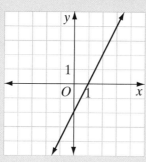

13.

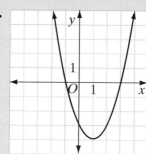

14.

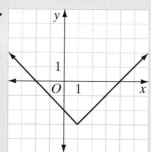

15.

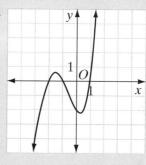

16.

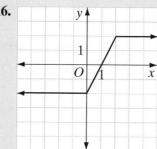

17.

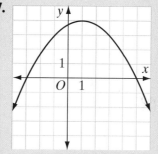

18. What are zeros of the function $f(x) = |x + 3| - 4$?

In 19–22, $p = \{(0, 6), (1, 5), (2, 4), (3, 3), (4, 2)\}$ and $q = \{(0, 1), (1, 3), (2, 5), (3, 7), (4, 9)\}$. Find:

19. $p + q$ **20.** $p - q$ **21.** pq **22.** $\frac{p}{q}$

In 23–30, for each function f and g, find $f(g(x))$ and $g(f(x))$.

23. $f(x) = 2x$ and $g(x) = x + 3$ **24.** $f(x) = |x|$ and $g(x) = 4x - 1$

25. $f(x) = x^2$ and $g(x) = x + 2$ **26.** $f(x) = \sqrt{x}$ and $g(x) = 5x - 3$

27. $f(x) = \frac{x}{2}$ and $g(x) = 2 + 3x$ **28.** $f(x) = \frac{5}{x}$ and $g(x) = 10x$

29. $f(x) = 2x + 7$ and $g(x) = \frac{x - 7}{2}$ **30.** $f(x) = x^3$ and $g(x) = \sqrt[3]{x}$

In 31–36, write the equation of each circle.

31.

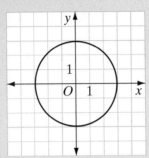

32.

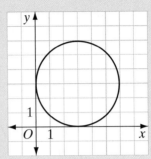

33.

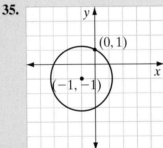

34.

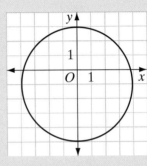

35.

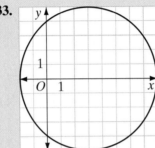

36.

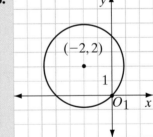

37. The graph of the circle
$(x - 2)^2 + (y + 1)^2 = 10$ and the line
$y = x - 1$ are shown in the diagram. Find
the common solutions of the equations and
check your answer.

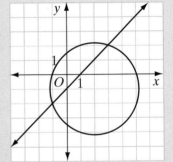

38. The graph of $y = x^2$ is shifted 4 units to the right and 2 units down. Write
the equation of the new function.

39. The graph of $y = |x|$ is stretched in the vertical direction by a factor of 3. Write an equation of the new function.

40. The graph of $y = 2x + 3$ is reflected in the x-axis. Write an equation of the new function.

41. The graph of $y = |x|$ is reflected in the x-axis and then moved 1 unit to the left and 3 units up. Write an equation of the new function.

In 42–45:

   a. Write an equation of the relation formed by interchanging the elements of the ordered pairs of each function.

   b. Determine if the relation written in **a** is an inverse function of the given function. Explain why or why not.

   c. If the given function does not have an inverse function, determine, if possible, a domain of the given function for which an inverse function exists.

42. $f(x) = 2x + 8$ **43.** $f(x) = \frac{3}{x}$

44. $f(x) = x^2$ **45.** $f(x) = \sqrt{x}$

Exploration

For this activity, use Styrofoam cones.

   The shapes obtained by cutting a cone are called **conic sections**. The equation of each of these curves is a relation.

1. What is the shape of the edge of the base of a cone? Cut the cone parallel to the base. What is the shape of the edges of the two cut surfaces?

2. Cut the cone at an angle so that the cut does not intersect the base of the cone but is not parallel to the base of the cone. What is the shape of the edges of the cut surfaces?

3. Cut the cone parallel to one slant edge of the cone. What is the shape of the curved portion of the edges of the cut surfaces?

4. Cut the cone perpendicular to the base of the cone but not through a diameter of the base. Repeat this cut on a second cone and place the two cones tip to tip. What is the shape of the edges of the cut surfaces?

5. Cut the cone perpendicular to the base of the cone through a diameter of the base. Repeat this cut on a second cone and place the two cones tip to tip. What is the shape of the edges of the cut surfaces?

CUMULATIVE REVIEW   CHAPTERS 1–4

Part I

Answer all questions in this part. Each correct answer will receive 2 credits. No partial credit will be allowed.

1. Which of the following is *not* a real number?
  (1) -1 (3) $\sqrt{-3}$
  (2) $\sqrt{3}$ (4) $0.101001000100001\ldots$

2. In simplest form, $2x(4x^2 - 5) - (7x^3 + 3x)$ is equal to
  (1) $x^3 - 7x$ (3) $x^3 + 3x - 5$
  (2) $x^3 - 13x$ (4) $x^3 - 3x - 10$

3. The factors of $3x^2 - 7x + 2$ are
  (1) $(3x - 2)(x - 1)$ (3) $(3x + 2)(x + 1)$
  (2) $(3x - 1)(x - 2)$ (4) $(3x + 1)(x + 2)$

4. The fraction $\frac{3}{\sqrt{3}}$ is equal to
  (1) $\frac{\sqrt{3}}{3}$ (2) $\sqrt{3}$ (3) $3\sqrt{3}$ (4) 3

5. The solution set of $|x + 2| < 5$ is
  (1) $\{x : x < 3\}$ (3) $\{x : -7 < x < 3\}$
  (2) $\{x : x < -7\}$ (4) $\{x : x < -7 \text{ or } x > 3\}$

6. The solution set of $x^2 - 5x = 6$ is
  (1) $\{3, -2\}$ (2) $\{-3, 2\}$ (3) $\{6, 1\}$ (4) $\{6, -1\}$

7. In simplest form, the sum $\sqrt{75} + \sqrt{27}$ is
  (1) $\sqrt{102}$ (2) $34\sqrt{3}$ (3) $8\sqrt{6}$ (4) $8\sqrt{3}$

8. Which of the following is *not* a one-to-one function?
  (1) $y = x + 2$ (3) $y = \frac{1}{x}$
  (2) $y = |x + 2|$ (4) $y = \frac{x}{2}$

9. For which of the following are x and y inversely proportional?
  (1) $y = 4x$ (2) $y = \frac{x}{4}$ (3) $y = x^2$ (4) $y = \frac{4}{x}$

10. The fraction $\frac{a - 3}{a^2 - 4}$ is undefined for
  (1) $a = 0$ (3) $a = 4$
  (2) $a = 3$ (4) $a = 2$ and $a = -2$

Part II

Answer all questions in this part. Each correct answer will receive 2 credits. Clearly indicate the necessary steps, including appropriate formula substitutions, diagrams, graphs, charts, etc. For all questions in this part, a correct numerical answer with no work shown will receive only 1 credit.

11. Factor completely: $2x^3 + 2x^2 - 2x - 2$.

12. Write $\dfrac{3 + \sqrt{5}}{3 - \sqrt{5}}$ as an equivalent fraction with a rational denominator.

Part III

Answer all questions in this part. Each correct answer will receive 4 credits. Clearly indicate the necessary steps, including appropriate formula substitutions, diagrams, graphs, charts, etc. For all questions in this part, a correct numerical answer with no work shown will receive only 1 credit.

13. Find the solution set of the inequality $x^2 - 2x - 15 < 0$.

14. Write in standard form the equation of the circle with center at $(2, -3)$ and radius 4.

Part IV

Answer all questions in this part. Each correct answer will receive 6 credits. Clearly indicate the necessary steps, including appropriate formula substitutions, diagrams, graphs, charts, etc. For all questions in this part, a correct numerical answer with no work shown will receive only 1 credit.

15. a. Sketch the graph of the polynomial function $y = 4x^2 - 8x - 5$.

    b. Write an equation of the axis of symmetry.

    c. What are the coordinates of the turning point?

    d. What are the zeros of the function?

16. Simplify the rational expression $\dfrac{2 + \dfrac{2}{a}}{1 - \dfrac{1}{a^2}}$ and list all values of a for which the fraction is undefined.

CHAPTER

5

QUADRATIC FUNCTIONS AND COMPLEX NUMBERS

Records of early Babylonian and Chinese mathematics provide methods for the solution of a quadratic equation. These methods are usually presented in terms of directions for the solution of a particular problem, often one derived from the relationship between the perimeter and the area of a rectangle. For example, a list of steps needed to find the length and width of a rectangle would be given in terms of the perimeter and the area of a rectangle.

The Arab mathematician Mohammed ibn Musa al-Khwarizmi gave similar steps for the solution of the relationship "a square and ten roots of the same amount to thirty-nine dirhems." In algebraic terms, x is the root, x^2 is the square, and the equation can be written as $x^2 + 10x = 39$. His solution used these steps:

1. Take half of the number of roots : $\frac{1}{2}$ of $10 = 5$
2. Multiply this number by itself: $5 \times 5 = 25$
3. Add this to 39: $25 + 39 = 64$
4. Now take the root of this number: $\sqrt{64} = 8$
5. Subtract from this number one-half the number of roots: $8 - \frac{1}{2}$ of $10 = 8 - 5 = 3$
6. This is the number which we sought.

In this chapter, we will derive a formula for the solution of any quadratic equation. The derivation of this formula uses steps very similar to those used by al-Khwarizmi.

5-1 REAL ROOTS OF A QUADRATIC EQUATION

The real roots of a polynomial function are the x-coordinates of the points at which the graph of the function intersects the x-axis. The graph of the polynomial function $y = x^2 - 2$ is shown at the right. The graph intersects the x-axis between -2 and -1 and between 1 and 2. The roots of $0 = x^2 - 2$ appear to be about -1.4 and 1.4. Can we find the exact values of these roots? Although we cannot factor $x^2 - 2$ over the set of integers, we can find the value of x^2 and take the square roots of each side of the equation.

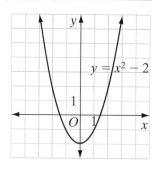

$$0 - x^2 - 2$$
$$2 = x^2$$
$$\pm\sqrt{2} = x$$

The roots of the function $y = x^2 - 2$ are the irrational numbers $-\sqrt{2}$ and $\sqrt{2}$.

The graph of the polynomial function $y = x^2 + 2x - 1$ is shown at the right. The graph shows that the function has two real roots, one between -3 and -2 and the other between 0 and 1. Can we find the exact values of the roots of the function $y = x^2 + 2x - 1$? We know that the equation $0 = x^2 + 2x - 1$ cannot be solved by factoring the right member over the set of integers. However, we can follow a process similar to that used to solve $0 = x^2 - 2$. We need to write the right member as a trinomial that is the square of a binomial. This process is called **completing the square**.

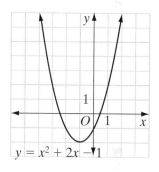

Recall that $(x + h)^2 = x^2 + 2hx + h^2$. To write a trinomial that is a perfect square using the terms $x^2 + 2x$, we need to let $2h = 2$. Therefore, $h = 1$ and $h^2 = 1$. The perfect square trinomial is $x^2 + 2x + 1$, which is the square of $(x + 1)$.

(1) Add 1 to both sides of the equation to isolate the terms in x:

$$0 = x^2 + 2x - 1$$
$$1 = x^2 + 2x$$

(2) Add the number needed to complete the square to both sides of the equation:

$$2 = x^2 + 2x + 1$$
$$2 = (x + 1)^2$$

$$\left(\tfrac{1}{2} \text{ coefficient of } x\right)^2 = \left(\tfrac{1}{2}(2)\right)^2 = 1^2 = 1$$

(3) Write the square root of both sides of the equation:

$$\pm\sqrt{2} = x + 1$$

(4) Solve for x:

$$-1 \pm \sqrt{2} = x$$

The roots of $y = x^2 + 2x - 1$ are $-1 - \sqrt{2}$ and $-1 + \sqrt{2}$.

We can use a calculator to approximate these roots to the nearest hundredth.

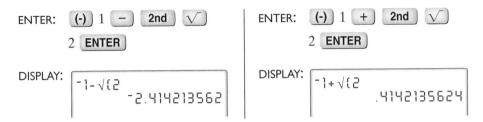

ENTER: (-) 1 − 2nd √

2 ENTER

DISPLAY:
```
-1-√{2
         -2.414213562
```

ENTER: (-) 1 + 2nd √

2 ENTER

DISPLAY:
```
-1+√{2
          .4142135624
```

To the nearest hundredth, the roots are -2.41 and 0.41. One root is between -3 and -2 and the other is between 0 and 1, as shown on the graph.

The process of completing the square can be illustrated geometrically. Let us draw the quadratic expression $x^2 + 2x$ as a square with sides of length x and a rectangle with dimensions $x \times 2$. Cut the rectangle in half and attach the two parts to the sides of the square. The result is a larger square with a missing corner. The factor $\left(\frac{1}{2} \text{ coefficient of } x\right)^2$ or 1^2 is the area of this missing corner. This is the origin of the expression "completing the square."

$$x^2 \quad + \quad 2x \quad \Rightarrow \quad (x+1)^2$$

Procedure

To solve a quadratic equation of the form $ax^2 + bx + c = 0$ where $a = 1$ by completing the square:

1. Isolate the terms in x on one side of the equation.

2. Add the square of one-half the coefficient of x or $\left(\frac{1}{2}b\right)^2$ to both sides of the equation.

3. Write the square root of both sides of the resulting equation and solve for x.

Note: The procedure outlined above only works when $a = 1$. If the coefficient a is not equal to 1, divide each term by a, and then proceed with completing the square using the procedure outlined above.

Not every second-degree polynomial function has real roots. Note that the graph of $y = x^2 - 6x + 10$ does not intersect the x-axis and therefore has no real roots. Can we find roots by completing the square?

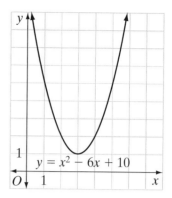

Use $\left(\frac{1}{2} \text{ coefficient of } x\right)^2 = \left(\frac{1}{2}(\ 6)\right)^2 - (-3)^2 = 9$ to complete the square.

$$0 = x^2 - 6x + 10$$
$$-10 = x^2 - 6x \qquad \text{Isolate the terms in } x.$$
$$-10 + 9 = x^2 - 6x + 9 \qquad \text{Add the square of one-half the}$$
$$-1 = (x - 3)^2 \qquad \text{coefficient of } x \text{ to complete square.}$$
$$\pm\sqrt{-1} = x - 3 \qquad \text{Take the square root of both sides.}$$
$$3 \pm \sqrt{-1} = x \qquad \text{Solve for } x.$$

Since there is no real number that is the square root of -1, this function has no real zeros. We will study a set of numbers that includes the square roots of negative numbers later in this chapter.

EXAMPLE 1

Find the exact values of the zeros of the function $f(x) = -x^2 - 4x + 2$.

Solution   *How to Proceed*

(1) Let $f(x) = 0$: $\qquad\qquad\qquad\qquad\qquad\qquad 0 = -x^2 - 4x + 2$

(2) Isolate the terms in x: $\qquad\qquad\qquad\qquad -2 = -x^2 - 4x$

(3) Multiply by -1: $\qquad\qquad\qquad\qquad\qquad\quad 2 = x^2 + 4x$

(4) Complete the square by adding $\qquad\qquad 4 + 2 = x^2 + 4x + 4$

$\left(\frac{1}{2} \text{ coefficient of } x\right)^2 = \left(\frac{1}{2}(4)\right)^2 = 2^2 = 4$ $\qquad 6 = x^2 + 4x + 4$

to both sides of the equation: $\qquad\qquad\qquad\quad 6 = (x + 2)^2$

(5) Take the square root of each side of the $\qquad \pm\sqrt{6} = x + 2$
equation:

(6) Solve for x: $\qquad\qquad\qquad\qquad\qquad\qquad -2 \pm \sqrt{6} = x$

Answer The zeros are $(-2 + \sqrt{6})$ and $(-2 - \sqrt{6})$.

EXAMPLE 2

Find the roots of the function $y = 2x^2 - x - 1$.

Solution

How to Proceed

(1) Let $y = 0$:

$$0 = 2x^2 - x - 1$$

(2) Isolate the terms in x:

$$1 = 2x^2 - x$$

(3) Divide both sides of the equation by the coefficient of x^2:

$$\tfrac{1}{2} = x^2 - \tfrac{1}{2}x$$

(4) Complete the square by adding

$$\tfrac{1}{16} + \tfrac{1}{2} = x^2 - \tfrac{1}{2}x + \left(-\tfrac{1}{4}\right)^2$$

$$\left(\tfrac{1}{2} \text{ coefficient of } x\right)^2 = \left(\tfrac{1}{2} \cdot -\tfrac{1}{2}\right)^2 = \tfrac{1}{16}$$

$$\tfrac{1}{16} + \tfrac{8}{16} = \left(x - \tfrac{1}{4}\right)^2$$

to both sides of the equation:

$$\tfrac{9}{16} = \left(x - \tfrac{1}{4}\right)^2$$

(5) Take the square root of both sides of the equation:

$$\pm\sqrt{\tfrac{9}{16}} = x - \tfrac{1}{4}$$

(6) Solve for x:

$$\tfrac{1}{4} \pm \tfrac{3}{4} = x$$

The roots of the function are:

$$\tfrac{1}{4} - \tfrac{3}{4} = -\tfrac{2}{4} = -\tfrac{1}{2} \quad \text{and} \quad \tfrac{1}{4} + \tfrac{3}{4} = \tfrac{4}{4} = 1$$

Answer $-\tfrac{1}{2}$ and 1

Note that since the roots are rational, the equation could have been solved by factoring over the set of integers.

$$0 = 2x^2 - x - 1$$
$$0 = (2x + 1)(x - 1)$$

$2x + 1 = 0$	$x - 1 = 0$
$2x = -1$	$x = 1$
$x = -\tfrac{1}{2}$	

Graphs of Quadratic Functions and Completing the Square

A quadratic function $y = a(x - h)^2 + k$ can be graphed in terms of the basic quadratic function $y = x^2$. In particular, the graph of $y = a(x - h)^2 + k$ is the graph of $y = x^2$ shifted h units horizontally and k units vertically. The coefficient a determines its orientation: when $a > 0$, the parabola opens upward; when

$a < 0$, the parabola opens downward. The coefficient a also determines how narrow or wide the graph is compared to $y = x^2$: when $|a| > 1$, the graph is narrower; when $|a| < 1$, the graph is wider.

We can use the process of completing the square to help us graph quadratic functions of the form $y = ax^2 + bx + c$, as the following examples demonstrate:

EXAMPLE 3

Graph $f(x) = x^2 + 4x + 9$.

Solution The square of one-half the coefficient of x is $\left(\frac{1}{2} \cdot 4\right)^2$ or 4.

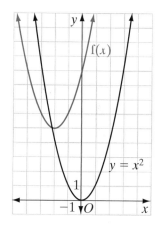

Add *and* subtract 4 to keep the function equivalent to the original and complete the square.

$$f(x) = x^2 + 4x + 9$$
$$= x^2 + 4x + 4 - 4 + 9$$
$$= (x^2 + 4x + 4) + 5$$
$$= (x + 2)^2 + 5$$

The graph of $f(x)$ is the graph of $y = x^2$ shifted 2 units to the left and 5 units up, as shown on the right. Note that this function has no real roots.

EXAMPLE 4

Graph $f(x) = 2x^2 + 4x + 1$.

Solution

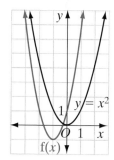

How to Proceed

(1) Factor 2 from the first two terms:

$$f(x) = 2x^2 + 4x + 1$$
$$= 2(x^2 + 2x) + 1$$

(2) Add *and* subtract

$$\left(\tfrac{1}{2}\text{ coefficient of }x\right)^2 = \left(\tfrac{1}{2} \cdot 2\right)^2 = 1$$

to keep the function equivalent to the original:

$$= 2(x^2 + 2x + 1 - 1) + 1$$
$$= 2(x^2 + 2x + 1) - 2 + 1$$
$$= 2(x^2 + 2x + 1) - 1$$

(3) Complete the square:

$$= 2(x + 1)^2 - 1$$

The graph of $f(x)$ is the graph of $y = x^2$ stretched vertically by a factor of 2, and shifted 1 unit to the left and 1 unit down as shown on the left. Note that this function has two real roots, one between -2 and -1 and the other between -1 and 0.

Exercises

Writing About Mathematics

1. To solve the equation given in step 2 of Example 2, Sarah multiplied each side of the equation by 8 and then added 1 to each side to complete the square. Show that $16x^2 - 8x - 8$ is the square of a binomial and will lead to the correct solution of $0 = 2x^2 - x - 1$.

2. Phillip said that the equation $0 = x^2 - 6x + 1$ can be solved by adding 8 to both sides of the equation. Do you agree with Phillip? Explain why or why not.

Developing Skills

In 3–8, complete the square of the quadratic expression.

3. $x^2 + 6x$ **4.** $x^2 - 8x$ **5.** $x^2 - 2x$

6. $x^2 - 12x$ **7.** $2x^2 - 4x$ **8.** $x^2 - 3x$

In 9–14: **a.** Sketch the graph of each function. **b.** From the graph, estimate the roots of the function to the nearest tenth. **c.** Find the exact irrational roots in simplest radical form.

9. $f(x) = x^2 - 6x + 4$ **10.** $f(x) = x^2 - 2x - 2$ **11.** $f(x) = x^2 + 4x + 2$

12. $f(x) = x^2 - 6x + 6$ **13.** $f(x) = x^2 - 2x - 1$ **14.** $f(x) = x^2 - 10x + 18$

In 9–26, solve each quadratic equation by completing the square. Express the answer in simplest radical form.

15. $x^2 - 2x - 2 = 0$ **16.** $x^2 + 6x + 4 = 0$ **17.** $x^2 - 4x + 1 = 0$

18. $x^2 + 2x - 5 = 0$ **19.** $x^2 - 6x + 2 = 0$ **20.** $x^2 - 8x + 4 = 0$

21. $2x^2 + 12x + 3 = 0$ **22.** $3x^2 - 6x - 1 = 0$ **23.** $2x^2 - 6x + 3 = 0$

24. $4x^2 - 20x + 9 = 0$ **25.** $\frac{1}{2}x^2 + x - 3 = 0$ **26.** $\frac{1}{4}x^2 + \frac{3}{4}x - \frac{3}{2} = 0$

27. a. Complete the square to find the roots of the equation $x^2 - 5x + 1 = 0$.

b. Write, to the nearest tenth, a rational approximation for the roots.

In 28–33, without graphing the parabola, describe the translation, reflection, and/or scaling that must be applied to $y = x^2$ to obtain the graph of each given function.

28. $f(x) = x^2 - 12x + 5$ **29.** $f(x) = x^2 + 2x - 2$ **30.** $f(x) = x^2 - 6x - 7$

31. $f(x) = x^2 + x + \frac{9}{4}$ **32.** $f(x) = -x^2 + x + 2$ **33.** $f(x) = 3x^2 + 6x + 3$

34. Determine the coordinates of the vertex and the equation of the axis of symmetry of $f(x) = x^2 + 8x + 5$ by writing the equation in the form $f(x) = (x - h)^2 + k$. Justify your answer.

Applying Skills

35. The length of a rectangle is 4 feet more than twice the width. The area of the rectangle is 38 square feet.

    a. Find the dimensions of the rectangle in simplest radical form.

    b. Show that the product of the length and width is equal to the area.

    c. Write, to the nearest tenth, rational approximations for the length and width.

36. One base of a trapezoid is 8 feet longer than the other base and the height of the trapezoid is equal to the length of the shorter base. The area of the trapezoid is 20 square feet.

    a. Find the lengths of the bases and of the height of the trapezoid in simplest radical form.

    b. Show that the area of the trapezoid is equal to one-half the height times the sum of the lengths of the bases.

    c. Write, to the nearest tenth, rational approximations for the lengths of the bases and for the height of the trapezoid.

37. Steve and Alice realized that their ages are consecutive odd integers. The product of their ages is 195. Steve is younger than Alice. Determine their ages by completing the square.

5-2 THE QUADRATIC FORMULA

We have solved equations by using the same steps to complete the square needed to find the roots or zeros of a function of the form $y = ax^2 + bx + c$, that is, to solve the quadratic equation $0 = ax^2 + bx + c$. We will apply those steps to the general equation to find a formula that can be used to solve any quadratic equation.

How to Proceed

(1) Let $y = 0$:

$$0 = ax^2 + bx + c$$

(2) Isolate the terms in x:

$$-c = ax^2 + bx$$

(3) Divide both sides of the equation by the coefficient of x^2:

$$\frac{-c}{a} = x^2 + \frac{b}{a}x$$

(4) Complete the square by adding

$$\left(\frac{1}{2} \text{ coefficient of } x\right)^2 = \left(\frac{1}{2} \cdot \frac{b}{a}\right)^2 = \frac{b^2}{4a^2}$$

to both sides of the equation:

$$\frac{b^2}{4a^2} + \frac{-c}{a} = x^2 + \frac{b}{a}x + \frac{b^2}{4a^2}$$

$$\frac{b^2}{4a^2} + \frac{-4ac}{4a^2} = x^2 + \frac{b}{a}x + \frac{b^2}{4a^2}$$

$$\frac{b^2 - 4ac}{4a^2} = \left(x + \frac{b}{2a}\right)^2$$

(5) Take the square root of each side of the equation:

$$\pm\sqrt{\frac{b^2 - 4ac}{4a^2}} = x + \frac{b}{2a}$$

(6) Solve for x:

$$\frac{-b}{2a} \pm \frac{\sqrt{b^2 - 4ac}}{2a} = x$$

$$\frac{-b \pm \sqrt{b^2 - 4ac}}{2a} = x$$

This result is called the **quadratic formula**.

▶ **When $a \neq 0$, the roots of $ax^2 + bx + c = 0$ are $x = \dfrac{-b \pm \sqrt{b^2 - 4ac}}{2a}$.**

EXAMPLE 1

Use the quadratic formula to find the roots of $2x^2 - 4x = 1$.

Solution   *How to Proceed*

(1) Write the equation in standard form:

$$2x^2 - 4x = 1$$
$$2x^2 - 4x - 1 = 0$$

(2) Determine the values of a, b, and c:

$$a = 2, b = -4, c = -1$$

(3) Substitute the values of a, b, and c in the quadratic formula:

$$x = \frac{-b \pm \sqrt{b^2 - 4ac}}{2a}$$

$$x = \frac{-(-4) \pm \sqrt{(-4)^2 - 4(2)(-1)}}{2(2)}$$

(4) Perform the computation:

$$x = \frac{4 \pm \sqrt{16 + 8}}{4}$$

$$x = \frac{4 \pm \sqrt{24}}{4}$$

(5) Write the radical in simplest form:

$$x = \frac{4 \pm \sqrt{4}\sqrt{6}}{4}$$

$$x = \frac{4 \pm 2\sqrt{6}}{4}$$

(6) Divide numerator and denominator by 2:

$$x = \frac{2 \pm \sqrt{6}}{2}$$

Answer $\frac{2 + \sqrt{6}}{2}$ and $\frac{2 - \sqrt{6}}{2}$

EXAMPLE 2

Use the quadratic formula to show that the equation $x^2 + x + 2 = 0$ has no real roots.

Solution For the equation $x^2 + x + 2 = 0$, $a = 1$, $b = 1$, and $c = 2$.

$$x = \frac{-b \pm \sqrt{b^2 - 4ac}}{2a} = \frac{-1 \pm \sqrt{1^2 - 4(1)(2)}}{2(1)} = \frac{-1 \pm \sqrt{1 - 8}}{2} = \frac{-1 \pm \sqrt{-7}}{2}$$

There is no real number that is the square root of a negative number. Therefore, $\sqrt{-7}$ is not a real number and the equation has no real roots.

EXAMPLE 3

One leg of a right triangle is 1 centimeter shorter than the other leg and the hypotenuse is 2 centimeters longer than the longer leg. What are lengths of the sides of the triangle?

Solution Let x = the length of the longer leg

$x - 1$ = the length of the shorter leg

$x + 2$ = the length of the hypotenuse

Use the Pythagorean Theorem to write an equation.

$$x^2 + (x - 1)^2 = (x + 2)^2$$
$$x^2 + x^2 - 2x + 1 = x^2 + 4x + 4$$
$$x^2 - 6x - 3 = 0$$

Use the quadratic formula to solve this equation: $a = 1, b = -6, c = -3$.

$$x = \frac{-b \pm \sqrt{b^2 - 4ac}}{2a}$$

$$x = \frac{-(-6) \pm \sqrt{(-6)^2 - 4(1)(-3)}}{2(1)}$$

$$= \frac{6 \pm \sqrt{36 + 12}}{2}$$

$$= \frac{6 \pm \sqrt{48}}{2} = \frac{6 \pm \sqrt{16}\sqrt{3}}{2} = \frac{6 \pm 4\sqrt{3}}{2} = 3 \pm 2\sqrt{3}$$

The roots are $3 + 2\sqrt{3}$ and $3 - 2\sqrt{3}$. Reject the negative root, $3 - 2\sqrt{3}$.

The length of the longer leg is $3 + 2\sqrt{3}$.

The length of the shorter leg is $3 + 2\sqrt{3} - 1$ or $2 + 2\sqrt{3}$.

The length of the hypotenuse is $3 + 2\sqrt{3} + 2$ or $5 + 2\sqrt{3}$.

Check

$$\left(3 + 2\sqrt{3}\right)^2 + \left(2 + 2\sqrt{3}\right)^2 \overset{?}{=} \left(5 + 2\sqrt{3}\right)^2$$

$$\left(9 + 12\sqrt{3} + 4 \cdot 3\right) + \left(4 + 8\sqrt{3} + 4 \cdot 3\right) \overset{?}{=} 25 + 20\sqrt{3} + 4 \cdot 3$$

$$37 + 20\sqrt{3} = 37 + 20\sqrt{3} \ ✔$$

Exercises

Writing About Mathematics

1. Noah said that the solutions of Example 1, $\frac{2 + \sqrt{6}}{2}$ and $\frac{2 - \sqrt{6}}{2}$, could have been written as $\frac{\overset{1}{2} + \sqrt{6}}{\underset{1}{2}} = 1 + \sqrt{6}$ and $\frac{\overset{1}{2} - \sqrt{6}}{\underset{1}{2}} = 1 - \sqrt{6}$. Do you agree with Noah? Explain why or why not.

2. Rita said that when a, b, and c are real numbers, the roots of $ax^2 + bx + c = 0$ are real numbers only when $b^2 \geq 4ac$. Do you agree with Rita? Explain why or why not.

Developing Skills

In 3–14, use the quadratic formula to find the roots of each equation. Irrational roots should be written in simplest radical form.

3. $x^2 + 5x + 4 = 0$

4. $x^2 + 6x - 7 = 0$

5. $x^2 - 3x + 1 = 0$

6. $x^2 - x - 4 = 0$

7. $x^2 + 5x - 2 = 0$

8. $x^2 - 8 = 0$

9. $x^2 - 3x = 0$

10. $x^2 + 2x = 4$

11. $3x^2 - 5x + 2 = 0$

12. $4x^2 - x - 1 = 0$

13. $5x + 1 = 2x^2$

14. $2x^2 = x + 4$

15. $x^2 - 6x + 3 = 0$

16. $4x^2 - 4x = 11$

17. $3x^2 = 4x + 2$

18. a. Sketch the graph of $y = x^2 + 6x + 2$.

b. From the graph, estimate the roots of the function to the nearest tenth.

c. Use the quadratic formula to find the exact values of the roots of the function.

d. Express the roots of the function to the nearest tenth and compare these values to your estimate from the graph.

Applying Skills

In 19–25, express each answer in simplest radical form. Check each answer.

19. The larger of two numbers is 5 more than twice the smaller. The square of the smaller is equal to the larger. Find the numbers.

20. The length of a rectangle is 2 feet more than the width. The area of the rectangle is 2 square feet. What are the dimensions of the rectangle?

21. The length of a rectangle is 4 centimeters more than the width. The measure of a diagonal is 10 centimeters. Find the dimensions of the rectangle.

22. The length of the base of a triangle is 6 feet more than the length of the altitude to the base. The area of the triangle is 18 square feet. Find the length of the base and of the altitude to the base.

23. The lengths of the bases of a trapezoid are $x + 10$ and $3x + 2$ and the length of the altitude is $2x$. If the area of the trapezoid is 40, find the lengths of the bases and of the altitude.

24. The altitude, $\overline{CD}$, to the hypotenuse, $\overline{AB}$, of right triangle ABC separates the hypotenuse into two segments, $\overline{AD}$ and $\overline{DB}$. If $AD = DB + 4$ and $CD = 12$ centimeters, find DB, AD, and AB. Recall that the length of the altitude to the hypotenuse of a right triangle is the mean proportional between the lengths of the segments into which the hypotenuse is separated, that is, $\frac{AD}{CD} = \frac{CD}{DB}$.

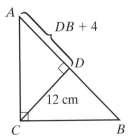

25. A parabola is symmetric under a line reflection. Each real root of the quadratic function $y = ax^2 + bx + c$ is the image of the other under a reflection in the axis of symmetry of the parabola.

a. What are the coordinates of the points at which the parabola whose equation is $y = ax^2 + bx + c$ intersects the x-axis?

b. What are the coordinates of the midpoint of the points whose coordinates were found in part **a**?

c. What is the equation of the axis of symmetry of the parabola $y = ax^2 + bx + c$?

d. The turning point of a parabola is on the axis of symmetry. What is the x-coordinate of the turning point of the parabola $y = ax^2 + bx + c$?

26. Gravity on the moon is about one-sixth of gravity on Earth. An astronaut standing on a tower 20 feet above the moon's surface throws a ball upward with a velocity of 30 feet per second. The height of the ball at any time t (in seconds) is $h(t) = -2.67t^2 + 30t + 20$. To the nearest tenth of a second, how long will it take for the ball to hit the ground?

27. Based on data from a local college, a statistician determined that the average student's grade point average (GPA) at this college is a function of the number of hours, h, he or she studies each week. The grade point average can be estimated by the function $G(h) = 0.006h^2 + 0.02h + 1.2$ for $0 \le h \le 20$.

a. To the nearest tenth, what is the average GPA of a student who does no studying?

b. To the nearest tenth, what is the average GPA of a student who studies 12 hours per week?

c. To the nearest tenth, how many hours of studying are required for the average student to achieve a 3.2 GPA?

28. Follow the steps given in the chapter opener to find a root of the equation $x^2 + \frac{b}{a}x + \frac{c}{a} = 0$. Compare your solution with the quadratic formula.

Hands-On Activity: Alternate Derivation of the Quadratic Formula
As you may have noticed, the derivation of the quadratic formula on page 193 uses fractions throughout (in particular, starting with step 3). We can use an alternate derivation of the quadratic equation that uses fractions only in the last step. This alternate derivation uses the following perfect square trinomial:

$$
\begin{aligned}
4a^2x^2 + 4abx + b^2 &= 4a^2x^2 + 2abx + 2abx + b^2 \\
&= 2ax(2ax + b) + b(2ax + b) \\
&= (2ax + b)(2ax + b) \\
&= (2ax + b)^2
\end{aligned}
$$

To write $ax^2 + bx$, the terms in x of the general quadratic equation, as this perfect square trinomial, we must multiply by $4a$ and add b^2. The steps for the derivation of the quadratic formula using this perfect square trinomial are shown below.

How to Proceed

(1) Isolate the terms in x:

$$ax^2 + bx + c = 0$$
$$ax^2 + bx = -c$$

(2) Multiply both sides of the equation by $4a$:

$$4a^2x^2 + 4abx = -4ac$$

(3) Add b^2 to both sides of the equation to complete the square:

$$4a^2x^2 + 4abx + b^2 = b^2 - 4ac$$
$$(2ax + b)^2 = b^2 - 4ac$$

(4) Now solve for x.

Does this procedure lead to the quadratic formula?

5-3 THE DISCRIMINANT

When a quadratic equation $ax^2 + bx + c = 0$ has rational numbers as the values of a, b, and c, the value of $b^2 - 4ac$, or the **discriminant**, determines the nature of the roots. Consider the following equations.

CASE 1 *The discriminant $b^2 - 4ac$ is a positive perfect square.*

$$2x^2 + 3x - 2 = 0 \text{ where } a = 2, b = 3, \text{ and } c = -2.$$

$$x = \frac{-b \pm \sqrt{b^2 - 4ac}}{2a}$$

$$x = \frac{-3 \pm \sqrt{3^2 - 4(2)(-2)}}{2(2)}$$

$$= \frac{-3 \pm \sqrt{9 + 16}}{4} = \frac{-3 \pm \sqrt{25}}{4} = \frac{-3 \pm 5}{4}$$

The roots are $\frac{-3 + 5}{4} = \frac{2}{4} = \frac{1}{2}$ and $\frac{-3 - 5}{4} = \frac{-8}{4} = -2$.

The roots are unequal rational numbers, and the graph of the corresponding quadratic function has two x-intercepts.

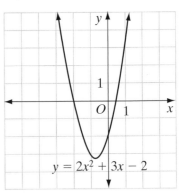

$$y = 2x^2 + 3x - 2$$

CASE 2 *The discriminant $b^2 - 4ac$ is a positive number that is not a perfect square.*

$$3x^2 + x - 1 = 0 \text{ where } a = 3, b = 1, \text{ and } c = -1.$$

$$x = \frac{-b \pm \sqrt{b^2 - 4ac}}{2a}$$

$$x = \frac{-1 \pm \sqrt{1^2 - 4(3)(-1)}}{2(3)}$$

$$= \frac{-1 \pm \sqrt{1 + 12}}{6} = \frac{-1 \pm \sqrt{13}}{6}$$

The roots are $\frac{-1 + \sqrt{13}}{6}$ and $\frac{-1 - \sqrt{13}}{6}$.

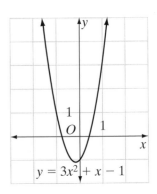

$$y = 3x^2 + x - 1$$

The roots are unequal irrational numbers, and the graph of the corresponding quadratic function has two x-intercepts.

CASE 3 *The discriminant $b^2 - 4ac$ is 0.*

$9x^2 - 6x + 1 = 0$ where $a = 9$, $b = -6$, and $c = 1$.

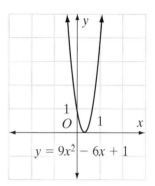

$$x = \frac{-b \pm \sqrt{b^2 - 4ac}}{2a}$$

$$x = \frac{-(-6) \pm \sqrt{(-6)^2 - 4(9)(1)}}{2(9)}$$

$$= \frac{6 \pm \sqrt{36 - 36}}{18} = \frac{6 \pm \sqrt{0}}{18} = \frac{6 \pm 0}{18}$$

The roots are $\frac{6 + 0}{18} = \frac{1}{3}$ and $\frac{6 - 0}{18} = \frac{1}{3}$.

$$y = 9x^2 - 6x + 1$$

The roots are equal rational numbers, and the graph of the corresponding quadratic function has one x-intercept. The root is a **double root**.

CASE 4 *The discriminant $b^2 - 4ac$ is a negative number.*

$2x^2 + 3x + 4 = 0$ where $a = 2$, $b = 3$, and $c = 4$.

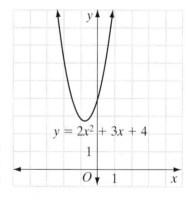

$$x = \frac{-b \pm \sqrt{b^2 - 4ac}}{2a}$$

$$x = \frac{-3 \pm \sqrt{3^2 - 4(2)(4)}}{2(2)}$$

$$= \frac{-3 \pm \sqrt{9 - 32}}{4} = \frac{-3 \pm \sqrt{-23}}{4}$$

$$y = 2x^2 + 3x + 4$$

This equation has no real roots because there is no real number equal to $\sqrt{-23}$, and the graph of the corresponding quadratic function has no x-intercepts.

In each case, when a, b, and c are rational numbers, the value of $b^2 - 4ac$ determined the nature of the roots and the x-intercepts. From the four cases given above, we can make the following observations.

Let $ax^2 + bx + c = 0$ be a quadratic equation and a, b, and c be rational numbers with $a \neq 0$:

When the discriminant $b^2 - 4ac$ is:	The roots of the equation are:	The number of x-intercepts of the function is:
> 0 and a perfect square	real, rational, and unequal	2
> 0 and not a perfect square	real, irrational, and unequal	2
$= 0$	real, rational, and equal	1
< 0	not real numbers	0

EXAMPLE I

Without solving the equation, find the nature of the roots of $3x^2 + 2x = 1$.

Solution Write the equation in standard form: $3x^2 + 2x - 1 = 0$.

Evaluate the discriminant of the equation when $a = 3, b = 2, c = -1$.

$$b^2 - 4ac = 2^2 - 4(3)(-1) = 4 + 12 = 16$$

Since a, b, and c are rational and the discriminant is greater than 0 and a perfect square, the roots are real, rational, and unequal. ∎

EXAMPLE 2

Show that it is impossible to draw a rectangle whose area is 20 square feet and whose perimeter is 10 feet.

Solution Let l and w be the length and width of the rectangle.

How to Proceed

(1) Use the formula for perimeter: $\qquad\qquad 2l + 2w = 10$

(2) Solve for l in terms of w: $\qquad\qquad\qquad l + w = 5$
$$l = 5 - w$$

(3) Use the formula for area: $\qquad\qquad\qquad lw = 20$

(4) Substitute $5 - w$ for l. Write the resulting quadratic equation in standard form:
$$(5 - w)(w) = 20$$
$$5w - w^2 = 20$$
$$-w^2 + 5w - 20 = 0$$
$$w^2 - 5w + 20 = 0$$

(5) Test the discriminant:
$$b^2 - 4ac = (-5)^2 - 4(1)(20)$$
$$= 25 - 80$$
$$= -55$$

Since the discriminant is negative, the quadratic equation has no real roots. There are no real numbers that can be the dimensions of this rectangle. ∎

EXAMPLE 3

A lamp manufacturer earns a weekly profit of P dollars according to the function $P(x) = -0.3x^2 + 50x - 170$ where x is the number of lamps sold. When $P(x) = 0$, the roots of the equation represent the level of sales for which the manufacturer will break even.

a. Interpret the meaning of the discriminant value of 2,296.

b. Use the discriminant to determine the number of break even points.

c. Use the discriminant to determine if there is a level of sales for which the weekly profit would be $3,000.

Solution **a.** Since the discriminant is greater than 0 and not a perfect square, the equation has two roots or break even points that are irrational and unequal. *Answer*

b. 2 (see part **a**) *Answer*

c. We want to know if the equation

$$3,000 = -0.3x^2 + 50x - 170$$

has at least one real root. Find the discriminant of this equation. The equation in standard form is $0 = -0.3x^2 + 50x - 3,170$.

$$b^2 - ac = 50^2 - 4(-0.3)(-3,170) = -1,304$$

The discriminant is negative, so there are no real roots. The company cannot achieve a weekly profit of $3,000. *Answer*

Exercises

Writing About Mathematics

1. a. What is the discriminant of the equation $x^2 + (\sqrt{5})x - 1 = 0$?

  b. Find the roots of the equation $x^2 + (\sqrt{5})x - 1 = 0$.

  c. Do the rules for the relationship between the discriminant and the roots of the equation given in this chapter apply to this equation? Explain why or why not.

2. Christina said that when a, b, and c are rational numbers and a and c have opposite signs, the quadratic equation $ax^2 + bx + c = 0$ must have real roots. Do you agree with Christina? Explain why or why not.

Developing Skills

In 3–8, each graph represents a quadratic function. Determine if the discriminant of the quadratic function is greater than 0, equal to 0, or less than 0.

3.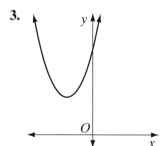

4.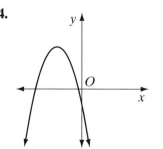

5.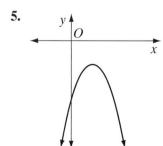

6.
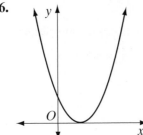

7.
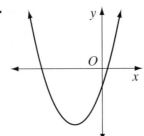

8.
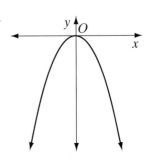

In 9–14: **a.** For each given value of the discriminant of a quadratic equation with rational coefficients, determine if the roots of the quadratic equation are (1) rational and unequal, (2) rational and equal, (3) irrational and unequal, or (4) not real numbers. **b.** Use your answer to part **a** to determine the number of x-intercepts of the graph of the corresponding quadratic function.

9. 9   **10.** 5   **11.** 0   **12.** 12   **13.** -4 **14.** 49

In 15–23: **a.** Find the value of the discriminant and determine if the roots of the quadratic equation are (1) rational and unequal, (2) rational and equal, (3) irrational and unequal, or (4) not real numbers. **b.** Use any method to find the real roots of the equation if they exist.

15. $x^2 - 12x + 36 = 0$ **16.** $2x^2 + 7x = 0$ **17.** $x^2 + 3x + 1 = 0$

18. $2x^2 - 8 = 0$ **19.** $4x^2 - x = 1$ **20.** $3x^2 = 5x - 3$

21. $4x - 1 = 4x^2$ **22.** $2x^2 - 3x - 5 = 0$ **23.** $3x - x^2 = 5$

24. When $b^2 - 4ac = 0$, is $ax^2 + bx + c$ a perfect square trinomial or a constant times a perfect square trinomial? Explain why or why not.

25. Find a value of c such that the roots of $x^2 + 2x + c = 0$ are:

    a. equal and rational.   **b.** unequal and rational.

    c. unequal and irrational.   **d.** not real numbers.

26. Find a value of b such that the roots of $x^2 + bx + 4 = 0$ are:

    a. equal and rational.   **b.** unequal and rational.

    c. unequal and irrational.   **d.** not real numbers.

Applying Skills

27. Lauren wants to fence off a rectangular flower bed with a perimeter of 30 yards and a diagonal length of 8 yards. Use the discriminant to determine if her fence can be constructed. If possible, determine the dimensions of the rectangle.

28. An open box is to be constructed as shown in the figure on the right. Is it possible to construct a box with a volume of 25 cubic feet? Use the discriminant to explain your answer.

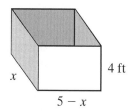

29. The height, in feet, to which of a golf ball rises when shot upward from ground level is described by the function $h(t) = -16t^2 + 48t$ where t is the time elapsed in seconds. Use the discriminant to determine if the golf ball will reach a height of 32 feet.

30. The profit function for a company that manufactures cameras is $P(x) = -x^2 + 350x - 15{,}000$. Under present conditions, can the company achieve a profit of \$20,000? Use the discriminant to explain your answer.

5-4 THE COMPLEX NUMBERS

The Set of Imaginary Numbers

We have seen that when we solve quadratic equations with rational coefficients, some equations have real, rational roots, some have real, irrational roots, and some do not have real roots.

For example:

$x^2 - 9 = 0$	$x^2 - 8 = 0$	$x^2 + 9 = 0$
$x^2 = 9$	$x^2 = 8$	$x^2 = -9$
$x = \pm\sqrt{9}$	$x = \pm\sqrt{8}$	$x = \pm\sqrt{-9}$
$x = \pm 3$	$x = \pm\sqrt{4}\sqrt{2}$	$x = \pm\sqrt{9}\sqrt{-1}$
	$x = \pm 2\sqrt{2}$	$x = \pm 3\sqrt{-1}$

The equations $x^2 - 9 = 0$ and $x^2 - 8 = 0$ have real roots, but the equation $x^2 + 9 = 0$ does not have real roots.

A number is an abstract idea to which we assign a meaning and a symbol. We can form a new set of numbers that includes $\sqrt{-1}$, to which we will assign the symbol i. We call i the unit of the **set of imaginary numbers**.

The numbers $+3\sqrt{-1}$ and $-3\sqrt{-1}$ that are the roots of the equation $x^2 + 9 = 0$ can be written as $3i$ and $-3i$.

DEFINITION

A number of the form $a\sqrt{-1} = ai$ is a **pure imaginary number** where a is a non-zero real number.

The idea of the square root of -1 as a number was not accepted by many early mathematicians. Rafael Bombelli (1526–1572) introduced the terms *plus of minus* and *minus of minus* to designate these numbers. For instance, he designated $2 + 3i$ as 2 *plus of minus* 3 and $2 - 3i$ as 2 *minus of minus* 3. The term *imaginary* was first used in 1637 by René Descartes, who used it in an almost derogatory sense, implying that somehow imaginary numbers had no significance in mathematics. However, the algebra of imaginary numbers was later developed and became an invaluable tool in real-world applications such as alternating current.

Some elements of the set of pure imaginary numbers are:

$$\sqrt{-25} = \sqrt{25}\sqrt{-1} = 5i$$
$$-\sqrt{-16} = -\sqrt{4}\sqrt{-1} = -2i$$
$$\sqrt{-10} = \sqrt{10}\sqrt{-1} = \sqrt{-1}\sqrt{10} = i\sqrt{10}$$
$$-\sqrt{-32} = -\sqrt{16}\sqrt{2}\sqrt{-1} = -4\sqrt{-1}\sqrt{2} = -4i\sqrt{2}$$

Note the order in which the factors of $i\sqrt{10}$ and $-4i\sqrt{2}$ are written. The factor that is a radical is written last to avoid confusion with $\sqrt{10i}$ and $-4\sqrt{2i}$.

EXAMPLE I ▬▬▬▬▬▬▬▬▬▬▬▬▬▬▬▬▬▬▬▬▬▬▬▬

Simplify: **a.** $\sqrt{-7}$ **b.** $\sqrt{-12}$

Solution   Write each imaginary number in terms of i and simplify.

a. $\sqrt{-7} = \sqrt{7}\sqrt{-1} = i\sqrt{7}$ **b.** $\sqrt{-12} = \sqrt{4}\sqrt{3}\sqrt{-1} = 2i\sqrt{3}$ ▨

Powers of i

The unit element of the set of pure imaginary numbers is $i = \sqrt{-1}$. Since $\sqrt{b}$ is one of the two equal factors of b, then $i = \sqrt{-1}$ is one of the two equal factors of -1 and $i \cdot i = -1$ or $i^2 = -1$. The box at the right shows the first four powers of i.

$$\boxed{\begin{aligned} i &= \sqrt{-1} \\ i^2 &= -1 \\ i^3 &= i^2 \cdot i = -1i = -i \\ i^4 &= i^2 \cdot i^2 = -1(-1) = 1 \end{aligned}}$$

In general, for any integer n:

$$i^{4n} = (i^4)^n = 1^n = 1$$

$$i^{4n+1} = (i^4)^n(i^1) = 1^n(i) = 1(i) = i$$

$$i^{4n+2} = (i^4)^n(i^2) = 1^n(-1) = 1(-1) = -1$$

$$i^{4n+3} = (i^4)^n(i^3) = 1^n(-i) = 1(-i) = -i$$

For any positive integer k, i^k is equal to i, -1, $-i$, or 1. We can use this relationship to evaluate any power of i.

EXAMPLE 2

Evaluate:

Answers

a. i^{14} $= i^{4(3)+2} = (i^4)^3 \cdot (i^2) = (1)^3(-1) = 1(-1) = -1$

b. i^7 $= i^7 = i^{4+3} = i^4 \cdot i^3 = (1)(-i) = -i$

Arithmetic of Imaginary Numbers

The rule for multiplying radicals, $\sqrt{a} \cdot \sqrt{b} = \sqrt{ab}$, is true if and only if a and b are non-negative numbers. When $a < 0$ or $b < 0$, $\sqrt{a} \cdot \sqrt{b} \neq \sqrt{ab}$. For example, it is *incorrect* to write: $\sqrt{-4} \times \sqrt{-9} = \sqrt{36} = 6$, but it is *correct* to write:

$$\sqrt{-4} \times \sqrt{-9} = \sqrt{4}\sqrt{-1} \times \sqrt{9}\sqrt{-1} = 2i \times 3i = 6i^2 = 6(-1) = -6$$

The distributive property of multiplication over addition is true for imaginary numbers.

$$\sqrt{-4} + \sqrt{-9}$$
$$= \sqrt{4}\sqrt{-1} + \sqrt{9}\sqrt{-1}$$
$$= 2i + 3i$$
$$= (2 + 3)i$$
$$= 5i$$

$$\sqrt{-8} + \sqrt{-50}$$
$$= \sqrt{4}\sqrt{-1}\sqrt{2} + \sqrt{25}\sqrt{-1}\sqrt{2}$$
$$= 2i\sqrt{2} + 5i\sqrt{2}$$
$$= (2 + 5)(i\sqrt{2})$$
$$= 7i\sqrt{2}$$

EXAMPLE 3

Find the sum of $\sqrt{-25} + \sqrt{-49}$.

Solution Write each imaginary number in terms of i and add similar terms.

$$\sqrt{-25} + \sqrt{-49} = \sqrt{25}\sqrt{-1} + \sqrt{49}\sqrt{-1} = 5i + 7i = 12i \ \textit{Answer}$$

The Set of Complex Numbers

The set of pure imaginary numbers is not closed under multiplication because the powers of i include both real numbers and imaginary numbers. For example, $i \cdot i = -1$. We want to define a set of numbers that includes both real numbers and pure imaginary numbers and that satisfies all the properties of the real numbers. This new set of numbers is called the set of *complex numbers* and is defined as the set of numbers of the form $a + bi$ where a and b are real numbers. When $a = 0$, $a + bi = 0 + bi = bi$, a pure imaginary number. When $b = 0$, $a + bi = a + 0i = a$, a real number. Therefore, both the set of pure imaginary numbers and the set of real numbers are contained in the set of complex numbers. A complex number that is *not* a real number is called an imaginary number.

DEFINITION _____

A **complex number** is a number of the form

$$a + bi$$

where a and b are real numbers and $i = \sqrt{-1}$.

Examples of complex numbers are:

$$2, \quad -4, \quad \tfrac{1}{3}, \quad \sqrt{2}, \quad \pi, \quad 4i, \quad -5i, \quad \tfrac{1}{4}i, \quad i\sqrt{5}, \quad \pi i,$$
$$2 - 3i, \quad -7 + i, \quad 8 + 3i, \quad 0 - i, \quad 5 + 0i$$

Complex Numbers and the Graphing Calculator

The TI family of graphing calculators can use complex numbers. To change your calculator to work with complex numbers, press ⬚MODE⬚, scroll down to the row that lists REAL, select $a + bi$, and then press ⬚ENTER⬚. Your calculator will now evaluate complex numbers. For example, with your calculator in $a + bi$ mode, write $6 + \sqrt{-9}$ as a complex number in terms of i.

ENTER: 6 ⬚+⬚ ⬚2nd⬚ ⬚√⬚ −9 ⬚)⬚ ⬚ENTER⬚

DISPLAY:

```
6+√(⁻9)
          6+3i
```

EXAMPLE 4 ▮▮▮▮▮▮▮▮▮▮▮▮▮▮▮▮▮▮▮▮▮▮▮▮▮▮▮▮▮▮▮▮▮▮▮▮▮▮▮

Evaluate each expression using a calculator:

a. $\sqrt{-25} - \sqrt{-16}$ **b.** $2 + \sqrt{-9} + 5\sqrt{-4}$

Calculator **a.** Press ⬚2nd⬚ ⬚√⬚ −25 ⬚)⬚ ⬚−⬚ ⬚2nd⬚ ⬚√⬚
Solution   −16 ⬚)⬚ ⬚ENTER⬚.

```
√(⁻25)-√(⁻16)
                i
```

b. Press 2 ⬚+⬚ ⬚2nd⬚ ⬚√⬚ −9 ⬚)⬚ ⬚+⬚ 5
⬚2nd⬚ ⬚√⬚ −4 ⬚)⬚ ⬚ENTER⬚.

```
2+√(⁻9)+5√(⁻4)
          2+13i
```

Answers **a.** i **b.** $2 + 13i$

What about powers of i? A calculator is not an efficient tool for evaluating an expression such as bi^n for large values of n. Large complex powers introduce rounding errors in the answer as shown in the calculator screen on the right.

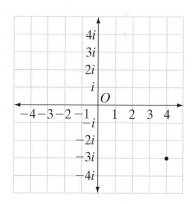

Graphing Complex Numbers

We are familiar with the one-to-one correspondence between the set of real numbers and the points on the number line. The complex numbers can be put in a one-to-one correspondence with the points of a plane. This correspondence makes it possible for a physical quantity, such as an alternating current that has both amplitude and direction, to be represented by a complex number.

On graph paper, draw a horizontal real number line. Through 0 on the real number line, draw a vertical line. This vertical line is the pure imaginary number line. On this line, mark points equidistant from one another such that the distance between them is equal to the unit distance on the real number line. Label the points above 0 as i, $2i$, $3i$, $4i$, and so on and the numbers below 0 as $-i$, $-2i$, $-3i$, $-4i$, and so on.

To find the point that corresponds to the complex number $4 - 3i$, move to 4 on the real number line and then, parallel to the imaginary number line, move down 3 units to a point on the same horizontal line as $3i$.

Hands-On Activity

On the complex plane, locate the points that correspond to $4 + 2i$ and $2 - 5i$. Draw a line segment from 0 to each of these points. Now draw a parallelogram with these two line segments as adjacent sides. What is the complex number that corresponds to the vertex of the parallelogram opposite 0? How is this complex number related to $4 + 2i$ and $2 - 5i$? Repeat the procedure given above for each of the following pairs of complex numbers.

1. $2 + 4i, 3 + i$ **2.** $-2 + 5i, -5 + 2i$ **3.** $4 - 2i, 2 - 2i$

4. $-4 - 2i, 3 - 5i$ **5.** $1 + 4i, 1 - 4i$ **6.** $-5 + 3i, -5 - 3i$

7. $-3 + 0i, 0 - 4i$ **8.** $4 + 0i, -4 + 3i$ **9.** $4, 2i$

Exercises

Writing About Mathematics

1. Pete said that $\sqrt{-2} \times \sqrt{-8} = \sqrt{16} = 4$. Do you agree with Pete? Explain why or why not.

2. Ethan said that the square of any pure imaginary number is a negative real number. Do you agree with Ethan? Justify your answer.

Developing Skills

In 3–18, write each number in terms of i.

3. $\sqrt{-4}$ 4. $\sqrt{-81}$ 5. $\sqrt{-9}$ 6. $-\sqrt{-36}$

7. $-\sqrt{-121}$ 8. $\sqrt{-8}$ 9. $\sqrt{-12}$ 10. $-\sqrt{-72}$

11. $5\sqrt{-27}$ 12. $-\frac{1}{2}\sqrt{-80}$ 13. $-\sqrt{-51}$ 14. $\sqrt{-500}$

15. $5 + \sqrt{-5}$ 16. $1 + \sqrt{-3}$ 17. $-4 - \sqrt{-24}$ 18. $-3 + 2\sqrt{-9}$

In 19–34, write each sum or difference in terms of i.

19. $\sqrt{-100} + \sqrt{-81}$ 20. $\sqrt{-25} - \sqrt{-4}$

21. $\sqrt{-144} + \sqrt{-1}$ 22. $\sqrt{-49} - \sqrt{-16}$

23. $\sqrt{-12} + \sqrt{-27} - \sqrt{-75}$ 24. $2\sqrt{-5} + \sqrt{-125}$

25. $\sqrt{4} + \sqrt{-4} + \sqrt{-36} - \sqrt{36}$ 26. $\sqrt{50} + \sqrt{-2} + \sqrt{200} + \sqrt{-50}$

27. $\sqrt{4} + \sqrt{-32} + \sqrt{-8}$ 28. $3 + \sqrt{-28} - 7 - \sqrt{-7}$

29. $-3 - \sqrt{-10} + 2 - \sqrt{-90}$ 30. $-2 + \sqrt{-16} + 7 - \sqrt{-49}$

31. $3 + \sqrt{-36} - 6 + \sqrt{-1}$ 32. $\frac{1}{7} + \sqrt{-\frac{1}{8}} + \frac{2}{7} - \sqrt{-\frac{1}{2}}$

33. $-\frac{1}{2} + \sqrt{-\frac{2}{3}} - \frac{1}{2} + \sqrt{-\frac{24}{9}}$ 34. $\sqrt{0.2025} + \sqrt{-0.09}$

In 35–43, write each number in simplest form.

35. $3i + 2i^3$ 36. $5i^2 + 2i^4$ 37. $i - 5i^3$

38. $2i^5 + 7i^7$ 39. $i + i^3 + i^5$ 40. $4i + 5i^8 + 6i^3 + 2i^4$

41. $i^2 + i^{12} + i^8$ 42. $i + i^2 + i^3 + i^4$ 43. i^{57}

In 44–51, locate the point that corresponds to each of the given complex numbers.

44. $2 + 4i$ 45. $-2 + 5i$ 46. $4 - 2i$ 47. $-4 - 2i$

48. $\frac{1}{2} + 4i$ 49. $0 + 3i$ 50. $-3 + 0i$ 51. $0 + 0i$

Applying Skills

52. Imaginary numbers are often used in electrical engineering. The *impedance Z* (measured in ohms) of a circuit measures the resistance of a circuit to alternating current (AC) electricity. For two AC circuits connected in series, the total impedance is the sum $Z_1 + Z_2$ of the individual impedances. Find the total impedance if $Z_1 = 5i$ ohms and $Z_2 = 8i$ ohms.

53. In certain circuits, the total impedance Z_T is given by the formula $Z_T = \frac{Z_1 Z_2}{Z_1 + Z_2}$. Find Z_T when $Z_1 = -3i$ and $Z_2 = 4i$.

Hands-On Activity: Multiplying Complex Numbers

Multiplication by i

In the complex plane, the image of 1 under a rotation of 90° about the origin is i, which is equivalent to multiplying 1 by i. Now rotate i by 90° about the origin. The image of i is -1, which models the product $i \cdot i = -1$. This result is true for *any* complex number. That is, multiplying any complex number $a + bi$ by i is equivalent to rotating that number by 90° about the origin. Use this graphical representation of multiplication by i to evaluate the following products:

1. $(2 + 3i) \cdot i$ **2.** $12i \cdot i$ **3.** $\left(\frac{1}{2} + \frac{3}{2}i\right) \cdot i$

4. $-5i \cdot i$ **5.** $-10 \cdot i$ **6.** $(-3 - 2i) \cdot i$

Multiplication by a real number

In the complex plane, the image of any number a under a dilation of k is ak, which is equivalent to multiplying a by k. The image any imaginary number ai under a dilation of k is aki, which is equivalent to multiplying ai by k. This result is true for *any* complex number. That is, multiplying any complex number $a + bi$ by a real number k is equivalent to a dilation of k. Use this graphical representation of multiplication by a real number to evaluate the following products:

1. $(2 + 4i) \cdot 4$ **2.** $(-5 - 5i) \cdot \frac{1}{2}$ **3.** $\left(\frac{3}{4} + 2i\right) \cdot 4$

4. $2i \cdot 2$ **5.** $-4i \cdot \frac{1}{2}$ **6.** $(-3 - 2i) \cdot 3i = (-3 - 2i) \cdot 3 \cdot i$

5-5 OPERATIONS WITH COMPLEX NUMBERS

Like the real numbers, the complex numbers are closed, commutative, and associative under addition and multiplication, and multiplication is distributive over addition.

Adding Complex Numbers

We can add complex numbers using the commutative, associative, and distributive properties that we use to add real numbers. For example:

$$(4 + 3i) + (2 + 5i) = 4 + (3i + 2) + 5i \qquad \text{Associative Property}$$
$$= 4 + (2 + 3i) + 5i \qquad \text{Commutative Property}$$
$$= (4 + 2) + (3i + 5i) \qquad \text{Associative Property}$$
$$= (4 + 2) + (3 + 5)i \qquad \text{Distributive Property}$$
$$= 6 + 8i \qquad \text{Substitution Property}$$

When we know that each of these steps leads to a correct sum, we can write:

$$(a + bi) + (c + di) = (a + c) + (b + d)i$$

Here are some other examples of the addition of complex numbers:

$$(8 - i) + (3 - 7i) = (8 + 3) + (-1 - 7)i = 11 - 8i$$
$$(3 + 2i) + (1 - 2i) = (3 + 1) + (2 - 2)i = 4 + 0i = 4$$
$$(-9 + i) + (9 + 6i) = (-9 + 9) + (1 + 6)i = 0 + 7i = 7i$$

Note that the sum of two complex numbers is always a complex number. That sum may be an element of one of the subsets of the complex numbers, that is, a real number or a pure imaginary number.

Since for any complex number $a + bi$:

$$(a + bi) + (0 + 0i) = (0 + 0i) + (a + bi) = (a + bi)$$

the real number $(0 + 0i)$ or 0 is the additive identity.

▶ **The additive identity of the complex numbers is the real number $(0 + 0i)$ or 0.**

Similarly, since for any complex number $a + bi$:

$$(a + bi) + (-a - bi) = [a + (-a)] + [b + (-b)]i = 0 + 0i = 0$$

the additive inverse of any complex number $(a + bi)$ is $(-a - bi)$.

▶ **The additive inverse of any complex number $(a + bi)$ is $(-a - bi)$.**

On some calculators, the key for i is the 2nd function of the decimal point key. When the calculator is in $a + bi$ mode, we can add complex numbers.

ENTER: 4 + 2 **2nd** *i* + 3 − 5 **2nd** *i* **ENTER**

DISPLAY:
```
4+2i+3-5i
            7-3i
```

EXAMPLE 1

Express the sum in $a + bi$ form: $(-3 + 7i) + (-5 + 2i)$

Solution $(-3 + 7i) + (-5 + 2i) = (-3 - 5) + (7 + 2)i = -8 + 9i$

Calculator Solution ENTER: `(` -3 `+` 7 `2nd` `i` `)` `+` `(` -5 `+`

2 `2nd` `i` `)` `ENTER`

DISPLAY:

```
(-3+7i)+(-5+2i)
                -8+9i
```

Answer $-8 + 9i$

Subtracting Complex Numbers

In the set of real numbers, $a - b = a + (-b)$, that is, $a - b$ is the sum of a and the additive inverse of b. We can apply the same rule to the subtraction of complex numbers.

EXAMPLE 2

Subtract: *Answers*

a. $(5 + 3i) - (2 + 8i)$ $= (5 + 3i) + (-2 - 8i) = (5 - 2) + (3 - 8)i$
                                $= 3 - 5i$

b. $(7 + 4i) - (-1 + 3i)$ $= (7 + 4i) + (1 - 3i) = (7 + 1) + (4 - 3)i$
                                $= 8 + 1i = 8 + i$

c. $(6 + i) - (3 + i)$ $= (6 + i) + (-3 - i) = (6 - 3) + (1 - 1)i$
                                $= 3 + 0i = 3$

d. $(1 + 2i) - (1 - 9i)$ $= (1 + 2i) + (-1 + 9i) = (1 - 1) + (2 + 9)i$
                                $= 0 + 11i = 11i$

Multiplying Complex Numbers

We can multiply complex numbers using the same principles that we use to multiply binomials. When simplifying each of the following products, recall that $i(i) = i^2 = -1$.

EXAMPLE 3

Multiply:

a. $(5 + 2i)(3 + 4i)$ **b.** $(3 - 5i)(2 + i)$

c. $(9 + 3i)(9 - 3i)$ **d.** $(2 + 6i)^2$

Solution   **a.** $(5 + 2i)(3 + 4i)$

$= 5(3 + 4i) + 2i(3 + 4i)$

$= 15 + 20i + 6i + 8i^2$

$= 15 + 26i + 8(-1)$

$= 15 + 26i - 8$

$= 7 + 26i$ *Answer*

b. $(3 - 5i)(2 + i)$

$= 3(2 + i) - 5i(2 + i)$

$= 6 + 3i - 10i - 5i^2$

$= 6 - 7i - 5(-1)$

$= 6 - 7i + 5$

$= 11 - 7i$ *Answer*

c. $(9 + 3i)(9 - 3i)$

$= 9(9 - 3i) + 3i(9 - 3i)$

$= 81 - 27i + 27i - 9i^2$

$= 81 + 0i - 9(-1)$

$= 81 + 9$

$= 90$ *Answer*

d. $(2 + 6i)^2$

$= (2 + 6i)(2 + 6i)$

$= 2(2 + 6i) + 6i(2 + 6i)$

$= 4 + 12i + 12i + 36i^2$

$= 4 + 24i + 36(-1)$

$= -32 + 24i$ *Answer*

In part **c** of Example 3, the product is a real number.

Since for any complex number $a + bi$:

$$(a + bi)(1 + 0i) = a(1 + 0i) + bi(1 + 0i) = a + 0i + bi + 0i = a + bi$$

the real number $1 + 0i$ or 1 is the multiplicative identity.

▶ **The multiplicative identity of the complex numbers is the real number $1 + 0i$ or 1.**

Does every non-zero complex number have a multiplicative inverse? In the set of rational numbers, the multiplicative inverse of a is $\frac{1}{a}$. We can say that in the set of complex numbers, the multiplicative inverse of $a + bi$ is $\frac{1}{a + bi}$. To express $\frac{1}{a + bi}$ in the form of a complex number, multiply numerator and denominator by the **complex conjugate** of $a + bi$, $a - bi$. The reason for doing so is that the product of a complex number and its conjugate is a real number. For example,

$$(1 + 2i)(1 - 2i) = 1(1 - 2i) + 2i(1 - 2i)$$

$$= 1 - 2i + 2i - 4i^2$$

$$= 1 - 4(-1)$$

$$= 5$$

In general,

$$(a + bi)(a - bi) = a(a - bi) + bi(a - bi)$$
$$= a^2 - abi + abi - b^2i^2$$
$$= a^2 - b^2(-1)$$
$$= a^2 + b^2$$

or

$$(a + bi)(a - bi) = a^2 + b^2$$

The product of a complex number $a + bi$ and its conjugate $a - bi$ is a real number, $a^2 + b^2$. Thus, if we the multiply numerator and denominator of $\frac{1}{a + bi}$ by the complex conjugate of the denominator, we will have an equivalent fraction with a denominator that is a real number.

$$\frac{1}{a + bi} \cdot \frac{a - bi}{a - bi} = \frac{a - bi}{a^2 + b^2} = \frac{a}{a^2 + b^2} - \frac{b}{a^2 + b^2}i$$

For example, the complex conjugate of $2 - 4i$ is $2 + 4i$ and the multiplicative inverse of $2 - 4i$ is

$$\frac{1}{2 - 4i} = \frac{1}{2 - 4i} \cdot \frac{2 + 4i}{2 - 4i} = \frac{2 + 4i}{2^2 + 4^2} = \frac{2}{2^2 + 4^2} + \frac{4i}{2^2 + 4^2}$$
$$= \frac{2}{20} + \frac{4}{20}i$$
$$= \frac{1}{10} + \frac{1}{5}i$$

We can check that $\frac{1}{10} + \frac{1}{5}i$ is the multiplicative inverse of $2 - 4i$ by multiplying the two numbers together. If $\frac{1}{10} + \frac{1}{5}i$ is indeed the multiplicative inverse, then their product will be 1.

$$(2 - 4i) \cdot \left(\frac{1}{10} + \frac{1}{5}i\right) \stackrel{?}{=} 1$$
$$2\left(\frac{1}{10} + \frac{1}{5}i\right) - 4i\left(\frac{1}{10} + \frac{1}{5}i\right) \stackrel{?}{=} 1$$
$$2\left(\frac{1}{10}\right) + 2\left(\frac{1}{5}i\right) - 4i\left(\frac{1}{10}\right) - 4i\left(\frac{1}{5}i\right) \stackrel{?}{=} 1$$
$$\frac{2}{10} + \frac{2}{5}i - \frac{4}{10}i - \frac{4}{5}i^2 \stackrel{?}{=} 1$$
$$\frac{1}{5} + \frac{2}{5}i - \frac{2}{5}i - \frac{4}{5}(-1) \stackrel{?}{=} 1$$
$$\frac{1}{5} + \frac{4}{5} = 1 ✔$$

▶ **For any non-zero complex number $a + bi$, its multiplicative inverse is**

$$\frac{1}{a + bi} \quad \text{or} \quad \frac{a}{a^2 + b^2} - \frac{b}{a^2 + b^2}i$$

EXAMPLE 4

Express the product in $a + bi$ form: $\left(\frac{1}{3} - \frac{1}{2}i\right)(12 + 3i)$

Solution

$$\left(\frac{1}{3} - \frac{1}{2}i\right)(12 + 3i) = \frac{1}{3}(12 + 3i) - \frac{1}{2}i(12 + 3i)$$

$$= 4 + i - 6i - \frac{3}{2}i^2$$

$$= 4 - 5i - \frac{3}{2}(-1)$$

$$= \frac{8}{2} + \frac{3}{2} - 5i$$

$$= \frac{11}{2} - 5i$$

Calculator ENTER: (1 ÷ 3 − 1 ÷ 2 **2nd** *i*) (12 + 3
Solution **2nd** *i*) MATH ENTER ENTER

DISPLAY:
```
(1/3-1/2i)(12+3i
)▶Frac
              11/2-5i
```

Answer $\frac{11}{2} - 5i$

Dividing Complex Numbers

We can write $(2 + 3i) \div (1 + 2i)$ as $\frac{2 + 3i}{1 + 2i}$. However, this is not a complex number of the form $a + bi$. Can we write $\frac{2 + 3i}{1 + 2i}$ as an equivalent fraction with a rational denominator? In Chapter 3, we found a way to write a fraction with an irrational denominator as an equivalent fraction with a rational denominator. We can use a similar procedure to write a fraction with a complex denominator as an equivalent fraction with a real denominator.

To write the quotient $\frac{2 + 3i}{1 + 2i}$ with a denominator that is a real number, multiply numerator and denominator of the fraction by the complex conjugate of the denominator. The complex conjugate of $1 + 2i$ is $1 - 2i$.

$$\frac{2 + 3i}{1 + 2i} \cdot \frac{1 - 2i}{1 - 2i} = \frac{2 - 4i + 3i - 6i^2}{1^2 - (2i)^2} = \frac{2 - i - 6i^2}{1 - 4i^2}$$

$$= \frac{2 - i - 6(-1)}{1 - 4(-1)}$$

$$= \frac{8 - i}{5}$$

$$= \frac{8}{5} - \frac{1}{5}i$$

The quotient $(2 + 3i) \div (1 + 2i) = \frac{2 + 3i}{1 + 2i} = \frac{8}{5} - \frac{1}{5}i$ is a complex number in $a + bi$ form.

EXAMPLE 5

Express the quotient in $a + bi$ form: $\dfrac{5 + 10i}{\frac{1}{2} - \frac{2}{3}i}$

Solution   *How to Proceed*

(1) Multiply numerator and denominator by 6 to express the denominator in terms of integers:

$$\frac{5 + 10i}{\frac{1}{2} - \frac{2}{3}i} = \frac{5 + 10i}{\frac{1}{2} - \frac{2}{3}i} \cdot \frac{6}{6} = \frac{30 + 60i}{3 - 4i}$$

(2) Multiply numerator and denominator of the fraction by the complex conjugate of the denominator, $3 + 4i$:

$$\frac{30 + 60i}{3 - 4i} \cdot \frac{3 + 4i}{3 + 4i} = \frac{90 + 120i + 180i + 240i^2}{3^2 + 4^2}$$

$$= \frac{90 + 120i + 180i + 240(-1)}{9 + 16}$$

(3) Replace i^2 by its equal, -1:

$$= \frac{90 + 120i + 180i - 240}{25}$$

(4) Combine like terms:

$$= \frac{-150 + 300i}{25}$$

(5) Divide numerator and denominator by 25:

$$= -6 + 12i \ \textit{Answer}$$

Calculator
Solution

ENTER: (5 + 10 2nd i) ÷ (1 ÷ 2 − 2 ÷ 3

2nd i) ENTER

DISPLAY:

```
(5+10i)/(1/2-2/3
i)
                -6+12i
```

Exercises

Writing About Mathematics

1. Tim said that the binomial $x^2 + 16$ can be written as $x^2 - 16i^2$ and factored over the set of complex numbers. Do you agree with Tim? Explain why or why not.

2. Joshua said that the product of a complex number and its conjugate is always a real number. Do you agree with Joshua? Explain why or why not.

Developing Skills

In 3–17, find each sum or difference of the complex numbers in $a + bi$ form.

3. $(6 + 7i) + (1 + 2i)$ **4.** $(3 - 5i) + (2 + i)$ **5.** $(5 - 6i) + (4 + 2i)$

6. $(-3 + 3i) - (1 + 5i)$ **7.** $(2 - 8i) - (2 + 8i)$ **8.** $(4 + 12i) + (-4 - 2i)$

9. $(1 + 9i) - (1 + 2i)$ **10.** $(10 - 12i) - (12 + 7i)$ **11.** $(0 + 3i) + (0 - 3i)$

12. $\left(\frac{1}{2} + \frac{1}{2}i\right) + \left(\frac{1}{4} - \frac{3}{4}i\right)$ **13.** $\left(\frac{2}{3} - \frac{1}{6}i\right) + (2 - i)$ **14.** $(1 + 0i) - \left(\frac{1}{5} - \frac{2}{5}i\right)$

15. $\left(-\frac{3}{2} + \frac{5}{3}i\right) - \left(\frac{9}{4} - \frac{1}{5}i\right)$ **16.** $\left(\frac{1}{4} + 12i\right) + \left(7 + \frac{i}{10}\right)$ **17.** $\left(\frac{3}{7} + \frac{1}{3}i\right) + \left(\frac{1}{6} - \frac{7}{8}i\right)$

In 18–25, write the complex conjugate of each number.

18. $3 + 4i$ **19.** $2 - 5i$ **20.** $-8 + i$ **21.** $-6 - 9i$

22. $\frac{1}{2} - 3i$ **23.** $-4 + \frac{1}{3}i$ **24.** $\frac{5}{3} - \frac{2}{3}i$ **25.** $\pi + 2i$

In 26–37, find each product.

26. $(1 + 5i)(1 + 2i)$ **27.** $(3 + 2i)(3 + 3i)$ **28.** $(2 - 3i)^2$

29. $(4 - 5i)(3 - 2i)$ **30.** $(7 + i)(2 + 3i)$ **31.** $(-2 - i)(1 + 2i)$

32. $(-4 - i)(-4 + i)$ **33.** $(-12 - 2i)(12 - 2i)$ **34.** $(5 - 3i)(5 + 3i)$

35. $(3 - i)\left(\frac{3}{10} + \frac{1}{10}i\right)$ **36.** $\left(\frac{1}{5} - \frac{1}{10}i\right)(4 + 2i)$ **37.** $\left(-\frac{1}{8} + \frac{1}{8}i\right)(4 + 4i)$

In 38–45, find the multiplicative inverse of each of the following in $a + bi$ form.

38. $1 + i$ **39.** $2 + 4i$ **40.** $-1 + 2i$ **41.** $3 - 3i$

42. $\frac{1}{2} - \frac{1}{4}i$ **43.** $2 - \frac{1}{2}i$ **44.** $\frac{5}{6} + 3i$ **45.** $9 + \pi i$

In 46–60, write each quotient in $a + bi$ form.

46. $(8 + 4i) \div (1 + i)$ **47.** $(2 + 4i) \div (1 - i)$ **48.** $(10 + 5i) \div (1 + 2i)$

49. $(5 - 15i) \div (3 - i)$ **50.** $\frac{2 + 2i}{4 - 2i}$ **51.** $\frac{10 - 5i}{2 + 6i}$

52. $\frac{8 + 2i}{1 + 3i}$ **53.** $\frac{3 + i}{3 - i}$ **54.** $\frac{5 - 2i}{5 + 2i}$

55. $\frac{12 + 3i}{3i}$ **56.** $\frac{8 + 6i}{2i}$ **57.** $\frac{\frac{1}{5} - \frac{1}{5}i}{3 - 4i}$

58. $\frac{1 + i}{\frac{1}{2} - \frac{1}{2}i}$ **59.** $\frac{3i}{\pi + \frac{1}{2}i}$ **60.** $\frac{\frac{1}{7} + i}{\frac{5}{i}}$

Applying Skills

61. Impedance is the resistance to the flow of current in an electric circuit measured in ohms. The impedance, Z, in a circuit is found by using the formula $Z = \frac{V}{I}$ where V is the voltage (measured in volts) and I is the current (measured in amperes). Find the impedance when $V = 1.8 - 0.4i$ volts and $I = -0.3i$ amperes.

62. Find the current that will flow when $V = 1.6 - 0.3i$ volts and $Z = 1.5 + 8i$ ohms using the formula $Z = \frac{V}{I}$.

5-6 COMPLEX ROOTS OF A QUADRATIC EQUATION

When we used the quadratic formula to find the roots of a quadratic equation, we found that some quadratic equations have no real roots. For example, the quadratic equation $x^2 + 4x + 5 = 0$ has no real roots since the graph of the corresponding quadratic function $f(x) = x^2 + 4x + 5$, shown on the right, has no x-intercepts. Can the roots of a quadratic equation be complex numbers?

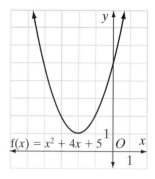

We will use the quadratic formula to find the roots of the equation $x^2 + 4x + 5 = 0$. For this equation, $a = 1$, $b = 4$, and $c = 5$. The value of the discriminant is

$$b^2 - 4ac = 4^2 - 4(1)(5) \qquad \text{or} \qquad -4$$

Therefore, the roots are not real numbers.

$$x = \frac{-b \pm \sqrt{b^2 - 4ac}}{2a} = \frac{-4 \pm \sqrt{4^2 - 4(1)(5)}}{2(1)} = \frac{-4 \pm \sqrt{-4}}{2}$$
$$= \frac{-4 \pm 2i}{2}$$
$$= -2 \pm i$$

The roots are the complex numbers $-2 + i$ and $-2 - i$.

Therefore, every quadratic equation with rational coefficients has two complex roots. Those roots may be two real numbers. Complex roots that are real may be rational or irrational and equal or unequal. Roots that are not real numbers are complex conjugates.

Note that complex roots that are not real are often referred to as *imaginary roots* and complex roots of the form $0 + bi = bi$ as *pure imaginary roots*. A quadratic equation with rational coefficients has roots that are pure imaginary only if $b = 0$ and $b^2 - 4ac < 0$.

Let $ax^2 + bx + c = 0$ be a quadratic equation and a, b, and c be rational numbers with $a \neq 0$:

When the discriminant $b^2 - 4ac$ is	The roots of the equation are	The number of x-intercepts of the function is:
> 0 and a perfect square	real, rational, and unequal	2
> 0 and not a perfect square	real, irrational, and unequal	2
$= 0$	real, rational, and equal	1
< 0	imaginary numbers	0

EXAMPLE 1

Find the complex roots of the equation $x^2 + 12 = 6x$.

Solution

How to Proceed

(1) Write the equation in standard form:

$$x^2 + 12 = 6x$$
$$x^2 - 6x + 12 = 0$$

(2) Determine the values of a, b, and c:

$$a = 1, b = -6, c = 12$$

(3) Substitute the values of a, b, and c into the quadratic formula:

$$x = \frac{-b \pm \sqrt{b^2 - 4ac}}{2a}$$

$$= \frac{-(-6) \pm \sqrt{(-6)^2 - 4(1)(12)}}{2(1)}$$

(4) Perform the computation:

$$= \frac{6 \pm \sqrt{36 - 48}}{2}$$

$$= \frac{6 \pm \sqrt{-12}}{2}$$

(5) Write $\sqrt{-12}$ in simplest form in terms of i:

$$= \frac{6 \pm \sqrt{4}\sqrt{3}\sqrt{-1}}{2}$$

$$= \frac{6 \pm 2i\sqrt{3}}{2}$$

$$= 3 \pm i\sqrt{3}$$

Answer The roots are $3 + i\sqrt{3}$ and $3 - i\sqrt{3}$.

EXAMPLE 2

For what values of c does the equation $2x^2 + 3x + c = 0$ have imaginary roots?

Solution

How to Proceed

(1) The equation has imaginary roots when $b^2 - 4ac < 0$. Substitute $a = 2$, $b = 3$:

$$b^2 - 4ac < 0$$
$$3^2 - 4(2)c < 0$$

(2) Isolate the term in c:

$$9 - 8c < 0$$
$$-8c < -9$$

(3) Solve the inequality. Dividing by a negative number reverses the order of the inequality:

$$\frac{-8c}{-8} > \frac{-9}{-8}$$
$$c > \frac{9}{8}$$

Answer The roots are imaginary when $c > \frac{9}{8}$.

Exercises

Writing About Mathematics

1. Emily said that when a and c are real numbers with the same sign and $b = 0$, the roots of the equation $ax^2 + bx + c = 0$ are pure imaginary. Do you agree with Emily? Justify your answer.

2. Noah said that if a, b, and c are rational numbers and $b^2 - 4ac < 0$, then the roots of the equation $ax^2 + bx + c = 0$ are complex conjugates. Do you agree with Noah? Justify your answer.

Developing Skills

In 3–14, use the quadratic formula to find the imaginary roots of each equation.

3. $x^2 - 4x + 8 = 0$ **4.** $x^2 + 6x + 10 = 0$

5. $x^2 - 4x + 13 = 0$ **6.** $2x^2 + 2x + 1 = 0$

7. $x^2 + 10x + 29 = 0$ **8.** $x^2 + 8x + 17 = 0$

9. $x^2 - 2x + 10 = 0$ **10.** $4x^2 - 4x + 5 = 0$

11. $4x^2 + 4x + 17 = 0$ **12.** $x^2 + 5 = 4x$

13. $4x - 7 = x^2$ **14.** $2x = x^2 + 3$

5-7 SUM AND PRODUCT OF THE ROOTS OF A QUADRATIC EQUATION

Writing a Quadratic Equation Given the Roots of the Equation

We know that when a quadratic equation is in standard form, the polynomial can often be factored in order to find the roots. For example, to find the roots of $2x^2 - 5x + 3 = 0$, we can factor the polynomial.

$$2x^2 - 5x + 3 = 0$$
$$(2x - 3)(x - 1) = 0$$

$$
\begin{array}{c|c}
2x - 3 = 0 & x - 1 = 0 \\
2x = 3 & x = 1 \\
x = \frac{3}{2} &
\end{array}
$$

We can work backward on each of these steps to write a quadratic equation with a given pair of factors. For example, write a quadratic equation whose roots are 3 and −5:

(1) Let x equal each of the roots: $\qquad$ $x = 3 \quad \mid \quad x = -5$

(2) Write equivalent equations with one side equal to 0: $\qquad$ $x - 3 = 0 \quad \mid \quad x + 5 = 0$

(3) If equals are multiplied by equals their products are equal: $\qquad$ $(x - 3)(x + 5) = 0(0)$
$$x^2 + 2x - 15 = 0$$

We can check to show that 3 and −5 are the roots of $x^2 + 2x - 15 = 0$:

$$\begin{array}{ll}
\textit{Check for } x = 3: & \textit{Check for } x = -5: \\
x^2 + 2x - 15 = 0 & x^2 + 2x - 15 = 0 \\
(3)^2 + 2(3) - 15 \stackrel{?}{=} 0 & (-5)^2 + 2(-5) - 15 \stackrel{?}{=} 0 \\
9 + 6 - 15 \stackrel{?}{=} 0 & 25 - 10 - 15 \stackrel{?}{=} 0 \\
0 = 0 \checkmark & 0 = 0 \checkmark
\end{array}$$

We can use the same procedure for the general case. Write the quadratic equation that has roots r_1 and r_2:

(1) Let x equal each of the roots: $\qquad$ $x = r_1 \quad \mid \quad x = r_2$

(2) Write equivalent equations with one side equal to 0: $\qquad$ $x - r_1 = 0 \quad \mid \quad x - r_2 = 0$

(3) If equals are multiplied by equals, their products are equal: $\qquad$ $(x - r_2)(x - r_1) = 0(0)$
$$x^2 - r_1 x - r_2 x + r_1 r_2 = 0$$

$$x^2 - (r_1 + r_2)x + r_1 r_2 = 0$$

In the example given above, $r_1 = 3$ and $r_2 = -5$. The equation with these roots is:

$$x^2 - (r_1 + r_2)x + r_1 r_2 = 0$$
$$x^2 - [3 + (-5)]x + 3(-5) = 0$$
$$x^2 + 2x - 15 = 0$$

EXAMPLE I

Write an equation with roots $\dfrac{1 + \sqrt{6}}{3}$ and $\dfrac{1 - \sqrt{6}}{3}$.

Solution $r_1 + r_2 = \dfrac{1 + \sqrt{6}}{3} + \dfrac{1 - \sqrt{6}}{3} = \dfrac{2}{3}$

$r_1 r_2 = \left(\dfrac{1 + \sqrt{6}}{3}\right)\left(\dfrac{1 - \sqrt{6}}{3}\right) = \dfrac{1^2 - (\sqrt{6})^2}{9} = \dfrac{1 - 6}{9} = \dfrac{-5}{9} = -\dfrac{5}{9}$

Therefore, the equation is $x^2 - \frac{2}{3}x - \frac{5}{9} = 0$ or $9x^2 - 6x - 5 = 0$.

Alternative Solution

(1) Let x equal each of the roots:

$$x = \left(\frac{1 + \sqrt{6}}{3}\right) \quad \bigg| \quad x = \left(\frac{1 - \sqrt{6}}{3}\right)$$

(2) Write equivalent equations with one side equal to 0:

$$x - \left(\frac{1 + \sqrt{6}}{3}\right) = 0 \quad \bigg| \quad x - \left(\frac{1 - \sqrt{6}}{3}\right) = 0$$

(3) If equals are multiplied by equals, their products are equal:

$$\left[x - \left(\frac{1 + \sqrt{6}}{3}\right)\right] \cdot \left[x - \left(\frac{1 - \sqrt{6}}{3}\right)\right] = 0$$

$$x^2 - \left(\frac{1 - \sqrt{6}}{3}\right)x - \left(\frac{1 + \sqrt{6}}{3}\right)x + \left(\frac{1 + \sqrt{6}}{3}\right)\left(\frac{1 - \sqrt{6}}{3}\right) = 0$$

$$x^2 - \frac{2}{3}x + \frac{1 - 6}{9} = 0$$

(4) Simplify:

$$x^2 - \frac{2}{3}x - \frac{5}{9} = 0$$

$$\text{or} \quad 9x^2 - 6x - 5 = 0$$

Answer $9x^2 - 6x - 5 = 0$

The Coefficients of a Quadratic Equation and Its Roots

Compare the equation $x^2 - (r_1 + r_2)x + r_1r_2 = 0$ derived previously with the standard form of the quadratic equation, $ax^2 + bx + c = 0$.

(1) Divide each side of the equation by a to write the general equation as an equivalent equation with the coefficient of x^2 equal to 1.

$$\frac{a}{a}x^2 + \frac{b}{a}x + \frac{c}{a} = \frac{0}{a}$$

$$x^2 + \frac{b}{a}x + \frac{c}{a} = 0$$

(2) Compare the equation in step 1 with $x^2 - (r_1 + r_2)x + r_1r_2 = 0$ by equating corresponding coefficients.

$$\frac{b}{a} = -(r_1 + r_2) \quad \text{or} \quad -\frac{b}{a} = r_1 + r_2$$

$$\frac{c}{a} = r_1r_2$$

We can conclude the following:

▶ **If r_1 and r_2 are the roots of $ax^2 + bx + c = 0$, then:**

$$-\frac{b}{a} \text{ is equal to the sum of the roots } r_1 + r_2.$$

$$\frac{c}{a} \text{ is equal to the product of the roots } r_1r_2.$$

Thus, we can find the sum and product of the roots of a quadratic equation without actually solving the equation.

Note that an equivalent equation of any quadratic equation can be written with the coefficient of x^2 equal to 1. For example, $3x^2 - 5x + 2 = 0$ and $x^2 - \frac{5}{3}x + \frac{2}{3} = 0$ are equivalent equations.

- The sum of the roots of $3x^2 - 5x + 2 = 0$ is $-\left(-\frac{5}{3}\right)$ or $\frac{5}{3}$.
- The product of the roots of $3x^2 - 5x + 2 = 0$ is $\frac{2}{3}$.

EXAMPLE 2

If one root of the equation $x^2 - 3x + c = 0$ is 5, what is the other root?

Solution The sum of the roots of the equation is $-\frac{b}{a} = -\frac{-3}{1} = 3$.

Let $r_1 = 5$. Then $r_1 + r_2 = 3$ or $r_2 = -2$.

Answer The other root is -2.

EXAMPLE 3

Find two numbers whose sum is 4 and whose product is 5.

Solution Let the numbers be the roots of the equation $ax^2 + bx + c = 0$.

The sum of the roots is $-\frac{b}{a} = 4$ and the product of the roots is $\frac{c}{a} = 5$.

We can choose any convenient value of a. Let $a = 1$.

Then:

$$-\frac{b}{1} = 4 \quad \text{or} \quad b = -4$$
$$\frac{c}{1} = 5 \quad \text{or} \quad c = 5$$

The equation is $x^2 - 4x + 5 = 0$.

Use the quadratic equation to find the roots.

$$x = \frac{-b \pm \sqrt{b^2 - 4ac}}{2a}$$

$$x = \frac{-(-4) \pm \sqrt{(-4)^2 - 4(1)(5)}}{2(1)} = \frac{4 \pm \sqrt{-4}}{2} = \frac{4 \pm 2i}{2} = 2 \pm i$$

The numbers are $(2 + i)$ and $(2 - i)$.

Check The sum $(2 + i) + (2 - i) = 4.$ ✔

The product $(2 + i)(2 - i) = 2^2 - i^2 = 4 - (-1) = 4 + 1 = 5.$ ✔

Answer $(2 + i)$ and $(2 - i)$

Exercises

Writing About Mathematics

1. The roots of a quadratic equation with rational coefficients are $p \pm \sqrt{q}$. Write the equation in standard form in terms of p and q.

2. Adrien said that if the roots of a quadratic equation are $\frac{1}{2}$ and $\frac{3}{4}$, the equation is $4x^2 - 5x + \frac{3}{2} = 0$. Olivia said that the equation is $8x^2 - 10x + 3 = 0$. Who is correct? Justify your answer.

Developing Skills

In 3–17, without solving each equation, find the sum and product of the roots.

3. $x^2 + x + 1 = 0$

4. $x^2 + 4x + 5 = 0$

5. $2x^2 - 3x - 2 = 0$

6. $5x^2 + 2x - 10 = 0$

7. $3x^2 - 6x + 4 = 0$

8. $-x^2 + 3x + 1 = 0$

9. $8x + 12 = x^2$

10. $4x^2 = 2x + 9$

11. $2x^2 - 8 = 5x$

12. $x^2 - \frac{1}{4} = 0$

13. $x^2 + 1 = 0$

14. $x^2 + 2x = 0$

15. $8x^2 - 9 = -6x$

16. $x^2 + 2x - 3 = 0$

17. $\frac{3x^2 + 3x + 5}{3} = 0$

In 18–27, one of the roots is given. Find the other root.

18. $x^2 + 15x + c = 0; -5$

19. $x^2 - 8x + c = 0; -3$

20. $2x^2 + 3x + c = 0; 1$

21. $-4x^2 - 5x + c = 0; 2$

22. $-6x^2 + 2x + c = 0; \frac{1}{2}$

23. $6x^2 - x + c = 0; -\frac{2}{3}$

24. $m^2 - 4m + n = 0; 3$

25. $z^2 + 2z + k = 0; \frac{3}{4}$

26. $x^2 + bx + 3 = 0; 1$

27. $-7w^2 + bw - 5 = 0; -1$

28. One root of the equation $-3x^2 + 9x + c = 0$ is $\sqrt{2}$.

    a. Find the other root.

    b. Find the value of c.

    c. Explain why the roots of this equation are not conjugates.

29. One root of the equation $-x^2 - 11x + c = 0$ is $\sqrt{3}$.

    a. Find the other root.

    b. Find the value of c.

    c. Explain why the roots of this equation are not conjugates.

In 30–43, write a quadratic equation with integer coefficients for each pair of roots.

30. $2, 5$ **31.** $4, 7$ **32.** $-3, 4$ **33.** $-2, -1$

34. $-3, 3$ **35.** $\frac{1}{2}, \frac{7}{2}$ **36.** $\frac{3}{4}, -\frac{3}{8}$ **37.** $1, 0$

38. $2 + \sqrt{3}, 2 - \sqrt{3}$ **39.** $\frac{1 + \sqrt{5}}{2}, \frac{1 - \sqrt{5}}{2}$ **40.** $\frac{-1 + \sqrt{3}}{3}, \frac{-1 - \sqrt{3}}{3}$

41. $3 + i, 3 - i$ **42.** $\frac{3 - 2i}{2}, \frac{3 + 2i}{2}$ **43.** $\frac{3}{2}i, -\frac{3}{2}i$

Applying Skills

44. One root of a quadratic equation is three more than the other. The sum of the roots is 15. Write the equation.

45. The difference between the roots of a quadratic equation is $4i$. The sum of the roots is 12. Write the equation.

46. Write a quadratic equation for which the sum of the roots is equal to the product of the roots.

47. Use the quadratic formula to prove that the sum of the roots of the equation $ax^2 + bx + c = 0$ is $-\frac{b}{a}$ and the product is $\frac{c}{a}$.

48. A root of $x^2 + bx + c = 0$ is an integer and b and c are integers. Explain why the root must be a factor of c.

5-8 SOLVING HIGHER DEGREE POLYNOMIAL EQUATIONS

The graph of the polynomial function $f(x) = x^3 - 3x^2 - 4x + 12$ is shown at the right. We know that the real roots of a polynomial function are the x-coordinates of the points at which the graph intersects the x-axis, that is, the values of x for which $y = f(x) = 0$. A polynomial function of degree three can have no more than three real roots. The graph appears to intersect the x-axis at $-2, 2$, and 3, but are these exact values of the roots? We know that we can find the rational roots of a polynomial function by factoring. We can also show that these are exact roots by substituting the roots in the equation to show that they satisfy the equation.

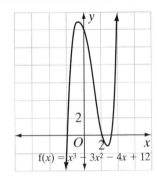
$f(x) = x^3 - 3x^2 - 4x + 12$

Solve by factoring:
Find a common binomial factor by first factoring the first two terms and then factoring the last two terms of the polynomial. Then factor out the common binomial factor.

$$f(x) = x^3 - 3x^2 - 4x + 12$$
$$0 = x^2(x - 3) - 4(x - 3)$$
$$0 = (x - 3)(x^2 - 4)$$

Now factor the difference of two squares.

$$0 = (x - 3)(x^2 - 4)$$
$$0 = (x - 3)(x - 2)(x + 2)$$

$x - 3 = 0$	$x - 2 = 0$	$x + 2 = 0$
$x = 3$	$x = 2$	$x = -2$

Solve by substitution:

$x = 3$: $f(x) = x^3 - 3x^2 - 4x + 12$
$$0 = 3^3 - 3(3)^2 - 4(3) + 12$$
$$0 = 27 - 27 - 12 + 12$$
$$0 = 0$$

$x = 2$: $f(x) = x^3 - 3x^2 - 4x + 12$
$$0 = 2^3 - 3(2)^2 - 4(2) + 12$$
$$0 = 8 - 12 - 8 + 12$$
$$0 = 0$$

$x = -2$: $f(x) = x^3 - 3x^2 - 4x + 12$
$$0 = (-2)^3 - 3(-2)^2 - 4(-2) + 12$$
$$0 = -8 - 12 + 8 + 12$$
$$0 = 0$$

Not every polynomial function has rational roots. The figure at the right shows the graph of the polynomial function $f(x) = x^3 + 2x^2 - 2x$. From the graph we see that this function appears to have a root between -3 and -2, a root at 0, and a root between 0 and 1. Can we verify that 0 is a root and can we find the exact values of the other two roots?

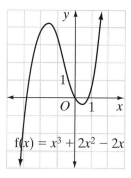

$f(x) = x^3 + 2x^2 - 2x$

(1) Let $f(x) = 0$. Factor the polynomial if possible. We can factor out the common factor x:

$$f(x) = x^3 + 2x^2 - 2x$$
$$0 = x^3 + 2x^2 - 2x$$
$$0 = x(x^2 + 2x - 2)$$

(2) The factor $x^2 + 2x - 2$ cannot be factored in the set of integers. Set each factor equal to 0. Use the quadratic formula to solve the quadratic equation:

$x = 0$ | $x^2 + 2x - 2 = 0$
$$x = \frac{-2 \pm \sqrt{2^2 - 4(1)(-2)}}{2(1)}$$
$$= \frac{-2 \pm \sqrt{12}}{2}$$
$$= \frac{-2 \pm 2\sqrt{3}}{2}$$
$$= -1 \pm \sqrt{3}$$

(3) The roots of $f(x) = x^3 + 2x^2 - 2x$ are 0, $-1 + \sqrt{3}$, and $-1 - \sqrt{3}$:

We can use a calculator to approximate the values of $-1 + \sqrt{3}$ and $-1 - \sqrt{3}$:

$$-1 - \sqrt{3} \approx -2.7 \text{ and } -1 + \sqrt{3} \approx 0.7$$

These are values between -3 and -2 and between 0 and 1 as seen on the graph on the previous page. The function has three real roots. One of the roots, 0, is rational and the other two roots, $-1 + \sqrt{3}$ and $-1 - \sqrt{3}$, are irrational roots.

Note that in factoring a polynomial of degree three or greater, we look for the following factors:

1. a common monomial factor.

2. the binomial factors of a trinomial.

3. a common binomial factor.

4. factors of the difference of two squares.

EXAMPLE I

Find three roots of the function $f(x) = x^3 - 2x^2 + 9x - 18$.

Solution   *How to Proceed*

(1) Try to factor the polynomial. There is no common monomial factor. Look for a common binomial factor:

$$x^3 - 2x^2 + 9x - 18$$
$$= x^2(x - 2) + 9(x - 2)$$
$$= (x - 2)(x^2 + 9)$$

(2) Let $f(x) = 0$:

$$f(x) = x^3 - 2x^2 + 9x - 18$$

(3) Set each factor equal to 0. and solve each equation for x:

$$0 = (x - 2)(x^2 + 9)$$

$$
\begin{array}{c|l}
x - 2 = 0 & x^2 + 9 = 0 \\
x = 2 & x^2 = -9 \\
 & x = \pm\sqrt{-9} \\
 & x = \pm\sqrt{9}\sqrt{-1} \\
 & x = \pm 3i
\end{array}
$$

Answer The roots are $2, 3i,$ and $-3i$.

EXAMPLE 2

Is 2 is a root of the function $f(x) = x^4 - 5x^2 + x + 2$?

Solution If 2 is a root of $f(x) = x^4 - 5x^2 + x + 2$, then $f(2) = 0$.

$$f(2) = 2^4 - 5(2)^2 + 2 + 2 = 16 - 20 + 2 + 2 = 0 ✔$$

Answer 2 is a root of $f(x) = x^4 - 5x^2 + x + 2$.

Exercises

Writing About Mathematics

1. Sharon said that if f(x) is a polynomial function and f(a) = 0, then a is a root of the function. Do you agree with Sharon? Explain why or why not.

2. Jordan said that if the roots of a polynomial function f(x) are r_1, r_2, and r_3, then the roots of g(x) = f(x − a) are $r_1 + a, r_2 + a$, and $r_3 + a$. Do you agree with Jordan? Explain why or why not.

Developing Skills

In 3–18, find all roots of each given function by factoring or by using the quadratic formula.

3. $f(x) = x^3 + 7x^2 + 10x$

4. $f(x) = 2x^3 + 2x^2 − 4x$

5. $f(x) = x^3 + 3x^2 + 4x + 12$

6. $f(x) = x^3 − x^2 + 3x − 3$

7. $f(x) = 2x^3 − 3x^2 − 2x + 3$

8. $f(x) = −2x^3 + 6x^2 − x + 3$

9. $f(x) = x^4 − 5x^2 + 4$

10. $f(x) = x^4 + 5x^2 + 4$

11. $f(x) = x^4 − 81$

12. $f(x) = 16x^4 − 1$

13. $f(x) = x^4 − 10x^2 + 9$

14. $f(x) = x^5 − x^4 − 2x^3$

15. $f(x) = x^3 − 18x$

16. $f(x) = (2x^2 + x − 1)(x^2 − 3x + 4)$

17. $f(x) = (x^2 − 1)(3x^2 + 2x + 1)$

18. $f(x) = x^3 + 2x^2 − x − 2$

In 19–28: **a.** Find f(a) for each given function. **b.** Is a a root of the function?

19. $f(x) = x^4 − 1$ and $a = 1$

20. $f(x) = x^3 + 4x$, and $a = −2$

21. $f(x) = 5x^2 + 4x + 1$ and $a = −1$

22. $f(x) = −x^3 + x − 24$ and $a = −3$

23. $f(x) = x^4 + x^2 + x + 1$ and $a = 0$

24. $f(x) = 2x^3 + 3x^2 − 1$ and $a = \frac{1}{2}$

25. $f(x) = x^3 − 3x^2 + x − 3$ and $a = i$

26. $f(x) = x^4 − 2x^2 + x$ and $a = \sqrt{3}$

27. $f(x) = x^3 − 2x + 3$ and $a = 2 + i$

28. $f(x) = −5x^3 + 5x^2 + 2x + 3$ and $a = \frac{3}{2}i$

Applying Skills

29. a. Verify by multiplication that $(x − 1)(x^2 + x + 1) = x^3 − 1$.

    b. Use the factors of $x^3 − 1$ to find the three roots of f(x) = $x^3 − 1$.

    c. If $x^3 − 1 = 0$, then $x^3 = 1$ and $x = \sqrt[3]{1}$. Use the answer to part **b** to write the three cube roots of 1. Explain your reasoning.

    d. Verify that each of the two imaginary roots of f(x) = $x^3 − 1$ is a cube root of 1.

30. a. Verify by multiplication that $(x + 1)(x^2 - x + 1) = x^3 + 1$.

  b. Use the factors of $x^3 + 1$ to find the three roots of $f(x) = x^3 + 1$.

  c. If $x^3 + 1 = 0$, then $x^3 = -1$ and $x = \sqrt[3]{-1}$. Use the answer to part **b** to write the three cube roots of -1. Explain your reasoning.

  d. Verify that each of the two imaginary roots of $f(x) = x^3 + 1$ is a cube root of -1.

31. Let $f(x) = x^3 + 3x^2 - 2x - 6$ and $g(x) = 2f(x) = 2x^3 + 6x - 4x - 12$ be two cubic polynomial functions.

  a. How does the graph of $f(x)$ compare with the graph of $g(x)$?

  b. How do the roots of $f(x)$ compare with the roots of $g(x)$?

  c. In general, if $p(x) = aq(x)$ and $a > 0$, how does the graph of $p(x)$ compare with the graph of $q(x)$?

  d. How do the roots of $p(x)$ compare with the roots of $q(x)$?

Hands-On Activity

The following activity will allow you to evaluate a function for any constant and test possible roots of a function. This process is called **synthetic substitution**. Let $f(x) = x^4 - 3x^3 + x^2 - 2x + 3$. Find $f(2)$:

(1) List the coefficients of the terms of the function and the constant to be tested as shown:

$$1 \quad -3 \quad 1 \quad -2 \quad 3 \quad \underline{|2}$$

(2) Bring down the first coefficient as shown and then multiply it by the constant 2 and write the product under the second coefficient:

$$\begin{array}{ccccc} 1 & -3 & 1 & -2 & 3 \quad \underline{|2} \\ & 2 \\ \hline 1 \end{array}$$

(3) Add the second coefficient and the product found in step 2:

$$\begin{array}{ccccc} 1 & -3 & 1 & -2 & 3 \quad \underline{|2} \\ & 2 \\ \hline 1 & -1 \end{array}$$

(4) Multiply this sum by the constant 2 and place the product under the next coefficient:

$$\begin{array}{ccccc} 1 & -3 & 1 & -2 & 3 \quad \underline{|2} \\ & 2 & -2 \\ \hline 1 & -1 \end{array}$$

(5) Add the coefficient and the product found in step 4:

$$\begin{array}{ccccc} 1 & -3 & 1 & -2 & 3 \quad \underline{|2} \\ & 2 & -2 \\ \hline 1 & -1 & -1 \end{array}$$

(6) Repeat steps 4 and 5 until a final sum is found using the last coefficient:

$$\begin{array}{ccccc} 1 & -3 & 1 & -2 & 3 \quad \underline{|2} \\ & 2 & -2 & -2 & -8 \\ \hline 1 & -1 & -1 & -4 & -5 \end{array}$$

The final number is -5. Therefore, $f(2) = -5$. Use a calculator to verify that this is true. Repeat the steps for $f(1)$. The final number should be 0. This means that $f(1) = 0$ and that 1 is a root of the function.

Repeat the process for the following function, using the numbers $-3, -2, -1, 1, 2,$ and 3 for each function. Find the roots of the function.

a. $f(x) = x^3 - 2x^2 - x + 2$

b. $f(x) = x^3 - 3x^2 - 4x + 12$

c. $f(x) = x^3 - 7x - 6 = x^3 + 0x^2 - 7x - 6$ (Include 0 in the list of coefficients.)

d. $f(x) = x^4 - 5x^2 + 4 = x^4 + 0x^3 - 5x^2 + 0x + 4$

5-9 SOLUTIONS OF SYSTEMS OF EQUATIONS AND INEQUALITIES

A system of equations is a set of two or more equations. The system is **consistent** if there exists at least one common solution in the set of real numbers. The solution set of a consistent system of equations can be found using a graphic method or an algebraic method.

Solving Quadratic-Linear Systems

A system that consists of a quadratic function whose graph is a parabola and a linear function whose graph is a straight line is a **quadratic-linear system** in two variables. In the set of real numbers, a quadratic-linear system may have two solutions, one solution, or no solutions. Each solution can be written as the coordinates of the points of intersection of the graph of the parabola and the graph of the line.

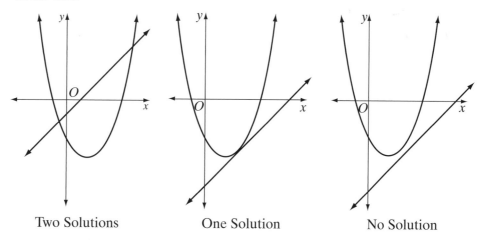

Two Solutions   One Solution   No Solution

Recall that for the function $y = ax^2 + bx + c$, the x-coordinate of the turning point and the equation of the axis of symmetry is $x = \frac{-b}{2a}$. When a is positive, the parabola opens upward and has a minimum value of y as shown in the figures above. When a is negative, the parabola opens downward and has a maximum value of y as shown in the following example.

We can use either a graphic method or an algebraic method to find the common solutions of $y = -x^2 + 6x - 3$ and $y - x = 1$.

Graphic Method

(1) Graph the parabola on the calculator. Choose a reasonable viewing window and enter $-x^2 + 6x - 3$ as Y_1.

ENTER: Y= CLEAR (-) X,T,Θ,n ^

2 + 6 X,T,Θ,n − 3 GRAPH

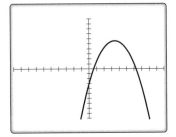

(2) On the same set of axes, draw the graph of $y - x = 1$. Write the equation in slope-intercept form: $y = x + 1$ and enter it as Y_2.

ENTER: Y= ▼ CLEAR X,T,Θ,n +

1 GRAPH

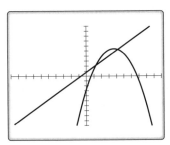

(3) The common solutions are the coordinates of the points of intersection of the graphs. We can find the intersection points by using the intersect function. Press 2nd CALC

5 ENTER ENTER to select both curves. When the calculator asks you for a guess, move the cursor near one of the intersection points using the arrow keys and then press ENTER . Repeat this process to find the other intersection point.

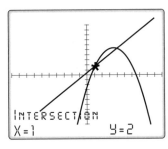

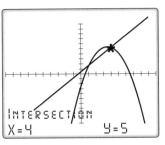

The common solutions are $(1, 2)$ and $(4, 5)$.

Note: The graphic method only gives exact values when the solutions are rational numbers. If the solutions are irrational, then the calculator will give approximations.

Algebraic Method

(1) Solve the linear equation for one of the variables in terms of the other. We will solve for y in terms of x:

$$y - x = 1$$
$$y = x + 1$$

(2) Replace y in the quadratic equation by its value in terms of x from the linear equation:

$$y = -x^2 + 6x - 3$$
$$x + 1 = -x^2 + 6x - 3$$

(3) Write the equation in standard form:

$$x^2 - 5x + 4 = 0$$

(4) Factor the trinomial:

$$(x - 4)(x - 1) = 0$$

(5) Set each factor equal to 0 and solve for x:

$$x - 4 = 0 \mid x - 1 = 0$$
$$x = 4 \qquad x = 1$$

(6) Use the simplest equation in x and y to find the corresponding value of y for each value of x:

$$y = x + 1 \mid y = x + 1$$
$$y = 4 + 1 \mid y = 1 + 1$$
$$y = 5 \qquad y = 2$$

The common solutions are $x = 4, y = 5$ and $x = 1, y = 2$.

EXAMPLE 1

For the given system of equations, compare the solution that can be read from the graph with the algebraic solution.

$$y = x^2 - 4$$
$$y = -2x$$

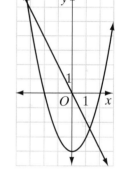

Solution   From the graph, the solutions appear to be $(-3.2, 6.5)$ and $(1.2, -2.5)$.

We can find exact values by using an algebraic solution.

(1) Substitute the value of y from the second equation in the first equation:

$$-2x = x^2 - 4$$

(2) Write the equation in standard form:

$$0 = x^2 + 2x - 4$$

(3) The polynomial $x^2 + 2x - 4$ cannot be factored in the set of integers. Use the quadratic formula with $a = 1, b = 2, c = -4$:

$$x = \frac{-b \pm \sqrt{b^2 - 4ac}}{2a}$$
$$= \frac{-2 \pm \sqrt{2^2 - 4(1)(-4)}}{2(1)}$$
$$= \frac{-2 \pm \sqrt{4 + 16}}{2}$$
$$= -1 \pm \sqrt{5}$$

(4) Use the linear equation to find y for each value of x:

$$y = -2x \mid y = -2x$$
$$y = -2(-1 - \sqrt{5}) \mid y = -2(-1 + \sqrt{5})$$
$$y = 2 + 2\sqrt{5} \mid y = 2 - 2\sqrt{5}$$

The exact solutions are $\left(-1 - \sqrt{5}, 2 + 2\sqrt{5}\right)$ and $\left(-1 + \sqrt{5}, 2 - 2\sqrt{5}\right)$. The rational approximations of these solutions, rounded to the nearest tenth, are the solutions that we found graphically.

EXAMPLE 2

Use an algebraic method to find the solution set of the equations:

$$y = 2x^2 - 4x - 5$$
$$3x - y = 1$$

Solution (1) Solve the linear equation for y:

$$3x - y = 1$$
$$-y = -3x + 1$$
$$y = 3x - 1$$

(2) Substitute for y in terms of x in the quadratic equation:

$$y = 2x^2 - 4x - 5$$
$$3x - 1 = 2x^2 - 4x - 5$$

(3) Write the equation in standard form and solve for x:

$$0 = 2x^2 - 7x - 4$$
$$0 = (2x + 1)(x - 4)$$

$2x + 1 = 0$	$x - 4 = 0$
$2x = -1$	$x = 4$
$x = -\frac{1}{2}$	

(4) Find the corresponding values of y using the linear equation solved for y:

$y = 3x - 1$	$y = 3x - 1$
$y = 3\left(-\frac{1}{2}\right) - 1$	$y = 3(4) - 1$
$y = -\frac{3}{2} - \frac{2}{2}$	$y = 12 - 1$
$y = -\frac{5}{2}$	$y = 11$

Answer $\left(-\frac{1}{2}, -\frac{5}{2}\right)$ and $(4, 11)$

EXAMPLE 3

The graph shows the circle whose equation is $(x - 2)^2 + (y + 1)^2 = 10$ and the line whose equation is $y = x - 1$.

a. Read the common solutions from the graph.

b. Check both solutions in both equations.

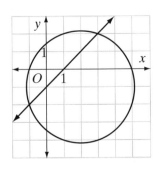

Solution **a.** The common solutions are $x = -1, y = -2$ and $x = 3, y = 2$.

b. *Check* $x = -1, y = -2$:

$$(x - 2)^2 + (y + 1)^2 = 10 \qquad\qquad y = x - 1$$
$$(-1 - 2)^2 + (-2 + 1)^2 \overset{?}{=} 10 \qquad -2 \overset{?}{=} -1 - 1$$
$$(-3)^2 + (-1)^2 \overset{?}{=} 10 \qquad\qquad -2 = -2 ✔$$
$$10 = 10 ✔$$

Check $x = 3, y = 2$:

$$(x - 2)^2 + (y + 1)^2 = 10 \qquad\qquad y = x - 1$$
$$(3 - 2)^2 + (2 + 1)^2 \overset{?}{=} 10 \qquad\quad 2 = 3 - 1$$
$$(1)^2 + (3)^2 \overset{?}{=} 10 \qquad\qquad 2 = 2 ✔$$
$$10 = 10 ✔$$

Solving Quadratic Inequalities

You already know how to graph a linear inequality in two variables. The process for graphing a quadratic inequality in two variables is similar.

EXAMPLE 3

Graph the inequality $y + 3 \geq x^2 + 4x$.

Solution

How to Proceed

(1) Solve the inequality for y:

$$y + 3 \geq x^2 + 4x$$
$$y \geq x^2 + 4x - 3$$

(2) Graph the corresponding quadratic function $y = x^2 + 4x - 3$. Since the inequality is $\geq$, use a solid curve to indicate that the parabola is part of the solution:

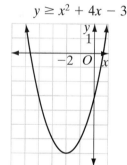

(3) The values of $y > x^2 + 4x - 3$ are those in the region above the parabola. Shade this region:

(4) *Check:* Use any convenient test point to verify that the correct region has been shaded. Try $(0, 0)$:

$$0 \overset{?}{\geq} 0^2 + 4(0) - 3$$
$$0 \geq -3 ✔$$

The test point satisfies the equation. The shaded region is the set of points whose coordinates make the inequality true.

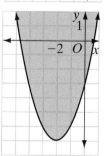

In general:

▶ **The solution set of the inequality $y > ax^2 + bx + c$ is the set of coordinates of the points above the graph of $y = ax^2 + bx + c$, which is drawn as a dotted line to show that it is not part of the solution set.**

▶ **The solution set of the inequality $y \geq ax^2 + bx + c$ is the set of coordinates of the points above the graph of $y = ax^2 + bx + c$, which is drawn as a solid line to show that it is part of the solution set.**

▶ **The solution set of the inequality $y < ax^2 + bx + c$ is the set of coordinates of the points below the graph of $y = ax^2 + bx + c$, which is drawn as a dotted line to show that it is not part of the solution set.**

▶ **The solution set of the inequality $y \leq ax^2 + bx + c$ is the set of coordinates of the points below the graph of $y = ax^2 + bx + c$, which is drawn as a solid line to show that it is part of the solution set.**

The following figures demonstrate some of these graphing rules:

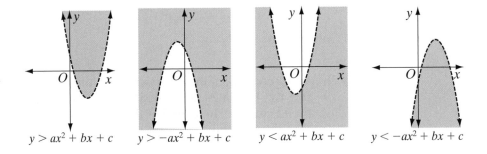

$y > ax^2 + bx + c$ $\quad$ $y > -ax^2 + bx + c$ $\quad$ $y < ax^2 + bx + c$ $\quad$ $y < -ax^2 + bx + c$

Graphs can also be used to find the solution of a quadratic inequality in one variable. At the right is the graph of the equation $y = x^2 + 4x - 5$. The shaded region is the graph of $y > x^2 + 4x - 5$. Let $y = 0$. The graph of $y = 0$ is the x-axis. Therefore, the solutions of $0 > x^2 + 4x - 5$ are the x-coordinates of the points common to the graph of $y > x^2 + 4x - 5$ and the x-axis. From the graph on the right, we can see that the solution of $0 > x^2 + 4x - 5$ or $x^2 + 4x - 5 < 0$ is the interval $-5 < x < 1$.

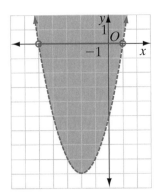

EXAMPLE 4

Solve the inequality $-x^2 + 3x + 4 \le 0$.

Solution

How to Proceed

(1) Let $y = 0$. The inequality
$-x^2 + 3x + 4 \le 0$ is equivalent to the
intersection of $-x^2 + 3x + 4 \le y$ and
$y = 0$. Graph the quadratic function
$y = -x^2 + 3x + 4$. Note that the graph
intersects the x-axis at -1 and 4. Shade
the area above the curve that is the graph
of $y > -x^2 + 3x + 4$.

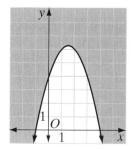

(2) The graph of $y = 0$ is the x-axis. The
solution of $-x^2 + 3x + 4 \le 0$ is
the set of x-coordinates of the
points common to the graph of
$y \ge -x^2 - 3x + 4$ and the x-axis.

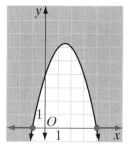

$x \le -1$ or $x \ge 4$

Answer $x \le -1$ or $x \ge 4$

Two Variable Inequalities and the Graphing Calculator

The graphing calculator can also be used to
graph quadratic inequalities in two variables.
For instance, to graph the inequality of
Example 3, first rewrite the inequality with the
x variable on the *right*: $y \ge x^2 + 4x - 3$. Enter
the quadratic function $x^2 + 4x - 3$ into Y_1. Now
move the cursor over the \ to the left of Y_1 using
the arrow keys. Since the inequality is $\ge$, press
[ENTER] until the cursor turns into ◥. Press
[GRAPH] to graph the inequality.

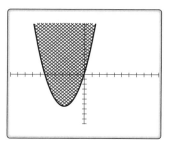

Note: For $\ge$ and $>$ inequalities, use ◥. For $\le$ and $<$ inequalities, use ◤.

Exercises

Writing About Mathematics

1. Explain the relationship between the solutions of $y > ax^2 + bx + c$ and the solutions of $0 > ax^2 + bx + c$.

2. Explain why the equations $y = x^2 + 2$ and $y = -2$ have no common solution in the set of real numbers.

Developing Skills

In 3–8, determine each common solution from the graph.

3.

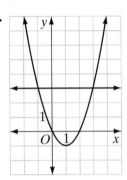

4.

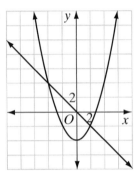

5.

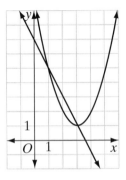

6.

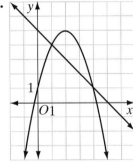

7.

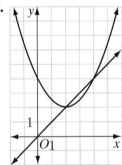

8.

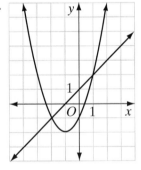

In 9–17, graph each system and determine the common solution from the graph.

9. $y = x^2 - 2x - 1$
$y = x + 3$

10. $y = x^2 + 2x$
$y = 2x + 1$

11. $-x^2 + 4x - y - 2 = 0$
$x + y = 4$

12. $y = -x^2 + 6x - 1$
$y = x + 3$

13. $y = 2x^2 + 2x + 3$
$y - x = 3$

14. $y = 2x^2 - 6x + 5$
$y = x + 2$

15. $x^2 - 4x - y + 4 = 0$
$y = \frac{4x + 7}{4}$

16. $\frac{x - 1}{y} = \frac{6}{x + 12}$
$y = x + 2$

17. $\frac{y}{x} = \frac{x + 7}{5}$
$y = 2x$

In 18–35, find each common solution algebraically. Express irrational roots in simplest radical form.

18. $y = x^2 - 2x$
   $y = 3x$

19. $y = x^2 + 4x$
   $2x - y = 1$

20. $y = x^2 + 2x + 3$
   $x + y = 1$

21. $y = x^2 - 8x + 6$
   $2x - y = 10$

22. $2x^2 - 3x + 3 - y = 0$
   $y - 2x = 1$

23. $y = x^2 - 4x + 5$
   $2y = x + 6$

24. $y = 2x^2 - 6x + 7$
   $y = x + 4$

25. $x^2 + x - y = 7$
   $\frac{1}{2}x = y + 2$

26. $y = 4x^2 - 6x - 10$
   $y = 25 - 2x$

27. $y = x^2 - 2x + 1$
   $y = -\frac{9x - 35}{2}$

28. $y = x^2 + 4x + 4$
   $y = 4x + 6$

29. $2x^2 + x - y + 1 = 0$
   $x - y + 7 = 0$

30. $y = x^2 + 2$
   $2x - y = -3$

31. $y = 2x^2 + 3x + 4$
   $y = 11x + 6$

32. $5x^2 + 3x - y = -1$
   $10x + 1 = y$

33. $x^2 + y^2 = 24$
   $x + y = 8$

34. $x^2 + y^2 = 20$
   $3x - y = 10$

35. $x^2 + y^2 = 16$
   $y = 2x$

36. Write the equations of the graphs shown in Exercise 4 and solve the system algebraically. ($a = 1$)

37. Write the equations of the graphs shown in Exercise 6 and solve the system algebraically. ($a = -1$)

In 38–43, match the inequality with its graph. The graphs are labeled (1) to (6).

(1)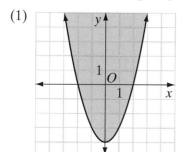

(2)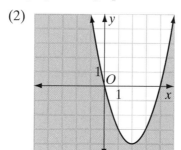

(3)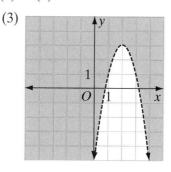

(4)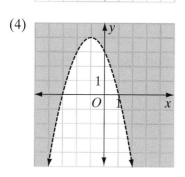

(5)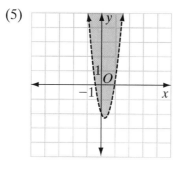

(6)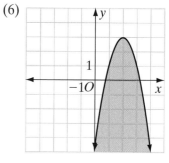

38. $x^2 \leq 4 + y$

39. $x^2 > -2x + 3 - y$

40. $y + 4x \leq x^2$

41. $y > -2(x - 2)^2 + 3$

42. $y \leq -2(x - 2)^2 + 3$

43. $4x^2 - 2x - y - 2 < 0$

In 44–51: **a.** Graph the given inequality. **b.** Determine if the given point is in the solution set.

44. $x^2 \le 2x + y$; $(5, 4)$

45. $x^2 + 5 \ge y$; $(-1, 3)$

46. $x(x - 13) > y$; $(-5, 130)$

47. $x^2 - 18 \ge y + 3x$; $(0, -18)$

48. $2(x - 3)^2 + 3 < y$; $\left(\frac{1}{2}, 6\right)$

49. $-4(x + 2)^2 - 5 \le y$; $\left(1, \frac{2}{3}\right)$

50. $4x^2 - 4x < 3 + y$; $\left(0, \frac{5}{3}\right)$

51. $6x^2 + x \le 2 - y$; $(10, 0)$

Applying Skills

52. The area of a rectangular rug is 48 square feet. The length of the rug is 2 feet longer than the width. What are the dimensions of the rug?

53. The difference in the lengths of the sides of two squares is 1 meter. The difference in the areas of the squares is 13 square meters. What are the lengths of the sides of the squares?

54. The perimeter of a rectangle is 24 feet. The area of the rectangle is 32 square feet. Find the dimensions of the rectangle.

55. The endpoints of a diameter of a circle are $(0, 0)$ and $(8, 4)$.

a. Write an equation of the circle and draw its graph.

b. On the same set of axes, draw the graph of $x + y = 4$.

c. Find the common solutions of the circle and the line.

d. Check the solutions in both equations.

56. a. On the same set of axes, sketch the graphs of $y = x^2 - 4x + 5$ and $y = 2x + 2$.

b. Does the system of equations $y = x^2 - 4x + 5$ and $y = 2x + 2$ have a common solution in the set of real numbers? Justify your answer.

c. Find the solution set.

57. a. On the same set of axes, sketch the graphs of $y = x^2 + 5$ and $y = 2x$.

b. Does the system of equations $y = x^2 + 5$ and $y = 2x$ have a common solution in the set of real numbers? Justify your answer.

c. Does the system of equations $y = x^2 + 5$ and $y = 2x$ have a common solution in the set of complex numbers? If so, find the solution.

58. The graphs of the given equations have three points of intersection. Use an algebraic method to find the three solutions of this system of equations:

$$y = x^3 - 2x + 1$$
$$y = 2x + 1$$

59. A soccer ball is kicked upward from ground level with an initial velocity of 52 feet per second. The equation $h(t) = -16t^2 + 52t$ gives the ball's height in feet after t seconds. To the nearest tenth of a second, during what period of time was the height of the ball at least 20 feet?

60. A square piece of cardboard measuring x inches by x inches is to be used to form an open box by cutting off 2-inch squares, as shown in the figure.

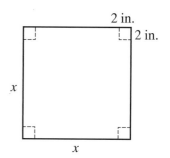

  a. Express the length of the sides of the base of the box in terms of x.

  b. Write a function $V(x)$ that represents the volume.

  c. If the volume of the box must be greater than 128 cubic inches, what are the minimum dimensions of the square cardboard that can be used?

61. The profit function, in thousands of dollars, for a company that makes graphing calculators is $P(x) = -5x^2 + 5{,}400x - 106{,}000$ where x is the number of calculators sold in the millions.

  a. Graph the profit function $P(x)$.

  b. How many calculators must the company sell in order to make a profit?

CHAPTER SUMMARY

The real **roots** of a polynomial function are the x-coordinates of the points where the graph of the function intersects the x-axis.

Any quadratic equation with rational roots can be solved by factoring over the set of integers.

Any quadratic equation can be solved by **completing the square**:

1. Write an equivalent equation with only the terms in x on one side.

2. Add the square of one-half the coefficient of x to both sides of the equation.

3. Take the square root of both sides of the equation.

4. Solve for x.

Any quadratic equation can be solved by using the **quadratic formula**:

$$x = \frac{-b \pm \sqrt{b^2 - 4ac}}{2a}$$

When a, b, and c are rational numbers, the value of the **discriminant**, $b^2 - 4ac$, determines the nature of the roots and the number of x-intercepts of the quadratic function.

When the discriminant $b^2 - 4ac$ is:	The roots of the equation are:	The number of x-intercepts of the function is:
> 0 and a perfect square	real, rational, and unequal	2
> 0 and not a perfect square	real, irrational, and unequal	2
$= 0$	real, rational, and equal	1
< 0	imaginary numbers	0

If r_1 and r_2 are the roots of a quadratic equation, the equation is

$$x^2 - (r_1 + r_2)x + r_1r_2 = 0$$

For the quadratic equation $ax^2 + bx + c = 0$:

- the sum of the roots is $-\frac{b}{a}$.
- the product of the roots is $\frac{c}{a}$.

A number of the form $a\sqrt{-1} = ai$ is a **pure imaginary number** when a is a non-zero real number.

For any integer n:

$$i^{4n} = 1 \qquad i^{4n+1} = i \qquad i^{4n+2} = -1 \qquad i^{4n+3} = -i$$

A **complex number** is a number of the form $a + bi$ where a and b are real numbers and $i = \sqrt{-1}$. A complex number that is not a real number is an imaginary number.

The identity element for addition is the real number $(0 + 0i)$ or 0.
The additive inverse of $(a + bi)$ is $(-a - bi)$.
The identity for multiplication is the real number $(1 + 0i)$ or 1.
The multiplicative inverse of $a + bi$ is $\frac{1}{a + bi}$ or $\frac{a}{a^2 + b^2} - \frac{b}{a^2 + b^2}i$.
The **complex conjugate** of $a + bi$ is $a - bi$.
A polynomial of degree three or greater can be factored using:

- a common monomial factor.
- the binomial factors of a trinomial.
- a common binomial factor.
- factors of the difference of two squares.

A system of equations is a set of two or more equations. The system is **consistent** if there exists at least one common solution in the set of real numbers. The solution set of a consistent system of equations can be found using a graphic method or an algebraic method.

The graph of a quadratic function can be used to estimate the solution of a quadratic inequality.

The solution set of the inequality $y > ax^2 + bx + c$ is the set of coordinates of the points above the graph of $y = ax^2 + bx + c$.

The solution set of the inequality $y < ax^2 + bx + c$ is the set of coordinates of the points below the graph of $y = ax^2 + bx + c$.

The solution set of $0 > ax^2 + bx + c$ is the set x-coordinates of points common to the graph of $y > ax^2 + bx + c$ and the x-axis.

VOCABULARY

5-1 Completing the square

5-2 Quadratic formula

5-3 Discriminant • Double root

5-4 i • Set of imaginary numbers • Pure imaginary number • Complex number

5-5 Complex conjugate

5-8 Synthetic substitution

5-9 Consistent • Quadratic-linear system

REVIEW EXERCISES

In 1–8, write each number in simplest form in terms of i.

1. $\sqrt{-1}$ **2.** $\sqrt{-16}$ **3.** $\sqrt{-9}$ **4.** $\sqrt{-12}$

5. $\sqrt{-4} + \sqrt{-25}$ **6.** $\sqrt{-18} + \sqrt{-32}$ **7.** $\sqrt{-64}\left(\sqrt{-\frac{1}{4}}\right)$ **8.** $\frac{\sqrt{-128}}{\sqrt{-12}}$

In 9–28, perform each indicated operation and express the result in $a + bi$ form.

9. $(2 + 3i) + (5 - 4i)$ **10.** $(1 + 2i) + (-1 + i)$

11. $(2 + 7i) + (2 - 7i)$ **12.** $(3 - 4i) + (-3 + 4i)$

13. $(1 + 2i) - (5 + 4i)$ **14.** $(8 - 6i) - (-2 - 2i)$

15. $(7 - 5i) - (7 + 5i)$ **16.** $(-2 + 3i) - (-2 - 3i)$

17. $(1 + 3i)(5 - 4i)$ **18.** $(2 + 6i)(3 - i)$

19. $(9 - i)(9 - i)$ **20.** $3i(4 - 2i)$

21. $\left(\frac{1}{2} - i\right)(2 + i)$ **22.** $\left(\frac{1}{5} - \frac{2}{5}i\right)(1 + 2i)$

23. $\frac{2 + 2i}{2i}$ **24.** $\frac{2 + 2i}{1 + i}$

25. $\frac{2 + 3i}{2 - 3i}$ **26.** $\frac{1 - i}{3 - i}$

27. $(1 + 4i)^2$ **28.** $(3 - 2i)^2$

In 29–36, find the roots of each equation by completing the square.

29. $x^2 - 7x + 1 = 0$ **30.** $x^2 - x - 12 = 0$

31. $x^2 + 4x + 5 = 0$ **32.** $x^2 - 6x - 10 = 0$

33. $x^2 - 6x + 10 = 0$ **34.** $\frac{x}{12} = \frac{5}{2x + 7}$

35. $3x^2 - 6x + 6 = 0$ **36.** $2x^2 + 3x - 2 = 0$

In 37–44, find the roots of each equation using the quadratic formula. Express irrational roots in simplest radical form.

37. $x^2 = x + 1$ **38.** $2x^2 - 2x = 1$

39. $5x^2 = 2x + 1$ **40.** $4x^2 - 12x + 13 = 0$

41. $x^3 - 2x^2 - 16x + 32 = 0$

42. $x^4 - 5x^2 + 4 = 0$

43. $0.1x^2 + 2x + 50 = 0$

44. $-3x^2 + \frac{1}{2}x + \frac{5}{3} = 0$

In 45–48, find the roots of the equation by any convenient method.

45. $x^3 - 6x^2 + 4x = 0$

46. $x^4 - 10x^2 + 9 = 0$

47. $3x^3 + 12x^2 - x - 4 = 0$

48. $(3x + 5)(x^2 + 5x - 6) = 0$

In 49–52, without graphing the parabola, describe the translation, reflection, and/or scaling that must be applied to $y = x^2$ to obtain the graph of each given function.

49. $y = x^2 + 2x + 2$

50. $y = x^2 + 3x + 10$

51. $y = 4x^2 - 6x + 3$

52. $y = -2x^2 + 5x - 10$

53. The graph on the right is the parabola $y = ax^2 + bx + c$ with a, b, and c rational numbers.

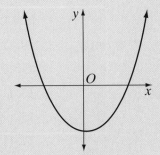

a. Describe the roots of the equation $0 = ax^2 + bx + c$ if the discriminant is 49.

b. Describe the roots of the equation $0 = ax^2 + bx + c$ if the discriminant is 32.

c. Can the discriminant be 0 or negative?

54. Let $h(x) = -20.5x^2 + 300.1x - 500$. Use the discriminant to determine if there exist real numbers, x, such that $h(x) = 325$.

In 55–60, find each common solution graphically.

55. $y = x^2 - 4$
$x + y = 2$

56. $y = -x^2 + 6x - 5$
$y = x - 1$

57. $y = 2x^2 - 8x$
$y = 2x$

58. $x^2 + y^2 = 25$
$x + y = 7$

59. $(x - 3)^2 + y^2 = 9$
$x + y = 6$

60. $(x + 1)^2 + (y - 2)^2 = 4$
$y = 3 - x$

In 61–70, find each common solution algebraically. Express irrational roots in simplest radical form.

61. $y = x^2 - 2x - 2$
$x + y = 4$

62. $y = x^2 - 8$
$2x - y = 0$

63. $x^2 + y = 3x$
$x - y = 3$

64. $\frac{x}{9} = \frac{8}{y}$
$y = 36 - 4x$

65. $x^2 = y + 5x$
$2x - y = 5$

66. $y = -x^2 + 2$
$x = y - 1$

67. $3x^2 - 2x - y = 0$
    $3x + 5 = 2y$

68. $y = x^2 - 3x + 3$
    $y = -x + 1$

69. $y = x^2 + 6x + 4$
    $y = 2x - 1$

70. $x^2 + y^2 = 5$
    $x - y - 1 = 0$

In 71–75, write each quadratic equation that has the given roots.

71. -3 and 5

72. $\frac{1}{2}$ and -4

73. $\sqrt{5}$ and $-\sqrt{5}$

74. $5 + 3\sqrt{2}$ and $5 - 3\sqrt{2}$

75. $6 + 2i$ and $6 - 2i$

76. For what value of c does the equation $x^2 - 6x + c = 0$ have equal roots?

77. For what values of b does $2x^2 + bx + 2 = 0$ have imaginary roots?

78. For what values of c does $x^2 - 3x + c = 0$ have real roots?

In 79 and 80: **a.** Graph the given inequality. **b.** Determine if the given point is in the solution set.

79. $y - (x + 2)^2 + 5 \geq 0; (2, 3)$

80. $-2x^2 + 3x < y; (0, \sqrt{5})$

81. The perimeter of a rectangle is 40 meters and the area is 97 square meters. Find the dimensions of the rectangle to the nearest tenth.

82. The manager of a theater is trying to determine the price to charge for tickets to a show. If the price is too low, there will not be enough money to cover expenses. If the price is too high, they may not get enough playgoers. The manager estimates that the profit, P, in hundreds of dollars per show, can be represented by $P = -(t - 12)^2 + 100$ where t is the price of a ticket in dollars. The manager is considering charging between \$0 and \$24.

   a. Graph the profit function for the given range of prices.

   b. The theater breaks even when profit is zero. For what ticket prices does the theater break even?

   c. What price results in maximum profit? What is the maximum profit?

83. Pam wants to make a scarf that is 20 inches longer than it is wide. She wants the area of the scarf to be more than 300 square inches. Determine the possible dimensions of the scarf.

Exploration

1. Find the three binomial factors of $x^3 - x^2 - 4x + 4$ by factoring.

2. One of these factors is $(x - 2)$. Write $x^3 - x^2 - 4x + 4$ as the product of $(x - 2)$ and the trinomial that is the product of the other two factors.

3. Use synthetic substitution (see the Hands-On Activity on page 228) to show that 2 is a root of $f(x) = x^3 - x^2 - 4x + 4$.

4. Compare the coefficients of the trinomial from step 2 with the first three numbers of the synthetic substitution in step 3.

In (1)–(4), test each of the given polynomial functions to see if the relationship above appears to be true.

a. Use synthetic substitution to find a root r. (Try integers that are factors of the constant term.)

b. Write $(x - r)$ as one factor and use the first three numbers from the bottom line of the synthetic substitution as the coefficients of a trinomial factor.

c. Multiply $(x - r)$ times the trinomial factor written in step 2. Is the product equal to the given polynomial?

d. Write the three factors of the given polynomial and the three roots of the given function.

(1) $f(x) = x^3 - 6x^2 + 11x - 6$ (3) $f(x) = x^3 - 5x^2 + 7x - 3$
(2) $f(x) = x^3 + 2x^2 - 5x - 6$ (4) $f(x) = x^3 - 4x^2 + 5x - 2$

CUMULATIVE REVIEW   CHAPTERS 1-5

Part I

Answer all questions in this part. Each correct answer will receive 2 credits. No partial credit will be allowed.

1. In simplest form, $3(1 - 2x)^2 - (x + 2)$ is equal to
(1) $36x^2 - 37x + 7$ (3) $12x^2 - 13x + 1$
(2) $36x^2 - 35x + 11$ (4) $12x^2 - 13x + 5$

2. The solution set of $x^2 - 2x - 8 = 0$ is
(1) $\{2, 4\}$ (3) $\{2, -4\}$
(2) $\{-2, 4\}$ (4) $\{-2, -4\}$

3. In simplest form, the fraction $\dfrac{1 - \frac{1}{2}}{2 + \frac{3}{4}}$ is equal to
(1) $\frac{1}{11}$ (2) $\frac{2}{11}$ (3) $\frac{3}{11}$ (4) $\frac{6}{11}$

4. The sum of $\left(2 - \sqrt{18}\right)$ and $\left(-4 + \sqrt{50}\right)$ is
(1) $-2 + 2\sqrt{2}$ (3) $-2 + 4\sqrt{2}$
(2) $-2 + \sqrt{32}$ (4) $-2 - 8\sqrt{2}$

5. Which of the following products is a rational number?

(1) $(2 + \sqrt{2})(2 + \sqrt{2})$

(2) $(2 + \sqrt{2})(2 - \sqrt{2})$

(3) $\sqrt{2}(2 + \sqrt{2})$

(4) $2(2 - \sqrt{2})$

6. If the graph of $g(x)$ is the graph of $f(x) = x^2$ moved 2 units to the right, then $g(x)$ is equal to

(1) $x^2 + 2$ (2) $x^2 - 2$ (3) $(x - 2)^2$ (4) $(x + 2)^2$

7. Which of the following is not a one-to-one function when the domain and range are the largest set of real numbers for which y is defined?

(1) $y = 2x + 3$ (2) $y = \frac{2}{x}$ (3) $y = |x|$ (4) $y = \sqrt{x}$

8. The equation of a circle with center at $(1, -1)$ and radius 2 is

(1) $(x - 1)^2 + (y + 1)^2 = 2$

(2) $(x - 1)^2 + (y + 1)^2 = 4$

(3) $(x + 1)^2 + (y - 1)^2 = 2$

(4) $(x + 1)^2 + (y - 1)^2 = 4$

9. The equation of the axis of symmetry of the graph of $y = 2x^2 - 4x + 7$ is

(1) $x = 1$ (2) $x = -1$ (3) $x = 2$ (4) $x = -2$

10. The graph of $f(x)$ is the graph of $f(x - 4)$ under the translation

(1) $T_{4,0}$ (2) $T_{-4,0}$ (3) $T_{0,4}$ (4) $T_{0,-4}$

Part II

Answer all questions in this part. Each correct answer will receive 2 credits. Clearly indicate the necessary steps, including appropriate formula substitutions, diagrams, graphs, charts, etc. For all questions in this part, a correct numerical answer with no work shown will receive only 1 credit.

11. Write the quotient $\frac{2 - i}{3 - i}$ in $a + bi$ form.

12. For what value(s) of x does $\frac{x - 3}{2} = \frac{2x + 1}{3}$?

Part III

Answer all questions in this part. Each correct answer will receive 4 credits. Clearly indicate the necessary steps, including appropriate formula substitutions, diagrams, graphs, charts, etc. For all questions in this part, a correct numerical answer with no work shown will receive only 1 credit.

13. Write $\frac{3 + \sqrt{5}}{3 - \sqrt{5}}$ with a rational denominator.

14. What are the roots of the function $f(x) = 3 - |2x|$?

Part IV

Answer all questions in this part. Each correct answer will receive 6 credits. Clearly indicate the necessary steps, including appropriate formula substitutions, diagrams, graphs, charts, etc. For all questions in this part, a correct numerical answer with no work shown will receive only 1 credit.

15. a. Sketch the graph of $y = -x^2 + 4x$.

    b. Shade the region that is the solution set of $y \leq -x^2 + 4x$.

16. Let $f(x) = 2x + 4$ and $g(x) = x^2$.

    a. Find $f \circ g(-3)$.

    b. Find an expression for $h(x) = f \circ g(x)$.

CHAPTER 6

SEQUENCES AND SERIES

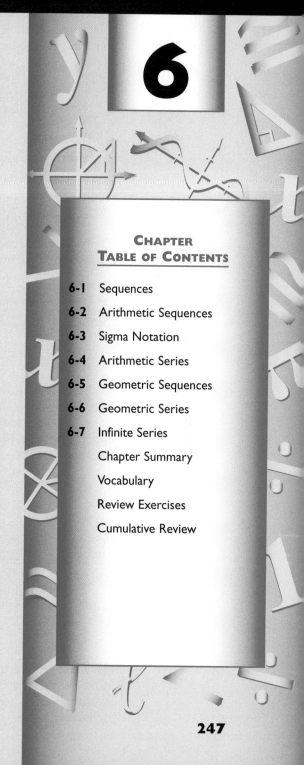

When the Grant family purchased a computer for $1,200 on an installment plan, they agreed to pay $100 each month until the cost of the computer plus interest had been paid. The interest each month was 1.5% of the unpaid balance. The amount that the Gant family still owed after each payment is a function of the number of months that have passed since they purchased the computer.

- At the end of month 1:
  they owed $(1,200 + 1,200 \times 0.015 - 100)$ dollars or
  $$(1,200 \times 1.015 - 100) = \$1,118.$$

- At the end of month 2:
  they owed $(1,118 + 1,118 \times 0.015 - 100)$ dollars or
  $$(1,118 \times 1.015 - 100) = \$1,034.77.$$

Each month, interest is added to the balance from the previous month and a payment of $100 is subtracted.

We can express the monthly payments with the function defined as $\{(n, f(n))\}$. Let the domain be the set of positive integers that represent the number of months since the initial purchase. Then:

$$f(1) = 1,118$$
$$f(2) = 1,034.77$$

In general, for positive integers n:
$$f(n) = f(n - 1) \times 1.015 - 100$$

This pattern continues until $(f(n - 1) \times 1.015)$ is between 0 and 100, since the final payment would be $(f(n - 1) \times 1.015)$ dollars.

In this chapter we will study sequential functions, such as the one described here, whose domain is the set of positive integers.

247

6-1 SEQUENCES

A ball is dropped from height of 16 feet. Each time that it bounces, it reaches a height that is half of its previous height. We can list the height to which the ball bounces in order until it finally comes to rest.

After Bounce	1	2	3	4	5	6	7
Height (ft)	8	4	2	1	$\frac{1}{2}$	$\frac{1}{4}$	$\frac{1}{8}$

The numbers $8, 4, 2, 1, \frac{1}{2}, \frac{1}{4}, \frac{1}{8}$ form a *sequence*. A **sequence** is a set of numbers written in a given order. We can list these heights as ordered pairs of numbers in which each height is paired with the number that indicates its position in the list. The set of ordered pairs would be:

$$\{(1, 8), (2, 4), (3, 2), (4, 1), (5, 0.5), (6, 0.25), (7, 0.125)\}$$

We associate each term of the sequence with the positive integer that specifies its position in the ordered set. Therefore, a sequence is a special type of function.

DEFINITION _____

A **finite sequence** is a function whose domain is the set of integers $\{1, 2, 3, \ldots, n\}$.

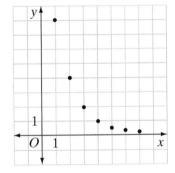

The function that lists the height of the ball after 7 bounces is shown on the graph at the right.

Often the sequence can continue without end. In this case, the domain is the set of positive integers.

DEFINITION _____

An **infinite sequence** is a function whose domain is the set of positive integers.

The terms of a sequence are often designated as $a_1, a_2, a_3, a_4, a_5, \ldots$. If the sequence is designated as the function f, then $f(1) = a_1$, $f(2) = a_2$, or in general:

$$f(n) = a_n$$

Most sequences are sets of numbers that are related by some pattern that can be expressed as a formula. The formula that allows any term of a sequence, except the first, to be computed from the previous term is called a **recursive definition**.

For example, the sequence that lists the heights to which a ball bounces when dropped from a height of 16 feet is 8, 4, 2, 1, 0.5, 0.25, 0.125, ... In this sequence, each term after the first is $\frac{1}{2}$ the previous term. Therefore, for each term after the first,

$$4 = \tfrac{1}{2}(8), 2 = \tfrac{1}{2}(4), 1 = \tfrac{1}{2}(2), 0.5 = \tfrac{1}{2}(1), 0.25 = \tfrac{1}{2}(0.5), 0.125 = \tfrac{1}{2}(0.25), \ldots.$$

For $n > 1$, we can write the recursive definition:

$$a_n = \tfrac{1}{2}a_{n-1}$$

Alternatively, we can write the recursive definition as:

$$a_{n+1} = \tfrac{1}{2}a_n \text{ for } n \geq 1$$

A rule that designates any term of a sequence can often be determined from the first few terms of the sequence.

EXAMPLE 1

a. List the next three terms of the sequence 2, 4, 8, 16, ...

b. Write a general expression for a_n.

c. Write a recursive definition for the sequence.

Solution **a.** It appears that each term of the sequence is a power of 2: $2^1, 2^2, 2^3, 2^4, \ldots$. Therefore, the next three terms should be $2^5, 2^6$, and 2^7 or 32, 64, and 128.

b. Each term is a power of 2 with the exponent equal to the number of the term. Therefore, $a_n = 2^n$.

c. Each term is twice the previous term. Therefore, for $n > 1$, $a_n = 2a_{n-1}$. Alternatively, for $n \geq 1$, $a_{n+1} = 2a_n$.

Answers **a.** 32, 64, 128

b. $a_n = 2^n$

c. For $n > 1$, $a_n = 2a_{n-1}$ or for $n \geq 1$, $a_{n+1} = 2a_n$.

EXAMPLE 2

Write the first three terms of the sequence $a_n = 3n - 1$.

Solution $\quad a_1 = 3(1) - 1 = 2 \qquad a_2 = 3(2) - 1 = 5 \qquad a_3 = 3(3) - 1 = 8$

Answer The first three terms of the sequence are 2, 5, and 8.

Sequences and the Graphing Calculator

We can use the sequence function on the graphing calculator to view a sequence for a specific range of terms. Evaluate a *sequence* in terms of *variable* from a *beginning term* to an *ending term*. That is:

seq(*sequence, variable, beginning term, ending term*)

For example, we can examine the sequence $a_n = n^2 + 2$ on the calculator. To view the first 20 terms of the sequence:

ENTER: `2nd`   `LIST`   `▶`   `5`   `X,T,Θ,n`   `x²`   `+` 2 `,`
`X,T,Θ,n`   `,` 1 `,` 20 `)`   `ENTER`

DISPLAY:
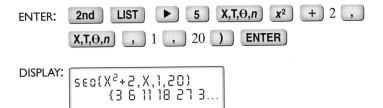

Note: We use the variable X instead of n to enter the sequence. We can also use the left and right arrow keys to examine the terms of the sequence.

Exercises

Writing About Mathematics

1. Nichelle said that sequence of numbers in which each term equals half of the previous term is a finite sequence. Randi said that it is an infinite sequence. Who is correct? Justify your answer.

2. a. Jacob said that if $a_n = 3n - 1$, then $a_{n+1} = a_n + 3$. Do you agree with Jacob? Explain why or why not.

b. Carlos said that if $a_n = 2^n$, then $a_{n+1} = 2^{n+1}$. Do you agree with Carlos? Explain why or why not.

Developing Skills

In 3–18, write the first five terms of each sequence.

3. $a_n = n$ **4.** $a_n = n + 5$ **5.** $a_n = 2n$ **6.** $a_n = \frac{1}{n}$

7. $a_n = \frac{n}{2}$ **8.** $a_n = 20 - n$ **9.** $a_n = 3^n$ **10.** $a_n = n^2$

11. $a_n = 2n + 3$ **12.** $a_n = 2n - 1$ **13.** $a_n = \frac{n}{n+1}$ **14.** $a_n = \frac{n+2}{n}$

15. $a_n = -n$ **16.** $a_n = 12 - 3n$ **17.** $a_n = \frac{4n}{3}$ **18.** $a_n = \frac{n}{2} + i$

In 19–30: **a.** Write an algebraic expression that represents a_n for each sequence. **b.** Find the ninth term of each sequence.

19. $2, 4, 6, 8, \ldots$ **20.** $3, 6, 9, 12, \ldots$ **21.** $1, 4, 7, 10, \ldots$

22. $3, 9, 27, 81, \ldots$ **23.** $12, 6, 3, 1.5, \ldots$ **24.** $7, 9, 11, 13, \ldots$

25. $10i, 8i, 6i, 4i, \ldots$ **26.** $\frac{1}{2}, \frac{1}{3}, \frac{1}{4}, \frac{1}{5}, \ldots$ **27.** $\frac{1}{2}, \frac{2}{3}, \frac{3}{4}, \frac{4}{5}, \ldots$

28. $2, 5, 10, 17, \ldots$ **29.** $1, -2, 3, -4, \ldots$ **30.** $1, \sqrt{2}, \sqrt{3}, 2, \ldots$

In 31–39, write the first five terms of each sequence.

31. $a_1 = 5, a_n = a_{n-1} + 1$ **32.** $a_1 = 1, a_{n+1} = 3a_n$ **33.** $a_1 = 1, a_n = 2a_{n-1} + 1$

34. $a_1 = -2, a_n = -2a_{n-1}$ **35.** $a_1 = 20, a_n = a_{n-1} - 4$ **36.** $a_1 = 4, a_{n+1} = a_n + n$

37. $a_2 = 36, a_n = \frac{1}{3}a_{n-1}$ **38.** $a_3 = 25, a_{n+1} = 2.5a_n$ **39.** $a_5 = \frac{1}{2}, a_n = \frac{1}{a_{n-1}}$

Applying Skills

40. Sean has started an exercise program. The first day he worked out for 30 minutes. Each day for the next six days, he increased his time by 5 minutes.

   a. Write the sequence for the number of minutes that Sean worked out for each of the seven days.

   b. Write a recursive definition for this sequence.

41. Sherri wants to increase her vocabulary. On Monday she learned the meanings of four new words. Each other day that week, she increased the number of new words that she learned by two.

   a. Write the sequence for the number of new words that Sherri learned each day for a week.

   b. Write a recursive definition for this sequence.

42. Julie is trying to lose weight. She now weighs 180 pounds. Every week for eight weeks, she was able to lose 2 pounds.

   a. List Julie's weight for each week.

   b. Write a recursive definition for this sequence.

43. January 1, 2008, was a Tuesday.

   a. List the dates for each Tuesday in January of that year.

   b. Write a recursive definition for this sequence.

44. Hui started a new job with a weekly salary of $400. After one year, and for each year that followed, his salary was increased by 10%. Hui left this job after six years.

   a. List the weekly salary that Hui earned each year.

   b. Write a recursive definition for this sequence.

45. One of the most famous sequences is the Fibonacci sequence. In this sequence, $a_1 = 1, a_2 = 1$, and for $n > 2$, $a_n = a_{n-2} + a_{n-1}$. Write the first ten terms of this sequence.

Hands-On Activity

The Tower of Hanoi is a famous problem that has challenged problem solvers throughout the ages. The tower consists of three pegs. On one peg there are a number of disks of different sizes, stacked according to size with the largest at the bottom. The task is to move the entire stack from one peg to the other side using the following rules:

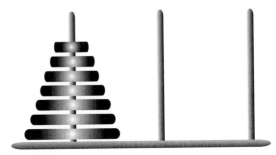

- Only one disk may be moved at a time.
- No disk may be placed on top of a smaller disk.

   Note that a move consists of taking the top disk from one peg and placing it on another peg.

   1. Use a stack of different-sized coins to model the Tower of Hanoi. What is the smallest number of moves needed if there are:

   a. 2 disks?

   b. 3 disks?

   c. 4 disks?

   d. n disks?

   2. Write a recursive definition for the sequence described in Exercise 1, **a–d**.

6-2 ARITHMETIC SEQUENCES

The set of positive odd numbers, 1, 3, 5, 7, . . . , is a sequence. The first term, a_1, is 1 and each term is 2 greater than the preceding term. The difference between consecutive terms is 2. We say that 2 is the **common difference** for the sequence. The set of positive odd numbers is an example an *arithmetic sequence*.

DEFINITION _____

An **arithmetic sequence** is a sequence such that for all n, there is a constant d such that $a_n - a_{n-1} = d$.

For an arithmetic sequence, the recursive formula is:

$$a_n = a_{n-1} + d$$

An arithmetic sequence is formed when each term after the first is obtained by adding the same constant to the previous term. For example, look at the first five terms of the sequence of positive odd numbers.

$$a_1 = 1 \qquad\qquad a_1 = 1$$
$$a_2 = a_1 + 2 = 1 + 2 = 3 \qquad a_2 = 1 + 1(2)$$
$$a_3 = a_2 + 2 = 3 + 2 = 5 \qquad a_3 = [1 + 1(2)] + 2 = 1 + 2(2)$$
$$a_4 = a_3 + 2 = 5 + 2 = 7 \qquad a_4 = [1 + 2(2)] + 2 = 1 + 3(2)$$
$$a_5 = a_4 + 2 = 7 + 2 = 9 \qquad a_5 = [1 + 3(2)] + 2 = 1 + 4(2)$$

For this sequence, $a_1 = 1$ and each term after the first is found by adding 2 to the preceding term. Therefore, for each term, 2 has been added to the first term one less time than the number of the term. For the second term, 2 has been added once; for the third term, 2 has been added twice; for the fourth term, 2 has been added three times. In general, for the nth term, 2 has been added $n - 1$ times.

Shown below is a general arithmetic sequence with a_1 as the first term and d as the common difference:

$$a_1, \; a_1 + d, \; a_1 + 2d, \; a_1 + 3d, \; a_1 + 4d, \; a_1 + 5d, \; \ldots, \; a_1 + (n - 1)d$$

If the first term of an arithmetic sequence is a_1 and the common difference is d, then for each term of the sequence, d has been added to a_1 one less time than the number of the term. Therefore, any term a_n of an arithmetic sequence can be evaluated with the formula

$$a_n = a_1 + (n - 1)d$$

where d is the common difference of the sequence.

EXAMPLE I

For the arithmetic sequence $100, 97, 94, 91, \ldots$, find:

a. the common difference.

b. the 20th term of the sequence.

Solution **a.** The common difference is the difference between any term and the previous term:

$$97 - 100 = -3 \qquad \text{or} \qquad 94 - 97 = -3 \qquad \text{or} \qquad 91 - 95 = -3$$

The common difference is -3.

b. $a_{20} = 100 + (20 - 1)(-3)$
$$= 100 + (19)(-3)$$
$$= 100 - 57$$
$$= 43$$

The 20th term can also be found by writing the sequence:

$$100, 97, 94, 91, 88, 85, 82, 79, 76, 73, 70, 67, 64, 61, 58, 55, 52, 49, 46, 43$$

Answers **a.** $d = -3$ **b.** $a_{20} = 43$

EXAMPLE 2

Scott is saving to buy a guitar. In the first week, he put aside $42 that he received for his birthday, and in each of the following weeks, he added $8 to his savings. He needs $400 for the guitar that he wants. In which week will he have enough money for the guitar?

Solution   Let $a_1 = 42$ and $d = 8$.

The value for n for which $a_n = 400$ is the week Scott will have enough money to buy the guitar.

$$a_n = a_1 + (n - 1)d$$
$$400 = 42 + (n - 1)(8)$$
$$400 = 42 + 8n - 8$$
$$400 = 34 + 8n$$
$$366 = 8n$$
$$45.75 = n$$

Scott adds money to his savings in $8 increments, so he will not have the needed $400 in savings until the 46th week.

Answer   46th week

EXAMPLE 3

The 4th term of an arithmetic sequence is 80 and the 12th term is 32.

a. What is the common difference?

b. What is the first term of the sequence?

Solution   **a.**   *How to Proceed*

(1) Use $a_n = a_1 + (n - 1)d$ to write two equation in two variables:

$$80 = a_1 + (4 - 1)d \ \rightarrow \ 80 = a_1 + 3d$$
$$32 = a_1 + (12 - 1)d \ \rightarrow \ \underline{-(32 = a_1 + 11d)}$$

(2) Subtract to eliminate a_1:

$$48 = -8d$$

(3) Solve for d:

$$-6 = d$$

b. Substitute in either equation to find a_1.

$$80 = a_1 + 3(-6)$$
$$80 = a_1 - 18$$
$$98 = a_1$$

Answers   **a.** $d = -6$ **b.** $a_1 = 98$

Arithmetic Means

We have defined the *arithmetic mean* of two numbers as their average, that is, the sum of the numbers divided by 2. For example, the arithmetic mean of 4 and 28 is $\frac{4 + 28}{2} = \frac{32}{2} = 16$. The numbers 4, 16, and 28 form an arithmetic sequence with a common difference of 12.

We can find three numbers between 4 and 28 that together with 4 and 28 form an arithmetic sequence: $4, a_2, a_3, a_4, 28$. This is a sequence of five terms.

1. Use the formula for a_5 to find the common difference:

$$a_5 = a_1 + (n - 1)d$$
$$28 = 4 + (5 - 1)d$$
$$28 = 4 + 4d$$
$$24 = 4d$$
$$6 = d$$

2. Now use the recursive formula, $a_n = a_{n-1} + d$, and $d = 6$, to write the sequence:

$$4, 10, 16, 22, 28$$

The numbers 10, 16, and 22 are three **arithmetic means** between 4 and 28. Looking at it another way, in *any* arithmetic sequence, any given term is the average of the term before it and the term after it. Thus, an *arithmetic mean* is any term of an arithmetic sequence.

EXAMPLE 4

Find five arithmetic means between 2 and 23.

Solution Five arithmetic means between 2 and 23 will form a sequence of seven terms. Use the formula for a_7 to find the common difference:

$$a_7 = a_1 + (n - 1)d$$
$$23 = 2 + (7 - 1)d$$
$$23 = 2 + 6d$$
$$21 = 6d$$
$$\tfrac{7}{2} = d$$

Evaluate $a_2, a_3, a_4, a_5,$ and a_6:

$$a_2 = 2 + (2 - 1)\tfrac{7}{2} \qquad a_3 = 2 + (3 - 1)\tfrac{7}{2} \qquad a_4 = 2 + (4 - 1)\tfrac{7}{2}$$
$$a_2 = \tfrac{11}{2} = 5\tfrac{1}{2} \qquad\qquad a_3 = 9 \qquad\qquad a_4 = \tfrac{25}{2} = 12\tfrac{1}{2}$$

$$a_5 = 2 + (5 - 1)\tfrac{7}{2} \qquad a_6 = 2 + (6 - 1)\tfrac{7}{2}$$
$$a_5 = 16 \qquad\qquad a_6 = \tfrac{39}{2} = 19\tfrac{1}{2}$$

Answer The five arithmetic means are $5\tfrac{1}{2}, 9, 12\tfrac{1}{2}, 16,$ and $19\tfrac{1}{2}$.

Exercises

Writing About Mathematics

1. Virginia said that Example 3 could have been solved without using equations. Since there are eight terms from a_4 to a_{12}, the difference between 80 and 32 has to be divided into eight parts. Each part is 6. Since the sequence is decreasing, the common difference is -6. Then using -6, work back from a_4 to a_1: 80, 86, 92, 98. Do you think that Virginia's solution is better than the one given in Example 3? Explain why or why not.

2. Pedro said that to form a sequence of five terms that begins with 2 and ends with 12, you should divide the difference between 12 and 2 by 5 to find the common difference. Do you agree with Pedro? Explain why or why not.

Developing Skills

In 3–8, determine if each sequence is an arithmetic sequence. If the sequence is arithmetic, find the common difference.

3. $2, 5, 8, 11, 14, \ldots$

4. $-3i, -1i, 1i, 3i, 5i, \ldots$

5. $1, 1, 2, 3, 5, 8, \ldots$

6. $20, 15, 10, 5, 0, \ldots$

7. $1, 2, 4, 8, 16, \ldots$

8. $1, 1.25, 1.5, 1.75, 2, \ldots$

In 9–14: **a.** Find the common difference of each arithmetic sequence. **b.** Write the nth term of each sequence for the given value of n.

9. $3, 6, 9, 12, \ldots, n = 8$

10. $2, 7, 12, 17, \ldots, n = 12$

11. $18, 16, 14, 12, \ldots, n = 10$

12. $\frac{1}{2}, 1, \frac{3}{2}, 2, \ldots, n = 7$

13. $-1, -3, -5, -7, \ldots, n = 10$

14. $2.1, 2.2, 2.3, 2.4, \ldots, n = 20$

15. Write the first six terms of the arithmetic sequence that has 12 for the first term and 42 for the sixth term.

16. Write the first nine terms of the arithmetic sequence that has 100 as the fifth term and 80 as the ninth term.

17. Find four arithmetic means between 3 and 18.

18. Find two arithmetic means between 1 and 5.

19. Write a recursive definition for an arithmetic sequence with a common difference of -3.

Applying Skills

20. On July 1, Mr. Taylor owed $6,000. On the 1st of each of the following months, he repaid $400. List the amount owed by Mr. Taylor on the 2nd of each month starting with July 2. Explain why the amount owed each month forms an arithmetic sequence.

21. Li is developing a fitness program that includes doing push-ups each day. On each day of the first week he did 20 push-ups. Each subsequent week, he increased his daily push-ups by 5. During which week did he do 60 push-ups a day?

    a. Use a formula to find the answer to the question.

    b. Write the arithmetic sequence to answer the question.

    c. Which method do you think is better? Explain you answer.

22. a. Show that a linear function whose domain is the set of positive integers is an arithmetic sequence.

   b. For the linear function $y = mx + b$, $y = a_n$ and $x = n$. Express a_1 and d of the arithmetic sequence in terms of m and b.

23. Leslie noticed that the daily number of e-mail messages she received over the course of two months form an arithmetic sequence. If she received 13 messages on day 3 and 64 messages on day 20:

   a. How many messages did Leslie receive on day 12?

   b. How many messages will Leslie receive on day 50?

6-3 SIGMA NOTATION

Ken wants to get more exercise so he begins by walking for 20 minutes. Each day for two weeks, he increases the length of time that he walks by 5 minutes. At the end of two weeks, the length of time that he has walked each day is given by the following arithmetic sequence:

$$20, 25, 30, 35, 40, 45, 50, 55, 60, 65, 70, 75, 80, 85$$

The total length of time that Ken has walked in two weeks is the sum of the terms of this arithmetic sequence:

$$20 + 25 + 30 + 35 + 40 + 45 + 50 + 55 + 60 + 65 + 70 + 75 + 80 + 85$$

This sum is called a *series*.

DEFINITION

A **series** is the indicated sum of the terms of a sequence.

The symbol Σ, which is the Greek letter sigma, is used to indicate a sum. The number of minutes that Ken walked on the nth day is $a_n = 20 + (n - 1)(5)$. We can write the sum of the number of minutes that Ken walked in **sigma notation**:

$$\sum_{n=1}^{14} a_n = \sum_{n=1}^{14} 20 + (n - 1)(5)$$

The "$n = 1$" below Σ is the value of n for the first term of the series, and the number above Σ is the value of n for the last term of the series. The symbol $\sum_{n=1}^{14} a_n$ can be read as "the sum of a_n for all integral values of n from 1 to 14."

In expanded form:

$$\sum_{n=1}^{14} a_n = a_1 + a_2 + a_3 + a_4 + a_5 + a_6 + a_7 + a_8 + a_9 + a_{10} + a_{11} + a_{12} + a_{13} + a_{14}$$

$$\sum_{n=1}^{14} 20 + (n-1)(5) = [20 + 0(5)] + [20 + 1(5)] + [20 + 2(5)] + \cdots + [20 + 13(5)]$$

$$= 20 + 25 + 30 + \cdots + 85$$

For example, we can indicate the sum of the first 50 positive even numbers as $\displaystyle\sum_{i=1}^{50} 2i.$

$$\sum_{i=1}^{50} 2i = 2(1) + 2(2) + 2(3) + 2(4) + \cdots + 2(50)$$

$$= 2 + 4 + 6 + 8 + \cdots + 100$$

Note that in this case, i was used to indicate the number of the term. Although any variable can be used, n, i, and k are the variables most frequently used. Be careful not to confuse the variable i with the imaginary number i.

The series $\displaystyle\sum_{i=1}^{50} 2i$ is an example of a **finite series** since it is the sum of a *finite* number of terms. An **infinite series** is the sum of an *infinite* number of terms of a sequence. We indicate that a series is infinite by using the symbol for infinity, ∞. For example, we can indicate the sum of *all* of the positive even numbers as:

$$\sum_{i=1}^{\infty} 2i = 2 + 4 + 6 + \cdots + 2i + \cdots$$

EXAMPLE 1

Write the sum given by $\displaystyle\sum_{k=1}^{7} (k + 5)$.

Solution $\displaystyle\sum_{k=1}^{7} (k + 5) = (1 + 5) + (2 + 5) + (3 + 5) + (4 + 5) + (5 + 5) + (6 + 5) + (7 + 5)$

$$= 6 + 7 + 8 + 9 + 10 + 11 + 12 \ \ Answer$$

EXAMPLE 2

Write the sum of the first 25 positive odd numbers in sigma notation.

Solution   The positive odd numbers are $1, 3, 5, 7, \ldots$.

The 1st positive odd number is 1 less than twice 1, the 2nd positive odd number is 1 less than twice 2, the 3rd positive odd number is 1 less than twice 3. In general, the nth positive odd number is 1 less than twice n or $a_n = 2n - 1$.

The sum of the first 25 odd numbers is $\displaystyle\sum_{n=1}^{25} (2n - 1)$. *Answer*

EXAMPLE 3

Use sigma notation to write the series $12 + 20 + 30 + 42 + 56 + 72 + 90 + 110$ in two different ways:

a. Express each term as a sum of two numbers, one of which is a square.

b. Express each term as a product of two numbers.

Solution **a.** The terms of this series can be written as $3^2 + 3, 4^2 + 4, 5^2 + 5, \ldots, 10^2 + 10$, or, in general, as $n^2 + n$ with n from 3 to 10.

The series can be written as $\displaystyle\sum_{n=3}^{10} (n^2 + n)$.

b. Write the series as

$$3(4) + 4(5) + 5(6) + 6(7) + 7(8) + 8(9) + 9(10) + 10(11).$$

The series is the sum of $n(n + 1)$ from $n = 3$ to $n = 10$.

The series can be written as $\displaystyle\sum_{n=3}^{10} n(n + 1)$.

Answers **a.** $\displaystyle\sum_{n=3}^{10} (n^2 + n)$ **b.** $\displaystyle\sum_{n=3}^{10} n(n + 1)$

EXAMPLE 4

Use sigma notation to write the sum of the reciprocals of the natural numbers.

Solution The reciprocals of the natural numbers are $1, \frac{1}{2}, \frac{1}{3}, \frac{1}{4}, \ldots, \frac{1}{n}$.

Since there is no largest natural number, this sequence has no last term. Therefore, the sum of the terms of this sequence is an infinite series. In sigma notation, the sum of the reciprocals of the natural numbers is:

$$\sum_{n=1}^{\infty} \frac{1}{n} \quad \textit{Answer}$$

Finite Series and the Graphing Calculator

The graphing calculator can be used to find the sum of a finite series. We use the sum(function along with the seq(function of the previous section to evaluate a series. For example, to evaluate $\displaystyle\sum_{k=1}^{37} \frac{1}{k(k + 2)}$ on the calculator:

STEP 1. Enter the sequence into the calculator and store it in list L_1.

ENTER: [2nd] [LIST] [▶] [5] [1] [÷] [(] [X,T,θ,n] [(]
[X,T,θ,n] [+] [2] [)] [)] [,] [X,T,θ,n] [,] [1] [,]
[37] [)] [STO▶] [2nd] [L1] [ENTER]

DISPLAY:

```
seq(1/(X(X+2)),X
,1,37)→L1
   {.3333333333 .1 ...
```

STEP 2. Use the sum(function to find the sum.

ENTER: [2nd] [LIST] [▶] [▶] [5] [2nd] [L1] [ENTER]

DISPLAY:

```
sum(L1
          .7240215924
```

The sum is approximately equal to 0.72.

Exercises

Writing About Mathematics

1. Is the series given in Example 3 equal to $\sum_{n=1}^{8}[(n+2)^2 + (n+2)]$? Justify your answer.

2. Explain why $\sum_{k=0}^{10}\frac{1}{k}$ is undefined.

Developing Skills

In 3–14: **a.** Write each arithmetic series as the sum of terms. **b.** Find each sum.

3. $\sum_{n=1}^{10} 3n$

4. $\sum_{k=1}^{5}(2k-2)$

5. $\sum_{k=1}^{4}k^2$

6. $\sum_{n=1}^{6}n^3$

7. $\sum_{k=1}^{10}(100-5k)$

8. $\sum_{n=5}^{10}(3n-3)$

9. $\sum_{n=2}^{5}(n^2+2i)$

10. $\sum_{h=1}^{10}(-1)^h h$

11. $\sum_{n=5}^{15}[4n-(n+1)]$

12. $\sum_{k=1}^{10}-ki$

13. $\sum_{k=3}^{5}(5-4k)$

14. $\sum_{n=0}^{5}(-2n)^{n+1}$

In 15–26, write each series in sigma notation.

15. $3 + 5 + 7 + 9 + 11 + 13 + 15$

16. $1 + 6 + 11 + 16 + 21 + 26 + 31 + 36$

17. $1^1 + 2^2 + 3^3 + 4^4 + 5^5$

18. $100 + 95 + 90 + 85 + \cdots + 5$

19. $3 + 6 + 9 + 12 + 15 + \cdots + 30$

20. $1 + \frac{1}{2} + \frac{1}{4} + \frac{1}{8} + \frac{1}{16} + \frac{1}{32}$

21. $\frac{1}{2} + \frac{2}{3} + \frac{3}{4} + \frac{4}{5} + \frac{5}{6} + \frac{6}{7} + \frac{7}{8} + \frac{8}{9} + \frac{9}{10}$

22. $\frac{1}{1!} + \frac{1}{2!} + \frac{1}{3!} + \frac{1}{4!} + \frac{1}{5!}$

23. $-\frac{1}{3} + \frac{2}{9} - \frac{3}{27} + \frac{4}{81} - \frac{5}{243}$

24. $\frac{1}{1 \times 2} + \frac{1}{2 \times 3} + \frac{1}{3 \times 4} + \frac{1}{4 \times 5} + \frac{1}{5 \times 6} + \frac{1}{6 \times 7}$

25. $1^2 + 2^2 + 3^2 + 4^2 + 5^2 + \cdots$

26. $\frac{1}{3} + \frac{2}{9} + \frac{3}{27} + \frac{4}{81} + \frac{5}{243} + \cdots$

Applying Skills

27. Show that $\displaystyle\sum_{i=1}^{n} ka_i = k \sum_{i=1}^{n} a_i$.

28. Show that $\displaystyle\sum_{i=1}^{n} (a_i + b_i) = \sum_{i=1}^{n} a_i + \sum_{i=1}^{n} b_i$.

29. In a theater, there are 20 seats in the first row. Each row has 3 more seats than the row ahead of it. There are 35 rows in the theater.

   a. Express the number of seats in the nth row of the theater in terms of n.

   b. Use sigma notation to represent the number of seats in the theater.

30. On Monday, Elaine spent 45 minutes doing homework. On the remaining four days of the school week, she spent 15 minutes longer doing homework than she had the day before.

   a. Express the number of minutes Elaine spent doing homework on the nth day of the school week.

   b. Use sigma notation to represent the total number of minutes Elaine spent doing homework from Monday to Friday.

31. Use the graphing calculator to evaluate the following series to the nearest *hundredth*:

   (1) $\displaystyle\sum_{n=1}^{50}\left(1 + \frac{1}{n}\right)$

   (2) $\displaystyle\sum_{k=1}^{18} \frac{5}{1 + k}$

   (3) $\displaystyle\sum_{n=1}^{20} \frac{(n - 1)(-1)^n}{n}$

6-4 ARITHMETIC SERIES

In the last section, we wrote the sequence of minutes that Ken walked each day for two weeks:

$$20, 25, 30, 35, 40, 45, 50, 55, 60, 65, 70, 75, 80, 85$$

Since the difference between each pair of consecutive times is a constant, 5, the sequence is an arithmetic sequence. The total length of time that Ken walked in two weeks is the sum of the terms of this sequence:

$$20 + 25 + 30 + 35 + 40 + 45 + 50 + 55 + 60 + 65 + 70 + 75 + 80 + 85$$

In general, if $a_1, a_1 + d, a_1 + 2d, \ldots, a_n + (n-1)d$ is an arithmetic sequence with n terms, then:

$$\sum_{i=1}^{n} [a_1 + (i-1)d] = a_1 + (a_1 + d) + (a_1 + 2d) + \cdots + [a_1 + (n-1)d]$$

This sum is called an *arithmetic series*.

DEFINITION _____

An **arithmetic series** is the indicated sum of the terms of an arithmetic sequence.

Once a given series is defined, we can refer to it simply as S (for sigma). S_n is called the **nth partial sum** and represents the sum of the first n terms of the sequence.

We can find the number of minutes that Ken walked in 14 days by adding the 14 numbers or by observing the pattern of this series. Begin by writing the sum first in the order given and then in reverse order.

$$S_{14} = 20 + 25 + 30 + 35 + 40 + 45 + 50 + 55 + 60 + 65 + 70 + 75 + 80 + 85$$
$$S_{14} = 85 + 80 + 75 + 70 + 65 + 60 + 55 + 50 + 45 + 40 + 35 + 30 + 25 + 20$$

Note that for this arithmetic series:

$$a_1 + a_{14} = a_2 + a_{13} = a_3 + a_{12} = a_4 + a_{11} = a_5 + a_{10} = a_6 + a_9 = a_7 + a_8$$

Add the sums together, combining corresponding terms. The sum of each pair is 105 and there are $\frac{14}{2}$ or 7 pairs.

$$S_{14} = 20 + 25 + 30 + 35 + 40 + 45 + 50 + 55 + 60 + 65 + 70 + 75 + 80 + 85$$
$$S_{14} = 85 + 80 + 75 + 70 + 65 + 60 + 55 + 50 + 45 + 40 + 35 + 30 + 25 + 20$$
$$\overline{2S_{14} = 105 + 105 + 105 + 105 + 105 + 105 + 105 + 105 + 105 + 105 + 105 + 105 + 105 + 105}$$

$2S_{14} = 14(105)$ ← Write the expression in factored form.

$S_{14} = 7(105)$ ← Divide both sides by 2.

$S_{14} = 735$ ← Simplify.

Therefore, the total number of minutes that Ken walked in 14 days is 7(105) or 735 minutes.

Does a similar pattern exist for every arithmetic series? Consider the general arithmetic series with n terms, $a_n = a_1 + (n - 1)d$. List the terms of the series in ascending order from a_1 to a_n and in descending order from a_n to a_1.

Ascending Order	Descending Order	Sum
a_1	$a_n = a_1 + (n - 1)d$	$a_1 + [a_1 + (n - 1)d] = 2a_1 + (n - 1)d$
$a_2 = a_1 + d$	$a_{n-1} = a_1 + (n - 2)d$	$[a_1 + d] + [a_1 + (n - 2)d] = 2a_1 + (n - 1)d$
$a_3 = a_1 + 2d$	$a_{n-2} = a_1 + (n - 3)d$	$[a_1 + 2d] + [a_1 + (n - 3)d] = 2a_1 + (n - 1)d$
$\vdots$	$\vdots$	$\vdots$
$a_{n-2} = a_1 + (n - 3)d$	$a_3 = a_1 + 2d$	$[a_1 + (n - 3)d] + [a_1 + 2d] = 2a_1 + (n - 1)d$
$a_{n-1} = a_1 + (n - 2)d$	$a_2 = a_1 + d$	$[a_1 + (n - 2)d] + [a_1 + d] = 2a_1 + (n - 1)d$
$a_n = a_1 + (n - 1)d$	a_1	$[a_1 + (n - 1)d] + a_1 = 2a_1 + (n - 1)d$

In general, for any arithmetic series with n terms there are $\frac{n}{2}$ pairs of terms whose sum is $a_1 + a_n$:

$$2S_n = n(a_1 + a_n) = n(2a_1 + (n - 1)d)$$
$$S_n = \tfrac{n}{2}(a_1 + a_n) = \tfrac{n}{2}(2a_1 + (n - 1)d)$$

EXAMPLE I

a. Write the sum of the first 15 terms of the arithmetic series $1 + 4 + 7 + \cdots$ in sigma notation.

b. Find the sum.

Solution **a.** For the related arithmetic sequence, $1, 4, 7, \ldots$, the common difference is 3. Therefore,

$$a_n = 1 + (n - 1)(3)$$
$$= 3n - 2$$

The series is written as $\displaystyle\sum_{n=1}^{15} 3n - 2$. *Answer*

b. Use the formula $S_n = \tfrac{n}{2}(2a_1 + (n - 1)d)$ with $a_1 = 1, n = 15$, and $d = 3$.

$$S_{15} = \tfrac{15}{2}[2(1) + (15 - 1)(3)]$$
$$= \tfrac{15}{2}[2 + 14(3)]$$
$$= \tfrac{15}{2}[2 + 42]$$
$$= \tfrac{15}{2}[44]$$
$$= 330 \ \text{\textit{Answer}}$$

Note: Part **b** can also be solved by using the formula $S_n = \frac{n}{2}(a_1 + a_n)$. First find a_{15}.

$$a_{15} = a_1 + (n - 1)d \qquad\qquad S_{15} = \frac{15}{2}(1 + 43)$$
$$= 1 + (15 - 1)(3) \qquad\qquad = \frac{15}{2}(44)$$
$$= 1 + 14(3) \qquad\qquad\qquad = 330$$
$$= 1 + 42$$
$$= 43$$

Notice that for this series there are $7\frac{1}{2}$ pairs. The first 7 numbers are paired with the last 7 numbers, and each pair has a sum of 44. The middle number, 22, is paired with itself, making half of a pair.

EXAMPLE 2

The sum of the first and the last terms of an arithmetic sequence is 80 and the sum of all the terms is 1,200. How many terms are in the sequence?

Solution
$$S_n = \frac{n}{2}(a_1 + a_n)$$
$$1{,}200 = \frac{n}{2}(80)$$
$$1{,}200 = 40n$$
$$30 = n$$

Answer   There are 30 terms in the sequence.

Exercises

Writing About Mathematics

1. Is there more than one arithmetic series such that the sum of the first and the last terms is 80 and the sum of the terms is 1,200? Justify your answer.

2. Is $1 + 1 + 2 + 3 + 5 + 8 + 13 + 21$ an arithmetic series? Justify your answer.

Developing Skills

In 3–8, find the sum of each series using the formula for the partial sum of an arithmetic series. Be sure to show your work.

3. $2 + 4 + 6 + 8 + 10 + 12$

4. $10 + 20 + 30 + 40 + 50 + 60$

5. $3 + 1 - 1 - 3 - 5 - 7 - 9 - 11 - 13$

6. $0i + 4i + 8i + 12i + 16i + 20i$

7. $0 + \frac{1}{3} + \frac{2}{3} + 1 + \frac{4}{3} + \frac{5}{3} + 2$

8. $\sqrt{2} + 2\sqrt{2} + 3\sqrt{2} + 4\sqrt{2} + \cdots + 15\sqrt{2}$

In 9–18, use the given information to **a.** write the series in sigma notation, and **b.** find the sum of the first n terms.

9. $a_1 = 3, a_n = 39, d = 4$

10. $a_1 = 24, a_n = 0, n = 6$

11. $a_1 = 24, a_n = 0, d = -6$

12. $a_1 = 10, d = 2, n = 14$

13. $a_1 = 2, d = \frac{1}{2}, n = 15$

14. $a_1 = 0, d = -2, n = 10$

15. $a_1 = \frac{1}{3}, d = \frac{1}{3}, n = 12$

16. $a_1 = 100, d = -5, n = 20$

17. $a_{10} = 50, d = 2.5, n = 10$

18. $a_5 = 15, d = 2, n = 12$

In 19–24: **a.** Write each arithmetic series as the sum of terms. **b.** Find the sum.

19. $\displaystyle\sum_{k=1}^{10} 2k$

20. $\displaystyle\sum_{k=2}^{8} (3 + k)$

21. $\displaystyle\sum_{n=0}^{9} (20 - 2n)$

22. $\displaystyle\sum_{i=0}^{19} (100 - 5i)$

23. $\displaystyle\sum_{n=1}^{25} -2n$

24. $\displaystyle\sum_{n=1}^{10} (-1 + 2n)$

Applying Skills

25. Madeline is writing a computer program for class. The first day she wrote 5 lines of code and each day, as she becomes more skilled in writing code, she writes one more line than the previous day. It takes Madeline 6 days to complete the program. How many lines of code did she write?

26. Jose is learning to cross-country ski. He began by skiing 1 mile the first day and each day he increased the distance skied by 0.2 mile until he reached his goal of 3 miles.

a. How many days did it take Jose to reach his goal?

b. How many miles did he ski from the time he began until the day he reached his goal?

27. Sarah wants to save for a special dress for the prom. The first month she saved $15 and each of the next five months she increased the amount that she saved by $2. What is the total amount Sarah saved over the six months?

28. In a theater, there are 20 seats in the first row. Each row has 3 more seats than the row ahead of it. There are 35 rows in the theater. Find the total number of seats in the theater.

29. On Monday, Enid spent 45 minutes doing homework. On the remaining four days of the school week she spent 15 minutes longer doing homework than she had the day before. Find the total number of minutes Enid spent doing homework from Monday to Friday.

30. Keegan started a job that paid $20,000 a year. Each year after the first, he received a raise of $600. What was the total amount that Keegan earned in six years?

31. A new health food store's net income was a loss of $2,300 in its first month, but its net income increased by $575 in each succeeding month for the next year. What is the store's net income for the year?

6-5 GEOMETRIC SEQUENCES

Pete reuses paper that is blank on one side to write phone messages. One day he took a stack of five sheets of paper and cut it into three parts and then cut each part into three parts. The number of pieces of paper that he had after each cut forms a sequence: 5, 15, 45. This sequence can be written as:

$$5, 5(3), 5(3)^2$$

Each term of the sequence is formed by multiplying the previous term by 3, or we could say that the ratio of each term to the previous term is a constant, 3. If Pete had continued to cut the pieces of paper in thirds, the terms of the sequence would be

$$5, 5(3), 5(3)^2, 5(3)^3, 5(3)^4, 5(3)^5, \ldots$$

This sequence is called a *geometric sequence*.

DEFINITION _____

A **geometric sequence** is a sequence such that for all n, there is a constant r such that $\dfrac{a_n}{a_{n-1}} = r$. The constant r is called the **common ratio**.

The recursive definition of a geometric sequence is:

$$a_n = a_{n-1}r$$

When written in terms of a_1 and r, the terms of a geometric sequence are:

$$a_1, \ a_2 = a_1r, \ a_3 = a_1r^2, \ a_4 = a_1r^3, \ \ldots$$

Each term after the first is obtained by multiplying the previous term by r. Therefore, each term is the product of a_1 times r raised to a power that is one less than the number of the term, that is:

$$a_n = a_1r^{n-1}$$

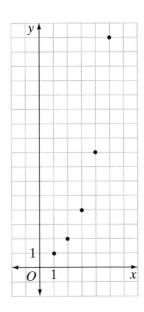

Since a sequence is a function, we can sketch the function on the coordinate plane. The geometric sequence $1, 1(2), 1(2)^2, 1(2)^3, 1(2)^4$ or $1, 2, 4, 8, 16$ can be written in function notation as $\{(1,1), (2, 2), (3, 4), (4, 8), (5, 16)\}$. Note that since the domain is the set of positive integers, the points on the graph are distinct points that are not connected by a curve.

Many common problems can be characterized by a geometric sequence. For example, if P dollars are invested at a yearly rate of 4%, then the value of the investment at the end of each year forms a geometric sequence:

Year 1: $P + 0.04P = P(1 + 0.04) = P(1.04)$

Year 2: $P(1.04) + 0.04(P(1.04)) = P(1.04)(1 + 0.04) = P(1.04)(1.04) = P(1.04)^2$

Year 3: $P(1.04)^2 + 0.04(P(1.04)^2) = P(1.04)^2(1 + 0.04) = P(1.04)^2(1.04) = P(1.04)^3$

Year 4: $P(1.04)^3 + 0.04(P(1.04)^3) = P(1.04)^3(1 + 0.04) = P(1.04)^3(1.04) = P(1.04)^4$

Year 5: $P(1.04)^4 + 0.04(P(1.04)^4) = P(1.04)^4(1 + 0.04) = P(1.04)^4(1.04) = P(1.04)^5$

The terms $P(1.04)$, $P(1.04)^2$, $P(1.04)^3$, $P(1.04)^4$, ... , $P(1.04)^n$ form a geometric sequence in which $a_1 = P(1.04)$ and $r = 1.04$. The nth term is $a_n = a_1 r^{n-1} = P(1.04)(1.04)^{n-1}$.

EXAMPLE 1

Is the sequence $4, 12, 36, 108, 324, \ldots$ a geometric sequence?

Solution In the sequence, $\frac{12}{4} = 3, \frac{36}{12} = 3, \frac{108}{36} = 3, \frac{324}{108} = 3$, the ratio of any term to the preceding term is a constant, 3, Therefore, $4, 12, 36, 108, 324, \ldots$ is a geometric sequence with $a_1 = 4$ and $r = 3$.

EXAMPLE 2

What is the 10th term of the sequence $4, 12, 36, 108, 324, \ldots$?

Solution The sequence $4, 12, 36, 108, 324, \ldots$ is a geometric sequence with $a_1 = 4$ and $r = 3$. Therefore,

$$a_n = a_1 r^{n-1}$$
$$a_{10} = 4(3)^9$$

Use a calculator for the computation.

ENTER: 4 $\boxed{\times}$ 3 $\boxed{\wedge}$ 9 $\boxed{\text{ENTER}}$

DISPLAY:
```
4*3^9
            78732
```

Answer 78,732

Geometric Means

In the proportion $\frac{4}{20} = \frac{20}{100}$, we say that 20 is the mean proportional between 4 and 100. These three numbers form a geometric sequence $4, 20, 100$. The mean proportional, 20, is also called the **geometric mean** between 4 and 100.

Between two numbers, there can be any number of geometric means. For example, to write three geometric means between 4 and 100, we want to form a geometric sequence 4, a_2, a_3, a_4, 100. In this sequence, $a_1 = 4$ and $a_5 = 100$.

$$a_n = a_1 r^{n-1}$$
$$a_5 = a_1 r^4$$
$$100 = 4r^4$$
$$25 = r^4$$
$$\sqrt[4]{25} = \sqrt[4]{r^4}$$
$$r = \pm\sqrt{5}$$

There are two possible values of r, $-\sqrt{5}$ and $+\sqrt{5}$. Therefore, there are two possible sequences and two possible sets of geometric means.

One sequence is 4, $4\sqrt{5}$, 20, $20\sqrt{5}$, 100 with $4\sqrt{5}$, 20, and $20\sqrt{5}$ three geometric means between 4 and 100.

The other sequence is 4, $-4\sqrt{5}$, 20, $-20\sqrt{5}$, 100 with $-4\sqrt{5}$, 20, and $-20\sqrt{5}$ three geometric means between 4 and 100.

Note that for each sequence, the ratio of each term to the preceding term is a constant.

$$\frac{4\sqrt{5}}{4} = \sqrt{5}$$

$$\frac{20}{4\sqrt{5}} = \frac{5}{\sqrt{5}} = \frac{5\sqrt{5}}{5} = \sqrt{5}$$

$$\frac{20\sqrt{5}}{20} = \sqrt{5}$$

$$\frac{100}{20\sqrt{5}} = \frac{5}{\sqrt{5}} = \frac{5\sqrt{5}}{5} = \sqrt{5}$$

$$\frac{-4\sqrt{5}}{4} = -\sqrt{5}$$

$$\frac{20}{-4\sqrt{5}} = \frac{5}{-\sqrt{5}} = \frac{5\sqrt{5}}{-5} = -\sqrt{5}$$

$$\frac{-20\sqrt{5}}{20} = -\sqrt{5}$$

$$\frac{100}{-20\sqrt{5}} = \frac{5}{-\sqrt{5}} = \frac{5\sqrt{5}}{-5} = -\sqrt{5}$$

EXAMPLE 3

Find four geometric means between 5 and 1,215.

Solution We want to find the missing terms in the sequence 5, a_2, a_3, a_4, a_5, 1,215. Use the formula to determine the common ratio r for $a_1 = 5$ and $a_6 = 1,215$.

$$a_n = a_1 r^{n-1}$$
$$a_6 = a_1 r^5$$
$$1,215 = 5r^5$$
$$243 = r^5$$
$$\sqrt[5]{243} = \sqrt[5]{r^5}$$
$$r = 3$$

The four geometric means are $5(3) = 15, 15(3) = 45, 45(3) = 135$, and $135(3) = 405$.

Answer 15, 45, 135, and 405

Exercises

Writing About Mathematics

1. Autumn said that the answer to Example 2 could have been found by entering 4 $\boxed{\times}$ 3 $\boxed{\text{ENTER}}$ on a calculator and then entering $\boxed{\times}$ 3 $\boxed{\text{ENTER}}$ eight times to display the sequence to the 9 terms after the first. Do you think that this is an easier way to find the 10th term? Explain your answer.

2. Sierra said that $8, 8\sqrt{2}, 16, 16\sqrt{2}, 32$ is a geometric sequence with three geometric means, $8\sqrt{2}, 16$, and $16\sqrt{2}$. Do you agree with Sierra? Justify your answer.

Developing Skills

In 3–14, determine whether each given sequence is geometric. If it is geometric, find r. If it is not geometric, explain why it is not.

3. $4, 8, 16, 32, 64, \ldots$

4. $1, 5, 25, 125, 625, \ldots$

5. $3, 6, 9, 12, \ldots$

6. $\frac{1}{2}, 2, 8, 32, \ldots$

7. $1, -3, 9, -27, 81, \ldots$

8. $36, 12, 4, \frac{4}{3}, \ldots$

9. $1, \frac{1}{3}, \frac{1}{9}, \frac{1}{27}, \ldots$

10. $\frac{5}{2}, 2, \frac{3}{2}, 1, \frac{1}{2}, \ldots$

11. $1, -10, 100, -1,000, 10,000, \ldots$

12. $1, 0.1, 0.01, 0.001, 0.0001, \ldots$

13. $0.05, -0.1, 0.2, -0.4, \ldots$

14. $a, a^2, a^3, a^4, \ldots$

In 15–26, write the first five terms of each geometric sequence.

15. $a_1 = 1, r = 6$

16. $a_1 = 40, r = \frac{1}{2}$

17. $a_1 = 2, r = 3$

18. $a_1 = \frac{1}{4}, r = -2$

19. $a_1 = 1, r = \sqrt{2}$

20. $a_1 = 10, a_2 = 30$

21. $a_1 = -1, a_2 = 4$

22. $a_1 = 100, a_3 = 1$

23. $a_1 = 1, a_3 = 16$

24. $a_1 = 1, a_3 = 2$

25. $a_1 = 1, a_4 = -8$

26. $a_1 = 81, a_5 = 1$

27. What is the 10th term of the geometric sequence $0.25, 0.5, 1, \ldots$?

28. What is the 9th term of the geometric sequence $125, 25, 5, \ldots$?

29. In a geometric sequence, $a_1 = 1$ and $a_5 = 16$. Find a_9.

30. The first term of a geometric sequence is 1 and the 4th term is 27. What is the 8th term?

31. In a geometric sequence, $a_1 = 2$ and $a_3 = 16$. Find a_6.

32. In a geometric sequence, $a_3 = 1$ and $a_7 = 9$. Find a_1.

33. Find two geometric means between 6 and 93.75.

34. Find three geometric means between 3 and $9\frac{13}{27}$.

35. Find three geometric means between 8 and 2,592.

Applying Skills

36. If $1,000 was invested at 6% annual interest at the beginning of 2001, list the geometric sequence that is the value of the investment at the beginning of each year from 2001 to 2010.

37. Al invested $3,000 in a certificate of deposit that pays 5% interest per year. What is the value of the investment at the end of each of the first four years?

38. In a small town, a census is taken at the beginning of each year. The census showed that there were 5,000 people living in the town at the beginning of 2001 and that the population decreased by 2% each year for the next seven years. List the geometric sequence that gives the population of the town from 2001 to 2008. (A decrease of 2% means that the population changed each year by a factor of 0.98.) Write your answer to the nearest integer.

39. It is estimated that the deer population in a park was increasing by 10% each year. If there were 50 deer in the park at the end of the first year in which a study was made, what is the estimated deer population for each of the next five years? Write your answer to the nearest integer.

40. A car that cost $20,000 depreciated by 20% each year. Find the value of the car at the end of each of the first four years. (A depreciation of 20% means that the value of the car each year was 0.80 times the value the previous year.)

41. A manufacturing company purchases a machine for $50,000. Each year the company estimates the depreciation to be 15%. What will be the estimated value of the machine after each of the first six years?

6-6 GEOMETRIC SERIES

DEFINITION

A **geometric series** is the indicated sum of the terms of a geometric sequence.

For example, 3, 12, 48, 192, 768, 3,072 is a geometric sequence with $r = 4$. The indicated sum of this sequence, $3 + 12 + 48 + 192 + 768 + 3,072$, is a geometric series.

In general, if $a_1, a_1r, a_1r^2, a_1r^3, \ldots, a_1r^{n-1}$ is a geometric sequence with n terms, then

$$\sum_{i=1}^{n} a_i r^{i-1} = a_1 + a_1r + a_1r^2 + a_1r^3 + \cdots + a_1r^{n-1}$$

is a geometric series.

Let the sum of these six terms be S_6. Now multiply S_6 by the *negative* of the common ratio, -4. This will result in a series in which each term but the last is the *opposite* of a term in S_6.

$$
\begin{aligned}
S_6 &= 3 + 12 + 48 + 192 + 3{,}072 \\
-4SS_6 &= \quad -12 - 48 - 192 - 3{,}072 - 12{,}288 \\
\hline
-3S_6 &= 3 - 12{,}288 \qquad\qquad \leftarrow \text{Add the sums.} \\
S_6 &= \frac{3 - 12{,}288}{-3} \qquad\quad \leftarrow \text{Divide by } -3. \\
&= -1 + 4{,}096 = 4{,}095 \quad \leftarrow \text{Simplify.}
\end{aligned}
$$

Thus, $S_6 = 4{,}095$. The pattern of a geometric series allows us to find a formula for the sum of the series. To S_n, add $-rS_n$:

$$
\begin{aligned}
S_n &= a_1 + a_1r + a_1r^2 + a_1r^3 + \cdots + a_1r^{n-1} \\
-rS_n &= \quad - a_1r - a_1r^2 - a_1r^3 - \cdots - a_1r^{n-1} - a_1r^n \\
\hline
S_n - rS_n &= a_1 - a_1r^n \\
S_n(1 - r) &= a_1(1 - r^n) \\
\\
S_n &= \frac{a_1(1 - r^n)}{1 - r}
\end{aligned}
$$

EXAMPLE I

Find the sum of the first 10 terms of the geometric series $2 + 1 + \frac{1}{2} + \cdots$.

Solution   For the series $2 + 1 + \frac{1}{2} + \cdots$, $a_1 = 2$ and $r = \frac{1}{2} = \frac{\frac{1}{2}}{1} = \frac{1}{2}$.

$$S_n = \frac{a_1(1 - r^n)}{1 - r}$$

$$S_{10} = \frac{2\left(1 - \left(\frac{1}{2}\right)^{10}\right)}{1 - \frac{1}{2}} = \frac{2\left(1 - \frac{1}{1{,}024}\right)}{\frac{1}{2}} = \frac{2\left(\frac{1{,}023}{1{,}024}\right)}{\frac{1}{2}} = 2\left(\frac{1{,}023}{1{,}024}\right) \times \frac{2}{1} = \frac{1{,}023}{256}$$

Answer $S_{10} = \frac{1{,}023}{256}$

EXAMPLE 2

Find the sum of five terms of the geometric series whose first term is 2 and whose fifth term is 162.

Solution   Use $a_5 = a_1 r^{5-1}$ to find r.

$$162 = 2r^4$$
$$81 = r^4$$
$$\sqrt[4]{81} = \sqrt[4]{r^4}$$
$$\pm 3 = r$$

$$S_n = \frac{a_1(1 - r^n)}{1 - r}$$

$$S_5 = \frac{2(1 - 3^5)}{1 - 3} = \frac{2(1 - 243)}{-2} = \frac{2(-242)}{-2} = 242$$

$$S_5 = \frac{2(1 - (-3)^5)}{1 - (-3)} = \frac{2(1 + 243)}{4} = \frac{2(244)}{4} = 122$$

Answer $S_5 = 242$ or $S_5 = 122$

Exercises

Writing About Mathematics

1. Casey said that the formula for the sum of a geometric series could be written as $S_n = \frac{a_1 - a_n r}{1 - r}$. Do you agree with Casey? Justify your answer.

2. Sherri said that Example 1 could have been solved by simply adding the ten terms of the series on a calculator. Do you think that this would have been a simpler way of finding the sum? Explain why or why not.

Developing Skills

In 3–14, find the sum of n terms of each geometric series.

3. $a_1 = 1, r = 2, n = 12$

4. $a_1 = 4, r = 3, n = 11$

5. $a_2 = 6, r = 4, n = 15$

6. $a_1 = 10, r = 10, n = 6$

7. $a_3 = 0.4, r = 2, n = 12$

8. $a_1 = 1, r = \frac{1}{3}, n = 10$

9. $5 + 10 + 20 + \cdots + a_n, n = 8$

10. $1 + 5 + 25 + \cdots + a_n, n = 10$

11. $\frac{1}{2} + \frac{1}{4} + \frac{1}{8} + \cdots + a_n, n = 6$

12. $2 - 8 + 32 - \cdots + a_n, n = 7$

13. $a_1 = 4, a_5 = 324, n = 9$

14. $a_1 = 1, a_8 = 128, n = 10$

In 15–22: **a.** Write each sum as a series. **b.** Find the sum of each series.

15. $3 + \sum_{n=1}^{5} 3(2)^n$

16. $1 + \sum_{n=1}^{5} \left(\frac{1}{3}\right)^n$

17. $10 + \sum_{n=1}^{5} 10\left(\frac{1}{2}\right)^n$

18. $-6 + \sum_{n=1}^{8} -6(4)^n$

19. $1 + \sum_{n=1}^{5}(-2)^n$

20. $1 + \sum_{n=1}^{5}\left(\frac{2}{3}\right)^n$

21. $100 + \sum_{k=1}^{6}100\left(\frac{1}{2}\right)^k$

22. $-81 + \sum_{k=1}^{6} - 81\left(-\frac{1}{3}\right)^k$

23. Find the sum of the first six terms of the series $1 + \sqrt{3} + 3 + 3\sqrt{3} + \ldots$.

24. Find the sum of the first eight terms of a series whose first term is 1 and whose eighth term is 625.

Applying Skills

25. A group of students are participating in a math contest. Students receive 1 point for their first correct answer, 2 points for their second correct answer, 4 points for their third correct answer, and so forth. What is the score of a student who answers 10 questions correctly?

26. Heidi deposited $400 at the beginning of each year for six years in an account that paid 5% interest. At the end of the sixth year, her first deposit had earned interest for six years and was worth $400(1.05)^6$ dollars, her second deposit had earned interest for five years and was worth $400(1.05)^5$ dollars, her third deposit had earned interest for four years and was worth $400(1.05)^4$ dollars. This pattern continues.

    a. What is the value of Heidi's sixth deposit at the end of the sixth year? Express your answer as a product and as a dollar value.

    b. Do the values of these deposits after six years form a geometric sequence? Justify your answer.

    c. What is the total value of Heidi's six deposits at the end of the sixth year?

27. A ball is thrown upward so that it reaches a height of 9 feet and then falls to the ground. When it hits the ground, it bounces to $\frac{1}{3}$ of its previous height. If the ball continues in this way, bouncing each time to $\frac{1}{3}$ of its previous height until it comes to rest when it hits the ground for the fifth time, find the total distance the ball has traveled, starting from its highest point.

28. If you start a job for which you are paid $1 the first day, $2 the second day, $4 the third day, and so on, how many days will it take you to become a millionaire?

6-7 INFINITE SERIES

Recall that an infinite series is a series that continues without end. That is, for a series $S_n = a_1 + a_2 + \cdots + a_n$, we say that n approaches infinity. We can also write $\sum_{n=1}^{\infty}a_n$. However, while there may be an infinite number of terms, a series can behave in only one of four ways. The series may increase without limit, decrease without limit, oscillate, or approach a limit.

CASE 1 *The series increases without limit.*

Consider an arithmetic series $S_n = \frac{n}{2}(2a_1 + (n-1)d)$.
If $a_1 = 1$ and $d = \frac{1}{2}$:

$$S_n = \frac{n}{2}\left[2(1) + (n-1)\left(\tfrac{1}{2}\right)\right]$$
$$= \frac{n}{2}\left[\tfrac{4}{2} + \tfrac{n}{2} - \tfrac{1}{2}\right]$$
$$= \frac{n}{2}\left[\tfrac{3+n}{2}\right] = \frac{3n + n^2}{4}$$

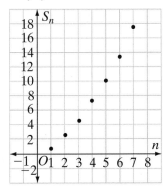

As the value of n approaches infinity, $\frac{3n + n^2}{4}$ increases without limit. This series has no limit. We can see this by graphing the function $S_n = \frac{3n + n^2}{4}$ for positive integer values.

CASE 2 *The series decreases without limit.*

For an arithmetic series $S_n = \frac{n}{2}(2a_1 + (n-1)d)$, if $a_1 = 1$ and $d = -1$:

$$S_n = \frac{n}{2}[2(1) + (n-1)(-1)]$$
$$= \frac{n}{2}[2 - n + 1] = \frac{n}{2}[3 - n]$$

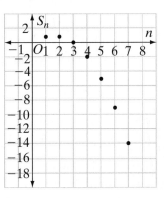

For this series, as the value of n approaches infinity, $\frac{n}{2}[3 - n]$ or $\frac{3n - n^2}{2}$ decreases without limit. This series also has no limit. We can see this by graphing the function $S_n = \frac{3n - n^2}{2}$ for positive integer values.

CASE 3 *The series approaches a limit.*

For a geometric series $S_n = \frac{a_1 - a_1 r^n}{1 - r}$, if $a_1 = 1$ and $r = \frac{1}{2}$:

$$S_n = \frac{1 - 1\left(\tfrac{1}{2}\right)^n}{1 - \tfrac{1}{2}} = \frac{1 - \left(\tfrac{1}{2}\right)^n}{\tfrac{1}{2}}$$
$$= 2 - 2\left(\tfrac{1}{2}\right)^n = 2 - \left(\tfrac{1}{2}\right)^{n-1}$$

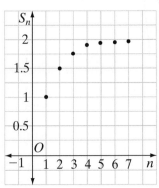

As n approaches infinity, $\left(\tfrac{1}{2}\right)^{n-1}$ approaches 0. Therefore, $S_n = 2 - \left(\tfrac{1}{2}\right)^{n-1}$ approaches $S_n = 2 - 0 = 2$ as n approaches infinity.

This series has a limit. The diagram on the bottom of page 274 shows how the sum of n terms approaches 2 as the number of terms, n, increases.

In general,

▶ **An infinite arithmetic series has no limit.**

▶ **An infinite geometric series has no limit when $|r| > 1$.**

▶ **An infinite geometric series has a finite limit when $|r| < 1$.**

When $|r| < 1$, r^n approaches 0. Therefore:

▶ $S_n = \dfrac{a_1 - a_1 r^n}{1 - r}$ **approaches** $\dfrac{a_1}{1 - r}$ **as n approaches infinity.**

EXAMPLE I

Find $\displaystyle\sum_{n=1}^{\infty} \left(\tfrac{1}{3}\right)^n$.

Solution

$$\sum_{n=1}^{\infty} \left(\tfrac{1}{3}\right)^n = \tfrac{1}{3} + \left(\tfrac{1}{3}\right)^2 + \left(\tfrac{1}{3}\right)^3 + \cdots$$

This is an infinite geometric series with $a_1 = \tfrac{1}{3}$ and $r = \tfrac{1}{3}$.

Using the formula $\dfrac{a_1 - a_1 r^n}{1 - r}$,

$$\sum_{n=1}^{\infty} \left(\tfrac{1}{3}\right)^n = \frac{\tfrac{1}{3} - \tfrac{1}{3}\left(\tfrac{1}{3}\right)^n}{1 - \tfrac{1}{3}}$$

As n approaches infinity, $\left(\tfrac{1}{3}\right)^n$ approaches 0. Therefore,

$$\sum_{n=1}^{\infty} \left(\tfrac{1}{3}\right)^n = \frac{\tfrac{1}{3}}{1 - \tfrac{1}{3}} = \tfrac{1}{3} \times \tfrac{3}{2} = \tfrac{1}{2} \quad \textit{Answer}$$

Series Containing Factorials

Series that are neither arithmetic nor geometric must be considered individually. Often the limit of a series can be found to be finite by comparing the terms of the series to the terms of another series with a known limit. One such series is $\displaystyle\sum_{k=1}^{n} \tfrac{1}{k!}$.

Recall *factorials* from previous courses. We write *n* **factorial** as follows:

DEFINITION

$n! = n(n - 1)(n - 2)(n - 3) \cdots (3)(2)(1)$

Consider the series $\sum_{k=1}^{n} \frac{1}{k!} = \frac{1}{1!} + \frac{1}{2!} + \frac{1}{3!} + \cdots + \frac{1}{n!}$. This series can be shown to have a limit by comparing its terms to the terms of the geometric series with $a_1 = 1$ and $r = \frac{1}{2}$. Each term, a_n $(n > 1)$, of this geometric series is $\frac{1}{2^{n-1}}$.

n	Term of Geometric Series	Term of Series to Be Tested	Comparison of Terms
1	1	$\frac{1}{1!} = 1$	$1 = 1$
2	$\frac{1}{2^1} = \frac{1}{2}$	$\frac{1}{2!} = \frac{1}{2 \times 1} = \frac{1}{2}$	$\frac{1}{2} = \frac{1}{2}$
3	$\frac{1}{2^2} = \frac{1}{4}$	$\frac{1}{3!} = \frac{1}{3 \times 2 \times 1} = \frac{1}{6}$	$\frac{1}{4} > \frac{1}{6}$
4	$\frac{1}{2^3} = \frac{1}{8}$	$\frac{1}{4!} = \frac{1}{4 \times 3 \times 2 \times 1} = \frac{1}{24}$	$\frac{1}{8} > \frac{1}{24}$
5	$\frac{1}{2^4} = \frac{1}{16}$	$\frac{1}{5!} = \frac{1}{5 \times 4 \times 3 \times 2 \times 1} = \frac{1}{120}$	$\frac{1}{16} > \frac{1}{120}$

We can conclude that $\frac{1}{n!} = \frac{1}{n(n-1)(n-2)\cdots(2)(1)} \leq \frac{1}{2^{n-1}}$. Therefore:

$$\frac{1}{1!} + \frac{1}{2!} + \frac{1}{3!} + \cdots + \frac{1}{n!} < 1 + \frac{1}{2} + \frac{1}{2^2} + \frac{1}{2^3} + \cdots + \frac{1}{2^{n-1}}$$

Previously, we found that for the geometric series with $a_1 = 1$ and $r = \frac{1}{2}$, the series approaches $\dfrac{1}{1 - \frac{1}{2}}$ or 2 as n approaches infinity. Therefore, $\sum_{n=1}^{\infty} \frac{1}{n!} < 2$. Thus, this series is bounded above by 2.

To find a lower bound, notice that the series $\frac{1}{1!} + \frac{1}{2!} + \frac{1}{3!} + \cdots + \frac{1}{n!}$ is the sum of positive terms. Therefore, for $n > 3$,

$$\frac{1}{1!} + \frac{1}{2!} < \frac{1}{1!} + \frac{1}{2!} + \frac{1}{3!} + \cdots + \frac{1}{n!}$$

$$\text{or} \quad \frac{3}{2} < \frac{1}{1!} + \frac{1}{2!} + \frac{1}{3!} + \cdots + \frac{1}{n!}$$

Putting it all together, what we have shown is that for the series $\sum_{n=1}^{\infty} \frac{1}{n!}$:

$$\frac{3}{2} < \frac{1}{1!} + \frac{1}{2!} + \frac{1}{3!} + \cdots + \frac{1}{n!} < 2$$

$$\text{or} \quad \frac{3}{2} < \sum_{n=1}^{\infty} \frac{1}{n!} < 2$$

The Number e

Let us add 1 to both sides of the inequality derived in the previous section. We find that as n approaches infinity,

$$1 + \frac{3}{2} < 1 + \frac{1}{1!} + \frac{1}{2!} + \frac{1}{3!} + \cdots + \frac{1}{n!} < 1 + 2$$

$$\text{or} \quad \frac{5}{2} < 1 + \frac{1}{1!} + \frac{1}{2!} + \frac{1}{3!} + \cdots + \frac{1}{n!} < 3$$

The infinite series $1 + \frac{1}{1!} + \frac{1}{2!} + \frac{1}{3!} + \cdots + \frac{1}{n!}$ is equal to an irrational number that is greater than $\frac{5}{2}$ and less than 3. We call this number e.

Eighteenth-century mathematicians computed this number to many decimal places and assigned to it the symbol e just as earlier mathematicians computed to many decimal places the ratio of the length of the circumference of a circle to the length of the diameter and assigned the symbol π to this ratio. Therefore, we say:

$$\sum_{n=0}^{\infty} \frac{1}{n!} = 1 + \frac{1}{1!} + \frac{1}{2!} + \frac{1}{3!} + \cdots + \frac{1}{(n-1)!} + \cdots = e$$

 This number has an important role in many different branches of mathematics. A calculator will give the value of e to nine decimal places.

ENTER: [**2nd**] [**eˣ**] 1 [**ENTER**]

DISPLAY:
```
e^(1
            2.718281828
```

EXAMPLE 2

Find, to the nearest hundredth, the value of $1 + e^2$.

Solution   Evaluate the expression on a calculator.

ENTER: 1 [**+**] [**2nd**] [**eˣ**] 2 [**ENTER**]

DISPLAY:
```
1+e^(2
            8.389056099
```

Answer   To the nearest hundredth, $1 + e^2 \approx 8.39$.

Exercises

Writing About Mathematics

1. Show that if the first term of an infinite geometric series is 1 and the common ratio is $\frac{1}{c}$, then the sum is $\frac{c}{c-1}$.

2. Cody said that since the calculator gives the value of e as 2.718281828, the value of e can be written as $2.71\overline{828}$, a repeating decimal and therefore a rational number. Do you agree with Cody? Explain why or why not.

Developing Skills

In 3–10: **a.** Write each series in sigma notation. **b.** Determine whether each sum increases without limit, decreases without limit, or approaches a finite limit. If the series has a finite limit, find that limit.

3. $1 + \frac{1}{3} + \frac{1}{9} + \cdots$

4. $2 + \frac{1}{2} + \frac{1}{8} + \frac{1}{32} + \cdots$

5. $2 + 4 + 6 + 8 + \cdots$

6. $5 + 1 + \frac{1}{5} + \frac{1}{25} + \cdots$

7. $5 + 1 - 3 - 7 - \cdots$

8. $6 + 3 + \frac{3}{2} + \frac{3}{4} + \cdots$

9. $\frac{1}{2!} + \frac{1}{3!} + \cdots + \frac{1}{(n+1)!} + \cdots$

10. $1 + 3 + 6 + 10 + \cdots$

In 11–16, an infinitely repeating decimal is an infinite geometric series. Find the rational number represented by each of the following infinitely repeating decimals.

11. $1.11111\ldots = 1 + 0.1 + 0.01 + 0.001 + \cdots$

12. $0.33333\ldots = 0.3 + 0.03 + 0.003 + 0.0003 + \cdots$

13. $0.444444\ldots$

14. $0.121212\ldots = 0.12 + 0.0012 + 0.000012 + \cdots$

15. $0.242424\ldots$

16. $0.126126126\ldots$

17. The sum of the infinite series $1 + \frac{1}{2} + \frac{1}{2^2} + \frac{1}{2^3} + \cdots + \frac{1}{2^n}$ is 2. Find values of n such that
$$2 - a_n < \frac{2^{26} - 1}{2^{25}}.$$

CHAPTER SUMMARY

A **sequence** is a set of numbers written in a given order. Each term of a sequence is associated with the positive integer that specifies its position in the ordered set. A **finite sequence** is a function whose domain is the set of integers $\{1, 2, 3, \ldots, n\}$. An **infinite sequence** is a function whose domain is the set of positive integers. The terms of a sequence are often designated as $a_1, a_2, a_3, \ldots$. The formula that allows any term of a sequence except the first to be computed from the previous term is called a **recursive definition**.

An **arithmetic sequence** is a sequence such that for all n, there is a constant d such that $a_{n+1} - a_n = d$. For an arithmetic sequence:

$$a_n = a_1 + (n - 1)d = a_{n-1} + d$$

A **geometric sequence** is a sequence such that for all n, there is a constant r such that $\frac{a_{n+1}}{a_n} = r$. For a geometric sequence:

$$a_n = a_1 r^{n-1} = a_{n-1} r$$

A **series** is the indicated sum of the terms of a sequence. The Greek letter Σ is used to indicate a sum defined for a set of consecutive integer.

If S_n represents the **nth partial sum**, the sum of the first n terms of a sequence, then

$$S_n = \sum_{k=1}^{n} a_k = a_1 + a_2 + a_3 + \cdots + a_n$$

For an arithmetic series:

$$S_n = \frac{n}{2}(a_1 + a_n) = \frac{n}{2}(2a_1 + (n - 1)d)$$

For a geometric series:

$$S_n = \frac{a_1(1 - r^n)}{1 - r}$$

For a geometric series, if $|r| < 1$ and n approaches infinity:

$$S_n = \frac{a_1}{1 - r} \quad \text{or} \quad \sum_{n=1}^{\infty} a_1 r^{n-1} = \frac{a_1}{1 - r}$$

As n approaches infinity:

$$\sum_{n=0}^{\infty} \frac{1}{n!} = 1 + \frac{1}{1!} + \frac{1}{2!} + \frac{1}{3!} + \cdots + \frac{1}{(n - 1)!} = e$$

The number e is an irrational number.

VOCABULARY

6-1 Sequence • Finite sequence • Infinite sequence • Recursive definition

6-2 Common difference • Arithmetic sequence • Arithmetic means

6-3 Series • Σ • Sigma notation • Finite series • Infinite series • ∞

6-4 Arithmetic series • S_n • nth partial sum

6-5 Geometric sequence • Common ratio • Geometric mean

6-6 Geometric series

6-7 n factorial • e

REVIEW EXERCISES

In 1–6: **a.** Write a recursive formula for a_n. **b.** Is the sequence arithmetic, geometric, or neither? **c.** If the sequence is arithmetic or geometric, write an explicit formula for a_n. **d.** Find a_{10}.

1. $1, 5, 9, 13, \ldots$

2. $3, 1, \frac{1}{3}, \frac{1}{9}, \ldots$

3. $12, 11, 10, 9, \ldots$

4. $1, 3, 6, 10, 15, \ldots$

5. $1i, 3i, 7i, 15i, 31i, \ldots$

6. $2, -6, 18, -54, 162, \ldots$

In 7–12, write each series in sigma notation.

7. $1 + 3 + 6 + 10 + 15 + 21 + 28 + 36$

8. $2 + 5 + 8 + 11 + 14 + 17 + 20$

9. $4 + 6 + 8 + 10 + 12 + 14$

10. $1^2 + 3^2 + 5^2 + 7^2 + 9^2 + 11^2$

11. $1 - 2 + 3 - 4 + 5 - 6 + 7$

12. $\frac{1}{2} + \frac{1}{2^2} + \frac{1}{2^3} + \frac{1}{2^4} + \cdots$

13. In an arithmetic sequence, $a_1 = 6$ and $d = 5$. Write the first five terms.

14. In an arithmetic sequence, $a_1 = 6$ and $d = 5$. Find a_{30}.

15. In an arithmetic sequence, $a_1 = 5$ and $a_4 = 23$. Find a_{12}.

16. In an arithmetic sequence, $a_3 = 0$ and $a_{10} = 70$. Find a_1.

17. In a geometric sequence, $a_1 = 2$ and $a_2 = 5$. Write the first five terms.

18. In a geometric sequence, $a_1 = 2$ and $a_2 = 5$. Find a_{10}.

19. In a geometric sequence, $a_1 = 2$ and $a_4 = 128$. Find a_6.

20. Write a recursive formula for the sequence $12, 20, 30, 42, 56, 70, 90, 110, \ldots$.

21. Write the first five terms of a sequence if $a_1 = 1$ and $a_n = 4a_{n-1} + 3$.

22. Find six arithmetic means between 1 and 36.

23. Find three geometric means between 1 and 625.

24. In an arithmetic sequence, $a_1 = 1, d = 8$. If $a_n = 89$, find n.

25. In a geometric sequence, $a_1 = 1$ and $a_5 = \frac{1}{4}$. Find r.

26. Write $\sum\limits_{n=1}^{4} 3^n$ as the sum of terms and find the sum.

27. Write $\sum\limits_{k=0}^{6} (12 - 3k)$ as the sum of terms and find the sum.

28. To the nearest hundredth, find the value of $3e^3$.

29. a. Write, in sigma notation, the series $3 + 1.5 + 0.75 + 0.375 + 0.1875 + \cdots$.

  b. Determine whether the series given in part **a** increases without limit, decreases without limit, or approaches a finite limit. If the series has a finite limit, find that limit.

30. A grocer makes a display of canned tomatoes that are on sale. There are 24 cans in the first (bottom) layer and, after the first, each layer contains three fewer cans than in the layer below. There are three cans in the top layer.

  a. How many layers of cans are in the display?

  b. If the grocer sold all of the cans in the display, how many cans of tomatoes did he sell?

31. A retail store pays cashiers $20,000 a year for the first year of employment and increases the salary by $500 each year.

  a. What is the salary of a cashier in the tenth year of employment?

  b. What is the total amount that a cashier earns in his or her first 10 years?

32. Ben started a job that paid $40,000 a year. Each year after the first, his salary was increased by 4%.

  a. What was Ben's salary in his eighth year of employment?

  b. What is the total amount that Ben earned in eight years?

33. The number of handshakes that are exchanged if every person in a room shakes hands once with every other person can be written as a sequence. Let a_{n-1} be the number of handshakes exchanged when there are $n - 1$ persons in the room. If one more persons enters the room so that there are n persons in the room, that person shakes hands with $n - 1$ persons, creating $n - 1$ additional handshakes.

  a. Write a recursive formula for a_n.

  b. If $a_1 = 0$, write the first 10 terms of the sequence.

  c. Write a formula for a_n in terms of n.

Exploration

The graphing calculator can be used to graph recursive sequences and examine the *convergence*, *divergence*, or *oscillation* of the sequence. A sequence **converges** if as n goes to infinity, the sequence approaches a limit. A sequence **diverges** if as n goes to infinity, the sequence increases or decreases without limit. A sequence **oscillates** if as n goes to infinity, the sequence neither converges nor diverges.

For example, to verify that the geometric sequence $a_n = a_{n-1}\left(\frac{1}{2}\right)$ with $a_1 = 5$ converges to 0:

STEP 1. Set the calculator to sequence graphing mode. Press $\boxed{\text{MODE}}$ and select Seq, the last option in the fourth row.

STEP 2. Enter the expression $a_{n-1}\left(\frac{1}{2}\right)$ as $u(n-1)\left(\frac{1}{2}\right)$ into $u(n)$. The function name u is found above the $\boxed{7}$ key.

ENTER: $\boxed{\text{Y=}}$ $\boxed{\text{2nd}}$ $\boxed{\text{u}}$ $\boxed{(}$ $\boxed{\text{X,T,}\theta\text{,}n}$

$\boxed{-}$ 1 $\boxed{)}$ $\boxed{\times}$ 1 $\boxed{\div}$ 2.

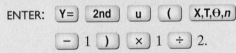

STEP 3. To set the initial value of $a_1 = 5$, set $u(n\text{Min}) = 5$.

ENTER: $\boxed{\blacktriangledown}$ 5

(To set more than one initial value, for example $a_1 = 5$ and $a_2 = 6$, set $u(n\text{Min}) = \{5, 6\}$.)

STEP 4. Graph the sequence up to $n = 25$.

Press $\boxed{\text{WINDOW}}$ $\boxed{\blacktriangledown}$ 25 to set

$n\text{Max} = 25$.

STEP 5. Finally, press ZOOM 0 to graph the sequence. You can use TRACE to explore the values of the sequence. We can see that the sequence approaches the value of 0 very quickly.

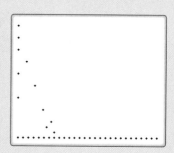

In 1–4: **a.** Graph each sequence up to the first 75 terms. **b.** Use the graph to make a conjecture as to whether the sequence converges, diverges, or oscillates.

1. $a_1 = 1, a_2 = 1, a_n = a_{n-2} + a_{n-1}$ (Fibonacci sequence)

2. $a_1 = 5, a_n = \frac{1}{a_{n-1}}$

3. $a_1 = 100, a_n = \frac{1}{2}a_{n-1} + 100$

4. $a_1 = 10, a_n = 0.95a_{n-1} + 500(0.75)^n$

CUMULATIVE REVIEW   CHAPTERS 1–6

Part I

Answer all questions in this part. Each correct answer will receive 2 credits. No partial credit will be allowed.

1. If the domain is the set of integers, then the solution set of $-2 \le x + 3 < 7$ is
(1) $\{-2, -1, 0, 1, 2, 3, 4, 5, 6, 7\}$
(2) $\{-2, -1, 0, 1, 2, 3, 4, 5, 6\}$
(3) $\{-5, -4, -3, -2, -1, 0, 1, 2, 3, 4\}$
(4) $\{-5, -4, -3, -2, -1, 0, 1, 2, 3\}$

2. The expression $(a + 3)^2 - (a + 3)$ is equal to
(1) $a^2 - a + 6$ (3) $a^2 + 5a + 12$
(2) $a^2 + 5a + 6$ (4) $a + 3$

3. In simplest form, $\frac{x}{3} + \frac{x-2}{4} - \frac{x-2}{8}$ is equal to
(1) $\frac{11x - 6}{24}$ (3) $\frac{11x + 6}{24}$
(2) $\frac{11x - 18}{24}$ (4) $\frac{x(x-2)^2}{24}$

4. The sum of $(3 - \sqrt{5}) + (2 - \sqrt{45})$ is

(1) $5 - 4\sqrt{5}$ (2) $5 - 3\sqrt{5}$ (3) $5 - 5\sqrt{2}$ (4) $\sqrt{2}$

5. The fraction $\dfrac{1}{1 - \sqrt{2}}$ is equal to

(1) $1 + \sqrt{2}$ (2) $-1 + \sqrt{2}$ (3) $-1 - \sqrt{2}$ (4) $1 - \sqrt{2}$

6. The quadratic equation $3x^2 - 7x = 3$ has roots that are
   (1) real, rational, and equal.
   (2) real, rational, and unequal.
   (3) real irrational and unequal.
   (4) imaginary.

7. A function that is one-to-one is
   (1) $y = 2x + 1$ (3) $y = x^4 + x$
   (2) $y = x^2 + 1$ (4) $y = |x|$

8. The function $y = x^2$ is translated 2 units up and 3 units to the left. The equation of the new function is
   (1) $y = (x - 3)^2 + 2$ (3) $y = (x + 3)^2 + 2$
   (2) $y = (x - 3)^2 - 2$ (4) $y = (x + 3)^2 - 2$

9. Which of the following is a real number?
   (1) i (2) i^2 (3) i^3 (4) $2i$

10. The sum of the roots of a quadratic equation is -4 and the product of the roots is 5. The equation could be
   (1) $x^2 - 4x + 5 = 0$ (3) $x^2 - 4x - 5 = 0$
   (2) $x^2 + 4x - 5 = 0$ (4) $x^2 + 4x + 5 = 0$

Part II

Answer all questions in this part. Each correct answer will receive 2 credits. Clearly indicated the necessary steps, including appropriate formula substitutions, diagrams, graphs, charts, etc. For all questions in this part, a correct numerical answer with no work will receive only 1 credit.

11. What number must be added to each side of the equation $x^2 + 3x = 0$ in order to make the left member the perfect square of a binomial?

12. Solve for x and check: $7 - \sqrt{x + 2} = 4$.

Part III

Answer all questions in this part. Each correct answer will receive 4 credits. Clearly indicated the necessary steps, including appropriate formula substitutions, diagrams, graphs, charts, etc. For all questions in this part, a correct numerical answer with no work will receive only 1 credit.

13. Express the fraction $\frac{(2 + i)^2}{i}$ in $a + bi$ form.

14. Graph the solution set of the inequality $2x^2 - 5x - 3 > 0$.

Part IV

Answer all questions in this part. Each correct answer will receive 6 credits. Clearly indicated the necessary steps, including appropriate formula substitutions, diagrams, graphs, charts, etc. For all questions in this part, a correct numerical answer with no work will receive only 1 credit.

15. a. Write the function that is the inverse function of $y = 3x + 1$.

   b. Sketch the graph of the function $y = 3x + 1$ and of its inverse.

   c. What are the coordinates of the point of intersection of the function and its inverse?

16. a. Write a recursive definition for the geometric sequence of five terms in which $a_1 = 3$ and $a_5 = 30{,}000$.

   b. Write the sum of the sequence in part **a** in sigma notation.

   c. Find the sum from part **b**.

CHAPTER

7

EXPONENTIAL FUNCTIONS

The use of exponents to indicate the product of equal factors evolved through many different notations. Here are some early methods of expressing a power using an exponent. In many cases, the variable was not expressed. Each is an example of how the polynomial $9x^4 + 10x^3 + 3x^2 + 7x + 4$ was written.

■ Joost Bürgi (1552–1632) used Roman numerals above the coefficient to indicate an exponent:

$$\overset{IV}{9} + \overset{III}{10} + \overset{II}{3} + \overset{I}{7} + 4$$

■ Adriaan van Roomen (1561–1615) used parentheses:

$$9(4) + 10(3) + 3(2) + 7(1) + 4$$

■ Pierre Hérigone (1580–1643) placed the coefficient before the variable and the exponent after:

$$9x4 + 10x3 + 3x2 + 7x1 + 4$$

■ James Hume used a raised exponent written as a Roman numeral:

$$9x^{iv} + 10x^{iii} + 3x^{ii} + 7x^{i} + 4$$

■ René Descartes (1596–1650) introduced the symbolism that is in common use today:

$$9x^4 + 10x^3 + 3x^3 + 7x + 4$$

7-1 LAWS OF EXPONENTS

When multiplying or dividing algebraic terms, we use the rules for exponents when the terms involve powers of a variable. For example:

$$a^3b^2(3a^2b) = 3a^5b^3$$

$$x^4y^5 \div x^3y^2 = xy^3$$

$$(c^3d^2)^2 = c^6d^4$$

These examples illustrate the following rules for exponents. If a and b are positive integers, then:

Multiplication: $x^a \cdot x^b = x^{a+b}$

Division: $x^a \div x^b = x^{a-b}$ or $\dfrac{x^a}{x^b} = x^{a-b}$ $(x \neq 0)$

Raising to a Power: $(x^a)^b = x^{ab}$

Power of a Product: $(xy)^a = x^a \cdot y^a$

Power of a Quotient: $\left(\dfrac{x}{y}\right)^a = \dfrac{x^a}{y^a}$

In the first example given above, we need the commutative and associative laws to group powers of the same base in order to apply the rules for exponents.

$$
\begin{aligned}
a^3b^2(3a^2b) &= 3a^2a^3b^2b && \text{Commutative property} \\
&= 3(a^2a^3)(b^2b) && \text{Associative property} \\
&= 3a^{2+3}b^{2+1} && x^a \cdot x^b = x^{a+b} \\
&= 3a^5b^3
\end{aligned}
$$

Note: When a variable has no exponent, the exponent is understood to be 1.

In the second example, we use the rule for the division of powers with like bases:

$$
\begin{aligned}
x^4y^5 \div x^3y^2 &= \frac{x^4y^5}{x^3y^2} \\
&= \frac{x^4}{x^3} \cdot \frac{y^5}{y^2} \\
&= x^{4-3}y^{5-2} && \frac{x^a}{x^b} = x^{a-b} \\
&= xy^3
\end{aligned}
$$

In the third example, we can use the rule for multiplying powers with like bases or the rule for raising a power to a power:

$$
\begin{aligned}
&(c^3d^2)^2 \\
&= (c^3d^2)(c^3d^2) && x^2 = x \cdot x \\
&= c^3c^3d^2d^2 && \text{Commutative property} \\
&= c^{3+3}d^{2+2} && x^a \cdot x^b = x^{a+b} \\
&= c^6d^4
\end{aligned}
$$

$$
\begin{aligned}
&(c^3d^2)^2 \\
&= (c^3)^2(d^2)^2 && (xy)^a = x^a \cdot y^a \\
&= c^{3(2)}d^{2(2)} && (x^a)^b = x^{ab} \\
&= c^6d^4
\end{aligned}
$$

EXAMPLE 1

Simplify: $\dfrac{a^4(3b)^7}{a(3b)^2}$

Solution $\dfrac{a^4(3b)^7}{a(3b)^2} = \dfrac{a^4}{a^1} \cdot \dfrac{(3b)^7}{(3b)^2} = a^{4-1}(3b)^{7-2} = a^3(3b)^5 = 3^5 a^3 b^5 = 243 a^3 b^5$ *Answer*

Alternative $\dfrac{a^4(3b)^7}{a(3b)^2} = \dfrac{a^4(3)^7(b)^7}{a^1(3)^2(b)^2} = a^{4-1}(3)^{7-2}(b)^{7-2} = a^3(3)^5(b)^5 = 243 a^3 b^5$ *Answer*
Solution

EXAMPLE 2

What is the value of the fraction $\dfrac{(-3)^4(2)^5}{(-3(2))^3}$?

Solution $\dfrac{(-3)^4(2)^5}{(-3(2))^3} = \dfrac{(-3)^4(2)^5}{(-3)^3(2)^3} = (-3)^{4-3}(2)^{5-3} = (-3)^1(2)^2 = (-3)(4) = -12$ *Answer*

Evaluating Powers

Note that the expression $(3a)^4$ is *not* equivalent to $3a^4$. We can see this by using the rule for multiplying powers with like bases or the rule for raising a power to a power. Let $a = 2$ or any number not equal to 0 or 1:

$$3a^4 = 3(a)(a)(a)(a) = 3(2)(2)(2)(2) = 48$$
$$(3a)^4 = (3a)(3a)(3a)(3a) = (3 \times 2)(3 \times 2)(3 \times 2)(3 \times 2) = (6)(6)(6)(6) = 1{,}296$$

or

$$(3a)^4 = (3)^4(a)^4 = (3)^4(2)^4 = (81)(16) = 1{,}296$$

Similarly, since $-a = -1a$, the expression $-a^4$ is *not* equivalent to $(-a)^4$. Let $a = 2$ (or any number not equal to 0 or 1):

$$-a^4 = -(a)(a)(a)(a) = -(2)(2)(2)(2) = -16$$
$$(-a)^4 = (-a)(-a)(-a)(-a) = (-2)(-2)(-2)(-2) = 16$$

Exercises

Writing About Mathematics

1. Randy said that $(2)^3(5)^2 = (10)^5$. Do you agree with Randy? Justify your answer.

2. Natasha said that $(2)^3(5)^3 = (10)^3$. Do you agree with Natasha? Justify your answer.

Developing Skills

In 3–26, simplify each expression. In each exercise, all variables are positive.

3. $x^3 \cdot x^4$

4. $y \cdot y^5$

5. $x^6 \div x^2$

6. $y^4 \div y$

7. $(x^5)^2$

8. $(2y^4)^3$

9. $10^2 \cdot 10^4$

10. $-2^6 \cdot 2^2$

11. $x^4 \cdot x^2 y^3$

12. $xy^5 \cdot xy^2$

13. $-(3x^3)^2$

14. $(-3x^3)^2$

15. $x^8 y^6 \div (x^3 y^5)$

16. $x^9 y^7 \div (x^8 y^7)$

17. $(x^2 y^3)^3 \cdot (x^2 y)$

18. $(-2x)^4 \cdot (2x^3)^2$

19. $\frac{(4x)^3}{4x^3}$

20. $\frac{3(x^3)^4 y^5}{3x^7}$

21. $\frac{-x^4 y^6}{(-x^3 y^4)}$

22. $\left(\frac{x^3 y^5}{(xy^2)^2} \right)^2$

23. $\frac{x^2 (y^3 z)^3}{(x^2 y)^2 z}$

24. $\left(\frac{2a^3}{a^2} \right)^5 \cdot b$

25. $\frac{4(ab)^2 c^5}{abc}$

26. $\frac{(a^x)^y b}{a^{xy}}$

Applying Skills

27. What is the value of n if $8^3 = 2^n$?

28. What is the value of a if $27^2 = 9^a$?

29. If $3^{a+1} = x$ and $3^a = y$, express y in terms of x.

30. If $25^{b+1} = x$ and $5^b = y$, express x in terms of y.

In 31–33, the formula $A = P(1 + r)^t$ expresses the amount A to which P dollars will increase if invested for t years at a rate of r per year.

31. Find A when $P = \$500$, $r = 0.04$ and $t = 5$.

32. Find the amount to which $2,400 will increase when invested at 5% for 10 years.

33. What is the minimum number of years that $1 must be in invested at 5% to increase to $2? (Use a calculator to try possible values of t.)

7-2 ZERO AND NEGATIVE EXPONENTS

Zero Exponents

We know that any nonzero number divided by itself is 1:

$$\frac{4}{4} = 1 \qquad \frac{3^4}{3^4} = \frac{81}{81} = 1 \qquad \frac{7^5}{7^5} = 1$$

In general, for $x \neq 0$ and n a positive integer:

$$\frac{x^n}{x^n} = 1$$

Can we apply the rule for the division of powers with like bases to these examples?

$$\frac{4}{4} = \frac{2^2}{2^2} = 2^{2-2} = 2^0 \qquad \frac{3^4}{3^4} = 3^{4-4} = 3^0 \qquad \frac{7^5}{7^5} = 7^{5-5} = 7^0$$

In general, for $x \neq 0$ and n a positive integer:

$$\frac{x^n}{x^n} = x^{n-n} = x^0$$

In order for the rule for the division of powers with like bases to be consistent with ordinary division, we must be able to show that for all $x \neq 0$, $x^0 = 1$.

Multiplication: $x^n \cdot x^0 = x^{n+0} = x^n$	$x^n \cdot x^0 = x^n \cdot 1 = x^n$
Division: $x^n \div x^0 = x^{n-0} = x^n$	$x^n \div x^0 = x^n \div 1 = x^n$
Raising to a Power: $(x^0)^n = x^{0 \cdot n} = x^0 = 1$	$(x^0)^n = 1^n = 1$
Power of a Product: $(xy)^0 = x^0 \cdot y^0 = 1 \cdot 1 = 1$	$(xy)^0 = 1$
Power of a Quotient: $\left(\frac{x}{y}\right)^0 = \frac{x^0}{y^0} = \frac{1}{1} = 1$	$\left(\frac{x}{y}\right)^0 = 1$

Therefore, it is reasonable to make the following definition:

DEFINITION

If $x \neq 0$, $x^0 = 1$.

EXAMPLE I

Find the value of $\frac{3x^2}{x^2}$ in two different ways.

Solution $\frac{3x^2}{x^2} = 3 \cdot \frac{x^2}{x^2} = 3(1) = 3$ or $\frac{3x^2}{x^2} = 3x^{2-2} = 3x^0 = 3(1) = 3$

Answer 3

Negative Exponents

When using the rule for the division of powers with like bases, we will allow the exponent of the denominator to be larger than the exponent of the numerator.

$$\frac{3^4}{3^6} = \frac{1 \times 3^4}{3^2 \times 3^4} = \frac{1}{3^2} \times \frac{3^4}{3^4} = \frac{1}{3^2} \times 1 = \frac{1}{3^2} \qquad\qquad \frac{3^4}{3^6} = 3^{4-6} = 3^{-2}$$

$$\frac{5}{5^4} = \frac{1 \times 5}{5^3 \times 5} = \frac{1}{5^3} \times \frac{5}{5} = \frac{1}{5^3} \times 1 = \frac{1}{5^3} \qquad\qquad \frac{5}{5^4} = 5^{1-4} = 5^{-3}$$

$$\frac{x^a}{x^{a+b}} = \frac{1 \cdot x^a}{x^b \cdot x^a} = \frac{1}{x^b} \cdot \frac{x^a}{x^a} = \frac{1}{x^b} \cdot 1 = \frac{1}{x^b} \qquad\qquad \frac{x^a}{x^{a+b}} = x^{a-(a+b)} = x^{-b}$$

This last example suggests that $x^{-b} = \frac{1}{x^b}$. Before we accept that $x^{-b} = \frac{1}{x^b}$, we must show that this is consistent with the rules for powers.

$$\text{Multiplication:} \quad 2^5 \cdot 2^{-3} = 2^{5+(-3)} = 2^2$$
$$2^5 \cdot 2^{-3} = 2^5 \cdot \frac{1}{2^3} = 2^2$$

$$\text{Division:} \quad 6^2 \div 6^{-5} = 6^{2-(-5)} = 6^7$$
$$6^2 \div 6^{-5} = 6^2 \div \frac{1}{6^5} = 6^2 \cdot 6^5 = 6^7$$

$$\text{Raising to a Power:} \quad (3^{-4})^{-2} = 3^{-4(-2)} = 3^8$$
$$(3^{-4})^{-2} = \left(\frac{1}{3^4}\right)^{-2} = \frac{1}{3^{-8}} = \frac{1}{\frac{1}{3^8}} = \frac{3^8}{1} = 3^8$$

$$\text{Power of a Product:} \quad (x^2 y^{-3})^4 = x^8 y^{-12} = x^8 \cdot \frac{1}{y^{12}} = \frac{x^8}{y^{12}}$$
$$(x^2 y^{-3})^4 = \left(x^2 \cdot \frac{1}{y^3}\right)^4 = (x^2)^4 \left(\frac{1}{y^3}\right)^4 = \frac{x^8}{y^{12}}$$

$$\text{Power of a Quotient:} \quad \left(\frac{x^3}{y^5}\right)^{-2} = \frac{x^{-6}}{y^{-10}} = \frac{\frac{1}{x^6}}{\frac{1}{y^{10}}} = \frac{y^{10}}{x^6}$$
$$\left(\frac{x^3}{y^5}\right)^{-2} = \frac{1}{\left(\frac{x^3}{y^5}\right)^2} = \frac{1}{\frac{x^6}{y^{10}}} = \frac{y^{10}}{x^6}$$

Therefore, it is reasonable to make the following definition:

DEFINITION

If $x \neq 0$, $x^{-n} = \frac{1}{x^n}$.

EXAMPLE 2

Show that for $x \neq 0$, $\frac{1}{x^{-n}} = x^n$.

Solution
$$\frac{1}{x^{-n}} = \frac{1}{\frac{1}{x^n}} = \frac{1}{\frac{1}{x^n}} \cdot \frac{x^n}{x^n} = \frac{x^n}{1} = x^n$$

EXAMPLE 3

Write $\frac{a^4 b^{-3}}{ab^{-2}}$ with only positive exponents.

Solution
$$\frac{a^4 b^{-3}}{ab^{-2}} = a^{4-1} b^{-3-(-2)} = a^3 b^{-1} = a^3 \cdot \frac{1}{b} = \frac{a^3}{b} \quad \textit{Answer}$$

EXAMPLE 4

Express as a fraction the value of $2x^0 + 3x^{-3}$ for $x = 5$.

Solution $\quad 2x^0 + 3x^{-3} = 2(5)^0 + 3(5)^{-3} = 2(1) + 3 \times \frac{1}{5^3} = 2 + \frac{3}{125} = \frac{250}{125} + \frac{3}{125} = \frac{253}{125}$

Answer $\frac{253}{125}$

Exercises

Writing About Mathematics

1. Kim said that $a^0 + a^0 = a^{0+0} = a^0 = 1$. Do you agree with Kim? Explain why or why not.

2. Tony said that $a^0 + a^0 = 2a^0 = 2$. Do you agree with Tony? Explain why or why not?

Developing Skills

In 3–10, write each expression as a rational number without an exponent.

3. 5^{-1} **4.** 4^{-2} **5.** 6^{-2} **6.** $\left(\frac{1}{2}\right)^{-1}$

7. $\left(\frac{1}{5}\right)^{-3}$ **8.** $\left(\frac{2}{3}\right)^{-1}$ **9.** $\frac{3^0}{4^{-2}}$ **10.** $\frac{(2 \cdot 5)^{-4}}{5^{-2}}$

In 11–22, find the value of each expression when $x \neq 0$.

11. 7^0 **12.** $(-5)^0$ **13.** x^0 **14.** -4^0

15. $(4x)^0$ **16.** $4x^0$ **17.** $-2x^0$ **18.** $(-2x)^0$

19. $\left(\frac{3}{4}\right)^0$ **20.** $\frac{3^0}{4}$ **21.** $\frac{3^0}{4^0}$ **22.** $\frac{3x^0}{(4x)^0}$

In 23–34, evaluate each function for the given value. Be sure to show your work.

23. $f(x) = x^{-3} \cdot x^4$; $f(1)$

24. $f(x) = x + x^{-5}$; $f(3)$

25. $f(x) = (2x)^{-6} \div x^3$; $f(-3)$

26. $f(x) = (x^{-7})^4$; $f(-6)$

27. $f(x) = \left(\frac{1}{x} + \frac{3}{2}\right)^{-2}$; $f(2)$

28. $f(x) = 10^x + 10^{-2x}$; $f(3)$

29. $f(x) = x^{-7} \div x^8$; $f\left(\frac{3}{4}\right)$

30. $f(x) = (3x^{-3} - 2x^{-3})^2$; $f(-2)$

31. $f(x) = x^8\left(x^{-2} + \frac{1}{x^3}\right)$; $f\left(\frac{1}{2}\right)$

32. $f(x) = \left(\frac{x^{-1}}{(2x)^{-2}}\right)^{-1}$; $f(8)$

33. $f(x) = \frac{1}{1 + \frac{2}{x^{-1}}}$; $f(-5)$

34. $f(x) = 4\left(\frac{1}{2}\right)^{-x} + 3\left(\frac{1}{2}\right)^{-x}$; $f(3)$

In 35–63, write each expression with only positive exponents and express the answer in simplest form. The variables are not equal to zero.

35. x^{-4} **36.** a^{-6} **37.** y^{-5} **38.** $2x^{-2}$

39. $7a^{-4}$ **40.** $-5y^{-8}$ **41.** $(2x)^{-2}$ **42.** $(3a)^{-4}$

43. $(4y)^{-3}$ **44.** $-(2x)^{-2}$ **45.** $-(3a)^{-4}$ **46.** $(-2x)^{-2}$

47. $\frac{1}{x^{-3}}$ **48.** $\frac{1}{y^{-7}}$ **49.** $\frac{3}{a^{-5}}$ **50.** $\frac{6}{a^{-4}}$

51. $\frac{9x^2}{a^{-3}}$ **52.** $\frac{(-x)^{-5}}{x^{-3}}$ **53.** $\frac{(2a)^{-1}}{2(a)^{-2}}$ **54.** $\frac{4y^{-3}}{2y^{-1}}$

55. $(xy^5z^{-2})^{-1}$

56. $(a^5b^{-5}c^{-4})^{-3}$

57. $\left(\frac{3m^{-3}}{2n^{-2}}\right)^{-3}$

58. $\left(\frac{6ab^4}{3x^{-3}y^{-4}}\right)^{-1}$

59. $\frac{x^5y^{-4}}{x^{-4}y^{-2}}$

60. $(3ab^{-2})(3a^{-1}b^3)$

61. $\frac{-64x^4a^{-2}}{2x^5b^{-4}}$

62. $\left(\frac{-49u^3v^4}{-7u^4v^7}\right)^{-1}$

63. $x^{-1} + x^{-5}$

In 64–75, write each quotient as a product without a denominator. The variables are not equal to zero.

64. $(xy) \div (xy^3)$

65. $(a^2b^3) \div (ab^5)$

66. $x^3 \div (x^3y^{\,4})$

67. $12ab \div 2ab^2$

68. $\frac{1}{a^{-3}}$

69. $\frac{3}{x^4}$

70. $\frac{8}{4a^3}$

71. $\frac{36}{9x^{-5}}$

72. $\frac{3a^0b^{-3}}{a^{-1}b^{-3}}$

73. $\frac{20x^0y^{-5}}{4x^{-1}y^5}$

74. $\frac{15x^{-2}y^2}{3xy^5}$

75. $\frac{25a^5b^{-3}}{5^0a^{-1}b}$

76. Find the value of $a^0 + (4a)^{-1} + 4a^{-2}$ if $a = 2$.

77. Find the value of $(-5a)^0 - 5a^{-2}$ if $a = 3$.

78. If $\left(\frac{3}{4}\right)^{-3} + \left(\frac{8}{3}\right)^2 = \frac{2^a}{3^b}$, find the values of a and of b.

79. Show that $3 \times 10^{-2} = \frac{3}{100}$.

7-3 FRACTIONAL EXPONENTS

We know the following:

Since $3^2 = 9$: $\quad \sqrt{9} = 3$ or $\sqrt{3^2}$

Since $5^3 = 125$: $\quad \sqrt[3]{125} = 5$ or $\sqrt[3]{5^3}$

Since $2^4 = 16$: $\quad \sqrt[4]{16} = 2$ or $\sqrt[4]{2^4}$

In general, for $x > 0$: $\quad \sqrt[n]{x^n} = x$

In other words, raising a positive number to the nth power and taking the nth root of the power are inverse operations. Is it possible to express the nth root as a power? Consider the following example:

(1) Let x be any positive real number: $\qquad \sqrt{x} \cdot \sqrt{x} = x$

(2) Assume that $\sqrt{x}$ can be expressed as x^a: $\qquad x^a \cdot x^a = x$

(3) Use the rule for multiplying powers with like bases: $\qquad x^{a+a} = x^1$

$$x^{2a} = x^1$$

$$2a = 1$$

$$a = \tfrac{1}{2}$$

(4) Therefore, we can write the following equality: $\qquad x^{\frac{1}{2}} \cdot x^{\frac{1}{2}} = x$

(5) Since for $x > 0$, $\sqrt{x}$ is one of the two equal factors of x, then: $\qquad \sqrt{x} = x^{\frac{1}{2}}$

We can apply this reasoning to any power and the corresponding root:

$$\underbrace{\sqrt[n]{x} \cdot \sqrt[n]{x} \cdot \sqrt[n]{x} \cdots \cdots \sqrt[n]{x}}_{n \text{ factors}} = x$$

Assume that $\sqrt[n]{x}$ can be expressed as x^a:

$$\underbrace{x^a \cdot x^a \cdot x^a \cdot x^a \cdots \cdots x^a}_{n \text{ factors}} = x$$

$$(x^a)^n = x^1$$

$$an = 1$$

$$a = \tfrac{1}{n}$$

For n a positive integer, $\sqrt[n]{x}$ is one of the n equal factors of x, and $x^{\frac{1}{n}}$ is one of the n equal factors of x. Therefore:

$$\sqrt[n]{x} = x^{\frac{1}{n}}$$

Before we accept this relationship, we must verify that operations with radicals and with fractional exponents give the same results.

Multiplication: $\quad \sqrt{8} \times \sqrt{2} = \sqrt{16} = 4$

$$(8)^{\frac{1}{2}} \times (2)^{\frac{1}{2}} = (2^3)^{\frac{1}{2}} \times (2)^{\frac{1}{2}}$$

$$= (2)^{\frac{3}{2}} \times (2)^{\frac{1}{2}}$$

$$= (2)^{\frac{3}{2}+\frac{1}{2}}$$

$$= (2)^{\frac{4}{2}}$$

$$= (2)^2 = 4$$

Division: $\dfrac{5}{\sqrt{5}} = \dfrac{5}{\sqrt{5}} \times \dfrac{\sqrt{5}}{\sqrt{5}} = \dfrac{5\sqrt{5}}{5} = \sqrt{5}$

$\dfrac{5^1}{5^{\frac{1}{2}}} = 5^{1-\frac{1}{2}} = 5^{\frac{1}{2}}$

Raising to a Power: $\sqrt[3]{2^6} = \sqrt[3]{64} = 4$

$(2^6)^{\frac{1}{3}} = 2^{6\left(\frac{1}{3}\right)} = 2^2 = 4$

Verifying that the laws for power of a product and power of a quotient also hold is left to the student. (See Exercise 83.) These examples indicate that it is reasonable to make the following definition:

DEFINITION

If $x \geq 0$ and n is a positive integer, $\sqrt[n]{x} = x^{\frac{1}{n}}$.

From this definition and the rules for finding the power of a power and the definition of negative powers, we can also write the following:

$$\sqrt[n]{x^m} = \left(\sqrt[n]{x}\right)^m = x^{\frac{m}{n}} \qquad \frac{1}{\sqrt[n]{x^m}} = \frac{1}{\left(\sqrt[n]{x}\right)^m} = x^{-\frac{m}{n}}$$

Note: When $x < 0$, these formulas may lead to invalid results.

For instance: $(-8)^{\frac{1}{3}} = \sqrt[3]{-8} = -2$ and $\dfrac{1}{3} = \dfrac{2}{6}$

However: $(-8)^{\frac{2}{6}} = \sqrt[6]{(-8)^2} = \sqrt[6]{64} = 2$

Thus, $(-8)^{\frac{1}{3}} \neq (-8)^{\frac{2}{6}}$ even though the exponents are equal.

 A calculator will evaluate powers with fractional exponents. For instance, to find $(125)^{\frac{2}{3}}$:

ENTER: 125 [^] [( ] 2 [÷] 3 [ )] [ENTER]

DISPLAY:
```
125^(2/3)
              25
```

We can also evaluate $(125)^{\frac{2}{3}}$ by writing it as $\sqrt[3]{125^2}$ or as $\left(\sqrt[3]{125}\right)^2$.

ENTER: [MATH] [4] 125 [x²]   ENTER: [( ] [MATH] [4] 125 [ )] [ )]

[ )] [ENTER]   [x²] [ENTER]

DISPLAY:
```
3√(125²)
              25
```
DISPLAY:
```
(3√(125))²
              25
```

The calculator shows that $(125)^{\frac{2}{3}} = \sqrt[3]{125^2} = \left(\sqrt[3]{125}\right)^2$.

EXAMPLE I

Find the value of $81^{-\frac{3}{4}}$.

Solution
$$81^{-\frac{3}{4}} = \frac{1}{81^{\frac{3}{4}}} = \frac{1}{\left(\sqrt[4]{81}\right)^3} = \frac{1}{(3)^3} = \frac{1}{27} \text{ Answer}$$

Calculator Solution

ENTER: 81 `^` `(-)` `(` 3 `÷` 4 `)` `MATH` `ENTER` `ENTER`

DISPLAY:
```
81^-(3/4)▶Frac
              1/27
```

Note that we used the Frac function from the `MATH` menu to convert the answer to a fraction.

EXAMPLE 2

Find the value of $2a^0 - (2a)^0 + a^{\frac{1}{3}}$ if $a = 64$.

Solution
$$\begin{aligned}
2a^0 - (2a)^0 + a^{\frac{1}{3}} &= 2(64)^0 - [2(64)]^0 + 64^{\frac{1}{3}} \\
&= 2(1) - 1 + \sqrt[3]{64} \\
&= 2 - 1 + 4 \\
&= 5 \text{ Answer}
\end{aligned}$$

Exercises

Writing About Mathematics

1. Use exponents to show that for $a > 0$, $\left(\sqrt[n]{a}\right)^0 = 1$.

2. Use exponents to show that for $a > 0$, $\sqrt{\sqrt{a}} = \sqrt[4]{a}$.

Developing Skills

In 3–37, express each power as a rational number in simplest form.

3. $4^{\frac{1}{2}}$ **4.** $9^{\frac{1}{2}}$ **5.** $100^{\frac{1}{2}}$ **6.** $8^{\frac{1}{3}}$

7. $125^{\frac{1}{3}}$ **8.** $216^{\frac{1}{3}}$ **9.** $32^{\frac{1}{5}}$ **10.** $(3 \times 12)^{\frac{1}{2}}$

11. $(2 \times 8)^{\frac{1}{4}}$ **12.** $5(81)^{\frac{1}{4}}$ **13.** $-4(1{,}000)^{\frac{1}{3}}$ **14.** $49^{\frac{3}{2}}$

15. $8^{\frac{5}{3}}$ **16.** $27^{\frac{4}{3}}$ **17.** $10{,}000^{\frac{3}{4}}$ **18.** $32^{\frac{4}{5}}$

19. $9^{-\frac{1}{2}}$ **20.** $8^{-\frac{1}{3}}$ **21.** $100^{-\frac{3}{2}}$ **22.** $125^{-\frac{2}{3}}$

23. $3^{\frac{1}{2}} \times 3^{\frac{3}{2}}$ **24.** $5^{\frac{1}{3}} \times 5^{\frac{2}{3}}$ **25.** $7^{\frac{3}{4}} \times 7^{\frac{5}{4}}$ **26.** $4 \times 4^{\frac{1}{2}}$

27. $32 \times 32^{\frac{1}{5}}$ **28.** $2^{\frac{1}{4}} \times 8^{\frac{1}{4}}$ **29.** $12^{\frac{5}{3}} \div 12^{\frac{2}{3}}$ **30.** $3^{\frac{7}{3}} \div 3^{\frac{1}{3}}$

31. $4^{\frac{2}{3}} \div 4^{\frac{1}{6}}$ **32.** $125^{\frac{2}{3}} \div 125^{\frac{1}{3}}$ **33.** $4^0 + 4^{-\frac{1}{2}}$ **34.** $9^{-2} + 9^{\frac{1}{2}}$

35. $2[(3)^{-2} + (4)^{-2}]^{-\frac{1}{2}}$ **36.** $\left(2.3 \times 10^{\frac{1}{3}}\right)\left(5.2 \times 10^{\frac{2}{3}}\right)$ **37.** $\dfrac{\left(2(3)^2 + \frac{1}{3^{-2}}\right)^{\frac{2}{3}}}{6\left(2 + \frac{1}{4}\right)^{-\frac{1}{2}}}$

In 38–57, write each radical expression as a power with positive exponents and express the answer in simplest form. The variables are positive numbers.

38. $\sqrt{7}$ **39.** $\sqrt{6}$ **40.** $\sqrt[3]{12}$ **41.** $\sqrt[3]{15}$

42. $\sqrt[4]{3}$ **43.** $\sqrt[5]{2^3}$ **44.** $\left(\sqrt[5]{9}\right)^4$ **45.** $\dfrac{1}{\left(\sqrt{5}\right)^3}$

46. $\sqrt{25a}$ **47.** $\sqrt{49x^2}$ **48.** $\sqrt{64a^3b^6}$ **49.** $\frac{1}{2}\sqrt{18a^6b^2}$

50. $\sqrt{9a^{-2}b^6}$ **51.** $\sqrt{\frac{3a}{4b}}$ **52.** $\sqrt[3]{27a^3}$ **53.** $\sqrt[4]{64x^5}$

54. $\dfrac{1}{\sqrt[5]{xyz^5}}$ **55.** $\sqrt{\frac{9a^{-2}}{4b^4}}$ **56.** $\sqrt[10]{\frac{w^{15}x^{20}}{y^5}}$ **57.** $\sqrt[8]{\sqrt[4]{a}} \cdot \sqrt[4]{b^7}$

In 58–73, write each power as a radical expression in simplest form. The variables are positive numbers.

58. $3^{\frac{1}{2}}$ **59.** $5^{\frac{1}{2}}$ **60.** $6^{\frac{1}{3}}$ **61.** $9^{\frac{1}{3}}$

62. $5^{\frac{3}{2}}$ **63.** $12^{\frac{5}{4}}$ **64.** $6^{\frac{5}{2}}$ **65.** $\dfrac{1}{5^{\frac{3}{2}}}$

66. $\left(x^{13}\right)^{\frac{1}{7}}$ **67.** $(25x^2y)^{\frac{1}{2}}$ **68.** $(50ab^4)^{\frac{1}{2}}$ **69.** $(16a^5b^6)^{\frac{1}{4}}$

70. $\dfrac{(x^5y^6)^{\frac{1}{7}}}{z^{-\frac{3}{7}}}$ **71.** $\dfrac{5^{\frac{1}{3}}a^{\frac{2}{3}}}{4^{\frac{1}{3}}}$ **72.** $\left(\dfrac{-32x^{10}}{y^4}\right)^{\frac{1}{5}}$ **73.** $\dfrac{8^{\frac{1}{6}}a^{\frac{5}{6}}b^{\frac{3}{6}}}{(27c^4)^{\frac{1}{6}}}$

In 74–82, write each expression as a power with positive exponents in simplest form.

74. $\left(\dfrac{2a^{\frac{1}{2}}}{3a^{\frac{1}{6}}}\right)^6$

75. $\left(\dfrac{x^2y}{3x^4b^2}\right)^{\frac{2}{3}}$

76. $\left(\dfrac{4a^4b^6}{25a^{-1}b}\right)^{\frac{1}{2}}$

77. $\left(\dfrac{8a^2z^6}{27x^9a^{-4}z^{-1}}\right)^{\frac{1}{3}}$

78. $\sqrt{x^2y}\cdot\sqrt{x^4y^3}$

79. $\dfrac{\sqrt[6]{a^5}}{\sqrt[5]{a^3}}$

80. $\dfrac{\sqrt[3]{11x^5y^4}}{\sqrt{2x^5y^2}}$

81. $\dfrac{\sqrt[5]{48xy^2}}{\sqrt[3]{6x^2y^4}}$

82. $\left(\sqrt{2xy^2}\right)\left(\sqrt[4]{16x^2y}\right)$

83. Verify that the laws for power of a product and power of a quotient are true for the following examples. In each example, evaluate the left side using the rules for radicals and the right side using the rules for fractional exponents:

a. $\left(\sqrt[3]{27}\cdot\sqrt[3]{3}\right)^2 \overset{?}{=} \left(27^{\frac{1}{3}}\cdot 3^{\frac{1}{3}}\right)^2$

b. $\left(\dfrac{\sqrt{3}}{\sqrt{9}}\right)^3 \overset{?}{=} \left(\dfrac{3^{\frac{1}{2}}}{9^{\frac{1}{2}}}\right)^3$

7-4 EXPONENTIAL FUNCTIONS AND THEIR GRAPHS

Two friends decide to start a chess club. At the first meeting, they are the only members of the club. To increase membership, they decide that at the next meeting, each will bring a friend. At the second meeting there are twice as many, or $2 \times 2 = 2^2$, members. Since they still want to increase membership, at the third meeting, each member brings a friend so that there are $2 \times 2^2 = 2^3$ members. If this pattern continues, then at the nth meeting, there will be 2^n members. We say that the membership has increased exponentially or is an example of **exponential growth**.

Often quantities that change exponentially decrease in size. For example, radioactive elements decrease by an amount that is proportional to the size of the sample. This is an example of **exponential decay**. The half-life of an element is the length of time required for the amount of a sample to decrease by one-half. If the weight of a sample is 1 gram and the half life is t years, then:

- After one period of t years, the amount present is $(1)\frac{1}{2} = 2^{-1}$ grams.
- After two periods of t years, the amount present is $\left(\frac{1}{2}\right)\left(\frac{1}{2}\right) = \frac{1}{4} = 2^{-2}$ grams.
- After three periods of t years, the amount present is $\left(\frac{1}{4}\right)\left(\frac{1}{2}\right) = \frac{1}{8} = 2^{-3}$ grams.
- After n periods of t years, the amount present is $\frac{1}{2^n} = 2^{-n}$ grams.

Each of the examples given above suggests a function of the form $f(x) = 2^x$. To study the function $f(x) = 2^x$, choose values of x, find the corresponding value of $f(x)$, and plot the points. Draw a smooth curve through these points to represent the function.

x	2^x	y	x	2^x	y
−3	$2^{-3} = \frac{1}{2^3}$	$\frac{1}{8}$	1	$2^1 = 2$	2
−2	$2^{-2} = \frac{1}{2^2}$	$\frac{1}{4}$	$\frac{3}{2}$	$2^{\frac{3}{2}} = \sqrt{2^3}$	$\sqrt{8} \approx 2.8$
−1	$2^{-1} = \frac{1}{2^1}$	$\frac{1}{2}$	2	$2^2 = 4$	4
0	$2^0 = 1$	1	$\frac{5}{2}$	$2^{\frac{5}{2}} = \sqrt{2^5}$	$\sqrt{32} \approx 5.7$
$\frac{1}{2}$	$2^{\frac{1}{2}} = \sqrt{2}$	$\sqrt{2} \approx 1.4$	3	$2^3 = 8$	8

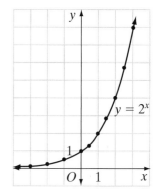

The function $y = 2^x$ is an increasing function. As we trace the graph from left to right, the values of y increase, that is, as x increases, y also increases.

If we reflect the graph of $y = 2^x$ over the y-axis, $(x, y) \to (-x, y)$ and the equation of the image is $y = 2^{-x}$, that is, $y = \frac{1}{2^x}$ or $y = \left(\frac{1}{2}\right)^x$. Compare the graph shown below with the values of $\left(\frac{1}{2}\right)^x$ shown in the table.

x	$\left(\frac{1}{2}\right)^x$	y	x	$\left(\frac{1}{2}\right)^x$	y
−3	$\left(\frac{1}{2}\right)^{-3} = 2^3$	8	1	$\left(\frac{1}{2}\right)^1$	$\frac{1}{2}$
−2	$\left(\frac{1}{2}\right)^{-2} = 2^2$	4	2	$\left(\frac{1}{2}\right)^2$	$\frac{1}{4}$
−1	$\left(\frac{1}{2}\right)^{-1} = 2^1$	2	3	$\left(\frac{1}{2}\right)^3$	$\frac{1}{8}$
0	$\left(\frac{1}{2}\right)^0$	1	4	$\left(\frac{1}{2}\right)^4$	$\frac{1}{16}$

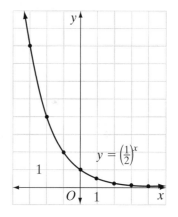

Note that the function $y = \left(\frac{1}{2}\right)^x$ is a decreasing function. As we trace the graph from left to right, the values of y decrease, that is, as x increases, y decreases.

An **exponential function** is a function of the form $y = b^x$ where b is a positive number not equal to 1. If $b > 1$, the function is an increasing function. If $0 < b < 1$, the function is a decreasing function.

We can use a graphing calculator to draw several exponential functions with $b > 1$. For example, let $y = \left(\frac{5}{2}\right)^x$, $y = 3^x$, and $y = 5^x$. First, set the **WINDOW** so that $Xmin = -3$, $Xmax = 3$, $Ymin = -1$, and $Ymax = 10$.

ENTER: Y= (5 ÷ 2) ^ X,T,Θ,n ENTER

3 ^ X,T,Θ,n ENTER

5 ^ X,T,Θ,n GRAPH

DISPLAY:

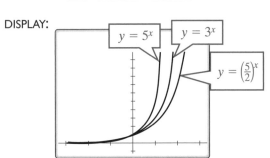

$y = 5^x$

$y = 3^x$

$y = \left(\frac{5}{2}\right)^x$

If we press **TRACE**, the calculator will display the values $x = 0, y = 1$ for the first graph. By pressing **▼**, the calculator will display this same set of values for the second function and again for the third. The point $(0, 1)$ is a point on every function of the form $y = b^x$. For each function, as x decreases through negative values, the values of y get smaller and smaller but are always positive. We say that as x approaches $-\infty$, y approaches 0. The x-axis or the line $y = 0$ is a **horizontal asymptote**.

EXAMPLE 1

a. Sketch the graph of $y = \left(\frac{3}{2}\right)^x$.

b. From the graph, estimate the value of $\left(\frac{3}{2}\right)^{1.2}$, the value of y when $x = 1.2$.

c. Sketch the graph of the image of the graph of $y = \left(\frac{3}{2}\right)^x$ under a reflection in the y-axis.

d. Write an equation of the graph drawn in part **c**.

Solution **a.**

x	y
-2	$\left(\frac{3}{2}\right)^{-2} = \left(\frac{2}{3}\right)^{2} = \frac{4}{9}$
-1	$\left(\frac{3}{2}\right)^{-1} = \left(\frac{2}{3}\right)^{1} = \frac{2}{3}$
0	$\left(\frac{3}{2}\right)^{0} = 1$
1	$\left(\frac{3}{2}\right)^{1} = \frac{3}{2}$
2	$\left(\frac{3}{2}\right)^{2} = \frac{9}{4}$
3	$\left(\frac{3}{2}\right)^{3} = \frac{27}{8}$

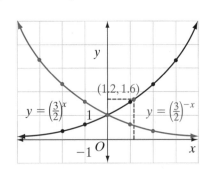

$y = \left(\frac{3}{2}\right)^x$

$y = \left(\frac{3}{2}\right)^{-x}$

$(1.2, 1.6)$

b. Locate the point 1.2 on the x-axis. Draw a vertical line to the graph. From that point on the graph draw a horizontal line to the y-axis. Read the approximate value of y from the graph. The value is approximately 1.6.

c. For each point (a, b) on the graph of $y = \left(\frac{3}{2}\right)^x$, locate the point $(-a, b)$ on the graph of the image.

d. The equation of the image is $y = \left(\frac{3}{2}\right)^{-x}$ or $y = \left(\frac{2}{3}\right)^x$.

Answers   **a.** Graph   **b.** 1.6   **c.** Graph   **d.** $y = \left(\frac{2}{3}\right)^x$

We can use the graphing calculator to find the values of the function used to graph $y = \left(\frac{3}{2}\right)^x$ in Example 1.

STEP 1. Enter the function into Y_1. Then change the graphing mode to split screen mode (G-T mode) by pressing MODE and selecting G-T. This will display the graph of the function alongside a table of values.

STEP 2. Press 2nd TBLSET to enter the TABLE SETUP screen. Change TblStart to −2 and make sure that △Tbl is set to 1. This tells the calculator to start the table of values at −2 and increase each successive x-value by 1 unit.

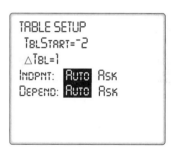

STEP 3. Finally, press GRAPH to plot the graph and the table of values. Press 2nd TABLE to use the table. Move through the table of values by pressing the up and down arrow keys.

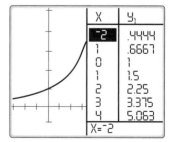

By tracing values on the graph of $y = \left(\frac{3}{2}\right)^x$, we can again see that as x approaches −∞, y approaches 0, and as x approaches ∞, y approaches ∞.

The Graph of $y = e^x$

In Section 6-7, we introduced the irrational number e. For many real-world applications, the number e is a convenient choice to use as the base. This is why the number e is also called the **natural base**, and an exponential function with e as the base is called a **natural exponential function**. Recall that

$$e = 2.718281828 \ldots$$

Note that e is an irrational number. The above pattern does *not* continue to repeat.

The figure on the left is a graph of the natural exponential function $y = e^x$. The figure on the right is a calculator screen of the same graph along with a table of selected values.

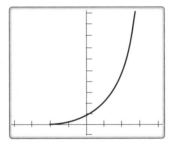

X	y_1
-3	.04979
-2	.13534
-1	.36788
0	1
1	2.7183
2	7.3891
3	20.086

X=-3

Exercises

Writing About Mathematics

1. Explain why $(0, 1)$ is a point on the graph of every function of the form $y = b^x$.

2. Explain why $y = b^x$ is *not* an exponential function for $b = 1$.

Developing Skills

In 3–6: **a.** Sketch the graph of each function. **b.** On the same set of axes, sketch the graph of the image of the reflection in the y-axis of the graph drawn in part **a**. **c.** Write an equation of the graph of the function drawn in part **b**.

3. $y = 4^x$

4. $y = 3^x$

5. $y = \left(\frac{7}{2}\right)^x$

6. $y = \left(\frac{3}{4}\right)^x$

7. a. Sketch the graph of $y = \left(\frac{5}{3}\right)^x$.

    b. From the graph of $y = \left(\frac{5}{3}\right)^x$, estimate the value of y, to the nearest tenth, when $x = 2.2$.

    c. From the graph of $y = \left(\frac{5}{3}\right)^x$, estimate the value of x, to the nearest tenth, when $y = 2.9$.

8. a. Sketch the graph of $f(x) = 2^x$.

   b. Sketch the graph of the image of $f(x) = 2^x$ under a reflection in the x-axis.

   c. Write an equation for the function whose graph was sketched in part **b**.

9. a. Sketch the graph of $f(x) = 1.2^x$.

   b. Sketch the graph of the image of $f(x) = 1.2^x$ under a reflection in the x-axis.

   c. Write an equation for the function whose graph was sketched in part **b**.

10. a. Make a table of values for e^x for integral values of x from -2 to 3.

   b. Sketch the graph of $f(x) = e^x$ by plotting points and joining them with a smooth curve:

   c. From the graph, estimate the value of $e^{\frac{1}{2}}$ and compare your answer to the value given by a calculator.

Applying Skills

11. The population of the United States can be modeled by the function $p(x) = 80.21e^{0.131x}$ where x is the number of decades (ten year periods) since 1900 and $p(x)$ is the population in millions.

   a. Graph $p(x)$ over the interval $0 \le x \le 15$.

   b. If the population of the United States continues to grow at this rate, predict the population in the years 2010 and 2020.

12. In 1986, the worst nuclear power plant accident in history occurred in the Chernobyl Nuclear Power Plant located in the Ukraine. On April 26, one of the reactors exploded, releasing large amounts of radioactive isotopes into the atmosphere. The amount of plutonium present after t years can be modeled by the function:

$$y = Pe^{-0.0000288t}$$

where P represents the amount of plutonium that is released.

   a. Graph this function over the interval $0 \le t \le 100{,}000$ and $P = 10$ grams.

   b. If 10 grams of the isotope plutonium-239 were released into the air, to the nearest hundredth, how many grams will be left after 10 years? After 100 years?

   c. Using the graph, approximate how long it will take for the 10 grams of plutonium-239 to decay to 1 gram.

13. a. Graph the functions $y = x^4$ and $y = 4^x$ on a graphing calculator using the following viewing windows:

   (1) $Xmin = 0$, $Xmax = 3$, $Ymin = 0$, $Ymax = 50$
   (2) $Xmin = 0$, $Xmax = 5$, $Ymin = 0$, $Ymax = 500$
   (3) $Xmin = 0$, $Xmax = 5$, $Ymin = 0$, $Ymax = 1{,}000$

   b. How many points of intersection can you find? Find the coordinates of these intersection points to the nearest tenth.

   c. Which function grows more rapidly for increasing values of x?

7-5 SOLVING EQUATIONS INVOLVING EXPONENTS

We know that to raise a power to a power, we multiply exponents. Therefore, for positive values of x and non-zero integer values of a:

$$(x^a)^{\frac{1}{a}} = x^{a\left(\frac{1}{a}\right)} = x^1 = x \qquad \left(x^{\frac{1}{a}}\right)^a = x^{\frac{1}{a}(a)} = x^1 = x$$

We can use this relationship to solve for x in an equation such as $x^{\frac{2}{3}} = 25$. To solve for x, we need to raise $x^{\frac{2}{3}}$ to the power that is the reciprocal of the exponent $\frac{2}{3}$. The reciprocal of $\frac{2}{3}$ is $\frac{3}{2}$.

$$x^{\frac{2}{3}} = 25$$
$$\left(x^{\frac{2}{3}}\right)^{\frac{3}{2}} = 25^{\frac{3}{2}}$$
$$x^1 = 25^{\frac{3}{2}}$$
$$x = 25^{\frac{3}{2}}$$

Note that $25^{\frac{3}{2}}$ means $\left(25^{\frac{1}{2}}\right)^3$, that is, the cube of the square root of 25.

$$x = \left(\sqrt{25}\right)^3 = 5^3 = 125$$

EXAMPLE I

Solve each equation and check: **a.** $2a^{-3} - 1 = 15$ **b.** $2\sqrt[3]{x^5} + 1 = 487$

Solution   *How to Proceed*

(1) Write the equation with only the variable term on one side of the equation:

a. $2a^{-3} - 1 = 15$
$2a^{-3} = 16$

b. $2\sqrt[3]{x^5} + 1 = 487$
$2\sqrt[3]{x^5} = 487$

(2) Divide both sides of the equation by the coefficient of the variable term:

$a^{-3} = 8$

$x^{\frac{5}{3}} = 243$

(3) Raise both sides of the equation to the power that is the reciprocal of the exponent of the variable:

$(a^{-3})^{-\frac{1}{3}} = 8^{-\frac{1}{3}}$
$a = 8^{-\frac{1}{3}}$

$\left(x^{\frac{5}{3}}\right)^{\frac{3}{5}} = 243^{\frac{3}{5}}$
$x = 243^{\frac{3}{5}}$

(4) Simplify the right side of the equation:

$a = \dfrac{1}{8^{\frac{1}{3}}}$
$= \dfrac{1}{\sqrt[3]{8}}$
$= \dfrac{1}{2}$

$x = 243^{\frac{3}{5}}$
$= \left(\sqrt[5]{243}\right)^3$
$= 3^3$
$= 27$

(5) *Check* the solution:

a.
$$2a^{-3} - 1 = 15$$
$$2\left(\tfrac{1}{2}\right)^{-3} - 1 \stackrel{?}{=} 15$$
$$2(2)^3 - 1 \stackrel{?}{=} 15$$
$$2(8) - 1 \stackrel{?}{=} 15$$
$$16 - 1 \stackrel{?}{=} 15$$
$$15 = 15 \checkmark$$

b.
$$2\sqrt[3]{x^5} + 1 = 487$$
$$2\sqrt[3]{27^5} + 1 \stackrel{?}{=} 487$$
$$2\sqrt[3]{(3^3)^5} + 1 \stackrel{?}{=} 487$$
$$2\sqrt[3]{3^{15}} + 1 \stackrel{?}{=} 486$$
$$2(3^5) + 1 \stackrel{?}{=} 487$$
$$487 = 487 \checkmark$$

Answers   **a.** $a = \tfrac{1}{2}$ **b.** $x = 27$

Exercises

Writing About Mathematics

1. Ethan said that to solve the equation $(x + 3)^{\frac{1}{2}} = 5$, the first step should be to square both sides of the equation. Do you agree with Ethan? Explain why or why not.

2. Chloe changed the equation $a^{-2} = 36$ to the equation $\frac{1}{a^2} = \frac{1}{36}$ and then took the square root of each side. Will Chloe's solution be correct? Explain why or why not.

Developing Skills

In 3–17 solve each equation and check.

3. $x^{\frac{1}{3}} = 4$

4. $a^{\frac{1}{5}} = 2$

5. $x^{\frac{2}{5}} = 9$

6. $b^{\frac{3}{2}} = 8$

7. $x^{-2} = 9$

8. $b^{-5} = \frac{1}{32}$

9. $2y^{-1} = 12$

10. $9a^{-\frac{3}{4}} = \frac{1}{3}$

11. $5x^{\frac{3}{4}} = 40$

12. $5x^{\frac{1}{2}} + 7 = 22$

13. $14 - 4b^{\frac{1}{3}} = 2$

14. $(2x)^{\frac{1}{2}} + 3 = 15$

15. $3a^3 = 81$

16. $x^5 = 3{,}125$

17. $z^{\frac{1}{2}} = \sqrt{81}$

In 18–23, solve for the variable in each equation. Express the solution to the nearest hundredth.

18. $x^{-3} = 24$

19. $y^{\frac{2}{3}} = 6$

20. $a^{-\frac{3}{4}} = 0.85$

21. $3z^3 + 2 = 27$

22. $5 + b^5 = 56$

23. $(3w)^9 + 2 = 81$

24. Solve for x and check: $\dfrac{x^{\frac{1}{3}}}{x^{\frac{2}{3}}} = 10$. Use the rule for the division of powers with like bases to simplify the left side of the equation.

Applying Skills

25. Show that if the area of one face of a cube is B, the volume of the cube is $B^{\frac{3}{2}}$.

26. If the area of one face of a cube is B and the volume of the cube is V, express B in terms of V.

7-6 SOLVING EXPONENTIAL EQUATIONS

Solving Exponential Equations With the Same Base

An **exponential equation** is an equation that contains a power with a variable exponent. For example, $2^{2x} = 8$ and $5^{x-1} = 0.04$ are exponential equations.

An exponential function $y = b^x$ is a one-to-one function since it is increasing for $b > 1$ and decreasing for $0 < b < 1$. Let $y_1 = b^{x_1}$ and $y_2 = b^{x_2}$. If $y_1 = y_2$, then $b^{x_1} = b^{x_2}$ and $x_1 = x_2$.

▶ **In general, if $b^p = b^q$, then $p = q$.**

We can use this fact to solve exponential equations that have the same base.

EXAMPLE I

Solve and check: $3^x = 3^{2x-2}$

Solution Since the bases are equal, the exponents must be equal.

$$3^x = 3^{2x-2} \qquad \qquad \textit{Check}$$
$$x = 2x - 2 \qquad \qquad 3^x = 3^{2x-2}$$
$$-x = -2 \qquad \qquad 3^2 \overset{?}{=} 3^{2(2)-2}$$
$$x = 2 \qquad \qquad 3^2 = 3^2 ✔$$

Answer $x = 2$

Solving Exponential Equations With Different Bases

How do we solve exponential equations such as $2^{2x} = 8$ or $5^{x-1} = 0.04$? One approach is, if possible, to write each term as a power of the same base. For example:

$$2^{2x} = 8 \qquad \qquad \qquad 5^{x-1} = 0.04$$
$$2^{2x} = 2^3 \qquad \qquad \qquad 5^{x-1} = \frac{4}{100}$$
$$2x = 3 \qquad \qquad \qquad 5^{x-1} = \frac{1}{25}$$
$$x = \frac{3}{2} \qquad \qquad \qquad 5^{x-1} = \frac{1}{5^2}$$
$$5^{x-1} = 5^{-2}$$
$$x - 1 = -2$$
$$x = -1$$

EXAMPLE 2

Solve and check: $4^a = 8^{a+1}$

Solution The bases, 4 and 8, can each be written as a power of 2: $4 = 2^2, 8 = 2^3$.

$4^a = 8^{a+1}$	*Check*
$(2^2)^a = (2^3)^{a+1}$	$4^a = 8^{a+1}$
$2^{2a} = 2^{3a+3}$	$4^{-3} \stackrel{?}{=} 8^{-3+1}$
$2a = 3a + 3$	$4^{-3} \stackrel{?}{=} 8^{-2}$
$-a = 3$	$\frac{1}{4^3} \stackrel{?}{=} \frac{1}{8^2}$
$a - -3$	$\frac{1}{64} = \frac{1}{64}$ ✔

Answer $a = -3$

EXAMPLE 3

Solve and check: $3 + 7^{x-1} = 10$

Solution Add -3 to each side of the equation to isolate the power.

$3 + 7^{x-1} = 10$	*Check*
$7^{x-1} = 7$	$3 + 7^{x-1} = 10$
$x - 1 = 1$	$3 + 7^{2-1} \stackrel{?}{=} 10$
$x = 2$	$3 + 7^1 \stackrel{?}{=} 10$
	$10 = 10$ ✔

Answer $x = 2$

Exercises

Writing About Mathematics

1. What value of a makes the equation $6^a = 1$ true? Justify your answer.

2. Explain why the equation $3^a = 5^{a-1}$ cannot be solved using the procedure used in this section.

Developing Skills

In 3–14, write each number as a power.

3. 9	**4.** 27	**5.** 25	**6.** 49
7. 1,000	**8.** 32	**9.** $\frac{1}{8}$	**10.** $\frac{1}{216}$
11. 0.001	**12.** 0.125	**13.** 0.81	**14.** 0.16

In 15–38, solve each equation and check.

15. $2^x = 16$

16. $3^x = 27$

17. $5^x = \frac{1}{5}$

18. $7^x = \frac{1}{49}$

19. $4^{x+2} = 4^{2x}$

20. $3^{x+1} = 3^{2x+3}$

21. $6^{3x} = 6^{x-1}$

22. $3^{x+2} = 9^x$

23. $25^x = 5^{x+3}$

24. $49^x = 7^{3x+1}$

25. $2^{2x+1} = 16^x$

26. $9^{x-1} = 27^x$

27. $100^x = 1{,}000^{x-1}$

28. $125^{x-1} = 25^x$

29. $6^{2-x} = \left(\frac{1}{36}\right)^2$

30. $\left(\frac{1}{4}\right)^x = 8^{1-x}$

31. $\left(\frac{1}{3}\right)^x = 9^{1-x}$

32. $(0.01)^{2x} = 100^{2-x}$

33. $(0.25)^{x-2} = 4^x$

34. $5^{x-1} = (0.04)^{2x}$

35. $4^x + 7 = 15$

36. $5 + 7^x = 6$

37. $e^{2x+2} = e^{x-1}$

38. $3^{x^2+2} = 3^6$

7-7 APPLICATIONS OF EXPONENTIAL FUNCTIONS

There are many situations in which an initial value A_0 is increased or decreased by adding or subtracting a percentage of A_0. For example, the value of an investment is the amount of money invested plus the interest, a percentage of the investment.

Let A_n be the value of the investment after n years if A_0 dollars are invested at rate r per year.

- After 1 year: $A_1 = A_0 + A_0 r = A_0(1 + r)$
  $$A_1 = A_0(1 + r)$$
- After 2 years: $A_2 = A_1 + A_1 r = A_0(1 + r) + A_0(1 + r)r$
  $$= A_0(1 + r)(1 + r) = A_0(1 + r)^2$$
  $$A_2 = A_0(1 + r)^2$$
- After 3 years: $A_3 = A_2 + A_2 r = A_0(1 + r)^2 + A_0(1 + r)^2 r$
  $$= A_0(1 + r)^2(1 + r) = A_0(1 + r)^3$$
  $$A_3 = A_0(1 + r)^3$$
- After 4 years: $A_4 = A_3 + A_3 r = A_0(1 + r)^3 + A_0(1 + r)^3 r$
  $$= A_0(1 + r)^3(1 + r) = A_0(1 + r)^4$$
  $$A_4 = A_0(1 + r)^4$$

We see that a pattern has been established. The values at the end of each year form a geometric sequence with the common ratio $(1 + r)$.

- After t years:

$$A_t = A_{t-1} + A_{t-1}r = A_0(1 + r)^{t-1} + A_0(1 + r)^{t-1}r = A_0(1 + r)^{t-1}(1 + r) = A_0(1 + r)^t$$

or

$$A = A_0(1 + r)^t$$

A calculator can be used to display the value of an investment year by year. The value of the investment after each year is $(1 + r)$ times the value of the investment from the previous year. If $100 is invested at a yearly rate of 5%, the value of the investment at the end of each of the first 5 years can be found by multiplying the value from the previous year by $(1 + 0.05)$ or 1.05.

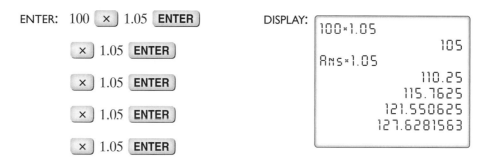

The value of the investment for each of the first five years is $105.00, $110.25, $115.76, $121.55, and $127.63.

EXAMPLE I

Amanda won $10,000 and decided to use it as a "vacation fund." Each summer, she withdraws an amount of money that reduces the value of the fund by 7.5% from the previous summer. How much will the fund be worth after the tenth withdrawal?

Solution Use the formula $A = A_0(1 + r)^t$ with $A_0 = 10{,}000$, $r = -0.075$ and $t = 10$.
$$A = 10{,}000(1 - 0.075)^{10} = 10{,}000(0.925)^{10}$$

ENTER: 10000 $\boxed{\times}$ 0.925 $\boxed{\wedge}$ 10 $\boxed{\text{ENTER}}$

DISPLAY:
```
10000*0.925^10
        4585.823414
```

Answer $4,585.82

  Calculator variables can be used to quickly evaluate an exponential model for different time periods. For instance, to find the value of the fund from Example 1 after the second and fifth withdrawals:

STEP 1. Let $Y_1 = 10{,}000(0.925)^x$.

STEP 2. Exit the Y= screen and let $X = 2$.

ENTER: 2 `STO▸` `X,T,θ,n` `ENTER`

STEP 3. Find the value of Y_1 when $X = 2$.

ENTER: `VARS` `▶` `ENTER` `ENTER`

`ENTER`

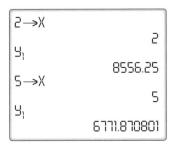

STEP 4. Repeat steps 2 and 3 to find the value of Y_1 when $X = 5$.

The value of the fund after the second withdrawal is $8,556.25 and after the fifth withdrawal, $6,771.87.

Other Types of Compounding Periods

The exponential function $A = A_0(1 + r)^t$, when applied to investments, represents interest *compounded annually* since interest is added once at the end of each year. However, this formula can be applied to any increase that takes place at a fixed rate for any specified period of time. For example, if an annual interest rate of 5% is paid monthly, then interest of $\frac{5}{12}\%$ is paid 12 times in one year. Or if an annual interest rate of 5% is paid daily, then interest of $\frac{5}{365}\%$ is paid 365 times in one year. Compare the value of an investment of $100 for these three cases.

Paid yearly for 1 year: $A = 100(1 + 0.05)^1 = 105$ or $105.00

Paid monthly for 1 year: $A = 100\left(1 + \frac{0.05}{12}\right)^{12} \approx 105.1161898$ or $105.12

Paid daily for 1 year: $A = 100\left(1 + \frac{0.05}{365}\right)^{365} \approx 105.1267496$ or $105.13

In general, if interest is compounded n times in one time period and the number of time periods is t, then:

$$A = A_0\left(1 + \frac{r}{n}\right)^{nt}$$

Note that as the number of times that the interest is compounded increases, the value of the investment for a fixed period of time increases.

What happens as we let n, the number of compoundings, increase without limit? Let $\frac{r}{n} = \frac{1}{k}$. Then $n = rk$.

$A = A_0\left(1 + \frac{r}{n}\right)^{nt}$

$A = A_0\left(1 + \frac{1}{k}\right)^{rkt}$ Substitute $\frac{r}{n} = \frac{1}{k}$ and $n = rk$.

$A = A_0\left[\left(1 + \frac{1}{k}\right)^k\right]^{rt}$

As n approaches infinity, k approaches infinity. It can be shown that $\left(1 + \frac{1}{k}\right)^k$ approaches e. The table on the right shows values of $\left(1 + \frac{1}{k}\right)^k$ rounded to eight decimal places for different values of k. Recall that e is an irrational number that is the sum of an infinite series:

$$1 + \tfrac{1}{1!} + \tfrac{1}{2!} + \tfrac{1}{3!} + \cdots \approx 2.718281828\ldots$$

Therefore, for very large values of n:

k	$\left(1 + \frac{1}{k}\right)^k$
1	2
10	2.59374246
100	2.70481383
1,000	2.71692393
100,000	2.71826823
1,000,000	2.71828047

$$A = A_0\left(1 + \tfrac{r}{n}\right)^{nt} = A_0\left(\left(1 + \tfrac{1}{k}\right)^k\right)^{rt} = A_0 e^{rt}$$

This formula can be applied to any change that takes place continuously at a fixed rate. Large populations of people or of animals can be said to increase continuously. If this happens at a fixed rate per year, then the size of the population in the future can be predicted.

EXAMPLE 2

In a state park, the deer population was estimated to be 2,000 and increasing continuously at a rate of 4% per year. If the increase continues at that rate, what is the expected deer population in 10 years?

Solution Use the formula $A = A_0 e^{rt}$. Let $A_0 = 2,000$, $r = 0.04$, and $t = 10$.

$$A = 2{,}000e^{0.04(10)} = 2{,}000e^{0.4}$$

ENTER: 2000 ×   2nd e^x 0.4)   ENTER

DISPLAY:
```
2000*e^(0.4)
       2983.649395
```

Answer The expected population is 2,984 deer.

Note: Since the answer in Example 2 is an approximation, it is probably more realistic to say that the expected population is 3,000 deer.

The formula $A = A_0 e^{rt}$ can also be used to study radioactive decay. Decay takes place continuously. The decay constant for a radioactive substance is negative to indicate a decrease.

EXAMPLE 3

The decay constant of radium is -0.0004 per year. How many grams will remain of a 50-gram sample of radium after 20 years?

Solution   Since change takes place continuously, use the formula $A = A_0 e^{rt}$. Let $A_0 = 50$, $r = -0.0004$ and $t = 20$.

$$A = 50e^{-0.0004(20)} = 50e^{-0.008}$$

ENTER: 50 ⌧×⌧ ⌧2nd⌧ ⌧eˣ⌧ -0.008 ⌧)⌧   ⌧ENTER⌧

DISPLAY:
```
50*e^(-0.008)
           49.60159574
```

Answer   Approximately 49.6 grams

Exercises

Writing About Mathematics

1. Show that the formula $A = A_0(1 + r)^n$ is equivalent to $A = A_0(2)^n$ when $r = 100\%$.

2. Explain why, if an investment is earning interest at a rate of 5% per year, the investment is worth more if the interest is compounded daily rather than if it is compounded yearly.

Developing Skills

In 3–10, find the value of x to the nearest hundredth.

3. $x = e^2$ **4.** $x = e^{1.5}$ **5.** $x = e^{-1}$ **6.** $xe^3 = e^4$

7. $12x = e$ **8.** $\frac{x}{e^3} = e^{-2}$ **9.** $x = e^3 e^5$ **10.** $x = e^3 + e^5$

In 11–16, use the formula $A = A_0\left(1 + \frac{r}{n}\right)^{nt}$ to find the missing variable to the nearest hundredth.

11. $A_0 = 50, r = 2\%, n = 12, t = 1$ **12.** $A = 400, r = 5\%, n = 4, t = 3$

13. $A = 100, A_0 = 25, n = 1, t = 2$ **14.** $A = 25, A_0 = 200, r = -50\%, n = 1$

15. $A = 250, A_0 = 10, n = 3, t = 1$ **16.** $A = 6, A_0 = 36, n = 1, t = 4$

Applying Skills

17. A bank offers certificates of deposit with variable compounding periods.

  a. Joe invested $1,000 at 6% per year compounded yearly. Find the values of Joe's investment at the end of each of the first five years.

  b. Sue invested $1,000 at 6% per year compounded monthly. Find the values of Sue's investment at the end of each of the first five years.

  c. Who had more money after the end of the fifth year?

  d. The *annual percentage yield* (APY) is the amount an investment actually increases during one year. Find the APY for Joe and Sue's certificates of deposit. Is the APY of each investment equal to 6%?

18. a. When Kyle was born, his grandparents invested $5,000 in a college fund that paid 4% per year, compounded yearly. What was the value of this investment when Kyle was ready for college at age 18? (Note that $r = 0.04$.)

  b. If Kyle's grandparents had invested the $5,000 in a fund that paid 4% compounded continuously, what would have been the value of the fund after 18 years?

19. A trust fund of $2.5 million was donated to a charitable organization. Once each year the organization spends 2% of the value of the fund so that the fund decreases by 2%. What will be the value of the fund after 25 years?

20. The decay constant of a radioactive element is -0.533 per minute. If a sample of the element weighs 50 grams, what will be its weight after 2 minutes?

21. The population of a small town decreased continually by 2% each year. If the population of the town is now 37,000, what will be the population 8 years from now if this trend continues?

22. A piece of property was valued at $50,000 at the end of 1990. Property values in the city where this land is located increase by 10% each year. The value of the land increases continuously. What is the property worth at the end of 2010?

23. A sample of a radioactive substance decreases continually at a rate of -0.04. If the weight of the sample is now 40 grams, what will be its weight in 7 days?

24. The number of wolves in a wildlife preserve is estimated to have increased continually by 3% per year. If the population is now estimated at 5,400 wolves, how many were present 10 years ago?

25. The amount of a certain medicine present in the bloodstream decreases at a rate of 10% per hour.

  a. Which is a better model to use for this scenario: $A = A_0(1 + r)^t$ or $A = A_0e^{rt}$? Explain your answer.

  b. Using both models, find the amount of medicine in the bloodstream after 10.5 hours if the initial dose was 200 milligrams.

CHAPTER SUMMARY

The rules for operations with powers with like bases can be extended to include zero, negative, and fractional exponents. If $x > 0$ and $y > 0$:

$$\text{\textit{Multiplication:}} \quad x^a \cdot x^b = x^{a+b}$$
$$\text{\textit{Division:}} \quad x^a \div x^b = x^{a-b} \quad \text{or} \quad \frac{x^a}{x^b} = x^{a+b} \quad (x \neq 0)$$
$$\text{\textit{Raising to a Power:}} \quad (x^a)^b = x^{ab}$$
$$\text{\textit{Power of a Product:}} \quad (xy)^a = x^a \cdot y^a$$
$$\text{\textit{Power of a Quotient:}} \quad \left(\frac{x}{y}\right)^a = \frac{x^a}{y^a}$$

If $x > 0$ and n is a positive integer:

$$\sqrt[n]{x} = x^{\frac{1}{n}}$$

In general, for $x > 0$ and m and n positive integers:

$$\sqrt[n]{x^m} = \left(\sqrt[n]{x}\right)^m = x^{\frac{m}{n}} \qquad \frac{1}{\sqrt[n]{x^m}} = \frac{1}{\left(\sqrt[n]{x}\right)^m} = x^{-\frac{m}{n}}$$

The graph of the exponential function, $y = b^x$:

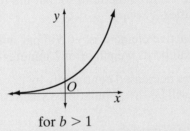

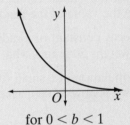

for $b > 1$ for $0 < b < 1$

To solve an equation of the form $x^{\frac{a}{b}} = c$, raise each side of the equation to the $\frac{b}{a}$ power so that the left member is x^1 or x.

An **exponential equation** is an equation that contains a power with a variable exponent. To solve an exponential equation, write each member with the same base and equate exponents.

The number e is also called the **natural base** and the function $y = e^x$ is called the **natural exponential function**.

If a quantity A_0 is increased or decreased by a rate r per interval of time, its value A after t intervals of time is $A = A_0(1 + r)^t$.

If a quantity A_0 is increased or decreased by a rate r per interval of time, compounded n times per interval, its value A after t intervals of time is:

$$A = A_0\left(1 + \frac{r}{n}\right)^{nt}$$

If the increase or decrease is continuous for t units of time, the formula becomes

$$A = A_0 e^{rt}$$

VOCABULARY

7-4 Exponential growth • Exponential decay • Exponential function •
Horizontal asymptote • Natural base • Natural exponential function

7-6 Exponential equation

REVIEW EXERCISES

In 1–12, evaluate each expression.

1. 3^0 **2.** 2^{-1} **3.** $8^{\frac{1}{3}}$ **4.** $5(3)^0$

5. $8(2)^3$ **6.** $5(0.1)^{-2}$ **7.** $12(9)^{\frac{1}{2}}$ **8.** $(2 \times 3)^{-2}$

9. $(9 \times 9)^{\frac{1}{4}}$ **10.** $100^{-\frac{1}{2}}$ **11.** $125^{\frac{2}{3}}$ **12.** $\left(12 \times \frac{4}{3}\right)^{-\frac{3}{2}}$

In 13–16, evaluate each function for the given value. Be sure to show your work.

13. $g(x) = x^{\frac{1}{4}} x^{\frac{1}{4}}; g(4)$ **14.** $g(x) = \left(\dfrac{1}{1 + \frac{1}{x^{-3}}}\right); g(3)$

15. $g(x) = 2^x + \left(\frac{1}{2^x}\right); g(4)$ **16.** $g(x) = \left(\frac{x^5}{x^7}\right)^{-2}; g(10)$

In 17–20, write each expression with only positive exponents and express the answer in simplest form. The variables are not equal to zero.

17. $(27x)^3 \cdot 3(x)^{-1}$ **18.** $\left(\frac{a^3bc}{abc^3}\right)^{-3}$

19. $\left(\frac{x^3}{b^{-2}}\right)^{-\frac{1}{5}}$ **20.** $\left(2x^2y^{-2}z^{\frac{3}{2}}\right)\left(3x^{-3}y^3z^{-1}\right)$

In 21–24, write each radical expression as a power with positive exponents and express the answer in simplest form. The variables are positive numbers.

21. $\dfrac{3}{\sqrt{25x}}$ **22.** $\sqrt[3]{\dfrac{81y^3}{3y^5}}$

23. $\sqrt[12]{64a^7b^{20}}$ **24.** $\sqrt[6]{32x^8y^3}$

In 25–28, write each power as a radical expression in simplest form. The variables are positive numbers.

25. $(2y)^{\frac{5}{2}}$ **26.** $(16x^8y)^{\frac{1}{2}}$

27. $\left(\frac{(a + 2)}{2^4}\right)^{\frac{3}{4}}$ **28.** $\left(\frac{a^{12}b^9c}{a^6b^3}\right)^{\frac{1}{4}}$

29. a. Sketch the graph of $y = 1.25^x$.

    b. On the same set of axes, sketch the graph of $y = 0.80^x$.

    c. Under what transformation is the graph of $y = 0.80^x$ the image of $y = 1.25^x$?

In 30–41, solve and check each equation.

30. $x^{-2} = 36$

31. $a^{\frac{1}{3}} = 4$

32. $b^{-\frac{1}{2}} = \frac{2}{3}$

33. $y^{-3} + 6 = 14$

34. $\frac{a^2}{a^{\frac{2}{3}}} + 7 = 9$

35. $4^x + 2 = 10$

36. $2(5)^{-x} = 50$

37. $7^x = 7^{2x-2}$

38. $2^{x+2} = 8^{x-2}$

39. $0.10^{-x} = 10^{2x+1}$

40. $3^{x+1} = 27^x$

41. $2 = 0.5^x$

42. Find the interest that has accrued on an investment of $500 if interest of 4% per year is compounded quarterly for a year.

43. The label on a prescription bottle directs the patient to take the medicine twice each day. The effective ingredient of the medicine decreases continuously at a rate of 25% per hour. If a dose of medicine containing 1 milligram of the effective ingredient is taken, how much is still present 12 hours later when the second dose is taken?

44. A sum of money that was invested at a fixed rate of interest doubled in value in 15 years. The interest was compounded yearly. Find the rate of interest to the nearest tenth of a percent.

45. A company records the value of a machine used for production at $25,000. As the machine ages, its value depreciates, that is, decreases in value. If the depreciation is estimated to be 20% of the value of the machine at the end of each year, what is the expected value of the machine after 6 years?

46. Determine the common solution of the system of equations:

$$15y = 27^x$$
$$5y = 3^x$$

Exploration

Prove that for all a:

a. $\dfrac{3^a - 3^{a-2}}{3^{a-1} + 3^a} = \dfrac{2}{3}$

b. $\dfrac{4^{a+1} + 4}{2^{a+5} \cdot 2^{a-1} + 16} = \dfrac{1}{4}$

CUMULATIVE REVIEW   CHAPTERS 1–7

Part I

Answer all questions in this part. Each correct answer will receive 2 credits. No partial credit will be allowed.

1. The expression $2x^3(4 - 3x) - 5x(x - 2)$ is equal to

(1) $6x^4 - 8x^3 + 5x^2 - 10x$

(3) $6x^4 + 8x^3 - 5x^2 + 10x$

(2) $-6x^4 + 8x^3 - 5x^2 + 10x$

(4) $-6x^4 + 8x^3 - 5x^2 - 10x$

2. If the domain is the set of integers, the solution set of $x^2 - 3x - 4 < 0$ is

(1) $\{-1, 0, 1, 2, 3, 4\}$ (3) $\{\ldots, -3, -2, -1, 4, 5, 6, \ldots\}$

(2) $\{0, 1, 2, 3\}$ (4) $\{\ldots, -4, -3, -2, 5, 6, 7, \ldots\}$

3. The fraction $\dfrac{\frac{1}{3} - 1}{\frac{1}{9} - 3}$ is equal to

(1) -1 (2) 1 (3) $\frac{3}{13}$ (4) $\frac{52}{27}$

4. An equation whose roots are -3 and 5 is

(1) $x^2 + 2x - 15 = 0$ (3) $x^2 + 2x + 15 = 0$

(2) $x^2 - 2x - 15 = 0$ (4) $x^2 - 2x + 15 = 0$

5. In simplest form, $\sqrt{12} + \sqrt{9} + \sqrt{27}$ is

(1) $5\sqrt{3} + 3$ (2) $6\sqrt{3}$ (3) $4\sqrt{3}$ (4) $\sqrt{48}$

6. The expression $\sqrt{-2}(\sqrt{-18}) + \sqrt{-25}$ is equal to

(1) 1 (2) $-6 + 5i$ (3) $6 + 5i$ (4) $11i$

7. The discriminant of a quadratic equation is 35. The roots are

(1) unequal rational numbers.   (3) equal rational numbers.

(2) unequal irrational numbers.   (4) imaginary numbers.

8. In the sequence $100, 10, 1, \ldots, a_{20}$ is equal to

(1) 10^{-20} (2) 10^{-19} (3) 10^{-18} (4) 10^{-17}

9. The factors of $x^3 - 4x^2 - x + 4$ are

(1) $(x - 4)(x - 1)(x + 1)$

(2) $(x - 2)(x - 1)(x + 1)(x + 2)$

(3) $(x + 4)(x - 1)(x + 1)$

(4) $(4 - x)(x - 1)(x + 1)$

10. The solution set of the equation $6 - \sqrt{7 - x} = 8$ is

(1) $\{3\}$ (2) $\{-3\}$ (3) $\{-3, 3\}$ (4) $\varnothing$

Part II

Answer all questions in this part. Each correct answer will receive 2 credits. Clearly indicate the necessary steps, including appropriate formula substitutions, diagrams, graphs, charts, etc. For all questions in this part, a correct numerical answer with no work shown will receive only 1 credit.

11. Write an expression for the nth term of the series $1 + 3 + 5 + 7 + \cdots$.

12. The roots of a quadratic equation are 2 and $\frac{5}{2}$. Write the equation.

Part III

Answer all questions in this part. Each correct answer will receive 4 credits. Clearly indicate the necessary steps, including appropriate formula substitutions, diagrams, graphs, charts, etc. For all questions in this part, a correct numerical answer with no work shown will receive only 1 credit.

13. Express $\dfrac{3 + \sqrt{-4}}{1 - \sqrt{-4}}$ in $a + bi$ form.

14. Find the solution set of the equation $1 + 27^{x+1} = 82$.

Part IV

Answer all questions in this part. Each correct answer will receive 6 credits. Clearly indicate the necessary steps, including appropriate formula substitutions, diagrams, graphs, charts, etc. For all questions in this part, a correct numerical answer with no work shown will receive only 1 credit.

15. What is the solution set of the inequality $x^2 - 7x - 12 < 0$?

16. A circle with center at $(2, 0)$ and radius 4 is intersected by a line whose slope is 1 and whose y-intercept is 2.

    a. Write the equation of the circle in center-radius form.

    b. Write the equation of the circle in standard form.

    c. Write the equation of the line.

    d. Find the coordinates of the points at which the line intersects the circle.

LOGARITHMIC FUNCTIONS

The heavenly bodies have always fascinated and challenged humankind. Our earliest records contain conclusions, some false and some true, that were believed about the relationships among the sun, the moon, Earth, and the other planets. As more accurate instruments for studying the heavens became available and more accurate measurements were possible, the mathematical computations absorbed a great amount of the astronomer's time.

A basic principle of mathematical computation is that it is easier to add than to multiply. John Napier (1550–1617) developed a system of logarithms that facilitated computation by using the principles of exponents to find a product by using addition. Henry Briggs (1560–1630) developed Napier's concept using base 10. Seldom has a new mathematical concept been more quickly accepted.

8-1 INVERSE OF AN EXPONENTIAL FUNCTION

In Chapter 7, we showed that any positive real number can be the exponent of a power by drawing the graph of the exponential function $y = b^x$ for $0 < b < 1$ or $b > 1$. Since $y = b^x$ is a one-to-one function, its reflection in the line $y = x$ is also a function. The function $x = b^y$ is the inverse function of $y = b^x$.

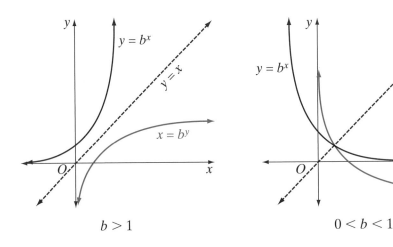

$b > 1$ $0 < b < 1$

The equation of a function is usually solved for y in terms of x. To solve the equation $x = b^y$ for y, we need to introduce some new terminology. First we will describe y in words:

$x = b^y$: "y is the exponent to the base b such that the power is x."

A **logarithm** is an exponent. Therefore, we can write:

$x = b^y$: "y is the *logarithm* to the base b of the power x."

The word *logarithm* is abbreviated as *log*. Look at the essential parts of this sentence:

$y = \log_b x$: "**y is** the **log**arithm to the **base b** of **x**."

The base b is written as a subscript to the word "log."

▶ **$x = b^y$ can be written as $y = \log_b x$.**

For example, let $b = 2$. Write pairs of values for $x = 2^y$ and $y = \log_2 x$.

$x = 2^y$	In Words	$y = \log_2 x$	(x, y)
$\frac{1}{2} = 2^{-1}$	-1 is the logarithm to the base 2 of $\frac{1}{2}$.	$-1 = \log_2 \frac{1}{2}$	$\left(\frac{1}{2}, -1\right)$
$1 = 2^0$	0 is the logarithm to the base 2 of 1.	$0 = \log_2 1$	$(1, 0)$
$\sqrt{2} = 2^{\frac{1}{2}}$	$\frac{1}{2}$ is the logarithm to the base 2 of $\sqrt{2}$.	$\frac{1}{2} = \log_2 \sqrt{2}$	$\left(\sqrt{2}, \frac{1}{2}\right)$
$2 = 2^1$	1 is the logarithm to the base 2 of 2.	$1 = \log_2 2$	$(2, 1)$
$4 = 2^2$	2 is the logarithm to the base 2 of 4.	$2 = \log_2 4$	$(4, 2)$
$8 = 2^3$	3 is the logarithm to the base 2 of 8.	$3 = \log_2 8$	$(8, 3)$

We say that $y = \log_b x$, with b a positive number not equal to 1, is a **logarithmic function**.

EXAMPLE I

Write the equation $x = 10^y$ for y in terms of x.

Solution $x = 10^y \leftarrow y$ is the exponent or logarithm to the base 10 of x.

$y = \log_{10} x$

Graphs of Logarithmic Functions

From our study of exponential functions in Chapter 7, we know that when $b > 1$ and when $0 < b < 1$, $y = b^x$ is defined for all real values of x. Therefore, the domain of $y = b^x$ is the set of real numbers. When $b > 1$, as the negative values of x get larger and larger in absolute value, the value of b^x gets smaller but is always positive. When $0 < b < 1$, as the positive values of x get larger and larger, the value of b^x gets smaller but is always positive. Therefore, the range of $y = b^x$ is the set of positive real numbers.

When we interchange x and y to form the inverse function $x = b^y$ or $y = \log_b x$:

▶ **The domain of $y = \log_b x$ is the set of positive real numbers.**

▶ **The range $y = \log_b x$ is the set of real numbers.**

▶ **The y-axis or the line $x = 0$ is a *vertical asymptote* of $y = \log_b x$.**

EXAMPLE 2

a. Sketch the graph of $f(x) = 2^x$.

b. Write the equation of $f^{-1}(x)$ and sketch its graph.

Solution   a. Make a table of values for $f(x) = 2^x$, plot the points, and draw the curve.

x	2^x	$f(x)$
−2	$2^{-2} = \frac{1}{2^2}$	$\frac{1}{4}$
−1	$2^{-1} = \frac{1}{2}$	$\frac{1}{2}$
0	2^0	1
1	2^1	2
2	2^2	4
3	2^3	8

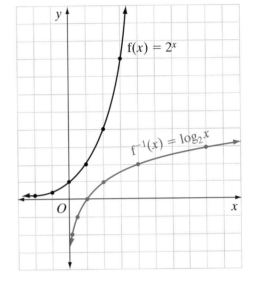

b. Let $f(x) = 2^x \rightarrow y = 2^x$.

To write $f^{-1}(x)$, interchange x and y.

$x = 2^y$ is written as $y = \log_2 x$. Therefore, $f^{-1}(x) = \log_2 x$.

To draw the graph, interchange x and y in each ordered pair or reflect the graph of $f(x)$ over the line $y = x$. Ordered pairs of $f^{-1}(x)$ include $\left(\frac{1}{4}, -2\right)$, $\left(\frac{1}{2}, -1\right)$, $(1, 0)$, $(2, 1)$, $(4, 2)$, and $(8, 3)$.

The function $y = \log_b x$ represents *logarithmic growth.* Quantities represented by a logarithmic function grow very slowly. For example, suppose that the time it takes for a computer to run a program could by modeled by the logarithmic function $y = \log_{10} x$ where x is the number of instructions of the computer program and y is the running time in milliseconds. The graph on the right shows the running time in the interval $1 < x < 1,000,000$. Note that

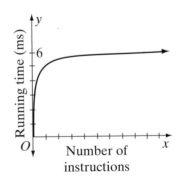

the graph increases from 0 to 5 for the first 100,000 instructions but only increases from 5 to 6 as the number of instructions increases from 100,000 to 1,000,000. (Each interval on the x-axis represents 100,000 instructions.)

Exercises

Writing About Mathematics

1. Peg said that $(1, 0)$ is always a point on the graph of $y = \log_b x$. Do you agree with Peg? Explain why or why not.

2. Sue said that if $x = b^{2y}$ for $b > 1$, then $y = \frac{1}{2} \log_b x$. Do you agree with Sue? Explain why or why not.

Developing Skills

In 3–10: **a.** For each $f(x)$, write an equation for $f^{-1}(x)$, the inverse function. **b.** Sketch the graph of $f(x)$ and of $f^{-1}(x)$.

3. $f(x) = 3^x$ 4. $f(x) = 5^x$ 5. $f(x) = \left(\frac{3}{2}\right)^x$ 6. $f(x) = \left(\frac{5}{2}\right)^x$

7. $f(x) = \left(\frac{1}{2}\right)^x$ 8. $f(x) = \left(\sqrt{2}\right)^x$ 9. $f(x) = \frac{1}{3^x}$ 10. $f(x) = -2^x$

In 11–22, solve each equation for y in terms of x.

11. $x = 6^y$ 12. $x = 10^y$ 13. $x = 8^y$ 14. $x = (0.1)^y$

15. $x = (0.2)^y$ 16. $x = 4^{-y}$ 17. $x = 12^{-y}$ 18. $x = \log_2 y$

19. $x = \log_5 y$ 20. $x = \log_{10} y$ 21. $x = \log_8 y$ 22. $x = \log_{0.1} y$

Applying Skills

23. Genevieve decided to organize a group of volunteers to help at a soup kitchen. Every week for the first three weeks, the number of volunteers tripled so that the number, $f(x)$, after x weeks is $f(x) = 3^x$.

    a. Write the ordered pairs of the function $f(x) = 3^x$ for $0 \le x \le 3$ and locate the pairs as points on a graph. The domain is the set of non-negative integers.

    b. Write the ordered pairs for $f^{-1}(x)$ and sketch the graph.

24. If money is invested at a rate of 5% compounded annually, then for each dollar invested, the amount of money in an account is $g(x)$, when $g(x) = 1.05^x$ after x years.

    a. Write the ordered pairs of the function g for $0 \le x \le 3$ and locate the pairs as points on a graph. The domain is the set of non-negative integers.

    b. Write the ordered pairs for $g^{-1}(x)$ and sketch the graph.

8-2 LOGARITHMIC FORM OF AN EXPONENTIAL EQUATION

An exponential equation and a logarithmic equation are two different ways of expressing the same relationship. For example, to change an exponential equation to a logarithmic equation, recall that a logarithm or log is an exponent. In the exponential equation $81 = 3^4$, the exponent or log is 4. The basic statement is:

$$\log = 4$$

Then write the base as a subscript of the word "log" to indicate the log to the base 3:

$$\log_3 = 4$$

Now add the value of the power, that is, log to the base 3 of the power 81:

$$\log_3 81 = 4$$

To change from a logarithmic equation to an exponential statement, we must again look for the exponent. For example, in the logarithmic equation $\log_{10} 0.01 = -2$, the basic statement is that the log or exponent is -2. The base is the number that is written as a subscript of the word "log":

$$10^{-2}$$

The number that follows the word "log" is the power:

$$10^{-2} = 0.01$$

Recall that $10^{-2} = \frac{1}{10^2} = \frac{1}{100} = 0.01$.

In general:

▶ $b^a = c$ **is equivalent to** $\log_b c = a.$

EXAMPLE I

Write $9^3 = 729$ in logarithmic form and express its meaning in words.

Solution   In the equation $9^3 = 729$, the logarithm (exponent) is 3:

$$\log = 3$$

Write the base, 9, as a subscript of "log":

$$\log_9 = 3$$

Write the power, 729, after "log":

$$\log_9 729 = 3$$

The logarithmic form of the equation means, "the exponent to the base 9 of the power 729 is 3." *Answer*

EXAMPLE 2

Write $\frac{3}{2} = \log_{25} 125$ in exponential form.

Solution The equation $\frac{3}{2} = \log_{25} 125$ says that $\frac{3}{2}$ is the exponent. The base is written as the subscript of "log." The power, $25^{\frac{3}{2}}$, is 125.

$$25^{\frac{3}{2}} = 125 \ \textit{Answer}$$

Note that $25^{\frac{3}{2}} = \left(25^{\frac{1}{2}}\right)^3 = \left(\sqrt{25}\right)^3 = 5^3 = 125$.

EXAMPLE 3

Find the value of a when $\log_8 a = \frac{1}{3}$.

Solution The equation $\log_8 a = \frac{1}{3}$ means that for the power a, the base is 8, and the exponent is $\frac{1}{3}$.

$$8^{\frac{1}{3}} = a$$
$$\sqrt[3]{8} = a$$
$$2 = a \ \textit{Answer}$$

EXAMPLE 4

Solve the following equation for x: $\log_3 \frac{1}{9} = \log_2 x$.

Solution Represent each side of the equation by y.

Let $y = \log_3 \frac{1}{9}$.	Let $y = \log_2 x$.
Then:	Then:
$3^y = \frac{1}{9}$	$-2 = \log_2 x$
$3^y = 3^{-2}$	$2^{-2} = x$
$y = -2$	$\frac{1}{4} = x$

Answer $x = \frac{1}{4}$

Exercises

Writing About Mathematics

1. If $\log_b c = a$, explain why $\log_b \frac{1}{c} = -a$.

2. If $\log_b c = a$, explain why $\log_b c^2 = 2a$.

Developing Skills

In 3–14, write each exponential equation in logarithmic form.

3. $2^4 = 16$ **4.** $5^3 = 125$ **5.** $64 = 8^2$

6. $12^0 = 1$ **7.** $216 = 6^3$ **8.** $10^{-1} = 0.1$

9. $5^{-3} = 0.008$ **10.** $4^{-2} = 0.0625$ **11.** $7^{-1} = \frac{1}{7}$

12. $64^{\frac{1}{3}} = 4$ **13.** $625^{\frac{3}{4}} = 125$ **14.** $0.001 = 100^{-\frac{3}{2}}$

In 15–26, write each logarithmic equation in exponential form.

15. $\log_{10} 100 = 2$ **16.** $\log_5 125 = 3$ **17.** $\log_4 16 = 2$

18. $7 = \log_2 128$ **19.** $5 = \log_3 243$ **20.** $\log_7 1 = 0$

21. $\log_{10} 0.001 = -3$ **22.** $\log_{100} 0.01 = -1$ **23.** $-2 = \log_5 0.04$

24. $\log_8 2 = \frac{1}{3}$ **25.** $\log_{49} 343 = \frac{3}{2}$ **26.** $-\frac{2}{5} = \log_{32} 0.25$

In 27–56, evaluate each logarithmic expression. Show all work.

27. $\log_8 8$ **28.** $5 \log_8 8$ **29.** $\log_6 216$

30. $4 \log_6 216$ **31.** $\log_{\frac{1}{2}} 2$ **32.** $8 \log_{\frac{1}{2}} 2$

33. $\log_4 \frac{1}{16}$ **34.** $3 \log_4 \frac{1}{16}$ **35.** $\log_3 729$

36. $\frac{1}{3} \log_3 729$ **37.** $\log_4 \frac{1}{64}$ **38.** $16 \log_4 \frac{1}{64}$

39. $\log_3 81$ **40.** $\log_2 16$ **41.** $\log_3 81 \cdot \log_2 16$

42. $\log_5 125$ **43.** $\log_{10} 1{,}000$ **44.** $\log_2 32$

45. $\log_{\frac{1}{2}} \frac{1}{4}$ **46.** $\frac{1}{2} \log_3 81$ **47.** $\log_{18} 324$

48. $\log_6 36$ **49.** $\log_{\frac{1}{3}} 27$ **50.** $\frac{6 \log_6 36}{\log_{\frac{1}{3}} 27}$

51. $\log_5 125 \cdot \log_{10} 1{,}000 \cdot 2 \log_2 32$ **52.** $\log_{\frac{1}{2}} \frac{1}{4} \cdot \log_3 81 \cdot \frac{1}{2} \log_{18} 324$

53. $\dfrac{\log_5 25 + 2\log_{10} 10}{\log_{16} 4}$

54. $\dfrac{2\log_{1.5} 2.25}{\log_4 64 - \log_{80} 80}$

55. $\dfrac{3\log_3 9 \cdot 4\log_8 8 \cdot \log_{13} 169}{6\log_2 256 + \log_{\frac{1}{2}} 8}$

56. $\dfrac{\log_3 27 + 8\log_{16} 2}{\log_8 512} \cdot \log_{1,000} 10$

In 57–68, solve each equation for the variable.

57. $\log_{10} x = 3$

58. $\log_2 32 = x$

59. $\log_5 625 = x$

60. $a = \log_4 16$

61. $\log_b 27 = 3$

62. $\log_b 64 = 6$

63. $\log_5 y = -2$

64. $\log_{25} c = -4$

65. $\log_{100} x = -\frac{1}{2}$

66. $\log_8 x = \frac{1}{2}$

67. $\log_{36} 6 = x$

68. $\log_b 1,000 = \frac{3}{2}$

69. If $f(x) = \log_3 x$, find f(81).

70. If $p(x) = \log_{25} x$, find p(5).

71. If $g(x) = \log_{10} x$, find g(0.001).

72. If $h(x) = \log_{32} x$, find h(8).

73. Solve for a: $\log_5 0.2 = \log_a 10$

74. Solve for x: $\log_{100} 10 = \log_{16} x$

Applying Skills

75. When \$1 is invested at 6% interest, its value, A, after t years is $A = 1.06^t$. Express t in terms of A.

76. R is the ratio of the population of a town n years from now to the population now. If the population has been decreasing by 3% each year, $R = 0.97^n$. Express n in terms of R.

77. The decay constant of radium is -0.0004/year. The amount of radium, A, present after t years in a sample that originally contained 1 gram of radium is $A = e^{-0.0004t}$.

   a. Express $-0.0004t$ in terms of A and e.

   b. Solve for t in terms of A.

8-3 LOGARITHMIC RELATIONSHIPS

Because a logarithm is an exponent, the rules for exponents can be used to derive the rules for logarithms. Just as the rules for exponents only apply to powers with like bases, the rules for logarithms will apply to logarithms with the same base.

Basic Properties of Logarithms

If $0 < b < 1$ or $b > 1$:

$$b^0 = 1 \leftrightarrow \log_b 1 = 0$$

$$b^1 = b \leftrightarrow \log_b b = 1$$

For example:

$$3^0 = 1 \leftrightarrow \log_3 1 = 1$$
$$3^1 = 3 \leftrightarrow \log_3 3 = 1$$

Logarithms of Products

If $0 < b < 1$ or $b > 1$:

$$b^x = c \leftrightarrow \log_b c = x$$
$$b^y = d \leftrightarrow \log_b d = y$$
$$b^{x+y} = cd \leftrightarrow \log_b cd = (x + y)$$

Therefore,

$$\log_b cd = \log_b c + \log_b d$$

▶ **The log of a product is the sum of the logs of the factors of the product.**

For example:

$$3^2 = 9 \leftrightarrow \log_3 9 = 2$$
$$3^4 = 81 \leftrightarrow \log_3 81 = 4$$
$$3^{2+4} = 9 \times 81 \leftrightarrow \log_3 (9 \times 81) = \log_3 9 + \log_3 81 = 2 + 4 = 6$$

EXAMPLE I

If $\log_5 125 = 3$ and $\log_5 25 = 2$, find $\log_5 (125 \times 25)$.

Solution
$$\log_b cd = \log_b c + \log_b d$$
$$\log_5 (125 \times 25) = \log_5 125 + \log_5 25$$
$$\log_5 (125 \times 25) = 3 + 2$$
$$\log_5 (125 \times 25) = 5 \ \textit{Answer}$$

Logarithms of Quotients

If $0 < b < 1$ or $b > 1$:

$$b^x = c \leftrightarrow \log_b c = x$$
$$b^y = d \leftrightarrow \log_b d = y$$
$$b^{x-y} = \frac{c}{d} \leftrightarrow \log_b \frac{c}{d} = (x - y)$$

Therefore,

$$\log_b \frac{c}{d} = \log_b c - \log_b d$$

▶ **The log of a quotient is the log of the dividend minus the log of the divisor.**

For example:

$$3^2 = 9 \leftrightarrow \log_3 9 = 2$$
$$3^4 = 81 \leftrightarrow \log_3 81 = 4$$
$$3^{2-4} = \frac{9}{81} \leftrightarrow \log_3\left(\frac{9}{81}\right) = \log_3 9 - \log_3 81 = 2 - 4 = -2$$

Do we get the same answer if the quotient is simplified?

$$\frac{9}{81} = \frac{1}{9}$$

$$\log_3 1 = 0 \qquad \log_3 9 = 2$$

$$\log_3\left(\frac{1}{9}\right) = \log_3 1 - \log_3 9 = 0 - 2 = -2 \checkmark$$

This is true because $3^{2-4} = 3^{-2} = \frac{1}{9}$.

EXAMPLE 2

Use logs to show that for $b > 0$, $b^0 = 1$.

Solution   Let $\log_b c = a$. Then,

$$\log_b c - \log_b c = \log_b\left(\frac{c}{c}\right)$$
$$a - a = \log_b 1$$
$$0 = \log_b 1$$
$$b^0 = 1$$

Logarithms of Powers

If $0 < b < 1$ or $b > 1$:

$$b^x = c \leftrightarrow \log_b c = x$$
$$(b^x)^a = b^{ax} = c^a \leftrightarrow \log_b c^a = ax$$

Therefore,

$$\log_b c^a = a \log_b c$$

▶ **The log of a power is the exponent times the log of the base.**

For example:

$$3^2 = 9 \leftrightarrow \log_3 9 = 2$$
$$(3^2)^3 = 3^{2\times3} = 9^3 \leftrightarrow \log_3 (9)^3 = 3 \log_3 9 = 3(2) = 6$$

EXAMPLE 3

Find $\log_2 8$: **a.** using the product rule, and **b.** using the power rule.

Solution   We know that $\log_2 2 = 1$.

a. Product rule: $\log_2 8 = \log_2 (2 \times 2 \times 2) = \log_2 2 + \log_2 2 + \log_2 2$
$$= 1 + 1 + 1$$
$$= 3$$

b. Power rule: $\log_2 8 = \log_2 (2^3) = 3 \log_2 2$
$$= 3(1)$$
$$= 3$$

Check $2^3 = 8 \leftrightarrow \log_2 8 = 3$

Answer   Using both the product rule and the power rule, $\log_2 8 = 3$.

EXAMPLE 4

Find the value of $\log_{12} 9 + \log_{12} 16$.

Solution   Use the rule for product of logs:
$$\log_{12} 9 + \log_{12} 16 = \log_{12} (9 \times 16) = \log_{12} 144$$
Since $12^2 = 144, \log_{12} 144 = 2$.

Answer $\log_{12} 9 + \log_{12} 16 = 2$

EXAMPLE 5

If $\log_{10} 3 = x$ and $\log_{10} 5 = y$, express $\log_{10} \frac{\sqrt{3}}{5^2}$ in terms of x and y.

Solution $$\log_{10} \frac{\sqrt{3}}{5^2} = \log_{10} \frac{3^{\frac{1}{2}}}{5^2} = \log_{10} 3^{\frac{1}{2}} - \log_{10} 5^2$$
$$= \tfrac{1}{2} \log_{10} 3 - 2 \log_{10} 5$$
$$= \tfrac{1}{2}x - 2y$$

Answer $\log_{10} \frac{\sqrt{3}}{5^2} = \tfrac{1}{2}x - 2y$

SUMMARY

	Logarithms
Multiplication	$\log_b cd = \log_b c + \log_b d$
Division	$\log_b \frac{c}{d} = \log_b c - \log_b d$
Logarithm of a Power	$\log_b c^a = a \log_b c$
Logarithm of 1	$\log_b 1 = 0$
Logarithm of the Base	$\log_b b = 1$

Exercises

Writing About Mathematics

1. Show that for all $a > 0$ and $a \neq 1$, $\log_a a^n = n$.

2. Terence said that $(\log_a b) \cdot (\log_a c) = \log_a bc$. Do you agree with Terence? Explain why or why not.

Developing Skills

In 3–14, use the table to the right and the properties of logarithms to evaluate each expression. Show all work.

3. 27×81　　　　　**4.** 243×27

5. $19{,}683 \div 729$　　　**6.** $6{,}561 \div 27$

7. 9^4　　　　　　　　　**8.** 243^2

9. $81^2 \times 9$　　　　　**10.** $\sqrt{6{,}561} \div 729$

11. $\sqrt[4]{243 \times 2{,}187}$　　**12.** $\sqrt{\frac{19{,}683}{2{,}187}}$

13. $81^3 \div \sqrt{729}$　　　**14.** $27 \times \sqrt[3]{\frac{729}{19{,}683}}$

$\log_3 \frac{1}{9} = -2$	$\log_3 243 = 5$
$\log_3 \frac{1}{3} = -1$	$\log_3 729 = 6$
$\log_3 1 = 0$	$\log_3 2{,}187 = 7$
$\log_3 3 = 1$	$\log_3 6{,}561 = 8$
$\log_3 9 = 2$	$\log_3 19{,}683 = 9$
$\log_3 27 = 3$	$\log_3 59{,}049 = 10$
$\log_3 81 = 4$	

In 15–20: **a.** Write each expression as a single logarithm. **b.** Find the value of each expression.

15. $\log_3 1 + \log_3 9$　　　　　　　　　**16.** $\log_3 27 + \log_3 81$

17. $\log_3 243 - \log_3 729$　　　　　　　**18.** $\log_3 6{,}561 - \log_3 243$

19. $\frac{1}{3} \log_3 2{,}187 + \frac{1}{6} \log_3 81$　　　　　**20.** $\log_3 9 - 2 \log_3 27 + \log_3 243$

In 21–23: **a.** Expand each expression as a difference, sum, and/or multiple of logarithms. **b.** Find the value of each expression.

21. $4 \log_3 \frac{9}{27}$　　　　　　　**22.** $\frac{1}{2} \log_3 3(243)$　　　　　　**23.** $\log_4 \sqrt{16^2}$

In 24–29, write each expression as a single logarithm.

24. $\log_e x + \log_e 10$

25. $\log_2 a + \log_2 b$

26. $4 \log_2 (x + 2)$

27. $\log_{10} y - 2 \log_{10} (y - 1)$

28. $\log_e x + 2 \log_e y - 2 \log_e z$

29. $\frac{1}{2} \log_3 x^{10} - \frac{2}{5} \log_3 x^5$

In 30–35, expand each expression using the properties of logarithms.

30. $\log_2 2ab$

31. $\log_3 \frac{10}{x}$

32. $\log_5 a^{-5}$

33. $\log_{10} (x + 1)^2$

34. $\log_4 \frac{x^6}{y^5}$

35. $\log_e \sqrt{x}$

In 36–47, write each expression in terms of A and B if $\log_2 x = A$ and $\log_2 y = B$.

36. $\log_2 xy$

37. $\log_2 x^2 y$

38. $\log_2 (xy)^3$

39. $\log_2 xy^3$

40. $\log_2 (x \div y)$

41. $\log_2 (x^2 \div y^3)$

42. $\log_2 \sqrt{xy}$

43. $\log_2 x\sqrt{y}$

44. $\log_2 \frac{\sqrt{x}}{y^3}$

45. $\log_2 \sqrt{\frac{x}{y}}$

46. $\log_2 x\sqrt{x}$

47. $\log_2 \sqrt[4]{y}$

In 48–53, solve each equation for the variable.

48. $\log_2 2^3 + \log_2 2^2 = \log_2 x$

49. $\log_2 16 + \log_2 2 = \log_2 x$

50. $\log_2 x - \log_2 8 = \log_2 4$

51. $\log_5 x + \log_5 x = \log_5 625$

52. $\log_b 64 - \log_b 16 = \log_4 16$

53. $\log_2 8 + \log_3 9 = \log_b 100,000$

8-4 COMMON LOGARITHMS

DEFINITION

A **common logarithm** is a logarithm to the base 10.

When the base is 10, the base need not be written as a subscript of the word "log."

$$\log_{10} 100 = \log 100$$

Before calculators, tables of common logarithms were used for computation. The slide rule, a mechanical device used for multiplication and division, was a common tool for engineers and scientists. Although logarithms are no longer needed for ordinary arithmetic computations, they make it possible for us to solve equations, particularly those in which the variable is an exponent.

The TI-83+/84+ graphing calculator has a [LOG] key that will return the common logarithm of any number. For example, to find $\log_{10} 8$ or log 8, enter the following sequence of keys.

ENTER: [LOG] 8 [ENTER]

DISPLAY:

```
Log(8
          .903089987
```

Therefore, $\log_{10} 8 \approx 0.903089987$ or $10^{0.903089987} \approx 8$.

To show that this last statement is true, store the value of log 8 as x, and then use [10ˣ], the [2nd] function of the [LOG] key.

ENTER: [STO▶] [X,T,θ,n] [ENTER] [2nd] [10ˣ] [X,T,θ,n] [ENTER]

DISPLAY:

```
Ans→X
          .903089987
10^(X
                   8
```

In the equation $10^{0.903089987} = 8$, the number 0.903089987 is the logarithm of 8 and 8 is the **antilogarithm** of 0.903089987. The logarithm is the exponent and the antilogarithm is the power. In general:

▶ For $\log_b x = y$, x is the **antilogarithm** of y.

EXAMPLE I

Find log 23.75.

Solution Use the [LOG] key of the calculator.

ENTER: [LOG] 23.75 [ENTER]

DISPLAY:

```
Log(23.75
          1.375663614
```

Answer log 23.75 ≈ 1.375663614

EXAMPLE 2

If log x = 2.87534, find x to the nearest tenth.

Solution Log x = 2.87534 can be written as $10^{2.87534} = x$.

Use the $\boxed{10^x}$ key to find the antilogarithm.

ENTER: $\boxed{\text{2nd}}$ $\boxed{10^x}$ 2.87534 $\boxed{\text{ENTER}}$

DISPLAY:

```
10^(2.87534
           750.4815156
```

Answer To the nearest tenth, log 750.5 = 2.87534 or $10^{2.87534}$ = 750.5.

EXAMPLE 3

Use logarithms and antilogarithms to find the value of $\frac{4.20^3 \times 0.781}{4.83}$ to the nearest tenth.

Solution First take the log of the expression.

$$\log \frac{4.20^3 \times 0.781}{4.83} = 3 \log 4.20 + \log 0.781 - \log 4.83$$

ENTER: 3 $\boxed{\text{LOG}}$ 4.2 $\boxed{\text{)}}$ $\boxed{+}$ $\boxed{\text{LOG}}$ .781 $\boxed{\text{)}}$ $\boxed{-}$ $\boxed{\text{LOG}}$ 4.83 $\boxed{\text{ENTER}}$

DISPLAY:

```
3LOG(4.2)+LOG(.7
81)-LOG(4.83
            1.078451774
```

Now store this value for x and find its antilogarithm.

ENTER: $\boxed{\text{STO▶}}$ $\boxed{\text{X,T,θ,}n}$ $\boxed{\text{ENTER}}$ $\boxed{\text{2nd}}$ $\boxed{10^x}$ $\boxed{\text{X,T,θ,}n}$ $\boxed{\text{ENTER}}$

DISPLAY:

```
Ans→X
            1.078451774
10^(X
            11.97986087
```

Answer To the nearest tenth, $\frac{4.20^3 \times 0.781}{4.83}$ = 12.0.

Exercises

Writing About Mathematics

1. Explain why log 80 = 1 + log 8.

2. Explain why log x is negative if $0 < x < 1$.

Developing Skills

In 3–14, find the common logarithm of each number to the nearest hundredth.

3. 3.75   **4.** 8.56   **5.** 47.88   **6.** 56.2

7. 562   **8.** 5,620   **9.** 0.342   **10.** 0.0759

11. 1   **12.** 10   **13.** 100   **14.** 0.1

In 15–23, evaluate each logarithm to the nearest hundredth.

15. log 1,024   **16.** log 80   **17.** log 0.002

18. log 9 + log 3   **19.** log 64 − log 16   **20.** 200 log $\frac{5}{2}$

21. $\frac{3 \log 4}{\log 5}$ **22.** $\frac{\log 100 - \frac{1}{2} \log 36}{\log 6}$ **23.** log $\frac{1}{2}$ · log 100 · log 300

In 24–35, for each given logarithm, find the antilogarithm, x. Write the answer to four decimal places.

24. log x = 0.5787   **25.** log x = 0.8297   **26.** log x = 1.3826

27. log x = 1.7790   **28.** log x = 2.2030   **29.** log x = 2.5619

30. log x = 4.8200   **31.** log x = −0.5373   **32.** log x = −0.05729

33. log x = −1.1544   **34.** log x = −3   **35.** log x = −4

In 36–47, if log 3 = x and log 5 = y, write each of the logs in terms of x and y.

36. log 15   **37.** log 9   **38.** log 25   **39.** log 45

40. log 75   **41.** log 27   **42.** log $\frac{1}{3}$ **43.** log $\frac{1}{5}$

44. log 0.04   **45.** log $\frac{9}{10}$ **46.** log $\frac{3}{5}$ **47.** log $\left(\frac{3}{5}\right)^2$

In 48–55, if log a = c, express each of the following in terms of c.

48. log a^2 **49.** log 10a **50.** log 100 a **51.** log $\frac{a}{10}$

52. log $\frac{100}{a}$ **53.** log $\frac{a^2}{10}$ **54.** log $\left(\frac{a}{10}\right)^2$ **55.** log $\sqrt{a}$

56. Write the following expression as a single logarithm: log (x^2 − 4) + 2 log 8 − log 6.

57. Write the following expression as a multiple, sum, and/or difference of logarithms: log $\sqrt{\frac{xy}{z}}$.

Applying Skills

58. The formula $t = \frac{\log K}{0.045 \log e}$ gives the time t (in years) that it will take an investment P that is compounded continuously at a rate of 4.5% to increase to an amount K times the original principal.

a. Use the formula to complete the table to three decimal places.

K	1	2	3	4	5	10	20	30
t								

b. Use the table to graph the function $t = \frac{\log K}{0.045 \log e}$.

c. If Paul invests $1,000 in a savings account that is compounded continuously at a rate of 4.5%, when will his investment double? triple?

59. The pH (hydrogen potential) measures the acidity or alkalinity of a solution. In general, acids have pH values less than 7, while alkaline solutions (bases) have pH values greater than 7. Pure water is considered neutral with a pH of 7. The pH of a solution is given by the formula pH $= -\log x$ where x represents the hydronium ion concentration of the solution. Find, to the nearest hundredth, the approximate pH of each of the following:

a. Blood: $x = 3.98 \times 10^{-8}$

b. Vinegar: $x = 6.4 \times 10^{-3}$

c. A solution with $x = 4.0 \times 10^{-5}$

8-5 NATURAL LOGARITHMS

DEFINITION

A **natural logarithm** is a logarithm to the base e.

Recall that e is an irrational constant that is approximately equal to 2.718281828. If $x = e^y$, then $y = \log_e x$. The expression $\log_e x$ is written as **ln** x. Therefore, $y = \ln x$ can be read as "y is the natural log of x" or as "y is the exponent to the base e of x" and is called the **natural logarithmic function**.

The TI-83+/84+ graphing calculator has a **LN** key that will return the natural logarithm of any number. For example, to find $\log_e 8$ or ln 8, enter the following sequence of keys.

ENTER: **LN** 8 **ENTER**

DISPLAY:
```
LN(8
          2.079441542
```

Therefore, ln 8 ≈ 2.079441542 or $e^{2.079441542} \approx 8$.

To show that this last statement is true, store the answer in your calculator and then use the e^x key, the 2nd function of the LN key.

ENTER: STO▸ X,T,Θ,*n* ENTER 2nd e^x X,T,Θ,*n* ENTER

DISPLAY:

```
Ans→X
             2.079441542
e^(X
                       8
```

EXAMPLE I

Find ln 23.75.

Solution Use the LN key of the calculator.

ENTER: LN 23.75 ENTER

DISPLAY:

```
ln(23.75
             3.16758253
```

Answer 3.16758253

EXAMPLE 2

If ln x = 2.87534, find x to the nearest tenth.

Solution Use the e^x key to find the antilogarithm.

ENTER: 2nd e^x 2.87534 ENTER

DISPLAY:

```
e^(2.87534
            17.73145179
```

Answer To the nearest tenth, ln 17.7 = 2.87534 or $e^{2.87534}$ = 17.7.

EXAMPLE 3

Use natural logs to find the value of $\frac{4.20^3 \times 0.781}{4.83}$ to the nearest tenth.

Solution $\ln \frac{4.20^3 \times 0.781}{4.83} = 3 \ln 4.20 + \ln 0.781 - \ln 4.83$

ENTER: 3 [LN] 4.2 [)] [+] [LN] .781 [)] [−] [LN] 4.83 [ENTER]

DISPLAY:
```
3LN(4.2)+LN(.781
)-LN(4.83
         2.483226979
```

Now store this value for *x* and find its antilogarithm.

ENTER: [STO▸] [X,T,θ,n] [ENTER] [2nd] [eˣ] [X,T,θ,n] [ENTER]

DISPLAY:
```
Ans→X
         2.483226979
e^(X
         11.97986087
```

Answer To the nearest tenth, $\frac{4.20^3 \times 0.781}{4.83} = 12.0$.

Exercises

Writing About Mathematics

1. Compare Example 3 in Section 8-4 with Example 3 in Section 8-5. Explain why the answers are the same.

2. For what value of *a* does $\log a = \ln a$? Justify your answer.

Developing Skills

In 3–14, find the natural logarithm of each number to the nearest hundredth.

3. 3.75	**4.** 8.56	**5.** 47.88	**6.** 56.2
7. 562	**8.** 5,620	**9.** 0.342	**10.** 0.0759
11. 1	**12.** e	**13.** e^2	**14.** $\frac{1}{e}$

In 15–20, evaluate each logarithm to the nearest hundredth.

15. $\ln \frac{1}{2}$ **16.** $\ln 5 + \ln 7$ **17.** $\frac{1}{2} \ln 3 - \frac{1}{2} \ln 1$

18. $\ln 1{,}000^2$ **19.** $\frac{\ln 6 - \ln e}{2 \ln 8}$ **20.** $\frac{\ln \sqrt{5}}{\ln 10}$

In 21–32, for each given logarithm, find x, the antilogarithm. Write the answer to four decimal places.

21. $\ln x = 0.5787$ **22.** $\ln x = 0.8297$ **23.** $\ln x = 1.3826$

24. $\ln x = 1.7790$ **25.** $\ln x = 2.2030$ **26.** $\ln x = 2.5619$

27. $\ln x = 4.8200$ **28.** $\ln x = -0.5373$ **29.** $\ln x = -0.05729$

30. $\ln x = -1.1544$ **31.** $\ln x = -1$ **32.** $\ln x = -2$

In 33–44, if $\ln 2 = x$ and $\ln 3 = y$, write each of the natural logs in terms of x and y.

33. $\ln 6$ **34.** $\ln 9$ **35.** $\ln 4$ **36.** $\ln 12$

37. $\ln 24$ **38.** $\ln 36$ **39.** $\ln \frac{1}{3}$ **40.** $\ln \frac{1}{2}$

41. $\ln \frac{1}{6}$ **42.** $\ln \frac{1}{36}$ **43.** $\ln \frac{2}{3}$ **44.** $\ln \left(\frac{2}{3} \right)^2$

In 45–52, if $\ln a = c$, express each of the following in terms of c.

45. $\ln a^2$ **46.** $\ln a^3$ **47.** $\ln a^{-1}$ **48.** $\ln \frac{1}{a}$

49. $\ln a^{-2}$ **50.** $\ln \frac{1}{a^2}$ **51.** $\ln a^{\frac{1}{2}}$ **52.** $\ln \sqrt{a}$

In 53–56, find each value of x to the nearest thousandth.

53. $e^x = 35$ **54.** $e^x = 217$ **55.** $e^x = 2$ **56.** $e^x = -2$

57. Write the following expression as a single logarithm: $\frac{1}{2} \ln x - \ln y + \ln z^3$.

58. Write the following expression as a multiple, sum, and/or difference of logarithms: $\ln \frac{e^2 x y^{\frac{1}{2}}}{z}$.

Hands-On Activity: The Change of Base Formula

You may have noticed that your calculator can only evaluate common and natural logarithms. For logarithms with other bases, derive a formula by using the following steps:

1. Let $x = \log_b y$.

2. If $x = \log_b y$, then $b^x = y$ and $\log b^x = \log y$. Use the rule for the logarithm of a power to rewrite the left side of this last equation and solve for x.

3. Write $\log_b y$ by equating the value of x from step 2 and the value of x from step 1. This equation expresses the logarithm to the base b of y in terms of the common logarithm of y and the common logarithm of b.

4. Repeat steps 2 and 3 using $\ln b^x = \ln y$.

The equation in step 4 expresses the logarithm to the base b of y in terms of the natural logarithm of y and the natural logarithm of b.

These steps lead to the following equations:

$$\log_b y = \frac{\log y}{\log b} \qquad\qquad \log_b y = \frac{\ln y}{\ln b}$$

In **a–h**, use the calculator to evaluate the following logarithms to the nearest hundredth using both common and natural logarithms.

a. $\log_5 30$ **b.** $\log_2 17$ **c.** $\log_4 5$ **d.** $\log_{3.5} 10$

e. $\log_9 0.2$ **f.** $\log_{\frac{1}{2}} 6$ **g.** $\log_4 (12 \times 80)$ **h.** $\log_5 \frac{15}{28}$

8-6 EXPONENTIAL EQUATIONS

In chapter 7, we solved exponential equations by writing each side of the equation to the same base. Often that is possible only by using logarithms.

Since $f(x) = \log_b x$ is a function:

▶ If $x_1 = x_2$, then $\log_b x_1 = \log_b x_2$.

For example, solve $8^x = 32$ for x. There are two possible methods.

METHOD 1	**METHOD 2**
Write each side of the equation to the base 2.	*Take the log of each side of the equation and solve for the variable.*
$8^x = 32$	$8^x = 32$
$(2^3)^x = 2^5$	$\log 8^x = \log 32$
$3x = 5$	$x \log 8 = \log 32$
$x = \frac{5}{3}$	$x = \frac{\log 32}{\log 8}$

ENTER: [LOG] 32 [)] [÷] [LOG]

8 [ENTER] [MATH] [ENTER]

[ENTER]

DISPLAY:
```
LOG(32)/LOG(8
            1.666666667
ANS▶FRAC
                    5/3
```

Check: $8^{\frac{5}{3}} = \left(\sqrt[3]{8}\right)^5 = 2^5 = 32$ ✔

For exponential equations with different bases, we use logarithms. For example, solve $5^x = 32$ for x. In this equation, there is no base of which both 5 and 32 are a power.

$$5^x = 32$$

$$\log 5^x = \log 32$$

$$x \log 5 = \log 32$$

$$x = \frac{\log 32}{\log 5}$$

ENTER: [LOG] 32 [ )] [÷]

[LOG] 5 [ENTER]

DISPLAY:

```
LOG(32)/LOG(5
          2.15338279
```

Check

ENTER: [STO▸] [X,T,Θ,n] [ENTER]

5 [^] [X,T,Θ,n] [ENTER]

DISPLAY:

```
Ans→X
              2.15338279
5^X
                      32
```

When solving an exponential equation for a variable, either common logs or natural logs can be used.

EXAMPLE 1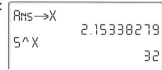

Solve for x to the nearest hundredth: $5.00(7.00)^x = 1{,}650$.

Solution   *How to Proceed*

(1) Write the equation: $5.00(7.00)^x = 1{,}650$

(2) Write the natural log of each side of the equation: $\ln 5.00(7.00)^x = \ln 1{,}650$

(3) Simplify the equation: $\ln 5.00 + \ln 7.00^x = \ln 1{,}650$

$\ln 5.00 + x \ln 7.00 = \ln 1{,}650$

(4) Solve the equation for x: $x \ln 7 = \ln 1{,}650 - \ln 5.00$

$$x = \frac{\ln 1{,}650 - \ln 5.00}{\ln 7.00}$$

(5) Use a calculator to compute x:

ENTER: [( ] [LN] 1650 [ )] [−]

[LN] 5 [ )] [ )] [÷] [LN]

7 [ENTER]

DISPLAY:

```
(LN(1650)-LN(5))
/LN(7
          2.980144102
```

Answer $x \approx 2.98$

EXAMPLE 2 ▰▰▰▰▰▰▰▰▰▰▰▰▰▰▰▰▰▰▰▰▰▰▰▰▰▰▰▰▰▰

If a $100 investment receives 5% interest each year, after how many years will the investment have doubled in value?

Solution Let A dollars represent the value of an investment of P dollars after t years at interest rate r.

Since interest is compounded yearly, $A = P(1 + r)^t$.

If the value of the investment doubles, then $A = 2P$.

$P = 100$, $A = 2P = 200$, and $(1 + r) = (1 + 0.05) = (1.05)$.

Solve for t:

$$200 = 100(1.05)^t$$
$$2 = (1.05)^t$$
$$\log 2 = t \log 1.05$$
$$\frac{\log 2}{\log (1.05)} = t$$

ENTER: [**LOG**] 2 [**)**] [**÷**] [**LOG**] 1.05 [**)**] [**ENTER**]

DISPLAY:
```
LOG(2)/LOG(1.05)
        14.20669908
```

The investment will have almost doubled after 14 years and will have doubled in value in the 15th year.

Answer 15 years

EXAMPLE 3 ▰▰▰▰▰▰▰▰▰▰▰▰▰▰▰▰▰▰▰▰▰▰▰▰▰▰▰▰▰▰

The element fermium has a decay constant of -0.00866/day. After how many days will 7.0 grams remain of a 10.0-gram sample?

Solution Since an element decays constantly, use the equation $A_n = A_0 e^{rt}$.

$A_0 = 10.0$, $A_n = 7.0$, $r = -0.00866$.

$$7.0 = 10.0\, e^{-0.00866t}$$
$$0.70 = e^{-0.00866t}$$
$$\ln 0.70 = \ln e^{-0.00866t}$$
$$\ln 0.70 = -0.00866t$$
$$\frac{\ln 0.70}{-0.00866} = t$$

ENTER: [LN] .7 [ )] [÷] [(-)] .00866 [ENTER]

DISPLAY:

```
LN(.7)/-.00866
            41.18648313
```

The sample will decay to 7.0 grams on the 41st day.

Answer   41 days

Exercises

Writing About Mathematics

1. Trisha said that the equation $5^x + 6 = 127$ could be solved by writing the logarithmic equation $x \log 5 + \log 6 = \log 127$. Do you agree with Trisha? Explain why or why not.

2. Melita said that the equation $4(3)^x = 72$ could be solved by writing the logarithmic equation $x \log 12 = \log 72$. Do you agree with Melita? Explain why or why not.

Developing Skills

In 3–14, solve each equation for the variable. Express each answer to the nearest hundredth.

3. $3^x = 12$

4. $2^b = 18$

5. $5^y = 100$

6. $10^x = 50$

7. $12^a = 254$

8. $6(3^x) = 532$

9. $7(2^b) = 815$

10. $5(10^y) = 1,200$

11. $(2 \times 8)^x = 0.25$

12. $(5 \times 7)^a = 0.585$

13. $12 + 9^x = 122$

14. $75 - 4^b = 20$

Applying Skills

15. When Rita was five, she had $1 in her piggy bank. The next year she doubled the amount that she had in her piggy bank to $2. She decided that each year she would double the amount in her piggy bank. How old will Rita be when she has at least $1,000 in her piggy bank?

16. An investment of $2,000 receives 5% interest annually. After how many years has the investment increased to at least $2,500?

17. When interest is compounded quarterly (4 times a year) at an annual rate of 6%, the rate of interest for each quarter is $\frac{0.06}{4}$, and the number of times that interest is added in t years is $4t$. After how many years will an investment of $100 compounded quarterly at 6% annually be worth at least $450? (Use the formula $A_n = A_0\left(1 + \frac{r}{n}\right)^{nt}$.)

18. After how many years will $100 invested at an annual rate of 6% compounded continuously be worth at least $450? (Use the formula $A_n = A_0 e^{rt}$.)

19. The decay constant of francium is -0.0315/minute.

    a. After how many minutes will 1.25 grams of francium remain of a 10.0-gram sample? Assume the exponential decay occurs continuously.

    b. What is the half-life of francium? (The half-life of an element is the length of time needed for half of a sample to decay. For example, it is the length of time for a sample of 10 grams to be reduced to 5 grams of the original element.)

20. The half-life of einsteinium is 276 days.

    a. To five decimal places, what is the decay constant of einsteinium? Assume the exponential decay occurs continuously.

    b. After how many days will 2.5 grams of einsteinium remain of a sample of 20 grams?

8-7 LOGARITHMIC EQUATIONS

We have solved equations involving exponents by equating the log of each side of the equation. An equation given in terms of logs can be solved by equating the antilog of each side of the equation.

Since $y = \log_b x$ is a one-to-one function:

▶ **If $\log_b x_1 = \log_b x_2$, then $x_1 = x_2$.**

EXAMPLE I

Solve for x and check: $\ln 12 - \ln x = \ln 3$.

Solution

How to Proceed

(1) Write the equation: $\ln 12 - \ln x = \ln 3$

(2) Solve for $\ln x$: $-\ln x = -\ln 12 + \ln 3$

                 $\ln x = \ln 12 - \ln 3$

(3) Simplify the right side of the equation: $\ln x = \ln \frac{12}{3}$

(4) Equate the antilog of each side of $x = \frac{12}{3}$
the equation: $x = 4$

Alternative Solution

How to Proceed

(1) Write the equation: $\ln 12 - \ln x = \ln 3$

(2) Simplify the left side of the equation: $\ln \frac{12}{x} = \ln 3$

(3) Equate the antilog of each side of the equation: $\frac{12}{x} = 3$

(4) Solve for x: $12 = 3x$

                 $4 = x$

Check	*Calculator check*

$$\ln 12 - \ln x = \ln 3$$

ENTER: [LN] 12 [ )] [−] [LN] 4 [ENTER]

$$\ln 12 - \ln 4 \stackrel{?}{=} \ln 3$$

[LN] 3 [ENTER]

$$\ln \tfrac{12}{4} \stackrel{?}{=} \ln 3$$

DISPLAY:

$$\ln 3 = \ln 3 ✔$$

```
LN(12)-LN(4
                1.098612289
LN(3
                1.098612289
```

Answer $x = 4$

Note that in both methods of solution, each side must be written as a single log before taking the antilog of each side.

EXAMPLE 2

Solve for x: $\log x + \log (x + 5) = \log 6$.

Solution

(1) Write the equation: $\qquad\qquad\qquad\qquad$ $\log x + \log (x + 5) = \log 6$

(2) Simplify the left side: $\qquad\qquad\qquad$ $\log [x(x + 5)] = \log 6$

(3) Equate the antilog of each side of the equation: $\qquad\qquad\qquad$ $x(x + 5) = 6$

(4) Solve the equation for x: $\qquad\qquad\qquad$ $x^2 + 5x = 6$

$$x^2 + 5x - 6 = 0$$

$$(x - 1)(x + 6) = 0$$

Reject the negative root. In the given equation, $\log x$ is only defined for positive values of x.

$x - 1 = 0$	$x + 6 = 0$
$x = 1$	$x = -6$ ✗

Check

$$\log x + \log (x + 5) = \log 6$$

$$\log 1 + \log (1 + 5) \stackrel{?}{=} \log 6$$

$$0 + \log 6 \stackrel{?}{=} \log 6$$

$$\log 6 = \log 6 ✔$$

Answer $x = 1$

EXAMPLE 3

Solve for b: $\log_b 8 = \log_4 64$.

Solution Let each side of the equation equal x.

Let $x = \log_4 64$.

$$4^x = 64$$
$$(2^2)^x = 2^6$$
$$2^{2x} = 2^6$$
$$2x = 6$$
$$x = 3$$

Let $x = \log_b 8$.

$$3 = \log_b 8$$
$$b^3 = 8$$
$$(b^3)^{\frac{1}{3}} = 8^{\frac{1}{3}}$$
$$b = \sqrt[3]{8}$$
$$b = 2$$

Check

$$\log_b 8 = \log_4 64$$
$$\log_2 8 \stackrel{?}{=} \log_4 64$$
$$3 = 3 \checkmark$$

Answer $b = 2$

Exercises

Writing About Mathematics

1. Randall said that the equation $\log x + \log 12 = \log 9$ can be solved by writing the equation $x + 12 = 9$. Do you agree with Randall? Explain why or why not.

2. Pritha said that before an equation such as $\log x = 1 + \log 5$ can be solved, 1 could be written as $\log 10$. Do you agree with Pritha? Explain why or why not.

Developing Skills

In 3–14, solve each equation for the variable and check.

3. $\log x + \log 8 = \log 200$

4. $\log x + \log 15 = \log 90$

5. $\ln x + \ln 18 = \ln 27$

6. $\log x - \log 5 = \log 6$

7. $\log x - \log 3 = \log 42$

8. $\ln x - \ln 24 = \ln 8$

9. $\log 8 - \log x = \log 2$

10. $\log (x + 3) = \log (x - 5) + \log 3$

11. $\log x + \log (x + 7) = \log 30$

12. $\log x + \log (x - 1) = \log 12$

13. $2 \log x = \log 25$

14. $3 \ln x + \ln 24 = \ln 3$

In 15–18, find x to the nearest hundredth.

15. $\log x - 2 = \log 5$

16. $\log x + \log (x + 2) = \log 3$

17. $2 \log x = \log (x - 1) + \log 5$

18. $2 \log x = \log (x + 3) + \log 2$

CHAPTER SUMMARY

A **logarithm** is an exponent: $x = b^y \leftrightarrow y = \log_b x$. The expression $y = \log_b x$ can be read as "y is the logarithm to the base b of x."

For $b > 0$ and $b \neq 1$, $f(x) = b^x$, then $f^{-1}(x) = \log_b x$, the **logarithmic function**. The domain of $f^{-1}(x) = \log_b x$ is the set of positive real numbers and the range is the set of real numbers. The y-axis is the **vertical asymptote** of $f^{-1}(x) = \log_b x$.

Because a logarithm is an exponent, the rules for logarithms are derived from the rules for exponents. If $b^x = c$ and $b^y = d$:

	Powers	**Logarithms**
Exponent of Zero	$b^0 = 1$	$\log_b 1 = 0$
Exponent of One	$b^1 = b$	$\log_b b = 1$
Products	$b^{x+y} = cd$	$\log_b cd = \log_b c + \log_b d$ $= x + y$
Quotients	$b^{x-y} = \frac{c}{d}$	$\log_b \frac{c}{d} = \log_b c - \log_b d$ $= x - y$
Powers	$(b^x)^a = b^{ax} = c^a$	$\log_b c^a = a \log_b c$ $= ax$

A **common logarithm** is a logarithm to the base 10. When no base is written, the logarithm is a common logarithm: $\log A = \log_{10} A$.

For $\log_b x = y$, x is the **antilogarithm** of y.

A **natural logarithm** is a logarithm to the base e. A natural logarithm is abbreviated ln: $\ln A = \log_e A$.

The rules for logarithms are used to simplify exponential and logarithmic equations:

- If $\log_b x_1 = \log_b x_2$, then $x_1 = x_2$.
- If $x_1 = x_2$, then $\log_b x_1 = \log_b x_2$.

VOCABULARY

8-1 Logarithm • Logarithm function • Vertical asymptote

8-4 Common logarithm • Antilogarithm

8-5 Natural logarithm • ln x • Natural logarithmic function

REVIEW EXERCISES

1. a. Sketch the graph of $f(x) = \log_3 x$.

   b. What is the domain of $f(x)$?

   c. What is the range of $f(x)$?

   d. Sketch the graph of $f^{-1}(x)$.

   e. Write an equation for $f^{-1}(x)$.

In 2–4, solve each equation for y in terms of x.

2. $x = \log_6 y$ **3.** $x = \log_{2.5} y$ **4.** $x = 82^y$

In 5–10, write each expression in logarithmic form.

5. $2^3 = 8$ **6.** $6^2 = 36$ **7.** $10^{-1} = 0.1$

8. $3^{\frac{1}{2}} = \sqrt{3}$ **9.** $4 = 8^{\frac{2}{3}}$ **10.** $\frac{1}{4} = 2^{-2}$

In 11–16, write each expression in exponential form.

11. $\log_3 81 = 4$ **12.** $\log_5 125 = 3$ **13.** $\log_4 8 = \frac{3}{2}$

14. $\log_7 \sqrt{7} = \frac{1}{2}$ **15.** $-1 = \log 0.1$ **16.** $0 = \ln 1$

In 17–20, evaluate each logarithmic expression. Show all work.

17. $3 \log_2 8$ **18.** $\frac{14}{9} \log_5 625$

19. $\log_3 \frac{1}{27} \div \log_9 \sqrt{3}$ **20.** $\dfrac{\log_2 256 \cdot \log_{861} 861^{\frac{1}{4}}}{4 \log_{13} \frac{1}{169}}$

21. If $f(x) = \log_3 x$, find $f(27)$.   **22.** If $f(x) = \log x$, find $f(0.01)$.

23. If $f(x) = \log_4 x$, find $f(32)$.   **24.** If $f(x) = \ln x$, find $f(e^4)$.

In 25–32, if $\log 5 = a$ and $\log 3 = b$, express each log in terms of a and b.

25. $\log 15$ **26.** $\log 25$ **27.** $\log 5(3)^2$ **28.** $\log \sqrt{45}$

29. $\log \frac{\sqrt{3}}{5}$ **30.** $\log \left(\frac{3}{5}\right)^2$ **31.** $\log \sqrt[3]{\frac{5}{3}}$ **32.** $\log \frac{25 \times \sqrt{3}}{9}$

33. If $b > 1$, then what is the value of $\log_b b$?

34. If $\log a = 0.5$, what is $\log (100a)$?

In 35–40, solve each equation for the variable. Show all work.

35. $x = \log_4 32$ **36.** $\log_b 27 = \frac{3}{2}$ **37.** $\log_6 x = -2$

38. $\log 0.001 = x$ **39.** $\log_{25} x = \frac{1}{4}$ **40.** $\log_b 16 = -2$

In 41–44: **a.** Write each expression as a single logarithm. **b.** Find the value of each expression.

41. $\frac{1}{4} \log_4 81 - \log_4 48$ **42.** $\log_{360} 5 + \log_{360} 12 + \log_{360} 6$

43. $\log_{0.5} 64 + \log_{0.5} 0.25 - 8 \log_{0.5} 2$ **44.** $\log_{1.5} \frac{3}{2} + \log_{1.5} 3 + \log_{1.5} \frac{1}{2}$

In 45–47: **a.** Expand each expression using the properties of logarithms. **b.** Evaluate each expression to the nearest hundredth.

45. $\ln \frac{42^2}{3}$ **46.** $\ln (14^2 \cdot 0.625)$ **47.** $\ln (0.25^4 \div (26 \cdot 3^{-5}))$

In 48–53, write an equation for A in terms of x and y.

48. $\log A = \log x + \log y$ **49.** $\log_2 A = \log_2 x - \log_2 y$

50. $\ln x = \ln y - \ln A$ **51.** $\log_5 A = \log_5 x + 3 \log_5 y$

52. $\log A = 2(\log x - \log y)$ **53.** $\ln x = \ln A - \frac{1}{3} \ln y$

In 54–56, if $\log x = 1.5$, find the value of each logarithm.

54. $\log 100x$ **55.** $\log x^2$ **56.** $\log \sqrt{x}$

57. If $\log x = \log 3 - \log (x - 2)$, find the value of x.

58. For what value of a does $1 + \log_3 a - \log_3 2 = \log_3 (a + 1)$?

59. If $2 \log (x + 1) = \log 5$, find x to the nearest hundredth.

60. In 2000, it was estimated that there were 50 rabbits in the local park and that the number of rabbits was growing at a rate of 8% per year. If the estimates are correct and the rabbit population continues to increase at the estimated rate, in what year will there be three times the number of rabbits, that is, 150 rabbits? (If A_0 = the number of rabbits present in 2000, A_n = the number of rabbits n years later and k = the annual rate of increase, $A_n = A_0 e^{kn}$.)

61. When Tobey was born, his parents invested $2,000 in a fund that paid an annual interest of 6%. How old will Tobey be when the investment is worth at least $5,000?

62. The population of a small town is decreasing at the rate of 2% per year. The town historian records the population at the end of each year. In 2000 ($n = 0$), the population was 5,300. If this decrease continued, what would have been the population, to the nearest hundred, in 2008 ($n = 8$)? (Use $P_n = P_0(1 + r)^n$ where P_0 is the population when $n = 0$, P_n is the population in n years, and r is the rate of change.)

Exploration

Using tables of logarithms was the common method of calculation before computers and calculators became everyday tools. These tables were relatively simple to use but their development was a complex process. John Napier spent many years developing these tables, often referred to as *Napier's bones*. Napier observed a relationship between arithmetic and geometric sequences. To distinguish the arithmetic sequence from the geometric sequence, we will let a_n be the terms of the arithmetic sequence and g_n be the terms of the geometric sequence.

For example, compare the arithmetic sequence with $a_1 = 0$ and $d = 1$ with the geometric sequence with $g_1 = 1$ and $r = 10$.

a_n	0	1	2	3	4	5	...
g_n	1	10	100	1,000	10,000	100,000	...

Note that for all n, $g_n = 10^{a_n}$ or $\log g_n = a_n$.

The arithmetic mean between a_1 and a_2, that is, between 0 and 1, is $\frac{0 + 1}{2} = 0.5$. The geometric mean between g_1 and g_2 is the mean proportional between the first and second terms of the sequence:

$$\frac{1}{x} = \frac{x}{10} \rightarrow x^2 = (1)(10) \text{ or } x = \sqrt{(1)(10)} \approx 3.16227766$$

For these two means, $3.16227766 = 10^{0.5}$ or $\log 3.16227766 = 0.5$.

Now use a recursive method. Use the results of the last step to find a new log. The arithmetic mean of 0 and 0.5 is $\frac{0 + 0.5}{2} = 0.25$. The geometric mean between 1 and 3.16227766 is 1.77827941. For these two means, $1.77827941 = 10^{0.25}$ or $\log 1.77827941 = 0.25$.

Complete the following steps. Use a calculator to find the geometric means.

a_n	g_n
0	1
1	10
(7)	(8)
(4)	(5)
(1)	(2)
2	100

STEP 1. Find the arithmetic mean of $a_2 = 1$ and $a_3 = 2$ and place it in slot (1) in the table on the left.

STEP 2. Find the geometric mean of $g_2 = 10$ and $g_3 = 100$ and place it in slot (2) in the table on the left.

STEP 3. Show that the log of the geometric mean found in step 2 is equal to the arithmetic mean found in step 1.

STEP 4. Repeat step 1 using $a_2 = 1$ and the mean in slot (1) and place it in slot (4).

STEP 5. Repeat step 2 using $g_2 = 10$ and the mean in slot (2) and place it in slot (5).

STEP 6. Show that the log of the geometric mean found in step 5 is equal to the arithmetic mean found in step 4.

STEP 7. Repeat step 1 using $a_2 = 1$ and the mean in slot (4) and place it in slot (7).

STEP 8. Repeat step 2 using $g_2 = 10$ and the mean in slot (5) and place it in slot (8).

STEP 9. Show that the log of the geometric mean found in step 8 is equal to the arithmetic mean found in step 7.

STEP 10. The geometric mean of 17.7827941 and 31.6227766 is 23.71373706. Find log 23.71373706 without using the LOG or LN keys of the calculator.

CUMULATIVE REVIEW   CHAPTERS 1–8

Part I

Answer all questions in this part. Each correct answer will receive 2 credits. No partial credit will be allowed.

1. If the range of f(x) is a subset of the set of real numbers, which of the following is not an element of the domain of $f(x) = \sqrt{9 - x^2}$?
(1) 0   (2) 1   (3) 3   (4) 4

2. An equation with imaginary roots is
(1) $x^2 - 2x - 5 = 0$ (3) $x^2 + 2x - 5 = 0$
(2) $x^2 - 2x + 5 = 0$ (4) $x^2 + 2x = 5$

3. The solution set of $5 - \sqrt{x + 1} = x$ is
(1) {3, 8}   (2) {−8, −3}   (3) {8}   (4) {3}

4. If f(x) = 2x + 3 and g(x) = x^2, then f(g(−2)) is equal to
(1) −4   (2) 1   (3) 4   (4) 11

5. Which of the following is an arithmetic sequence?
(1) 1, 3, 9, 27, 81   (3) 1, 3, 6, 12, 24
(2) 1, 2, 4, 7, 11   (4) 1, 4, 7, 10, 13

6. The expression $(3 - i)^2$ is equivalent to
(1) 8   (2) 8 − 6i   (3) 10   (4) 8 + 6i

7. The 9th term of the sequence 13, 9, 5, 1, . . . is
(1) −36   (2) −32   (3) −23   (4) −19

8. The fraction $\frac{2 + \sqrt{3}}{2 - \sqrt{3}}$ is equivalent to
(1) $11\sqrt{3}$ (2) $7 - 4\sqrt{3}$ (3) $7 + 4\sqrt{3}$ (4) $\frac{7 + 4\sqrt{3}}{7}$

9. The sum $2(3)^{-2} + 6^{-1}$ is equal to

(1) $\frac{7}{36}$ (2) $\frac{7}{18}$ (3) $\frac{1}{6^3}$ (4) $2(3)^{-3}$

10. $\displaystyle\sum_{i=0}^{3} 2^i i$ is equal to

(1) 24   (2) 26   (3) 34   (4) 90

Part II

Answer all questions in this part. Each correct answer will receive 2 credits. Clearly indicate the necessary steps, including appropriate formula substitutions, diagrams, graphs, charts, etc. For all questions in this part, a correct numerical answer with no work shown will receive only 1 credit.

11. Find all the zeros of the function $f(x) = 4x^3 - x$.

12. The sum of the roots of a quadratic equation is 10 and the product is 34. Write the equation and find the roots.

Part III

Answer all questions in this part. Each correct answer will receive 4 credits. Clearly indicate the necessary steps, including appropriate formula substitutions, diagrams, graphs, charts, etc. For all questions in this part, a correct numerical answer with no work shown will receive only 1 credit.

13. Find, to the nearest hundredth, the value of x such that $5^{3x} = 1,000$.

14. Express the sum of $\sqrt{200} + \sqrt{50} + 2\sqrt{8}$ in simplest radical form.

Part IV

Answer all questions in this part. Each correct answer will receive 6 credits. Clearly indicate the necessary steps, including appropriate formula substitutions, diagrams, graphs, charts, etc. For all questions in this part, a correct numerical answer with no work shown will receive only 1 credit.

15. a. Write $\displaystyle\sum_{n=1}^{8} 30(0.2)^{n-1}$ as the sum of terms.

b. Find $\displaystyle\sum_{n=1}^{8} 30(0.2)^{n-1}$ using a formula for the sum of a series.

16. Carbon-14 has a decay constant of -0.000124 when time is in years. A sample of wood that is estimated to have originally contained 15 grams of carbon-14 now contains 9.25 grams. Estimate, to the nearest hundred years, the age of the wood.

TRIGONOMETRIC FUNCTIONS

A triangle is the most basic shape in our study of mathematics. The word **trigonometry** means the measurement of triangles. The study of the sun, Earth, and the other planets has been furthered by an understanding of the ratios of the sides of similar triangles. Eratosthenes (276–194 B.C.) used similar right triangles to estimate the circumference of Earth to about 25,000 miles. If we compare this to the best modern estimate, 24,902 miles, we see that even though his methods involved some inaccuracies, his final results were remarkable.

Although in the history of mathematics, the applications of trigonometry are based on the right triangle, the scope of trigonometry is much greater than that. Today, trigonometry is critical to fields ranging from computer art to satellite communications.

9-1 TRIGONOMETRY OF THE RIGHT TRIANGLE

The right triangle is an important geometric figure that has many applications in the world around us. As we begin our study of trigonometry, we will recall what we already know about right triangles.

A right triangle has one right angle and two complementary acute angles. The side opposite the right angle is the longest side of the triangle and is called the **hypotenuse**. Each side opposite an acute angle is called a **leg**.

In the diagram, $\triangle ABC$ is a right triangle with $\angle C$ the right angle and $\overline{AB}$ the
hypotenuse. The legs are $\overline{BC}$, which is opposite $\angle A$, and $\overline{AC}$, which is opposite $\angle B$. The
leg $\overline{AC}$ is said to be adjacent to $\angle A$ and the
leg $\overline{BC}$ is also said to be adjacent to $\angle B$.
The acute angles, $\angle A$ and $\angle B$, are complementary.

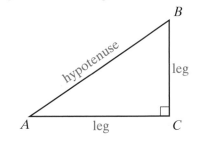

When the corresponding angles
of two triangles are congruent, the
triangles are **similar**. In the diagram,
$\triangle ABC$ and $\triangle DEF$ are right triangles with right angles $\angle C$ and $\angle F$.
If $\angle A \cong \angle D$, then $\angle B \cong \angle E$
because complements of congruent
angles are congruent. Therefore,
the corresponding angles of $\triangle ABC$
and $\triangle DEF$ are congruent and
$\triangle ABC \sim \triangle DEF$.

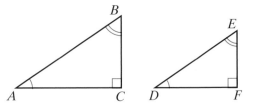

Two triangles are similar if and only if the lengths of their corresponding sides are proportional. For similar triangles $\triangle ABC$ and $\triangle DEF$, we can write:

$$\frac{BC}{EF} = \frac{AC}{DF} = \frac{AB}{DE}$$

In any proportion, we can interchange the means of the proportion to write a new proportion:

- Since $\frac{BC}{EF} = \frac{AB}{DE}$, we can write $\frac{BC}{AB} = \frac{EF}{DE}$.
- Since $\frac{AC}{DF} = \frac{AB}{DE}$, we can write $\frac{AC}{AB} = \frac{DF}{DE}$.
- Since $\frac{BC}{EF} = \frac{AC}{DF}$, we can write $\frac{BC}{AC} = \frac{EF}{DF}$.

The ratio of any two sides of one triangle is equal to the ratio of the corresponding sides of a similar triangle.

Let $\triangle ABC$ be a right triangle with a right angle at C. Any other right trian-
gle that has an acute angle congruent to $\angle A$ is similar to $\triangle ABC$. In $\triangle ABC$, $\overline{BC}$
is the leg that is opposite $\angle A$, $\overline{AC}$ is the leg that is adjacent to $\angle A$, and $\overline{AB}$ is
the hypotenuse. We can write three ratios of sides of $\triangle ABC$. The value of each
of these ratios is a function of the measure of $\angle A$, that is, a set of ordered pairs

whose first element is the measure of $\angle A$ and whose second element is one and only one real number that is the value of the ratio.

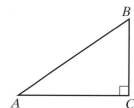

$$\text{sine of } \angle A = \frac{BC}{AB} = \frac{\text{length of the leg opposite } \angle A}{\text{length of the hypotenuse}}$$

$$\text{cosine of } \angle A = \frac{AC}{AB} = \frac{\text{length of the leg adjacent to } \angle A}{\text{length of the hypotenuse}}$$

$$\text{tangent of } \angle A = \frac{BC}{AC} = \frac{\text{length of the leg opposite } \angle A}{\text{length of the leg adjacent to } \angle A}$$

We name each of these functions by using the first three letters of the name of the function just as we named the absolute value function by using the first three letters of the word "absolute." We also abbreviate the description of the sides of the triangle used in each ratio.

$$\sin A = \frac{\text{opp}}{\text{hyp}} \qquad \cos A = \frac{\text{adj}}{\text{hyp}} \qquad \tan A = \frac{\text{opp}}{\text{adj}}$$

Note: A word or sentence can be helpful in remembering these ratios. Use the first letter of the symbols in each ratio to make up a memory aid such as "SohCahToa," which comes from:

$$\text{Sin } A = \frac{\text{Opp}}{\text{Hyp}} \qquad \text{Cos } A = \frac{\text{Adj}}{\text{Hyp}} \qquad \text{Tan } A = \frac{\text{Opp}}{\text{Adj}}$$

EXAMPLE I

In right triangle PQR, $\angle R$ is a right angle, $PQ = 25$, $QR = 24$, and $PR = 7$. Find the exact value of each trigonometric ratio.

a. $\sin P$ **b.** $\cos P$ **c.** $\tan P$ **d.** $\sin Q$ **e.** $\cos Q$ **f.** $\tan Q$

Solution The hypotenuse is $\overline{PQ}$ and $PQ = 25$.

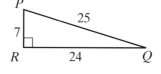

The side opposite $\angle P$ and the side adjacent to $\angle Q$ is $\overline{QR}$ and $QR = 24$.

The side opposite $\angle Q$ and the side adjacent to $\angle P$ is $\overline{PR}$ and $PR = 7$.

a. $\sin P = \frac{\text{opp}}{\text{hyp}} = \frac{24}{25}$ **b.** $\cos P = \frac{\text{adj}}{\text{hyp}} = \frac{7}{25}$ **c.** $\tan P = \frac{\text{opp}}{\text{adj}} = \frac{24}{7}$

d. $\sin Q = \frac{\text{opp}}{\text{hyp}} = \frac{7}{25}$ **e.** $\cos Q = \frac{\text{adj}}{\text{hyp}} = \frac{24}{25}$ **f.** $\tan Q = \frac{\text{opp}}{\text{adj}} = \frac{7}{24}$

EXAMPLE 2

An altitude to a side of an equilateral triangle bisects that side forming two congruent right triangles. One of the acute angles of each right triangle is an angle of the equilateral triangle and has a measure of 60°. Show that $\sin 60° = \frac{\sqrt{3}}{2}$, $\cos 60° = \frac{1}{2}$, and $\tan 60° = \sqrt{3}$.

Solution Let s be the length of the equal sides, $\overline{AC}$ and $\overline{CB}$, of equilateral triangle ABC and $\overline{CM}$ the altitude to $\overline{AB}$. The midpoint of $\overline{AB}$ is M and $AM = \frac{s}{2}$.

Use the Pythagorean Theorem to find CM:

$$(AC)^2 = (AM)^2 + (CM)^2$$
$$s^2 = \left(\frac{s}{2}\right)^2 + (CM)^2$$
$$\frac{4s^2}{4} - \frac{s^2}{4} = (CM)^2$$
$$\frac{3s^2}{4} = (CM)^2$$
$$\frac{s\sqrt{3}}{2} = CM$$

Use the ratio of sides to find sine, cosine, and tangent:

The measure of $\angle A$ is 60, hyp $= AC = s$, opp $= CM = \frac{s\sqrt{3}}{2}$, and adj $= AM = \frac{s}{2}$.

$$\sin 60° = \frac{\text{opp}}{\text{hyp}} = \frac{CM}{AC} = \frac{\frac{s\sqrt{3}}{2}}{s} = \frac{\sqrt{3}}{2}$$

$$\cos 60° = \frac{\text{adj}}{\text{hyp}} = \frac{AM}{AC} = \frac{\frac{s}{2}}{s} = \frac{1}{2}$$

$$\tan 60° = \frac{\text{opp}}{\text{adj}} = \frac{CM}{AM} = \frac{\frac{s\sqrt{3}}{2}}{\frac{s}{2}} = \sqrt{3}$$

Note that the sine, cosine, and tangent values are independent of the value of s and therefore independent of the lengths of the sides of the triangle.

Exercises

Writing About Mathematics

1. In any right triangle, the acute angles are complementary. What is the relationship between the sine of the measure of an angle and the cosine of the measure of the complement of that angle? Justify your answer.

2. Bebe said that if A is the measure of an acute angle of a right triangle, $0 < \sin A < 1$. Do you agree with Bebe? Justify your answer.

Developing Skills

In 3–10, the lengths of the sides of $\triangle ABC$ are given. For each triangle, $\angle C$ is the right angle and $m\angle A < m\angle B$. Find: **a.** $\sin A$ **b.** $\cos A$ **c.** $\tan A$.

3. 6, 8, 10   **4.** 5, 12, 13   **5.** 11, 60, 61   **6.** 8, 17, 15

7. 16, 30, 34   **8.** $2\sqrt{5}, 2, 4$ **9.** $\sqrt{2}, 3, \sqrt{7}$ **10.** $6, 3\sqrt{5}, 9$

11. Two of the answers to Exercises 3–10 are the same. What is the relationship between the triangles described in these two exercises? Justify your answer.

12. Use an isosceles right triangle with legs of length 3 to find the exact values of $\sin 45°$, $\cos 45°$, and $\tan 45°$.

13. Use an equilateral triangle with sides of length 4 to find the exact values of $\sin 30°$, $\cos 30°$, and $\tan 30°$.

Applying Skills

14. A 20-foot ladder leaning against a vertical wall reaches to a height of 16 feet. Find the sine, cosine, and tangent values of the angle that the ladder makes with the ground.

15. An access ramp reaches a doorway that is 2.5 feet above the ground. If the ramp is 10 feet long, what is the sine of the angle that the ramp makes with the ground?

16. The bed of a truck is 5 feet above the ground. The driver of the truck uses a ramp 13 feet long to load and unload the truck. Find the sine, cosine, and tangent values of the angle that the ramp makes with the ground.

17. A 20-meter line is used to keep a weather balloon in place. The sine of the angle that the line makes with the ground is $\frac{3}{4}$. How high is the balloon in the air?

18. From a point on the ground that is 100 feet from the base of a building, the tangent of the angle of elevation of the top of the building is $\frac{5}{4}$. To the nearest foot, how tall is the building?

19. From the top of a lighthouse 75 feet high, the cosine of the angle of depression of a boat out at sea is $\frac{4}{5}$. To the nearest foot, how far is the boat from the base of the lighthouse?

9-2 ANGLES AND ARCS AS ROTATIONS

Many machines operate by turning a wheel. As a wheel turns, each spoke from the center of the wheel moves through a central angle and any point on the rim of the wheel moves through an arc whose degree measure is equal to that of the central angle. The angle and the arc can have any real number as their measure.

For example, the diagram shows a rotation of three right angles or 270° as point A moves to point B. The rotation is counterclockwise (opposite to the direction in which the hands of a clock move). A counterclockwise rotation is said to be a positive rotation. Therefore, $m\angle AOB = m\overset{\frown}{AXB} = 270$.

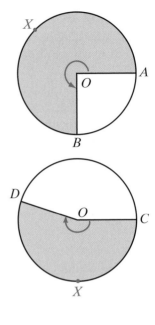

When a wheel turns in a clockwise direction (the same direction in which the hands of a clock turn), the rotation is said to be negative. For example, the diagram shows a rotation of $-200°$ as point C moves to point D. Therefore, $m\angle COD = m\overset{\frown}{CXD} = -200$. The ray at which an angle of rotation begins is the **initial side** of the angle. Here, $\overrightarrow{OC}$ is the initial side of $\angle COD$. The ray at which an angle of rotation ends is the **terminal side** of the angle. Here, $\overrightarrow{OD}$ is the terminal side of $\angle COD$.

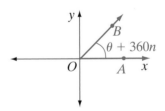

In order to study rotations through angles and arcs, and the lengths of line segments associated with these angles and arcs, we compare the rotations with central angles and arcs of a circle in the coordinate plane. An angle is in **standard position** when its vertex is at the origin and its initial side is the nonnegative ray of the x-axis.

We classify angles in standard position according to the quadrant in which the terminal side lies. The Greek symbol θ (theta) is often used to represent angle measure.

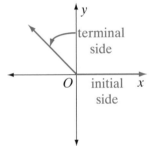

- If $0 < \theta < 90$ and $m\angle AOB = \theta + 360n$ for any integer n, then $\angle AOB$ is a first-quadrant angle.

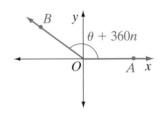

- If $90 < \theta < 180$ and $m\angle AOB = \theta + 360n$ for any integer n, then $\angle AOB$ is a second-quadrant angle.

- If $180 < \theta < 270$ and m$\angle AOB = \theta + 360n$
  for any integer n, then $\angle AOB$ is a
  third-quadrant angle.

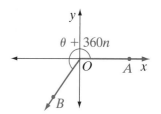

- If $270 < \theta < 360$ and m$\angle AOB = \theta + 360n$
  for any integer n, then $\angle AOB$ is a
  fourth-quadrant angle.

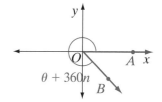

An angle in standard position whose terminal side lies on either the x-axis
or the y-axis is called a **quadrantal angle**.

▶ **The degree measure of a quadrantal angle is a multiple of 90.**

The following figures show examples of quadrantal angles:

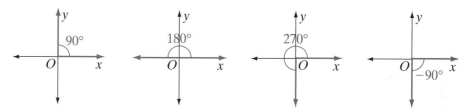

Coterminal Angles

When D on the positive ray of the x-axis is rotated
counterclockwise about the origin 270° to point D', a
quadrantal angle in standard position, $\angle DOD'$, is
formed. When point D on the positive ray of the x-axis
is rotated clockwise 90° to point D'', $\overrightarrow{OD'}$ and $\overrightarrow{OD''}$
coincide. We say that $\angle DOD''$ and $\angle DOD'$ are *coter-minal angles* because they have the same terminal
side. The measure of $\angle DOD'$ is 270° and the measure
of $\angle DOD''$ is $-90°$. The measure of $\angle DOD''$ and the
measure of $\angle DOD'$ differ by 360° or a complete rotation. The measure of $\angle DOD''$, -90, can be written as
$270 + 360(-1)$ or m$\angle DOD'$ plus a multiple of 360.

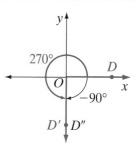

DEFINITION _____

Angles in standard position that have the same terminal side are **coterminal angles**.

If $\angle AOB$ and $\angle AOC$ are angles in standard position with the same terminal side, then m$\angle AOB$ = m$\angle AOC$ + 360n for some integer n.

EXAMPLE I

An angle in standard form with a measure of 500° lies in what quadrant?

Solution Find a coterminal angle of the 500° angle with a measure between 0° and 360°.

The measures of coterminal angles can be found by adding or subtracting multiples of 360°, the measure of a complete rotation.

$$500 - 360 = 140$$

Therefore, angles of 500° and 140° are coterminal.

An angle of 140° is in quadrant II.

An angle of 500° is a second-quadrant angle. *Answer*

EXAMPLE 2

Find the measures of five angles that are coterminal with an angle of 120°.

Solution If angles are coterminal, then their degree measures differ by a multiple of 360°. The measures of angles coterminal with an angle of 120° are of the form 120 + 360n for any integer n. Five of the infinitely many coterminal angle measures are given below:

$$120 + 360(1) = 480 \qquad 120 + 360(2) = 840 \qquad 120 + 360(3) = 1{,}200$$
$$120 + 360(-1) = -240 \qquad 120 + 360(-2) = -600$$

Angles whose measures are −600°, −240°, 480°, 840°, and 1,200° are all coterminal with an angle of 120°. *Answer*

Exercises

Writing About Mathematics

1. Is an angle of 810° a quadrantal angle? Explain why or why not.

2. Huey said that if the sum of the measures of two angles in standard position is a multiple of 360, then the angles are coterminal. Do you agree with Huey? Explain why or why not.

Developing Skills

In 3–7, draw each angle in standard position.

3. 45°   **4.** 540°   **5.** −180°   **6.** −120°   **7.** 110°

In 8–17, name the quadrant in which an angle of each given measure lies.

8. 25°   **9.** 150°   **10.** 200°   **11.** 300°   **12.** −75°

13. −200°   **14.** −280°   **15.** −400°   **16.** 750°   **17.** 1,050°

In 18–27, for each given angle, find a coterminal angle with a measure of θ such that 0 ≤ θ < 360.

18. 390°   **19.** 412°   **20.** 1,000°   **21.** −10°   **22.** −85°

23. –270°   **24.** −500°   **25.** 540°   **26.** 360°   **27.** 980°

Applying Skills

28. Do the wheels of a car move in the clockwise or counterclockwise direction when the car is moving to the right of a person standing at the side of the car?

29. To remove the lid of a jar, should the lid be turned clockwise or counterclockwise?

30. a. To insert a screw, should the screw be turned clockwise or counterclockwise?

   b. The thread spirals six and half times around a certain screw. How many degrees should the screw be turned to insert it completely?

31. The blades of a windmill make one complete rotation per second. How many rotations do they make in one minute?

32. An airplane propeller rotates 750 times per minute. How many times will a point on the edge of the propeller rotate in 1 second?

33. The Ferris wheel at the county fair takes 2 minutes to complete one full rotation.

   a. To the nearest second, how long does it take the wheel to rotate through an angle of 260°?

   b. How many minutes will it take for the wheel to rotate through an angle of 1,125°?

34. The measure of angle *POA* changes as *P* is rotated around the origin. The ratio of the change in the measure of the angle to the time it takes for the measure to change is called the **angular speed** of point *P*. For example, if a ceiling fan rotates 30 times per minute, its angular speed in degrees per minute is:

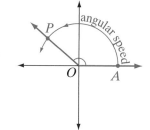

$$30(360°) = 10,800° \text{ per minute}$$

Find the angular speed in degrees per second of a tire rotating:

   a. 3 times per minute

   b. 90 times per minute

   c. 600 times per minute

9-3 THE UNIT CIRCLE, SINE, AND COSINE

Wheels, machinery, and other items that rotate suggest that trigonometry must apply to angles that have measures greater than a right angle and that have negative as well as positive measures. We want to write definitions of sine, cosine, and tangent that include the definitions that we already know but go beyond them. We will use a circle in the coordinate plane to do this.

A circle with center at the origin and radius 1 is called the **unit circle** and has the equation $x^2 + y^2 = 1$. Let θ be the measure of a central angle in standard position.

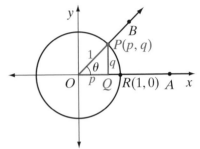

Let $\angle AOB$ be an angle in standard position with its terminal side in the first quadrant and $m\angle AOB = \theta$. Let $P(p, q)$ be the point at which the terminal side intersects the unit circle and Q be the point at which the vertical line through P intersects the x-axis. Triangle POQ is a right triangle with right angle at Q, $m\angle QOP = m\angle AOB = \theta$, $PQ = q$, $OQ = p$, and $OP = 1$.

$$\cos \theta = \frac{\text{adj}}{\text{hyp}} = \frac{OQ}{OP} = \frac{p}{1} = p$$

$$\sin \theta = \frac{\text{opp}}{\text{hyp}} = \frac{PQ}{OP} = \frac{q}{1} = q$$

Therefore, the coordinates of P are $(\cos \theta, \sin \theta)$. We have shown that for a first-quadrant angle in standard position, the cosine of the angle measure is the x-coordinate of the point at which the terminal side of the angle intersects the unit circle and the sine of the angle measure is the y-coordinate of the same point. Since P is in the first quadrant, p and q are positive numbers and $\cos \theta$ and $\sin \theta$ are positive numbers.

We can use the relationship between the point P on the unit circle and angle POR in standard position to define the sine and cosine functions for *any* angle, not just first-quadrant angles. In particular, notice that $\angle POR$ with measure θ determines the y-coordinate of P for any value of θ. This y-value is defined to be $\sin \theta$. Similarly, $\angle POR$ with measure θ determines the x-coordinate of P for any value of θ. This x-value is defined to be $\cos \theta$.

Let us consider angle $\angle POR$ in each quadrant.

CASE 1 *First-quadrant angles*

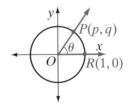

$\angle POR$ is a first-quadrant angle.
The coordinates of P are (p, q).
$\cos \theta = p$ is positive.
$\sin \theta = q$ is positive.

CASE 2 *Second-quadrant angles*

∠POR is a second-quadrant angle.
The coordinates of P are (p, q).
cos θ = p is negative.
sin θ = q is positive.

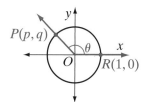

CASE 3 *Third-quadrant angles*

∠POR is a third-quadrant angle.
The coordinates of P are (p, q).
cos θ = p is negative.
sin θ = q is negative.

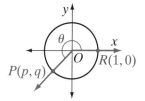

CASE 4 *Fourth-quadrant angles*

∠POR is a fourth-quadrant angle.
The coordinates of P are (p, q).
cos θ = p is positive.
sin θ = q is negative.

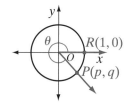

CASE 5 *Quadrantal angles*

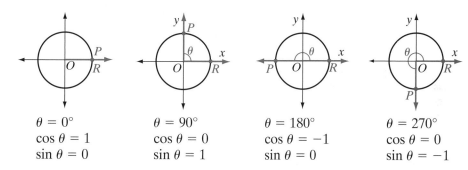

θ = 0°
cos θ = 1
sin θ = 0

θ = 90°
cos θ = 0
sin θ = 1

θ = 180°
cos θ = −1
sin θ = 0

θ = 270°
cos θ = 0
sin θ = −1

These examples allow us to use the unit circle to define the sine and cosine functions for the measure of any angle or rotation.

DEFINITION

Let ∠POQ be an angle in standard position and P be the point where the terminal side of the angle intersects the unit circle. Let m∠POQ = θ. Then:

- The **sine function** is the set of ordered pairs (θ, sin θ) such that sin θ is the y-coordinate of P.
- The **cosine function** is the set of ordered pairs (θ, cos θ) such that cos θ is the x-coordinate of P.

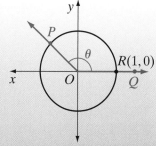

EXAMPLE I

If $P\left(-\frac{1}{2}, \frac{\sqrt{3}}{2}\right)$ is a point on the unit circle and on the terminal side of an angle in standard position whose measure is θ, find: **a.** $\sin \theta$ **b.** $\cos \theta$

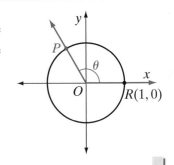

Solution   **a.** $\sin \theta = y$-coordinate of $P = \frac{\sqrt{3}}{2}$. *Answer*

b. $\cos \theta = x$-coordinate of $P = -\frac{1}{2}$. *Answer*

EXAMPLE 2

Angle AOB is an angle in standard position with measure θ. If $\theta = 72°$ and A is a point on the unit circle and on the terminal side of $\angle AOB$, find the coordinates of A to the nearest hundredth.

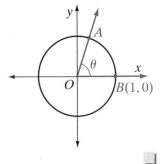

Solution   The coordinates of $A = (\cos 72°, \sin 72°)$.

Since $\cos 72° \approx 0.3091$ and $\sin 72° \approx 0.9515$, the coordinates of A to the nearest hundredth are $(0.31, 0.95)$. *Answer*

EXAMPLE 3

Find $\sin (-450°)$ and $\cos (-450°)$.

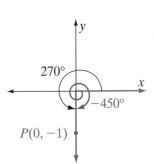

Solution   An angle in standard position whose measure is -450 has the same terminal side as an angle of $-450 + 2(360)$ or $270°$. Let P be the point where the terminal side of an angle of $270°$ intersects the unit circle. The coordinates of P are $(0, -1)$.

$$\sin (-450°) = \sin 270° = -1$$
$$\cos (-450°) = \cos (-450°) = 0$$

Therefore, $\sin (-450°) = -1$ and $\cos (-450°) = 0$. *Answer*

EXAMPLE 4

In the diagram $\triangle OPT$ is an equilateral triangle with O at the origin, P on the unit circle, $T(-1, 0)$, and $R(1, 0)$. Let $\theta = m\angle ROP$. Find:
a. θ **b.** $\cos \theta$ **c.** $\sin \theta$

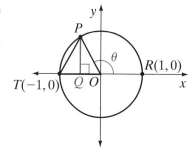

Solution **a.** Since $\angle ROP$ and $\angle TOP$ are a linear pair of angles, they are supplementary. The measure of an angle of an equilateral triangle is $60°$. Therefore, $m\angle TOP = 60$ and $m\angle ROP = 120$ or $\theta = 120$.

b. The value of $\cos \theta$ or $\cos 120$ is equal to the x-coordinate of P. Draw $\overline{PQ}$, the altitude from P in $\triangle OPT$. In an equilateral triangle, the altitude bisects the side to which it is drawn. Since $OT = 1$, $OQ = \frac{1}{2}$. The x-coordinate of P is negative and equal to $-OQ$. Therefore, $\cos \theta = -OQ = -\frac{1}{2}$.

c. The value of $\sin \theta$ or $\sin 120$ is equal to the y-coordinate of P. Triangle OQP is a right triangle with $OP = 1$.

$$OP^2 = OQ^2 + PQ^2$$
$$1^2 = \left(\tfrac{1}{2}\right)^2 + PQ^2$$
$$1 = \tfrac{1}{4} + PQ^2$$
$$\tfrac{4}{4} - \tfrac{1}{4} = PQ^2$$
$$\tfrac{3}{4} = PQ^2$$
$$\tfrac{\sqrt{3}}{2} = PQ$$

The y-coordinate of P is positive and equal to PQ. Therefore, $\sin \theta = PQ = \frac{\sqrt{3}}{2}$.

Answers **a.** $120°$ **b.** $\cos 120° = -\frac{1}{2}$ **c.** $\sin 120° = \frac{\sqrt{3}}{2}$

The values given here are exact values. Compare these values with the rational approximations given on a calculator. Press **MODE** on your calculator. In the third line, DEGREE should be highlighted.

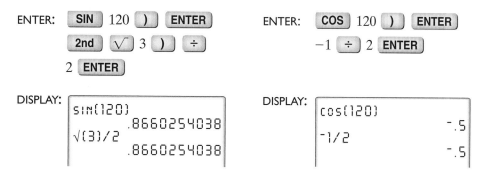

ENTER: **SIN** 120 **)** **ENTER**
        2nd **√** 3 **)** **÷**
        2 **ENTER**

ENTER: **COS** 120 **)** **ENTER**
        −1 **÷** 2 **ENTER**

DISPLAY:
```
sin(120)
            .8660254038
√(3)/2
            .8660254038
```

DISPLAY:
```
cos(120)
                     -.5
-1/2
                     -.5
```

Exercises

Writing About Mathematics

1. If P is the point at which the terminal side of an angle in standard position intersects the unit circle, what are the largest and smallest values of the coordinates of P? Justify your answer.

2. Are the sine function and the cosine function one-to-one functions? Justify your answer.

Developing Skills

In 3–10, the terminal side of $\angle ROP$ in standard position intersects the unit circle at P. If m$\angle ROP$ is θ, find: **a.** sin θ **b.** cos θ **c.** the quadrant of $\angle ROP$

3. $P\left(\frac{3}{5}, \frac{4}{5}\right)$

4. $P(0.6, -0.8)$

5. $P\left(-\frac{\sqrt{3}}{2}, \frac{1}{2}\right)$

6. $P\left(\frac{\sqrt{5}}{5}, -\frac{2\sqrt{5}}{5}\right)$

7. $P\left(-\frac{5}{13}, -\frac{12}{13}\right)$

8. $P\left(\frac{24}{25}, -\frac{7}{25}\right)$

9. $P\left(\frac{\sqrt{2}}{2}, \frac{\sqrt{2}}{2}\right)$

10. $P\left(-\frac{9}{41}, \frac{40}{41}\right)$

In 11–14, for each of the following function values, find θ if $0° \le \theta < 360°$.

11. $\sin \theta = 1$,
$\cos \theta = 0$

12. $\sin \theta = -1$,
$\cos \theta = 0$

13. $\sin \theta = 0$,
$\cos \theta = 1$

14. $\sin \theta = 0$,
$\cos \theta = -1$

In 15–22, for each given angle in standard position, determine to the nearest tenth the coordinates of the point where the terminal side intersects the unit circle.

15. $90°$

16. $-180°$

17. $81°$

18. $137°$

19. $229°$

20. $312°$

21. $540°$

22. $-45°$

23. If $P\left(\frac{2}{3}, y\right)$ is a point on the unit circle and on the terminal side of an angle in standard position with measure θ, find: **a.** y **b.** sin θ **c.** cos θ

Applying Skills

24. Let $\angle ROP$ be an angle in standard position, P the point at which the terminal side intersects the unit circle and m$\angle ROP = \theta$.

   a. Under the dilation D_a, the image of $A(x, y)$ is $A'(ax, ay)$. What are the coordinates P' of the image of P under the dilation D_5?

   b. What are the coordinates of P'', the image of P under the dilation D_{-5}?

   c. Express m$\angle ROP'$ and m$\angle ROP''$ in terms of θ.

25. Under a reflection in the y-axis, the image of $A(x, y)$ is $A'(-x, y)$. The measure of $\angle ROP = \theta$ and $P(\cos \theta, \sin \theta)$ is a point on the terminal side of $\angle ROP$. Let P' be the image of P and R' be the image of R under a reflection in the y-axis.

   a. What are the coordinates of P'?

   b. Express the measure of $\angle R'OP'$ in terms of θ.

   c. Express the measure of $\angle ROP'$ in terms of θ.

Hands-On Activity

Use graph paper, a protractor, a ruler, and a pencil, or a computer program for this activity.

1. Let the side of each square of the grid represent 0.1. Draw a circle with center at the origin and radius 1.

2. Draw the terminal side of an angle of 20° in standard position. Label the point where the terminal side intersects the circle P.

3. Estimate the coordinates of P to the nearest hundredth.

4. Use a calculator to find sin 20° and cos 20°.

5. Compare the numbers obtained from the calculator with the coordinates of P.

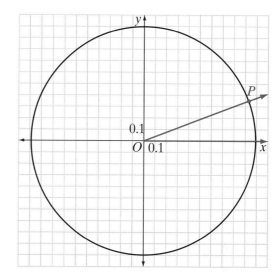

6. Repeat steps 2 through 5, replacing 20° with the following values: 70°, 100°, 165°, 200°, 250°, 300°, 345°.

In each case, are the values of the sine and the cosine approximately equal to the coordinates of P?

Hands-On Activity: Finding Sine and Cosine Using *Any* Point on the Plane

For any $\angle ROP$ in standard position, point $P(x, y)$ is on the terminal side, $OP = r = \sqrt{x^2 + y^2}$, and m$\angle ROP = \theta$. The trigonometric function values of θ are:

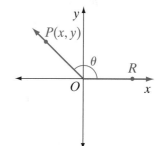

$$\sin \theta = \frac{y}{r} \qquad \cos \theta = \frac{x}{r}$$

Note that the point P is no longer restricted to the unit circle.

In (1)–(8), the given point is on the terminal side of angle θ. Find r, sin θ, and cos θ.

(1) $(3, 4)$
(2) $(-5, 12)$
(3) $(8, -15)$
(4) $(-2, -7)$

(5) $(1, 7)$
(6) $(-1, 7)$
(7) $(1, -2)$
(8) $(-3, -3)$

9-4 THE TANGENT FUNCTION

We have used the unit circle and an angle in standard position to write a general definition for the sine and cosine functions of an angle of any measure. We can write a similar definition for the tangent function in terms of a line *tangent* to the unit circle.

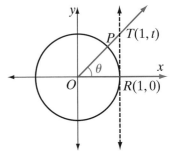

We will begin with an angle in the first quadrant. Angle *ROP* is an angle in standard position and *P* is the point at which the terminal side of ∠*ROP* intersects the unit circle. Let m∠*ROP* = θ. At *R*, the point at which the unit circle intersects the positive ray of the *x*-axis, construct a line tangent to the circle and let *T* be the point at which the terminal side intersects this tangent line. Recall that a line tangent to the circle is perpendicular to the radius drawn to the point at which the tangent line intersects the circle. Since $\overline{OR}$ is a portion of the *x*-axis, $\overline{RT}$ is a vertical line. The *x*-coordinate of *T* is the same as the *x*-coordinate of *R*. The coordinates of *T* are $(1, t)$. Since ∠*ORT* is a right angle and △*ORT* is a right triangle, $OR = 1$ and $RT = t$.

$$\tan \theta = \frac{\text{opp}}{\text{adj}} = \frac{RT}{OR} = \frac{t}{1} = t$$

Therefore, the coordinates of *T* are $(1, \tan \theta)$. We have shown that for a first-quadrant angle in standard position with measure θ, tan θ is equal to the *y*-coordinate of the point where the terminal side of the angle intersects the line that is tangent to the unit circle at the point $(1, 0)$. Since *T* is in the first quadrant, *t* is a positive number and tan θ is a positive number.

We can use this relationship between the point *T* and the tangent to the unit circle to define the tangent function for angles other than first-quadrant angles. In particular, ∠*ROP* with measure θ determines the *y*-coordinate of *T* for any value of θ. This *y*-value is defined to be tan θ.

Let us consider angle ∠*ROP* in each quadrant.

CASE 1 *First-quadrant angles*

∠*ROP* is a first-quadrant angle.
The coordinates of *T* are $(1, t)$.
tan θ = *t* is positive.

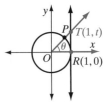

CASE 2 *Second-quadrant angles*

∠*ROP* is a second-quadrant angle.
Extend the terminal side through *O*.
The coordinates of *T* are $(1, t)$.
tan θ = *t* is negative.

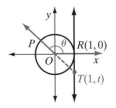

CASE 3 *Third-quadrant angles*

$\angle ROP$ is a third-quadrant angle.
Extend the terminal side through O.
The coordinates of T are $(1, t)$.
$\tan \theta = t$ is positive.

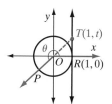

CASE 4 *Fourth-quadrant angles*

$\angle ROP$ is a fourth-quadrant angle.
The coordinates of T are $(1, t)$.
$\tan \theta = t$ is negative.

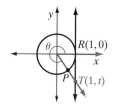

CASE 5 *Quadrantal angles*

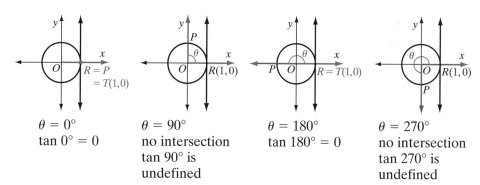

$\theta = 0°$
$\tan 0° = 0$

$\theta = 90°$
no intersection
$\tan 90°$ is
undefined

$\theta = 180°$
$\tan 180° = 0$

$\theta = 270°$
no intersection
$\tan 270°$ is
undefined

We have extended the definition of the tangent function on the unit circle for any angle or rotation for which $\tan \theta$ is defined and to determine those angles or rotations for which $\tan \theta$ is not defined.

DEFINITION

Let T be the point where the terminal side of an angle in standard position intersects the line that is tangent to the unit circle at $R(1, 0)$. Let $m\angle ROP = \theta$. Then:

- The **tangent function** is a set of ordered pairs $(\theta, \tan \theta)$ such that $\tan \theta$ is the y-coordinate of T.

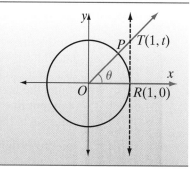

EXAMPLE 1

Line l is tangent to the unit circle at $A(1,0)$. If $T(1, -\sqrt{5})$ is a point on l, what is $\tan \angle AOT$?

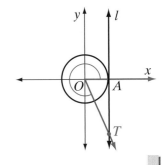

Solution $\tan \angle AOT = y$-coordinate of T
$$= -\sqrt{5} \ \textit{Answer}$$

EXAMPLE 2

Angle AOT is an angle in standard position with measure θ. If $\theta = 242°$ and T is a point on the line tangent to the unit circle at $A(1,0)$, find the coordinates of T to the nearest hundredth.

Solution The coordinates of $T = (1, \tan 242°)$.

Since $\tan 242° \approx 1.8807$, the coordinates of T to the nearest hundredth are $(1, 1.88)$. *Answer*

EXAMPLE 3

Angle AOB is an angle in standard position and $m\angle AOB = \theta$. If $\sin \theta < 0$ and $\tan \theta < 0$, in what quadrant is $\angle AOB$?

Solution When the sine of an angle is negative, the angle could be in quadrants III or IV. When the tangent of an angle is negative, the angle could be in quadrants II or IV. Therefore, when both the sine and the tangent of the angle are negative, the angle must be in quadrant IV. *Answer*

EXAMPLE 4

In the diagram, $\overleftrightarrow{RT}$ is tangent to the unit circle at $R(1,0)$. The terminal ray of $\angle ROP$ intersects the unit circle at $P(p, q)$ and the line that contains the terminal ray of $\angle ROP$ intersects the tangent line at $T(1, t)$. The vertical line from P intersects the x-axis at $Q(p, 0)$.

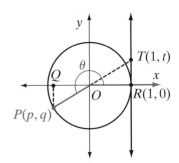

a. Prove that if $m\angle ROP = \theta$, then $\tan \theta = \frac{\sin \theta}{\cos \theta}$.

b. If the coordinates of P are $\left(-\frac{2\sqrt{2}}{3}, -\frac{1}{3}\right)$, find $\sin \theta, \cos \theta$, and $\tan \theta$.

Solution **a.** In the diagram, $\overline{RT}$ and $\overline{PQ}$ are vertical segments perpendicular to the x-axis. Therefore, $\triangle ROT$ and $\triangle QOP$ are right triangles. The vertical angles $\angle ROT$ and $\angle QOP$ are congruent. Therefore, $\triangle ROT \sim \triangle QOP$ by AA$\sim$. The ratios of corresponding sides of similar triangles are in proportion.

(1) Write a proportion using the lengths of the legs of the right triangles. The lengths of the legs are positive numbers. $RT = |t|$, $QP = |q|$, and $OQ = |p|$:

$$\frac{RT}{OR} = \frac{QP}{OQ}$$

$$\frac{|t|}{1} = \frac{|q|}{|p|}$$

(2) Since t is a positive number, $|t| = t$. Since p and q are negative numbers, $|p| = -p$ and $|q| = -q$:

$$\frac{t}{1} = \frac{-q}{-p}$$

$$t = \frac{q}{p}$$

(3) Replace t, q, and p with the function values that they represent:

$$\tan \theta = \frac{\sin \theta}{\cos \theta}$$

b. The coordinates of P are $\left(-\frac{2\sqrt{2}}{3}, -\frac{1}{3} \right)$, $\cos \theta = -\frac{2\sqrt{2}}{3}$, $\sin \theta = -\frac{1}{3}$, and:

$$\tan \theta = \frac{\sin \theta}{\cos \theta} = \frac{-\frac{1}{3}}{-\frac{2\sqrt{2}}{3}} = \frac{1}{2\sqrt{2}} = \frac{1}{2\sqrt{2}} \times \frac{\sqrt{2}}{\sqrt{2}} = \frac{\sqrt{2}}{4} \; Answer$$

What we have established in the last example is that for a given value of θ, the tangent function value is equal to the ratio of the sine function value to the cosine function value.

$$\mathbf{\tan \theta = \frac{\sin \theta}{\cos \theta}}$$

EXAMPLE 5

Use the properties of coterminal angles to find $\tan (-300°)$.

Solution Coterminal angles are angles in standard position that have the same terminal side.

$$-300 + 360 = 60$$

Therefore, a $-300°$ angle is coterminal with a $60°$ angle.

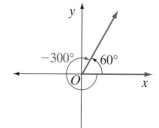

$$\sin (-300°) = \sin 60° = \frac{\sqrt{3}}{2}$$

$$\cos (-300°) = \cos 60° = \frac{1}{2}$$

$$\tan (-300°) = \frac{\cos (-300°)}{\sin (-300°)} = \frac{\frac{\sqrt{3}}{2}}{\frac{1}{2}} = \sqrt{3} \; Answer$$

Exercises

Writing About Mathematics

1. a. What are two possible measures of θ if $0° < \theta < 360°$ and $\sin \theta = \cos \theta$? Justify your answer.

b. What are two possible measures of θ if $0° < \theta < 360°$ and $\tan \theta = 1$? Justify your answer.

2. What is the value of $\cos \theta$ when $\tan \theta$ is undefined? Justify your answer.

Developing Skills

In 3–11, P is the point at which the terminal side of an angle in standard position intersects the unit circle. The measure of the angle is θ. For each point P, the x-coordinate and the quadrant is given. Find: **a.** the y-coordinate of P **b.** $\cos \theta$ **c.** $\sin \theta$ **d.** $\tan \theta$

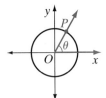

3. $\left(\frac{3}{5}, y\right)$, first quadrant

4. $\left(\frac{5}{13}, y\right)$, fourth quadrant

5. $\left(-\frac{6}{10}, y\right)$, second quadrant

6. $\left(-\frac{1}{4}, y\right)$, third quadrant

7. $\left(-\frac{\sqrt{3}}{2}, y\right)$, second quadrant

8. $\left(\frac{\sqrt{5}}{3}, y\right)$, first quadrant

9. $\left(-\frac{\sqrt{2}}{2}, y\right)$, third quadrant

10. $\left(\frac{1}{5}, y\right)$, fourth quadrant

11. $\left(\frac{3}{4}, y\right)$, first quadrant

12. $\left(-\frac{\sqrt{7}}{3}, y\right)$, second quadrant

In 13–20, P is a point on the terminal side of an angle in standard position with measure θ and on a circle with center at the origin and radius r. For each point P, find: **a.** r **b.** $\cos \theta$ **c.** $\sin \theta$ **d.** $\tan \theta$

13. $(7, 24)$

14. $(8, 15)$

15. $(-1, -1)$

16. $(-3, -4)$

17. $(-3, 6)$

18. $(-2, 6)$

19. $(4, -4)$

20. $(9, -3)$

In 21–26, if θ is the measure of $\angle AOB$, an angle in standard position, name the quadrant in which the terminal side of $\angle AOB$ lies.

21. $\sin \theta > 0, \cos \theta > 0$

22. $\sin \theta < 0, \cos \theta > 0$

23. $\sin \theta < 0, \cos \theta < 0$

24. $\sin \theta > 0, \tan \theta > 0$

25. $\tan \theta < 0, \cos \theta < 0$

26. $\sin \theta < 0, \tan \theta > 0$

27. When $\sin \theta = -1$, find a value of: **a.** $\cos \theta$ **b.** $\tan \theta$ **c.** θ

28. When $\tan \theta = 0$, find a value of: **a.** $\sin \theta$ **b.** $\cos \theta$ **c.** θ

29. When $\tan \theta$ is undefined, find a value of: **a.** $\sin \theta$ **b.** $\cos \theta$ **c.** θ

Applying Skills

30. Angle *ROP* is an angle in standard position with
    m∠*ROP* = θ, *R*(1, 0) a point on the initial side of ∠*ROP*,
    and *P*(*p*, *q*) the point at which the terminal side of ∠*ROP*
    intersects the unit circle.

    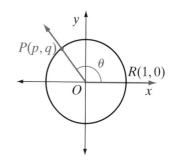

    a. What is the domain of cos θ and of sin θ?

    b. What is the range of cos θ and of sin θ?

    c. Is tan θ defined for all angle measures?

    d. What is the domain of tan θ?

    e. What is the range of tan θ?

31. Use the definitions of sin θ and cos θ based on the unit circle to prove that $\sin^2 \theta + \cos^2 \theta = 1$.

32. Show that if ∠*ROP* is an angle in standard position and m∠*ROP* = θ, then the slope of
    $\overleftrightarrow{PO}$ = tan θ.

Hands-On Activity

Use graph paper, protractor, ruler, and
pencil, or geometry software for this
activity.

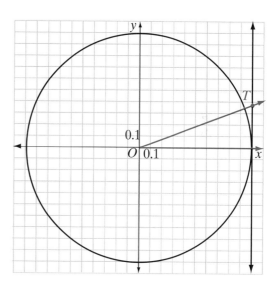

1. On graph paper, let the side of each
   square represent 0.1. Draw a circle with
   center at the origin and radius 1 and the
   line tangent to the circle at (1, 0).

2. Draw the terminal side of an angle of 20°.
   Label point *T*, the point at which the terminal side intersects the tangent.

3. Estimate the *y*-coordinate of *T* to the
   nearest hundredth.

4. Use a calculator to find tan 20.

5. Compare the number obtained from the
   calculator with the *y*-coordinates of *T*.

6. Repeat steps 2 through 5, replacing 20° with the following values: 70°, 100°, 165°, 200°, 250°,
   300°, 345°.

In each case, are the values of the tangent approximately equal to the *y*-coordinate of *T*?

9-5 THE RECIPROCAL TRIGONOMETRIC FUNCTIONS

We have defined three trigonometric functions: sine, cosine, and tangent. There are three other trigonometric functions that can be defined in terms of $\sin \theta$, $\cos \theta$, and $\tan \theta$. These functions are the **reciprocal functions**.

The Secant Function

DEFINITION

The **secant function** is the set of ordered pairs $(\theta, \sec \theta)$ such that for all θ for which $\cos \theta \neq 0$, $\sec \theta = \frac{1}{\cos \theta}$.

We know that $\cos \theta = 0$ for $\theta = 90, \theta = 270$, and for any value of θ that differs from 90 or 270 by 360. Since $270 = 90 + 180$, $\cos \theta = 0$ for all values of θ such that $\theta = 90 + 180n$ where n is an integer. Therefore, $\sec \theta$ is defined for the degree measures of angles, θ, such that $\theta \neq 90 + 180n$.

The secant function values are the reciprocals of the cosine function values. We know that $-1 \leq \cos \theta \leq 1$. The reciprocal of a positive number less than or equal to 1 is a positive number greater than or equal to 1. For example, the reciprocal of $\frac{3}{4}$ is $\frac{4}{3}$ and the reciprocal of $\frac{2}{9}$ is $\frac{9}{2}$. Also, the reciprocal of a negative number greater than or equal to -1 is a negative number less than or equal to -1. For example, the reciprocal of $-\frac{7}{8}$ is $-\frac{8}{7}$ and the reciprocal of $-\frac{1}{5}$ is -5. Therefore:

$$-1 \leq \cos \theta < 0 \rightarrow \sec \theta \leq -1$$
$$0 < \cos \theta \leq 1 \rightarrow \sec \theta \geq 1$$

▶ **The set of secant function values is the set of real numbers that are less than or equal to -1 or greater than or equal to 1, that is, $\{x : x \geq 1 \text{ or } x \leq -1\}$.**

The Cosecant Function

DEFINITION

The **cosecant function** is the set of ordered pairs $(\theta, \csc \theta)$ such that for all θ for which $\sin \theta \neq 0$, $\csc \theta = \frac{1}{\sin \theta}$.

We know that $\sin \theta = 0$ for $\theta = 0$, for $\theta = 180$, and for any value of θ that differs from 0 or 180 by 360. Therefore, $\sin \theta = 0$ for all values of θ such that $\theta = 180n$ for all integral values of n. Therefore, $\csc \theta$ is defined for the degree measures of angles, θ, such that $\theta \neq 180n$.

The cosecant function values are the reciprocals of the sine function values. We know that $-1 \leq \sin \theta \leq 1$. Therefore, the cosecant function values is the same set of values as the secant function values. Therefore:

$$-1 \leq \sin \theta < 0 \;\rightarrow\; \csc \theta \leq -1$$

$$0 < \sin \theta \leq 1 \;\rightarrow\; \csc \theta \geq 1$$

▶ **The set of cosecant function values is the set of real number that are less than or equal to -1 or greater than or equal to 1, that is, $\{x : x \geq 1 \text{ or } x \leq -1\}$.**

The Cotangent Function

DEFINITION

The **cotangent function** is the set of ordered pairs $(\theta, \cot \theta)$ such that for all θ for which $\tan \theta$ is defined and not equal to 0, $\cot \theta = \frac{1}{\tan \theta}$ and for all θ for which $\tan \theta$ is undefined, $\cot \theta = 0$.

We know that $\tan \theta = 0$ for $\theta = 0$, for $\theta = 180$, and for any value of θ that differs from 0 or 180 by 360. Therefore, $\tan \theta = 0$ for all values of θ such that $\theta = 180n$, when n is an integer. Therefore, $\cot \theta$ is defined for the degree measures of angles, θ, such that $\theta \neq 180n$.

The cotangent function values are the real numbers that are 0 and the reciprocals of the nonzero tangent function values. We know that the set of tangent function values is the set of all real numbers. The reciprocal of any nonzero real number is a nonzero real number. Therefore, the set of cotangent function values is the set of real numbers. Therefore:

$$\tan \theta < 0 \;\rightarrow\; \cot \theta < 0$$

$$\tan \theta > 0 \;\rightarrow\; \cot \theta > 0$$

$$\tan \theta \text{ is undefined} \;\rightarrow\; \cot \theta = 0$$

▶ **The set of cotangent function values is the set of real numbers.**

Function Values in the Right Triangle

In right triangle ABC, $\angle C$ is a right angle, $\overline{AB}$ is the hypotenuse, $\overline{BC}$ is the side opposite $\angle A$, and $\overline{AC}$ is the side adjacent to $\angle A$. We can express the three reciprocal function values in terms of these sides.

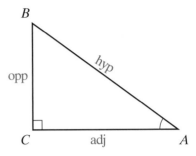

$$\sin A = \frac{\text{opp}}{\text{hyp}} = \frac{BC}{AB} \qquad \csc A = \frac{1}{\sin A} = \frac{1}{\frac{\text{opp}}{\text{hyp}}} = \frac{\text{hyp}}{\text{opp}} = \frac{AB}{BC}$$

$$\cos A = \frac{\text{adj}}{\text{hyp}} = \frac{AC}{AB} \qquad \sec A = \frac{1}{\cos A} = \frac{1}{\frac{\text{adj}}{\text{hyp}}} = \frac{\text{hyp}}{\text{adj}} = \frac{AB}{AC}$$

$$\tan A = \frac{\text{opp}}{\text{adj}} = \frac{BC}{AC} \qquad \cot A = \frac{1}{\tan A} = \frac{1}{\frac{\text{opp}}{\text{adj}}} = \frac{\text{adj}}{\text{opp}} = \frac{AC}{BC}$$

EXAMPLE 1

The terminal side of an angle in standard position intersects the unit circle at $P\left(\frac{5}{7}, -\frac{2\sqrt{6}}{7}\right)$. If the measure of the angle is θ, find the six trigonometric function values.

Solution If the coordinates of the point at which the terminal side of the angle intersects the unit circle are $\left(\frac{5}{7}, -\frac{2\sqrt{6}}{7}\right)$, then:

$$\sin \theta = -\frac{2\sqrt{6}}{7} \qquad\qquad \csc \theta = -\frac{7}{2\sqrt{6}} = -\frac{7}{2\sqrt{6}} \times \frac{\sqrt{6}}{\sqrt{6}} = -\frac{7\sqrt{6}}{12}$$

$$\cos \theta = \frac{5}{7} \qquad\qquad \sec \theta = \frac{7}{5}$$

$$\tan \theta = \frac{\sin \theta}{\cos \theta} = \frac{-\frac{2\sqrt{6}}{7}}{\frac{5}{7}} = -\frac{2\sqrt{6}}{5} \qquad \cot \theta = \frac{1}{\tan \theta} = -\frac{5}{2\sqrt{6}} = -\frac{5}{2\sqrt{6}} \times \frac{\sqrt{6}}{\sqrt{6}} = \frac{5\sqrt{6}}{12}$$

EXAMPLE 2

Show that $(\tan \theta)(\csc \theta) = \sec \theta$.

Solution Write $\tan \theta$ and $\csc \theta$ in terms of $\sin \theta$ and $\cos \theta$.

$$(\tan \theta)(\csc \theta) = \frac{\sin \theta}{\cos \theta}\left(\frac{1}{\sin \theta}\right) = \frac{1}{\cos \theta} = \sec \theta$$

Exercises

Writing About Mathematics

1. Explain why sec θ cannot equal 0.5.

2. When tan θ is undefined, cot θ is defined to be equal to 0. Use the fact that $\tan \theta = \frac{\sin \theta}{\cos \theta}$ and cot $\theta = \frac{1}{\tan \theta}$ to explain why it is reasonable to define cot 90 = 0.

Developing Skills

In 3–10, the terminal side of $\angle ROP$ in standard position intersects the unit circle at P. If m$\angle ROP$ is θ, find: **a.** sin θ **b.** cos θ **c.** tan θ **d.** sec θ **e.** csc θ **f.** cot θ

3. $P(0.6, 0.8)$ **4.** $P(0.96, -0.28)$ **5.** $P\left(-\frac{1}{6}, \frac{\sqrt{35}}{6}\right)$ **6.** $P\left(-\frac{1}{2}, \frac{\sqrt{3}}{2}\right)$

7. $P\left(\frac{2\sqrt{2}}{3}, \frac{1}{3}\right)$ **8.** $P\left(-\frac{\sqrt{5}}{3}, -\frac{2}{3}\right)$ **9.** $P\left(-\frac{\sqrt{7}}{5}, \frac{3\sqrt{2}}{5}\right)$ **10.** $P\left(-\frac{2\sqrt{10}}{7}, -\frac{3}{7}\right)$

In 11–18, P is a point on the terminal side of an angle in standard position with measure θ and on a circle with center at the origin and radius r. For each point P, find: **a.** r **b.** csc θ **c.** sec θ **d.** cot θ

11. $(3, 4)$ **12.** $(1, 4)$ **13.** $(-3, -3)$ **14.** $(-5, -5)$

15. $(-6, 6)$ **16.** $(-4, 8)$ **17.** $(9, -9)$ **18.** $(9, -3)$

19. If sin $\theta = 0$, find all possible values of: **a.** cos θ **b.** tan θ **c.** sec θ

20. If cos $\theta = 0$, find all possible values of: **a.** sin θ **b.** cot θ **c.** csc θ

21. If tan θ is undefined, find all possible values of: **a.** cos θ **b.** sin θ **c.** cot θ

22. If sec θ is undefined, find all possible values of sin θ.

23. What is the smallest positive value of θ such that cos $\theta = 0$?

Applying Skills

24. Grace walked within range of a cell phone tower. As soon as her cell phone received a signal, she looked up at the tower. The cotangent of the angle of elevation of the top of the tower is $\frac{1}{10}$. If the top of the tower is 75 feet above the ground, to the nearest foot, how far is she from the cell phone tower?

25. A pole perpendicular to the ground is braced by a wire 13 feet long that is fastened to the ground 5 feet from the base of the pole. The measure of the angle the wire makes with the ground is θ. Find the value of:

     a. sec θ **b.** csc θ **c.** cot θ

26. An airplane travels at an altitude of 6 miles. At a point on the ground, the measure of the angle of elevation to the airplane is θ. Find the distance to the plane from the point on the ground when:

     a. csc $\theta = 2$ **b.** sec $\theta = \frac{5}{3}$ **c.** cot $= \sqrt{3}$

27. The equation $\sin^2 \theta + \cos^2 \theta = 1$ is true for all θ.

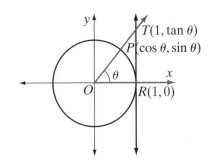

a. Use the given equation to prove that $\tan^2 \theta + 1 = \sec^2 \theta$.

b. Is the equation in part **a** true for all θ? Justify your answer.

c. In the diagram, if $OR = 1$ and $RT = \tan \theta$, the length of what line segment is equal to $\sec \theta$?

28. Show that $\cot \theta = \frac{\cos \theta}{\sin \theta}$ for all values of θ for which $\sin \theta \neq 0$.

9-6 FUNCTION VALUES OF SPECIAL ANGLES

Equilateral triangles and isosceles right triangles occur frequently in our study of geometry and in the applications of geometry and trigonometry. The angles associated with these triangles are multiples of 30 and 45 degrees. Although a calculator will give rational approximations for the trigonometric function values of these angles, we often need to know their exact values.

The Isosceles Right Triangle

The measure of an acute angle of an isosceles right triangle is 45 degrees. If the measure of a leg is 1, then the length of the hypotenuse can be found by using the Pythagorean Theorem.

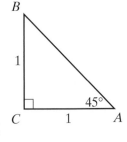

$c^2 = a^2 + b^2$

$AB^2 = 1^2 + 1^2$

$AB^2 = 1 + 1$

$AB = \sqrt{2}$

Therefore:

$\sin 45° = \frac{\text{opp}}{\text{hyp}} = \frac{1}{\sqrt{2}} = \frac{1}{\sqrt{2}} \times \frac{\sqrt{2}}{\sqrt{2}} = \frac{\sqrt{2}}{2}$

$\cos 45° = \frac{\text{adj}}{\text{hyp}} = \frac{1}{\sqrt{2}} = \frac{1}{\sqrt{2}} \times \frac{\sqrt{2}}{\sqrt{2}} = \frac{\sqrt{2}}{2}$

$\tan 45° = \frac{\text{opp}}{\text{adj}} = \frac{1}{1} = 1$

The Equilateral Triangle

An equilateral triangle is separated into two congruent right triangles by an altitude from any vertex. In the diagram, $\overline{CD}$ is altitude from C, $\triangle ACD \cong \triangle BCD$, $\angle CDA$ is a right angle, $m\angle A = 60$, and $m\angle ACD = 30$. If $AC = 2$, $AD = 1$.

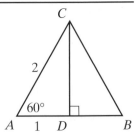

$$c^2 = a^2 + b^2$$
$$2^2 = 1^2 + CD^2$$
$$4 = 1 + CD^2$$
$$3 = CD^2$$
$$\sqrt{3} = CD$$

Therefore:

$$\sin 30° = \frac{\text{opp}}{\text{hyp}} = \frac{1}{2}$$

$$\cos 30° = \frac{\text{adj}}{\text{hyp}} = \frac{\sqrt{3}}{2}$$

$$\tan 30° = \frac{\text{opp}}{\text{adj}} = \frac{1}{\sqrt{3}} = \frac{1}{\sqrt{3}} \times \frac{\sqrt{3}}{\sqrt{3}} = \frac{\sqrt{3}}{3}$$

$$\sin 60° = \frac{\text{opp}}{\text{hyp}} = \frac{\sqrt{3}}{2}$$

$$\cos 60° = \frac{\text{adj}}{\text{hyp}} = \frac{1}{2}$$

$$\tan 60° = \frac{\text{opp}}{\text{adj}} = \frac{\sqrt{3}}{1} = \sqrt{3}$$

The $0°, 30°, 45°, 60°$, and $90°$ angles and their multiples occur frequently and it is useful to memorize the exact trigonometric function values summarized below:

θ	0°	30°	45°	60°	90°
$\sin \theta$	0	$\frac{1}{2}$	$\frac{\sqrt{2}}{2}$	$\frac{\sqrt{3}}{2}$	1
$\cos \theta$	1	$\frac{\sqrt{3}}{2}$	$\frac{\sqrt{2}}{2}$	$\frac{1}{2}$	0
$\tan \theta$	0	$\frac{\sqrt{3}}{3}$	1	$\sqrt{3}$	undefined

EXAMPLE 1

Use a counterexample to show that $\sin 2\theta \neq 2 \sin \theta$.

Solution   Let $\theta = 30°$ and $2\theta = 3(30°) = 60°$.

$\sin 2\theta = \sin 60° = \frac{\sqrt{3}}{2}$ and $2 \sin \theta = 2 \sin 30° = 2\left(\frac{1}{2}\right) = 1$.

Therefore, there exists at least one value of θ for which $\sin 2\theta \neq 2 \sin \theta$.

EXAMPLE 2

What are the lengths of the diagonals of a rhombus if the measure of a side is 10 centimeters and the measure of an angle is 120 degrees?

Solution   The diagonals of a rhombus are perpendicular, bisect each other, and bisect the angles of the rhombus. Therefore, the diagonals of this rhombus separate it into four congruent right triangles.

Let $ABCD$ be the given rhombus with E the midpoint of $\overline{AC}$ and $\overline{BD}$, $m\angle DAB = 120$ and $AB = 10$ centimeters. Therefore, $\angle AEB$ is a right angle and $m\angle EAB = 60$.

$$\cos 60 = \frac{AE}{AB}$$
$$\frac{1}{2} = \frac{AE}{10}$$
$$2AE = 10$$
$$AE = 5$$

$$\sin 60 = \frac{BE}{AB}$$
$$\frac{\sqrt{3}}{2} = \frac{BE}{10}$$
$$2BE = 10\sqrt{3}$$
$$BE = 5\sqrt{3}$$

Answers $AC = 2AE = 2(5) = 10$
$BD = 2BE = 2(5\sqrt{3}) = 10\sqrt{3}$

EXAMPLE 3

What is the exact value of sec 30°?

Solution $\sec 30° = \dfrac{1}{\cos 30°} = \dfrac{1}{\frac{\sqrt{3}}{2}} = \dfrac{2}{\sqrt{3}} = \dfrac{2}{\sqrt{3}} \times \dfrac{\sqrt{3}}{\sqrt{3}} = \dfrac{2\sqrt{3}}{3}$ *Answer*

Exercises

Writing About Mathematics

1. R is the point $(1, 0)$, P' is the point on a circle with center at the origin, O, and radius r, and $m\angle ROP' = \theta$. Alicia said that the coordinates of P' are $(r \cos \theta, r \sin \theta)$. Do you agree with Alicia? Explain why or why not.

2. Hannah said that if $\cos \theta = a$, then $\sin \theta = \pm\sqrt{1 - a^2}$. Do you agree with Hannah? Explain why or why not.

Developing Skills

In 3–44, find the exact value.

3. $\cos 30°$ **4.** $\sin 30°$ **5.** $\csc 30°$ **6.** $\tan 30°$ **7.** $\cot 30°$

8. $\cos 60°$ **9.** $\sec 60°$ **10.** $\sin 60°$ **11.** $\csc 60°$ **12.** $\tan 60°$

13. $\cot 60°$ **14.** $\cos 45°$ **15.** $\sec 45°$ **16.** $\sin 45°$ **17.** $\csc 45°$

18. $\tan 45°$ **19.** $\cot 45°$ **20.** $\cos 180°$ **21.** $\sec 180°$ **22.** $\sin 180°$

23. $\csc 180°$ **24.** $\tan 180°$ **25.** $\cot 180°$ **26.** $\cos 270°$ **27.** $\sec 270°$

28. $\sin 270°$ **29.** $\csc 270°$ **30.** $\tan 270°$ **31.** $\cot 270°$ **32.** $\sin 450°$

33. $\sin 0° + \cos 0° + \tan 0°$ **34.** $\sin 45° + \cos 60°$ **35.** $\sin 90° + \cos 0° + \tan 45°$

36. $(\cos 60°)^2 + (\sin 60°)^2$ **37.** $(\sec 45°)^2 - (\tan 45°)^2$ **38.** $(\sin 30°)(\cos 60°)$

39. $(\tan 45°)(\cot 45°)$ **40.** $(\sin 45°)(\cos 45°)(\tan 45°)$ **41.** $(\sin 30°)(\sec 60°)$

42. $\dfrac{\tan 30°}{\cos 60°}$ **43.** $\dfrac{\sin 45°}{\cos 45°}$ **44.** $\dfrac{\sin 30°}{\csc 30°}$

Applying Skills

45. A diagonal path across a rectangular field makes an angle of 30 degrees with the longer side of the field. If the length of the path is 240 feet, find the exact dimensions of the field.

46. Use a counterexample to show that $\sin A + \sin B = \sin (A + B)$ is false.

47. Use a counterexample to show that $A < B$ implies $\cos A < \cos B$ is false.

9-7 FUNCTION VALUES FROM THE CALCULATOR

We used our knowledge of an equilateral triangle and of an isosceles right triangle to find the trigonometric function values for some special angles. How can we find the trigonometric function values of any number?

Before calculators and computers were common tools, people who worked with trigonometric functions used tables that supplied the function values. Now these function values are stored in most calculators.

Compare the values given by a calculator to the exact values for angles of 30°, 45° and 60°. On your calculator, press **MODE** . The third line of that menu lists RADIAN and DEGREE. These are the two common angle measures. In a later chapter, we will work with radians. For now, DEGREE should be highlighted on your calculator.

We know that $\cos 30° = \frac{\sqrt{3}}{2}$. When you enter cos 30° into your calculator, the display will show an approximate decimal value.

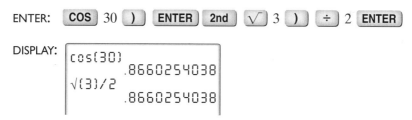

These two displays tell us that 0.8660254038 is a rational approximation of the exact irrational value of cos 30°.

A degree, like an hour, is divided into 60 minutes. The measure of an angle written as 37° 45′ is read 37 degrees, 45 minutes, which is equivalent to $37\frac{45}{60}$ degrees. Use the following calculator key sequence to find the tangent of an angle with this measure.

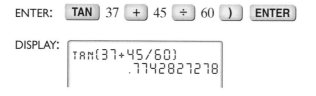

EXAMPLE 1

Find sin 70° to four decimal places.

Solution   Use a calculator.

ENTER: **SIN**   70   **)**   **ENTER**

DISPLAY:

```
sin(70)
         .9396926208
```

Round the decimal value. Since the digit following the fourth decimal place is greater than 4, add 1 to the digit in the fourth decimal place.

Answer   sin 70° ≈ 0.9397

EXAMPLE 2

Find sec 54° to four decimal places.

Solution $\sec 54° = \dfrac{1}{\cos 54°}$

ENTER: 1   **÷**   **COS**   54   **)**   **ENTER**

DISPLAY:

```
1/cos(54)
        1.701301617
```

Round the decimal value. Since the digit following the fourth decimal place is less than 5, drop that digit and all digits that follow.

Answer   sec 54° ≈ 1.7013

EXAMPLE 3

Find sin 215° to four decimal places.

Solution   ENTER: **SIN**   215   **)**   **ENTER**

DISPLAY:

```
sin(215)
        -.5735764364
```

Recall that when the degree measure of an angle is greater than 180 and less than 270, the angle is a third-quadrant angle. The y-coordinate of the intersection of the terminal side with the unit circle is negative. Therefore, the sine of the angle is negative.

Answer   −0.5736

Degree Measures of Angles

If sin θ = 0.9205, then θ is the measure of an angle whose sine is 0.9205. In a circle, the degree measure of a central angle is equal to the degree measure of the intercepted arc. Therefore, we can also say that θ is the measure of an arc whose sine is 0.9205. This can be written in symbols.

$$\sin \theta = 0.9205 \rightarrow \theta = \text{the angle whose sine is } 0.9205$$
$$\sin \theta = 0.9205 \rightarrow \theta = \text{the arc whose sine is } 0.9205$$

The words "the arc whose sine is" are abbreviated as **arcsine**. We further shorten arcsine to arcsin.

$$\sin \theta = 0.9205 \rightarrow \theta = \arcsin 0.9205$$

On a calculator, the symbol for arcsin is "$\sin^{-1}$."

$$\sin \theta = 0.9205 \rightarrow \theta = \sin^{-1} 0.9205$$

We can also write:

$$\cos \theta = 0.3907 \rightarrow \theta = \arccos 0.3907 = \cos^{-1} 0.3907$$
$$\tan \theta = 2.3559 \rightarrow \theta = \arctan 2.3559 = \tan^{-1} 2.3559$$

Arccos and $\cos^{-1}$ are abbreviations for **arccosine**. Similarly, arctan and $\tan^{-1}$ are abbreviations for **arctangent**.

EXAMPLE 4

Find θ to the nearest degree if cos θ = 0.8988 and 0° < θ < 180°.

Solution   cos θ = 0.8988 → θ = arccos 0.8988 = $\cos^{-1}$ 0.8988

ENTER: [**2nd**] [**COS⁻¹**] 0.8988 [**)**]   [**ENTER**]

DISPLAY:
```
cos⁻¹(0.8988)
          25.99922183
```

To the nearest degree, θ = 26°. *Answer*

EXAMPLE 5

Find θ to the nearest minute if $\tan \theta = 3.5782$ and $-90° < \theta < 90°$.

Solution The calculator will return the value of θ as a whole number of degrees followed by a decimal value. To change the decimal part of the number to minutes, multiply by 60.

ENTER: [2nd] [TAN⁻¹] 3.5782 [ )] [ENTER]

DISPLAY:
```
TAN⁻¹(3.5782)
          74.38590988
```

Now subtract 74 from this answer and multiply the difference by 60.

ENTER: [( ] [2nd] [ANS] [−] 74 [ )] [×] 60 [ENTER]

DISPLAY:
```
TAN⁻¹(3.5782)
          74.38590988
(Ans-74)*60
          23.15459294
```

To the nearest minute, $\theta = 74° 23'$. *Answer*

Exercises

Writing About Mathematics

1. Explain why the calculator displays an error message when [TAN] 90 is entered.

2. Explain why the calculator displays the same value for sin 400° as for sin 40°.

Developing Skills

In 3–38, find each function value to four decimal places.

3. sin 28°	**4.** cos 35°	**5.** tan 78°	**6.** cos 100°
7. sin 170°	**8.** tan 200°	**9.** tan 20°	**10.** cos 255°
11. cos 75°	**12.** sin 280°	**13.** sin 80°	**14.** tan 375°
15. tan 15°	**16.** cos 485°	**17.** cos 125°	**18.** sin (−10°)
19. sin 350°	**20.** sin 190°	**21.** cos 18° 12′	**22.** sin 57° 40′

23. tan 88° 30′   **24.** sin 105° 50′   **25.** tan 172° 18′   **26.** cos 205° 12′

27. sin 205° 12′   **28.** tan 266° 27′   **29.** sec 72°   **30.** csc 15°

31. cot 63°   **32.** sec 100°   **33.** csc 125°   **34.** cot 165°

35. csc 245°   **36.** cot 254°   **37.** sec 307°   **38.** csc 347°

In 39–50, find the smallest positive value of θ to the nearest degree.

39. $\sin \theta = 0.3455$ **40.** $\cos \theta = 0.4383$ **41.** $\tan \theta = 0.2126$ **42.** $\cos \theta = 0.7660$

43. $\tan \theta = 0.7000$ **44.** $\sin \theta = 0.9990$ **45.** $\cos \theta = 0.9990$ **46.** $\tan \theta = 1.8808$

47. $\sin \theta = 0.5446$ **48.** $\cos \theta = 0.5446$ **49.** $\tan \theta = 1.0355$ **50.** $\tan \theta = 12.0000$

In 51–58, find the smallest positive value of θ to the nearest minute.

51. $\sin \theta = 0.2672$ **52.** $\cos \theta = 0.2672$ **53.** $\sin \theta = 0.9692$ **54.** $\cos \theta = 0.9692$

55. $\sin \theta = 0.6534$ **56.** $\cos \theta = 0.6534$ **57.** $\tan \theta = 7.3478$ **58.** $\tan \theta = 0.0892$

Applying Skills

59. A ramp that is 12 feet long is used to reach a doorway that is 3.5 feet above the level ground. Find, to the nearest degree, the measure the ramp makes with the ground.

60. The bed of a truck is 4.2 feet above the ground. In order to unload boxes from the truck, the driver places a board that is 12 feet long from the bed of the truck to the ground. Find, to the nearest minute, the measure the board makes with the ground.

61. Three roads intersect to enclose a small triangular park. A path that is 72 feet long extends from the intersection of two of the roads to the third road. The path is perpendicular to that road at a point 65 feet from one of the other intersections and 58 feet from the third. Find, to the nearest ten minutes, the measures of the angles at which the roads intersect.

62. The terminal side of an angle in standard position intersects the unit circle at the point $(-0.8, 0.6)$.

a. In what quadrant does the terminal side of the angle lie?

b. Find, to the nearest degree, the smallest positive measure of the angle.

63. The terminal side of an angle in standard position intersects the unit circle at the point $(0.28, -0.96)$.

a. In what quadrant does the terminal side of the angle lie?

b. Find, to the nearest degree, the smallest positive measure of the angle.

9-8 REFERENCE ANGLES AND THE CALCULATOR

In Section 9-7, we used the calculator to find the function values of angles. We know that two or more angles can have the same function value. For example, when the degree measures of two angles differ by 360, these angles, in standard position, have the same terminal side and therefore the same function values. In addition to angles whose measures differ by a multiple of 360, are there other angles that have the same function values?

Second-Quadrant Angles

In the diagram, $R(1, 0)$ and $P(a, b)$ are points on the unit circle. Under a reflection in the y-axis, the image of $P(a, b)$ is $P'(-a, b)$ and the image of $R(1, 0)$ is $R'(-1, 0)$. Since angle measure is preserved under a line reflection, $m\angle ROP = m\angle R'OP'$.

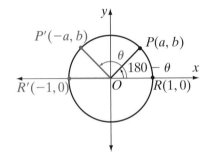

The rays $\overrightarrow{OR}$ and $\overrightarrow{OR'}$ are opposite rays and $\angle ROP'$ and $\angle R'OP'$ are supplementary. Therefore, $\angle ROP'$ and $\angle ROP$ are supplementary.

$$m\angle ROP' + m\angle ROP = 180$$
$$m\angle ROP = 180 - m\angle ROP'$$

- If $m\angle ROP' = \theta$, then $m\angle ROP = 180 - \theta$.
- If $m\angle ROP' = \theta$, then $\sin \theta = b$ and $\cos \theta = -a$.
- If $m\angle ROP = (180 - \theta)$, then $\sin (180 - \theta) = b$ and $\cos (180 - \theta) = a$.

Therefore:

$$\sin \theta = \sin (180 - \theta) \qquad \cos \theta = -\cos (180 - \theta)$$

Since $\tan \theta = \frac{\sin \theta}{\cos \theta}$, then also:

$$\tan \theta = \frac{\sin (180 - \theta)}{-\cos (180 - \theta)} = -\tan (180 - \theta)$$

Let θ be the measure of a second-quadrant angle. Then there exists a first-quadrant angle with measure $180 - \theta$ such that:

$$\sin \theta = \sin (180 - \theta)$$

$$\cos \theta = -\cos (180 - \theta)$$

$$\tan \theta = -\tan (180 - \theta)$$

The positive acute angle $(180 - \theta)$ is the **reference angle of the second-quadrant angle**. When drawn in standard position, the reference angle is in the first quadrant.

EXAMPLE 1

Find the exact values of sin 120°, cos 120°, and tan 120°.

Solution The measure of the reference angle for an angle of 120° is 180° − 120° = 60°.

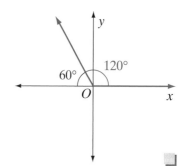

$$\sin 120° = \sin 60° = \frac{\sqrt{3}}{2}$$

$$\cos 120° = -\cos 60° = -\frac{1}{2}$$

$$\tan 120° = -\tan 60° = -\sqrt{3}$$

Third-Quadrant Angles

In the diagram, $R(1, 0)$ and $P(a, b)$ are two points on the unit circle. Under a reflection in the origin, the image of $R(1,0)$ is $R'(-1,0)$ and the image of $P(a, b)$ is $P'(-a, -b)$. Since angle measure is preserved under a point reflection, $m\angle ROP = m\angle R'OP'$.

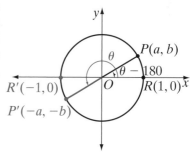

$$m\angle ROP' = 180 + m\angle R'OP'$$
$$= 180 + m\angle ROP$$

$$m\angle ROP = m\angle ROP' - 180$$

- If $m\angle ROP' = \theta$, then $m\angle ROP = \theta - 180$.
- If $m\angle ROP' = \theta$, then $\sin \theta = -b$ and $\cos \theta = -a$.
- If $m\angle ROP = (\theta - 180)$, then $\sin (\theta - 180) = b$ and $\cos (\theta - 180) = a$.

Therefore:

$$\sin \theta = -\sin (\theta - 180) \qquad \cos \theta = -\cos (\theta - 180)$$

Since $\tan \theta = \frac{\sin \theta}{\cos \theta}$, then:

$$\tan \theta = \frac{-\sin (\theta - 180)}{-\cos (\theta - 180)} = \tan (\theta - 180)$$

Let θ be the measure of a third-quadrant angle. Then there exists a first-quadrant angle with measure $\theta - 180$ such that:

$$\mathbf{\sin \theta = -\sin (\theta - 180)}$$

$$\mathbf{\cos \theta = -\cos (\theta - 180)}$$

$$\mathbf{\tan \theta = \tan (\theta - 180)}$$

The positive acute angle $(\theta - 180)$ is the **reference angle of the third-quadrant angle**. When drawn in standard position, the reference angle is in the first quadrant.

EXAMPLE 2

Express sin 200°, cos 200°, and tan 200° in terms of a function value of an acute angle.

Solution   An angle of 200° is in the third quadrant.

For a third-quadrant angle with measure θ, the measure of the reference angle is $\theta - 180$.

The measure of the reference angle for an angle of 200° is 200° − 180° = 20°.

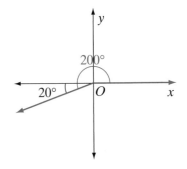

$$\sin 200° = -\sin 20°$$
$$\cos 200° = -\cos 20°$$
$$\tan 200° = \tan 20°$$

ENTER: [SIN] 200 [ )] [ENTER]

[COS] 200 [ )] [ENTER]

[TAN] 200 [ )] [ENTER]

ENTER: [(-)] [SIN] 20 [ )] [ENTER]

[(-)] [COS] 20 [ )] [ENTER]

[TAN] 20 [ )] [ENTER]

DISPLAY:
```
sin(200)
            -.3420201433
cos(200)
            -.9396926208
TAN(200)
             .3639702343
```

DISPLAY:
```
-sin(20)
            -.3420201433
-cos(20)
            -.9396926208
TAN(20)
             .3639702343
```

Answer   sin 200° = −sin 20°, cos 200° = −cos 20°, tan 200° = tan 20°

EXAMPLE 3

If $180 < \theta < 270$ and sin $\theta = -0.5726$, what is the value of θ to the nearest degree?

Solution   If $m\angle A = \theta$, $\angle A$ is a third-quadrant angle. The measure of the reference angle is $\theta - 180$ and sin $(\theta - 180) = 0.5726$.

ENTER: [**2nd**] [**SIN⁻¹**] 0.5726 [**)**] [**ENTER**]

DISPLAY:

```
SIN⁻¹(0.5726)
           34.9317314
```

To the nearest degree, $(\theta - 180) = 35$. Therefore:

$$\theta = 35 + 180 = 215° \quad \textit{Answer}$$

Fourth-Quadrant Angles

In the diagram, $R(1, 0)$ and $P(a, b)$ are points on the unit circle. Under a reflection in the x-axis, the image of $P(a, b)$ is $P'(a, -b)$ and the image of $R(1, 0)$ is $R(1, 0)$. Since angle measure is preserved under a line reflection, the measure of the acute angle $\angle ROP$ is equal to the measure of the acute angle $\angle ROP'$.

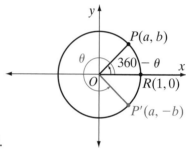

- If $m\angle ROP' = \theta$, then $m\angle ROP = 360 - \theta$.
- If $m\angle ROP' = \theta$, then $\sin \theta = -b$ and $\cos \theta = a$.
- If $m\angle ROP = (360 - \theta)$, then $\sin (360 - \theta) = b$ and $\cos (360 - \theta) = a$.

Therefore:

$$\sin \theta = -\sin (360 - \theta) \qquad \cos \theta = \cos (360 - \theta)$$

Since $\tan \theta = \frac{\sin \theta}{\cos \theta}$, then:

$$\tan \theta = \frac{-\sin (360 - \theta)}{\cos (360 - \theta)} = -\tan (360 - \theta)$$

Let θ be the measure of a fourth-quadrant angle. Then there exists a first-quadrant angle with measure $360 - \theta$ such that:

$$\sin \theta = -\sin (360 - \theta)$$

$$\cos \theta = \cos (360 - \theta)$$

$$\tan \theta = -\tan (360 - \theta)$$

The positive acute angle $(360 - \theta)$ is the **reference angle of the fourth-quadrant angle**. When drawn in standard position, the reference angle is in the first quadrant.

EXAMPLE 4 ▰▰▰▰▰▰▰▰▰▰▰▰▰▰▰▰▰▰▰▰▰▰▰▰▰▰▰▰

If $\sin \theta = -0.7424$, find, to the nearest degree, two positive values of θ that are less than 360°.

Solution   Find the reference angle by finding arcsin 0.7424.

ENTER: **2nd**   **SIN⁻¹**   .7424   **)**   **ENTER**

DISPLAY:
```
SIN⁻¹(.7424)
          47.93626204
```

To the nearest degree, the measure of the reference angle is 48°. The sine is negative in the third quadrant and in the fourth quadrant.

In the third quadrant:	In the fourth quadrant:
$48 = \theta - 180$	$48 = 360 - \theta$
$\theta = 228°$	$\theta = 360 - 48$
	$\theta = 312°$

Alternative θ = the angle whose sine is -0.7424 and arcsin $-0.7424 = \sin^{-1} -0.7424$.
Solution

ENTER: **2nd**   **SIN⁻¹**   **(-)**   .7424   **)**   **ENTER**

DISPLAY:
```
SIN⁻¹(-.7424)
         -47.93626204
```

There are many angles whose sine is -0.7424. The calculator returns the measure of the angle with the smallest absolute value. When the sine of the angle is negative, this is an angle with a negative degree measure, that is, an angle formed by a clockwise rotation. To find the positive measure of an angle with the same terminal side, add 360: $\theta = -48 + 360 = 312°$. This is a fourth-quadrant angle whose reference angle is $360 - 312 = 48$. There is also a third-quadrant angle whose sine is -0.7424 and whose reference angle is 48. In the third quadrant:

$$48 = \theta + 180$$
$$\theta = 228°$$

Answer   228° and 312°

SUMMARY

Let θ be the measure of an angle $90° < \theta < 360°$.

	Second Quadrant	**Third Quadrant**	**Fourth Quadrant**
Reference Angle	$180 - \theta$	$\theta - 180$	$360 - \theta$
sin θ	$\sin(180 - \theta)$	$-\sin(\theta - 180)$	$-\sin(360 - \theta)$
cos θ	$-\cos(180 - \theta)$	$-\cos(\theta - 180)$	$\cos(360 - \theta)$
tan θ	$-\tan(180 - \theta)$	$\tan(\theta - 180)$	$-\tan(360 - \theta)$

Exercises

Writing About Mathematics

1. Liam said that if $0 < \theta < 90$, then when the degree measure of a fourth-quadrant angle is $-\theta$, the degree measure of the reference angle is θ. Do you agree with Liam? Explain why or why not.

2. Sammy said that if a negative value is entered for $\sin^{-1}$, $\cos^{-1}$, or $\tan^{-1}$, the calculator will return a negative value for the measure of the angle. Do you agree with Sammy? Explain why or why not.

Developing Skills

In 3–7, for each angle with the given degree measure: **a.** Draw the angle in standard position. **b.** Draw its reference angle as an acute angle formed by the terminal side of the angle and the x-axis. **c.** Draw the reference angle in standard position. **d.** Give the measure of the reference angle.

3. $120°$ **4.** $250°$ **5.** $320°$ **6.** $-45°$ **7.** $405°$

In 8–17, for each angle with the given degree measure, find the measure of the reference angle.

8. $100°$ **9.** $175°$ **10.** $210°$ **11.** $250°$ **12.** $285°$

13. $310°$ **14.** $95°$ **15.** $290°$ **16.** $-130°$ **17.** $505°$

In 18–27, express each given function value in terms of a function value of a positive acute angle (the reference angle).

18. $\sin 215°$ **19.** $\cos 95°$ **20.** $\tan 255°$ **21.** $\cos 312°$ **22.** $\tan 170°$

23. $\sin 285°$ **24.** $\cos 245°$ **25.** $\tan 305°$ **26.** $\sin -56°$ **27.** $\sin 500°$

In 28–43, for each function value, if $0° \leq \theta < 360°$, find, to the nearest degree, two values of θ.

28. $\sin \theta = 0.3420$ **29.** $\cos \theta = 0.6283$ **30.** $\tan \theta = 0.3240$ **31.** $\tan \theta = 1.4281$

32. $\sin \theta = 0.8090$ **33.** $\sin \theta = -0.0523$ **34.** $\cos \theta = -0.3090$ **35.** $\cos \theta = 0.9205$

36. $\tan \theta = -9.5141$ **37.** $\sin \theta = 0.2419$ **38.** $\cos \theta = -0.7431$ **39.** $\sin \theta = -0.1392$

40. $\tan \theta = -0.1405$ **41.** $\sin \theta = 0$ **42.** $\cos \theta = 0$ **43.** $\tan \theta = 0$

In $\triangle ABC$ with a right angle at C, $\overline{BC}$ is the leg that is opposite $\angle A$, $\overline{AC}$ is the leg that is adjacent to $\angle A$, and $\overline{AB}$ is the hypotenuse.

$$\sin A = \frac{\text{opp}}{\text{hyp}} \qquad \cos A = \frac{\text{adj}}{\text{hyp}} \qquad \tan A = \frac{\text{opp}}{\text{adj}}$$

An angle is in **standard position** when its vertex is at the origin and its **initial side** is the nonnegative ray of the x-axis. The measure of an angle in standard position is positive when the rotation from the initial side to the **terminal side** is in the counterclockwise direction. The measure of an angle in standard position is negative when the rotation from the initial side to the terminal side is in the clockwise direction.

We classify angles in standard position according to the quadrant in which the terminal side lies.

- If $0 < \theta < 90$ and $m\angle AOB = \theta + 360n$ for any integer n, then $\angle AOB$ is a first-quadrant angle.
- If $90 < \theta < 180$ and $m\angle AOB = \theta + 360n$ for any integer n, then $\angle AOB$ is a second-quadrant angle.
- If $180 < \theta < 270$ and $m\angle AOB = \theta + 360n$ for any integer n, then $\angle AOB$ is a third-quadrant angle.
- If $270 < \theta < 360$ and $m\angle AOB = \theta + 360n$ for any integer n, then $\angle AOB$ is a fourth-quadrant angle.

An angle in standard position whose terminal side lies on either the x-axis or the y-axis is called a **quadrantal angle**. The measure of a quadrantal angle is a multiple of 90.

Angles in standard position that have the same terminal side are **coterminal angles**.

A circle with center at the origin and radius 1 is the **unit circle** and has the equation $x^2 + y^2 = 1$.

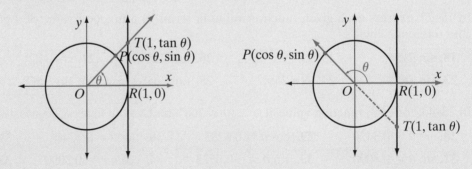

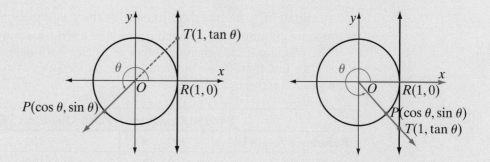

Let $P(p, q)$ be the point at which the terminal side of an angle in standard position intersects the unit circle, and $T(1, t)$ be the point at which the line tangent to the circle at $R(1, 0)$ intersects the terminal side of the angle. Let $m\angle ROP = \theta$.

- The **sine function** is a set of ordered pairs $(\theta, \sin \theta)$ such that $\sin \theta = q$.
- The **cosine function** is a set of ordered pairs $(\theta, \cos \theta)$ such that $\cos \theta = p$.
- The **tangent function** is the set of ordered pairs $(\theta, \tan \theta)$ such that $\tan \theta = t$.
- The **secant function** is the set of ordered pairs $(\theta, \sec \theta)$ such that for all θ for which $\cos \theta \neq 0$, $\sec \theta = \frac{1}{\cos \theta}$.
- The **cosecant function** is the set of ordered pairs $(\theta, \csc \theta)$ such that for all θ for which $\sin \theta \neq 0$, $\csc \theta = \frac{1}{\sin \theta}$.
- The **cotangent function** is the set of ordered pairs $(\theta, \cot \theta)$ such that for all θ for which $\tan \theta$ is defined and not equal to 0, $\cot \theta = \frac{1}{\tan \theta}$ and for all θ for which $\tan \theta$ is undefined, $\cot \theta = 0$.

The secant, cosecant, and cotangent functions are called **reciprocal functions**.

The equilateral triangle and the isosceles right triangle make it possible to find exact values for angles of 30°, 45°, and 60°.

θ	0°	30°	45°	60°	90°
$\sin \theta$	0	$\frac{1}{2}$	$\frac{\sqrt{2}}{2}$	$\frac{\sqrt{3}}{2}$	1
$\cos \theta$	1	$\frac{\sqrt{3}}{2}$	$\frac{\sqrt{2}}{2}$	$\frac{1}{2}$	0
$\tan \theta$	0	$\frac{\sqrt{3}}{3}$	1	$\sqrt{3}$	undefined

The rational approximation of the values of the sine, cosine, and tangent functions can be displayed by most calculators.

If $\sin \theta = a$, then $\theta = \textbf{arcsin } a$ or $\theta = \sin^{-1} a$, read as "θ is the angle whose measure is a." We can also write $\tan \theta = a$ as $\theta = \textbf{arctan } a = \tan^{-1} a$ and $\cos \theta = a$ as $\theta = \textbf{arccos } a = \cos^{-1} a$.

A minute is $\frac{1}{60}$ of a degree. To express the measure of an angle in degrees and minutes, multiply the decimal part of the degree measure by 60.

The trigonometric function values of angles with degree measure greater than 90 or less than 0 can be found from their values at corresponding acute angles called **reference angles**.

	Second Quadrant	Third Quadrant	Fourth Quadrant
Reference Angle	$180 - \theta$	$\theta - 180$	$360 - \theta$
sin θ	$\sin(180 - \theta)$	$-\sin(\theta - 180)$	$-\sin(360 - \theta)$
cos θ	$-\cos(180 - \theta)$	$-\cos(\theta - 180)$	$\cos(360 - \theta)$
tan θ	$-\tan(180 - \theta)$	$\tan(\theta - 180)$	$-\tan(360 - \theta)$

VOCABULARY

9-1 Hypotenuse • Leg • Similar triangles • Sine • Cosine • Tangent

9-2 Initial side • Terminal side • Standard position • θ (theta) • Quadrantal angle • Coterminal angle • Angular speed

9-3 Unit circle • Sine function • Cosine function

9-4 Tangent function

9-5 Reciprocal function • Secant function • Cosecant function • Cotangent function

9-7 Arcsine • Arccosine • Arctangent

9-8 Reference angle

REVIEW EXERCISES

In 1–5: **a.** Draw each angle in standard position. **b.** Draw its reference angle as an acute angle formed by the terminal side of the angle and the x-axis. **c.** Draw the reference angle in standard position. **d.** Give the measure of the reference angle.

1. 220°   **2.** 300°   **3.** 145°   **4.** −100°   **5.** 600°

6. For each given angle, find a coterminal angle with a measure of θ such that $0° \le \theta < 360°$: **a.** 505° **b.** −302°

In 7–10, the terminal side of $\angle ROP$ in standard position intersects the unit circle at P. If m$\angle ROP$ is θ, find:

a. the quadrant of $\angle ROP$ **b.** $\sin \theta$ **c.** $\cos \theta$ **d.** $\tan \theta$ **e.** $\csc \theta$ **f.** $\sec \theta$ **g.** $\cot \theta$

7. $P(0.8, -0.6)$ **8.** $P\left(\frac{3}{4}, \frac{\sqrt{7}}{4}\right)$ **9.** $P\left(-\frac{2\sqrt{6}}{5}, \frac{1}{5}\right)$ **10.** $P\left(-\frac{2}{3}, -\frac{\sqrt{5}}{3}\right)$

In 11–14, the given point is on the terminal side of angle θ in standard position. Find: **a.** the quadrant of the angle **b.** sin θ **c.** cos θ **d.** tan θ **e.** csc θ **f.** sec θ **g.** cot θ

11. $(12, 9)$ **12.** $(-100, 0)$ **13.** $(-8, -6)$ **14.** $(9, -13)$

15. For each given angle in standard position, determine, to the nearest tenth, the coordinates of the point where the terminal side intersects the unit circle: **a.** 405° **b.** 79°

In 16–27, find each exact function value.

16. cos 30°   **17.** sin 60°   **18.** tan 45°   **19.** tan 120°

20. cos 135°   **21.** sin 240°   **22.** cos 330°   **23.** sin 480°

24. sec 45°   **25.** csc 30°   **26.** cot 60°   **27.** sec 120°

In 28–35, find each function value to four decimal places.

28. cos 50°   **29.** tan 80°   **30.** sin 110°   **31.** tan 230°

32. cos 187°   **33.** sin $(-24°)$ **34.** cos $(-230°)$ **35.** tan 730°

In 36–43, express each given function value in terms of a function value of a positive acute angle (the reference angle).

36. cos 100°   **37.** sin 300°   **38.** cos 280°   **39.** tan 210°

40. sin 150°   **41.** cos 255°   **42.** cos $(-40°)$ **43.** tan $(-310°)$

In 44–51, for each function value, if $0 \leq \theta < 360$, find, to the nearest degree, two values of θ.

44. sin $\theta = 0.3747$ **45.** cos $\theta = 0.9136$

46. tan $\theta = 1.376$ **47.** sin $\theta = -0.7000$

48. tan $\theta = -0.5775$ **49.** cos $\theta = -0.8192$

50. sec $\theta = 1.390$ **51.** csc $\theta = 3.072$

In 52–55, $0 \leq \theta < 360$.

52. For what two values of θ is tan θ undefined?

53. For what two values of θ is cot θ undefined?

54. For what two values of θ is sec θ undefined?

55. For what two values of θ is csc θ undefined?

56. In the diagram, m∠*ROP* is θ. Express each of the following in terms of a coordinate of *R*, *P*, or *T*.

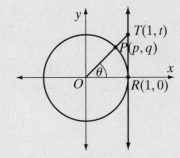

a. sin θ   **b.** cos θ   **c.** tan θ

d. Use similar triangles to show that $\tan \theta = \frac{\sin \theta}{\cos \theta}$.

57. A road rises 630 feet every mile. Find to the nearest ten minutes the angle that the road makes with the horizontal.

58. From the top of a building that is 56 feet high, the angle of depression to the base of an adjacent building is 72°. Find, to the nearest foot, the distance between the buildings.

Exploration

In the diagram, $\overrightarrow{OP}$ intersects the unit circle at *P*. The line that is tangent to the circle at *R*(1, 0) intersects $\overrightarrow{OP}$ at *T*. The line that is tangent to the circle at *Q*(0,1) intersects $\overrightarrow{OP}$ at *S*. The measure of ∠*ROP* is θ. Show that *OT* = sec θ, *OS* = csc θ, and *QS* = cot θ. (*Hint:* Use similar triangles.)

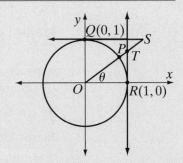

CUMULATIVE REVIEW   CHAPTERS 1–9

Part I

Answer all questions in this part. Each correct answer will receive 2 credits. No partial credit will be allowed.

1. In simplest form, $(2x + 1)^2 - (x + 2)^2$ is equal to

(1) $3x^2 - 3$ (3) $3x^2 + 5$

(2) $3x^2 + 8x + 5$ (4) $3x^2 + 8x - 3$

2. The solution set of $x^2 - 5x = 6$ is

(1) {2, 3}   (2) {−2, −3}   (3) {6, −1}   (4) {−6, 1}

3. In simplest form, the fraction $\dfrac{a - \frac{1}{a}}{a + 1}$ is equal to

(1) $-\frac{1}{a}$ (2) $\frac{a - 1}{a}$ (3) $\frac{a + 1}{a}$ (4) $a - 1$

4. The sum of $\left(3 + \sqrt{12}\right)$ and $\left(-5 + \sqrt{27}\right)$ is

(1) 16   (2) $-2 + \sqrt{39}$ (3) $-2 + 13\sqrt{3}$ (4) $-2 + 5\sqrt{3}$

5. Which of the following products is a rational number?

(1) $\left(10 + \sqrt{10}\right)\left(10 + \sqrt{10}\right)$ (3) $\sqrt{10}\left(2 + \sqrt{10}\right)$

(2) $\left(10 + \sqrt{10}\right)\left(10 - \sqrt{10}\right)$ (4) $10\left(10 - \sqrt{10}\right)$

6. The fraction $\dfrac{1 - \sqrt{3}}{1 + \sqrt{3}}$ is equal to

(1) $-\frac{1}{2}$ (2) 2   (3) $2 - \sqrt{3}$ (4) $-2 + \sqrt{3}$

7. Which of the following is a one-to-one function when the domain is the set of real numbers?

(1) $y = x - 5$ (3) $y = x^2 - 2x + 5$

(2) $x^2 + y^2 = 9$ (4) $y = |x - 4|$

8. The sum of the roots of the equation $2x^2 - 5x + 3 = 0$ is

(1) 5   (2) -5 (3) $\frac{5}{2}$ (4) $-\frac{5}{2}$

9. When x and y vary inversely

(1) xy equals a constant.   (3) $x + y$ equals a constant.

(2) $\frac{x}{y}$ equals a constant.   (4) $x - y$ equals a constant.

10. The expression $2 \log a + \frac{1}{3} \log b$ is equivalent to

(1) $\log \frac{1}{3}a^2 b$ (3) $\log \left(a^2 + \sqrt[3]{b}\right)$

(2) $\log \frac{2}{3}ab$ (4) $\log a^2\left(\sqrt[3]{b}\right)$

Part II

Answer all questions in this part. Each correct answer will receive 2 credits. Clearly indicate the necessary steps, including appropriate formula substitutions, diagrams, graphs, charts, etc. For all questions in this part, a correct numerical answer with no work shown will receive only 1 credit.

11. Express the product $(3 - 2i)(-1 + i)$ in $a + bi$ form.

12. Find the solution set of $|2x - 4| < 3$.

Part III

Answer all questions in this part. Each correct answer will receive 4 credits. Clearly indicate the necessary steps, including appropriate formula substitutions, diagrams, graphs, charts, etc. For all questions in this part, a correct numerical answer with no work shown will receive only 1 credit.

13. Solve for x: $3 + (x + 3)^{\frac{1}{2}} = x$.

14. Write the equation of the circle if the center of the circle is $C(2, 1)$ and one point on the circle is $A(4, 0)$.

Part IV

Answer all questions in this part. Each correct answer will receive 6 credits. Clearly indicate the necessary steps, including appropriate formula substitutions, diagrams, graphs, charts, etc. For all questions in this part, a correct numerical answer with no work shown will receive only 1 credit.

15. What are the roots of the equation $x^2 + 4 = 6x$?

16. Let $f(x) = x^2 - x$ and $g(x) = 5x + 7$.

a. Find $f \circ g(-2)$.

b. Write $h(x) = f \circ g(x)$ as a polynomial in simplest form.

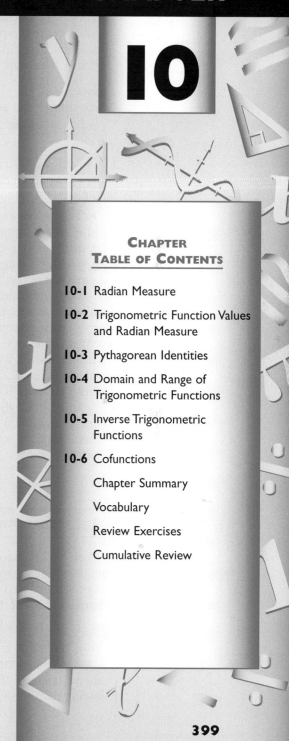

MORE TRIGONOMETRIC FUNCTIONS

The relationships among the lengths of the sides of an isosceles right triangle or of the right triangles formed by the altitude to a side of an equilateral triangle make it possible for us to find exact values of the trig functions of angles of 30°, 45°, and their multiples. How are the trigonometric function values for other angles determined? Before calculators and computers were readily available, handbooks that contained tables of values were a common tool of mathematicians. Now calculators or computers will return these values. How were the values in these tables or those displayed by a calculator determined?

In more advanced math courses, you will learn that trigonometric function values can be approximated by evaluating an infinite series to the required number of decimal places.

$$\sin x = x - \frac{x^3}{3!} + \frac{x^5}{5!} - \frac{x^7}{7!} + \cdots$$

To use this formula correctly, x must be expressed in a unit of measure that has no dimension. This unit of measure, which we will define in this chapter, is called a *radian*.

10-1 RADIAN MEASURE

A line segment can be measured in different units of measure such as feet, inches, centimeters, and meters. Angles and arcs can also be measured in different units of measure. We have measured angles in degrees, but another frequently used unit of measure for angles is called a *radian*.

DEFINITION _____

A **radian** is the unit of measure of a central angle that intercepts an arc equal in length to the radius of the circle.

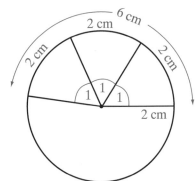

In a circle of radius 1 centimeter, a central angle whose measure is 1 radian intercepts an arc whose length is 1 centimeter.

The radian measure of any central angle in a circle of radius 1 is equal to the length of the arc that it intercepts. In a circle of radius 1 centimeter, a central angle of 2 radians intercepts an arc whose length is 2 centimeters, and a central angle of 3 radians intercepts an arc whose length is 3 centimeters.

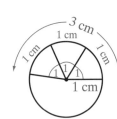

In a circle of radius 2 centimeters, a central angle whose measure is 1 radian intercepts an arc whose measure is 2 centimeters, a central angle of 2 radians intercepts an arc whose measure is 4 centimeters, and a central angle of 3 radians intercepts an arc whose measure is 6 centimeters.

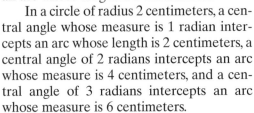

In general, the radian measure θ of a central angle of a circle is the length of the intercepted arc, s, divided by the radius of the circle, r.

$$\theta = \frac{s}{r}$$

Relationship Between Degrees and Radians

When the diameter of a circle is drawn, the straight angle whose vertex is at the center of the circle intercepts an arc that is a semicircle. The length of the semicircle is one-half the circumference of the circle or $\frac{1}{2}(2\pi r) = \pi r$. Therefore, the radian measure of a straight angle is $\frac{\pi r}{r} = \pi$. Since the degree measure of a straight angle is 180°, π radians and 180 degrees are measures of the same angle. We can write the following relationship:

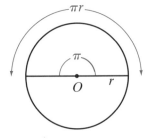

$$\pi \text{ radians} = 180 \text{ degrees}$$

We will use this relationship to change the degree measure of any angle to radian measure.

Changing Degrees to Radians

The measure of any angle can be expressed in degrees or in radians. We can do this by forming the following proportion:

$$\frac{\textbf{measure in degrees}}{\textbf{measure in radians}} = \frac{\textbf{180}}{\pi}$$

This proportion can be used to find the radian measure of an angle whose degree measure is known. For example, to find θ, the radian measure of an angle of 30°, substitute 30 for "measure in degrees" and θ for "measure in radians" in the proportion. Then use the fact that the product of the means is equal to the product of the extremes to solve for θ.

$$\frac{30}{\theta} = \frac{180}{\pi}$$
$$180\theta = 30\pi$$
$$\theta = \frac{30\pi}{180}$$
$$\theta = \frac{\pi}{6}$$

The radian measure of an angle of 30° is $\frac{\pi}{6}$.

 We know that $\sin 30° = \frac{1}{2}$. Use a calculator to compare this value to $\sin \theta$ when θ is $\frac{\pi}{6}$ radians.

Begin by changing the calculator to radian mode.

ENTER: [MODE] [▼] [▼]

[ENTER]

DISPLAY:

```
NORMAL    SCI ENG
FLOAT   0123456789
RADIAN     DEGREE
```

Then enter the expression.

ENTER: [SIN] [2nd] [π] [÷] 6 [ )]

[MATH] [ENTER] [ENTER]

DISPLAY:

```
SIN(π/6)▶FRAC
              1/2
```

The calculator confirms that $\sin 30° = \sin \frac{\pi}{6}$.

Changing Radians to Degrees

When the radian measure of an angle is known, we can use either of the two methods, shown below, to find the degree measure of the angle.

METHOD 1 *Proportion*

Use the same proportion that was used to change from degrees to radians. For example, to convert an angle of $\frac{\pi}{2}$ radians to degrees, let θ = degree measure of the angle.

(1) Write the proportion that relates degree measure to radian measure:

$$\frac{\text{measure in degrees}}{\text{measure in radians}} = \frac{180}{\pi}$$

(2) Substitute θ for the degree measure and $\frac{\pi}{2}$ for the radian measure:

$$\frac{\theta}{\left(\frac{\pi}{2}\right)} = \frac{180}{\pi}$$

(3) Solve for θ:

$$\pi\theta = \frac{\pi}{2}(180)$$
$$\pi\theta = 90\pi$$
$$\theta = 90$$

The measure of the angle is 90°.

METHOD 2 *Substitution*

Since we know that π radians = 180°, substitute 180° for π radians and simplify the expression. For example:

$$\frac{\pi}{2} \text{ radians} = \frac{1}{2}(\pi \text{ radians}) = \frac{1}{2}(180°) = 90°$$

Radian measure is a ratio of the length of an arc to the length of the radius of a circle. These ratios must be expressed in the same unit of measure. Therefore, the units cancel and the ratio is a number that has no unit of measure.

EXAMPLE 1

The radius of a circle is 4 centimeters. A central angle of the circle intercepts an arc of 12 centimeters.

a. What is the radian measure of the angle?

b. Is the angle acute, obtuse, or larger than a straight angle?

Solution **a.** *How to Proceed*

(1) Write the rule that states that the radian measure $\qquad \theta = \frac{s}{r}$
of a central angle, θ, is equal to the length s of the
intercepted arc divided by the length r of a radius:

(2) Substitute the given values: $\qquad \theta = \frac{12}{4}$

(3) Simplify the fraction: $\qquad \theta = 3$

b. Since we know that π radians $= 180°$, then 1 radian $= \frac{180°}{\pi}$.

3 radians $= 3\left(\frac{180°}{\pi}\right) \approx 171.88°$. Therefore, the angle is obtuse.

Answers **a.** 3 radians $\qquad$ **b.** Obtuse

EXAMPLE 2

The radius of a circle is 8 inches. What is the length of
the arc intercepted by a central angle of the circle if the
measure of the angle is 2.5 radians?

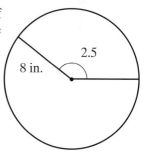

Solution $\qquad\qquad \theta = \frac{s}{r}$

$\qquad\qquad 2.5 = \frac{s}{8}$

$\qquad\qquad s = 2.5(8) = 20$

Answer 20 inches

EXAMPLE 3

What is the radian measure of an angle of 45°?

Solution *How to Proceed*

(1) Write the proportion: $\qquad \dfrac{\text{measure in degrees}}{\text{measure in radians}} = \dfrac{180}{\pi}$

(2) Substitute x for the radian
measure and 45 for the
degree measure: $\qquad \dfrac{45}{x} = \dfrac{180}{\pi}$

(3) Solve for x: $\qquad 180x = 45\pi$

$\qquad\qquad x = \frac{45}{180}\pi$

(4) Express x in simplest form: $\qquad x = \frac{1}{4}\pi$

Answer $\frac{1}{4}\pi$ or $\frac{\pi}{4}$

EXAMPLE 4

The radian measure of an angle is $\frac{5\pi}{6}$.

a. What is the measure of the angle in degrees?

b. What is the radian measure of the reference angle?

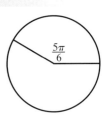

Solution **a.** **METHOD 1:** *Proportions*

Let θ = degree measure of the angle.

$$\frac{\text{measure in degrees}}{\text{measure in radians}} = \frac{180}{\pi}$$

$$\frac{\theta}{\frac{5\pi}{6}} = \frac{180}{\pi}$$

$$\pi\theta = 180\left(\frac{5\pi}{6}\right)$$

$$\pi\theta = 150\pi$$

$$\theta = 150$$

METHOD 2: *Substitution*

Substitute 180° for π radians, and simplify.

$$\frac{5\pi}{6} \text{ radians} = \frac{5}{6}(\pi \text{ radians})$$

$$= \frac{5}{6}(180°)$$

$$= 150°$$

b. Since $\frac{5\pi}{6}$ is a second-quadrant angle, its reference angle is equal to $180° - 150°$ or, in radians:

$$180° - 150° \rightarrow \pi - \frac{5\pi}{6} = \frac{\pi}{6}$$

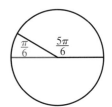

Answers **a.** $150°$ **b.** $\frac{\pi}{6}$

Exercises

Writing About Mathematics

1. Ryan and Rebecca each found the radian measure of a central angle by measuring the radius of the circle and the length of the intercepted arc. Ryan used inches and Rebecca used centimeters when making their measurements. If Ryan and Rebecca each measured accurately, will the measures that they obtain for the angle be equal? Justify your answer.

2. If a wheel makes two complete revolutions, each spoke on the wheel turns through an angle of how many radians? Explain your answer.

Developing Skills

In 3–12, find the radian measure of each angle whose degree measure is given.

3. 30°   **4.** 90°   **5.** 45°   **6.** 120°   **7.** 160°

8. 135°   **9.** 225°   **10.** 240°   **11.** 270°   **12.** 330°

In 13–22, find the degree measure of each angle whose radian measure is given.

13. $\frac{\pi}{3}$ **14.** $\frac{\pi}{9}$ **15.** $\frac{\pi}{10}$ **16.** $\frac{2\pi}{5}$ **17.** $\frac{10\pi}{9}$

18. $\frac{3\pi}{2}$ **19.** 3π **20.** $\frac{11\pi}{6}$ **21.** $\frac{7\pi}{2}$ **22.** 1

In 23–27, for each angle with the given radian measure: **a.** Give the measure of the angle in degrees. **b.** Give the measure of the reference angle in radians. **c.** Draw the angle in standard position and its reference angle as an acute angle formed by the terminal side of the angle and the *x*-axis.

23. $\frac{\pi}{3}$ **24.** $\frac{7\pi}{36}$ **25.** $\frac{10\pi}{9}$ **26.** $-\frac{7\pi}{18}$ **27.** $\frac{25\pi}{9}$

In 28–37, θ is the radian measure of a central angle that intercepts an arc of length s in a circle with a radius of length r.

28. If $s = 6$ and $r = 1$, find θ.   **29.** If $\theta = 4.5$ and $s = 9$, find r.

30. If $\theta = 2.5$ and $r = 10$, find s.   **31.** If $r = 2$ and $\theta = 1.6$, find s.

32. If $r = 2.5$ and $s = 15$, find θ.   **33.** If $s = 16$ and $\theta = 0.4$, find r.

34. If $r = 4.2$ and $s = 21$, find θ.   **35.** If $r = 6$ and $\theta = \frac{2\pi}{3}$, find s.

36. If $s = 18$ and $\theta = \frac{6\pi}{5}$, find r.   **37.** If $\theta = 6\pi$ and $r = 1$, find s.

38. Circle O has a radius of 1.7 inches. What is the length, in inches, of an arc intercepted by a central angle whose measure is 2 radians?

39. In a circle whose radius measures 5 feet, a central angle intercepts an arc of length 12 feet. Find the radian measure of the central angle.

40. The central angle of circle O has a measure of 4.2 radians and it intercepts an arc whose length is 6.3 meters. What is the length, in meters, of the radius of the circle?

41. Complete the following table, expressing degree measures in radian measure in terms of π.

Degrees	30°	45°	60°	90°	180°	270°	360°
Radians							

Applying Skills

42. The pendulum of a clock makes an angle of 2.5 radians as its tip travels 18 feet. What is the length of the pendulum?

43. A wheel whose radius measures 16 inches is rotated. If a point on the circumference of the wheel moves through an arc of 12 feet, what is the measure, in radians, of the angle through which a spoke of the wheel travels?

44. The wheels on a bicycle have a radius of 40 centimeters. The wheels on a cart have a radius of 10 centimeters. The wheels of the bicycle and the wheels of the cart all make one complete revolution.

    a. Do the wheels of the bicycle rotate through the same angle as the wheels of the cart? Justify your answer.

    b. Does the bicycle travel the same distance as the cart? Justify your answer.

45. Latitude represents the measure of a central angle with vertex at the center of the earth, its initial side passing through a point on the equator, and its terminal side passing through the given location. (See the figure.) Cities A and B are on a north-south line. City A is located at 30°N and City B is located at 52°N. If the radius of the earth is approximately 6,400 kilometers, find d, the distance between the two cities along the circumference of the earth. Assume that the earth is a perfect sphere.

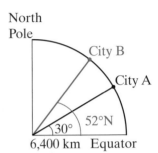

10-2 TRIGONOMETRIC FUNCTION VALUES AND RADIAN MEASURE

Using Radians to Find Trigonometric Function Values

Hands-On Activity

In previous chapters, we found the exact trigonometric function values of angles of 0°, 30°, 45°, 60°, 90°, 180°, and 270°. Copy and complete the table below to show measures of these angles in radians and the corresponding function values. Let θ be the measure of an angle in standard position.

θ in Degrees	0°	30°	45°	60°	90°	180°	270°
θ in Radians							
sin θ							
cos θ							
tan θ							

    We can use these function values to find the exact function value of any angle whose radian measure is a multiple of $\frac{\pi}{6}$ or $\frac{\pi}{4}$.

EXAMPLE 1

Find the exact value of $\tan \frac{2\pi}{3}$.

| *Solution* | **METHOD 1** | **METHOD 2** |

	METHOD 1	**METHOD 2**
	The radian measure $\frac{2\pi}{3}$ is equivalent to the degree measure $\frac{2(180°)}{3} = 120°$. An angle of 120° is a second-quadrant angle whose tangent is negative. The reference angle is $180° - 120° = 60°$. $$\tan \frac{2\pi}{3} = \tan 120° = -\tan 60° = -\sqrt{3}$$	Since $\frac{\pi}{2} < \frac{2\pi}{3} < \pi$, an angle of $\frac{2\pi}{3}$ is a second-quadrant angle. The reference angle of an angle of $\frac{2\pi}{3}$ radians is $\pi - \frac{2\pi}{3} = \frac{\pi}{3}$. The tangent of a second-quadrant angle is negative. $$\tan \frac{2\pi}{3} = -\tan \frac{\pi}{3} = -\sqrt{3}$$

Answer $-\sqrt{3}$

The graphing calculator will return the trigonometric function values of angles entered in degree or radian measure. Before entering an angle measure, be sure that the calculator is in the correct mode. For example, to find $\tan \frac{\pi}{8}$ to four decimal places, use the following sequence of keys and round the number given by the calculator. Be sure that your calculator is set to radian mode.

ENTER: TAN 2nd π ÷ 8   DISPLAY:

```
TAN(π/8)
         .4142135624
```

) ENTER

Therefore, $\tan \frac{\pi}{8} \approx 0.4142$.

EXAMPLE 2

Find the value of $\cos \theta$ to four decimal places when $\theta = 2.75$ radians.

Solution ENTER: COS 2.75) ENTER   DISPLAY:

```
cos(2.75)
       -.9243023786
```

Note that the cosine is negative because the angle is a second-quadrant angle.

Answer -0.9243

Finding Angle Measures in Radians

When the calculator is in radian mode, it will return the decimal value of the radian measure whose function value is entered. The measure will *not* be in terms of π.

For instance, we know that tan 45° = 1. Since 45 = $\frac{180}{4}$, an angle of 45° is an angle of $\frac{\pi}{4}$ radians. Use a calculator to find θ in radians when tan θ = 1. Your calculator should be in radian mode.

ENTER: [2nd] [TAN⁻¹] 1 [ )] [ENTER]   DISPLAY:
```
TAN⁻¹(1)
         .7853981634
```

The value 0.7853981634 returned by the calculator is approximately equal to $\frac{\pi}{4}$. We can verify this by dividing π by 4 and comparing the answers.

The exact radian measure of the angle is the irrational number $\frac{\pi}{4}$. The rational approximation of the radian measure of the angle is 0.7853981634.

EXAMPLE 3

Find, in radians, the smallest positive value of θ if tan θ = −1.8762. Express the answer to the nearest hundredth.

Solution   The calculator should be in radian mode.

ENTER: [2nd] [TAN⁻¹] −1.8762 [ )]   DISPLAY:
```
TAN⁻¹(-1.8762)
         -1.081104612
```
[ENTER]

Tangent is negative in the second and fourth quadrants. The calculator has returned the measure of the negative rotation in the fourth quadrant. The angle in the second quadrant with a positive measure is formed by the opposite ray and can be found by adding π, the measure of a straight angle, to the given measure.

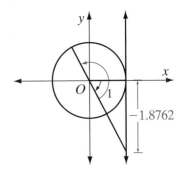

ENTER: [2nd] [ANS] [+] [2nd]   DISPLAY:
```
ANS+π
         2.060488041
```
[π] [ENTER]

Check   Verify this answer by finding tan 2.060488041.

ENTER: [TAN] [2nd] [ANS] [ )]   DISPLAY:
```
TAN(ANS)
         -1.8762
```
[ENTER]

Answer 2.06

EXAMPLE 4

Find, to the nearest ten-thousandth, a value of θ in radians if csc $\theta = 2.369$.

Solution If csc $\theta = 2.369$, then sin $\theta = \frac{1}{2.369}$.

ENTER: [2nd] [SIN⁻¹] 1 [÷] 2.369   DISPLAY:
$$\boxed{\begin{array}{l} \text{SIN}^{-1}(1/2.369) \\ \qquad\qquad .4357815508 \end{array}}$$

[ )] [ENTER]

Rounded to four decimal places, $\theta = 0.4358$. *Answer*

Exercises

Writing About Mathematics

1. Alexia said that when θ is a second-quadrant angle whose measure is in radians, the measure of the reference angle in radians is $\pi - \theta$. Do you agree with Alexia? Explain why or why not.

2. Diego said that when θ is the radian measure of an angle, the angle whose radian measure is $2\pi n + \theta$ is an angle with the same terminal side for all integral values of n. Do you agree with Diego? Explain why or why not.

Developing Skills

In 3–12, find the exact function value of each of the following if the measure of the angle is given in radians.

3. sin $\frac{\pi}{4}$ **4.** tan $\frac{\pi}{3}$ **5.** cos $\frac{\pi}{2}$ **6.** tan $\frac{\pi}{6}$ **7.** cos $\frac{2\pi}{3}$

8. sin $\frac{4\pi}{3}$ **9.** tan $\frac{5\pi}{4}$ **10.** sec $\frac{\pi}{3}$ **11.** csc π **12.** cot $\frac{\pi}{4}$

In 13–24, find, to the nearest ten-thousandth, the radian measure θ of a first-quadrant angle with the given function value.

13. sin $\theta = 0.2736$ **14.** cos $\theta = 0.5379$ **15.** tan $\theta = 3.726$

16. cos $\theta = 0.9389$ **17.** sin $\theta = 0.8267$ **18.** cos $\theta = 0.8267$

19. tan $\theta = 1.5277$ **20.** cot $\theta = 1.5277$ **21.** sec $\theta = 5.232$

22. cot $\theta = 0.3276$ **23.** csc $\theta = 2.346$ **24.** cot $\theta = 0.1983$

25. If f$(x) = $ sin $\left(\frac{1}{3}x\right)$, find f$\left(\frac{\pi}{2}\right)$.   **26.** If f$(x) = $ cos $2x$, find f$\left(\frac{3\pi}{4}\right)$.

27. If f$(x) = $ sin $2x + $ cos $3x$, find f$\left(\frac{\pi}{4}\right)$.   **28.** If f$(x) = $ tan $5x - $ sin $2x$, find f$\left(\frac{\pi}{6}\right)$.

Applying Skills

29. The unit circle intersects the x-axis at $R(1, 0)$ and the terminal side of $\angle ROP$ at P. What are the coordinates of P if m$\overset{\frown}{RP} = 4.275$?

30. The *x*-axis intersects the unit circle at $R(1,0)$ and a circle of radius 3 centered at the origin at $A(3,0)$. The terminal side of $\angle ROP$ intersects the unit circle at P and the circle of radius 3 at B. The measure of $\overset{\frown}{RP}$ is 2.50.

a. What is the radian measure of $\angle ROP$?

b. What is the radian measure of $\angle AOB$?

c. What are the coordinates of P?

d. What are the coordinates of B?

e. In what quadrant do P and B lie? Justify your answer.

31. The wheels of a cart that have a radius of 12 centimeters move in a counterclockwise direction for 20 meters.

a. What is the radian measure of the angle through which each wheel has turned?

b. What is sine of the angle through which the wheels have turned?

32. A supporting cable runs from the ground to the top of a tree that is in danger of falling down. The tree is 18 feet tall and the cable makes an angle of $\frac{2\pi}{9}$ with the ground. Determine the length of the cable to the nearest tenth of a foot.

33. An airplane climbs at an angle of $\frac{\pi}{15}$ with the ground. When the airplane has reached an altitude of 500 feet:

a. What is the distance in the air that the airplane has traveled?

b. What is the horizontal distance that the airplane has traveled?

Hands-On Activity 1: The Unit Circle and Radian Measure

In Chapter 9, we explored trigonometric function values on the unit circle with regard to degree measure. In the following Hands-On Activity, we will examine trigonometric function values on the unit circle with regard to *radian measure*.

1. Recall that for any point P on the unit circle, $\overrightarrow{OP}$ is the terminal side of an angle in standard position. If the measure of this angle is θ, the coordinates of P are $(\cos \theta, \sin \theta)$. Use your knowledge of the *exact* cosine and sine values for $0°, 30°, 45°, 60°$, and $90°$ angles to find the coordinates of the first-quadrant points.

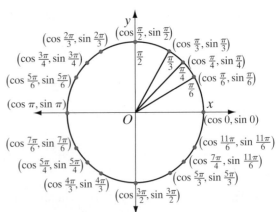

2. The cosine and sine values of the other angles marked on the unit circle can be found by relating them to the points in the first quadrant or along the positive axes. For example, the point $\left(\cos \frac{2\pi}{3}, \sin \frac{2\pi}{3} \right)$ is the image of the point $\left(\cos \frac{\pi}{3}, \sin \frac{\pi}{3} \right)$ under a reflection in the y-axis. Thus, if $\left(\cos \frac{\pi}{3}, \sin \frac{\pi}{3} \right) = (a, b)$, then

$$\left(\cos \frac{2\pi}{3}, \sin \frac{2\pi}{3} \right) = (-a, b).$$

Determine the *exact* coordinates of the points shown on the unit circle to find the function values.

Hands-On Activity 2: Evaluating the Sine and Cosine Functions

The functions $\sin x$ and $\cos x$ can be represented by the following series when x is in radians:

$$\sin x = x - \frac{x^3}{3!} + \frac{x^5}{5!} - \frac{x^7}{7!} + \cdots$$
$$\cos x = 1 - \frac{x^2}{2!} + \frac{x^4}{4!} - \frac{x^6}{6!} + \cdots$$

1. Write the next two terms for each series given above.

2. Use the first six terms of the series for $\sin x$ to approximate $\sin \frac{\pi}{4}$ to four decimal places.

3. Use the first six terms of the series for $\cos x$ to approximate $\cos \frac{\pi}{3}$ to four decimal places.

10-3 PYTHAGOREAN IDENTITIES

The unit circle is a circle of radius 1 with center at the origin. Therefore, the equation of the unit circle is $x^2 + y^2 = 1^2$ or $x^2 + y^2 = 1$. Let P be any point on the unit circle and $\overrightarrow{OP}$ be the terminal side of $\angle ROP$, an angle in standard position whose measure is θ. The coordinates of P are $(\cos \theta, \sin \theta)$. Since P is a point on the circle whose equation is $x^2 + y^2 = 1$, and $x = \cos \theta$ and $y = \sin \theta$:

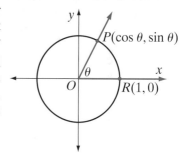

$$(\cos \theta)^2 + (\sin \theta)^2 = 1$$

In order to emphasize that it is the cosine value and the sine value that are being squared, we write $(\cos \theta)^2$ as $\cos^2 \theta$ and $(\sin \theta)^2$ as $\sin^2 \theta$. Therefore, the equation is written as:

$$\cos^2 \theta + \sin^2 \theta = 1$$

This equation is called an *identity*.

DEFINITION

An **identity** is an equation that is true for all values of the variable for which the terms of the variable are defined.

Because the identity $\cos^2 \theta + \sin^2 \theta = 1$ is based on the Pythagorean theorem, we refer to it as a **Pythagorean identity**.

EXAMPLE 1

Verify that $\cos^2 \frac{\pi}{3} + \sin^2 \frac{\pi}{3} = 1$.

Solution   We know that $\cos \frac{\pi}{3} = \frac{1}{2}$ and $\sin \frac{\pi}{3} = \frac{\sqrt{3}}{2}$. Therefore:

$$\cos^2 \theta + \sin^2 \theta = 1$$
$$\left(\tfrac{1}{2}\right)^2 + \left(\tfrac{\sqrt{3}}{2}\right)^2 = 1$$
$$\tfrac{1}{4} + \tfrac{3}{4} = 1$$
$$1 = 1 \checkmark$$

We can write two related Pythagorean identities by dividing both sides of the equation by the same expression.

Divide by $\cos^2 \theta$:	*Divide by $\sin^2 \theta$:*
Recall that $\tan \theta = \frac{\sin \theta}{\cos \theta}$ and	Recall that $\cot \theta = \frac{1}{\tan \theta} = \frac{\cos \theta}{\sin \theta}$ and
$\sec \theta = \frac{1}{\cos \theta}$.	$\csc \theta = \frac{1}{\sin \theta}$.
For all values of θ for which $\cos \theta \neq 0$:	For all values of θ for which $\sin \theta \neq 0$:
$\cos^2 \theta + \sin^2 \theta = 1$	$\cos^2 \theta + \sin^2 \theta = 1$
$\frac{\cos^2 \theta}{\cos^2 \theta} + \frac{\sin^2 \theta}{\cos^2 \theta} = \frac{1}{\cos^2 \theta}$	$\frac{\cos^2 \theta}{\sin^2 \theta} + \frac{\sin^2 \theta}{\sin^2 \theta} = \frac{1}{\sin^2 \theta}$
$1 + \left(\frac{\sin \theta}{\cos \theta}\right)^2 = \left(\frac{1}{\cos \theta}\right)^2$	$\left(\frac{\cos \theta}{\sin \theta}\right)^2 + 1 = \left(\frac{1}{\sin \theta}\right)^2$
$1 + \tan^2 \theta = \sec^2 \theta$	
$1 + \tan^2 \theta = \sec^2 \theta$	**$\cot^2 \theta + 1 = \csc^2 \theta$**

EXAMPLE 2

If $\cos \theta = \frac{1}{3}$ and θ is in the fourth quadrant, use an identity to find:

a. $\sin \theta$ **b.** $\tan \theta$ **c.** $\sec \theta$ **d.** $\csc \theta$ **e.** $\cot \theta$

Solution   Since θ is in the fourth quadrant, cosine and its reciprocal, secant, are positive; sine and tangent and their reciprocals, cosecant and cotangent, are negative.

a. $\cos^2 \theta + \sin^2 \theta = 1$

$\left(\frac{1}{3}\right)^2 + \sin^2 \theta = 1$

$\frac{1}{9} + \sin^2 \theta = 1$

$\sin^2 \theta = \frac{8}{9}$

$\sin \theta = -\sqrt{\frac{8}{9}}$

$\sin \theta = -\frac{2\sqrt{2}}{3}$

b. $\tan \theta = \frac{\sin \theta}{\cos \theta} = \frac{-\frac{2\sqrt{2}}{3}}{\frac{1}{3}} = -2\sqrt{2}$

c. $\sec \theta = \frac{1}{\cos \theta} = \frac{1}{\frac{1}{3}} = 3$

d. $\csc \theta = \frac{1}{\sin \theta} = \frac{1}{-\frac{2\sqrt{2}}{3}} = -\frac{3}{2\sqrt{2}} = -\frac{3\sqrt{2}}{4}$

e. $\cot \theta = \frac{1}{\tan \theta} = \frac{1}{-2\sqrt{2}} = -\frac{\sqrt{2}}{4}$

Answers **a.** $-\frac{2\sqrt{2}}{3}$ **b.** $-2\sqrt{2}$ **c.** 3 **d.** $-\frac{3\sqrt{2}}{4}$ **e.** $-\frac{\sqrt{2}}{4}$

EXAMPLE 3

Write $\tan \theta + \cot \theta$ in terms of $\sin \theta$ and $\cos \theta$ and write the sum as a single fraction in simplest form.

Solution

$$\tan \theta + \cot \theta = \frac{\sin \theta}{\cos \theta} + \frac{\cos \theta}{\sin \theta}$$

$$= \frac{\sin \theta}{\cos \theta} \times \frac{\sin \theta}{\sin \theta} + \frac{\cos \theta}{\sin \theta} \times \frac{\cos \theta}{\cos \theta}$$

$$= \frac{\sin^2 \theta}{\cos \theta \sin \theta} + \frac{\cos^2 \theta}{\cos \theta \sin \theta}$$

$$= \frac{\sin^2 \theta + \cos^2 \theta}{\cos \theta \sin \theta}$$

$$= \frac{1}{\cos \theta \sin \theta} \quad Answer$$

SUMMARY

Reciprocal Identities	Quotient Identities	Pythagorean Identities
$\sec \theta = \frac{1}{\cos \theta}$	$\tan \theta = \frac{\sin \theta}{\cos \theta}$	$\cos^2 \theta + \sin^2 \theta = 1$
$\csc \theta = \frac{1}{\sin \theta}$	$\cot \theta = \frac{\cos \theta}{\sin \theta}$	$1 + \tan^2 \theta = \sec^2 \theta$
$\cot \theta = \frac{1}{\tan \theta}$		$\cot^2 \theta + 1 = \csc^2 \theta$

Exercises

Writing About Mathematics

1. Emma said that the answer to Example 3 can be simplified to $(\sec \theta)(\csc \theta)$. Do you agree with Emma? Justify your answer.

2. Ethan said that the equations $\cos^2 \theta = 1 - \sin^2 \theta$ and $\sin^2 \theta = 1 - \cos^2 \theta$ are identities. Do you agree with Ethan? Explain why or why not.

Developing Skills

In 3–14, for each given function value, find the remaining five trigonometric function values.

3. $\sin \theta = \frac{1}{5}$ and θ is in the second quadrant.

4. $\cos \theta = \frac{3}{4}$ and θ is in the first quadrant.

5. $\cos \theta = -\frac{3}{4}$ and θ is in the third quadrant.

6. $\sin \theta = -\frac{2}{3}$ and θ is in the fourth quadrant.

7. $\sin \theta = \frac{2}{3}$ and θ is in the second quadrant.

8. $\tan \theta = -2$ and θ is in the second quadrant.

9. $\tan \theta = 4$ and θ is in the third quadrant.

10. $\sec \theta = -8$ and θ is in the second quadrant.

11. $\cot \theta = \frac{5}{3}$ and θ is in the third quadrant.

12. $\csc \theta = \frac{5}{4}$ and θ is in the second quadrant.

13. $\sin \theta = \frac{4}{5}$ and θ is in the second quadrant.

14. $\cot \theta = -6$ and θ is in the fourth quadrant.

In 15–22, write each given expression in terms of sine and cosine and express the result in simplest form.

15. $(\csc \theta)(\sin \theta)$

16. $(\sin \theta)(\cot \theta)$

17. $(\tan \theta)(\cos \theta)$

18. $(\sec \theta)(\cot \theta)$

19. $\sec \theta + \tan \theta$

20. $\frac{\sec \theta}{\csc \theta}$

21. $\csc^2 \theta - \frac{\cot \theta}{\tan \theta}$

22. $\sec \theta \, (1 + \cot \theta) - \csc \theta \, (1 + \tan \theta)$

10-4 DOMAIN AND RANGE OF TRIGONOMETRIC FUNCTIONS

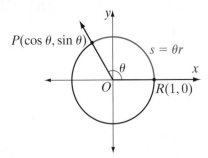

In Chapter 9, we defined the trigonometric functions for degree measures on the unit circle. Using arc length, we can define the trigonometric functions on the set of real numbers. Recall that the function values of sine and cosine are given by the coordinates of P, that is, $P(\cos \theta, \sin \theta)$. The length of the arc intercepted by the angle is given by

$$\theta = \frac{s}{r} \quad \text{or} \quad s = \theta r$$

However, since the radius of the unit circle is 1, the length of the arc s is equal to θ. In other words, the length of the arc s is equivalent to the measure of the angle θ in radians, a real number. Thus, any trigonometric function of θ is a function on the set of real numbers.

The Sine and Cosine Functions

We have defined the sine function and the cosine function in terms of the measure of angle formed by a rotation. The measure is positive if the rotation is in the counterclockwise direction and negative if the rotation is in the clockwise rotation. The rotation can continue indefinitely in both directions.

▶ **The domain of the sine function and of the cosine function is the set of real numbers.**

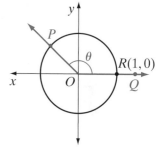

To find the range, let P be any point on the unit circle and $\overrightarrow{OP}$ be the terminal side of $\angle ROP$, an angle in standard position with measure θ. The coordinates of P are (cos θ, sin θ). Since the unit circle has a radius of 1, as P moves on the circle, its coordinates are elements of the interval $[-1, 1]$. Therefore, the range of cos θ and sin θ is the set of real numbers $[-1, 1]$.

To see this algebraically, consider the identity $\cos^2 \theta + \sin^2 \theta = 1$. Subtract $\cos^2 \theta$ from both sides of this identity and solve for sin θ. Similarly, subtract $\sin^2 \theta$ from both sides of this identity and solve for cos θ.

$$\cos^2 \theta + \sin^2 \theta = 1$$
$$\sin^2 \theta = 1 - \cos^2 \theta$$
$$\sin \theta = \pm\sqrt{1 - \cos^2 \theta}$$

$$\cos^2 \theta + \sin^2 \theta = 1$$
$$\cos^2 \theta = 1 - \sin^2 \theta$$
$$\cos \theta = \pm\sqrt{1 - \sin^2 \theta}$$

The value of sin θ will be a real number if and only if $1 - \cos^2 \theta \geq 0$, that is, if and only if $|\cos \theta| \leq 1$ or $-1 \leq \cos \theta \leq 1$.

The value of cos θ will be a real number if and only if $1 - \sin^2 \theta \geq 0$, that is, if and only if $|\sin \theta| \leq 1$ or $-1 \leq \sin \theta \leq 1$.

▶ **The range of the sine function and of the cosine function is the set of real numbers $[-1, 1]$, that is, the set of real numbers from -1 to 1, including -1 and 1.**

The Secant Function

The secant function is defined for all real numbers such that cos $\theta \neq 0$. We know that cos $\theta = 0$ when $\theta = \frac{\pi}{2}$, when $\theta = \frac{3\pi}{2}$, and when θ differs from either of these values by a complete rotation, 2π.

Therefore, $\cos \theta = 0$ when θ equals:

$\frac{\pi}{2}$	$\frac{3\pi}{2} = \frac{\pi}{2} + \pi$
$\frac{\pi}{2} + 2\pi$	$\frac{3\pi}{2} + 2\pi = \frac{\pi}{2} + 3\pi$
$\frac{\pi}{2} + 4\pi$	$\frac{3\pi}{2} + 4\pi = \frac{\pi}{2} + 5\pi$
$\frac{\pi}{2} + 6\pi$	$\frac{3\pi}{2} + 6\pi = \frac{\pi}{2} + 7\pi$
$\vdots$	$\vdots$

We can continue with this pattern, which can be summarized as $\frac{\pi}{2} + n\pi$ for all integral values of n. The domain of the secant function is the set of all real numbers except those for which $\cos \theta = 0$.

▶ **The domain of the secant function is the set of real numbers except $\frac{\pi}{2} + n\pi$ for all integral values of n.**

The secant function is defined as $\sec \theta = \frac{1}{\cos \theta}$. Therefore:

When $0 < \cos \theta \leq 1$,

$$\frac{0}{\cos \theta} < \frac{\cos \theta}{\cos \theta} \leq \frac{1}{\cos \theta} \qquad \text{or} \qquad 0 < 1 \leq \sec \theta$$

When $-1 \leq \cos \theta < 0$,

$$\frac{-1}{\cos \theta} \geq \frac{\cos \theta}{\cos \theta} > \frac{0}{\cos \theta} \qquad \text{or} \qquad -\sec \theta \geq 1 > 0$$

$$\text{or} \qquad \sec \theta \leq -1 < 0$$

The union of these two inequalities defines the range of the secant function.

▶ **The range of the secant function is the set of real numbers $(-\infty, -1] \cup [1, \infty)$, that is, the set of real numbers greater than or equal to 1 or less than or equal to -1.**

The Cosecant Function

The cosecant function is defined for all real numbers such that $\sin \theta \neq 0$. We know that $\sin \theta = 0$ when $\theta = 0$, when $\theta = \pi$, and when θ differs from either of these values by a complete rotation, 2π. Therefore, $\sin \theta = 0$ when θ equals:

0	π
$0 + 2\pi = 2\pi$	$\pi + 2\pi = 3\pi$
$0 + 4\pi = 4\pi$	$\pi + 4\pi = 5\pi$
$0 + 6\pi = 6\pi$	$\pi + 6\pi = 7\pi$
$\vdots$	$\vdots$

We can continue with this pattern, which can be summarized as $n\pi$ for all integral values of n. The domain of the cosecant function is the set of all real numbers except those for which $\sin \theta = 0$.

▶ The domain of the cosecant function is the set of real numbers except $n\pi$ for all integral values of n.

The range of the cosecant function can be found using the same procedure as was used to find the range of the secant function.

▶ The range of the cosecant function is the set of real numbers $(-\infty, -1] \cup [1, \infty)$, that is, the set of real numbers greater than or equal to 1 or less than or equal to -1.

The Tangent Function

We can use the identity $\tan \theta = \frac{\sin \theta}{\cos \theta}$ to determine the domain of the tangent function. Since $\sin \theta$ and $\cos \theta$ are defined for all real numbers, $\tan \theta$ is defined for all real numbers for which $\cos \theta \neq 0$, that is, for all real numbers except $\frac{\pi}{2} + n\pi$.

▶ The domain of the tangent function is the set of real numbers except $\frac{\pi}{2} + n\pi$ for all integral values of n.

To find the range of the tangent function, recall that the tangent function is defined on the line tangent to the unit circle at $R(1, 0)$. In particular, let T be the point where the terminal side of an angle in standard position intersects the line tangent to the unit circle at $R(1, 0)$. If $m\angle ROT = \theta$, then the coordinates of T are $(1, \tan \theta)$. As the point T moves on the tangent line, its y-coordinate can take on any real-number value. Therefore, the range of $\tan \theta$ is the set of real numbers.

To see this result algebraically, solve the identity $1 + \tan^2 \theta = \sec^2 \theta$ for $\tan^2 \theta$.

$$1 + \tan^2 \theta = \sec^2 \theta$$
$$\tan^2 \theta = \sec^2 \theta - 1$$

The range of the secant function is $\sec \theta \geq 1$ or $\sec \theta \leq -1$. Therefore, $\sec^2 \theta \geq 1$. If we subtract 1 from each side of this inequality:

$$\sec^2 \theta - 1 \geq 0$$
$$\tan^2 \theta \geq 0$$
$$\tan \theta \geq 0 \text{ or } \tan \theta \leq 0$$

▶ The range of the tangent function is the set of all real numbers.

The Cotangent Function

We can find the domain and range of the cotangent function by a procedure similar to that used for the tangent function.

We can use the identity $\cot \theta = \frac{\cos \theta}{\sin \theta}$ to determine the domain of the cotangent function. Since $\sin \theta$ and $\cos \theta$ are defined for all real numbers, $\cot \theta$ is defined for all real numbers for which $\sin \theta \neq 0$, that is, for all real numbers except $n\pi$.

▶ **The domain of the cotangent function is the set of real numbers except $n\pi$ for all integral values of n.**

To find the range of the cotangent function, solve the identity $1 + \cot^2 \theta = \csc^2 \theta$ for $\cot^2 \theta$.

$$1 + \cot^2 \theta = \csc^2 \theta$$
$$\cot^2 \theta = \csc^2 \theta - 1$$

The range of the cosecant function is $\csc \theta \geq 1$ or $\csc \theta \leq -1$. Therefore, $\csc^2 \theta \geq 1$. If we subtract 1 from each side of this inequality:

$$\csc^2 \theta - 1 \geq 0$$
$$\cot^2 \theta \geq 0$$
$$\cot \theta \geq 0 \text{ or } \cot \theta \leq 0$$

▶ **The range of the cotangent function is the set of all real numbers.**

EXAMPLE I

Explain why that is no value of θ such that $\cos \theta = 2$.

Solution The range of the cosine function is $-1 \leq \cos \theta \leq 1$. Therefore, there is no value of θ for which $\cos \theta$ is greater than 1.

SUMMARY

Function	Domain (n is an integer)	Range
Sine	All real numbers	$[-1, 1]$
Cosine	All real numbers	$[-1, 1]$
Tangent	All real numbers except $\frac{\pi}{2} + n\pi$	All real numbers
Cotangent	All real numbers except $n\pi$	All real numbers
Secant	All real numbers except $\frac{\pi}{2} + n\pi$	$(-\infty, -1] \cup [1, \infty)$
Cosecant	All real numbers except $n\pi$	$(-\infty, -1] \cup [1, \infty)$

Exercises

Writing About Mathematics

1. Nathan said that cot θ is undefined for the values of θ for which csc θ is undefined. Do you agree with Nathan? Explain why or why not.

2. Jonathan said that cot θ is undefined for $\frac{\pi}{2}$ because tan θ is undefined for $\frac{\pi}{2}$. Do you agree with Nathan? Explain why or why not.

Developing Skills

In 3–18, for each function value, write the value or tell why it is undefined. Do not use a calculator.

3. $\sin \frac{\pi}{2}$ **4.** $\cos \frac{\pi}{2}$ **5.** $\tan \frac{\pi}{2}$ **6.** $\sec \frac{\pi}{2}$

7. $\csc \frac{\pi}{2}$ **8.** $\cot \frac{\pi}{2}$ **9.** $\tan \pi$ **10.** $\cot \pi$

11. $\sec \frac{3\pi}{2}$ **12.** $\csc \frac{7\pi}{2}$ **13.** $\tan 0$ **14.** $\cot 0$

15. $\tan \left(-\frac{\pi}{2} \right)$ **16.** $\csc \left(-\frac{9\pi}{2} \right)$ **17.** $\sec \left(-\frac{9\pi}{2} \right)$ **18.** $\cot (-8\pi)$

19. List five values of θ for which sec θ is undefined.

20. List five values of θ for which cot θ is undefined.

10-5 INVERSE TRIGONOMETRIC FUNCTIONS

Inverse Sine Function

We know that an angle of 30° is an angle of $\frac{\pi}{6}$ radians and that an angle of 5(30°) or 150° is an angle of $5\left(\frac{\pi}{6}\right)$ or $\frac{5\pi}{6}$. Since 30 = 180 − 150, an angle of 30° is the reference angle for an angle of 150°. Therefore:

$$\sin 30° = \sin 150° = \tfrac{1}{2}$$

and $\quad \sin \frac{\pi}{6} = \sin \frac{5\pi}{6} = \tfrac{1}{2}$

Two ordered pairs of the sine function are $\left(\frac{\pi}{6}, \frac{1}{2}\right)$ and $\left(\frac{5\pi}{6}, \frac{1}{2}\right)$. Therefore, the sine function is *not* a one-to-one function. Functions that are not one-to-one do not have an inverse function. When we interchange the elements of the ordered pairs of the sine function, the set of ordered pairs is a relation that is not a function.

$\{(a, b) : b = \sin a\}$ is a function.

$\{(b, a) : a = \arcsin b\}$ is *not* a function.

For every value of θ such that $0 \leq \theta \leq \frac{\pi}{2}$, there is exactly one nonnegative value of $\sin \theta$. For every value of θ such that $-\frac{\pi}{2} \leq \theta < 0$, that is, for every fourth-quadrant angle, there is exactly one negative value of $\sin \theta$.

Therefore, if we **restrict the domain** of the sine function to $-\frac{\pi}{2} \leq \theta \leq \frac{\pi}{2}$, the function is one-to-one and has an inverse function. We designate the **inverse sine function** by *arcsin* or *sin^{-1}*.

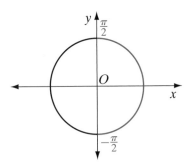

Sine Function with a Restricted Domain
$y = \sin x$
Domain $= \left\{x : -\frac{\pi}{2} \leq x \leq \frac{\pi}{2}\right\}$
Range $= \{y : -1 \leq y \leq 1\}$

Inverse Sine Function
$y = \arcsin x$ or $y = \sin^{-1} x$
Domain $= \{x : -1 \leq x \leq 1\}$
Range $= \left\{y : -\frac{\pi}{2} \leq y \leq \frac{\pi}{2}\right\}$

Note that when we use the $\boxed{\text{SIN}^{-1}}$ key on the graphing calculator with a number in the interval $[-1, 1]$, the response will be an number in the interval $\left[-\frac{\pi}{2}, \frac{\pi}{2}\right]$.

Inverse Cosine Function

The cosine function, like the sine function, is *not* a one-to-one function and does not have an inverse function. We can also restrict the domain of the cosine function to form a one-to-one function that has an inverse function.

For every value of θ such that $0 \leq \theta \leq \frac{\pi}{2}$, there is exactly one nonnegative value of $\cos \theta$. For every value of θ such that $\frac{\pi}{2} < \theta \leq \pi$, that is, for every second-quadrant angle, there is exactly one negative value of $\cos \theta$. Therefore, if we restrict the domain of the cosine function to $0 \leq \theta \leq \pi$, the function is one-to-one and has an inverse function. We designate the **inverse cosine function** by *arccos* or *cos^{-1}*.

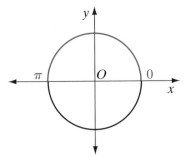

Cosine Function with a Restricted Domain
$y = \cos x$
Domain $= \{x : 0 \leq x \leq \pi\}$
Range $= \{y : -1 \leq y \leq 1\}$

Inverse Cosine Function
$y = \arccos x$ or $y = \cos^{-1} x$
Domain $= \{x : -1 \leq x \leq 1\}$
Range $= \{y : 0 \leq y \leq \pi\}$

Note that when we use the COS⁻¹ key on the graphing calculator with a number in the interval $[-1, 1]$, the response will be an number in the interval $[0, \pi]$.

Inverse Tangent Function

The tangent function, like the sine and cosine functions, is *not* a one-to-one function and does not have an inverse function. We can also restrict the domain of the tangent function to form a one-to-one function that has an inverse function.

For every of θ such that $0 \le \theta < \frac{\pi}{2}$, there is exactly one nonnegative value of tan θ. For every value of θ such that $-\frac{\pi}{2} < \theta < 0$, that is, for every fourth-quadrant angle, there is exactly one negative value of tan θ. Therefore, if we restrict the domain of the tangent function to $-\frac{\pi}{2} < \theta < \frac{\pi}{2}$, the function is one-to-one and has an inverse function. We designate the **inverse tangent function** by *arctan* or *tan⁻¹*.

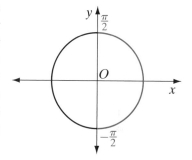

Tangent Function with a Restricted Domain
$y = \tan x$
Domain $= \left\{x : -\frac{\pi}{2} < x < \frac{\pi}{2}\right\}$
Range $= \{y : y \text{ is a real number}\}$

Inverse Tangent Function
$y = \arctan x \text{ or } y = \tan^{-1} x$
Domain $= \{x : x \text{ is a real number}\}$
Range $= \left\{y : -\frac{\pi}{2} < y < \frac{\pi}{2}\right\}$

Note that when we use the TAN⁻¹ key on the graphing calculator with any real number, the response will be an number in the interval $\left(-\frac{\pi}{2}, \frac{\pi}{2}\right)$.

These same restrictions can be used to define inverse functions for the secant, cosecant, and cotangent functions, which will be left to the student. (See Exercises 40–42.)

EXAMPLE I

Express the range of the arcsine function in degrees.

Solution Expressed in radians, the range of the arcsine function is $\left[-\frac{\pi}{2}, \frac{\pi}{2}\right]$, that is, $-\frac{\pi}{2} \le \arcsin \theta \le \frac{\pi}{2}$. Since an angle of $\frac{\pi}{2}$ radians is an angle of $90°$ and an angle of $-\frac{\pi}{2}$ radians is an angle of $-90°$, the range of the arcsine function is:

$$[-90°, 90°] \quad \text{or} \quad -90° \le \theta \le 90° \; \textit{Answer}$$

EXAMPLE 2

Find $\sin (\cos^{-1} (-0.5))$.

Solution   Let $\theta = \cos^{-1} (-0.5)$. Then θ is a second-quadrant angle whose reference angle, r, is $r = \pi - \theta$ in radians or $r = 180 - \theta$ in degrees.

In radians:	In degrees:
$\cos r = 0.5$	$\cos r = 0.5$
$r = \frac{\pi}{3}$	$r = 60°$
$\pi - \theta = \frac{\pi}{3}$	$180° - \theta = 60°$
$\frac{2\pi}{3} = \theta$	$120° = \theta$
$\sin \theta = \sin \frac{2\pi}{3} = \frac{\sqrt{3}}{2}$	$\sin \theta = \sin 120° = \frac{\sqrt{3}}{2}$

Alternative   To find $\sin (\cos^{-1}(-0.5))$, find $\sin \theta$ when $\theta = \cos^{-1} (-0.5)$. If $\theta = \cos^{-1}$
Solution (-0.5), then $\cos \theta = -0.5$. For the function $y = \cos^{-1} x$, when x is negative, $\frac{\pi}{2} < y \le \pi$. Therefore, θ is the measure of a second-quadrant angle and $\sin \theta$ is positive.

Use the Pythagorean identity:

$$\sin^2 \theta + \cos^2 \theta = 1$$
$$\sin^2 \theta + (-0.5)^2 = 1$$
$$\sin^2 \theta + 0.25 = 1$$
$$\sin^2 \theta = 0.75$$
$$\sin \theta = \sqrt{0.75} = \sqrt{0.25}\sqrt{3} = 0.5\sqrt{3}$$

Check   ENTER: [SIN] [2nd] [COS⁻¹] [(-)]

[.5] [)] [)] [ENTER]

[2nd] [√] 3 [)] [÷] 2

[ENTER]

DISPLAY:

```
sin(cos⁻¹(-.5))
            .8660254038
√(3)/2
            .8660254038
```

Answer $\frac{\sqrt{3}}{2}$

EXAMPLE 3

Find the exact value of $\sec \left(\arcsin \frac{-3}{5} \right)$ if the angle is a third-quadrant angle.

Solution   Let $\theta = \left(\arcsin \frac{-3}{5} \right)$. This can be written as $\sin \theta = \left(\frac{-3}{5} \right)$. Then:

$$\sec \left(\arcsin \frac{-3}{5} \right) = \sec \theta = \frac{1}{\cos \theta}$$

We know the value of $\sin \theta$. We can use a Pythagorean identity to find $\cos \theta$:

$$\cos \theta = \pm\sqrt{1 - \sin^2 \theta}$$

$$\frac{1}{\cos \theta} = \frac{1}{\pm\sqrt{1 - \sin^2 \theta}} = \frac{1}{\pm\sqrt{1 - \left(-\frac{3}{5}\right)^2}} = \frac{1}{\pm\sqrt{1 - \frac{9}{25}}} = \frac{1}{\pm\sqrt{\frac{16}{25}}} = \frac{1}{\pm\frac{4}{5}} = \pm\frac{5}{4}$$

An angle in the third quadrant has a negative secant function value.

Answer $\sec\left(\arcsin \frac{-3}{5}\right) = -\frac{5}{4}$

Calculator Solution ENTER: 1 [÷] [COS] [2nd] [SIN⁻¹]
-3 [÷] 5 [)] [)] [MATH]
[ENTER] [ENTER]

DISPLAY:
```
1/cos(sin⁻¹(-3/5)
)▶Frac
                5/4
```

The calculator returns the value $\frac{5}{4}$. The calculator used the function value of $\left(\arcsin \frac{-3}{5}\right)$, a fourth-quadrant angle for which the secant function value is positive. For the given third-quadrant angle, the secant function value is negative.

Answer $\sec\left(\arcsin \frac{-3}{5}\right) = -\frac{5}{4}$

Exercises

Writing About Mathematics

1. Nicholas said that the restricted domain of the cosine function is the same as the restricted domain of the tangent function. Do you agree with Nicholas? Explain why or why not.

2. Sophia said that the calculator solution to Example 2 could have been found with the calculator set to either degree or radian mode. Do you agree with Sophia? Explain why or why not.

Developing Skills

In 3–14, find each value of θ: **a.** in degrees **b.** in radians

3. $\theta = \arcsin \frac{1}{2}$

4. $\theta = \arctan 1$

5. $\theta = \arccos 1$

6. $\theta = \arctan (-1)$

7. $\theta = \arccos \left(-\frac{1}{2}\right)$

8. $\theta = \arcsin \left(-\frac{1}{2}\right)$

9. $\theta = \arctan (-\sqrt{3})$

10. $\theta = \arcsin \left(-\frac{\sqrt{3}}{2}\right)$

11. $\theta = \arccos (-1)$

12. $\theta = \arcsin 1$

13. $\theta = \arctan 0$

14. $\theta = \arccos 0$

In 15–23, use a calculator to find each value of θ to the nearest degree.

15. $\theta = \arcsin 0.6$ **16.** $\theta = \arccos(-0.6)$ **17.** $\theta = \arctan 4.4$

18. $\theta = \arctan(-4.4)$ **19.** $\theta = \arccos 0.9$ **20.** $\theta = \arccos(-0.9)$

21. $\theta = \arcsin 0.72$ **22.** $\theta = \arcsin(-0.72)$ **23.** $\theta = \arctan(-17.3)$

In 24–32, find the exact value of each expression.

24. $\sin(\arctan 1)$ **25.** $\cos(\arctan 0)$ **26.** $\tan(\arccos 1)$

27. $\cos(\arccos(-1))$ **28.** $\tan\left(\arcsin\left(-\frac{1}{2}\right)\right)$ **29.** $\cos\left(\arcsin\left(-\frac{\sqrt{3}}{2}\right)\right)$

30. $\tan\left(\arccos\left(-\frac{\sqrt{2}}{2}\right)\right)$ **31.** $\sin\left(\arccos\left(-\frac{\sqrt{2}}{2}\right)\right)$ **32.** $\cos\left(\arcsin\left(-\frac{\sqrt{2}}{2}\right)\right)$

In 33–38, find the exact radian measure θ of an angle with the smallest absolute value that satisfies the equation.

33. $\sin\theta = \frac{\sqrt{2}}{2}$ **34.** $\cos\theta = \frac{\sqrt{3}}{2}$ **35.** $\tan\theta = 1$

36. $\sec\theta = -1$ **37.** $\csc\theta = -\sqrt{2}$ **38.** $\cot\theta = \sqrt{3}$

39. In **a–c** with the triangles labeled as shown, use the inverse trigonometric functions to express θ in terms of x.

a.

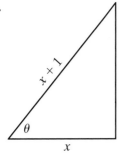

b.

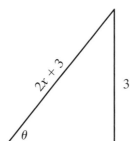

c.

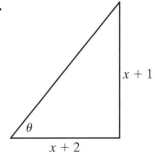

Applying Skills

40. a. Restrict the domain of the secant function to form a one-to-one function that has an inverse function. Justify your answer.

b. Is the restricted domain found in **a** the same as the restricted domain of the cosine function?

c. Find the range of the restricted secant function.

d. Find the domain of the inverse secant function, that is, the arcsecant function.

e. Find the range of the arcsecant function.

41. a. Restrict the domain of the cosecant function to form a one-to-one function that has an inverse function. Justify your domain.

b. Is the restricted domain found in **a** the same as the restricted domain of the sine function?

c. Find the range of the restricted cosecant function.

d. Find the domain of the inverse cosecant function, that is, the arccosecant function.

e. Find the range of the arccosecant function.

42. a. Restrict the domain of the cotangent function to form a one-to-one function that has an inverse function. Justify your domain.

b. Is the restricted domain found in **a** the same as the restricted domain of the tangent function?

c. Find the range of the restricted cotangent function.

d. Find the domain of the inverse cotangent function, that is, the arccotangent function.

e. Find the range of the arccotangent function.

43. Jennifer lives near the airport. An airplane approaching the airport flies at a constant altitude of 1 mile toward a point, P, above Jennifer's house. Let θ be the measure of the angle of elevation of the plane and d be the horizontal distance from P to the airplane.

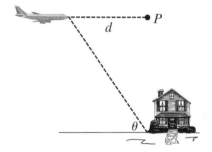

a. Express θ in terms of d.

b. Find θ when $d = 1$ mile and when $d = 0.5$ mile.

10-6 COFUNCTIONS

Two acute angles are complementary if the sum of their measures is 90°. In the diagram, $\angle ROP$ is an angle in standard position whose measure is θ and (p, q) are the coordinates of P, the intersection of $\overrightarrow{OP}$ with the unit circle. Under a reflection in the line $y = x$, the image of (x, y) is (y, x). Therefore, under a reflection in the line $y = x$, the image of $P(p, q)$ is $P'(q, p)$, the image of $R(1, 0)$ is $R'(0, 1)$, and the image of $O(0, 0)$ is $O(0, 0)$. Under a line reflection, angle measure is preserved. Therefore,

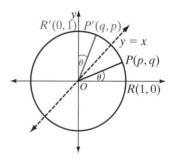

$$m\angle ROP = m\angle R'OP' = \theta$$
$$m\angle ROP' = 90° - m\angle R'OP' = 90° - \theta$$

The two angles, $\angle ROP$ and $\angle ROP'$ are in standard position. Therefore,

$\cos \theta = p$	$\sin \theta = q$
$\sin (90° - \theta) = p$	$\cos (90° - \theta) = q$
$\cos \theta = \sin (90° - \theta)$	$\sin \theta = \cos (90° - \theta)$

The sine of an acute angle is equal to the cosine of its complement. The sine and cosine functions are called **cofunctions**. There are two other pairs of cofunctions in trigonometry.

$\tan \theta = \dfrac{\sin \theta}{\cos \theta} = \dfrac{q}{p}$	$\cot \theta = \dfrac{\cos \theta}{\sin \theta} = \dfrac{p}{q}$
$\cot (90° - \theta) = \dfrac{\cos (90° - \theta)}{\sin (90° - \theta)} = \dfrac{q}{p}$	$\tan (90° - \theta) = \dfrac{\sin (90° - \theta)}{\cos (90° - \theta)} = \dfrac{p}{q}$
$\tan \theta = \cot (90° - \theta)$	$\cot \theta = \tan (90° - \theta)$

The tangent and cotangent functions are cofunctions.

$\sec \theta = \dfrac{1}{\cos \theta} = \dfrac{1}{p}$	$\csc \theta = \dfrac{1}{\sin \theta} = \dfrac{1}{q}$
$\csc (90° - \theta) = \dfrac{1}{\sin (90° - \theta)} = \dfrac{1}{p}$	$\sec (90° - \theta) = \dfrac{1}{\cos (90° - \theta)} = \dfrac{1}{q}$
$\sec \theta = \csc (90° - \theta)$	$\csc \theta = \sec (90° - \theta)$

The secant and cosecant functions are cofunctions. Cofunctions represented in radians are left to the student. (See Exercise 25.)

In any right triangle ABC, if $\angle C$ is the right angle, then $\angle A$ and $\angle B$ are complementary angles. Therefore, $m\angle A = 90 - m\angle B$ and $m\angle B = 90 - m\angle A$.

▶ **Sine and cosine are cofunctions: $\sin A = \cos B$ and $\sin B = \cos A$.**

▶ **Tangent and cotangent are cofunctions: $\tan A = \cot B$ and $\tan B = \cot A$.**

▶ **Secant and cosecant are cofunctions: $\sec A = \csc B$ and $\sec B = \csc A$.**

EXAMPLE I

If $\tan 63.44° = 2.00$, find a value of θ such that $\cot \theta = 2.00$.

Solution If θ is the degree measure of a first-quadrant angle, then:

$$\tan (90° - \theta) = \cot \theta$$
$$\tan 63.44° = \cot \theta$$

Therefore, $\tan (90° - \theta) = \tan 63.44$ and $63.44 = (90° - \theta)$.

$$63.44° = 90° - \theta$$
$$\theta = 90° - 63.44°$$
$$= 26.56°$$

Answer $\theta = 26.56°$

EXAMPLE 2

Express sin 100° as the function value of an acute angle that is less than 45°.

Solution An angle whose measure is 100° is a second-quadrant angle whose reference angle is 180° − 100° or 80°.

Therefore, sin 100° = sin 80° = cos (90° − 80°) = cos 10°.

Answer sin 100° = cos 10°

Exercises

Writing About Mathematics

1. Mia said that if you know the sine value of each acute angle, then you can find any trigonometric function value of an angle of any measure. Do you agree with Mia? Explain why or why not.

2. If $\sin A = \frac{1}{2}$, is $\cos A = \frac{\sqrt{3}}{2}$ always true? Explain why or why not.

Developing Skills

In 3–22: **a.** Rewrite each function value in terms of its cofunction. **b.** Find, to four decimal places, the value of the function value found in **a**.

3. sin 65°	**4.** cos 80°	**5.** tan 54°	**6.** sin 86°
7. csc 48°	**8.** sec 75°	**9.** cot 57°	**10.** cos 70°
11. sin 110°	**12.** tan 95°	**13.** cos 130°	**14.** sec 125°
15. sin 230°	**16.** cos 255°	**17.** tan 237°	**18.** csc 266°
19. cos 300°	**20.** sin 295°	**21.** cot 312°	**22.** sec 285°

23. If $\sin \theta = \cos (20 + \theta)$, what is the value of θ?

24. For what value of x does $\tan (x + 10) = \cot (40 + x)$?

25. Complete the following table of cofunctions for radian values.

Cofunctions (degrees)		Cofunctions (radians)	
$\cos \theta = \sin (90° - \theta)$	$\sin \theta = \cos (90° - \theta)$		
$\tan \theta = \cot (90° - \theta)$	$\cot \theta = \tan (90° - \theta)$		
$\sec \theta = \csc (90° - \theta)$	$\csc \theta = \sec (90° - \theta)$		

In 26–33: **a.** Rewrite each function value in terms of its cofunction. **b.** Find the exact value of the function value found in **a.**

26. $\sin \frac{\pi}{3}$

27. $\cos \frac{\pi}{4}$

28. $\tan \frac{\pi}{6}$

29. $\sec \frac{2\pi}{3}$

30. $\csc \frac{5\pi}{6}$

31. $\cot \pi$

32. $\sin \left(-\frac{\pi}{4}\right)$

33. $\cos \frac{8\pi}{3}$

34. $\tan \left(-\frac{5\pi}{3}\right)$

CHAPTER SUMMARY

A **radian** is the unit of measure of a central angle that intercepts an arc equal in length to the radius of the circle.

In general, the radian measure θ of a central angle is the length of the intercepted arc, s, divided by the radius of the circle, r.

$$\theta = \frac{s}{r}$$

The length of the semicircle is πr. Therefore, the radian measure of a straight angle is $\frac{\pi r}{r} = \pi$. This leads to the relationship:

$$\pi \text{ radians} = 180 \text{ degrees}$$

The measure of any angle can be expressed in degrees or in radians. We can do this by forming the following proportion:

$$\frac{\text{measure in degrees}}{\text{measure in radians}} = \frac{180}{\pi}$$

When the calculator is in radian mode, it will return the decimal value of the radian measure whose function value is entered. The measure will not be in terms of π.

An **identity** is an equation that is true for all values of the variable for which the terms of the variable are defined.

Reciprocal Identities	Quotient Identities	Pythagorean Identities
$\sec \theta = \frac{1}{\cos \theta}$	$\tan \theta = \frac{\sin \theta}{\cos \theta}$	$\cos^2 \theta + \sin^2 \theta = 1$
$\csc \theta = \frac{1}{\sin \theta}$	$\cot \theta = \frac{\cos \theta}{\sin \theta}$	$1 + \tan^2 \theta = \sec^2 \theta$
$\cot \theta = \frac{1}{\tan \theta}$		$\cot^2 \theta + 1 = \csc^2 \theta$

The domains and ranges of the six basic trigonometric functions are as follows.

Function	Domain (n is an integer)	Range
Sine	All real numbers	$[-1, 1]$
Cosine	All real numbers	$[-1, 1]$
Tangent	All real numbers except $\frac{\pi}{2} + n\pi$	All real numbers
Cotangent	All real numbers except $n\pi$	All real numbers
Secant	All real numbers except $\frac{\pi}{2} + n\pi$	$(-\infty, -1] \cup [1, \infty)$
Cosecant	All real numbers except $n\pi$	$(-\infty, -1] \cup [1, \infty)$

The **inverse trigonometric functions** are defined for **restricted domains**.

Sine Function with a Restricted Domain
$y = \sin x$
Domain $= \left\{x : -\frac{\pi}{2} \leq x \leq \frac{\pi}{2}\right\}$
Range $= \{y : -1 \leq y \leq 1\}$

Inverse Sine Function
$y = \arcsin x$ or $y = \sin^{-1} x$
Domain $= \{x : -1 \leq x \leq 1\}$
Range $= \left\{y : -\frac{\pi}{2} \leq y \leq \frac{\pi}{2}\right\}$

Cosine Function with a Restricted Domain
$y = \cos x$
Domain $= \{x : 0 \leq x \leq \pi\}$
Range $= \{y : -1 \leq y \leq 1\}$

Inverse Cosine Function
$y = \arccos x$ or $y = \cos^{-1} x$
Domain $= \{x : -1 \leq x \leq 1\}$
Range $= \{y : 0 \leq y \leq \pi\}$

Tangent Function with a Restricted Domain
$y = \tan x$
Domain $= \left\{x : -\frac{\pi}{2} < x < \frac{\pi}{2}\right\}$
Range $= \{y : y \text{ is a real number}\}$

Inverse Tangent Function
$y = \arctan x$ or $y = \tan^{-1} x$
Domain $= \{x : x \text{ is a real number}\}$
Range $= \left\{y : -\frac{\pi}{2} < y < \frac{\pi}{2}\right\}$

In any right triangle ABC, if $\angle C$ is the right angle, then $\angle A$ and $\angle B$ are complementary angles. Therefore, $m\angle A = 90 - m\angle B$ and $m\angle B = 90 - m\angle A$.

- Sine and cosine are cofunctions: $\sin A = \cos B$ and $\sin B = \cos A$.
- Tangent and cotangent are cofunctions: $\tan A = \cot B$ and $\tan B = \cot A$.
- Secant and cosecant are cofunctions: $\sec A = \csc B$ and $\sec B = \csc A$.

10-1 Radian

10-3 Identity • Pythagorean identity

10-5 Restricted domain • Inverse sine function • Inverse cosine function • Inverse tangent function

10-6 Cofunction

REVIEW EXERCISES

In 1–4, express each degree measure in radian measure in terms of π.

1. 75°   **2.** 135°   **3.** 225°   **4.** $-60°$

In 5–8, express each radian measure in degree measure.

5. $\frac{\pi}{4}$ **6.** $\frac{2\pi}{5}$ **7.** $\frac{7\pi}{6}$ **8.** $-\frac{\pi}{8}$

9. What is the radian measure of a right angle?

In 10–17, in a circle of radius r, the measure of a central angle is θ and the length of the arc intercepted by the angle is s.

10. If $r = 5$ cm and $s = 5$ cm, find θ.

11. If $r = 5$ cm and $\theta = 2$, find s.

12. If $r = 2$ in. and $s = 3$ in., find θ.

13. If $s = 8$ cm and $\theta = 2$, find r.

14. If $s = 9$ cm and $r = 3$ mm, find θ.

15. If $s = \pi$ cm and $\theta = \frac{\pi}{4}$, find r.

16. If $\theta = 3$ and $s = 7.5$ cm, find r.

17. If $\theta = \frac{\pi}{5}$ and $r = 10$ ft, find s.

18. For what values of θ $(0° \leq \theta \leq 360°)$ is each of the following undefined?

    a. $\tan \theta$ **b.** $\cot \theta$ **c.** $\sec \theta$ **d.** $\csc \theta$

19. What is the domain and range of each of the following functions?

    a. $\arcsin \theta$ **b.** $\arccos \theta$ **c.** $\arctan \theta$

In 20–27, find the exact value of each trigonometric function.

20. $\sin \frac{\pi}{6}$ **21.** $\cos \frac{\pi}{4}$ **22.** $\tan \frac{\pi}{3}$ **23.** $\sin \frac{\pi}{2}$

24. $\cos \frac{2\pi}{3}$ **25.** $\tan \frac{3\pi}{4}$ **26.** $\sin \frac{11\pi}{6}$ **27.** $\cos \left(\frac{-7\pi}{6} \right)$

In 28–33, find the exact value of each expression.

28. $\tan \frac{3\pi}{4} + \cot \frac{3\pi}{4}$

29. $2 \csc \frac{\pi}{2} + 5 \cot \frac{\pi}{2}$

30. $2 \tan \frac{2\pi}{3} - 3 \sec \frac{5\pi}{6}$

31. $\tan^2 \left(\frac{7\pi}{6}\right) - \sec^2 \left(\frac{7\pi}{6}\right)$

32. $\cot^2 \left(\frac{\pi}{3}\right) + \sec^2 \left(\frac{\pi}{4}\right)$

33. $3 \cot \left(\frac{\pi}{4}\right) \cdot 2 \csc^2 \left(\frac{\pi}{4}\right)$

In 34–39, find the exact value of each expression.

34. $\sin \left(\arcsin -\frac{\sqrt{3}}{2}\right)$

35. $\tan \left(\operatorname{arccot} -\sqrt{3}\right)$

36. $\cot \left(\arccos -1\right)$

37. $\sec \left(\arcsin \frac{\sqrt{2}}{2}\right)$

38. $\csc \left(\operatorname{arccot} -1\right)$

39. $\cos \left(\operatorname{arcsec} 2\right)$

Exploration

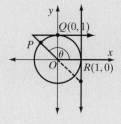

In this exploration, we will continue the work begun in Chapter 9.

Let $\overrightarrow{OP}$ intersect the unit circle at P. The line that is tangent to the circle at $R(1, 0)$ intersects $\overleftrightarrow{OP}$ at T. The line that is tangent to the circle at $Q(0, 1)$ intersects $\overleftrightarrow{OP}$ at S. The measure of $\angle ROP$ is θ.

1. $\angle ROP$ is a second-quadrant angle as shown to the left. Locate points S and T. Show that $-OT = \sec \theta$, $OS = \csc \theta$, and $-QS = \cot \theta$.

2. Draw $\angle ROP$ as a third-quadrant angle. Let $R(1, 0)$ and $Q(0, 1)$ be fixed points. Locate points S and T. Show that $-OT = \sec \theta$, $-OS = \csc \theta$, and $QS = \cot \theta$.

3. Draw $\angle ROP$ as a fourth-quadrant angle. Let R and Q be defined as before. Locate points S and T. Show that $OT = \sec \theta$, $-OS = \csc \theta$, and $-QS = \cot \theta$.

CUMULATIVE REVIEW   CHAPTERS 1–10

Part I

Answer all questions in this part. Each correct answer will receive 2 credits. No partial credit will be allowed.

1. If $\sec \theta = \sqrt{5}$, then $\cos \theta$ equals

(1) $-\sqrt{5}$ (2) $\frac{2\sqrt{5}}{5}$ (3) $\frac{\sqrt{5}}{5}$ (4) $-\frac{\sqrt{5}}{5}$

2. Which of the following is an irrational number?

(1) $0.\overline{34}$ (2) $\sqrt{121}$ (3) $3 + 2i$ (4) 2π

3. Which of the following is a function from $\{1, 2, 3, 4, 5\}$ to $\{1, 2, 3, 4, 5\}$ that is one-to-one and onto?

    (1) $\{(1, 2), (2, 3), (3, 4), (4, 5), (5, 5)\}$ (3) $\{(1, 3), (2, 3), (3, 3), (4, 3), (5, 3)\}$

    (2) $\{(1, 5), (2, 4), (3, 3), (4, 2), (5, 1)\}$ (4) $\{(1, 1), (2, 4), (3, 1), (4, 2), (5, 1)\}$

4. When written with a rational denominator, $\dfrac{1}{3 + \sqrt{5}}$ is equal to

    (1) $\dfrac{3 + \sqrt{5}}{14}$ (2) $\dfrac{3 - \sqrt{5}}{-2}$ (3) $\dfrac{3 - \sqrt{5}}{4}$ (4) $\dfrac{3 + \sqrt{5}}{8}$

5. The fraction $\dfrac{1 - \frac{1}{a}}{1 - \frac{1}{a^2}}$ is equal to

    (1) $\dfrac{a^2}{a - 1}$ (2) $\dfrac{a}{a - 1}$ (3) $\dfrac{a^2}{a + 1}$ (4) $\dfrac{a}{a + 1}$

6. The inverse function of the function $y = 2x + 1$ is

    (1) $y = \frac{1}{2}x + 1$ (3) $y = \dfrac{x - 1}{2}$

    (2) $y = \dfrac{x + 1}{2}$ (4) $y = -2x - 1$

7. The exact value of $\cos \frac{2\pi}{3}$ is

    (1) $\dfrac{\sqrt{3}}{2}$ (2) $-\dfrac{\sqrt{3}}{2}$ (3) $\dfrac{1}{2}$ (4) $-\dfrac{1}{2}$

8. The series $-2 + 2 + 6 + 10 + \cdots$ written in sigma notation is

    (1) $\displaystyle\sum_{k=1}^{\infty} -2k$ (3) $\displaystyle\sum_{k=1}^{\infty} (k + 4k)$

    (2) $\displaystyle\sum_{k=1}^{\infty} (-2 + 2k)$ (4) $\displaystyle\sum_{k=1}^{\infty} (-2 + 4(k - 1))$

9. What are the zeros of the function $y = x^3 + 3x^2 + 2x$?

    (1) 0 only   (3) 0, 1, and 2

    (2) $-2, -1$, and 0   (4) The function has no zeros.

10. The product $\left(\sqrt{-8}\right)\left(\sqrt{-2}\right)$ is equal to

    (1) $4i$ (2) 4   (3) -4 (4) $3i\sqrt{2}$

Part II

Answer all questions in this part. Each correct answer will receive 2 credits. Clearly indicate the necessary steps, including appropriate formula substitutions, diagrams, graphs, charts, etc. For all questions in this part, a correct numerical answer with no work shown will receive only 1 credit.

11. Express the roots of $x^2 - 6x + 13 = 0$ in $a + bi$ form.

12. If 280 is the measure of an angle in degrees, what is the measure of the angle in radians?

Part III

Answer all questions in this part. Each correct answer will receive 4 credits. Clearly indicate the necessary steps, including appropriate formula substitutions, diagrams, graphs, charts, etc. For all questions in this part, a correct numerical answer with no work shown will receive only 1 credit.

13. What is the equation of a circle if the endpoints of the diameter are $(-2, 4)$ and $(6, 0)$?

14. Given $\log_{\frac{1}{3}} \frac{1}{9} - 2 \log_{\frac{1}{3}} \frac{1}{27} + \log_{\frac{1}{3}} \frac{1}{243}$:

a. Write the expression as a single logarithm.

b. Evaluate the logarithm.

Part IV

Answer all questions in this part. Each correct answer will receive 6 credits. Clearly indicate the necessary steps, including appropriate formula substitutions, diagrams, graphs, charts, etc. For all questions in this part, a correct numerical answer with no work shown will receive only 1 credit.

15. If $\sec \theta = \sqrt{3}$ and θ is in the fourth quadrant, find $\cos \theta$, $\sin \theta$, and $\tan \theta$ in simplest radical form.

16. Find graphically the common solutions of the equations:

$$y = x^2 - 4x + 1$$
$$y = x - 3$$

CHAPTER

11

GRAPHS OF TRIGONOMETRIC FUNCTIONS

Music is an integral part of the lives of most people. Although the kind of music they prefer will differ, all music is the effect of sound waves on the ear. Sound waves carry the energy of a vibrating string or column of air to our ears. No matter what vibrating object is causing the sound wave, the frequency of the wave (that is, the number of waves per second) creates a sensation that we call the pitch of the sound. A sound wave with a high frequency produces a high pitch while a sound wave with a lower frequency produces a lower pitch. When the frequencies of two sounds are in the ratio of 2 : 1, the sounds differ by an octave and produce a pleasing combination. In general, music is the result of the mixture of sounds that are mathematically related by whole-number ratios of their frequencies.

Sound is just one of many physical entities that are transmitted by waves. Light, radio, television, X-rays, and microwaves are others. The trigonometric functions that we will study in this chapter provide the mathematical basis for the study of waves.

11-1 GRAPH OF THE SINE FUNCTION

The sine function is a set of ordered pairs of real numbers. Each ordered pair can be represented as a point of the coordinate plane. The domain of the sine function is the set of real numbers, that is, every real number is a first element of one pair of the function.

To sketch the graph of the sine function, we will plot a portion of the graph using the subset of the real numbers in the interval $0 \leq x \leq 2\pi$. We know that

$$\sin \tfrac{\pi}{6} = \tfrac{1}{2} = 0.5$$

and that $\frac{\pi}{6}$ is the measure of the reference angle for angles with measures of $\frac{5\pi}{6}, \frac{7\pi}{6}, \frac{11\pi}{6}, \ldots$. We also know that

$$\sin \tfrac{\pi}{3} = \tfrac{\sqrt{3}}{2} = 0.866025 \ldots$$

and that $\frac{\pi}{3}$ is the measure of the reference angle for angles with measures of $\frac{2\pi}{3}, \frac{4\pi}{3}, \frac{5\pi}{3}, \ldots$. We can round the rational approximation of $\sin \frac{\pi}{3}$ to two decimal places, 0.87.

x	0	$\frac{\pi}{6}$	$\frac{\pi}{3}$	$\frac{\pi}{2}$	$\frac{2\pi}{3}$	$\frac{5\pi}{6}$	π	$\frac{7\pi}{6}$	$\frac{4\pi}{3}$	$\frac{3\pi}{2}$	$\frac{5\pi}{3}$	$\frac{11\pi}{6}$	2π
sin x	0	0.5	0.87	1	0.87	0.5	0	−0.5	−0.87	−1	−0.87	−0.5	0

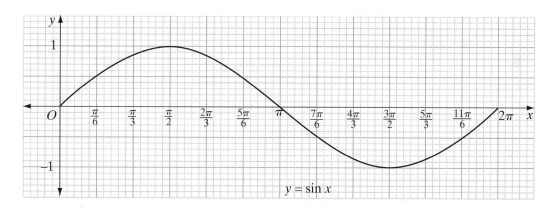

$y = \sin x$

On the graph, we plot the points whose coordinates are given in the table. Through these points, we draw a smooth curve. Note how x and y change.

- As x increases from 0 to $\frac{\pi}{2}$, y increases from 0 to 1.
- As x increases from $\frac{\pi}{2}$ to π, y decreases from 1 to 0.
- As x increases from π to $\frac{3\pi}{2}$, y continues to decrease from 0 to −1.
- As x increases from $\frac{3\pi}{2}$ to 2π, y increases from −1 to 0.

When we plot a larger subset of the domain of the sine function, this pattern is repeated. For example, add to the points given above the point whose x-coordinates are in the interval $-2\pi \le x \le 0$.

x	-2π	$-\frac{11\pi}{6}$	$-\frac{5\pi}{3}$	$-\frac{3\pi}{2}$	$-\frac{4\pi}{3}$	$-\frac{7\pi}{6}$	$-\pi$	$-\frac{5\pi}{6}$	$-\frac{2\pi}{3}$	$-\frac{\pi}{2}$	$-\frac{\pi}{3}$	$-\frac{\pi}{6}$	0
$\sin x$	0	0.5	0.87	1	0.87	0.5	0	-0.5	-0.87	-1	-0.87	-0.5	0

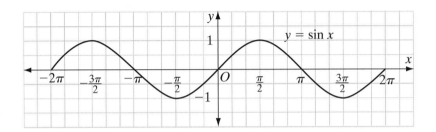

Each time we increase or decrease the value of the x-coordinates by a multiple of 2π, the basic sine curve is repeated. Each portion of the graph in an interval of 2π is one **cycle** of the sine function.

The graph of the function $y = \sin x$ is its own image under the translation $T_{2\pi,0}$. The function $y = \sin x$ is called a **periodic function** with a **period** of 2π because for every x in the domain of the sine function, $\sin x = \sin(x + 2\pi)$.

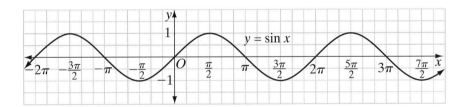

▶ **The period of the sine function $y = \sin x$ is 2π.**

Each cycle of the sine curve can be separated into four quarters. In the first quarter, the sine curve increases from 0 to the maximum value of the function. In the second quarter, it decreases from the maximum value to 0. In the third quarter, it decreases from 0 to the minimum value, and in the fourth quarter, it increases from the minimum value to 0.

A graphing calculator will display the graph of the sine function.

STEP 1. Put the calculator in radian mode.

ENTER: MODE ▼ ▼ ENTER

STEP 2. Enter the equation for the sine function.

ENTER: Y= SIN X,T,θ,*n* ENTER

STEP 3. To display one cycle of the curve, let the window include values from 0 to 2π for x and values slightly smaller than -1 and larger than 1 for y. Use the following viewing window: $Xmin = 0$, $Xmax = 2\pi$, $Xscl = \frac{\pi}{6}$, $Ymin = -1.5$, $Ymax = 1.5$. (Note: $Xscl$ changes the scale of the x-axis.)

ENTER: WINDOW 0 ENTER 2 2nd π ENTER 2nd

π ÷ 6 ENTER −1.5 ENTER 1.5 ENTER

STEP 4. Finally, graph the sin curve by pressing GRAPH . To display more than one cycle of the curve, change $Xmin$ or $Xmax$ of the window.

ENTER: WINDOW −2 2nd π ENTER 4 2nd π

ENTER GRAPH

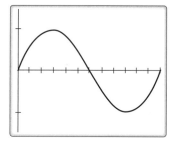 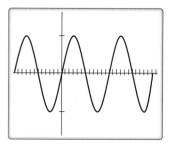

EXAMPLE I

In the interval $-2\pi \le x \le 0$, for what values of x does $y = \sin x$ increase and for what values of x does $y = \sin x$ decrease?

Solution The graph shows that $y = \sin x$ increases in the interval $-2\pi \le x \le -\frac{3\pi}{2}$ and in the interval $-\frac{\pi}{2} \le x \le 0$ and decreases in the interval $-\frac{3\pi}{2} \le x \le -\frac{\pi}{2}$.

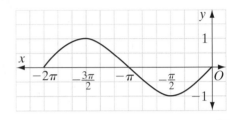

The Graph of the Sine Function and the Unit Circle

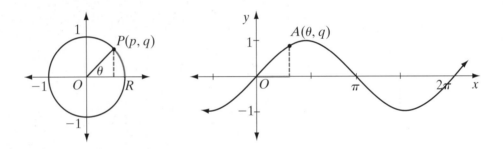

Recall from Chapter 9 that if $\angle ROP$ is an angle in standard position with measure θ and $P(p, q)$ is a point on the unit circle, then $(p, q) = (\cos \theta, \sin \theta)$ and $A(\theta, q)$ is a point on the graph of $y = \sin x$. Note that the x-coordinate of A on the graph of $y = \sin x$ is θ, the length of $\overset{\frown}{RP}$.

Compare the graph of the unit circle and the graph of $y = \sin x$ in the figures below for different values of θ.

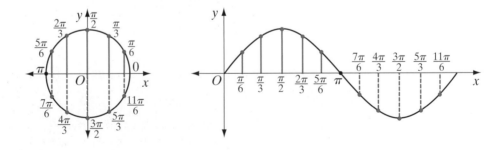

Hands-On Activity: Unwrapping the Unit Circle

We can use the graphing calculator to explore the unit circle and its relationship to the sine and cosine functions.

STEP 1. Press **MODE**. Select RADIAN mode, PAR graphing mode, and SIMUL graphing mode.

STEP 2. Press **WINDOW** to enter the window screen. Use the following viewing window:

$Tmin = 0$, $Tmax = 2\pi$, $Tstep = 0.1$, $Xmin = -1$, $Xmax = 2\pi$, $Xscl = \frac{\pi}{6}$, $Ymin = -2.5$, $Ymax = 2.5$

```
WINDOW
 Tmin=0
 Tmax=6.2831853…
 Tstep=.1
 Xmin=-1
 Xmax=6.2831853…
 Xscl=.52359877…
↓Ymin=-2.5
```

```
WINDOW
↑Tstep=.1
 Xmin=-1
 Xmax=6.2831853…
 Xscl=.52359877…
 Ymin=-2.5
 Ymax=2.5
 Yscl=1
```

Use the ▲ and ▼ arrow keys to display *Ymin* and *Ymax*.

STEP 3. Recall from Chapter 9 that a point P on the unit circle has coordinates $(\cos\theta, \sin\theta)$ where θ is the measure of the standard angle with terminal side through P. We can define a function on the graphing calculator that consists of the set of ordered pairs $(\cos\theta, \sin\theta)$. Its graph will be the unit circle.

```
PLOT1 PLOT2 PLOT3
\X1T ＝cos(T)
 Y1T ＝sin(T)
\X2T ＝T
 Y2T ＝sin(T)
\X3T =
 Y3T =
\X4T =
```

ENTER: **Y=** **COS** **X,T,θ,n** **)** **▼** **SIN** **X,T,θ,n** **)**

This key sequence defines the function consisting of the set of ordered pairs $(\cos T, \sin T)$. The variable T represents θ on the graphing calculator.

STEP 4. Similarly, we can define a function consisting of the set of ordered pairs $(\theta, \sin\theta)$.

ENTER: **▼** **X,T,θ,n** **▼** **SIN** **X,T,θ,n** **)**

STEP 5. Press GRAPH to watch the unit circle "unwrap" into the sine function. As the two functions are plotted, press ENTER to pause and resume the animation. You will see the unit circle swept out by point P (represented by a dot). Simultaneously, the graph of $y = \sin \theta$ will be plotted to the right of the y-axis by a point R. Notice that as point P is rotated about the unit circle, the x-coordinate of R is equal to θ, the length of the arc from the positive ray of the x-axis to P. The y-coordinate of R is equal to the y-coordinate of P.

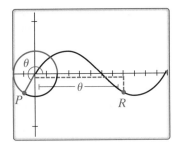

Exercises

Writing About Mathematics

1. Is the graph of $y = \sin x$ symmetric with respect to a reflection in the origin? Justify your answer.

2. Is the graph of $y = \sin x$ symmetric with respect to the translation $T_{-2\pi,0}$? Justify your answer.

Developing Skills

3. Sketch the graph of $y = \sin x$ in the interval $0 \leq x \leq 4\pi$.

    a. In the interval $0 \leq x \leq 4\pi$, for what values of x is the graph of $y = \sin x$ increasing?

    b. In the interval $0 \leq x \leq 4\pi$, for what values of x is the graph of $y = \sin x$ decreasing?

    c. How many cycles of the graph of $y = \sin x$ are in the interval $0 \leq x \leq 4\pi$?

4. What is the maximum value of y on the graph of $y = \sin x$?

5. What is the minimum value of y on the graph of $y = \sin x$?

6. What is the period of the sine function?

7. Is the sine function one-to-one? Justify your answer.

8. a. Point P is a point on the unit circle. The y-coordinate of P is $\sin \frac{\pi}{3}$. What is the x-coordinate of P?

    b. Point A is a point on the graph $y = \sin x$. The y-coordinate of A is $\sin \frac{\pi}{3}$. What is the x-coordinate of A?

Applying Skills

9. A function f is **odd** if and only if $f(x) = -f(-x)$ for all x in the domain of the function. Note that a function is odd if it is symmetric with respect to the origin. In other words, the function is its own image under a reflection about the origin.

a. Draw a unit circle and any first-quadrant angle ROP in standard position, with point P on the unit circle. Let m$\angle ROP = \theta$.

b. On the same set of axes, draw an angle in standard position with measure $-\theta$. What is the relationship between θ and $-\theta$? Between $\sin \theta$ and $-\sin (-\theta)$?

c. Repeat steps **a** and **b** for second-, third-, and fourth-quadrant angles. Does $\sin \theta = -\sin (-\theta)$ for second-, third-, and fourth-quadrant angles? Justify your answer.

d. Does $\sin \theta = -\sin (-\theta)$ for quadrantal angles? Explain.

e. Do parts **a–d** show that $y = \sin x$ is an odd function? Justify your answer.

10. City firefighters are told that they can use their 25-foot long ladder provided the measure of the angle that the ladder makes with the ground is at least 15° and no more than 75°.

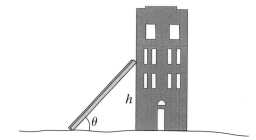

a. If θ represents the measure of the angle that the ladder makes with the ground *in radians*, what is a reasonable set of values for θ? Explain.

b. Express as a function of θ, the height h of the point at which the ladder will rest against a building.

c. Graph the function from part **b** using the set of values for θ from part **a** as the domain of the function.

d. What is the highest point that the ladder is allowed to reach?

11. In later courses, you will learn that the sine function can be written as the sum of an infinite sequence. In particular, for x in radians, the sine function can be approximated as the finite series:

$$\sin x \approx x - \tfrac{x^3}{3!} + \tfrac{x^5}{5!}$$

a. Graph $Y_1 = \sin x$ and $Y_2 = x - \tfrac{x^3}{3!} + \tfrac{x^5}{5!}$ on the graphing calculator. For what values of x does Y_2 seem to be a good approximation for Y_1?

b. The next term of the sine approximation is $-\tfrac{x^7}{7!}$. Repeat part **a** using Y_1 and $Y_3 = x - \tfrac{x^3}{3!} + \tfrac{x^5}{5!} - \tfrac{x^7}{7!}$. For what values of x does Y_3 seem to be a good approximation for Y_1?

c. Use Y_2 and Y_3 to find approximations to the sine function values below. Which function gives a better approximation? Is this what you expected? Explain.

(1) $\sin \tfrac{\pi}{6}$ (2) $\sin \tfrac{\pi}{4}$ (3) $\sin \pi$

11-2 GRAPH OF THE COSINE FUNCTION

The cosine function, like the sine function, is a set of ordered pairs of real numbers. Each ordered pair can be represented as a point of the coordinate plane. The domain of the cosine function is the set of real numbers, that is, every real number is a first element of one pair of the function.

To sketch the graph of the cosine function, we plot a portion of the graph using a subset of the real numbers in the interval $0 \leq x \leq 2\pi$. We know that

$$\cos \frac{\pi}{6} = \frac{\sqrt{3}}{2} = 0.866025 \ldots$$

and that $\frac{\pi}{6}$ is the measure of the reference angle for angles with measures of $\frac{5\pi}{6}$, $\frac{7\pi}{6}, \frac{11\pi}{6}, \ldots$. We can round the rational approximation of $\cos \frac{\pi}{6}$ to two decimal places, 0.87.

We also know that

$$\cos \frac{\pi}{3} = \frac{1}{2} = 0.5$$

and that $\frac{\pi}{3}$ is the measure of the reference angle for angle with measures of $\frac{2\pi}{3}$, $\frac{4\pi}{3}, \frac{5\pi}{3}, \ldots$.

x	0	$\frac{\pi}{6}$	$\frac{\pi}{3}$	$\frac{\pi}{2}$	$\frac{2\pi}{3}$	$\frac{5\pi}{6}$	π	$\frac{7\pi}{6}$	$\frac{4\pi}{3}$	$\frac{3\pi}{2}$	$\frac{5\pi}{3}$	$\frac{11\pi}{6}$	2π
cos x	1	0.87	0.5	0	−0.5	−0.87	−1	−0.87	−0.5	0	0.5	0.87	1

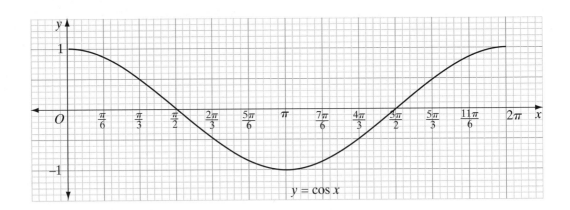

$y = \cos x$

On the graph, we plot the points whose coordinates are given in the table. Through these points, we draw a smooth curve. Note how x and y change.

- As x increases from 0 to $\frac{\pi}{2}$, y decreases from 1 to 0.
- As x increases from $\frac{\pi}{2}$ to π, y decreases from 0 to −1.

- As x increases from π to $\frac{3\pi}{2}$, y increases from -1 to 0.

- As x increases from $\frac{3\pi}{2}$ to 2π, y increases from 0 to 1.

When we plot a larger subset of the domain of the cosine function, this pattern is repeated. For example, add to the points given above the point whose x-coordinates are in the interval $-2\pi \le x \le 0$.

x	-2π	$-\frac{11\pi}{6}$	$-\frac{5\pi}{3}$	$-\frac{3\pi}{2}$	$-\frac{4\pi}{3}$	$-\frac{7\pi}{6}$	$-\pi$	$-\frac{5\pi}{6}$	$-\frac{2\pi}{3}$	$-\frac{\pi}{2}$	$-\frac{\pi}{3}$	$-\frac{\pi}{6}$	0
cos x	1	0.87	0.5	0	-0.5	-0.87	-1	-0.87	-0.5	0	0.5	0.87	1

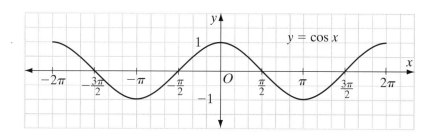

Each time we change the value of the x-coordinates by a multiple of 2π, the basic cosine curve is repeated. Each portion of the graph in an interval of 2π is one cycle of the cosine function.

The graph of the function $y = \cos x$ is its own image under the translation $T_{2\pi,0}$. The function $y = \cos x$ is a periodic function with a period of 2π because for every x in the domain of the cosine function, $\cos x = \cos (x + 2\pi)$.

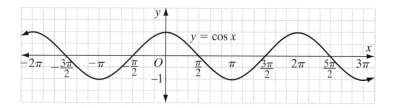

▶ **The period of the cosine function $y = \cos x$ is 2π.**

Each cycle of the cosine curve can be separated into four quarters. In the first quarter, the cosine curve decreases from the maximum value of the function to 0. In the second quarter, it decreases from 0 to the minimum value. In the third quarter, it increases from the minimum value to 0, and in the fourth quarter, it increases from 0 to the maximum value.

A graphing calculator will display the graph of the cosine function.

STEP 1. Put the calculator in radian mode.

STEP 2. Enter the equation for the cosine function.

ENTER: `Y=` `COS` `X,T,θ,n` `ENTER`

```
PLOT1 PLOT2 PLOT3
\Y₁■cos(X
\Y₂=
```

STEP 3. To display one cycle of the curve, let the window include values from 0 to 2π for x and values slightly smaller than -1 and larger than 1 for y. Use the following viewing window: $Xmin = 0$, $Xmax = 2\pi$, $Xscl = \frac{\pi}{6}$, $Ymin = -1.5$, $Ymax = 1.5$.

```
WINDOW
  Xmin=0
  Xmax=6.2831853...
  Xscl=.52359877...
  Ymin=-1.5
  Ymax=1.5
  Yscl=1
  Xres=1
```

ENTER: `WINDOW` 0 `ENTER` 2 `2nd` `π` `ENTER`

`2nd` `π` `÷` 6 `ENTER` -1.5 `ENTER` 1.5 `ENTER`

STEP 4. Finally, graph the sin curve by pressing `GRAPH`. To display more than one cycle of the curve, change *Xmin* or *Xmax* of the window.

ENTER: `WINDOW` -2 `2nd` `π` `ENTER` 4 `2nd` `π`

`ENTER` `GRAPH`

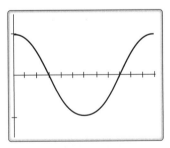

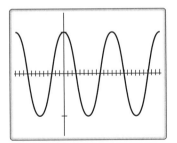

EXAMPLE 1

For what values of x in the interval $0 \leq x \leq 2\pi$ does $y = \cos x$ have a maximum value and for what values of x does it have a minimum value?

Solution The graph shows that $y = \cos x$ has a maximum value, 1, at $x = 0$ and at $x = 2\pi$ and has a minimum value, -1, at $x = \pi$.

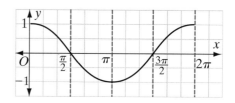

The Graph of the Cosine Function and the Unit Circle

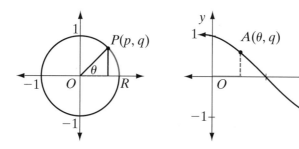

If $\angle ROP$ is an angle in standard position with measure θ and $P(p, q)$ is a point on the unit circle, then $(p, q) = (\cos \theta, \sin \theta)$ and $A(\theta, p)$ is a point on the graph of $y = \cos x$. Note that the x-coordinate of A on the graph of $y = \cos x$ is θ, the length of $\overset{\frown}{RP}$.

Compare the graph of the unit circle and the graph of $y = \cos x$ in the figures below for different values of θ.

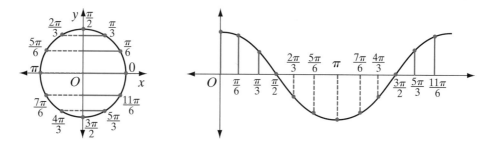

Exercises

Writing About Mathematics

1. Is the graph of $y = \cos x$ its own image under a reflection in the y-axis? Justify your answer.

2. Is the graph of $y = \cos x$ its own image under the translation $T_{-2\pi,0}$? Justify your answer.

Developing Skills

3. Sketch the graph of $y = \cos x$ in the interval $0 \leq x \leq 4\pi$.

   a. In the interval $0 \leq x \leq 4\pi$, for what values of x is the graph of $y = \cos x$ increasing?

   b. In the interval $0 \leq x \leq 4\pi$, for what values of x is the graph of $y = \cos x$ decreasing?

   c. How many cycles of the graph of $y = \cos x$ are in the interval $0 \leq x \leq 4\pi$?

4. What is the maximum value of y on the graph of $y = \cos x$?

5. What is the minimum value of y on the graph of $y = \cos x$?

6. What is the period of the cosine function?

7. Is the cosine function one-to-one? Justify your answer.

Applying Skills

8. A function f is **even** if and only if $f(x) = f(-x)$ for all x in the domain of the function. Note that a function is even if it is symmetric with respect to the y-axis. In other words, a function is even if it is its own image under a reflection about the y-axis.

   a. Draw a unit circle and any first-quadrant angle ROP in standard position, with point P on the unit circle. Let $m\angle ROP = \theta$.

   b. On the same set of axes, draw an angle in standard position with measure $-\theta$. What is the relationship between θ and $-\theta$? Between $\cos \theta$ and $\cos (-\theta)$?

   c. Repeat steps **a** and **b** for second-, third-, and fourth-quadrant angles. Does $\cos \theta = \cos (-\theta)$ for second-, third-, and fourth-quadrant angles? Justify your answer.

   d. Does $\cos \theta = \cos (-\theta)$ for quadrantal angles? Explain.

   e. Do parts **a–d** show that $y = \cos x$ is an even function? Justify your answer.

9. A wheelchair user brings along a 6-foot long portable ramp to get into a van. For safety and ease of wheeling, the ramp should make a 5- to 10-degree angle with the ground.

   a. Let θ represent the measure of the angle that the ramp makes with the ground *in radians*. Express, as a function of θ, the distance d between the foot of ramp and the base of the van on which the ramp sits.

   b. What is the domain of the function from part **a**?

   c. Graph the function from part **a** using the domain found in part **b**.

   d. What is the smallest safe distance from the foot of the ramp to the base of the van?

10. In later courses, you will learn that the cosine function can be written as the sum of an infinite sequence. In particular, for x in radians, the cosine function can be approximated by the finite series:

$$\cos x \approx 1 - \frac{x^2}{2!} + \frac{x^4}{4!}$$

   a. Graph $Y_1 = \cos x$ and $Y_2 = 1 - \frac{x^2}{2!} + \frac{x^4}{4!}$ on the graphing calculator. For what values of x does Y_2 seem to be a good approximation for Y_1?

   b. The next term of the cosine approximation is $-\frac{x^6}{6!}$. Repeat part **a** using Y_1 and $Y_3 = 1 - \frac{x^2}{2!} + \frac{x^4}{4!} - \frac{x^6}{6!}$. For what values of x does Y_3 seem to be a good approximation for Y_1?

c. Use Y_2 and Y_3 to find approximations to the cosine function values below. Which function gives a better approximation? Is this what you expected? Explain.

(1) $\cos -\frac{\pi}{6}$ (2) $\cos -\frac{\pi}{4}$ (3) $\cos -\pi$

11-3 AMPLITUDE, PERIOD, AND PHASE SHIFT

In Chapter 4 we saw that the functions $a\mathrm{f}(x)$, $\mathrm{f}(ax)$, and $\mathrm{f}(x) + a$ are transformations of the function $\mathrm{f}(x)$. Each of these transformations can be applied to the sine function and the cosine function.

Amplitude

How do the functions $y = 2\sin x$ and $y = \frac{1}{2}\sin x$ compare with the function $y = \sin x$? We will make a table of values and sketch the curves. In the following table, approximate values of irrational values of the sine function are used.

x	0	$\frac{\pi}{6}$	$\frac{\pi}{3}$	$\frac{\pi}{2}$	$\frac{2\pi}{3}$	$\frac{5\pi}{6}$	π	$\frac{7\pi}{6}$	$\frac{4\pi}{3}$	$\frac{3\pi}{2}$	$\frac{5\pi}{3}$	$\frac{11\pi}{6}$	2π
$\sin x$	0	0.5	0.87	1	0.87	0.5	0	-0.5	-0.87	-1	-0.87	-0.5	0
$2\sin x$	0	1	1.73	2	1.73	1	0	-1	-1.73	-2	-1.73	-1	0
$\frac{1}{2}\sin x$	0	0.25	0.43	0.5	0.43	0.25	0	-0.25	-0.43	-0.5	-0.43	-0.25	0

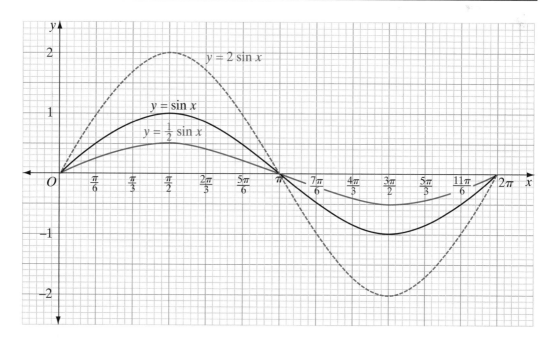

When we use these values to sketch the curves $y = \sin x$, $y = 2 \sin x$, and $y = \frac{1}{2} \sin x$, we see that $y = 2 \sin x$ is the function $y = \sin x$ stretched in the vertical direction and $y = \frac{1}{2} \sin x$ is the function $y = \sin x$ compressed in the vertical direction as expected.

- For $y = \sin x$, the maximum function value is 1 and the minimum function value is -1.

- For $y = 2 \sin x$, the maximum function value is 2 and the minimum function value is -2.

- For $y = \frac{1}{2} \sin x$, the maximum function value is $\frac{1}{2}$ and the minimum function value is $-\frac{1}{2}$.

In general:

▶ **For the function $y = a \sin x$, the maximum function value is $|a|$ and the minimum function value is $-|a|$.**

This is also true for the function $y = a \cos x$.

▶ **For the function $y = a \cos x$, the maximum function value is $|a|$ and the minimum function value is $-|a|$.**

The **amplitude** of a periodic function is the absolute value of one-half the difference between the maximum and minimum y-values.

- For $y = \sin x$, the amplitude is $\left| \frac{1 - (-1)}{2} \right| = 1$.

- For $y = 2 \sin x$, the amplitude is $\left| \frac{2 - (-2)}{2} \right| = 2$.

- For $y = \frac{1}{2} \sin x$, the amplitude is $\left| \frac{\frac{1}{2} - \left(-\frac{1}{2}\right)}{2} \right| = \frac{1}{2}$.

In general:

▶ **For $y = a \sin x$ and $y = a \cos x$, the amplitude is $\left| \frac{a - (-a)}{2} \right| = |a|$.**

EXAMPLE I

For the function $y = 3 \cos x$:

a. What are the maximum and minimum values of the function?

b. What is the range of the function?

c. What is the amplitude of the function?

Solution The range of the function $y = \cos x$ is $-1 \le y \le 1$.

The function $y = 3 \cos x$ is the function $y = \cos x$ stretched by a factor of 3 in the vertical direction.

When $x = 0$, $\cos 0 = 1$, the maximum value, and $y = 3 \cos 0 = 3(1) = 3$.

When $x = \pi$, $\cos \pi = -1$, the minimum value, and $y = 3 \cos \pi = 3(-1) = -3$.

a. The maximum value of the function is 3 and the minimum value is -3. *Answer*

b. The range of $y = 3 \cos x$ is $-3 \le y \le 3$. *Answer*

c. The amplitude of the function is $\left| \frac{3 - (-3)}{3} \right| = 3$. *Answer*

EXAMPLE 2

Describe the relationship between the graph of $y = -4 \sin x$ and the graph of $y = \sin x$ and sketch the graphs.

Solution
- The function $g(x) = 4f(x)$ is the function $f(x)$ stretched by the factor 4 in the vertical direction.
- The function $-g(x)$ is the function $g(x)$ reflected in the x-axis.

  Apply these rules to the sine function.
- The function $y = -4 \sin x$ is the function $y = \sin x$ stretched by the factor 4 in the vertical direction and reflected in the x-axis.
  - The amplitude of $y = -4 \sin x = \dfrac{4 - (-4)}{2} = 4$.

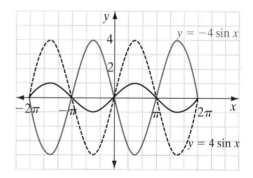

Period

The function $g(x) = f(ax)$ is the function $f(x)$ stretched or compressed by a factor of a in the horizontal direction. Compare the graphs of $y = \sin x$, $y = \sin 2x$, and $y = \sin \frac{1}{2}x$. Consider the maximum, zero, and minimum values of y for one cycle of the graph of $y = \sin x$.

	zero	maximum	zero	minimum	zero
$y = \sin x$	$x = 0$	$x = \frac{\pi}{2}$	$x = \pi$	$x = \frac{3\pi}{2}$	$x = 2\pi$
$y = \sin 2x$	$2x = 0$	$2x = \frac{\pi}{2}$	$2x = \pi$	$2x = \frac{3\pi}{2}$	$2x = 2\pi$
	$x = 0$	$x = \frac{\pi}{4}$	$x = \frac{\pi}{2}$	$x = \frac{3\pi}{4}$	$x = \pi$
$y = \sin \frac{1}{2}x$	$\frac{1}{2}x = 0$	$\frac{1}{2}x = \frac{\pi}{2}$	$\frac{1}{2}x = \pi$	$\frac{1}{2}x = \frac{3\pi}{2}$	$\frac{1}{2}x = 2\pi$
	$x = 0$	$x = \pi$	$x = 2\pi$	$x = 3\pi$	$x = 4\pi$

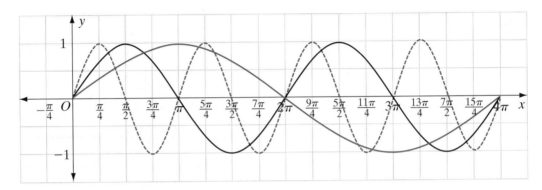

The graph shows the functions $y = \sin x$ (—), $y = \sin 2x$ (---), and $y = \sin \frac{1}{2}x$ (—) in the interval $0 \le x \le 4\pi$. The graph of $y = \sin 2x$ is the graph of $y = \sin x$ compressed by the factor $\frac{1}{2}$ in the horizontal direction. The graph of $y = \sin \frac{1}{2}x$ is the graph of $y = \sin x$ stretched by the factor 2 in the horizontal direction.

- For $y = \sin x$, there is one complete cycle in the interval $0 \le x \le 2\pi$.
- For $y = \sin 2x$, there is one complete cycle in the interval $0 \le x \le \pi$.
- For $y = \sin \frac{1}{2}x$, there is one complete cycle in the interval $0 \le x \le 4\pi$.

The difference between the x-coordinates of the endpoints of the interval for one cycle of the graph is the period of the graph.

- The period of $y = \sin x$ is 2π.
- The period of $y = \sin 2x$ is π.
- The period of $y = \sin \frac{1}{2}x$ is 4π.

In general:

▶ **The period of $y = \sin bx$ and $y = \cos bx$ is $\left|\frac{2\pi}{b}\right|$.**

EXAMPLE 3

a. Use a calculator to sketch the graph of $y = \cos 2x$ in the interval $0 \le x \le 2\pi$.

b. What is the period of $y = \cos 2x$?

Solution **a.** With the calculator in radian mode:

ENTER: Y=   COS   2   X,T,θ,n   DISPLAY:

    ENTER   WINDOW   0   ENTER

2   2nd   π   ENTER   2nd

π   ÷   6   ENTER   −1.5

ENTER   1.5   ENTER   GRAPH

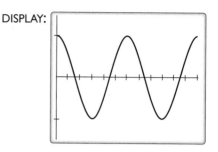

b. The graph shows that there are two cycles of the function in the 2π interval. The period of the graph is $\frac{2\pi}{2} = \pi$. *Answer*

Phase Shift

The graph of $f(x + c)$ is the graph of $f(x)$ moved $|c|$ units to the right when c is negative or $|c|$ units to the left when c is positive. The horizontal translation of a trigonometric function is called a **phase shift**. Compare the graph of $y = \sin x$ and the graph of $y = \sin\left(x + \frac{\pi}{2}\right)$.

x	0	$\frac{\pi}{2}$	π	$\frac{3\pi}{2}$	2π	$\frac{5\pi}{2}$	3π	$\frac{7\pi}{2}$	4π
sin x	0	1	0	−1	0	1	0	−1	0
$x + \frac{\pi}{2}$		$\frac{\pi}{2}$	π	$\frac{3\pi}{2}$	2π	$\frac{5\pi}{2}$	3π	$\frac{7\pi}{2}$	4π	$\frac{9\pi}{2}$
$\sin\left(x + \frac{\pi}{2}\right)$		1	0	−1	0	1	0	−1	0	1

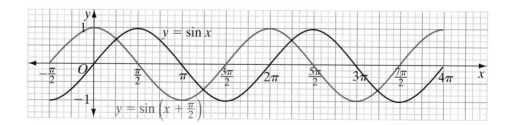

The graph of $y = \sin\left(x + \frac{\pi}{2}\right)$ is the graph of $y = \sin x$ moved $\frac{\pi}{2}$ units to the left.

EXAMPLE 4

Sketch the graph of $y = 3 \sin 2\left(x + \frac{\pi}{6}\right)$.

Solution

How to Proceed

(1) The phase shift is $-\frac{\pi}{6}$. Locate a point on the x-axis that is $\frac{\pi}{6}$ to the left of the origin. This point, $\left(-\frac{\pi}{6}, 0\right)$, is the starting point of one cycle of the sine curve:

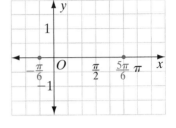

(2) The period is $\frac{2\pi}{2} = \pi$. Locate a point π units to the right of the point in step 1. This point, $\left(\frac{5\pi}{6}, 0\right)$, is the upper endpoint of one cycle of the curve:

(3) Divide the interval for one cycle into four parts of equal length along the x-axis:

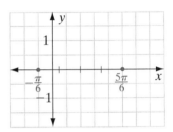

(4) The amplitude of the curve is 3:

(5) The y-coordinates of the sine curve increase from 0 to the maximum in the first quarter of the cycle, decrease from the maximum to 0 in the second quarter, decrease from 0 to the minimum in the third quarter, and increase from the minimum to 0 in the fourth quarter. Sketch the curve:

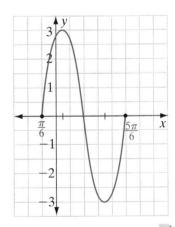

SUMMARY

For the graphs of $y = a \sin b(x + c)$ and $y = a \cos b(x + c)$:

1. The amplitude is $|a|$.

- The maximum value of the function is $|a|$ and the minimum value is $-|a|$.

2. The period of the function is $\left|\frac{2\pi}{b}\right|$.

- There are $|b|$ cycles in a 2π interval.

3. The phase shift is $-c$.

- The graph of $y = \sin (x + c)$ or $y = \cos (x + c)$ is the graph of $y = \sin x$ or $y = \cos x$ shifted c units in the horizontal direction.
    - If c is positive, the graph is shifted $|c|$ units to the left.
    - If c is negative, the graph is shifted $|c|$ units to the right.

4. The graph of $y = -a \sin b(x + c)$ or $y = -a \cos b(x + c)$ is a reflection in the x-axis of $y = a \sin b(x + c)$ or $y = a \cos b(x + c)$.

5. The domain of the function is the set of real numbers.

6. The range of the function is the interval $[-a, a]$.

Exercises

Writing About Mathematics

1. Is the graph of $y = \sin 2\left(x + \frac{\pi}{2}\right)$ the same as the graph of $y = \sin 2\left(x - \frac{\pi}{2}\right)$? Justify your answer.

2. Is the graph of $y = \cos 2\left(x + \frac{\pi}{4}\right)$ the same as the graph of $y = \cos \left(2x + \frac{\pi}{4}\right)$? Justify your answer.

Developing Skills

In 3–10, find the amplitude of each function.

3. $y = \sin x$ **4.** $y = 2 \cos x$ **5.** $y = 5 \cos x$ **6.** $y = 3 \sin x$

7. $y = \frac{3}{4} \sin x$ **8.** $y = \frac{1}{2} \cos x$ **9.** $y = 0.6 \cos x$ **10.** $y = \frac{1}{8} \sin x$

In 11–18, find the period of each function.

11. $y = \sin x$ **12.** $y = \cos x$ **13.** $y = \cos 3x$ **14.** $y = \sin 2x$

15. $y = \cos \frac{1}{2}x$ **16.** $y = \sin \frac{1}{3}x$ **17.** $y = \sin 1.5x$ **18.** $y = \cos 0.75x$

In 19–26, find the phase shift of each function.

19. $y = \cos \left(x + \frac{\pi}{2}\right)$ **20.** $y = \cos \left(x - \frac{\pi}{2}\right)$ **21.** $y = \sin \left(x + \frac{\pi}{3}\right)$

22. $y = \sin \left(x - \frac{\pi}{4}\right)$ **23.** $y = \cos \left(x - \frac{\pi}{6}\right)$ **24.** $y = \sin 2\left(x + \frac{3\pi}{4}\right)$

25. $y = \sin 2(x + \pi)$ **26.** $y = \cos (2x - \pi)$

In 27–38, sketch one cycle of each function.

27. $y = \sin x$ **28.** $y = \cos x$ **29.** $y = \sin 2x$

30. $y = \sin \frac{1}{2}x$ **31.** $y = \cos 3x$ **32.** $y = 3 \cos x$

33. $y = 4 \sin 3x$ **34.** $y = \frac{1}{2} \cos \frac{1}{3}x$ **35.** $y = -\sin 2x$

36. $y = -\cos \frac{1}{2}x$ **37.** $y = \sin\left(x + \frac{\pi}{2}\right)$ **38.** $y = \frac{1}{2} \cos\left(x - \frac{\pi}{4}\right)$

Applying Skills

39. Show that the graph of $y = \sin x$ is the graph of $y = \cos\left(x - \frac{\pi}{2}\right)$.

40. Electromagnetic radiation emitted by a radio signal can be described by the formula

$$e = 0.014 \cos (2\pi ft)$$

where e is in volts, the frequency f is in kilohertz (kHz), and t is time.

a. Graph two cycles of e for $f = 10$ kHz.

b. What is the voltage when $t = 2$ seconds?

41. As stated in the Chapter Opener, sound can be thought of as vibrating air. Simple sounds can be modeled by a function h(t) of the form

$$h(t) = \sin (2\pi ft)$$

where the frequency f is in kilohertz (kHz) and t is time.

a. The frequency of "middle C" is approximately 0.261 kHz. Graph two cycles of h(t) for middle C.

b. The frequency of C_3, or the C note that is one octave lower than middle C, is approximately 0.130 kHz. On the same set of axes, graph two cycles of h(t) for C_3.

c. Based on the graphs from parts **a** and **b**, the periods of each function appear to be related in what way?

42. In 2008, the temperature (in Fahrenheit) of a city can be modeled by:

$$f(t) = 71.3 + 12.1 \sin (0.5t - 1.4)$$

where t represents the number of months that have passed since the first of the year. (For example, $t = 4$ represents the temperature at May 1.)

a. Graph f(t) in the interval $[0, 11]$.

b. Is it reasonable to extend this model to the year 2009? Explain.

Hands-On Activity

1. Sketch the graph of $y = 2 \sin x$ in the interval $[0, 2\pi]$.

2. Sketch the graph of the image of $y = 2 \sin x$ under the translation $T_{0,3}$.

3. Write an equation of the graph drawn in step 2.

4. What are the maximum and minimum values of y for the image of $y = 2 \sin x$ under the translation $T_{0,3}$?

5. The amplitude of the image of $y = 2 \sin x$ under the translation $T_{0,3}$ is 2. How can the maximum and minimum values be used to find the amplitude?

6. Repeat steps 2 through 5 for the translation $T_{0,-4}$.

11-4 WRITING THE EQUATION OF A SINE OR COSINE GRAPH

Each of the graphs that we have studied in this chapter have had an equation of the form $y = a \sin b(x + c)$ or $y = a \cos b(x + c)$. Each of these graphs has a maximum value that is the amplitude, $|a|$, and a minimum, $-|a|$. The positive number b is the number of cycles in an interval of 2π, and $\frac{2\pi}{b}$ is the period or the length of one cycle. For $y = a \sin b(x + c)$, a basic cycle begins at $y = 0$, and for $y = a \cos b(x + c)$, a basic cycle begins at $y = a$. The basic cycle of the graph closest to the origin is contained in the interval $-c \leq x \leq -c + \frac{2\pi}{b}$ and the phase shift is $-c$. Using these values, we can write an equation of the graph.

$$y = a \sin (bx + c)$$

1. Identify the maximum and minimum values of y for the function. Find a.

$$a = \frac{\text{maximum} - \text{mininum}}{2}$$

2. Identify one basic cycle of the sine graph that begins at $y = 0$, increases to the maximum value, decreases to 0, continues to decrease to the minimum value, and then increases to 0. Determine the x-coordinates of the endpoints of this cycle. Write in interval notation the domain of one cycle, $x_0 \leq x \leq x_1$ or $[x_0, x_1]$.

3. The period or the length of one cycle is $\frac{2\pi}{b} = x_1 - x_0$. Find b using this formula.

4. The value of c is the opposite of the lower endpoint of the interval of the basic cycle: $c = -x_0$.

$$y = a \cos (bx + c)$$

1. Identify the maximum and minimum values of y for the function. Find a.

$$a = \frac{\text{maximum} - \text{mininum}}{2}$$

2. Identify one basic cycle of the graph that begins at the maximum value, decreases to 0, continues to decrease to the minimum value, increases to 0, and then increases to the maximum value. Find the x-coordinates of the endpoints of this cycle. Write in interval notation the domain of one cycle, $x_0 \leq x \leq x_1$ or $[x_0, x_1]$.

3. The period or the length of one cycle is $\frac{2\pi}{b} = x_1 - x_0$. Find b using this formula.

4. The value of c is the opposite of the lower endpoint of the interval of the basic cycle: $c = -x_0$.

EXAMPLE 1

Write an equation of the graph below in the form $y = a \cos bx$.

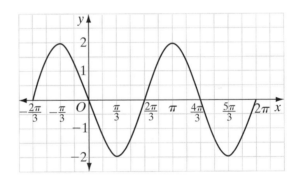

Solution (1) The maximum y value is 2 and the minimum y value is -2.
$$a = \frac{2 - (-2)}{2} = 2$$

(2) There is one cycle of the curve in the interval $-\frac{\pi}{3} \leq x \leq \pi$.

(3) The period is the difference between the endpoints of the interval for one cycle. The period is $\pi - \left(-\frac{\pi}{3}\right)$ or $\frac{4\pi}{3}$. Therefore:
$$\frac{2\pi}{b} = \frac{4\pi}{3}$$
$$4\pi b = 6\pi$$
$$b = \frac{3}{2}$$

(4) The value of c is $-\left(-\frac{\pi}{3}\right)$ or $\frac{\pi}{3}$.

The phase shift is $-\frac{\pi}{3}$.

The equation of the curve is $y = 2 \cos \frac{3}{2}\left(x + \frac{\pi}{3}\right)$. *Answer*

EXAMPLE 2

Write the equation of the graph from Example 1 as a sine function.

Solution (1) The maximum y value is 2 and the minimum y value is -2.

$$a = \frac{2 - (-2)}{2} = 2$$

(2) There is one cycle of the curve in the interval $-\frac{2\pi}{3} \le x \le \frac{2\pi}{3}$.

(3) The period is the difference between the endpoints of the interval for one cycle. The period is $\frac{2\pi}{3} - \left(-\frac{2\pi}{3}\right)$ or $\frac{4\pi}{3}$. Therefore:

$$\frac{2\pi}{b} = \frac{4\pi}{3}$$
$$4\pi b = 6\pi$$
$$b = \frac{3}{2}$$

(4) The value of c is $-\left(-\frac{2\pi}{3}\right)$ or $\frac{2\pi}{3}$.

The phase shift is $-\frac{2\pi}{3}$.

The equation of the curve is $y = 2 \sin \frac{3}{2}\left(x + \frac{2\pi}{3}\right)$. *Answer*

Note: The equations of the sine function and the cosine function differ only in the phase shift.

To determine the phase shift, we usually choose the basic cycle with its lower endpoint closest to zero. For this curve, the interval for the sine function, $\left[\frac{2\pi}{3}, 2\pi\right]$, with the lower endpoint $\frac{2\pi}{3}$ could also have been chosen. The equation could also have been written as $y = 2 \sin \frac{3}{2}\left(x - \frac{2\pi}{3}\right)$.

Exercises

Writing About Mathematics

1. Tyler said that one cycle of a cosine curve has a maximum value at $\left(\frac{\pi}{4}, 5\right)$ and a minimum value at $\left(\frac{5\pi}{4}, -5\right)$. The equation of the curve is $y = 5 \cos\left(2x - \frac{\pi}{2}\right)$. Do you agree with Tyler? Explain why or why not.

2. Is the graph of $y = \sin 2(x + \pi)$ the same as the graph of $y = \sin 2x$? Explain why or why not.

Developing Skills

In 3–14, for each of the following, write the equation of the graph as: **a.** a sine function **b.** a cosine function. In each case, choose the function with the smallest absolute value of the phase shift.

3.

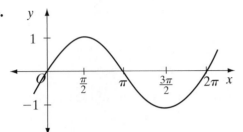

4.

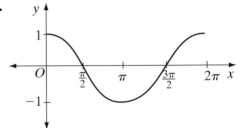

5.

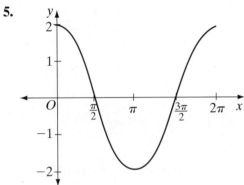

6.

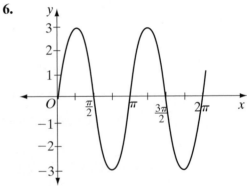

7.

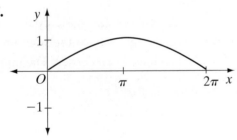

8.

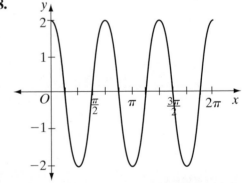

9.

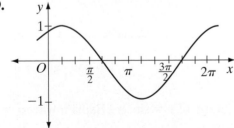

10.

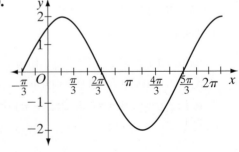

11.

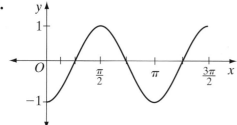

12.

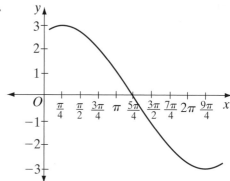

13.

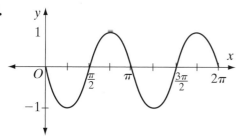

14.

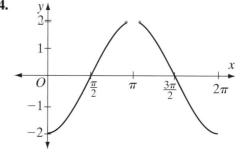

Applying Skills

15. Motion that can be described by a sine or cosine function is called **simple harmonic motion**. During the day, a buoy in the ocean oscillates in simple harmonic motion. The **frequency** of the oscillation is equal to the reciprocal of the period. The distance between its high point and its low point is 1.5 meters. It takes the buoy 5 seconds to move between its low point and its high point, or 10 seconds for one complete oscillation from high point to high point. Let h(t) represent the height of the buoy as a function of time t.

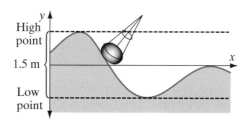

    a. What is the amplitude of h(t)?

    b. What is the period of h(t)?

    c. What is the frequency of h(t)?

    d. If h(0) represents the maximum height of the buoy, write an expression for h(t).

    e. Is there a value of t for which h(t) = 1.5 meters? Explain.

11-5 GRAPH OF THE TANGENT FUNCTION

We can use the table shown below to draw the graph of $y = \tan x$. The values of x are given at intervals of $\frac{\pi}{6}$ from -2π to 2π. The values of $\tan x$ are the approximate decimal values displayed by a calculator, rounded to two decimal places. No value is listed for those values of x for which $\tan x$ is undefined.

x	-2π	$-\frac{11\pi}{6}$	$-\frac{5\pi}{3}$	$-\frac{3\pi}{2}$	$-\frac{4\pi}{3}$	$-\frac{7\pi}{6}$	$-\pi$	$-\frac{5\pi}{6}$	$-\frac{2\pi}{3}$	$-\frac{\pi}{2}$	$-\frac{\pi}{3}$	$-\frac{\pi}{6}$
$\tan x$	0	0.58	1.73	—	-1.73	-0.58	0	0.58	1.73	—	-1.73	-0.58

x	0	$\frac{\pi}{6}$	$\frac{\pi}{3}$	$\frac{\pi}{2}$	$\frac{2\pi}{3}$	$\frac{5\pi}{6}$	π	$\frac{7\pi}{6}$	$\frac{4\pi}{3}$	$\frac{3\pi}{2}$	$\frac{5\pi}{3}$	$\frac{11\pi}{6}$	2π
$\tan x$	0	0.58	1.73	—	-1.73	-0.58	0	0.58	1.73	—	-1.73	-0.58	0

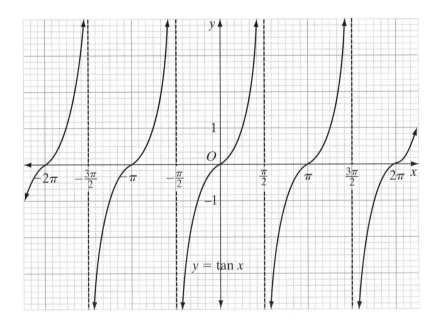

The graph of the tangent function is a curve that increases through negative values of $\tan x$ to 0 and then continues to increase through positive values. At odd multiples of $\frac{\pi}{2}$, the graph is discontinuous and then repeats the same pattern. Since there is one complete cycle of the curve in the interval from $x = -\frac{\pi}{2}$ to $x = \frac{\pi}{2}$, the period of the curve is $\frac{\pi}{2} - \left(-\frac{\pi}{2}\right) = \pi$. The curve is its own image under the transformation $T_{\pi,0}$.

The graph shows a vertical line at $x = \frac{\pi}{2}$ and at every value of x that is an odd multiple of $\frac{\pi}{2}$. These lines are vertical asymptotes. As x approaches $\frac{\pi}{2}$ or any odd multiple of $\frac{\pi}{2}$ from the left, y increases; that is, y approaches infinity. As x

approaches $\frac{\pi}{2}$ or any odd multiple of $\frac{\pi}{2}$ from the right, y decreases; that is, y approaches negative infinity.

Compare the graph shown above with the graph displayed on a graphing calculator.

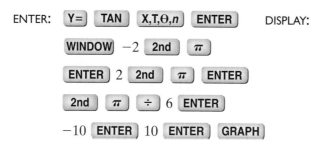 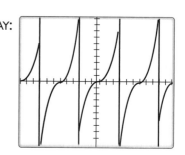

When in connected mode, the calculator displays a line connecting the points on each side of the values of x for which $\tan x$ is undefined. The lines that appear to be vertical lines at $-\frac{3\pi}{2}$, $-\frac{\pi}{2}$, $\frac{\pi}{2}$, and $\frac{3\pi}{2}$ are not part of the graph. On the table given by the graphing calculator, the y-values associated with these x-values are given as ERROR.

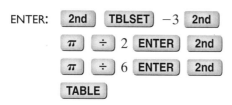 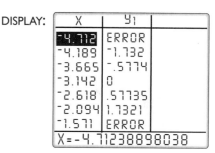

EXAMPLE I

Sketch one cycle of the graph of $y = \tan\left(x - \frac{\pi}{4}\right)$.

Solution The graph of $y = \tan\left(x - \frac{\pi}{4}\right)$ is the graph of $y = \tan x$ with a phase shift of $\frac{\pi}{4}$. Since there is one cycle of $y = \tan x$ in the interval $-\frac{\pi}{2} < x < \frac{\pi}{2}$, there will be one cycle of $y = \tan\left(x + \frac{\pi}{4}\right)$ in the interval $-\frac{\pi}{2} + \frac{\pi}{4} < x < \frac{\pi}{2} + \frac{\pi}{4}$, that is, in the interval $-\frac{\pi}{4} < x < \frac{3\pi}{4}$.

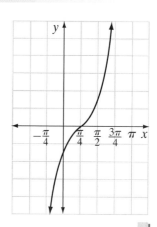

SUMMARY

For the graph of $y = \tan x$:

1. There is no amplitude.
   - The graph has no maximum or minimum values.
2. The period of the function is π.
3. The domain of the function is $\{x : x \neq \frac{\pi}{2} + n\pi \text{ for } n \text{ an integer}\}$.
   - For integral values of n, the graph has vertical asymptotes at $x = \frac{\pi}{2} + n\pi$.
4. The range of the function is {Real numbers} or $(-\infty, \infty)$.

Exercises

Writing About Mathematics

1. List at least three ways in which the graph of the tangent function differs from the graph of the sine function and the cosine function.

2. Does $y = \tan x$ have a maximum and a minimum value? Justify your answer.

Developing Skills

3. Sketch the graph of $y = \tan x$ from $x = -\frac{3\pi}{2}$ to $x = \frac{3\pi}{2}$.
   - **a.** What is the period of $y = \tan x$?
   - **b.** What is the domain of $y = \tan x$?
   - **c.** What is the range of $y = \tan x$?

4. **a.** Sketch the graph of $y = \tan x$ from $x = 0$ to $x = 2\pi$.
   - **b.** On the same set of axes, sketch the graph of $y = \cos x$ from $x = 0$ to $x = 2\pi$.
   - **c.** For how many pairs of values does $\tan x = \cos x$ in the interval $[0, 2\pi]$?

5. **a.** Sketch the graph of $y = \tan x$ from $x = -\frac{\pi}{2}$ to $x = \frac{\pi}{2}$.
   - **b.** Sketch the graph of $y = \tan(-x)$ from $x = -\frac{\pi}{2}$ to $x = \frac{\pi}{2}$.
   - **c.** Sketch the graph of $y = -\tan x$ from $x = -\frac{\pi}{2}$ to $x = \frac{\pi}{2}$.
   - **d.** How does the graph of $y = \tan(-x)$ compare with the graph of $y = -\tan x$?

Applying Skills

6. The volume of a cone is given by the formula

$$V = \tfrac{1}{3}Bh$$

where B is the area of the base and h is the height of the cone.

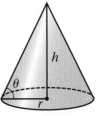

   - **a.** Find a formula for h in terms of r and θ when r is the radius of the base and θ is the measure of the angle that the side of the cone makes with the base.
   - **b.** Use part **a** to write a formula for the volume of the cone in terms of r and θ.

7. Recall from your geometry course that a polygon is *circumscribed* about a circle if each side of the polygon is tangent to the circle. Since each side is tangent to the circle, the radius of the circle is perpendicular to each side at the point of tangency. We will use the tangent function to examine the formula for the perimeter of a circumscribed regular polygon.

a. Let square $ABCD$ be circumscribed about circle O. A radius of the circle, $\overline{OP}$, is perpendicular to $\overline{AB}$ at P.

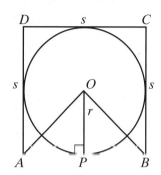

(1) In radians, what is the measure of $\angle AOB$?

(2) Let m$\angle AOP = \theta$. If θ is equal to one-half the measure of $\angle AOB$, find θ.

(3) Write an expression for AP in terms of $\tan \theta$ and r, the radius of the circle.

(4) Write an expression for $AB = s$ in terms of $\tan \theta$ and r.

(5) Use part (4) to write an expression for the perimeter in terms of r and the number of sides, n.

b. Let regular pentagon $ABCDE$ be circumscribed about circle O. Repeat part **a** using pentagon $ABCDE$.

c. Do you see a pattern in the formulas for the perimeter of the square and of the pentagon? If so, make a conjecture for the formula for the perimeter of a circumscribed regular polygon in terms of the radius r and the number of sides n.

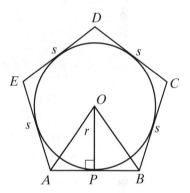

11-6 GRAPHS OF THE RECIPROCAL FUNCTIONS

The Cosecant Function

The cosecant function is defined in terms of the sine function: $\csc x = \frac{1}{\sin x}$. To graph the cosecant function, we can use the reciprocals of the sine function values. The reciprocal of 1 is 1 and the reciprocal of a positive number less than 1 is a number greater than 1. The reciprocal of -1 is -1 and the reciprocal of a negative number greater than -1 is a number less than -1. The reciprocal of 0 is undefined. Reciprocal values of the sine function exist for $-1 \leq \sin x < 0$, and for $0 < \sin x \leq 1$. Therefore:

$$-\infty < \csc x \leq -1 \qquad 1 \leq \csc x < \infty$$

(Recall that the symbol ∞ is called infinity and is used to indicate that a set of numbers has no upper bound. The symbol $-\infty$ indicates that the set of numbers has no lower bound.) For values of x that are multiples of π, $\sin x = 0$ and $\csc x$ is undefined. For integral values of n, the vertical lines on the graph at $x = n\pi$ are asymptotes.

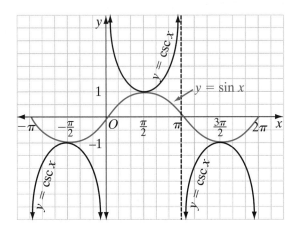

The Secant Function

The secant function is defined in terms of the cosine function: $\sec x = \frac{1}{\cos x}$. To graph the secant function, we can use the reciprocals of the cosine function values. Reciprocal values of the cosine function exist for $-1 \le \cos x < 0$, and for $0 < \cos x \le 1$. Therefore:

$$-\infty < \sec x \le -1 \qquad 1 \le \sec x < \infty$$

For values of x that are odd multiples of $\frac{\pi}{2}$, $\cos x = 0$ and $\sec x$ is undefined. For integral values of n, the vertical lines on the graph at $x = \frac{\pi}{2} + n\pi$ are asymptotes.

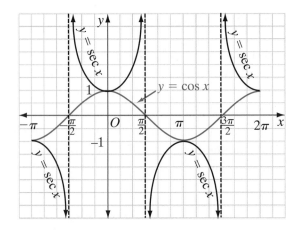

The Cotangent Function

The cotangent function is defined in terms of the tangent function: $\cot x = \frac{1}{\tan x}$.

To graph the cotangent function, we can to use the reciprocals of the tangent function values. For values of x that are multiples of π, $\tan x = 0$ and $\cot x$ is undefined. For values of x for which $\tan x$ is undefined, $\cot x = 0$. For integral values of n, the vertical lines on the graph at $x = n\pi$ are asymptotes.

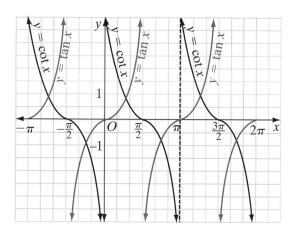

EXAMPLE 1

Use a calculator to sketch the graphs of $y = \sin x$ and $y = \csc x$ for $-2\pi < x < 2\pi$.

Solution There is no key for the cosecant function on the TI graphing calculator. Therefore, the function must be entered in terms of the sine function.

ENTER: Y= SIN X,T,θ,n)

ENTER Y= 1 ÷ SIN

X,T,θ,n) GRAPH

DISPLAY:

SUMMARY

	y = csc x	y = sec x	y = cot x
Amplitude:	none	none	none
Maximum:	$+\infty$	$+\infty$	$+\infty$
Minimum:	$-\infty$	$-\infty$	$-\infty$
Period:	2π	2π	π
Domain:	$\{x : x \neq n\pi\}$	$\left\{x : x \neq \frac{\pi}{2} + n\pi\right\}$	$\{x : x \neq n\pi\}$
Range:	$(-\infty, -1] \cup [1, \infty)$	$(-\infty, -1] \cup [1, \infty)$	$(-\infty, \infty)$

Exercises

Writing About Mathematics

1. If tan x increases for all values of x for which it is defined, explain why cot x decreases for all values of x for which it is defined.

2. In the interval $0 \leq x \leq \pi$, cos x decreases. Describe the change in sec x in the same interval.

Developing Skills

In 3–10, match each graph with its function.

(1)
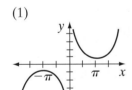

(2)
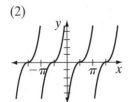

(3)
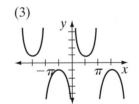

(4)
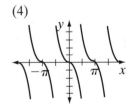

(5)
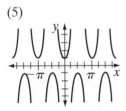

(6)
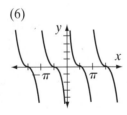

(7)
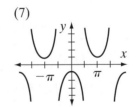

(8)
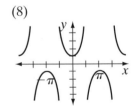

3. $y = \csc x$ **4.** $y = \sec x$ **5.** $y = \cot x$ **6.** $y = -\sec x$

7. $y = \csc \frac{x}{2}$ **8.** $y = \sec 2x$ **9.** $y = -\cot x$ **10.** $y = \cot\left(x + \frac{\pi}{2}\right)$

11. a. Sketch the graphs of $y = \sin x$ and $y = \csc x$ for $-2\pi \leq x \leq 2\pi$.

     b. Name four values of x in the interval $-2\pi \leq x \leq 2\pi$ for which $\sin x = \csc x$.

12. a. Sketch the graphs of $y = \cos x$ and $y = \sec x$ for $-2\pi \le x \le 2\pi$.

   b. Name four values of x in the interval $-2\pi \le x \le 2\pi$ for which $\cos x = \sec x$.

13. a. Sketch the graphs of $y = \tan x$ and $y = \cot x$ for $-\pi \le x \le \pi$.

   b. Name four values of x in the interval $-\pi \le x \le \pi$ for which $\tan x = \cot x$.

14. List two values of x in the interval $-2\pi \le x \le 2\pi$ for which $\sec x$ is undefined.

15. List two values of x in the interval $-2\pi \le x \le 2\pi$ for which $\csc x$ is undefined.

16. List two values of x in the interval $-2\pi \le x \le 2\pi$ for which $\tan x$ is undefined.

17. List two values of x in the interval $-2\pi \le x \le 2\pi$ for which $\cot x$ is undefined.

18. The graphs of which two trigonometric functions have an asymptote at $x = 0$?

19. The graphs of which two trigonometric functions have an asymptote at $x = \frac{\pi}{2}$?

20. Using the graphs of each function, determine whether each function is even, odd, or neither.

   a. $y = \tan x$ **b.** $y = \csc x$

   c. $y = \sec x$ **d.** $y = \cot x$

Applying Skills

21. A rotating strobe light casts its light on the ceiling of the community center as shown in the figure. The light is located 10 feet from the ceiling.

   a. Express a, the distance from the light to its projection on the ceiling, as a function of θ and a reciprocal trigonometric function.

   b. Complete the following table, listing each value to the nearest tenth.

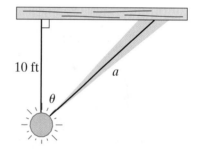

θ	$\frac{\pi}{18}$	$\frac{\pi}{9}$	$\frac{\pi}{6}$	$\frac{2\pi}{9}$
a				

   c. The given values of θ in part **b** increase in equal increments. For these values of θ, do the distances from the light to its projection on the ceiling increase in equal increments? Explain.

   d. The maximum value of θ, in radians, before the light stops shining on the ceiling is $\frac{4\pi}{9}$. To the nearest tenth, how wide is the ceiling?

11-7 GRAPHS OF INVERSE TRIGONOMETRIC FUNCTIONS

The trigonometric functions are not one-to-one functions. By restricting the domain of each function, we were able to write inverse functions in Chapter 10.

Inverse of the Sine Function

The graph of $y = \sin x$ is shown below on the left. When the graph of $y = \sin x$ is reflected in the line $y = x$, the image, $x = \sin y$ or $y = \arcsin x$, is a relation that is not a function. The interval $-\frac{\pi}{2} \leq x \leq \frac{\pi}{2}$ includes all of the values of $\sin x$ from -1 to 1.

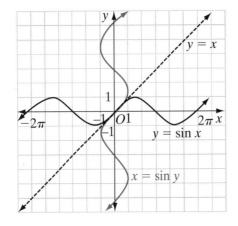
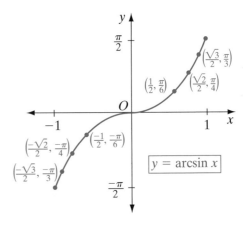

If we restrict the domain of the sine function to $-\frac{\pi}{2} \leq x \leq \frac{\pi}{2}$, that subset of the sine function is a one-to-one function and has an inverse function. When we reflect that subset over the line $y = x$, the image is the function $y = \arcsin x$ or $y = \sin^{-1} x$.

Sine Function with a Restricted Domain
$y = \sin x$
Domain $= \left\{ x : -\frac{\pi}{2} \leq x \leq \frac{\pi}{2} \right\}$
Range $= \{ y : -1 \leq y \leq 1 \}$

Inverse Sine Function
$y = \arcsin x$
Domain $= \{ x : -1 \leq x \leq 1 \}$
Range $= \left\{ y : -\frac{\pi}{2} \leq y \leq \frac{\pi}{2} \right\}$

Inverse of the Cosine Function

The graph of $y = \cos x$ is shown on the top of page 469. When the graph of $y = \cos x$ is reflected in the line $y = x$, the image, $x = \cos y$ or $y = \arccos x$, is a relation that is not a function. The interval $0 \leq x \leq \pi$ includes all of the values of $\cos x$ from -1 to 1.

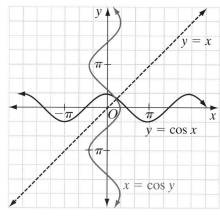

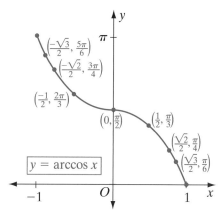

If we restrict the domain of the cosine function to $0 \leq x \leq \pi$, that subset of the cosine function is a one-to-one function and has an inverse function. When we reflect that subset over the line $y = x$, the image is the function $y = \arccos x$ or $y = \cos^{-1} x$.

Cosine Function with a Restricted Domain
$y = \cos x$
Domain $= \{x : 0 \leq x \leq \pi\}$
Range $= \{y : -1 \leq y \leq 1\}$

Inverse Cosine Function
$y = \arccos x$
Domain $= \{x : -1 \leq x \leq 1\}$
Range $= \{y : 0 \leq y \leq \pi\}$

Inverse of the Tangent Function

The graph of $y = \tan x$ is shown below on the left. When the graph of $y = \tan y$ is reflected in the line $y = x$, the image, $x = \tan y$ or $y = \arctan x$, is a relation that is not a function. The interval $-\frac{\pi}{2} < x < \frac{\pi}{2}$ includes all of the values of $\tan x$ from $-\infty$ to ∞.

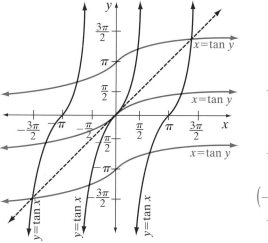

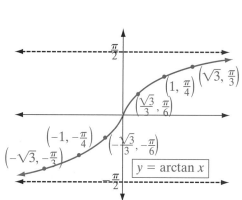

If we restrict the domain of the tangent function to $-\frac{\pi}{2} < x < \frac{\pi}{2}$, that subset of the tangent function is a one-to-one function and has an inverse function. When we reflect that subset over the line $y = x$, the image is the function $y = \arctan x$ or $y = \tan^{-1} x$.

Tangent Function with a Restricted Domain
$y = \tan x$
Domain $= \left\{x : -\frac{\pi}{2} < x < \frac{\pi}{2}\right\}$
Range $= \{y : y \text{ is a real number}\}$

Inverse Tangent Function
$y = \arctan x$
Domain $= \{x : x \text{ is a real number}\}$
Range $= \left\{y : -\frac{\pi}{2} < y < \frac{\pi}{2}\right\}$

If we know that $\sin x = 0.5738$, a calculator will give the value of x in the restricted domain. We can write $x = \arcsin 0.5770$. On the calculator, the **SIN⁻¹** key is used for the arcsin function.

To write the value of x in degrees, change the calculator to degree mode.

ENTER: **MODE** **▼** **▼** **►**
ENTER **CLEAR** **2nd**
SIN⁻¹ 0.5738 **)** **ENTER**

DISPLAY:
```
SIN⁻¹(0.5738)
        35.01563871
```

In degrees, $\sin 35° \approx 0.5738$. This is the value of x for the restricted domain of the sine function. There is also a second quadrant angle whose reference angle is $35°$. That angle is $180° - 35°$ or $145°$. These two measures, $35°$ and $145°$, and any measures that differ from one of these by a complete rotation, $360°$, are a value of x. Therefore, if $\sin x = 0.5738$,

$$x = 35 + 360n \quad \text{or} \quad x = 145 + 360n$$

for all integral values of n.

EXAMPLE I

Find, in radians, *all* values of θ such that $\tan \theta = -1.200$ in the interval $0 \le \theta \le 2\pi$. Express the answer to four decimal places.

Solution If $\tan \theta = -1.200$, $\theta = \arctan - 1.200$.

Set the calculator to radian mode. Then use the **TAN⁻¹** key.

ENTER: **2nd** **TAN⁻¹** −1.200 **)**
ENTER

DISPLAY:
```
TAN⁻¹(-1.200)
       -.8760580506
```

To four decimal places, $\theta = -0.8761$. This is the value of θ in the restricted domain of the tangent function. An angle of -0.8761 radians is a fourth-quadrant angle. The fourth-quadrant angle with the same terminal side and a measure between 0 and 2π is $-0.8761 + 2\pi \approx 5.4070$. The tangent function values are negative in the second and fourth quadrants. There is a second-quadrant angle with the same tangent value. The first-quadrant reference angle has a measure of 0.8761 radians. The second quadrant angle with a reference angle of 0.8761 radians has a measure of $\pi - 0.8761 \approx 2.2655$.

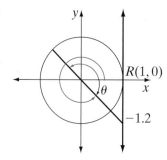

Answer $x \approx 2.2655$ and $x \approx 5.4070$

Exercises

Writing About Mathematics

1. Show that if $\arcsin -\frac{1}{2} = x$, then the measure of the reference angle for x is $30°$.

2. Is $\arctan 1 = 220°$ a true statement? Justify your answer.

Developing Skills

In 3–14, find each exact value in degrees.

3. $y = \arcsin \frac{1}{2}$

4. $y = \arccos \frac{1}{2}$

5. $y = \arctan 1$

6. $y = \arctan \sqrt{3}$

7. $y = \arcsin(-1)$

8. $y = \arccos 0$

9. $y = \arcsin\left(-\frac{\sqrt{3}}{2}\right)$

10. $y = \arccos\left(-\frac{\sqrt{2}}{2}\right)$

11. $y = \arctan(-1)$

12. $y = \arcsin\left(-\frac{\sqrt{2}}{2}\right)$

13. $y = \arctan 0$

14. $y = \arccos(-1)$

In 15–26, find each exact value in radians, expressing each answer in terms of π.

15. $y = \arcsin 1$

16. $y = \arccos 1$

17. $y = \arctan 1$

18. $y = \arcsin\left(\frac{\sqrt{3}}{2}\right)$

19. $y = \arcsin\left(-\frac{\sqrt{3}}{2}\right)$

20. $y = \arccos \frac{1}{2}$

21. $y = \arccos\left(-\frac{1}{2}\right)$

22. $y = \arctan \sqrt{3}$

23. $y = \arctan\left(-\sqrt{3}\right)$

24. $y = \arctan 0$

25. $y = \arccos 0$

26. $y = \arcsin 0$

In 27–32, for each of the given inverse trigonometric function values, find the exact function value.

27. sin (arccos 1)   **28.** cos (arcsin 1)   **29.** tan (arctan 1)

30. sin $\left(\arccos -\frac{\sqrt{3}}{2} \right)$ **31.** sin (arctan -1)   **32.** cos $\left(\arccos -\frac{1}{2} \right)$

33. a. On the same set of axes, sketch the graph of $y = \arcsin x$ and of its inverse function.

    b. What are the domain and range of each of the functions graphed in part **a**?

34. a. On the same set of axes, sketch the graph of $y = \arccos x$ and of its inverse function.

    b. What are the domain and range of each of the functions graphed in part **a**?

35. a. On the same set of axes, sketch the graph of $y = \arctan x$ and of its inverse function.

    b. What are the domain and range of each of the functions graphed in part **a**?

Applying Skills

36. A television camera 100 meters from the starting line is filming a car race, as shown in the figure. The camera will follow car number 2.

    a. Express θ as a function of d, the distance of the car to the starting line.

    b. Find θ when the car is 50 meters from the starting line.

    c. If the finish line is 300 meters away from the starting line, what is the maximum value of θ to the nearest minute?

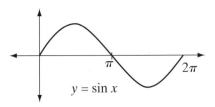

11-8 SKETCHING TRIGONOMETRIC GRAPHS

One cycle of the basic sine curve and of the basic cosine curve are shown to the right. In the previous sections, we have seen how the values of a, b, and c change these curves without changing the fundamental shape of a cycle of the graph. For $y = a \sin b(x + c)$ and for $y = a \cos b(x + c)$:

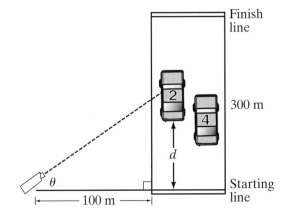

- $|a|$ = amplitude
- $|b|$ = number of cycles in a 2π interval
- $\frac{2\pi}{|b|}$ = period of the graph
- $-c$ = phase shift

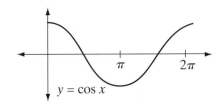

EXAMPLE I

a. Sketch two cycles of the graph of $y = 2 \sin \left(x - \frac{\pi}{4} \right)$ without using a calculator.

b. On the same set of axes, sketch one cycle of the graph of $y = \cos \frac{1}{2}x$ without using a calculator.

c. In the interval $0 \le x \le 4\pi$, how many points do the two curves have in common?

Solution

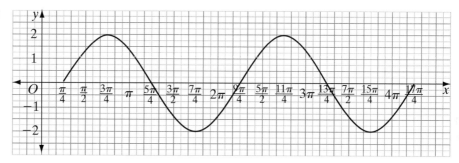

a. For the function $y = 2 \sin \left(x - \frac{\pi}{4} \right)$, $a = 2$, $b = 1$, and $c = -\frac{\pi}{4}$. Therefore, one cycle begins at $x = \frac{\pi}{4}$. There is one complete cycle in the 2π interval, that is from $\frac{\pi}{4}$ to $\frac{9\pi}{4}$. Divide this interval into four equal intervals and sketch one cycle of the sine curve with a maximum of 2 and a minimum of -2.

There will be a second cycle in the interval from $\frac{9\pi}{4}$ to $\frac{17\pi}{4}$. Divide this interval into four equal intervals and sketch one cycle of the sine curve with a maximum of 2 and a minimum of -2.

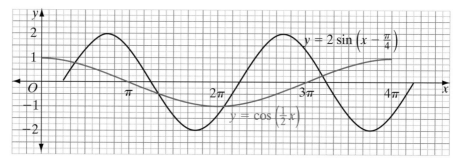

b. For the function $y = \cos \frac{1}{2}x$, $a = 1$, $b = \frac{1}{2}$, and $c = 0$. Therefore, one cycle begins at the origin, $x = 0$. There is one-half of a complete cycle in the 2π interval, so the interval for one cycle is $\frac{2\pi}{\frac{1}{2}} = 4\pi$. Divide the interval from 0 to 4π into four equal intervals and sketch one cycle of the cosine curve with a maximum of 1 and a minimum of -1.

c. The curves have four points in common. *Answer*

Exercises

Writing About Mathematics

1. Calvin said that the graph of $y = \tan\left(x - \frac{\pi}{4}\right)$ has asymptotes at $x = \frac{3\pi}{4} + n\pi$ for all integral values of n. Do you agree with Calvin? Explain why or why not.

2. Is the graph of $y = \sin\left(2x - \frac{\pi}{4}\right)$ the graph of $y = \sin 2x$ moved $\frac{\pi}{4}$ units to the right? Explain why or why not.

Developing Skills

In 3–14, sketch one cycle of the graph.

3. $y = 2 \sin x$

4. $y = 3 \sin 2x$

5. $y = \cos 3x$

6. $y = 2 \sin \frac{1}{2}x$

7. $y = 4 \cos 2x$

8. $y = 3 \sin\left(x - \frac{\pi}{3}\right)$

9. $y = \cos 2\left(x + \frac{\pi}{6}\right)$

10. $y = 4 \sin(x - \pi)$

11. $y = \tan x$

12. $y = \tan\left(x - \frac{\pi}{2}\right)$

13. $y = -2 \sin x$

14. $y = -\cos x$

In 15–20, for each of the following, write the equation of the graph as: **a.** a sine function **b.** a cosine function. In each case, choose the function with the smallest absolute value of the phase shift.

15.

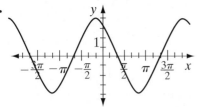

16.

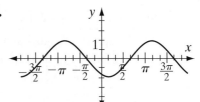

17.

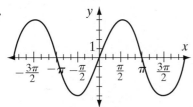

18.

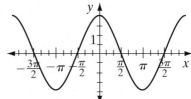

19.

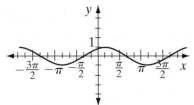

20.

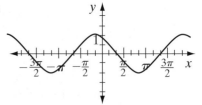

21. a. On the same set of axes, sketch the graphs of $y = 2 \sin x$ and $y = \cos x$ in the interval $0 \le x \le 2\pi$.

b. How many points do the graphs of $y = 2 \sin x$ and $y = \cos x$ have in common in the interval $0 \le x \le 2\pi$?

22. a. On the same set of axes, sketch the graphs of $y = \tan x$ and $y = \cos\left(x + \frac{\pi}{2}\right)$ in the interval $-\frac{\pi}{2} \le x \le \frac{3\pi}{2}$.

  b. How many points do the graphs of $y = \tan x$ and $y = \cos\left(x + \frac{\pi}{2}\right)$ have in common in the interval $-\frac{\pi}{2} \le x \le \frac{3\pi}{2}$?

23. a. On the same set of axes, sketch the graphs of $y = \sin 3x$ and $y = 2\cos 2x$ in the interval $0 \le x \le 2\pi$.

  b. How many points do the graphs of $y = \sin 3x$ and $y = 2\cos 2x$ have in common in the interval $0 \le x \le 2\pi$?

CHAPTER SUMMARY

	Domain (n an integer)	Range	Period	$\left(0, \frac{\pi}{2}\right)$	$\left(\frac{\pi}{2}, \pi\right)$	$\left(\pi, \frac{3\pi}{2}\right)$	$\left(\frac{3\pi}{2}, 2\pi\right)$
$y = \sin x$	All real numbers	$[-1, 1]$	2π	increase	decrease	decrease	increase
$y = \cos x$	All real numbers	$[-1, 1]$	2π	decrease	decrease	increase	increase
$y = \tan x$	$x \ne \frac{\pi}{2} + n\pi$	All real numbers	π	increase	increase	increase	increase
$y = \csc x$	$x \ne n\pi$	$(-\infty, -1] \cup [1, \infty)$	2π	decrease	increase	increase	decrease
$y = \sec x$	$x \ne \frac{\pi}{2} + n\pi$	$(-\infty, -1] \cup [1, \infty)$	2π	increase	increase	decrease	decrease
$y = \cot x$	$x \ne n\pi$	All real numbers	π	decrease	decrease	decrease	decrease

For $y = a \sin b(x + c)$ or $y = a \cos b(x + c)$:

- The amplitude, $|a|$, is the maximum value of the function, and $-|a|$ is the minimum value.

- The number, $|b|$, is the number of cycles in the 2π interval, the period, $\frac{2\pi}{|b|}$, is the length of the interval for one cycle, and the frequency, $\frac{|b|}{2\pi}$, is the reciprocal of the period.

- The phase shift is $-c$. If c is positive, the graph is shifted $|c|$ units to the left. If c is negative, the graph is shifted $|c|$ units to the right.

The graphs of the trigonometric functions are periodic curves. Each graph of $y = a \sin b(x + c)$ or of $y = a \cos b(x + c)$ is its own image under the translation $(x, y) \to \left(x, y + \frac{2\pi}{|b|}\right)$. The graph of $y = \tan x$ is its own image under the translation $(x, y) \to (x, y + \pi)$.

When the domain of a trigonometric function is restricted to a subset for which the function is one-to-one, the function has an inverse function.

Function	Restricted Domain	Inverse Function
$y = \sin x$	$-\frac{\pi}{2} \leq x \leq \frac{\pi}{2}$	$y = \arcsin x$ or $y = \sin^{-1} x$
$y = \cos x$	$0 \leq x \leq \pi$	$y = \arccos x$ or $y = \cos^{-1} x$
$y = \tan x$	$-\frac{\pi}{2} < x < \frac{\pi}{2}$	$y = \arctan x$ or $y = \tan^{-1} x$

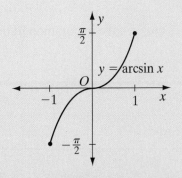

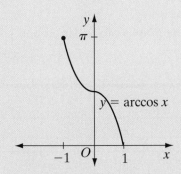

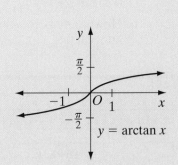

VOCABULARY

11-1 Cycle • Periodic function • Period • Odd function

11-2 Even function

11-3 Amplitude • Phase shift

11-4 Simple harmonic motion • Frequency

REVIEW EXERISES

In 1–6, for each function, state: **a.** the amplitude **b.** the period **c.** the frequency **d.** the domain **e.** the range. **f.** Sketch one cycle of the graph.

1. $y = 2 \sin 3x$ **2.** $y = 3 \cos \frac{1}{2}x$ **3.** $y = \tan x$

4. $y = \cos 2\left(x - \frac{\pi}{3}\right)$ **5.** $y = \sin(x + \pi)$ **6.** $y = -2 \cos x$

In 7–10, for each graph, write the equation in the form: **a.** $y = a \sin b(x + c)$ **b.** $y = a \cos b(x + c)$. In each case, choose one cycle with its lower endpoint closest to zero to find the phase shift.

7.

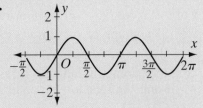

8.

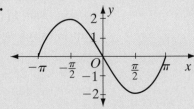

9.

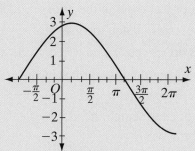

10.

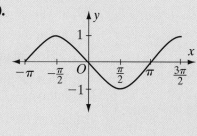

In 11–14, match each graph with its function.

(1)

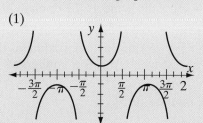

(2)

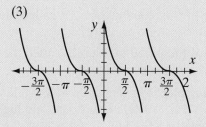

(3)

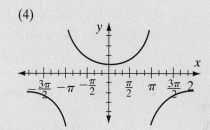

(4)

11. $y = \sec \frac{x}{2}$

12. $y = \csc \left(x + \frac{\pi}{2} \right)$

13. $y = -\cot \left(x + \frac{\pi}{2} \right)$

14. $y = -\tan \left(x - \frac{\pi}{2} \right)$

In 15–20, find, in radians, the exact value of y for each trigonometric function.

15. $y = \arcsin \frac{1}{2}$

16. $y = \sin^{-1} 1$

17. $y = \arctan \frac{\sqrt{3}}{3}$

18. $y = \arccos \left(-\frac{\sqrt{3}}{2} \right)$

19. $y = \cos^{-1} \frac{\sqrt{2}}{2}$

20. $y = \arctan \left(-\frac{\sqrt{3}}{3} \right)$

21. a. What is the restricted domain for which $y = \sin x$ is a one-to-one function?

b. What is the domain of the function $y = \arcsin x$?

c. What is the range of the function $y = \arcsin x$?

d. Sketch the graph of the function $y = \arcsin x$.

22. a. Sketch the graph of $y = \cos x$ in the interval $-2\pi \le x \le 2\pi$.

   b. On the same set of axes, sketch the graph of $y = \csc x$.

   c. How many points do $y = \cos x$ and $y = \csc x$ have in common?

23. What are the equations of the asymptotes of the graph of $y = \tan x$ in the interval $-2\pi < x < 2\pi$?

24. a. On the same set of axes, sketch the graph of $y = 2 \cos x$ and $y = \sin \frac{1}{2}x$ in the interval $-\pi \le x \le \pi$.

   b. From the graph, determine the number of values of x for which $\cos x = \sin \frac{1}{2}x$ in the interval $-\pi \le x \le \pi$.

25. Is the domain of $y = \csc x$ the same as the domain of $y = \sin x$? Explain why or why not.

26. The function $p(t) = 85 + 25 \sin (2\pi t)$ approximates the blood pressure of Mr. Avocado while at rest where $p(t)$ is in milligrams of mercury (mmHg) and t is in seconds.

   a. Graph $p(t)$ in the interval $[0, 3]$.

   b. Find the period of $p(t)$.

   c. Find the amplitude of $p(t)$.

   d. The higher value of the blood pressure is called the *systolic pressure*. Find Mr. Avocado's systolic pressure.

   e. The lower value of the blood pressure is called the *diastolic pressure*. Find Mr. Avocado's diastolic pressure.

27. The water at a fishing pier is 11 feet deep at low tide and 20 feet deep at high tide. On a given day, low tide is at 6 A.M. and high tide is at 1 P.M. Let $h(t)$ represent the height of the tide as a function of time t.

   a. What is the amplitude of $h(t)$?

   b. What is the period of $h(t)$?

   c. If $h(0)$ represents the height of the tide at 6 A.M., write an expression for $h(t)$.

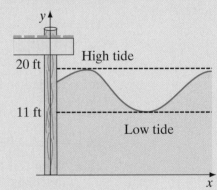

Exploration

Natural phenomena often occur in a cyclic pattern that can be modeled by a sine or cosine function. For example, the time from sunrise to sunset for any given latitude is a maximum at the beginning of summer and a minimum at the beginning of winter. If we plot this difference at weekly intervals for a year, beginning with the first day of summer, the curve will closely resemble a cosine curve after the translation $T_{0,d}$. (A translation $T_{0,d}$ moves the graph of a function d units in the vertical direction.) The equation of the cosine curve can then be written as $y = a \cos b(x + c) + d$.

STEP 1. Research in the library or on the Internet to find the time of sunrise and sunset at weekly intervals. Let the week of June 21 be week 0 and the week of June 21 for the next year be week 52. Round the time from sunrise to sunset to the nearest quarter hour. For example, let 14 hours 12 minutes be $14\frac{1}{4}$ hours and 8 hours 35 minutes be $8\frac{1}{2}$ hours.

STEP 2. Plot the data.

STEP 3. What is the amplitude that most closely approximates the data?

STEP 4. What is the period that most closely approximates the data?

STEP 5. Let d equal the average of the maximum and minimum values of the data or:

$$d = \frac{\text{maximum} + \text{minimum}}{2}$$

Find an approximate value for d.

STEP 6. Write a cosine function of the form $y = a \cos b(x + c) + d$ that approximates the data.

CUMULATIVE REVIEW   CHAPTERS 1–11

Part I

Answer all questions in this part. Each correct answer will receive 2 credits. No partial credit will be allowed.

1. The sum $\sqrt{-25} + \sqrt{-9}$ is equal to
(1) $\sqrt{-34}$ (2) $-8i$ (3) $8i$ (4) $34i$

2. The solution set of $|2x + 2| - 4 = 0$ is
(1) $\varnothing$ (2) $\{1\}$ (3) $\{1, -1\}$ (4) $\{1, -3\}$

3. In radians, $225°$ is equivalent to
(1) $\frac{\pi}{4}$ (2) $\frac{3\pi}{4}$ (3) $\frac{5\pi}{4}$ (4) $\frac{7\pi}{4}$

4. Which of the following is a geometric sequence?
(1) $1, 2, 4, 7, 11, \ldots$ (3) $1, 1, 2, 3, 5, \ldots$
(2) $1, 2, 3, 4, 5, \ldots$ (4) $1, 2, 4, 8, 16, \ldots$

5. Which of the following functions is one-to-one?
(1) $f(x) = 2x^2$ (3) $f(x) = |2x|$
(2) $f(x) = 2^x$ (4) $f(x) = 2 \tan x$

6. When written with a rational denominator, $\dfrac{3}{2 - \sqrt{2}}$ is equal to

(1) $\dfrac{3(2 + \sqrt{2})}{2}$

(3) $3(1 + \sqrt{2})$

(2) $\dfrac{3(2 - \sqrt{2})}{2}$

(4) $\dfrac{2 + \sqrt{2}}{2}$

7. The solution set of $2x^2 + 5x - 3 = 0$ is

(1) $\left\{\tfrac{1}{2}, 3\right\}$ (2) $\left\{\tfrac{1}{2}, -3\right\}$ (3) $\left\{-\tfrac{1}{2}, 3\right\}$ (4) $\left\{-\tfrac{1}{2}, -3\right\}$

8. If $g(x) = x^2$ and $f(x) = 2x + 1$, then $g(f(x))$ equals

(1) $(2x + 1)^2$

(3) $(2x + 1)(x^2)$

(2) $2x^2 + 1$

(4) $x^2 + 2x + 1$

9. If $\theta = \arcsin\left(-\dfrac{\sqrt{3}}{2}\right)$, then in radians, θ is equal to

(1) $-\dfrac{\pi}{3}$ (2) $-\dfrac{\pi}{6}$ (3) $\dfrac{\pi}{3}$ (4) $\dfrac{2\pi}{3}$

10. The coordinates of the center of the circle $(x + 3)^2 + (y - 2)^2 = 9$ are

(1) $(3, -2)$

(3) $(-3, 2)$

(2) $(3, 2)$

(4) $(-3, -2)$

Part II

Answer all questions in this part. Each correct answer will receive 2 credits. Clearly indicate the necessary steps, including appropriate formula substitutions, diagrams, graphs, charts, etc. For all questions in this part, a correct numerical answer with no work shown will receive only 1 credit.

11. Solve for x and graph the solution set on the number line:

$$|2x - 5| < 7$$

12. Find, to the nearest degree, all values of θ in the interval $0 \le \theta \le 360$ for which $\tan \theta = -1.54$.

Part III

Answer all questions in this part. Each correct answer will receive 4 credits. Clearly indicate the necessary steps, including appropriate formula substitutions, diagrams, graphs, charts, etc. For all questions in this part, a correct numerical answer with no work shown will receive only 1 credit.

13. Express the roots of $x^3 - 8x^2 + 25x = 0$ in simplest form.

14. Find the value of $\displaystyle\sum_{n=0}^{5} 3(2)^{n-1}$.

Part IV

Answer all questions in this part. Each correct answer will receive 6 credits. Clearly indicate the necessary steps, including appropriate formula substitutions, diagrams, graphs, charts, etc. For all questions in this part, a correct numerical answer with no work shown will receive only 1 credit.

15. *Given: ABCDEFGH* is a cube with sides of length 1.

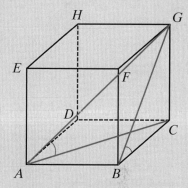

  a. Find the exact measure of $\angle GBC$, the angle formed by diagonal $\overline{BG}$ and side $\overline{BC}$.

  b. Find, to the nearest degree, the measure of $\angle GAC$, the angle formed by diagonals $\overline{AG}$ and $\overline{AC}$.

16. Find the solution set of the following system of equations algebraically.

$$y = 2x^2 - 3x - 5$$
$$2x - y = 7$$

TRIGONOMETRIC IDENTITIES

When a busy street passes through the business center of a town, merchants want to insure maximum parking in order to make stopping to shop convenient. The town planners must decide whether to allow parallel parking (parking parallel to the curb) on both sides of the street or angle parking (parking at an angle with the curb). The size of angle the parking space makes with the curb determines the amount of road space needed for parking and may limit parking to only one side of the street. Problems such as this require the use of function values of the parking angles and illustrate one way trigonometric identities help us to solve problems.

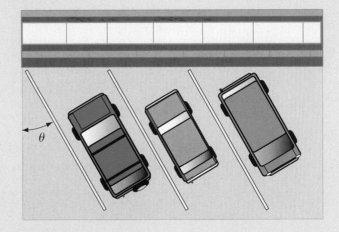

12-1 BASIC IDENTITIES

Throughout our study of mathematics, we have used the solution of equations to solve problems. The domain of an equation is the set of numbers for which each side of the equation is defined. When the solution set is a proper subset of the domain, the equation is a *conditional equation*. When the solution set is the domain of the equation, the equation is an *identity*.

Recall that an identity is an equation that is true for all possible replacements of the variable.

In the following examples, the domain is the set of real numbers.

Algebraic Equations

Conditional equation:

$x^2 + 3x - 10 = 0$

Solution set: $\{-5, 2\}$

Identity:

$3x + 12 = 3(x + 4)$

Solution set: {Real numbers}

Trigonometric Equations

Conditional equation:

$\sin \theta = 1$

Solution set: $\left\{\frac{\pi}{2} + 2\pi n, n = \text{an integer}\right\}$

Identity:

$\sin^2 \theta + \cos^2 \theta = 1$

Solution set: {Real numbers}

When stated in the form of an equation, a property of the real numbers or an application of a property of the real numbers is an algebraic identity. For example, the additive identity property can be expressed as an algebraic identity: $a + 0 = a$ is true for all real numbers.

When we defined the six trigonometric functions, we proved relationships that are true for all values of θ for which the function is defined. There are eight basic trigonometric identities.

Pythagorean Identities	Reciprocal Identities	Quotient Identities
$\cos^2 \theta + \sin^2 \theta = 1$	$\sec \theta = \frac{1}{\cos \theta}$	$\tan \theta = \frac{\sin \theta}{\cos \theta}$
$1 + \tan^2 \theta = \sec^2 \theta$	$\csc \theta = \frac{1}{\sin \theta}$	$\cot \theta = \frac{\cos \theta}{\sin \theta}$
$\cot^2 \theta + 1 = \csc^2 \theta$	$\cot \theta = \frac{1}{\tan \theta}$	

Each of these identities is true for all values of θ for which both sides of the identity are defined. For example, $\cos^2 \theta + \sin^2 \theta = 1$ is true for all real numbers and $1 + \tan^2 \theta = \sec^2 \theta$ is true for all real numbers except $\theta = \frac{\pi}{2} + n\pi$ when n is an integer.

We can use the eight basic identities to write other equations that are true for all replacements of the variable for which the function values exist.

EXAMPLE 1

Use the basic identities to show that $\tan \theta \csc \theta = \sec \theta$ for all values of θ for which each side of the equation is defined.

Solution   Each of the functions in the given equation can be written in terms of $\sin \theta$, $\cos \theta$, or both.

(1) Use the basic identities to write each side of the identity in terms of $\sin \theta$ and $\cos \theta$:

$$\tan \theta \csc \theta \overset{?}{=} \sec \theta$$

$$\left(\frac{\sin \theta}{\cos \theta}\right)\left(\frac{1}{\sin \theta}\right) \overset{?}{=} \frac{1}{\cos \theta}$$

(2) Divide a numerator and a denominator of the left side of the equation by $\sin \theta$:

$$\left(\frac{\overset{1}{\cancel{\sin \theta}}}{\cos \theta}\right)\left(\frac{1}{\underset{1}{\cancel{\sin \theta}}}\right) \overset{?}{=} \frac{1}{\cos \theta}$$

Note: If $\csc \theta$ is defined, $\sin \theta \neq 0$.

$$\frac{1}{\cos \theta} = \frac{1}{\cos \theta} \checkmark$$

EXAMPLE 2

Use the Pythagorean identities to write:

a. $\sin \theta$ in terms of $\cos \theta$.

b. $\cos \theta$ in terms of $\sin \theta$.

Solution   **a.** $\cos^2 \theta + \sin^2 \theta = 1$

$\sin^2 \theta = 1 - \cos^2 \theta$

$\sin \theta = \pm\sqrt{1 - \cos^2 \theta}$

b. $\cos^2 \theta + \sin^2 \theta = 1$

$\cos^2 \theta = 1 - \sin^2 \theta$

$\cos \theta = \pm\sqrt{1 - \sin^2 \theta}$

Note: For a given value of θ, the sign of $\cos \theta$ or of $\sin \theta$ depends on the quadrant in which the terminal side of the angle lies:

- When θ is a first-quadrant angle, $\sin \theta = \sqrt{1 - \cos^2 \theta}$ and $\cos \theta = \sqrt{1 - \sin^2 \theta}$.

- When θ is a second-quadrant angle, $\sin \theta = \sqrt{1 - \cos^2 \theta}$ and $\cos \theta = -\sqrt{1 - \sin^2 \theta}$.

- When θ is a third-quadrant angle, $\sin \theta = -\sqrt{1 - \cos^2 \theta}$ and $\cos \theta = -\sqrt{1 - \sin^2 \theta}$.

- When θ is a fourth-quadrant angle, $\sin \theta = -\sqrt{1 - \cos^2 \theta}$ and $\cos \theta = \sqrt{1 - \sin^2 \theta}$.

Exercises

Writing About Mathematics

1. If we know the value of sin θ, is it possible to find the other five trigonometric function values? If not, what other information is needed?

2. a. Explain how the identities $1 + \tan^2 \theta = \sec^2 \theta$ and $\cot^2 \theta + 1 = \csc^2 \theta$ can be derived from the identity $\cos^2 \theta + \sin^2 \theta = 1$.

b. The identity $\cos^2 \theta + \sin^2 \theta = 1$ is true for all real numbers. Are the identities $1 + \tan^2 \theta = \sec^2 \theta$ and $\cot^2 \theta + 1 = \csc^2 \theta$ also true for all real numbers? Explain your answer.

Developing Skills

In 3–14, write each expression as a single term using sin θ, cos θ, or both.

3. $\tan \theta$

4. $\cot \theta$

5. $\sec \theta$

6. $\csc \theta$

7. $\cot \theta \sec \theta$

8. $\tan^2 \theta + 1$

9. $\cot^2 \theta + 1$

10. $\tan \theta \sec \theta \cot \theta$

11. $\frac{1}{\sec \theta \csc \theta}$

12. $\frac{\tan \theta}{\cot \theta} + \tan \theta \cot \theta$

13. $\frac{1}{\tan \theta} + \cot \theta$

14. $\sec \theta + \frac{1}{\csc \theta}$

12-2 PROVING AN IDENTITY

The eight basic identities are used to prove other identities. To prove an identity means to show that the two sides of the equation are always equivalent. It is generally more efficient to work with the more complicated side of the identity and show, by using the basic identities and algebraic principles, that the two sides are the same.

> **Tips for Proving an Identity**
>
> To prove an identity, use one or more of the following tips:
>
> **1.** Work with the more complicated side of the equation.
>
> **2.** Use basic identities to rewrite unlike functions in terms of the same function.
>
> **3.** Remove parentheses.
>
> **4.** Find common denominators to add fractions.
>
> **5.** Simplify complex fractions and reduce fractions to lowest terms.

EXAMPLE 1

Prove that $\sec \theta \sin \theta = \tan \theta$ is an identity.

Solution   Write the left side of the equation in terms of $\sin \theta$ and $\cos \theta$.

$$\sec \theta \sin \theta \overset{?}{=} \tan \theta$$

$$\frac{1}{\cos \theta} \sin \theta \overset{?}{=} \tan \theta$$

$$\frac{\sin \theta}{\cos \theta} \overset{?}{=} \tan \theta$$

$$\tan \theta = \tan \theta \checkmark$$

Proof begins with what is known and proceeds to what is to be proved. Although we have written the proof in Example 1 by starting with what is to be proved and ending with what is obviously true, the proof of this identity really begins with the obviously true statement:

$$\tan \theta = \tan \theta; \text{ therefore, } \sec \theta \sin \theta = \tan \theta.$$

EXAMPLE 2

Prove that $\sin \theta \, (\csc \theta - \sin \theta) = \cos^2 \theta$.

Solution   Use the distributive property to simplify the left side.

$$\sin \theta \, (\csc \theta - \sin \theta) \overset{?}{=} \cos^2 \theta$$

$$\sin \theta \csc \theta - \sin^2 \theta \overset{?}{=} \cos^2 \theta$$

$$\sin \theta \left(\frac{1}{\sin \theta}\right) - \sin^2 \theta \overset{?}{=} \cos^2 \theta$$

$$1 - \sin^2 \theta \overset{?}{=} \cos^2 \theta$$

$$(\cos^2 \theta + \sin^2 \theta) - \sin^2 \theta \overset{?}{=} \cos^2 \theta \qquad \textit{Use the Pythagorean identity}$$

$$\cos^2 \theta = \cos^2 \theta \checkmark \qquad \textit{cos}^2 \, \theta + \textit{sin}^2 \, \theta = 1.$$

The Pythagorean identity $\cos^2 \theta + \sin^2 \theta = 1$ can be rewritten as $\cos^2 \theta = 1 - \sin^2 \theta$. The second to last line of the proof is often omitted and the left side, $1 - \sin^2 \theta$, replaced by $\cos^2 \theta$.

EXAMPLE 3

Prove the identity $1 - \sin \theta = \frac{\cos^2 \theta}{1 + \sin \theta}$.

Solution   For this identity, it appears that we need to multiply both sides of the equation by $(1 + \sin \theta)$ to clear the denominator. However, in proving an identity we perform only operations that change the form but not the value of that side of the equation.

Here we will work with the right side because it is more complicated and multiply by $\frac{1 - \sin \theta}{1 - \sin \theta}$, a fraction equal to 1.

$$1 - \sin \theta \overset{?}{=} \frac{\cos^2 \theta}{1 + \sin \theta}$$

$$1 - \sin \theta \overset{?}{=} \frac{\cos^2 \theta}{1 + \sin \theta} \times \frac{1 - \sin \theta}{1 - \sin \theta}$$

$$1 - \sin \theta \overset{?}{=} \frac{\cos^2 \theta \, (1 - \sin \theta)}{1 - \sin^2 \theta}$$

$$1 - \sin \theta \overset{?}{=} \frac{\cos^2 \theta \, (1 - \sin \theta)}{\cos^2 \theta}$$

$$1 - \sin \theta = 1 - \sin \theta \; ✔$$

EXAMPLE 4

Prove the identity $\frac{\cot^2 \theta}{\csc \theta + 1} + 1 = \csc \theta$.

Solution   In this identity we will work with the left side.

How to Proceed

(1) Write the given equation:

$$\frac{\cot^2 \theta}{\csc \theta + 1} + 1 \overset{?}{=} \csc \theta$$

(2) Write 1 as a fraction with the same denominator as the given fraction:

$$\frac{\cot^2 \theta}{\csc \theta + 1} + \frac{\csc \theta + 1}{\csc \theta + 1} \overset{?}{=} \csc \theta$$

(3) Add the fractions:

$$\frac{\cot^2 \theta + \csc \theta + 1}{\csc \theta + 1} \overset{?}{=} \csc \theta$$

(4) Use the identity $\cot^2 \theta + 1 = \csc^2 \theta$:

$$\frac{\csc^2 \theta + \csc \theta}{\csc \theta + 1} \overset{?}{=} \csc \theta$$

(5) Factor the numerator:

$$\frac{\csc \theta \, (\csc \theta + 1)}{\csc \theta + 1} \overset{?}{=} \csc \theta$$

(6) Divide the numerator and denominator by $(\csc \theta + 1)$:

$$\csc \theta = \csc \theta \; ✔$$

Exercises

Writing About Mathematics

1. Is $\sin \theta = \sqrt{1 - \cos^2 \theta}$ an identity? Explain why or why not.

2. Cory said that in Example 3, $1 - \sin \theta = \frac{\cos^2 \theta}{1 + \sin \theta}$ could have been shown to be an identity by multiplying the left side by $\frac{1 + \sin \theta}{1 + \sin \theta}$. Do you agree with Cory? Explain why or why not.

Developing Skills

In 3–26, prove that each equation is an identity.

3. $\sin \theta \csc \theta \cos \theta = \cos \theta$

4. $\tan \theta \sin \theta \cos \theta = \sin^2 \theta$

5. $\cot \theta \sin \theta \cos \theta = \cos^2 \theta$

6. $\sec \theta (\cos \theta - \cot \theta) = 1 - \csc \theta$

7. $\csc \theta (\sin \theta + \tan \theta) = 1 + \sec \theta$

8. $1 - \frac{\cos \theta}{\sec \theta} = \sin^2 \theta$

9. $1 - \frac{\sin \theta}{\csc \theta} = \cos^2 \theta$

10. $\sin \theta (\csc \theta - \sin \theta) = \cos^2 \theta$

11. $\cos \theta (\sec \theta - \cos \theta) = \sin^2 \theta$

12. $\frac{\tan \theta}{\sec \theta} = \sin \theta$

13. $\frac{\cot \theta}{\csc \theta} = \cos \theta$

14. $\frac{\csc \theta}{\sec \theta} = \cot \theta$

15. $\frac{\sec \theta}{\csc \theta} = \tan \theta$

16. $\frac{1}{\sin \theta \cos \theta} - \frac{\cos \theta}{\sin \theta} = \tan \theta$

17. $\frac{1}{\sin \theta \cos \theta} - \frac{\sin \theta}{\cos \theta} = \cot \theta$

18. $\frac{\sin^2 \theta}{1 + \cos \theta} = 1 - \cos \theta$

19. $\frac{\cos^2 \theta}{1 + \sin \theta} = 1 - \sin \theta$

20. $\sec \theta \csc \theta = \tan \theta + \cot \theta$

21. $\frac{\tan^2 \theta}{\sec \theta - 1} - 1 = \sec \theta$

22. $\cos \theta + \frac{\sin^2 \theta}{1 + \cos \theta} = 1$

23. $\sin \theta + \frac{\cos^2 \theta}{1 + \sin \theta} = 1$

24. $\frac{\sec \theta}{\cos \theta} - \tan^2 \theta = 1$

25. $\frac{\csc \theta}{\sin \theta} - \cot^2 \theta = 1$

26. $\frac{\cos \theta}{\sec \theta} + \frac{\sin \theta}{\csc \theta} = 1$

27. For what values of θ is the identity $\frac{\cos \theta}{\sec \theta} + \frac{\sin \theta}{\csc \theta} = 1$ undefined?

12-3 COSINE (A − B)

We can prove that $\cos (A - B) = \cos A - \cos B$ is *not* an identity by finding one pair of values of A and B for which each side of the equation is defined and the equation is false. For example, if, in degree measure, $A = 90°$ and $B = 60°$,

$$\cos (90° - 60°) \overset{?}{=} \cos 90° - \cos 60°$$

$$\cos 30° \overset{?}{=} 0 - \tfrac{1}{2}$$

$$\frac{\sqrt{3}}{2} \neq -\tfrac{1}{2} \; ✗$$

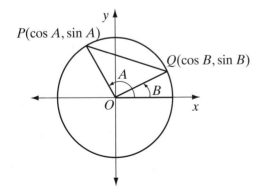

In order to write an identity that expresses $\cos (A - B)$ in terms of function values of A and B, we will use the relationship between the unit circle and the sine and cosine of an angle.

Let A and B be any two angles in standard position. The terminal side of $\angle A$ intersects the unit circle at $P(\cos A, \sin A)$ and the terminal side of $\angle B$ intersects the unit circle at $Q(\cos B, \sin B)$. Use the distance formula to express PQ in terms of $\sin A$, $\cos A$, $\sin B$, and $\cos B$:

$$PQ^2 = (\cos A - \cos B)^2 + (\sin A - \sin B)^2$$
$$= (\cos^2 A - 2 \cos A \cos B + \cos^2 B) + (\sin^2 A - 2 \sin A \sin B + \sin^2 B)$$
$$= (\cos^2 A + \sin^2 A) + (\cos^2 B + \sin^2 B) - 2 \cos A \cos B - 2 \sin A \sin B$$
$$= 1 + 1 - 2(\cos A \cos B + \sin A \sin B)$$
$$= 2 - 2(\cos A \cos B + \sin A \sin B)$$

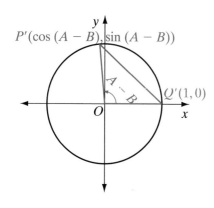

P'(cos (A − B), sin (A − B))

Q'(1, 0)

Now rotate $\triangle OQP$ through an angle of $-B$, that is, an angle of B units in the clockwise direction so that the image of P is P' and the image of Q is Q'. Q' is a point on the x-axis whose coordinates are $(1, 0)$. Angle $Q'OP'$ is an angle in standard position whose measure is $(A - B)$. Therefore, the coordinates of P' are $(\cos (A - B), \sin(A - B))$. Use the distance formula to find $P'Q'$.

$$(P'Q')^2 = (\cos (A - B) - 1)^2 + (\sin(A - B) - 0)^2$$
$$= \cos^2 (A - B) - 2 \cos (A - B) + 1 + \sin^2(A - B)$$
$$= [\cos^2 (A - B) + \sin^2(A - B)] - 2 \cos (A - B) + 1$$
$$= 1 - 2 \cos (A - B) + 1$$
$$= 2 - 2 \cos (A - B)$$

Distance is preserved under a rotation. Therefore,

$$(P'Q')^2 = (PQ)^2$$
$$2 - 2 \cos (A - B) = 2 - 2(\cos A \cos B + \sin A \sin B)$$
$$-2 \cos (A - B) = -2(\cos A \cos B + \sin A \sin B)$$

$$\mathbf{\cos (A - B) = \cos A \cos B + \sin A \sin B}$$

At the beginning of this section we showed that:

$$\cos (90° - 60°) \neq \cos 90° - \cos 60°$$

Does the identity that we proved make it possible to find $\cos (90° - 60°)$? We can check.

$$\cos (A - B) = \cos A \cos B + \sin A \sin B$$
$$\cos (90° - 60°) \overset{?}{=} \cos 90° \cos 60° + \sin 90° \sin 60°$$
$$\cos 30° \overset{?}{=} 0\left(\tfrac{1}{2}\right) + 1\left(\tfrac{\sqrt{3}}{2}\right)$$
$$\tfrac{\sqrt{3}}{2} \overset{?}{=} 0 + \tfrac{\sqrt{3}}{2}$$
$$\tfrac{\sqrt{3}}{2} = \tfrac{\sqrt{3}}{2} ✔$$

EXAMPLE 1

Use $(60° - 45°) = 15°$ to find the exact value of $\cos 15°$.

Solution

$$\cos (A - B) = \cos A \cos B + \sin A \sin B$$
$$\cos (60° - 45°) = \cos 60° \cos 45° + \sin 60° \sin 45°$$
$$\cos 15° = \left(\tfrac{1}{2}\right)\left(\tfrac{\sqrt{2}}{2}\right) + \left(\tfrac{\sqrt{3}}{2}\right)\left(\tfrac{\sqrt{2}}{2}\right)$$
$$\cos 15° = \tfrac{\sqrt{2}}{4} + \tfrac{\sqrt{6}}{4}$$
$$\cos 15° = \tfrac{\sqrt{2} + \sqrt{6}}{4} \quad \textit{Answer}$$

Note: Example 1 shows that $\cos 15°$ is an *irrational* number.

Cosine of (90° − B)

The cofunction relationship between cosine and sine can be proved using $\cos (A - B)$. Use the identity for $\cos (A - B)$ to express $\cos (90° - B)$ in terms of a function of B.

Let $A = 90°$.

$$\cos (A - B) = \cos A \cos B + \sin A + \sin B$$
$$\cos (90° - B) = \cos 90° \cos B + \sin 90° \sin B$$
$$\cos (90° - B) = 0 \cos B + 1 \sin B$$
$$\cos (90° - B) = \sin B$$

This is an identity, a statement true for all values of B.

EXAMPLE 2

Use the identity $\cos (90° - B) = \sin B$ to find $\sin (90° - B)$.

Solution   Let $B = (90° - A)$.

$$\cos (90° - B) = \sin B$$
$$\cos (90° - (90° - A)) = \sin (90° - A)$$
$$\cos (90° - 90° + A) = \sin (90° - A)$$
$$\cos A = \sin (90° - A) \quad \textit{Answer}$$

EXAMPLE 3

Given that A and B are second-quadrant angles, $\sin A = \tfrac{1}{3}$, and $\sin B = \tfrac{1}{5}$, find $\cos (A - B)$.

Solution   Use the identity $\cos^2 \theta = 1 - \sin^2 \theta$ to find $\cos A$ and $\cos B$.

$$\cos^2 A = 1 - \sin^2 A \qquad\qquad \cos^2 B = 1 - \sin^2 B$$

$$\cos^2 A = 1 - \left(\tfrac{1}{3}\right)^2 \qquad\qquad \cos^2 B = 1 - \left(\tfrac{1}{5}\right)^2$$

$$\cos^2 A = 1 - \tfrac{1}{9} \qquad\qquad\qquad \cos^2 B = 1 - \tfrac{1}{25}$$

$$\cos^2 A = \tfrac{8}{9} \qquad\qquad\qquad\quad \cos^2 B = \tfrac{24}{25}$$

$$\cos A = \pm\sqrt{\tfrac{8}{9}} \qquad\qquad\qquad \cos B = \pm\sqrt{\tfrac{24}{25}}$$

A is in the second quadrant. $\qquad\qquad$ B is in the second quadrant.

$$\cos A = -\tfrac{2\sqrt{2}}{3} \qquad\qquad\qquad \cos B = -\tfrac{2\sqrt{6}}{5}$$

Use the identity for the cosine of the difference of two angles.

$$\cos (A - B) = \cos A \cos B + \sin A \sin B$$

$$\cos (A - B) = \left(-\tfrac{2\sqrt{2}}{3}\right)\left(-\tfrac{2\sqrt{6}}{5}\right) + \left(\tfrac{1}{3}\right)\left(\tfrac{1}{5}\right)$$

$$\cos (A - B) = \tfrac{4\sqrt{12}}{15} + \tfrac{1}{15}$$

$$\cos (A - B) = \tfrac{4\sqrt{12} + 1}{15}$$

$$\cos (A - B) = \tfrac{8\sqrt{3} + 1}{15}$$

Note that since $\cos (A - B)$ is positive, $(A - B)$ must be a first- or fourth-quadrant angle.

Answer $\cos (A - B) = \tfrac{8\sqrt{3} + 1}{15}$

SUMMARY
We have proved the following identities:

$$\cos (A - B) = \cos A \cos B + \sin A \sin B$$

$$\cos (90° - B) = \sin B$$

$$\sin (90° - A) = \cos A$$

Exercises

Writing About Mathematics

1. Are the equations $\sin \theta = \cos (90° - \theta)$ and $\cos \theta = \sin (90° - \theta)$ true for all real numbers or only for values of θ in the interval $0 < \theta < 90°$?

2. Emily said that, without finding the values on a calculator, she knows that $\sin 100° = \cos (-10°)$. Do you agree with Emily? Explain why or why not.

Developing Skills

In 3–17, find the exact value of cos $(A - B)$ for each given pair of values.

3. $A = 180°, B = 60°$ **4.** $A = 180°, B = 45°$ **5.** $A = 180°, B = 30°$

6. $A = 270°, B = 60°$ **7.** $A = 270°, B = 30°$ **8.** $A = 60°, B = 90°$

9. $A = 30°, B = 90°$ **10.** $A = 90°, B = 60°$ **11.** $A = 60°, B = 270°$

12. $A = 45°, B = 270°$ **13.** $A = 30°, B = 270°$ **14.** $A = 360°, B = 60°$

15. $A = \pi, B = \frac{2\pi}{3}$ **16.** $A = \frac{\pi}{6}, B = \frac{4\pi}{3}$ **17.** $A = \frac{3\pi}{4}, B = \frac{7\pi}{4}$

Applying Skills

In 18–20, show all work.

18. a. Find the exact value of cos 15° by using cos $(45° - 30°)$.

    b. Use the value of cos 15° found in **a** to find cos 165° by using cos $(180° - 15°)$.

    c. Use the value of cos 15° found in **a** to find cos 345° by using cos $(360° - 15°)$.

    d. Use cos A = sin $(90° - A)$ to find the exact value of sin 75°.

19. a. Find the exact value of cos 120° by using cos $(180° - 60°)$.

    b. Find the exact value of sin 120° by using $\cos^2 \theta + \sin^2 \theta = 1$ and the value of cos 120° found in **a**.

    c. Find the exact value of cos 75° by using cos $(120° - 45°)$.

    d. Use the value of cos 75° found in **c** to find cos 105° by using cos $(180° - 75°)$.

    e. Use the value of cos 75° found in **c** to find cos 285° by using cos $(360° - 75°)$.

    f. Find the exact value of sin 15°.

20. a. Find the exact value of cos 210° by using cos $(270° - 60°)$.

    b. Find the exact value of sin 210° by using $\cos^2 \theta + \sin^2 \theta = 1$ and the value of cos 210° found in **a**.

    c. Find the exact value of cos 165° by using cos $(210° - 45°)$.

    d. Use the value of cos 165° found in **c** to find cos $(-15°)$ by using cos $(165° - 180°)$.

    e. Use the value of cos $(-15°)$ found in **d** to find cos 195° by using cos $(180° - (-15°))$.

    f. Use the value of cos $(-15°)$ found in **d** to find the exact value of sin 105°.

21. A telephone pole is braced by two wires that are both fastened to the ground at a point 15 feet from the base of the pole. The shorter wire is fastened to the pole 15 feet above the ground and the longer wire 20 feet above the ground.

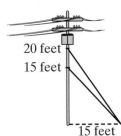

    a. What is the measure, in degrees, of the angle that the shorter wire makes with the ground?

b. Let θ be the measure of the angle that the longer wire makes with the ground. Find $\sin \theta$ and $\cos \theta$.

c. Find the cosine of the angle between the wires where they meet at the ground.

d. Find, to the nearest degree, the measure of the angle between the wires.

12-4 COSINE (A + B)

We know that $\cos (A + B)$ can be written as $\cos (A - (-B))$. Therefore,

$$\cos (A + B) = \cos (A - (-B)) = \cos A \cos (-B) + \sin A \sin (-B)$$

We would like to write this identity in term of $\cos B$ and $\sin B$. Therefore, we want to find the relationship between $\cos B$ and $\cos (-B)$ and between $\sin A$ and $\sin (-A)$. In the exercises of Sections 11-1 and 11-2, we showed that $y = \sin x$ is an odd function and $y = \cos x$ is an even function. That is, for all values of x, $-\sin x = \sin (-x)$ and $\cos x = \cos (-x)$. We can establish these results graphically on the unit circle as the following hands-on activity demonstrates.

Hands-On Activity

1. Draw a first-quadrant angle in standard position. Let the point of intersection of the initial side with the unit circle be Q and the intersection of the terminal side with the unit circle be P. Let the measure of the angle be θ; $m\angle QOP = \theta$.

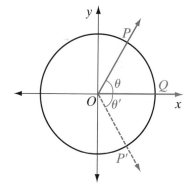

2. Reflect $\angle QOP$ in the x-axis. The image of $\angle QOP$ is $\angle QOP'$ and $m\angle QOP' = -\theta$.

3. Express $\sin \theta$, $\cos \theta$, $\sin (-\theta)$, and $\cos (-\theta)$ as the lengths of line segments. Show that $\cos (-\theta) = \cos \theta$ and that $\sin (-\theta) = -\sin \theta$.

4. Repeat steps 1 through 3 for a second-quadrant angle.

5. Repeat steps 1 through 3 for a third-quadrant angle.

6. Repeat steps 1 through 3 for a fourth-quadrant angle.

An algebraic proof can be used to prove the relationships $\cos (-\theta) = \cos \theta$ and $\sin (-\theta) = -\sin \theta$.

Proof of cos $(-\theta)$ = cos θ

In the identity for $\cos{(A - B)}$, let $A = 0$ and $B = \theta$.

$$\cos{(A - B)} = \cos A \cos B + \sin A \sin B$$
$$\cos{(0 - \theta)} = \cos 0 \cos \theta + \sin 0 \sin \theta$$
$$\cos{(-\theta)} = 1 \cos \theta + 0 \sin \theta$$
$$\cos{(-\theta)} = \cos \theta + 0$$
$$\cos{(-\theta)} = \cos \theta \ ✔$$

Proof of sin $(-\theta)$ = $-$sin θ

In the identity $\sin A = \cos{(90° - A)}$, let $A = -\theta$.

$$\sin{(-\theta)} = \cos{(90° - (-\theta))}$$
$$\sin{(-\theta)} = \cos{(90° + \theta)}$$
$$\sin{(-\theta)} = \cos{(\theta + 90°)}$$
$$\sin{(-\theta)} = \cos{(\theta - (-90°))}$$
$$\sin{(-\theta)} = \cos \theta \cos{(-90°)} + \sin \theta \sin{(-90°)}$$
$$\sin{(-\theta)} = (\cos \theta)(0) + (\sin \theta)(-1)$$
$$\sin{(-\theta)} = 0 - \sin \theta$$
$$\sin{(-\theta)} = -\sin \theta \ ✔$$

Proof of the identity for cos $(A + B)$

$$\cos{(A + B)} = \cos{(A - (-B))}$$
$$\cos{(A + B)} = \cos A \cos{(-B)} + \sin A \sin{(-B)}$$
$$\cos{(A + B)} = \cos A \cos B + \sin A \,(-\sin B)$$

$$\mathbf{\cos{(A + B)} = \cos A \cos B - \sin A \sin B}$$

EXAMPLE I

Find the exact value of $\cos 105° = \cos{(45° + 60°)}$ using identities.

Solution

$$\cos{(A + B)} = \cos A \cos B - \sin A \sin B$$
$$\cos{(45° + 60°)} = \cos 45° \cos 60° - \sin 45° \sin 60°$$
$$\cos 105° = \left(\tfrac{\sqrt{2}}{2}\right)\left(\tfrac{1}{2}\right) - \left(\tfrac{\sqrt{2}}{2}\right)\left(\tfrac{\sqrt{3}}{2}\right)$$
$$\cos 105° = \tfrac{\sqrt{2}}{4} - \tfrac{\sqrt{6}}{4}$$
$$\cos 105° = \tfrac{\sqrt{2} - \sqrt{6}}{4} \ \textit{Answer}$$

EXAMPLE 2

Show that $\cos (\pi + \theta) = -\cos \theta$.

Solution   We are now working in radians.

$$\cos (A + B) = \cos A \cos B - \sin A \sin B$$
$$\cos (\pi + \theta) = \cos \pi \cos \theta - \sin \pi \sin \theta$$
$$\cos (\pi + \theta) = -1 \cos \theta - 0 \sin \theta$$
$$\cos (\pi + \theta) = -\cos \theta \checkmark$$

SUMMARY

We have proved the following identities:

$$\cos (-\theta) = \cos \theta$$

$$\sin (-\theta) = -\sin \theta$$

$$\cos (A + B) = \cos A \cos B - \sin A \sin B$$

Exercises

Writing About Mathematics

1. Maggie said that $\cos (A + B) + \cos (A - B) = \cos 2A$. Do you agree with Maggie? Justify your answer.

2. Germaine said $\cos (A + B) + \cos (A - B) = 2 \cos A \cos B$. Do you agree with Germaine? Justify your answer.

Developing Skills

In 3–17, find the exact value of $\cos (A + B)$ for each given pair of values.

3. $A = 90°, B = 60°$ **4.** $A = 90°, B = 45°$ **5.** $A = 90°, B = 30°$

6. $A = 180°, B = 60°$ **7.** $A = 180°, B = 30°$ **8.** $A = 180°, B = 45°$

9. $A = 270°, B = 30°$ **10.** $A = 270°, B = 60°$ **11.** $A = 270°, B = 45°$

12. $A = 60°, B = 60°$ **13.** $A = 60°, B = 270°$ **14.** $A = 45°, B = 270°$

15. $A = \frac{\pi}{2}, B = \frac{2\pi}{3}$ **16.** $A = \frac{\pi}{3}, B = \frac{4\pi}{3}$ **17.** $A = \frac{\pi}{4}, B = \frac{\pi}{6}$

Applying Skills

In 18–20, show all work.

18. a. Find the exact value of cos 75° by using cos (45° + 30°).

  b. Use the value of cos 75° found in **a** to find cos 255° by using cos (180° + 75°).

  c. Use cos A = sin (90° − A) to find the exact value of sin 15°.

19. a. Find the exact value of cos 120° by using cos (60° + 60°).

  b. Find the exact value of sin 120° by using cos² θ + sin² θ = 1 and the value of cos 120° found in **a**.

  c. Find the exact value of cos 165° by using cos (120° + 45°).

  d. Use the value of cos 165° found in **c** to find cos 345° by using cos (180° + 165°).

20. a. Find the exact value of cos 315° by using cos (270° + 45°).

  b. Find the exact value of sin 315° by using cos² θ + sin² θ = 1 and the value of cos 315° found in **a**.

  c. Find the exact value of cos 345° by using cos (315° + 30°).

  d. Explain why cos 405° = cos 45°.

21. An engineer wants to determine CD, the exact height of a building. To do this, he first locates B on $\overline{CD}$, a point 30 feet above C at the foot of the building. Then he locates A, a point on the ground 40 feet from C. From A, the engineer then finds that the angle of elevation of D is 45° larger than θ, the angle of elevation of B.

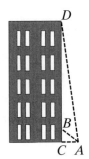

  a. Find AB, sin θ, and cos θ.

  b. Use sin θ and cos θ found in **a** to find the exact value of cos (θ + 45°).

  c. Use the value of cos (θ + 45°) found in **b** to find AD.

  d. Find CD, the height of the building.

12-5 SINE (A − B) AND SINE (A + B)

We can use the cofunction identity sin θ = cos (90° − θ) and the identities for cos (A − B) and cos (A + B) to derive identities for sin (A − B) and sin (A + B).

Sine of (A − B)

Let $\theta = A - B$.

$$\sin \theta = \cos(90° - \theta)$$
$$\sin(A - B) = \cos(90° - (A - B))$$
$$= \cos(90° - A + B)$$
$$= \cos((90° - A) + B)$$
$$= \cos(90° - A)\cos B - \sin(90° - A)\sin B$$
$$= \sin A \cos B - \cos A \sin B$$

$$\sin(A - B) = \sin A \cos B - \cos A \sin B$$

Sine of (A + B)

The derivation of an identity for $\sin(A + B)$ is similar.

Let $\theta = A + B$.

$$\sin \theta = \cos(90° - \theta)$$
$$\sin(A + B) = \cos(90° - (A + B))$$
$$= \cos(90° - A - B)$$
$$= \cos((90° - A) - B)$$
$$= \cos(90° - A)\cos B + \sin(90° - A)\sin B$$
$$= \sin A \cos B + \cos A \sin B$$

$$\sin(A + B) = \sin A \cos B + \cos A \sin B$$

EXAMPLE I

Use $\sin(45° - 30°)$ to find the exact value of $\sin 15°$.

Solution

$$\sin(A - B) = \sin A \cos B - \cos A \sin B$$
$$\sin(45° - 30°) = \sin 45° \cos 30° - \cos 45° \sin 30°$$
$$\sin 15° = \left(\frac{\sqrt{2}}{2}\right)\left(\frac{\sqrt{3}}{2}\right) - \left(\frac{\sqrt{2}}{2}\right)\left(\frac{1}{2}\right)$$
$$= \frac{\sqrt{6}}{4} - \frac{\sqrt{2}}{4}$$
$$= \frac{\sqrt{6} - \sqrt{2}}{4} \quad \textit{Answer}$$

EXAMPLE 2 ▰▰▰▰▰▰▰▰▰▰▰▰▰▰▰▰▰▰▰▰▰▰▰▰▰▰▰▰▰▰▰▰▰▰

Use $\sin(60° + 45°)$ to find the exact value of $\sin 105°$.

Solution
$$\sin(A + B) = \sin A \cos B + \cos A \sin B$$
$$\sin(60° + 45°) = \sin 60° \cos 45° + \cos 60° \sin 45°$$
$$\sin 105° = \left(\frac{\sqrt{3}}{2}\right)\left(\frac{\sqrt{2}}{2}\right) + \left(\frac{1}{2}\right)\left(\frac{\sqrt{2}}{2}\right)$$
$$= \frac{\sqrt{6}}{4} + \frac{\sqrt{2}}{4}$$
$$= \frac{\sqrt{6} + \sqrt{2}}{4} \quad \textit{Answer}$$

EXAMPLE 3 ▰▰▰▰▰▰▰▰▰▰▰▰▰▰▰▰▰▰▰▰▰▰▰▰▰▰▰▰▰▰▰▰▰▰

Show that $\sin(\pi + \theta) = -\sin\theta$.

Solution   We are now working in radians.
$$\sin(A + B) = \sin A \cos B + \cos A \sin B$$
$$\sin(\pi + \theta) = \sin\pi \cos\theta + \cos\pi \sin\theta$$
$$\sin(\pi + \theta) = 0\cos\theta + (-1)\sin\theta$$
$$\sin(\pi + \theta) = 0 - \sin\theta$$
$$\sin(\pi + \theta) = -\sin\theta \checkmark$$

SUMMARY

We have proved the following identities:

$$\textbf{sin}\,(A - B) = \textbf{sin}\,A\,\textbf{cos}\,B - \textbf{cos}\,A\,\textbf{sin}\,B$$

$$\textbf{sin}\,(A + B) = \textbf{sin}\,A\,\textbf{cos}\,B + \textbf{cos}\,A\,\textbf{sin}\,B$$

Exercises

Writing About Mathematics

1. William said that $\sin(A + B) + \sin(A - B) = \sin 2A$. Do you agree with William? Justify your answer.

2. Freddy said that $\sin(A + B) + \sin(A - B) = 2\sin A \cos B$. Do you agree with Freddy? Justify your answer.

Developing Skills

In 3–17, find the exact value of sin $(A - B)$ and of sin $(A + B)$ for each given pair of values.

3. $A = 180°, B = 60°$

4. $A = 180°, B = 45°$

5. $A = 180°, B = 30°$

6. $A = 270°, B = 60°$

7. $A = 270°, B = 30°$

8. $A = 60°, B = 90°$

9. $A = 30°, B = 90°$

10. $A = 90°, B = 60°$

11. $A = 60°, B = 270°$

12. $A = 45°, B = 270°$

13. $A = 30°, B = 270°$

14. $A = 360°, B = 60°$

15. $A = \frac{3\pi}{2}, B = 2\pi$

16. $A = \frac{2\pi}{3}, B = \frac{\pi}{6}$

17. $A = \frac{\pi}{3}, B = \frac{5\pi}{4}$

Applying Skills

In 18–20, show all work.

18. a. Find the exact value of sin 15° by using sin (45° − 30°).

   b. Use the value of sin 15° found in **a** to find sin 165° by using sin (180° − 15°).

   c. Use the value of sin 15° found in **a** to find sin 345° by using sin (360° − 15°).

   d. Use the value of sin 15° found in **a** to find sin 195° by using sin (180° + 15°).

19. a. Find the exact value of sin 120° by using sin (180° − 60°).

   b. Find the exact value of cos 120° by using cos (180° − 60°).

   c. Find the exact value of sin 75° by using sin (120° − 45°).

   d. Use the value of sin 75° found in **c** to find sin 105° by using sin (180° − 75°).

   e. Use the value of sin 75° found in **c** to find sin 285° by using sin (360° − 75°).

20. a. Find the exact value of sin 210° by using sin (270° − 60°).

   b. Find the exact value of cos 210° by using cos (270° − 60°).

   c. Find the exact value of sin 165° by using sin (210° − 45°).

   d. Use the value of sin 165° found in **c** to find sin (−15°) by using sin (165° − 180°).

   e. Use the value of sin (−15°) found in **d** to find sin 195° by using sin (180° − (−15°)).

   f. Use the value of sin (−15°) found in **d** to find the exact value of sin 105°.

21. A telephone pole is braced by two wires that are both fastened to the ground at the same point. The shorter wire is 30 feet long and is fastened to the pole 10 feet above the foot of the pole. The longer wire is 33 feet long and is fastened to the pole 17 feet above the foot of the pole.

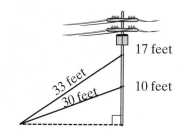

   a. If the measure of the angle that the longer wire makes with the ground is x and the measure of the angle that the shorter wire makes with the ground is y, find sin x, cos x, sin y, and cos y.

   b. Find the exact value of sin $(x - y)$, the sine of the angle between the two wires.

   c. Find to the nearest degree the measure of the angle between the two wires.

22. Two boats leave the same dock to cross a river that is 500 feet wide. The first boat leaves the dock at an angle of θ with the shore and travels 1,300 feet to reach a point downstream on the opposite shore of the river. The second boat leaves the dock at an angle of $(\theta + 30°)$ with the shore.

a. Find $\sin \theta$ and $\cos \theta$.

b. Find $\sin (\theta + 30°)$.

c. Find the distance that the second boat travels to reach the opposite shore.

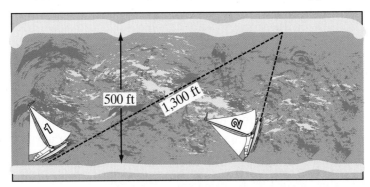

23. The coordinates of any point in the coordinate plane can be written as $A(r \cos a, r \sin a)$ when a is the measure of the angle between the positive ray of the x-axis and $\overrightarrow{OA}$ and $r = OA$. Under a rotation of θ about the origin, the coordinates of A', the image of A, are $(r \cos (a + \theta), r \sin (a + \theta))$. Find the coordinates of A', the image of $A(6, 8)$, under a rotation of $45°$ about the origin.

12-6 TANGENT (A − B) AND TANGENT (A + B)

We can use the identity $\tan \theta = \frac{\sin \theta}{\cos \theta}$ and the identities for $\sin (A + B)$ and $\cos (A + B)$ to write identities for $\tan (A + B)$ and $\tan (A - B)$.

Tangent of (A + B)

$$\tan (A + B) = \frac{\sin (A + B)}{\cos (A + B)} = \frac{\sin A \cos B + \cos A \sin B}{\cos A \cos B - \sin A \sin B}$$

We would like to write this identity for $\tan (A + B)$ in terms of $\tan A$ and $\tan B$. We can do this by dividing each term of the numerator and each term of the denominator by $\cos A \cos B$. When we do this we are dividing by a fraction equal to 1 and therefore leaving the value of the expression unchanged.

$$\tan (A + B) = \frac{\frac{\sin A \cos B}{\cos A \cos B} + \frac{\cos A \sin B}{\cos A \cos B}}{\frac{\cos A \cos B}{\cos A \cos B} - \frac{\sin A \sin B}{\cos A \cos B}} = \frac{\frac{\sin A \cancel{\cos B}}{\cos A \cancel{\cos B}} + \frac{\cancel{\cos A} \sin B}{\cancel{\cos A} \cos B}}{\frac{\cancel{\cos A \cos B}}{\cancel{\cos A \cos B}} - \frac{\sin A \sin B}{\cos A \cos B}}$$

$$\tan (A + B) = \frac{\tan A + \tan B}{1 - \tan A \tan B}$$

Tangent of $(A - B)$

The identity for $\tan (A - B)$ can be derived in a similar manner.

$$\tan (A - B) = \frac{\sin (A - B)}{\cos (A - B)} = \frac{\sin A \cos B - \cos A \sin B}{\cos A \cos B + \sin A \sin B}$$

$$\tan (A - B) = \frac{\frac{\sin A \cos B}{\cos A \cos B} - \frac{\cos A \sin B}{\cos A \cos B}}{\frac{\cos A \cos B}{\cos A \cos B} + \frac{\sin A \sin B}{\cos A \cos B}} = \frac{\frac{\sin A \cos B}{\cos A \cos B} - \frac{\cos A \sin B}{\cos A \cos B}}{\frac{\cos A \cos B}{\cos A \cos B} + \frac{\sin A \sin B}{\cos A \cos B}}$$

$$\tan (A - B) = \frac{\tan A - \tan B}{1 + \tan A \tan B}$$

These identities are true for all replacements of A and B for which $\cos A \neq 0$ and $\cos B \neq 0$, and for which $\tan (A + B)$ or $\tan (A - B)$ are defined.

EXAMPLE 1

Use $\tan 2\pi = 0$ and $\tan \frac{\pi}{4} = 1$ to show that $\tan \frac{7\pi}{4} = -1$.

Solution

$$\tan (A - B) = \frac{\tan A - \tan B}{1 + \tan A \tan B}$$

$$\tan \left(2\pi - \frac{\pi}{4}\right) = \frac{\tan 2\pi - \tan \frac{\pi}{4}}{1 + \tan 2\pi \tan \frac{\pi}{4}}$$

$$\tan \frac{7\pi}{4} = \frac{0 - 1}{1 + (0)(1)}$$

$$\tan \frac{7\pi}{4} = -1 \ ✔$$

EXAMPLE 2

Use $(45° + 120°) = 165°$ to find the exact value of $\tan 165°$.

Solution $\sin 45° = \frac{\sqrt{2}}{2}$ and $\cos 45° = \frac{\sqrt{2}}{2}$, so $\tan 45° = 1$.

$\sin 120° = \frac{\sqrt{3}}{2}$ and $\cos 120° = -\frac{1}{2}$, so $\tan 120° = -\sqrt{3}$.

$$\tan (A + B) = \frac{\tan A + \tan B}{1 - \tan A \tan B}$$

$$\tan (45° + 120°) = \frac{\tan 45° + \tan 120°}{1 - \tan 45° \tan 120°}$$

$$\tan 165° = \frac{1 + (-\sqrt{3})}{1 - (1)(-\sqrt{3})}$$

$$= \frac{1 - \sqrt{3}}{1 + \sqrt{3}} \times \frac{1 - \sqrt{3}}{1 - \sqrt{3}}$$

$$\tan 165° = \frac{4 - 2\sqrt{3}}{-2} = -2 + \sqrt{3}$$

Alternative Write the identity tan $(A + B)$ in terms of sine and cosine.
Solution

$$\tan (A + B) = \frac{\sin A \cos B + \cos A \sin B}{\cos A \cos B - \sin A \sin B}$$

$$\tan (45° + 120°) = \frac{\sin 45° \cos 120° + \cos 45° \sin 120°}{\cos 45° \cos 120° - \sin 45° \sin 120°}$$

$$\tan 165° = \frac{\left(\frac{\sqrt{2}}{2}\right)\left(-\frac{1}{2}\right) + \left(\frac{\sqrt{2}}{2}\right)\left(\frac{\sqrt{3}}{2}\right)}{\left(\frac{\sqrt{2}}{2}\right)\left(-\frac{1}{2}\right) - \left(\frac{\sqrt{2}}{2}\right)\left(\frac{\sqrt{3}}{2}\right)}$$

$$= \frac{-\frac{\sqrt{2}}{4} + \frac{\sqrt{6}}{4}}{-\frac{\sqrt{2}}{4} - \frac{\sqrt{6}}{4}} \times \frac{\frac{4}{\sqrt{2}}}{\frac{4}{\sqrt{2}}}$$

$$\tan 165° = \frac{-1 + \sqrt{3}}{-1 - \sqrt{3}} = -2 + \sqrt{3}$$

Answer $\tan 165° = -2 + \sqrt{3}$ or $\sqrt{3} - 2$

SUMMARY

We have proved the following identities:

$$\tan (A + B) = \frac{\tan A + \tan B}{1 - \tan A \tan B}$$

$$\tan (A - B) = \frac{\tan A - \tan B}{1 + \tan A \tan B}$$

Exercises

Writing About Mathematics

1. Explain why the identity $\tan (A + B) = \frac{\tan A + \tan B}{1 - \tan A \tan B}$ is not valid when A or B is equal to $\frac{\pi}{2} + n\pi$ for any integer n.

2. Explain why $\frac{\tan A + \tan B}{1 - \tan A \tan B}$ is undefined when $A = \frac{\pi}{6}$ and $B = \frac{\pi}{3}$.

Developing Skills

In 3–17, find the exact value of tan $(A + B)$ and of tan $(A - B)$ for each given pair of values.

3. $A = 45°, B = 30°$ **4.** $A = 45°, B = 60°$ **5.** $A = 60°, B = 60°$

6. $A = 180°, B = 30°$ **7.** $A = 180°, B = 45°$ **8.** $A = 180°, B = 60°$

9. $A = 120°, B = 30°$ **10.** $A = 120°, B = 45°$ **11.** $A = 120°, B = 60°$

12. $A = 120°, B = 120°$ **13.** $A = 240°, B = 120°$ **14.** $A = 360°, B = 60°$

15. $A = \pi, B = \frac{\pi}{3}$ **16.** $A = \frac{5\pi}{6}, B = \frac{5\pi}{6}$ **17.** $A = \frac{\pi}{3}, B = \frac{\pi}{4}$

Applying Skills

18. Prove that $\tan(180° + \theta) = \tan \theta$.

19. Find $\tan(A + B)$ if $\tan A = 3$ and $\tan B = -\frac{1}{2}$.

20. Find $\tan(A - B)$ if $\tan A = \frac{3}{4}$ and $\tan B = -8$.

21. Find $\tan(A + B)$ if A is in the second quadrant, $\sin A = 0.6$, and $\tan B = 4$.

22. If $A = \arctan 2$ and $B = \arctan(-2)$, find $\tan(A - B)$.

23. If $A = \arctan\left(-\frac{2}{3}\right)$ and $B = \arctan \frac{2}{3}$, find $\tan(A + B)$.

24. In the diagram, $ABCD$, $BEFC$, and $EGHF$ are congruent squares with $AD = 1$. Let $m\angle GAH = x$, $m\angle GBH = y$, and $m\angle GEH = z$.

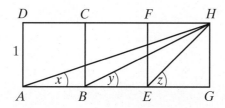

a. Find $\tan(x + y)$.

b. Does $x + y = z$? Justify your answer.

25. A tower that is 20 feet tall stands at the edge of a 30-foot cliff. From a point on level ground that is 20 feet from a point directly below the tower at the base of the cliff, the measure of the angle of elevation of the top of the tower is x and the measure of the angle of elevation of the foot of the tower is y.

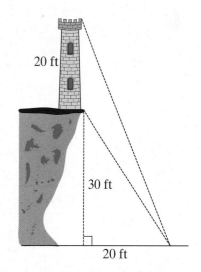

a. Find the exact value of $\tan(x - y)$, the tangent of the angle between the lines of sight to the foot and top of the tower.

b. Find to the nearest degree the measure of the angle between the lines of sight to the foot and the top of the tower.

26. Two boats leave a dock to cross a river that is 80 meters wide. The first boat travels to a point that is 100 meters downstream from a point directly opposite the starting point, and the second boat travels to a point that is 200 meters downstream from a point directly opposite the starting point.

a. Let *x* be the measure of the angle between the river's edge and the path of the first boat and *y* be the measure of the angle between the river's edge and the path of the second boat. Find tan *x* and tan *y*.

b. Find the tangent of the measure of the angle between the paths of the boats.

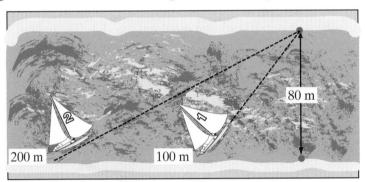

12-7 FUNCTIONS OF 2*A*

The identities for sin $(A + B)$, cos $(A + B)$, and tan $(A + B)$ are true when $A = B$. These identities can be used to find the function values of 2*A*. We often call the identities used to find the function values of twice an angle **double-angle formulas**.

Sine of 2A

In the identity for sin $(A + B)$, let $B = A$.

$$\sin (A + B) = \sin A \cos B + \cos A \sin B$$
$$\sin (A + A) = \sin A \cos A + \cos A \sin A$$
$$\textbf{sin } \mathbf{(2A) = 2 \sin A \cos A}$$

Cosine of 2A

In the identity for cos $(A + B)$, let $B = A$.

$$\cos (A + B) = \cos A \cos B - \sin A \sin B$$
$$\cos (A + A) = \cos A \cos A - \sin A \sin A$$
$$\textbf{cos } \mathbf{(2A) = \cos^2 A - \sin^2 A}$$

The identity for cos 2A can be written in two other ways:

$$\cos (2A) = 2 \cos^2 A - 1$$

$$\cos (2A) = 1 - 2 \sin^2 A$$

Proofs of the first of these identities are given in Example 2. The proof of the second identity is left to the student in Exercise 23.

Tangent of 2A

In the identity for tan $(A + B)$, let $B = A$.

$$\tan (A + B) = \frac{\tan A + \tan B}{1 - \tan A \tan B}$$

$$\tan (A + A) = \frac{\tan A + \tan A}{1 - \tan A \tan A}$$

$$\tan (2A) = \frac{2 \tan A}{1 - \tan^2 A}$$

EXAMPLE I

If $\tan \theta = -\frac{\sqrt{7}}{3}$ and θ is a second-quadrant angle, find:

a. sec θ **b.** cos θ **c.** sin θ

d. sin 2θ **e.** cos 2θ **f.** tan 2θ

g. In what quadrant does 2θ lie?

Solution **a.** Use the identity $1 + \tan^2 \theta = \sec^2 \theta$ to find sec θ. Since θ is a second-quadrant angle, sec θ is negative.

$$1 + \left(-\frac{\sqrt{7}}{3}\right)^2 = \sec^2 \theta$$

$$1 + \frac{7}{9} = \sec^2 \theta$$

$$\frac{16}{9} = \sec^2 \theta$$

$$-\frac{4}{3} = \sec \theta \quad \textit{Answer}$$

b. Use $\sec \theta = \frac{1}{\cos \theta}$ to find cos θ.

$$-\frac{4}{3} = \frac{1}{\cos \theta}$$

$$-\frac{3}{4} = \cos \theta \quad \textit{Answer}$$

c. Use $\cos^2 \theta + \sin^2 \theta = 1$ to find sin θ. Since θ is a second-quadrant angle, sin θ will be positive.

$$\left(-\frac{3}{4}\right)^2 + \sin^2 \theta = 1$$

$$\frac{9}{16} + \sin^2 \theta = 1$$

$$\sin^2 \theta = \frac{7}{16}$$

$$\sin \theta = \frac{\sqrt{7}}{4} \quad \textit{Answer}$$

d. $\sin 2\theta = 2 \sin \theta \cos \theta$

$$= 2\left(\frac{\sqrt{7}}{4}\right)\left(-\frac{3}{4}\right)$$

$$= -\frac{3\sqrt{7}}{8} \quad \textit{Answer}$$

e. $\cos 2\theta = \cos^2 \theta - \sin^2 \theta$

$$= \left(-\tfrac{3}{4}\right)^2 - \left(\tfrac{\sqrt{7}}{4}\right)^2$$

$$= \tfrac{9}{16} - \tfrac{7}{16}$$

$$= \tfrac{2}{16} = \tfrac{1}{8} \ \textit{Answer}$$

f. $\tan 2\theta = \dfrac{2\tan\theta}{1 - \tan^2\theta}$

$$= \dfrac{2\left(-\tfrac{\sqrt{7}}{3}\right)}{1 - \left(-\tfrac{\sqrt{7}}{3}\right)^2}$$

$$= \dfrac{-\tfrac{2\sqrt{7}}{3}}{\tfrac{2}{9}} = -\tfrac{6\sqrt{7}}{2} = -3\sqrt{7} \ \textit{Answer}$$

g. Since sin 2θ and tan 2θ are negative and cos 2θ is positive, 2θ must be a fourth-quadrant angle. *Answer*

EXAMPLE 2

Prove the identity cos $2\theta = 2 \cos^2 \theta - 1$.

Solution   First write the identity for cos 2θ that we have already proved. Then substitute for sin^2 θ using the Pythagorean identity sin^2 $\theta = 1 - \cos^2 \theta$.

$$\cos 2\theta = \cos^2 \theta - \sin^2 \theta$$
$$= \cos^2 \theta - (1 - \cos^2 \theta)$$
$$= \cos^2 \theta - 1 + \cos^2 \theta$$
$$= 2 \cos^2 \theta - 1 ✔$$

Alternative Solution   To the right side of the identity for cos 2θ, add 0 in the form cos^2 θ − cos^2 θ.

$$\cos 2\theta = \cos^2 \theta - \sin^2 \theta$$
$$= \cos^2 \theta - \sin^2 \theta + \cos^2 \theta - \cos^2 \theta$$
$$= \cos^2 \theta + \cos^2 \theta - (\sin^2 \theta + \cos^2 \theta)$$
$$= 2 \cos^2 \theta - 1 ✔$$

SUMMARY
We have proved the following identities:

$$\sin 2A = 2 \sin A \cos A$$

$$\cos 2A = \cos^2 A - \sin^2 A$$

$$\cos 2A = 2 \cos^2 A - 1$$

$$\cos 2A = 1 - 2 \sin^2 A$$

$$\tan 2A = \dfrac{2 \tan A}{1 - \tan^2 A}$$

Exercises

Writing About Mathematics

1. Does $\cos 2\theta = \sin 2(90° - \theta)$? Justify your answer.

2. Does $\tan 2\theta = \frac{\sin 2\theta}{\cos 2\theta}$? Justify your answer.

Developing Skills

In 3–8, for each value of θ, use double-angle formulas to find **a.** $\sin 2\theta$, **b.** $\cos 2\theta$, **c.** $\tan 2\theta$. Show all work.

3. $\theta = 30°$

4. $\theta = 225°$

5. $\theta = 330°$

6. $\theta = \frac{\pi}{4}$

7. $\theta = \frac{7\pi}{6}$

8. $\theta = \frac{5\pi}{3}$

In 9–20, for each given function value, find **a.** $\sin 2\theta$, **b.** $\cos 2\theta$, **c.** $\tan 2\theta$, **d.** the quadrant in which 2θ lies. Show all work.

9. $\tan \theta = \frac{3}{5}$, θ in the first quadrant

10. $\tan \theta = -\frac{\sqrt{11}}{5}$ in the second quadrant

11. $\sin \theta = \frac{2\sqrt{6}}{7}$ in the first quadrant

12. $\sin \theta = -0.5$ in the third quadrant

13. $\cos \theta = \frac{2\sqrt{10}}{7}$ in the first quadrant

14. $\cos \theta = \frac{-2\sqrt{5}}{5}$ in the second quadrant

15. $\tan \theta = \frac{12}{5}$ in the third quadrant

16. $\sin \theta = -\frac{\sqrt{2}}{3}$ in the third quadrant

17. $\csc \theta = \sqrt{5}$ in the second quadrant

18. $\sec \theta = \frac{\sqrt{13}}{2}$ in the fourth quadrant

19. $\cot \theta = -\frac{1}{3}$ in the second quadrant

20. $\tan \theta = 2$ in the third quadrant

Applying Skills

21. Prove the identity: $\cot \theta = \frac{\sin 2\theta}{1 - \cos 2\theta}$.

22. Prove the identity: $\frac{\cos 2\theta}{\sin \theta} + \frac{\sin 2\theta}{\cos \theta} = \csc \theta$.

23. Show that $\cos 2\theta = 1 - 2\sin^2 \theta$.

24. Show that $\csc 2\theta = \frac{1}{2} \sec \theta \csc \theta$.

25. a. Derive an identity for $\sin 4A$ in terms of the functions of $2A$.

    b. Derive an identity for $\cos 4A$ in terms of the functions of $2A$.

    c. Derive an identity for $\tan 4A$ in terms of the functions of $2A$.

    Hint: Let $4A = 2\theta$.

26. A park in the shape of a rectangle, *ABCD*, is crossed by two paths: $\overline{AC}$, a diagonal, and $\overline{AE}$, which intersects $\overline{BC}$ at *E*. The measure of $\angle BAC$ is twice the measure of $\angle BAE$, $AB = 10$ miles and $AE = 12$ miles.

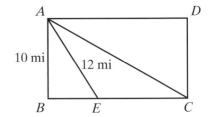

  a. Let m$\angle BAE = \theta$. Find the exact value of cos θ.

  b. Express m$\angle BAC$ in terms of θ and find the exact value of the cosine of $\angle BAC$.

  c. Find the exact value of *AC*.

27. The image of $A(4, 0)$ under a rotation of θ about the origin is $A'(\sqrt{10}, \sqrt{6})$. What are the coordinates of A'', the image of A' under the same rotation?

12-8 FUNCTIONS OF $\frac{1}{2}A$

Just as there are identities to find the function values of $2A$, there are identities to find cos $\frac{1}{2}A$, sin $\frac{1}{2}A$, and tan $\frac{1}{2}A$. We often call the identities used to find the function values of half an angle **half-angle formulas**.

Cosine of $\frac{1}{2}A$

We can use the identity for cos 2θ to write an identity for $\frac{1}{2}A$. Begin with the identity for cos 2θ written in terms of cos θ. Then solve for cos θ.

$$\cos 2\theta = 2\cos^2 \theta - 1$$
$$1 + \cos 2\theta = 2\cos^2 \theta$$
$$\frac{1 + \cos 2\theta}{2} = \cos^2 \theta$$
$$\pm\sqrt{\frac{1 + \cos 2\theta}{2}} = \cos \theta$$
$$\cos \theta = \pm\sqrt{\frac{1 + \cos 2\theta}{2}}$$

Let $2\theta = A$ and $\theta = \frac{1}{2}A$.

$$\cos \tfrac{1}{2}A = \pm\sqrt{\frac{1 + \cos A}{2}}$$

Sine of $\frac{1}{2}A$

Begin with the identity for cos 2θ, this time written in terms of sin θ. Then solve for sin θ.

$$\cos 2\theta = 1 - 2 \sin^2 \theta$$
$$2 \sin^2 \theta = 1 - \cos 2\theta$$
$$\sin^2 \theta = \frac{1 - \cos 2\theta}{2}$$
$$\sin \theta = \pm\sqrt{\frac{1 - \cos 2\theta}{2}}$$

Let $2\theta = A$ and $\theta = \frac{1}{2}A$.

$$\sin \tfrac{1}{2}A = \pm\sqrt{\frac{1 - \cos A}{2}}$$

Tangent of $\frac{1}{2}A$

Use the identity $\tan \theta = \frac{\sin \theta}{\cos \theta}$. Let $\theta = \frac{1}{2}A$. Then substitute in the values of $\sin \frac{1}{2}A$ and $\cos \frac{1}{2}A$.

$$\tan \tfrac{1}{2}A = \frac{\sin \frac{1}{2}A}{\cos \frac{1}{2}A}$$

$$\tan \tfrac{1}{2}A = \frac{\pm\sqrt{\frac{1 - \cos A}{2}}}{\pm\sqrt{\frac{1 + \cos A}{2}}}$$

$$\tan \tfrac{1}{2}A = \frac{\pm\sqrt{\frac{1 - \cos A}{2}}}{\pm\sqrt{\frac{1 + \cos A}{2}}} \times \frac{\frac{\sqrt{2}}{1}}{\frac{\sqrt{2}}{1}}$$

$$\tan \tfrac{1}{2}A = \pm\frac{\sqrt{1 - \cos A}}{\sqrt{1 + \cos A}}$$

When we use the identities for the function values of $(A + B), (A - B)$, and $2A$, the sign of the function value is a result of the computation. When we use the identities for the function values of $\frac{1}{2}A$, the sign of the function value must be chosen according to the quadrant in which $\frac{1}{2}A$ lies.

For example, if A is a third-quadrant angle such that $180° < A < 270°$, then $90° < \frac{1}{2}A < 135°$. Therefore, $\frac{1}{2}A$ is a second-quadrant angle. The sine value of $\frac{1}{2}A$ is positive and the cosine and tangent values of $\frac{1}{2}A$ are negative.

If A is a third-quadrant angle and $540° < A < 630°$, then $270° < \frac{1}{2}A < 315°$. Therefore, $\frac{1}{2}A$ is a fourth-quadrant angle. The cosine value of $\frac{1}{2}A$ is positive and the sine and tangent values of $\frac{1}{2}A$ are negative.

EXAMPLE 1

If $180° < A < 270°$ and $\sin A = -\frac{\sqrt{5}}{3}$, find:

a. $\sin \frac{1}{2}A$ **b.** $\cos \frac{1}{2}A$ **c.** $\tan \frac{1}{2}A$

Solution   The identities for the function values of $\frac{1}{2}A$ are given in terms of $\cos A$. Use $\cos A = \pm\sqrt{1 - \sin^2 A}$ to find $\cos A$. Since A is a third-quadrant angle, $\cos A$ is negative.

$$\cos A = -\sqrt{1 - \left(\frac{\sqrt{5}}{3}\right)^2} = -\sqrt{\frac{9}{9} - \frac{5}{9}} = -\sqrt{\frac{4}{9}} = -\frac{2}{3}$$

If $180° < A < 270°$, then $90° < \frac{1}{2}A < 135°$. Therefore, $\frac{1}{2}A$ is a second-quadrant angle: $\sin \frac{1}{2}A$ is positive and $\cos \frac{1}{2}A$ and $\tan \frac{1}{2}A$ are negative. Use the half-angle identities to find the function values of $\frac{1}{2}A$.

a. $\sin \frac{1}{2}A = \sqrt{\frac{1 - \cos A}{2}}$ **b.** $\cos \frac{1}{2}A = -\sqrt{\frac{1 + \cos A}{2}}$

$\qquad\quad = \sqrt{\dfrac{1 - \left(-\frac{2}{3}\right)}{2}}$ $\qquad\quad = -\sqrt{\dfrac{1 + \left(-\frac{2}{3}\right)}{2}}$

$\qquad\quad = \sqrt{\frac{5}{6}}$ $\qquad\qquad\quad = -\sqrt{\frac{1}{6}}$

$\qquad\quad = \frac{\sqrt{30}}{6}$ $\qquad\qquad\quad = -\frac{\sqrt{6}}{6}$

c. $\tan \frac{1}{2}A = -\sqrt{\frac{1 - \cos A}{1 + \cos A}}$

$\qquad\quad = -\sqrt{\dfrac{1 - \left(-\frac{2}{3}\right)}{1 + \left(-\frac{2}{3}\right)}}$

$\qquad\quad = -\sqrt{\frac{5}{1}}$

$\qquad\quad = -\sqrt{5}$

Answers   **a.** $\sin \frac{1}{2}A = \frac{\sqrt{30}}{6}$ **b.** $\cos \frac{1}{2}A = -\frac{\sqrt{6}}{6}$ **c.** $\tan \frac{1}{2}A = -\sqrt{5}$

EXAMPLE 2

Show that $\tan \frac{1}{2}A = \pm\frac{\sin A}{1 + \cos A}$.

Solution

$$\tan \tfrac{1}{2}A = \pm\sqrt{\dfrac{1 - \cos A}{1 + \cos A}}$$

$$= \pm\sqrt{\dfrac{1 - \cos A}{1 + \cos A} \times \dfrac{1 + \cos A}{1 + \cos A}}$$

$$= \pm\sqrt{\dfrac{1 - \cos^2 A}{(1 + \cos A)^2}}$$

$$= \pm\sqrt{\dfrac{\sin^2 A}{(1 + \cos A)^2}}$$

$$= \pm\dfrac{\sin A}{1 + \cos A} ✔$$

SUMMARY

We have proved the following identities:

$$\cos \tfrac{1}{2}A = \pm\sqrt{\dfrac{1 + \cos A}{2}}$$

$$\sin \tfrac{1}{2}A = \pm\sqrt{\dfrac{1 - \cos A}{2}}$$

$$\tan \tfrac{1}{2}A = \pm\sqrt{\dfrac{1 - \cos A}{1 + \cos A}}$$

Exercises

Writing About Mathematics

1. Karla said that if $\cos A$ is positive, then $-\frac{\pi}{2} < A < \frac{\pi}{2}$, $-\frac{\pi}{4} < \frac{1}{2}A < \frac{\pi}{4}$, and $\cos \frac{1}{2}A$ is positive. Do you agree with Karla? Explain why or why not.

2. In Example 1, can $\tan \frac{1}{2}A$ be found by using $\dfrac{\sin \frac{1}{2}A}{\cos \frac{1}{2}A}$? Explain why or why not.

Developing Skills

In 3–8, for each value of θ, use half-angle formulas to find **a.** $\sin \frac{1}{2}\theta$ **b.** $\cos \frac{1}{2}\theta$ **c.** $\tan \frac{1}{2}\theta$. Show all work.

3. $\theta = 480°$ **4.** $\theta = 120°$ **5.** $\theta = 300°$

6. $\theta = 2\pi$ **7.** $\theta = \frac{7\pi}{2}$ **8.** $\theta = \frac{3\pi}{2}$

In 9–14, for each value of $\cos A$, find **a.** $\sin \frac{1}{2}A$ **b.** $\cos \frac{1}{2}A$ **c.** $\tan \frac{1}{2}A$. Show all work.

9. $\cos A = \frac{3}{4}, 0° < A < 90°$ **10.** $\cos A = -\frac{5}{12}, 90° < A < 180°$

11. $\cos A = -\frac{1}{9}, 180° < A < 270°$ **12.** $\cos A = \frac{1}{8}, 270° < A < 360°$

13. $\cos A = -\frac{1}{5}, 450° < A < 540°$ **14.** $\cos A = \frac{7}{9}, 360° < A < 450°$

15. If $\sin A = \frac{24}{25}$ and $90° < A < 180°$, find: **a.** $\sin \frac{1}{2}A$ **b.** $\cos \frac{1}{2}A$ **c.** $\tan \frac{1}{2}A$

16. If $\sin A = -\frac{4}{5}$ and $180° < A < 270°$, find: **a.** $\sin \frac{1}{2}A$ **b.** $\cos \frac{1}{2}A$ **c.** $\tan \frac{1}{2}A$

17. If $\sin A = -\frac{24}{25}$ and $540° < A < 630°$, find: **a.** $\sin \frac{1}{2}A$ **b.** $\cos \frac{1}{2}A$ **c.** $\tan \frac{1}{2}A$

18. If $\tan A = 3$ and $180° < A < 270°$, find: **a.** $\sin \frac{1}{2}A$ **b.** $\cos \frac{1}{2}A$ **c.** $\tan \frac{1}{2}A$

Applying Skills

19. Show that $\tan \frac{1}{2}A = \pm \frac{1 - \cos A}{\sin A}$.

20. Use $\cos A = \cos \frac{\pi}{4} = \frac{\sqrt{2}}{2}$ to show that the exact value of $\tan \frac{\pi}{8} = \sqrt{2} - 1$.

21. Use $\cos A = \cos 30° = \frac{\sqrt{3}}{2}$ to show that the exact value of $\tan 15° = 2 - \sqrt{3}$.

22. Use $\cos A = \cos 30° = \frac{\sqrt{3}}{2}$ to show that the exact value of $\sin 15° = \frac{\sqrt{2 - \sqrt{3}}}{2}$.

23. a. Derive an identity for $\sin \frac{1}{4}A$ in terms of $\cos \frac{1}{2}A$.

   b. Derive an identity for $\cos \frac{1}{4}A$ in terms of $\cos \frac{1}{2}A$.

   c. Derive an identity for $\tan \frac{1}{4}A$ in terms of $\cos \frac{1}{2}A$.

    Hint: Let $\frac{1}{4}A = \frac{1}{2}\theta$.

24. The top of a billboard that is mounted on a base is 60 feet above the ground. At a point 25 feet from the foot of the base, the measure of the angle of elevation to the top of the base is one-half the measure of the angle of elevation to the top of the billboard.

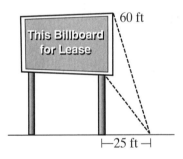

   a. Let θ be the measure of the angle of elevation to the top of the billboard. Find the $\cos \theta$ and $\tan \frac{1}{2}\theta$.

   b. Find the height of the base and the height of the billboard.

Hands-On Activity:
Graphical Support for the Trigonometric Identities

We can use the graphing calculator to provide support that an identity is true. Treat each side of the identity as a function and graph each function on the same set of axes. If the graphs of the functions coincide, then we have provided graphical support that the identity is true. Note that *support* is not the same as *proof*. In order to prove an identity, we need to use the algebraic methods from this chapter or similar algebraic methods.

For example, to provide support for the identity $\sin^2 \theta + \cos^2 \theta = 1$, treat each side as a function. In your graphing calculator, enter $Y_1 = \sin^2 \theta + \cos^2 \theta$ and $Y_2 = 1$. Graph both functions in the interval $-2\pi \le \theta \le 2\pi$. As the graph shows, both functions appear to coincide.

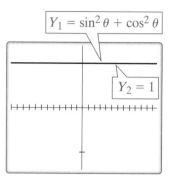

For 1–3, explore each equation on the graphing calculator. Do these equations appear to be identities?

1. $\cot \theta \sin 2\theta = 1 + \cos 2\theta$

2. $\sin^3 \theta = \dfrac{3 \sin \theta - \sin 3\theta}{4}$

3. $\tan^2 \theta \cos^2 \theta = 1 - \cos^2 \theta$

4. Kevin tried to provide graphical support for the identity $\cos \tfrac{1}{2}A = \pm\sqrt{\dfrac{1 + \cos A}{2}}$ by graphing the functions $Y_1 = \cos\left(\dfrac{X}{2}\right)$, $Y_2 = \sqrt{\dfrac{1 + \cos X}{2}}$, $Y_3 = -\sqrt{\dfrac{1 + \cos X}{2}}$ in the interval $0 \le X \le 2\pi$. The graphs did *not* coincide, as shown on the right. Explain to Kevin what he did wrong.

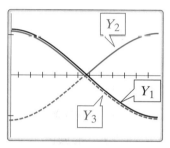

CHAPTER SUMMARY

A trigonometric identity can be proved by using the basic identities to change one side of the identity into the form of the other side.

The function values of $(A \pm B)$, $2A$, and $\tfrac{1}{2}A$, can be written in terms of the function values of A and of B, using the following identities:

Sums of Angle Measures	**Differences of Angle Measures**
$\cos (A + B) = \cos A \cos B - \sin A \sin B$	$\cos (A - B) = \cos A \cos B + \sin A \sin B$
$\sin (A + B) = \sin A \cos B + \cos A \sin B$	$\sin (A - B) = \sin A \cos B - \cos A \sin B$
$\tan (A + B) = \dfrac{\tan A + \tan B}{1 - \tan A \tan B}$	$\tan (A - B) = \dfrac{\tan A - \tan B}{1 + \tan A \tan B}$

Double-Angle Formulas	**Half-Angle Formulas**
$\sin (2A) = 2 \sin A \cos A$	$\sin \tfrac{1}{2}A = \pm\sqrt{\dfrac{1 - \cos A}{2}}$
$\cos (2A) = \cos^2 A - \sin^2 A$	$\cos \tfrac{1}{2}A = \pm\sqrt{\dfrac{1 + \cos A}{2}}$
$\cos (2A) = 2 \cos^2 A - 1$	$\tan \tfrac{1}{2}A = \pm\sqrt{\dfrac{1 - \cos A}{1 + \cos A}}$
$\cos (2A) = 1 - 2 \sin^2 A$	
$\tan (2A) = \dfrac{2 \tan A}{1 - \tan^2 A}$	

VOCABULARY

12-7 Double-angle formulas

12-8 Half-angle formulas

REVIEW EXERCISES

In 1–6, prove each identity.

1. $\sec \theta = \csc \theta \tan \theta$

2. $\cos \theta \cot \theta + \sin \theta = \csc \theta$

3. $2 \sin^2 \theta = 1 - \cos 2\theta$

4. $\tan \theta + \dfrac{1}{\csc \theta} = \dfrac{1 + \cos \theta}{\cot \theta}$

5. $\dfrac{\sin 2\theta + \sin \theta}{\cos 2\theta + \cos \theta + 1} = \tan \theta$

6. $\dfrac{\sin (A + B) + \sin (A - B)}{\sin (A + B) - \sin (A - B)} = \tan A \cot B$

In 7–21, $\sin A = -\frac{7}{25}$, $\sin B = -\frac{3}{5}$, and both A and B are third-quadrant angles. Find each function value.

7. $\cos A$

8. $\cos B$

9. $\tan A$

10. $\tan B$

11. $\sin (A + B)$

12. $\sin (A - B)$

13. $\cos (A + B)$

14. $\cos (A - B)$

15. $\tan (A + B)$

16. $\tan (A - B)$

17. $\sin 2A$

18. $\cos 2B$

19. $\tan 2A$

20. $\sin \frac{1}{2}A$

21. $\cos \frac{1}{2}A$

22. If $\cos A = 0.2$ and A and B are complementary angles, find $\cos B$.

23. If $\sin A = 0.6$ and A and B are supplementary angles, find $\cos B$.

In 24–32, θ is the measure of an acute angle and $\sin \theta = \frac{\sqrt{7}}{4}$. Find each function value.

24. $\cos \theta$

25. $\tan \theta$

26. $\sin 2\theta$

27. $\cos 2\theta$

28. $\tan 2\theta$

29. $\sin \frac{1}{2}\theta$

30. $\cos \frac{1}{2}\theta$

31. $\tan \frac{1}{2}\theta$

32. $\cos (2\theta + \theta)$

In 33–41, $360° < \theta < 450°$ and $\tan \theta = \frac{1}{7}$. Find, in simplest radical form, each function value.

33. $\sec \theta$

34. $\cos \theta$

35. $\sin \theta$

36. $\cot \theta$

37. $\sin 2\theta$

38. $\cos 2\theta$

39. $\tan 2\theta$

40. $\sin \frac{1}{2}\theta$

41. $\cos \frac{1}{2}\theta$

42. Show that if A and B are complementary, $\cos A \cos B = \sin A \sin B$.

Exploration

In this activity, you will derive the **triple-angle formulas**.

$$\sin(3A) = 3\sin A - 4\sin^3 A$$

$$\cos(3A) = 4\cos^3 A - 3\cos A$$

$$\tan(3A) = \frac{3\tan A - \tan^3 A}{1 - 3\tan^2 A}$$

 1. Examine the graphs of these equations on the graphing calculator in the interval $0 \le A \le 2\pi$. Explain why these equations appear to be identities.

2. Use $2A + A = 3A$ to prove that each equation is an identity.

CUMULATIVE REVIEW   CHAPTERS 1-12

Part I

Answer all questions in this part. Each correct answer will receive 2 credits. No partial credit will be allowed.

1. The solution set of $2 - \sqrt{x+3} = 6$ is
(1) $\{1\}$ (2) $\{-19\}$ (3) $\{13\}$ (4) $\varnothing$

2. The radian measure of an angle of $240°$ is
(1) $\frac{4\pi}{3}$ (2) $\frac{2\pi}{3}$ (3) $\frac{5\pi}{6}$ (4) $\frac{7\pi}{6}$

3. If $3^x = 27^{\frac{2}{3}}$, then x is equal to
(1) 1   (2) 2   (3) 3   (4) 4

4. When expressed in $a + bi$ form, $(12 + \sqrt{-9}) - (3 - \sqrt{4})$ is equal to
(1) $10 + 3i$ (2) $9 + 5i$ (3) $11 + 3i$ (4) $9 + i$

5. $\sum\limits_{n=0}^{3} 2^n$ is equal to
(1) 8   (2) 9   (3) 14   (4) 15

6. The fraction $\frac{3 + \sqrt{7}}{3 - \sqrt{7}}$ is equal to
(1) $8 + 3\sqrt{7}$ (3) $8 + \sqrt{7}$
(2) 8 (4) $8 + \frac{1}{2}\sqrt{7}$

7. If $f(x) = (x + 2)^2$ and $g(x) = x - 1$, then $g(f(x)) =$

(1) $(x + 2)^2 - 1$ (3) $x^2 + 5x + 3$

(2) $(x + 1)^2$ (4) $(2x + 1)^2$

8. The solution set of $2x^2 - 5x = 3$ is

(1) $\left\{\frac{1}{2}, -3\right\}$ (3) $\left\{\frac{3}{2}, 1\right\}$

(2) $\left\{-\frac{1}{2}, 3\right\}$ (4) $\left\{-\frac{3}{2}, -1\right\}$

9. When $\log x = 2 \log A - \frac{1}{2} \log B$, x is equal to

(1) $2A - \frac{1}{2}B$ (3) $\frac{A^2}{\sqrt{B}}$

(2) $A^2 - \sqrt{B}$ (4) $\frac{2A}{\frac{1}{2}B}$

10. If $y = \arccos 0$, then y is equal to

(1) 0 (2) 1 (3) $\frac{\pi}{2}$ (4) π

Part II

Answer all questions in this part. Each correct answer will receive 2 credits. Clearly indicate the necessary steps, including appropriate formula substitutions, diagrams, graphs, charts, etc. For all questions in this part, a correct numerical answer with no work shown will receive only 1 credit.

11. Write the multiplicative inverse of $1 + i$ in $a + bi$ form.

12. For what values of c and a does $x^2 + 5x + c = (x + a)^2$? Justify your answer.

Part III

Answer all questions in this part. Each correct answer will receive 4 credits. Clearly indicate the necessary steps, including appropriate formula substitutions, diagrams, graphs, charts, etc. For all questions in this part, a correct numerical answer with no work shown will receive only 1 credit.

13. Solve $x^2 + 3x - 10 \geq 0$ for x and graph the solution set on a number line.

14. Write the equation of a circle if the endpoints of a diameter of the circle are $(0, -1)$ and $(2, 5)$.

Part IV

Answer all questions in this part. Each correct answer will receive 6 credits. Clearly indicate the necessary steps, including appropriate formula substitutions, diagrams, graphs, charts, etc. For all questions in this part, a correct numerical answer with no work shown will receive only 1 credit.

15. a. Sketch the graph of $f(x) = \sin x$ in the interval $-2\pi \leq x \leq 2\pi$.

b. On the same set of axes, sketch the graph of $g(x) = f\left(x + \frac{\pi}{4}\right) + 3$ and write an equation for $g(x)$.

16. The first term of a geometric sequence is 1 and the fifth term is 9.

a. What is the common ratio of the sequence?

b. Write the first eight terms of the sequence.

c. Write the sum of the first eight terms of the sequence using sigma notation.

13

TRIGONOMETRIC EQUATIONS

The triangle is a rigid figure, that is, its shape cannot be changed without changing the lengths of its sides. This fact makes the triangle a basic shape in construction. The theorems of geometry give us relationships among the measures of the sides and angles of triangle. Precisely calibrated instruments enable surveyors to obtain needed measurements. The identities and formulas of trigonometry enable architects and builders to formulate plans needed to construct the roads, bridges, and buildings that are an essential part of modern life.

13-1 FIRST-DEGREE TRIGONOMETRIC EQUATIONS

A **trigonometric equation** is an equation whose variable is expressed in terms of a trigonometric function value. To solve a trigonometric equation, we use the same procedures that we used to solve algebraic equations. For example, in the equation $4 \sin \theta + 5 = 7$, $\sin \theta$ is multiplied by 4 and then 5 is added. Thus, to solve for $\sin \theta$, first add the opposite of 5 and then divide by 4.

$$\begin{array}{r} 4 \sin \theta + 5 = 7 \\ -5 = -5 \\ \hline 4 \sin \theta = 2 \\ \frac{4 \sin \theta}{4} = \frac{2}{4} \\ \sin \theta = \frac{1}{2} \end{array}$$

We know that $\sin 30° = \frac{1}{2}$, so one value of θ is $30°$; $\theta_1 = 30$. We also know that since $\sin \theta$ is positive in the second quadrant, there is a second-quadrant angle, θ_2, whose sine is $\frac{1}{2}$. Recall the relationship between an angle in any quadrant to the acute angle called the reference angle. The following table compares the degree measures of θ from $-90°$ to $360°$, the radian measures of θ from $-\frac{\pi}{2}$ to 2π, and the measure of its reference angle.

	Fourth Quadrant	First Quadrant	Second Quadrant	Third Quadrant	Fourth Quadrant
Angle	$-90° < \theta < 0°$ $-\frac{\pi}{2} < \theta < 0$	$0° < \theta < 90°$ $0 < \theta < \frac{\pi}{2}$	$90° < \theta < 180°$ $\frac{\pi}{2} < \theta < \pi$	$180° < \theta < 270°$ $\pi < \theta < \frac{3\pi}{2}$	$270° < \theta < 360°$ $\frac{3\pi}{2} < \theta < 2\pi$
Reference Angle	$-\theta$ $-\theta$	θ θ	$180° - \theta$ $\pi - \theta$	$\theta - 180°$ $\theta - \pi$	$360° - \theta$ $2\pi - \theta$

The reference angle for the second-quadrant angle whose sine is $\frac{1}{2}$ has a degree measure of $30°$ and $\theta_2 = 180° - 30°$ or $150°$. Therefore, $\sin \theta_2 = \frac{1}{2}$. For $0° \leq \theta < 360°$, the solution set of $4 \sin \theta + 5 = 7$ is $\{30°, 150°\}$. In radian measure, the solution set is $\{\frac{\pi}{6}, \frac{5\pi}{6}\}$.

In the example given above, it was possible to give the exact value of θ that makes the equation true. Often it is necessary to use a calculator to find an approximate value. Consider the solution of the following equation.

$$5 \cos \theta + 7 = 3$$
$$5 \cos \theta = -4$$
$$\cos \theta = -\frac{4}{5}$$
$$\theta = \arccos\left(-\frac{4}{5}\right)$$

When we use a calculator to find θ, the calculator will return the value of the function $y = \arccos x$ whose domain is $0° \leq x \leq 180°$ in degree measure or $0 \leq x \leq \pi$ in radian measure.

In degree measure:

ENTER: [2nd] [COS⁻¹] [(-)] 4 [÷]   DISPLAY:

5 [)] [ENTER]

```
cos⁻¹(-4/5)
           143.1301024
```

To the nearest degree, one value of θ is 143°. In addition to this second-quadrant angle, there is a third-quadrant angle such that $\cos \theta = -\frac{4}{5}$. To find this third-quadrant angle, find the reference angle for θ.

Let R be the measure of the reference angle of the second-quadrant angle. That is, R is the acute angle such that $\cos \theta = -\cos R$.

$$R = 180° - \theta = 180° - 143° = 37°$$

The measure of the third-quadrant angle is:

$$\theta = R + 180°$$
$$\theta = 37° + 180°$$
$$\theta = 217°$$

For $0° \leq \theta \leq 360°$, the solution set of $5 \cos \theta + 7 = 3$ is $\{143°, 217°\}$. If the value of θ can be any angle measure, then for all integral values of n, $\theta = 143 + 360n$ or $\theta = 217 + 360n$.

Procedure

To solve a linear trigonometric equation:

1. Solve the equation for the function value of the variable.

2. Use a calculator or your knowledge of the exact function values to write one value of the variable to an acceptable degree of accuracy.

3. If the measure of the angle found in step 2 is not that of a quadrantal angle, find the measure of its reference angle.

4. Use the measure of the reference angle to find the degree measures of each solution in the interval $0° \leq \theta < 360°$ or the radian measures of each solution in the interval $0 \leq \theta < 2\pi$.

5. Add $360n$ (n an integer) to the solutions in degrees found in steps 2 and 4 to write all possible solutions in degrees. Add $2\pi n$ (n an integer) to the solutions in radians found in steps 2 and 4 to write all possible solutions in radians.

The following table will help you find the locations of the angles that satisfy trigonometric equations. The values in the table follow from the definitions of the trigonometric functions on the unit circle.

	Sign of a and b $(0 < |a| < 1, b \neq 0)$	
	+	**−**
$\sin \theta = a$	Quadrants I and II	Quadrants III and IV
$\cos \theta = a$	Quadrants I and IV	Quadrants II and III
$\tan \theta = b$	Quadrants I and III	Quadrants II and IV

EXAMPLE I

Find the solution set of the equation $7 \tan \theta = 2\sqrt{3} + \tan \theta$ in the interval $0° \leq \theta < 360°$.

Solution

How to Proceed

(1) Solve the equation for $\tan \theta$:

$$7 \tan \theta = 2\sqrt{3} + \tan \theta$$
$$6 \tan \theta = 2\sqrt{3}$$
$$\tan \theta = \frac{\sqrt{3}}{3}$$

(2) Since $\tan \theta$ is positive, θ_1 can be a first-quadrant angle:

$$\theta_1 = 30°$$

(3) Since θ is a first-quadrant angle, $R = \theta$:

$$R = 30°$$

(4) Tangent is also positive in the third quadrant. Therefore, there is a third-quadrant angle such that $\tan \theta = \frac{\sqrt{3}}{3}$. In the third quadrant, $\theta_2 = 180° + R$:

$$\theta_2 = 180° + R$$
$$\theta_2 = 180° + 30° = 210°$$

Answer The solution set is $\{30°, 210°\}$.

EXAMPLE 2

Find, to the nearest hundredth, all possible solutions of the following equation in radians:

$$3(\sin A + 2) = 3 - \sin A$$

Solution

How to Proceed

(1) Solve the equation for $\sin A$:

$$3(\sin A + 2) = 3 - \sin A$$
$$3 \sin A + 6 = 3 - \sin A$$
$$4 \sin A = -3$$
$$\sin A = -\tfrac{3}{4}$$

(2) Use a calculator to find one value of A (be sure that the calculator is in RADIAN mode):

ENTER: [2nd] [SIN⁻¹] [(-)] 3 [÷] 4 [ )]

[ENTER]

DISPLAY:

```
SIN⁻¹(-3/4)
        -.848062079
```

One value of A is -0.848.

(3) Find the reference angle:

$$R = -A = -(-0.848) = 0.848$$

(4) Sine is negative in quadrants III and IV. Use the reference angle to find a value of A in each of these quadrants:

In quadrant III: $A_1 = \pi + 0.848 \approx 3.99$
In quadrant IV: $A_2 = 2\pi - 0.848 \approx 5.44$

(5) Write the solution set:

$\{3.99 + 2\pi n, 5.44 + 2\pi n\}$ *Answer* ▮

Note: When $\sin^{-1}\left(-\tfrac{3}{4}\right)$ was entered, the calculator returned the value in the interval $-\tfrac{\pi}{2} \le \theta \le \tfrac{\pi}{2}$, the range of the inverse of the sine function. This is the measure of a fourth-quadrant angle that is a solution of the equation. However, solutions are usually given as angle measures in radians between 0 and 2π plus multiples of 2π. Note that the value returned by the calculator is $5.43 + 2\pi(-1) \approx -0.85$.

EXAMPLE 3

Find all possible solutions to the following equation in degrees:

$$\tfrac{1}{2}(\sec \theta + 3) = \sec \theta + \tfrac{5}{2}$$

Solution (1) Solve the equation for $\sec\theta$:

$$\tfrac{1}{2}(\sec\theta + 3) = \sec\theta + \tfrac{5}{2}$$
$$\sec\theta + 3 = 2\sec\theta + 5$$
$$-2 = \sec\theta$$

(2) Rewrite the equation in terms of $\cos\theta$:

$$\cos\theta = -\tfrac{1}{2}$$

(3) Use a calculator to find one value of θ:

ENTER: 2nd   COS⁻¹   −1   ÷   2   )

       ENTER

DISPLAY:

```
cos⁻¹(-1/2)
                    120
```

Cosine is negative in quadrant II so $\theta_1 = 120°$.

(4) Find the reference angle:

$$R = 180 - 120 = 60°$$

(5) Cosine is also negative in quadrant III. Use the reference angle to find a value of θ_2 in quadrant III:

In quadrant III: $\theta_2 = 180 + 60 = 240°$

(6) Write the solution set:

$\{120 + 360n, 240 + 360n\}$ *Answer* ◻

Trigonometric Equations and the Graphing Calculator

Just as we used the graphing calculator to approximate the irrational solutions of quadratic-linear systems in Chapter 5, we can use the graphing calculator to approximate the irrational solutions of trigonometric equations. For instance,

$$3(\sin A + 2) = 3 - \sin A$$

from Example 2 can be solved using the intersect feature of the calculator.

STEP 1. Treat each side of the equation as a function.

     Enter $Y_1 = 3(\sin X + 2)$ and $Y_2 = 3 - \sin X$ into the Y= menu.

STEP 2. Using the following viewing window:

$$Xmin = 0, Xmax = 2\pi, Xscl = \tfrac{\pi}{6},$$
$$Ymin = 0, Ymax = 10$$

and with the calculator set to radian mode, GRAPH the functions.

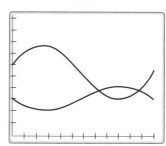

STEP 3. The solutions are the *x*-coordinates of the intersection points of the graphs. We can find the intersection points by using the intersect function. Press `2nd` `CALC` 5 `ENTER` `ENTER` to select both curves. When the calculator asks you for a guess, move the cursor near one of the intersection points using the arrow keys and then press `ENTER`. Repeat this process to find the other intersection point.

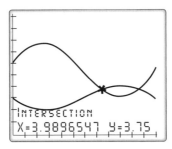 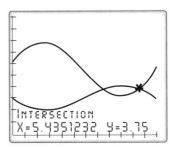

As before, the solutions in the interval $0 \leq \theta < 2\pi$ are approximately 3.99 and 5.44. The solution set is $\{3.99 + 2\pi n, 5.44 + 2\pi n\}$.

Exercises

Writing About Mathematics

1. Explain why the solution set of the equation $2x + 4 = 8$ is $\{2\}$ but the solution set of the equation $2 \sin x + 4 = 8$ is $\{\ \}$, the empty set.

2. Explain why $2x + 4 = 8$ has only one solution in the set of real numbers but the equation $2 \tan x + 4 = 8$ has infinitely many solutions in the set of real numbers.

Developing Skills

In 3–8, find the exact solution set of each equation if $0° \leq \theta < 360°$.

3. $2 \cos \theta - 1 = 0$

4. $3 \tan \theta + \sqrt{3} = 0$

5. $4 \sin \theta - 1 = 2 \sin \theta + 1$

6. $5(\cos \theta + 1) = 5$

7. $3(\tan \theta - 2) = 2 \tan \theta - 7$

8. $\sec \theta + \sqrt{2} = 2\sqrt{2}$

In 9–14, find the exact values for θ in the interval $0 \leq \theta < 2\pi$.

9. $3 \sin \theta - \sqrt{3} = \sin \theta$

10. $5 \cos \theta + 3 = 3 \cos \theta + 5$

11. $\tan \theta + 12 = 2 \tan \theta + 11$

12. $\sin \theta + \sqrt{2} = \frac{\sqrt{2}}{2}$

13. $3 \csc \theta + 5 = \csc \theta + 9$

14. $4(\cot \theta + 1) = 2(\cot \theta + 2)$

In 15–20, find, to the nearest degree, the measure of an acute angle for which the given equation is true.

15. $\sin \theta + 3 = 5 \sin \theta$

16. $3 \tan \theta - 1 = \tan \theta + 9$

17. $5 \cos \theta + 1 = 8 \cos \theta$

18. $4(\sin \theta + 1) = 6 - \sin \theta$

19. $\csc \theta - 1 = 3 \csc \theta - 11$

20. $\cot \theta + 8 = 3 \cot \theta + 2$

In 21–24, find, to the nearest tenth, the degree measures of all θ in the interval $0° \leq \theta < 360°$ that make the equation true.

21. $8 \cos \theta = 3 - 4 \cos \theta$

22. $5 \sin \theta - 1 = 1 - 2 \sin \theta$

23. $\tan \theta - 4 = 3 \tan \theta + 4$

24. $2 - \sec \theta = 5 + \sec \theta$

In 25–28, find, to the nearest hundredth, the radian measures of all θ in the interval $0 \leq \theta < 2\pi$ that make the equation true.

25. $10 \sin \theta + 1 = 3 - 2 \sin \theta$

26. $9 - 2 \cos \theta = 8 - 4 \cos \theta$

27. $15 \tan \theta - 7 = 5 \tan \theta - 3$

28. $\cot \theta - 6 = 2 \cot \theta + 2$

Applying Skills

29. The voltage E (in volts) in an electrical circuit is given by the function

$$E = 20 \cos (\pi t)$$

where t is time in seconds.

a. Graph the voltage E in the interval $0 \leq t \leq 2$.

b. What is the voltage of the electrical circuit when $t = 1$?

c. How many times does the voltage equal 12 volts in the first two seconds?

d. Find, to the nearest hundredth of a second, the times in the first two seconds when the voltage is equal to 12 volts.

(1) Let $\theta = \pi t$. Solve the equation $20 \cos \theta = 12$ in the interval $0 \leq \theta < 2\pi$.

(2) Use the formula $\theta = \pi t$ and your answers to part (1) to find t when $0 \leq \theta < 2\pi$ and the voltage is equal to 12 volts.

30. A water balloon leaves the air cannon at an angle of θ with the ground and an initial velocity of 40 feet per second. The water balloon lands 30 feet from the cannon. The distance d traveled by the water balloon is given by the formula

$$d = \tfrac{1}{32} v^2 \sin 2\theta$$

where v is the initial velocity.

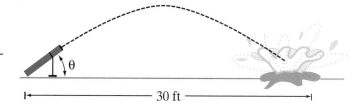

a. Let $x = 2\theta$. Solve the equation $30 = \frac{1}{32}(40)^2 \sin\ x$ to the nearest tenth of a degree.

b. Use the formula $x = 2\theta$ and your answer to part **a** to find the measure of the angle that the cannon makes with the ground.

31. It is important to understand the underlying mathematics before using the calculator to solve trigonometric equations. For example, Adrian tried to use the intersect feature of his graphing calculator to find the solutions of the equation $\cot\theta = \sin\left(\theta - \frac{\pi}{2}\right)$ in the interval $0 \le \theta \le \pi$ but got an error message. Follow the steps that Adrian used to solve the equation:

(1) Enter $Y_1 = \frac{1}{\tan\ X}$ and $Y_2 = \sin\left(X - \frac{\pi}{2}\right)$ into the $\boxed{\textsf{Y=}}$ menu.

(2) Use the following viewing window to graph the equations:

$$Xmin = 0,\ Xmax = \pi,\ Xscl = \tfrac{\pi}{6},\ Ymin = -5,\ Ymax = 5$$

(3) The curves seem to intersect at $\left(\frac{\pi}{2}, 0\right)$. Press $\boxed{\textsf{2nd}}$ $\boxed{\textsf{CALC}}$ 5 $\boxed{\textsf{ENTER}}$ $\boxed{\textsf{ENTER}}$ to select both curves. When the calculator asks for a guess, move the cursor near the intersection point using the arrow keys and then press $\boxed{\textsf{ENTER}}$.

a. Why does the calculator return an error message?

b. Is $\theta = \frac{\pi}{2}$ a solution to the equation? Explain.

13-2 USING FACTORING TO SOLVE TRIGONOMETRIC EQUATIONS

We know that the equation $3x^2 - 5x - 2 = 0$ can be solved by factoring the left side and setting each factor equal to 0. The equation $3\tan^2\theta - 5\tan\theta - 2 = 0$ can be solved for $\tan\theta$ in a similar way.

$$3x^2 - 5x - 2 = 0 \qquad\qquad 3\tan^2\theta - 5\tan\theta - 2 = 0$$
$$(3x + 1)(x - 2) = 0 \qquad\qquad (3\tan\theta + 1)(\tan\theta - 2) = 0$$

$3x + 1 = 0$	$x - 2 = 0$	$3\tan\theta + 1 = 0$	$\tan\theta - 2 = 0$
$3x = -1$	$x = 2$	$3\tan\theta = -1$	$\tan\theta = 2$
$x = -\frac{1}{3}$		$\tan\theta = -\frac{1}{3}$	

In the solution of the algebraic equation, the solution is complete. The solution set is $\left\{-\frac{1}{3}, 2\right\}$. In the solution of the trigonometric equation, we must now find the values of θ.

There are two values of θ in the interval $0° \le \theta < 360°$ for which $\tan\theta = 2$, one in the first quadrant and one in the third quadrant. The calculator will display the measure of the first-quadrant angle, which is also the reference angle for the third-quadrant angle.

ENTER: 2nd TAN⁻¹ 2) ENTER   DISPLAY:

$$TAN^{-1}(2)$$
$$63.43494882$$

To the nearest tenth of a degree, the measure of θ in the first quadrant is 63.4. This is also the reference angle for the third-quadrant angle.

In quadrant I: $\theta_1 = 63.4°$
In quadrant III: $\theta_2 = 180 + R = 180 + 63.4 = 243.4°$

There are two values of θ in the interval $0° \leq \theta < 360°$ for which $\tan \theta = -\frac{1}{3}$, one in the second quadrant and one in the fourth quadrant.

ENTER: 2nd TAN⁻¹ −1 ÷ 3)   DISPLAY:
ENTER

$$TAN^{-1}(-1/3)$$
$$-18.43494882$$

The calculator will display the measure of a fourth-quadrant angle, which is negative. To the nearest tenth of a degree, one measure of θ in the fourth quadrant is -18.4. The opposite of this measure, 18.4, is the measure of the reference angle for the second- and fourth-quadrant angles.

In quadrant II: $\theta_3 = 180 - 18.4 = 161.6°$
In quadrant IV: $\theta_4 = 360 - 18.4 = 341.6°$

The solution set of $3 \tan^2 \theta - 5 \tan \theta - 2 = 0$ is

$$\{63.4°, 161.6°, 243.4°, 341.6°\}$$

when $0° \leq \theta < 360°$.

EXAMPLE I

Find all values of θ in the interval $0 \leq \theta < 2\pi$ for which $2 \sin \theta - 1 = \frac{3}{\sin \theta}$.

Solution   *How to Proceed*

(1) Multiply both sides of the equation by $\sin \theta$:

$$2 \sin \theta - 1 = \frac{3}{\sin \theta}$$
$$2 \sin^2 \theta - \sin \theta = 3$$

(2) Write an equivalent equation with 0 as the right side:

$$2 \sin^2 \theta - \sin \theta - 3 = 0$$

(3) Factor the left side:

$$(2 \sin \theta - 3)(\sin \theta + 1) = 0$$

(4) Set each factor equal to 0 and solve for $\sin \theta$:

$$2 \sin \theta - 3 = 0 \quad | \quad \sin \theta + 1 = 0$$
$$2 \sin \theta = 3 \quad \quad \quad \sin \theta = -1$$
$$\sin \theta = \frac{3}{2}$$

(5) Find all possible values of θ:

There is no value of θ such that $\sin \theta > 1$. For $\sin \theta = -1, \theta = \frac{3\pi}{2}$.

Answer $\theta = \frac{3\pi}{2}$

EXAMPLE 2

Find the solution set of $4 \sin^2 A - 1 = 0$ for the degree measures of A in the interval $0° \le A < 360°$.

Solution

METHOD 1	**METHOD 2**

METHOD 1

Factor the left side.

$$4 \sin^2 A - 1 = 0$$

$$(2 \sin A - 1)(2 \sin A + 1) = 0$$

$2 \sin A - 1 = 0$	$2 \sin A + 1 = 0$
$2 \sin A = 1$	$2 \sin A = -1$
$\sin A = \frac{1}{2}$	$\sin A = -\frac{1}{2}$

METHOD 2

Solve for $\sin^2 A$ and take the square root of each side of the equation.

$$4 \sin^2 A - 1 = 0$$

$$4 \sin^2 A = 1$$

$$\sin^2 A = \frac{1}{4}$$

$$\sin A = \pm\frac{1}{2}$$

If $\sin A = \frac{1}{2}$, $A = 30°$ or $A = 150°$. If $\sin A = -\frac{1}{2}$, $A = 210°$ or $A = 330°$.

Answer $\{30°, 150°, 210°, 330°\}$

Note: The graphing calculator does not use the notation $\sin^2 A$, so we must enter the square of the trig function as $(\sin A)^2$ or enter $\sin (A)^2$. For example, to check the solution $A = 30°$ for Example 2:

ENTER: 4 $($ $\boxed{\text{SIN}}$ 30 $)$ $)$ $\boxed{x^2}$ $\boxed{-}$ 1 $\boxed{\text{ENTER}}$

4 $\boxed{\text{SIN}}$ 30 $)$ $\boxed{x^2}$ $\boxed{-}$ 1 $\boxed{\text{ENTER}}$

DISPLAY:

```
4(sin(30))2-1
                    0
4sin(30)2-1
                    0
```

Factoring Equations with Two Trigonometric Functions

To solve an equation such as $2 \sin \theta \cos \theta + \sin \theta = 0$, it is convenient to rewrite the left side so that we can solve the equation with just one trigonometric function value. In this equation, we can rewrite the left side as the product of two factors. Each factor contains one function.

$$2 \sin \theta \cos \theta + \sin \theta = 0$$
$$\sin \theta (2 \cos \theta + 1) = 0$$

$\sin \theta = 0$	$2 \cos \theta + 1 = 0$
$\theta_1 = 0°$	$2 \cos \theta = -1$
$\theta_2 = 180°$	$\cos \theta = -\frac{1}{2}$

Since $\cos 60° = \frac{1}{2}$, $R = 60°$.

In quadrant II: $\theta_3 = 180 - 60 = 120°$
In quadrant III: $\theta_4 = 180 + 60 = 240°$

The solution set is $\{0°, 120°, 180°, 240°\}$.

EXAMPLE 3

Find, in radians, all values of θ in the interval $0 \leq \theta < 2\pi$ that are in the solution set of:

$$\sec \theta \csc \theta + \sqrt{2} \csc \theta = 0$$

Solution Factor the left side of the equation and set each factor equal to 0.

$$\sec \theta \csc \theta + \sqrt{2} \csc \theta = 0$$
$$\csc \theta (\sec \theta + \sqrt{2}) = 0$$

$\csc \theta = 0$ ✗	$\sec \theta + \sqrt{2} = 0$
No solution	$\sec \theta = -\sqrt{2}$
	$\cos \theta = -\frac{1}{\sqrt{2}} = -\frac{\sqrt{2}}{2}$

Since $\cos \frac{\pi}{4} = \frac{\sqrt{2}}{2}$, $R = \frac{\pi}{4}$.

In quadrant II: $\theta = \pi - \frac{\pi}{4} = \frac{3\pi}{4}$
In quadrant III: $\theta = \pi + \frac{\pi}{4} = \frac{5\pi}{4}$

Answer $\left\{\frac{3\pi}{4}, \frac{5\pi}{4}\right\}$.

Exercises

Writing About Mathematics

1. Can the equation $\tan \theta + \sin \theta \tan \theta = 1$ be solved by factoring the left side of the equation? Explain why or why not.

2. Can the equation $2(\sin \theta)(\cos \theta) + \sin \theta + 2 \cos \theta + 1 = 0$ be solved by factoring the left side of the equation? Explain why or why not.

Developing Skills

In 3–8, find the exact solution set of each equation if $0° \le \theta < 360°$.

3. $2 \sin^2 \theta + \sin \theta - 1 = 0$

4. $3 \tan^2 \theta = 1$

5. $\tan^2 \theta - 3 = 0$

6. $2 \sin^2 \theta - 1 = 0$

7. $6 \cos^2 \theta + 5 \cos \theta - 4 = 0$

8. $2 \sin \theta \cos \theta + \cos \theta = 0$

In 9–14, find, to the nearest tenth of a degree, the values of θ in the interval $0° \le \theta < 360°$ that satisfy each equation.

9. $\tan^2 \theta - 3 \tan \theta + 2 = 0$

10. $3 \cos^2 \theta - 4 \cos \theta + 1 = 0$

11. $9 \sin^2 \theta - 9 \sin \theta + 2 = 0$

12. $25 \cos^2 \theta - 4 = 0$

13. $\tan^2 \theta + 4 \tan \theta - 12 = 0$

14. $\sec^2 \theta - 7 \sec \theta + 12 = 0$

In 15–20, find, to the nearest hundredth of a radian, the values of θ in the interval $0 \le \theta < 2\pi$ that satisfy the equation.

15. $\tan^2 \theta - 5 \tan \theta + 6 = 0$

16. $4 \cos^2 \theta - 3 \cos \theta = 1$

17. $5 \sin^2 \theta + 2 \sin \theta = 0$

18. $3 \sin^2 \theta + 7 \sin \theta + 2 = 0$

19. $\csc^2 \theta - 6 \csc \theta + 8 = 0$

20. $2 \cot^2 \theta - 13 \cot \theta + 6 = 0$

21. Find the smallest positive value of θ such that $4 \sin^2 \theta - 1 = 0$.

22. Find, to the nearest hundredth of a radian, the value of θ such that $\sec \theta = \frac{5}{\sec \theta}$ and $\frac{\pi}{2} < \theta < \pi$.

23. Find two values of A such that $(\sin A)(\csc A) = -\sin A$.

13-3 USING THE QUADRATIC FORMULA TO SOLVE TRIGONOMETRIC EQUATIONS

Not all quadratic equations can be solved by factoring. It is often useful or necessary to use the quadratic formula to solve a second-degree trigonometric equation.

The trigonometric equation $2 \cos^2 \theta - 4 \cos \theta + 1 = 0$ is similar in form to the algebraic equation $2x^2 - 4x + 1 = 0$. Both are quadratic equations that cannot be solved by factoring over the set of integers but can be solved by using the quadratic formula with $a = 2$, $b = -4$, and $c = 1$.

Algebraic equation:	*Trigonometric equation:*
$2x^2 - 4x + 1 = 0$	$2\cos^2\theta - 4\cos\theta + 1 = 0$

$$x = \frac{-b \pm \sqrt{b^2 - 4ac}}{2a} \qquad \cos\theta = \frac{-b \pm \sqrt{b^2 - 4ac}}{2a}$$

$$x = \frac{-(-4) \pm \sqrt{(-4)^2 - 4(2)(1)}}{2(2)} \qquad \cos\theta = \frac{-(-4) \pm \sqrt{(-4)^2 - 4(2)(1)}}{2(2)}$$

$$x = \frac{4 \pm \sqrt{16 - 8}}{4} \qquad \cos\theta = \frac{4 \pm \sqrt{16 - 8}}{4}$$

$$x = \frac{4 \pm \sqrt{8}}{4} \qquad \cos\theta = \frac{4 \pm \sqrt{8}}{4}$$

$$x = \frac{2 \pm \sqrt{2}}{2} \qquad \cos\theta = \frac{2 \pm \sqrt{2}}{2}$$

There are no differences between the two solutions up to this point. However, for the algebraic equation, the solution is complete. There are two values of x that make the equation true: $x = \frac{2 + \sqrt{2}}{2}$ or $x = \frac{2 - \sqrt{2}}{2}$.

For the trigonometric equation, there appear to be two values of $\cos\theta$. Can we find values of θ for each of these two values of $\cos\theta$?

CASE 1 $\cos\theta = \frac{2 + \sqrt{2}}{2} \approx \frac{1 + 1.414}{2} = 1.207$

There is no value of θ such that $\cos\theta > 1$.

CASE 2 $\cos\theta = \frac{2 - \sqrt{2}}{2} \approx \frac{2 - 1.414}{2} = 0.293$

There are values of θ in the first quadrant and in the fourth quadrant such that $\cos\theta$ is a positive number less than 1. Use a calculator to approximate these values to the nearest degree.

ENTER: [2nd] [COS⁻¹] [(] [2] [−]   DISPLAY: `cos⁻¹((2-√(2))/2)` `72.96875154`

[2nd] [√] [2] [)] [)]

[÷] [2] [)] [ENTER]

To the nearest degree, the value of θ in the first quadrant is 73°. This is also the value of the reference angle. Therefore, in the fourth quadrant, $\theta = 360° - 73°$ or 287°.

In the interval $0° \le \theta < 360°$, the solution set of $2\cos^2\theta - 4\cos\theta + 1 = 0$ is:

$$\{73°, 287°\}$$

EXAMPLE 1

a. Use three different methods to solve $\tan^2 \theta - 1 = 0$ for $\tan \theta$.

b. Find all possible values of θ in the interval $0 \leq \theta < 2\pi$.

Solution **a.**

METHOD 1: FACTOR

$$\tan^2 \theta - 1 = 0$$
$$(\tan \theta + 1)(\tan \theta - 1) = 0$$

$\tan \theta + 1 = 0 \quad | \quad \tan \theta - 1 = 0$

$\tan \theta = -1 \quad | \quad \tan \theta = 1$

METHOD 2: SQUARE ROOT

$$\tan^2 \theta - 1 = 0$$
$$\tan^2 \theta = 1$$
$$\tan \theta = \pm 1$$

METHOD 3: QUADRATIC FORMULA

$$\tan^2 \theta - 1 = 0$$
$$a = 1, b = 0, c = -1$$
$$\tan \theta = \frac{-0 \pm \sqrt{0^2 - 4(1)(-1)}}{2(1)} = \frac{\pm \sqrt{4}}{2} = \frac{\pm 2}{2} = \pm 1$$

b. When $\tan \theta = 1$, θ is in quadrant I or in quadrant III.

In quadrant I: if $\tan \theta = 1$, $\theta = \frac{\pi}{4}$

This is also the measure of the reference angle.

In quadrant III: $\theta = \pi + \frac{\pi}{4} = \frac{5\pi}{4}$

When $\tan \theta = -1$, θ is in quadrant II or in quadrant IV.

One value of θ is $-\frac{\pi}{4}$. The reference angle is $\frac{\pi}{4}$.

In quadrant II: $\theta = \pi - \frac{\pi}{4} = \frac{3\pi}{4}$

In quadrant IV: $\theta = 2\pi - \frac{\pi}{4} = \frac{7\pi}{4}$

Answer $\left\{ \frac{\pi}{4}, \frac{3\pi}{4}, \frac{5\pi}{4}, \frac{7\pi}{4} \right\}$

EXAMPLE 2

Find, to the nearest degree, all possible values of B such that:

$$3 \sin^2 B + 3 \sin B - 2 = 0$$

Solution

$$a = 3, b = 3, c = -2$$

$$\sin B = \frac{-3 \pm \sqrt{(3)^2 - 4(3)(-2)}}{2(3)} = \frac{-3 \pm \sqrt{9 + 24}}{6} = \frac{-3 \pm \sqrt{33}}{6}$$

CASE 1 Let $\sin B = \frac{-3 + \sqrt{33}}{6}$.

Since $\frac{-3 + \sqrt{33}}{6} \approx 0.46$ is a number between -1 and 1, it is in the range of the sine function.

ENTER: [2nd] [SIN⁻¹] [( ] [−3] [+]   DISPLAY: sin⁻¹((⁻3+√(33))
   [2nd] [√] 33 [ )] [ )]   /6)   27.22120768
   [÷] 6 [ )] [ENTER]

The sine function is positive in the first and second quadrants.

In quadrant I: $B = 27°$

In quadrant II: $B = 180 - 27 = 153°$

CASE 2 Let $\sin B = \dfrac{-3 - \sqrt{33}}{6}$.

Since $\dfrac{-3 - \sqrt{33}}{6} \approx -1.46$ is not a number between -1 and 1, it is not in the range of the sine function. There are no values of B such that $\sin B = \dfrac{-3 - \sqrt{33}}{6}$. ✗

Calculator Solution   Enter $Y_1 = 3 \sin^2 X + 3 \sin X - 2$ into the [Y=] menu.

ENTER: [Y=] 3 [SIN] [X,T,θ,n] [ )] [x²] [+] 3 [SIN] [X,T,θ,n] [ )] [−] 2

With the calculator set to DEGREE mode, graph the function in the following viewing window:

$Xmin = 0$, $Xmax = 360$, $Xscl = 30$, $Ymin = -5$, $Ymax = 5$

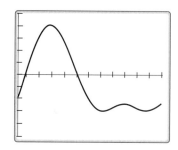

The solutions are the x-coordinates of the x-intercepts, that is, the roots of Y_1. Use the zero function of your graphing calculator to find the roots. Press [2nd] [CALC] [2]. Use the arrows to enter a left bound to the left of one of the zero values, a right bound to the right of the zero value, and a guess near the zero value. The calculator will display the coordinates of the point at which the graph intersects the x-axis. Repeat to find the other root.

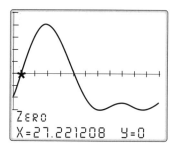
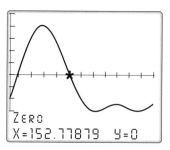

As before, we find that the solutions in the interval $0° \leq \theta < 360°$ are approximately $27°$ and $153°$.

Answer $B = 27° + 360n$ or $B = 153° + 360n$ for integral values of n.

Exercises

Writing About Mathematics

1. The discriminant of the quadratic equation $\tan^2 \theta + 4 \tan \theta + 5 = 0$ is -4. Explain why the solution set of this equation is the empty set.

2. Explain why the solution set of $2 \csc^2 \theta - \csc \theta = 0$ is the empty set.

Developing Skills

In 3–14, use the quadratic formula to find, to the nearest degree, all values of θ in the interval $0° \leq \theta < 360°$ that satisfy each equation.

3. $3 \sin^2 \theta - 7 \sin \theta - 3 = 0$

4. $\tan^2 \theta - 2 \tan \theta - 5 = 0$

5. $7 \cos^2 \theta - 1 = 5 \cos \theta$

6. $9 \sin^2 \theta + 6 \sin \theta = 2$

7. $\tan^2 \theta + 3 \tan \theta + 1 = 0$

8. $8 \cos^2 \theta - 7 \cos \theta + 1 = 0$

9. $2 \cot^2 \theta + 3 \cot \theta - 4 = 0$

10. $\sec^2 \theta - 2 \sec \theta - 4 = 0$

11. $3 \csc^2 \theta - 2 \csc \theta = 2$

12. $2 \tan \theta (\tan \theta + 1) = 3$

13. $3 \cos \theta + 1 = \frac{1}{\cos \theta}$

14. $\frac{\sin \theta}{2} = \frac{3}{\sin \theta + 2}$

15. Find all radian values of θ in the interval $0 \leq \theta < 2\pi$ for which $\frac{\sin \theta}{1} = \frac{1}{2 \sin \theta}$.

16. Find, to the nearest hundredth of a radian, all values of θ in the interval $0 \leq \theta < 2\pi$ for which $\frac{\cos \theta}{3} = \frac{1}{3 \cos \theta + 1}$.

13-4 USING SUBSTITUTION TO SOLVE TRIGONOMETRIC EQUATIONS INVOLVING MORE THAN ONE FUNCTION

When an equation contains two different functions, it may be possible to factor in order to write two equations, each with a different function. We can also use identities to write an equivalent equation with one function.

The equation $\cos^2 \theta + \sin \theta = 1$ cannot be solved by factoring. We can use the identity $\cos^2 \theta + \sin^2 \theta = 1$ to change the equation to an equivalent equation in $\sin \theta$ by replacing $\cos^2 \theta$ with $1 - \sin^2 \theta$.

$$\cos^2 \theta + \sin \theta = 1$$
$$1 - \sin^2 \theta + \sin \theta = 1$$
$$-\sin^2 \theta + \sin \theta = 0$$
$$\sin \theta (-\sin \theta + 1) = 0$$

$\sin \theta = 0$	$-\sin \theta + 1 = 0$
$\theta_1 = 0°$	$1 = \sin \theta$
$\theta_2 = 180°$	$\theta_3 = 90°$

Check $\theta_1 = 0°$	*Check* $\theta_2 = 180°$	*Check* $\theta_3 = 90°$
$\cos^2\theta + \sin\theta = 1$	$\cos^2\theta + \sin\theta = 1$	$\cos^2\theta + \sin\theta = 1$
$\cos^2 0° + \sin 0° \overset{?}{=} 1$	$\cos^2 180° + \sin 180° \overset{?}{=} 1$	$\cos^2 90° + \sin 90° \overset{?}{=} 1$
$1^2 + 0 \overset{?}{=} 1$	$(-1)^2 + 0 \overset{?}{=} 1$	$0^2 + 1 \overset{?}{=} 1$
$1 = 1$ ✔	$1 = 1$ ✔	$1 = 1$ ✔

In the interval $0° \le \theta < 360°$, the solution set of $\cos^2\theta + \sin\theta = 1$ is $\{0°, 90°, 180°\}$.

Is it possible to solve the equation $\cos^2\theta + \sin\theta = 1$ by writing an equivalent equation in terms of $\cos\theta$? To do so we must use an identity to write $\sin\theta$ in terms of $\cos\theta$. Since $\sin^2\theta = 1 - \cos^2\theta$, $\sin\theta = \pm\sqrt{1 - \cos^2\theta}$.

(1) Write the equation: $\qquad\qquad\qquad\qquad \cos^2\theta + \sin\theta = 1$

(2) Replace $\sin\theta$ with $\qquad\qquad\quad \cos^2\theta \pm\sqrt{1 - \cos^2\theta} = 1$
$\pm\sqrt{1 - \cos^2\theta}$:

(3) Isolate the radical: $\qquad\qquad\quad \pm\sqrt{1 - \cos^2\theta} = 1 - \cos^2\theta$

(4) Square both sides of $\qquad\qquad\qquad 1 - \cos^2\theta = 1 - 2\cos^2\theta + \cos^4\theta$
the equation:

(5) Write an equivalent
equation with the right $\qquad\qquad\quad \cos^2\theta - \cos^4\theta = 0$
side equal to 0:

(6) Factor the left side: $\qquad\qquad\quad \cos^2\theta\,(1 - \cos^2\theta) = 0$

(7) Set each factor equal to 0:

$\cos^2\theta = 0$	$1 - \cos^2\theta = 0$
$\cos\theta = 0$	$1 = \cos^2\theta$
$\theta_1 = 90°$	$\pm 1 = \cos\theta$
$\theta_2 = 270°$	$\theta_3 = 0°$
	$\theta_4 = 180°$

This approach uses more steps than the first. In addition, because it involves squaring both sides of the equation, an extraneous root, 270°, has been introduced. Note that 270° is a root of the equation $\cos^2\theta - \cos^4\theta = 0$ but is *not* a root of the given equation.

$$\cos^2\theta + \sin\theta = 1$$
$$\cos^2 270° + \sin 270° \overset{?}{=} 1$$
$$(0)^2 + (-1) \overset{?}{=} 1$$
$$0 - 1 \neq 1 \text{ ✗}$$

Any of the eight basic identities or the related identities can be substituted in a given equation.

EXAMPLE I

Find all values of A in the interval $0° \leq A < 360°$ such that $2 \sin A + 1 = \csc A$.

Solution

How to Proceed

(1) Write the equation:

$$2 \sin A + 1 = \csc A$$

(2) Replace $\csc A$ with $\frac{1}{\sin A}$:

$$2 \sin A + 1 = \frac{1}{\sin A}$$

(3) Multiply both sides of the equation by $\sin A$:

$$2 \sin^2 A + \sin A = 1$$

(4) Write an equivalent equation with 0 as the right side:

$$2 \sin^2 A + \sin A - 1 = 0$$

(5) Factor the left side:

$$(2 \sin A - 1)(\sin A + 1) = 0$$

(6) Set each factor equal to 0 and solve for $\sin A$:

$2 \sin A - 1 = 0$	$\sin A + 1 = 0$
$2 \sin A = 1$	$\sin A = -1$
$\sin A = \frac{1}{2}$	$A = 270°$
$A = 30°$	
$or\ A = 150°$	

Answer $A = 30°$ or $A = 150°$ or $A = 270°$

Often, more than one substitution is necessary to solve an equation.

EXAMPLE 2

If $0 \leq \theta < 2\pi$, find the solution set of the equation $2 \sin \theta = 3 \cot \theta$.

Solution

How to Proceed

(1) Write the equation:

$$2 \sin \theta = 3 \cot \theta$$

(2) Replace $\cot \theta$ with $\frac{\cos \theta}{\sin \theta}$:

$$2 \sin \theta = 3\left(\frac{\cos \theta}{\sin \theta}\right)$$

(3) Multiply both sides of the equation by $\sin \theta$:

$$2 \sin^2 \theta = 3 \cos \theta$$

(4) Replace $\sin^2 \theta$ with $1 - \cos^2 \theta$:

$$2(1 - \cos^2 \theta) = 3 \cos \theta$$

(5) Write an equivalent equation in standard form:

$$2 - 2 \cos^2 \theta = 3 \cos \theta$$

$$2 - 2 \cos^2 \theta - 3 \cos \theta = 0$$

$$2 \cos^2 \theta + 3 \cos \theta - 2 = 0$$

(6) Factor and solve for $\cos \theta$:

$(2 \cos \theta - 1)(\cos \theta + 2) = 0$

(7) Find all values of θ in the given interval:

$$2 \cos \theta - 1 = 0 \quad | \quad \cos \theta + 2 = 0$$
$$2 \cos \theta = 1 \quad | \quad \cos \theta = -2 \; \textbf{✗}$$
$$\cos \theta = \tfrac{1}{2} \quad | \quad \text{No solution}$$
$$\theta = \tfrac{\pi}{3}$$
$$or \; \theta = \tfrac{5\pi}{3}$$

Answer $\left\{\tfrac{\pi}{3}, \tfrac{5\pi}{3}\right\}$

The following identities from Chapter 10 will be useful in solving trigonometric equations:

Reciprocal Identities	Quotient Identities	Pythagorean Identities
$\sec \theta = \frac{1}{\cos \theta}$	$\tan \theta = \frac{\sin \theta}{\cos \theta}$	$\cos^2 \theta + \sin^2 \theta = 1$
$\csc \theta = \frac{1}{\sin \theta}$	$\cot \theta = \frac{\cos \theta}{\sin \theta}$	$1 + \tan^2 \theta = \sec^2 \theta$
$\cot \theta = \frac{1}{\tan \theta}$		$\cot^2 \theta + 1 = \csc^2 \theta$

Exercises

Writing About Mathematics

1. Sasha said that $\sin \theta + \cos \theta = 2$ has no solution. Do you agree with Sasha? Explain why or why not.

2. For what values of θ is $\sin \theta = \sqrt{1 - \cos^2 \theta}$ true?

Developing Skills

In 3–14, find the exact values of θ in the interval $0° \leq \theta < 360°$ that satisfy each equation.

3. $2 \cos^2 \theta - 3 \sin \theta = 0$

4. $4 \cos^2 \theta + 4 \sin \theta - 5 = 0$

5. $\csc^2 \theta - \cot \theta - 1 = 0$

6. $2 \sin \theta + 1 = \csc \theta$

7. $2 \sin^2 \theta + 3 \cos \theta - 3 = 0$

8. $3 \tan \theta = \cot \theta$

9. $2 \cos \theta = \sec \theta$

10. $\sin \theta = \csc \theta$

11. $\tan \theta = \cot \theta$

12. $2 \cos^2 \theta = \sin \theta + 2$

13. $\cot^2 \theta = \csc \theta + 1$

14. $2 \sin^2 \theta - \tan \theta \cot \theta = 0$

Applying Skills

15. An engineer would like to model a piece for a factory machine on his computer. As shown in the figure, the machine consists of a link fixed to a circle at point A. The other end of the link is fixed to a slider at point B. As the circle rotates, point B slides back and forth between the two ends of the slider (C and D). The movement is restricted so that θ, the measure of $\angle AOD$, is in the interval $-45° \leq \theta \leq 45°$. The motion of point B can be described mathematically by the formula

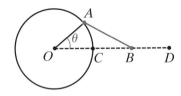

$$CB = r(\cos \theta - 1) + \sqrt{l^2 - r^2 \sin^2 \theta}$$

where r is the radius of the circle and l is the length of the link. Both the radius of the circle and the length of the link are 2 inches.

a. Find the exact value of CB when: (1) $\theta = 30°$ (2) $\theta = 45°$.

b. Find the exact value(s) of θ when $CB = 2$ inches.

c. Find, to the nearest hundredth of a degree, the value(s) of θ when $CB = 1.5$ inches.

13-5 USING SUBSTITUTION TO SOLVE TRIGONOMETRIC EQUATIONS INVOLVING DIFFERENT ANGLE MEASURES

If an equation contains function values of two different but related angle measures, we can use identities to write an equivalent equation in terms of just one variable. For example: Find the value(s) of θ such that $\sin 2\theta - \sin \theta = 0$.

Recall that $\sin 2\theta = 2 \sin \theta \cos \theta$. We can use this identity to write the equation in terms of just one variable, θ, and then use any convenient method to solve the equation.

(1) Write the equation:

$$\sin 2\theta - \sin \theta = 0$$

(2) For $\sin 2\theta$, substitute its equal, $2 \sin \theta \cos \theta$:

$$2 \sin \theta \cos \theta - \sin \theta = 0$$

(3) Factor the left side:

$$\sin \theta (2 \cos \theta - 1) = 0$$

(4) Set each factor equal to zero and solve for θ:

$\sin \theta = 0$	$2 \cos \theta - 1 = 0$
$\theta = 0°$	$2 \cos \theta = 1$
or $\theta = 180°$	$\cos \theta = \frac{1}{2}$
	$\theta = 60°$
	or $\theta = 300°$

The degree measures $0°, 60°, 180°,$ and $300°$ are all of the values in the interval $0° \leq \theta < 360°$ that make the equation true. Any values that differ from these values by a multiple of $360°$ will also make the equation true.

EXAMPLE 1

Find, to the nearest degree, the roots of $\cos 2\theta - 2 \cos \theta = 0$.

Solution

How to Proceed

(1) Write the given equation:

$$\cos 2\theta - 2 \cos \theta = 0$$

(2) Use an identity to write $\cos 2\theta$ in terms of $\cos \theta$:

$$2 \cos^2 \theta - 1 - 2 \cos \theta = 0$$

(3) Write the equation in standard form:

$$2 \cos^2 \theta - 2 \cos \theta - 1 = 0$$

(4) The equation cannot be factored over the set of integers. Use the quadratic formula:

$$\cos \theta = \frac{-(-2) \pm \sqrt{(-2)^2 - 4(2)(-1)}}{2(2)}$$

$$= \frac{2 \pm \sqrt{12}}{4}$$

$$= \frac{1 \pm \sqrt{3}}{2}$$

(5) When we use a calculator to approximate the value of $\arccos \frac{1 - \sqrt{3}}{2}$, the calculator will return the value $111°$. Cosine is negative in both the second and the third quadrants. Therefore, there is both a second-quadrant and a third-quadrant angle such that $\cos \theta = \frac{1 - \sqrt{3}}{2}$:

```
cos⁻¹((1-√(3))/2)
            111.4707014
```

$\theta_1 = 111°$

$R = 180° - 111° = 69°$

$\theta_2 = 180° + 69° = 249°$

(6) When we use a calculator to approximate the value of $\arccos \frac{1 + \sqrt{3}}{2}$, the calculator will return an error message because $\frac{1 + \sqrt{3}}{2} > 1$ is not in the domain of arccosine:

```
ERR:DOMAIN
1 Quit
2:Goto
```

Answer To the nearest degree, $\theta = 111°$ or $\theta = 249°$.

EXAMPLE 2

Find, to the nearest degree, the values of θ in the interval $0° \leq \theta \leq 360°$ that are solutions of the equation $\sin(90° - \theta) + 2\cos\theta = 2$.

Solution   Use the identity $\sin(90° - \theta) = \cos\theta$.

$$\sin(90° - \theta) + 2\cos\theta = 2$$
$$\cos\theta + 2\cos\theta = 2$$
$$3\cos\theta = 2$$
$$\cos\theta = \tfrac{2}{3}$$

To the nearest degree, a calculator returns the value of θ as $48°$.

In quadrant I, $\theta = 48°$ and in quadrant IV, $\theta = 360° - 48° = 312°$.

Answer $\theta = 48°$ or $\theta = 312°$

The basic trigonometric identities along with the cofunction, double-angle, and half-angle identities will be useful in solving trigonometric equations:

Cofunction Identities	Double-Angle Identities	Half-Angle Identities
$\cos\theta = \sin(90° - \theta)$	$\sin(2\theta) = 2\sin\theta\cos\theta$	$\sin\tfrac{1}{2}\theta = \pm\sqrt{\dfrac{1 - \cos\theta}{2}}$
$\sin\theta = \cos(90° - \theta)$	$\cos(2\theta) = \cos^2\theta - \sin^2\theta$	$\cos\tfrac{1}{2}\theta = \pm\sqrt{\dfrac{1 + \cos\theta}{2}}$
$\tan\theta = \cot(90° - \theta)$	$\tan(2\theta) = \dfrac{2\tan\theta}{1 - \tan^2\theta}$	$\tan\tfrac{1}{2}\theta = \pm\sqrt{\dfrac{1 - \cos\theta}{1 + \cos\theta}}$
$\cot\theta = \tan(90° - \theta)$		
$\sec\theta = \csc(90° - \theta)$		
$\csc\theta = \sec(90° - \theta)$		

Note that in radians, the right sides of the cofunction identities are written in terms of $\tfrac{\pi}{2} - \theta$.

Exercises

Writing About Mathematics

1. Isaiah said that if the equation $\cos 2x + 2\cos^2 x = 2$ is divided by 2, an equivalent equation is $\cos x + \cos^2 x = 1$. Do you agree with Isaiah? Explain why or why not.

2. Aaron solved the equation $2\sin\theta\cos\theta = \cos\theta$ by first dividing both sides of the equation by $\cos\theta$. Aaron said that for $0 \leq \theta \leq 2\pi$, the solution set is $\left\{\frac{\pi}{6}, \frac{5\pi}{6}\right\}$. Do you agree with Aaron? Explain why or why not.

Developing Skills

In 3–10, find the exact values of θ in the interval $0° \leq \theta \leq 360°$ that make each equation true.

3. $\sin 2\theta - \cos \theta = 0$

4. $\cos 2\theta + \sin^2 \theta = 1$

5. $\sin 2\theta + 2 \sin \theta = 0$

6. $\tan 2\theta = \cot \theta$

7. $\cos 2\theta + 2 \cos^2 \theta = 2$

8. $\sin \frac{1}{2}\theta = \cos \theta$

9. $3 - 3 \sin \theta - 2 \cos^2 \theta = 0$

10. $3 \cos 2\theta - 4 \cos^2 \theta + 2 = 0$

In 11–18, find all radian measures of θ in the interval $0 \leq \theta \leq 2\pi$ that make each equation true. Express your answers in terms of π when possible; otherwise, to the nearest hundredth.

11. $\cos 2\theta = 2 \cos \theta - 2 \cos^2 \theta$

12. $2 \sin 2\theta + \sin \theta = 0$

13. $5 \sin^2 \theta - 4 \sin \theta + \cos 2\theta = 0$

14. $\cos \theta = 3 \sin 2\theta$

15. $3 \sin 2\theta = \tan \theta$

16. $\sin \left(\frac{\pi}{2} - \theta \right) + \cos^2 \theta = \frac{1}{4}$

17. $2 \cos^2 \theta + 3 \sin \theta - 2 \cos 2\theta = 1$

18. $(2 \sin \theta \cos \theta)^2 + 4 \sin 2\theta - 1 = 0$

Applying Skills

19. Martha swims 90 meters from point A on the north bank of a stream to point B on the opposite bank. Then she makes a right angle turn and swims 60 meters from point B to point C, another point on the north bank. If $m\angle CAB = \theta$, then $m\angle ACB = 90° - \theta$.

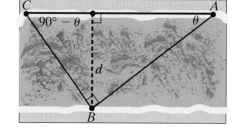

a. Let d be the width of the stream, the length of the perpendicular distance from B to $\overline{AC}$. Express d in terms of $\sin \theta$.

b. Express d in terms of $\sin (90° - \theta)$.

c. Use the answers to **a** and **b** to write an equation. Solve the equation for θ.

d. Find d, the width of the stream.

20. A pole is braced by two wires of equal length as shown in the diagram. One wire, $\overline{AB}$, makes an angle of θ with the ground, and the other wire, $\overline{CD}$, makes an angle of 2θ with the ground. If $FD = 1.75FB$, find, to the nearest degree, the measure of θ:

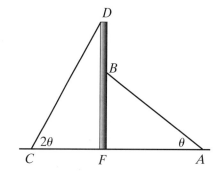

a. Let $AB = CD = x$, $FB = y$, and $FD = 1.75y$. Express $\sin \theta$ and $\sin 2\theta$ in terms of x and y.

b. Write an equation that expresses a relationship between $\sin \theta$ and $\sin 2\theta$ and solve for θ to the nearest degree.

CHAPTER SUMMARY

A **trigonometric equation** is an equation whose variable is expressed in terms of a trigonometric function value.

The following table compares the degree measures of θ from $-90°$ to $360°$, the radian measures of θ from $-\frac{\pi}{2}$ to 2π, and the measure of its reference angle.

	Fourth Quadrant	First Quadrant	Second Quadrant	Third Quadrant	Fourth Quadrant
Angle	$-90° < \theta < 0°$ $-\frac{\pi}{2} < \theta < 0$	$0° < \theta < 90°$ $0 < \theta < \frac{\pi}{2}$	$90° < \theta < 180°$ $\frac{\pi}{2} < \theta < \theta$	$180° < \theta < 270°$ $\pi < \theta < \frac{3\pi}{2}$	$270° < \theta < 360°$ $\frac{3\pi}{2} < \theta < 2\pi$
Reference Angle	$-\theta$ $-\theta$	θ θ	$180° - \theta$ $\pi - \theta$	$\theta - 180°$ $\theta - \pi$	$360° - \theta$ $2\pi - \theta$

To solve a trigonometric equation:

1. If the equation involves more than one variable, use identities to write the equation in terms of one variable.

2. If the equation involves more than one trigonometric function of the same variable, separate the functions by factoring or use identities to write the equation in terms of one function of one variable.

3. Solve the equation for the function value of the variable. Use factoring or the quadratic formula to solve a second-degree equation.

4. Use a calculator or your knowledge of the exact function values to write one value of the variable to an acceptable degree of accuracy.

5. If the measure of the angle found in step 4 is not that of a quadrantal angle, find the measure of its reference angle.

6. Use the measure of the reference angle to find the degree measures of each solution in the interval $0° \leq \theta < 360°$ or the radian measures of each solution in the interval $0 \leq \theta < 2\pi$.

7. Add $360n$ (n an integer) to the solutions in degrees found in steps 4 and 6 to write all possible solutions in degrees. Add $2\pi n$ (n an integer) to the solutions in radians found in steps 2 and 4 to write all possible solutions in radians.

VOCABULARY

13-1 Trigonometric equation

REVIEW EXERCISES

In 1–10, find the exact values of x in the interval $0° \leq x \leq 360°$ that make each equation true.

1. $2 \cos x + 1 = 0$

2. $\sqrt{3} - \sin x = \sin x + \sqrt{12}$

3. $2 \sec x = 2 + \sec x$

4. $2 \cos^2 x + \cos x - 1 = 0$

5. $\cos x \sin x + \sin x = 0$

6. $\tan x - 3 \cot x = 0$

7. $2 \cos x - \sec x = 0$

8. $\sin^2 x - \cos^2 x = 0$

9. $2 \tan x = 1 - \tan^2 x$

10. $\cos^3 x - \frac{3}{4} \cos x = 0$

In 11–22, find, to the nearest hundredth, all values of θ in the interval $0 \leq \theta < 2\pi$ that make each equation true.

11. $7 \sin \theta + 3 = 1$

12. $5(\cos \theta - 1) = 6 + \cos \theta$

13. $4 \sin^2 \theta - 3 \sin \theta = 1$

14. $3 \cos^2 \theta - \cos \theta - 2 = 0$

15. $\tan^2 \theta - 4 \tan \theta - 1 = 0$

16. $\sec^2 \theta - 10 \sec \theta + 20 = 0$

17. $\tan 2\theta = 4 \tan \theta$

18. $2 \sin 2\theta + \cos \theta = 0$

19. $\frac{\sin 2\theta}{1 + \cos 2\theta} = 4$

20. $3 \cos 2\theta + \cos \theta + 2 = 0$

21. $2 \tan^2 \theta + 6 \tan \theta = 20$

22. $\cos 2\theta - \cos^2 \theta + \cos \theta + \frac{1}{4} = 0$

23. Explain why the solution set of $\tan \theta - \sec \theta = 0$ is the empty set.

24. In $\triangle ABC$, $\text{m}\angle A = \theta$ and $\text{m}\angle B = 2\theta$. The altitude from C intersects $\overline{AB}$ at D and $AD : DB = 5 : 2$.

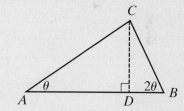

a. Write $\tan \theta$ and $\tan 2\theta$ as ratios of the sides of $\triangle ADC$ and $\triangle BDC$, respectively.

b. Solve the equation for $\tan 2\theta$ found in **a** for CD.

c. Substitute the value of CD found in **b** into the equation of $\tan \theta$ found in **a**.

d. Solve for θ.

e. Find the measures of the angles of the triangle to the nearest degree.

Exploration

For (1)–(6): **a.** Use your knowledge of geometry and trigonometry to express the area, A, of the shaded region in terms of θ. **b.** Find the measure of θ when $A = 0.5$ square unit or explain why there is no possible value of θ. Give the exact value when possible; otherwise, to the nearest hundredth of a radian.

(1)

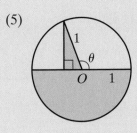

$$0 < \theta < \tfrac{\pi}{2}$$

(2)

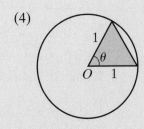

$$0 < \theta < \tfrac{\pi}{2}$$

(3)

$$0 < \theta < \tfrac{\pi}{2}$$

(4)

$$0 < \theta < \pi$$

(5)

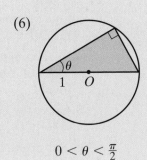

$$\tfrac{\pi}{2} < \theta \leq \pi$$

(6)

$$0 < \theta < \tfrac{\pi}{2}$$

CUMULATIVE REVIEW   CHAPTERS 1–13

Part I

Answer all questions in this part. Each correct answer will receive 2 credits. No partial credit will be allowed.

1. The expression $2(5)^0 + 3(27)^{-\frac{1}{3}}$ is equal to

(1) -27 (2) $2 + \frac{\sqrt{3}}{9}$ (3) 3 (4) $2\frac{1}{9}$

2. The sum of $\frac{3}{4}a^2 - \frac{1}{2}a$ and $a - \frac{1}{2}a^2$ is

(1) $\frac{1}{4}a^2 + \frac{1}{2}a$ (3) $\frac{5}{4}a^2 - \frac{3}{2}a$

(2) $2a^3$ (4) $\frac{7}{4}a^2 - a$

3. The fraction $\dfrac{4}{1 - \sqrt{5}}$ is equivalent to

(1) $1 + \sqrt{5}$ (3) $-1 + \sqrt{5}$

(2) $1 - \sqrt{5}$ (4) $-1 - \sqrt{5}$

4. The complex number $i^{12} + i^{10}$ can be written as

(1) 0 (2) 1 (3) 2 (4) $1 + i$

5. $\displaystyle\sum_{n=1}^{4} \left[(-1)^n \frac{n}{2} \right]$ is equal to

(1) -5 (2) 1 (3) 2 (4) 5

6. When the roots of a quadratic equation are real and irrational, the discriminant must be

(1) zero.

(2) a positive number that is a perfect square.

(3) a positive number that is not a perfect square.

(4) a negative number.

7. When $f(x) = x^2 + 1$ and $g(x) = 2x$, then $g(f(x))$ equals

(1) $4x^2 + 1$ (3) $4x^2 + 2$

(2) $2x^2 + 2x + 1$ (4) $2x^2 + 2$

8. If $\log x = 2 \log a - \frac{1}{3} \log b$, then x equals

(1) $2a - \frac{1}{3}b$ (3) $\dfrac{a^2}{\frac{1}{3}b}$

(2) $a^2 - \sqrt[3]{b}$ (4) $\dfrac{a^2}{\sqrt[3]{b}}$

9. What number must be added to the binomial $x^2 + 5x$ in order to change it into a trinomial that is a perfect square?

(1) $\frac{5}{2}$ (2) $\frac{25}{4}$ (3) $\frac{25}{2}$ (4) 25

10. If f(x) is a function from the set of real numbers to the set of real numbers, which of the following functions is one-to-one and onto?

(1) f(x) = 2x − 1   (3) f(x) = |2x − 1|

(2) f(x) = x^2 (4) f(x) = −x^2 + 2x

Part II

Answer all questions in this part. Each correct answer will receive 2 credits. Clearly indicate the necessary steps, including appropriate formula substitutions, diagrams, graphs, charts, etc. For all questions in this part, a correct numerical answer with no work shown will receive only 1 credit.

11. Sketch the graph of the inequality $y \leq -x^2 - 2x + 3$.

12. When f(x) = 4x − 2, find f^{-1}(x), the inverse of f(x).

Part III

Answer all questions in this part. Each correct answer will receive 4 credits. Clearly indicate the necessary steps, including appropriate formula substitutions, diagrams, graphs, charts, etc. For all questions in this part, a correct numerical answer with no work shown will receive only 1 credit.

13. Write the first six terms of the geometric function whose first term is 2 and whose fourth term is 18.

14. The endpoints of a diameter of a circle are $(-2, 5)$ and $(4, -1)$. Write an equation of the circle.

Part IV

Answer all questions in this part. Each correct answer will receive 6 credits. Clearly indicate the necessary steps, including appropriate formula substitutions, diagrams, graphs, charts, etc. For all questions in this part, a correct numerical answer with no work shown will receive only 1 credit.

15. If csc θ = 3, and cos θ < 0, find sin θ, cos θ, tan θ, cot θ and sec θ.

16. Solve for x and write the roots in $a + bi$ form: $\frac{6}{x} - 2 = \frac{5}{x^2}$.

TRIGONOMETRIC APPLICATIONS

An ocean is a vast expanse that can be life-threatening to a person who experiences a disaster while boating. In order for help to arrive on time, it is necessary that the coast guard or a ship in the area be able to make an exact identification of the location. A distress signal sent out by the person in trouble can be analyzed by those receiving the signal from different directions. In this chapter we will derive the formulas that can be used to determine distances and angle measures when sufficient information is available.

14-1 SIMILAR TRIANGLES

In the study of geometry, we learned that if two triangles are similar, the corresponding angles are congruent and the corresponding sides are in proportion. If certain pairs of corresponding angles and sides are congruent or proportional, then the triangles must be similar. The following pairs of congruent corresponding angles and proportional corresponding sides are sufficient to prove triangles similar.

1. Two angles (AA~)

2. Three sides (SSS~)

3. Two sides and the included angles (SAS~)

4. Hypotenuse and one leg of a right triangle (HL~)

We can use similar triangles to write the coordinates of a point in terms of its distance from the origin and the sine and cosine of an angle in standard position.

Let $B(x_1, y_1)$ be a point in the first quadrant of the coordinate plane with $OB = b$ and θ the measure of the angle formed by $\overrightarrow{OB}$ and the positive ray of the x-axis. Let $P(\cos \theta, \sin \theta)$ be the point at which $\overrightarrow{OB}$ intersects the unit circle. Let $Q(\cos \theta, 0)$ be the point at which a perpendicular line from P intersects the x-axis and $A(x_1, 0)$ be the point at which a perpendicular line from B intersects the x-axis. Then $\triangle OPQ \sim \triangle OBA$ by AA~. Therefore:

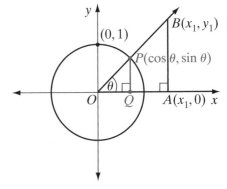

$$\frac{OA}{OQ} = \frac{OB}{OP} \qquad\qquad \frac{AB}{QP} = \frac{OB}{OP}$$

$$\frac{|x_1 - 0|}{\cos \theta} = \frac{b}{1} \qquad\qquad \frac{|y_1 - 0|}{\sin \theta} = \frac{b}{1}$$

$$\frac{x_1}{\cos \theta} = \frac{b}{1} \qquad\qquad \frac{y_1}{\sin \theta} = \frac{b}{1}$$

$$x_1 = b \cos \theta \qquad\qquad y_1 = b \sin \theta$$

Therefore, if B is a point b units from the origin of the coordinate plane and $\overrightarrow{OB}$ is the terminal side of an angle in standard position whose measure is θ, then the coordinates of B are $(b \cos \theta, b \sin \theta)$. This statement can be shown to be true for any point in the coordinate plane.

Let $\angle AOB$ be an angle in standard position whose measure is θ. If the terminal side, $\overrightarrow{OB}$, intersects the unit circle at P, then the coordinates of P are $(\cos\theta, \sin\theta)$. The diagrams below show $\angle AOB$ in each quadrant.

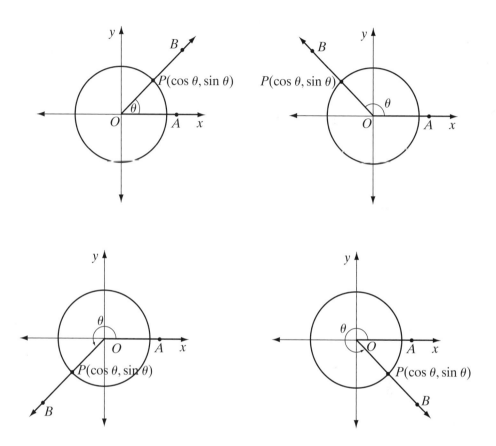

A dilation of b with center at the origin will stretch each segment whose endpoint is the origin by a factor of b. Under the dilation D_b, the image of (x, y) is (bx, by). If B is the image of $P(\cos\theta, \sin\theta)$ under the dilation D_b, then the coordinates of B are $(b\cos\theta, b\sin\theta)$. Therefore, $(b\cos\theta, b\sin\theta)$ are the coordinates of a point b units from the origin on the terminal ray of an angle in standard position whose measure is θ.

Any triangle can be positioned on the coordinate plane so that each vertex is identified by the coordinates of a point in the plane. The coordinates of the vertices can be expressed in terms of the trigonometric function values of angles in standard position.

EXAMPLE I

Point S is 12 units from the origin and $\overrightarrow{OS}$ makes an angle of 135° with the positive ray of the x-axis. What are the exact coordinates of S?

Solution   The coordinates of S are $(12 \cos 135°, 12 \sin 135°)$.

Since $135° = 180° - 45°$,

$$\cos 135° = -\cos 45° = -\frac{\sqrt{2}}{2} \text{ and}$$

$$\sin 135° = \sin 45° = \frac{\sqrt{2}}{2}.$$

Therefore, the coordinates of S are

$$\left(12 \times \frac{-\sqrt{2}}{2}, 12 \times \frac{\sqrt{2}}{2}\right) = (-6\sqrt{2}, 6\sqrt{2}). \text{ Answer}$$

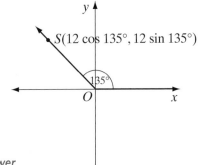

EXAMPLE 2

The coordinates of A are $(-5.30, -8.48)$.

a. Find OA to the nearest hundredth.

b. Find, to the nearest degree, the measure of the angle in standard position whose terminal side is $\overrightarrow{OA}$.

Solution   **a.** Let $C(-5.30, 0)$ be the point at which a vertical line from A intersects the x-axis. Then $\overline{OA}$ is the hypotenuse of right $\triangle OAC$, $OC = |-5.30 - 0| = 5.30$ and $AC = |-8.48 - 0| = 8.48$. Using the Pythagorean Theorem:

$$OA^2 = OC^2 + AC^2$$

$$OA^2 = 5.30^2 + 8.48^2$$

$$OA = \sqrt{5.30^2 + 8.48^2} \quad \text{We reject the negative root.}$$

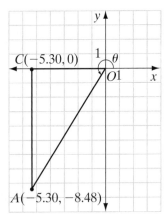

ENTER: [2nd] [√] 5.30 [x²]   DISPLAY:
         [+] 8.48 [x²] [)]

$$\boxed{\begin{array}{l} \sqrt{(5.30^2 + 8.48^2)} \\ \hspace{3em} 10.00002 \end{array}}$$

[ENTER]

Write the measure of OA to the nearest hundredth: $OA = 10.00$.

b. Let θ be the measure of the angle in standard position with terminal side $\overrightarrow{OA}$. A is in the third quadrant. We can use either coordinate to find the measure of θ, a third-quadrant angle.

$$(OA \cos \theta, OA \sin \theta) = (-5.30, -8.48)$$

$10.00 \cos \theta = -5.30$	$10.00 \sin \theta = -8.48$
$\cos \theta = -0.530$	$\sin \theta = -0.848$

ENTER: **2nd** **COS⁻¹** **(-)** 0.530   ENTER: **2nd** **SIN⁻¹** **(-)** 0.848

) **ENTER**   **)** **ENTER**

DISPLAY:
```
cos⁻¹(-0.530)
         122.0054548
```
     DISPLAY:
```
sin⁻¹(-.848)
        -57.99480014
```

The calculator returns the measure of a second-quadrant angle. Use this measure to find the reference angle to the nearest degree.

$$R = 180 - 122 = 58°$$

The calculator returns the measure of a fourth-quadrant angle. Use this measure to find the reference angle to the nearest degree.

$$R = -(-58°) = 58°$$

Use the reference angle to find the measure of the third-quadrant angle.

$$\theta = 180 + 58 = 238°$$

Answers   **a.** $OA = 10.00$ **b.** $\theta = 238°$

Exercises

Writing About Mathematics

1. In Example 2, is it possible to find the measure of θ without first finding OA? Justify your answer.

2. In what quadrant is a point whose coordinates in radian measure are $\left(2 \cos \frac{4\pi}{3}, 2 \sin \frac{4\pi}{3}\right)$? Justify your answer.

Developing Skills

In 3–14, write in simplest radical form the coordinates of each point A if A is on the terminal side of an angle in standard position whose degree measure is θ.

3. $OA = 4, \theta = 45°$ **4.** $OA = 2, \theta = 30°$ **5.** $OA = 6, \theta = 90°$

6. $OA = 8, \theta = 120°$ **7.** $OA = 15, \theta = 135°$ **8.** $OA = 0.5, \theta = 180°$

9. $OA = 9, \theta = 150°$ **10.** $OA = 25, \theta = 210°$ **11.** $OA = 12, \theta = 270°$

12. $OA = \sqrt{2}, \theta = 225°$ **13.** $OA = \sqrt{3}, \theta = 300°$ **14.** $OA = 2, \theta = -60°$

In 15–23, the coordinates of a point are given. **a.** Find the distance of the point from the origin. Express approximate distances to the nearest hundredth. **b.** Find the measure, to the nearest degree, of the angle in standard position whose terminal side contains the given point.

15. $(6, 8)$ **16.** $(-5, 12)$ **17.** $(0, 7)$

18. $(12, -9)$ **19.** $(15, 0)$ **20.** $(-8, -12)$

21. $(24, 7)$ **22.** $(6, -10)$ **23.** $(-8, 8)$

In 24–29, for each $\triangle ORS$, O is the origin, R is on the positive ray of the x-axis and $\overline{PS}$ is the altitude from S to $\overrightarrow{OR}$. **a.** Find the exact coordinates of R and S. **b.** Find the exact area of $\triangle ORS$.

24. $OR = 5, m\angle ROS = \frac{\pi}{3}, OS = 3$ **25.** $OR = 12, m\angle ROS = \frac{\pi}{2}, OS = 8$

26. $OR = 8, m\angle ROS = \frac{3\pi}{4}, OS = 8$ **27.** $OR = 20, OS = RS, PS = 10$

28. $OR = 9, \triangle ORS$ is equilateral   **29.** $OR = 7, m\angle ROS = \frac{\pi}{6}, PS = 8$

14-2 LAW OF COSINES

When we know the measures of two sides and the included angle of a triangle (SAS), the size and shape of the triangle are determined. Therefore, we should be able to find the measure of the third side of the triangle. In order to derive a formula to do this, we will position the triangle in the coordinate plane with one endpoint at the origin and one angle in standard position. As shown in the diagrams, the angle in standard position can be either acute, obtuse, or right.

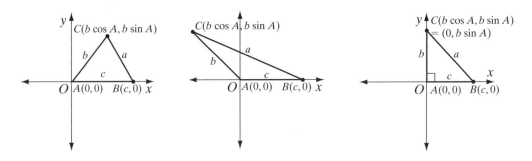

Let $\triangle ABC$ be a triangle with $AB = c$, $BC = a$, and $CA = b$. The coordinates of A are $(0, 0)$, of B are $(c, 0)$, and of C are $(b \cos A, b \sin A)$. If b, c, and $m\angle A$ are known measures, then the coordinates of each vertex are known. We can find a, the length of the third side of the triangle, by using the distance formula.

The distance between the two points $P(x_1, y_1)$ and $Q(x_2, y_2)$ is given by the formula:

$$PQ^2 = (x_2 - x_1)^2 + (y_2 - y_1)^2$$

Let $P(x_1, y_1) = B(c, 0)$ and $Q(x_2, y_2) = C(b \cos A, b \sin A)$.

$$
\begin{aligned}
BC^2 &= (b \cos A - c)^2 + (b \sin A - 0)^2 \\
&= b^2 \cos^2 A - 2bc \cos A + c^2 + b^2 \sin^2 A \\
&= b^2 \cos^2 A + b^2 \sin^2 A + c^2 - 2bc \cos A \\
&= b^2 (\cos^2 A + \sin^2 A) + c^2 - 2bc \cos A \\
&= b^2(1) + c^2 - 2bc \cos A \\
&= b^2 + c^2 - 2bc \cos A
\end{aligned}
$$

If we let $BC = a$, we can write:

$$a^2 = b^2 + c^2 - 2bc \cos A$$

This formula is called the **Law of Cosines**. The law of cosines for $\triangle ABC$ can be written in terms of the measures of any two sides and the included angle.

$$a^2 = b^2 + c^2 - 2bc \cos A$$

$$b^2 = a^2 + c^2 - 2ac \cos B$$

$$c^2 = a^2 + b^2 - 2ab \cos C$$

We can rewrite the Law of Cosines in terms of the letters that represent the vertices of any triangle. For example, in $\triangle DEF$, side $\overline{DE}$ is opposite $\angle F$ so we let $DE = f$, side $\overline{EF}$ is opposite $\angle D$ so we let $EF = d$, and side $\overline{DF}$ is opposite $\angle E$ so we let $DF = e$. We can use the Law of Cosines to write a formula for the square of the measure of each side of $\triangle DEF$ terms of the measures of the other two sides and the included angle.

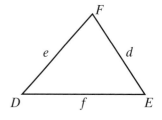

$$d^2 = e^2 + f^2 - 2ef \cos D$$
$$e^2 = d^2 + f^2 - 2df \cos E$$
$$f^2 = d^2 + e^2 - 2de \cos F$$

EXAMPLE I

In $\triangle ABC$, $AB = 8$, $AC = 10$, and $\cos A = \frac{1}{8}$. Find BC.

Solution $AB = c = 8$, $AC = b = 10$, and $\cos A = \frac{1}{8}$.

Use the Law of Cosines to find $BC = a$.

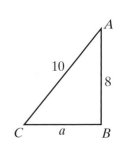

$$a^2 = b^2 + c^2 - 2bc \cos A$$
$$a^2 = 10^2 + 8^2 - 2(10)(8)\left(\tfrac{1}{8}\right)$$
$$a^2 = 100 + 64 - 20$$
$$a^2 = 144$$
$$a = \pm 12$$

Since a is the length of a line segment, a is a positive number.

Answer $BC = 12$

EXAMPLE 2

The diagonals of a parallelogram measure 12 centimeters and 22 centimeters and intersect at an angle of 143 degrees. Find the length of the longer sides of the parallelogram to the nearest tenth of a centimeter.

Solution Let the diagonals of parallelogram $PQRS$ intersect at T. The diagonals of a parallelogram bisect each other. If $PR = 12$, then $PT = 6$ and if $QS = 22$, then $QT = 11$. Let PQ be the longer side of the parallelogram, the side opposite the larger angle at which the diagonals intersect. Therefore, $\mathrm{m}\angle PTQ = 143$. Write the Law of Cosines for $PQ^2 = t^2$ in terms of $QT = p = 11$, $PT = q = 6$, and $\cos T = \cos 143°$.

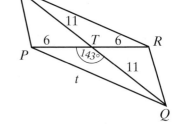

$$t^2 = p^2 + q^2 - 2pq \cos T$$
$$= 11^2 + 6^2 - 2(11)(6) \cos 143°$$
$$= 121 + 36 - 132 \cos 143°$$
$$t = \sqrt{157 - 132 \cos 143°}$$

Note that $\cos 143°$ is negative so $-132 \cos 143°$ is positive. Using a calculator, we find that $t \approx 16.19937923$.

Answer To the nearest tenth, $PQ = 16.2$ cm.

Exercises

Writing About Mathematics

1. Explain how the Law of Cosines can be used to show that in an obtuse triangle, the side opposite an obtuse angle is the longest side of the triangle.

2. Explain the relationship between the Law of Cosines and the Pythagorean Theorem.

Developing Skills

3. In $\triangle MAR$, express m^2 in terms of a, r, and cos M.

4. In $\triangle NOP$, express p^2 in terms of n, o, and cos P.

5. In $\triangle ABC$, if $a = 3, b - 5$, and cos $C = \frac{1}{5}$, find the exact value of c.

6. In $\triangle DEF$, if $e = 8, f = 3$, and cos $D = \frac{3}{4}$, find the exact value of d.

7. In $\triangle HIJ$, if $h = 10, j = 7$, and cos $I = 0.6$, find the exact value of i.

In 8–13, find the exact value of the third side of each triangle.

8. In $\triangle ABC, b = 4, c = 4$, and m$\angle A = \frac{\pi}{3}$.

9. In $\triangle PQR, p = 6, q = \sqrt{2}$, and m$\angle R = \frac{\pi}{4}$.

10. In $\triangle DEF, d = \sqrt{3}, e = 5$, and m$\angle F = \frac{\pi}{6}$.

11. In $\triangle ABC, a = 6, b = 4$, and m$\angle C = \frac{2\pi}{3}$.

12. In $\triangle RST, RS = 9, ST = 9\sqrt{3}$, and m$\angle S = \frac{5\pi}{6}$.

13. In $\triangle ABC, AB = 2\sqrt{2}, BC = 4$, and m$\angle B = \frac{3\pi}{4}$.

In 14–19, find, to the nearest tenth, the measure of the third side of each triangle.

14. In $\triangle ABC, b = 12.4, c = 8.70$, and m$\angle A = 23$.

15. In $\triangle PQR, p = 126, q = 214$, and m$\angle R = 42$.

16. In $\triangle DEF, d = 3.25, e = 5.62$, and m$\angle F = 58$.

17. In $\triangle ABC, a = 62.5, b = 44.7$, and m$\angle C = 133$.

18. In $\triangle RST, RS = 0.375, ST = 1.29$, and m$\angle S = 167$.

19. In $\triangle ABC, AB = 2.35, BC = 6.24$, and m$\angle B = 115$.

Applying Skills

20. Ann and Bill Bekebrede follow a familiar triangular path when they take a walk. They walk from home for 0.52 mile along a straight road, turn at an angle of 95°, walk for another 0.46 mile, and then return home.

   a. Find, to the nearest hundredth of a mile, the length of the last portion of their walk.

   b. Find, to the nearest hundredth of a mile, the total distance that they walk.

21. When two forces act on an object, the resultant force is the single force that would have produced the same result. When the magnitudes of the two forces are represented by the lengths of two sides of a parallelogram, the resultant can be represented by the length of the diagonal of the parallelogram. If forces of 12 pounds and 18 pounds act at an angle of 75°, what is the magnitude of the resultant force to the nearest hundredth pound?

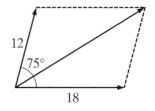

22. A field is in the shape of a parallelogram. The lengths of two adjacent sides are 48 meters and 65 meters. The measure of one angle of the parallelogram is 100°.

   a. Find, to the nearest meter, the length of the longer diagonal.

   b. Find, to the nearest meter, the length of the shorter diagonal.

23. A pole is braced by two wires that extend from the top of the pole to the ground. The lengths of the wires are 16 feet and 18 feet and the measure of the angle between the wires is 110°. Find, to the nearest foot, the distance between the points at which the wires are fastened to the ground.

24. Two points A and B are on the shoreline of Lake George. A surveyor is located at a third point C some distance from both points. The distance from A to C is 180.0 meters and the distance from B to C is 120.0 meters. The surveyor determines that the measure of $\angle ACB$ is 56.3°. To the nearest tenth of a meter, what is the distance from A to B?

25. Two sailboats leave a dock at the same time sailing on courses that form an angle of 112° with each other. If one boat sails at 10.0 knots and the other sails at 12.0 knots, how many nautical miles apart are the boats after two hours? (nautical miles = knots × time) Round to the nearest tenth.

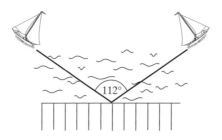

26. Use the Law of Cosines to prove that if the angle between two congruent sides of a triangle measures 60°, the triangle is equilateral.

14-3 USING THE LAW OF COSINES TO FIND ANGLE MEASURE

The measures of three sides of a triangle determine the size and shape of the triangle. If we know the measures of three sides of a triangle, we can use the Law of Cosines to find the measure of any angle of the triangle. For example, in $\triangle ABC$, if $a = 7$, $b = 5$, and $c = 8$, use the Law of Cosines to find $\cos A$.

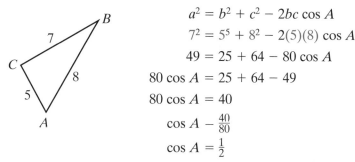

$$a^2 = b^2 + c^2 - 2bc \cos A$$
$$7^2 = 5^5 + 8^2 - 2(5)(8) \cos A$$
$$49 = 25 + 64 - 80 \cos A$$
$$80 \cos A = 25 + 64 - 49$$
$$80 \cos A = 40$$
$$\cos A - \tfrac{40}{80}$$
$$\cos A = \tfrac{1}{2}$$

Since A is an angle of a triangle, $0° < A < 180°$. Therefore, $A = 60°$.

The steps used to solve for $\cos A$ in terms of the measures of the sides can be applied to the general formula of the Law of Cosines to express the cosine of any angle of the triangle in terms of the lengths of the sides.

$$a^2 = b^2 + c^2 - 2bc \cos A$$
$$2bc \cos A = b^2 + c^2 - a^2$$
$$\cos A = \tfrac{b^2 + c^2 - a^2}{2bc}$$

This formula can be rewritten in terms of the cosine of any angle of $\triangle ABC$.

$$\cos A = \tfrac{b^2 + c^2 - a^2}{2bc}$$

$$\cos B = \tfrac{a^2 + c^2 - b^2}{2ac}$$

$$\cos C = \tfrac{a^2 + b^2 - c^2}{2ab}$$

EXAMPLE I

In $\triangle ABC$, $a = 12$, $b = 8$, $c = 6$. Find $\cos C$.

Solution

How to Proceed

(1) Write the Law of Cosines in terms of $\cos C$: $\cos C = \tfrac{a^2 + b^2 - c^2}{2ab}$

(2) Substitute the given values: $\cos C = \tfrac{12^2 + 8^2 - 6^2}{2(12)(8)}$

(3) Perform the computation. Reduce the fractional value of $\cos C$ to lowest terms: $\cos C = \tfrac{144 + 64 - 36}{192}$
$$= \tfrac{172}{192}$$
$$= \tfrac{43}{48}$$

Answer $\cos C = \tfrac{43}{48}$

EXAMPLE 2

Find, to the nearest degree, the measure of the largest angle of $\triangle DEF$ if $DE = 7.5$, $EF = 9.6$, and $DF = 13.5$.

Solution The largest angle of the triangle is opposite the longest side. The largest angle is $\angle E$, the angle opposite the longest side, DF. Let $DE = f = 7.5$, $EF = d = 9.6$, and $DF = e = 13.5$. Write the formula in terms of $\cos E$.

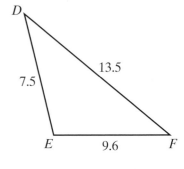

$$\cos E = \frac{d^2 + f^2 - e^2}{2df}$$

$$= \frac{9.6^2 + 7.5^2 - 13.5^2}{2(9.6)(7.5)}$$

$$\approx -0.235$$

Therefore, $E = \arccos(-0.235)$. Use a calculator to find the arccosine:

ENTER: [2nd] [COS⁻¹] [(-)] 0.235   DISPLAY:
       [ )] [ENTER]

```
cos⁻¹(-0.235)
        103.5916228
```

Answer $m\angle E = 104$

Exercises

Writing About Mathematics

1. Explain how the Law of Cosines can be used to show that 4, 7, and 12 cannot be the measures of the sides of a triangle.

2. Show that if $\angle C$ is an obtuse angle, $a^2 + b^2 < c^2$.

Developing Skills

3. In $\triangle TUV$, express $\cos T$ in terms of t, u, and v.

4. In $\triangle PQR$, express $\cos Q$ in terms of p, q, and r.

5. In $\triangle KLM$, if $k = 4$, $l = 5$, and $m = 8$, find the exact value of $\cos M$.

6. In $\triangle XYZ$, if $x = 1$, $y = 2$, and $z = \sqrt{5}$, find the exact value of $\cos Z$.

In 7–12, find the cosine of each angle of the given triangle.

7. In $\triangle ABC$, $a = 4$, $b = 6$, $c = 8$.   **8.** In $\triangle ABC$, $a = 12$, $b = 8$, $c = 8$.

9. In $\triangle DEF$, $d = 15$, $e = 12$, $f = 8$.   **10.** In $\triangle PQR$, $p = 2$, $q = 4$, $r = 5$.

11. In $\triangle MNP$, $m = 16$, $n = 15$, $p = 8$.   **12.** In $\triangle ABC$, $a = 5$, $b = 12$, $c = 13$.

In 13–18, find, to the nearest degree, the measure of each angle of the triangle with the given measures of the sides.

13. 12, 20, 22   **14.** 9, 10, 15   **15.** 30, 35, 45   **16.** 11, 11, 15   **17.** 32, 40, 38   **18.** 7, 24, 25

Applying Skills

19. Two lighthouses are 12 miles apart along a straight shore. A ship is 15 miles from one lighthouse and 20 miles from the other. Find, to the nearest degree, the measure of the angle between the lines of sight from the ship to each lighthouse.

20. A tree is braced by wires 4.2 feet and 4.7 feet long that are fastened to the tree at the same point and to the ground at points 7.8 feet apart. Find, to the nearest degree, the measure of the angle between the wires at the tree.

21. A kite is in the shape of a quadrilateral with two pair of congruent adjacent sides. The lengths of two sides are 20.0 inches and the lengths of the other two sides are 35.0 inches. The two shorter sides meet at an angle of 115°.

　a. Find the length of the diagonal between the points at which the unequal sides meet. Write the length to the nearest tenth of an inch.

　b. Using the answer to part **a**, find, to the nearest degree, the measure of the angle at which the two longer sides meet.

22. A beam 16.5 feet long supports a roof with rafters each measuring 12.4 feet long. What is the measure of the angle at which the rafters meet?

16.5 ft

23. A walking trail is laid out in the shape of a triangle. The lengths of the three paths that make up the trail are 2,500 meters, 2,000 meters, and 1,800 meters. Determine, to the nearest degree, the measure of the greatest angle of the trail.

24. Use the formula $\cos C = \frac{a^2 + b^2 - c^2}{2ab}$ to show that the measure of each angle of an equilateral triangle is 60°.

14-4 AREA OF A TRIANGLE

When the measures of two sides and the included angle of a triangle are known, the size and shape of the triangle is determined. Therefore, it is possible to use these known values to find the area of the triangle. Let $\triangle ABC$ be any triangle. If we know the measures of $AB = c$, $AC = b$, and the included angle, $\angle A$, we can find the area of the triangle.

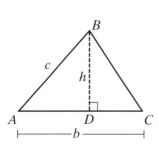
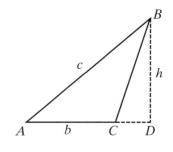

In $\triangle ABC$ let $\angle A$ be an acute angle, $\overline{BD}$ be the altitude from B to $\overleftrightarrow{AC}$, and $BD = h$. In right $\triangle ABD$, $\sin A = \frac{\text{opp}}{\text{hyp}} = \frac{BD}{AB} = \frac{h}{c}$ or $h = c \sin A$. Therefore:

$$\text{Area of } \triangle ABC = \tfrac{1}{2}bh = \tfrac{1}{2}bc \sin A$$

What if $\angle A$ is an obtuse angle?

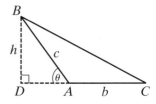

Let $\angle A$ be an obtuse angle of $\triangle ABC$, $\overline{BD}$ be the altitude from B to $\overleftrightarrow{AC}$ and $BD = h$. In right $\triangle ABD$, let $m\angle DAB = \theta$. Then $\sin \theta = \frac{\text{opp}}{\text{hyp}} = \frac{BD}{AB} = \frac{h}{c}$ or $h = c \sin \theta$. Therefore:

$$\text{Area of } \triangle ABC = \tfrac{1}{2}bh = \tfrac{1}{2}bc \sin \theta$$

Since $\angle DAB$ and $\angle BAC$ are adjacent angles whose sum is a straight angle, $m\angle DAB = 180 - m\angle BAC$ and $\sin \theta = \sin A$. Therefore, the area of $\triangle ABC$ is again equal to $\frac{1}{2}bc \sin A$. Thus, for any angle, we have shown that:

$$\textbf{Area of } \triangle ABC = \tfrac{1}{2}bh = \tfrac{1}{2}bc \sin A$$

The area of a triangle is equal to one-half the product of the measures of two sides of the triangle times the sine of the measure of the included angle. This formula can be written in terms of any two sides and the included angle.

$$\text{Area } \triangle ABC = \tfrac{1}{2}bc \sin A = \tfrac{1}{2}ac \sin B = \tfrac{1}{2}ab \sin C$$

Triangles in the Coordinate Plane

When a triangle is drawn in the coordinate plane, the area formula follows easily. Let $\triangle ABC$ be any triangle. Place the triangle with $A(0, 0)$ at the origin and $C(b, 0)$ on the positive ray of the x-axis, and $\angle A$ an angle in standard position.

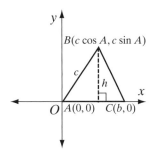 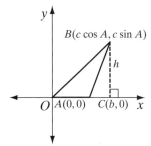 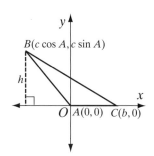

From Section 14-1, we know that the coordinates of B are $(c \cos A, c \sin A)$. For each triangle, h is the length of the perpendicular from B to the x-axis and $h = c \sin A$. Therefore,

$$\text{Area } \triangle ABC = \tfrac{1}{2}bh = \tfrac{1}{2}bc \sin A$$

EXAMPLE 1

Find the area of $\triangle DEF$ if $DE = 14$, $EF = 9$, and $m\angle E = 30$.

Solution $DE = f = 14$, $EF = d = 9$, and $m\angle E = 30$.

$$\text{Area of } \triangle DEF = \tfrac{1}{2}df \sin E = \tfrac{1}{2}(9)(14)\left(\tfrac{1}{2}\right) = \tfrac{63}{2} = 31\tfrac{1}{2} \ \textit{Answer}$$

EXAMPLE 2

The adjacent sides of parallelogram $ABCD$ measure 12 and 15. The measure of one angle of the parallelogram is 135°. Find the area of the parallelogram.

Solution The diagonal of a parallelogram separates the parallelogram into two congruent triangles. Draw diagonal $\overline{BD}$. In $\triangle DAB$, $DA = b = 12$, $AB = d = 15$, and $m\angle A = 135$.

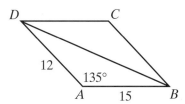

$$\text{Area of } \triangle DAB = \tfrac{1}{2}bd \sin A = \tfrac{1}{2}(12)(15) \sin 135° = \tfrac{1}{2}(12)(15)\left(\tfrac{\sqrt{2}}{2}\right) = 45\sqrt{2}$$

$$\text{Area of } \triangle DBC = \text{Area of } \triangle DAB = 45\sqrt{2}$$

$$\text{Area of parallelogram } ABCD = \text{Area of } \triangle DBC + \text{Area of } \triangle DAB = 90\sqrt{2}$$

Answer $90\sqrt{2}$ square units

Note: The same answer to Example 2 is obtained if we use adjacent angle B or D. Consecutive angles of a parallelogram are supplementary. If $m\angle A = 135$, then $m\angle B = 180 - 135 = 45$. Opposite sides of a parallelogram are congruent. If $DA = 12$, then $BC = 12$. Draw diagonal $\overline{AC}$. In $\triangle ABC$, $BC = a = 12$, $AB = c = 15$, and $m\angle B = 45$.

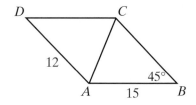

Area of $\triangle ABC = \frac{1}{2}ac \sin B = \frac{1}{2}(12)(15) \sin 45° = \frac{1}{2}(12)(15)\left(\frac{\sqrt{2}}{2}\right) = 45\sqrt{2}$

Area of $\triangle CDA$ = Area of $\triangle ABC = 45\sqrt{2}$

Area of parallelogram $ABCD$ = Area of $\triangle ABC$ + Area of $\triangle CDA = 90\sqrt{2}$

EXAMPLE 3

Three streets intersect in pairs enclosing a small triangular park. The measures of the distances between the intersections are 85.5 feet, 102 feet, and 78.2 feet. Find the area of the park to the nearest ten square feet.

Solution Let A, B, and C be the intersections of the streets, forming $\triangle ABC$. Use the Law of Cosines to find the measure of any angle, for example, $\angle A$. Then use the formula for the area of a triangle in terms of the measures of two sides and an angle.

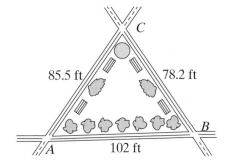

$$\cos A = \frac{b^2 + c^2 - a^2}{2bc} = \frac{85.5^2 + 102^2 - 78.2^2}{2(85.5)(102)} \approx 0.6650$$

Use a calculator to find the measure of $\angle A$.

$$m\angle A = \cos^{-1} 0.6650 \approx 48.31$$

$$\text{Area of } \triangle ABC = \frac{1}{2}bc \sin A = \frac{1}{2}(85.5)(102)(\sin 48.31°) \approx 3,256$$

Answer The area of the park is approximately 3,260 square feet.

Exercises

Writing About Mathematics

1. Rosa found the area of parallelogram $ABCD$ by using $(AB)(BC)(\sin B)$. Riley found the area of parallelogram $ABCD$ by using $(AB)(BC)(\sin A)$. Explain why Rosa and Riley both got the correct answer.

2. Jessica said that the area of rhombus $PQRS$ is $(PQ)^2(\sin P)$. Do you agree with Jessica? Explain why or why not.

Developing Skills

In 3–8, find the area of each $\triangle ABC$.

3. $b = 3, c = 8, \sin A = \frac{1}{4}$

4. $a = 12, c = 15, \sin B - \frac{1}{3}$

5. $b = 9, c = 16, \sin A = \frac{5}{6}$

6. $a = 24, b = 12, \sin C = \frac{3}{4}$

7. $b = 7, c = 8, \sin A = \frac{3}{5}$

8. $a = 10, c = 8, \sin B = \frac{3}{10}$

In 9–14, find the area of each triangle to the nearest tenth.

9. In $\triangle ABC, b = 14.6, c = 12.8, m\angle A = 56$.

10. In $\triangle ABC, a = 326, c = 157, m\angle B = 72$.

11. In $\triangle DEF, d = 5.83, e = 5.83, m\angle F = 48$.

12. In $\triangle PQR, p = 212, q = 287, m\angle R = 124$.

13. In $\triangle RST, t = 15.7, s = 15.7, m\angle R = 98$.

14. In $\triangle DEF, e = 336, f = 257, m\angle D = 122$.

15. Find the exact value of the area of an equilateral triangle if the length of one side is 40 meters.

16. Find the exact value of the area of an isosceles triangle if the measure of a leg is 12 centimeters and the measure of the vertex angle is 45 degrees.

17. Find the area of a parallelogram if the measures of two adjacent sides are 40 feet and 24 feet and the measure of one angle of the parallelogram is 30 degrees.

Applying Skills

18. A field is bordered by two pairs of parallel roads so that the shape of the field is a parallelogram. The lengths of two adjacent sides of the field are 2 kilometers and 3 kilometers, and the length of the shorter diagonal of the field is 3 kilometers.

   a. Find the cosine of the acute angle of the parallelogram.

   b. Find the exact value of the sine of the acute angle of the parallelogram.

   c. Find the exact value of the area of the field.

   d. Find the area of the field to the nearest integer.

19. The roof of a shed consists of four congruent isosceles triangles. The length of each equal side of one triangular section is 22.0 feet and the measure of the vertex angle of each triangle is 75°. Find, to the nearest square foot, the area of one triangular section of the roof.

20. A garden is in the shape of an isosceles trapezoid. The lengths of the parallel sides of the garden are 30 feet and 20 feet, and the length of each of the other two sides is 10 feet. If a base angle of the trapezoid measures 60°, find the exact area of the garden.

21. In $\triangle ABC$, $\text{m}\angle B = 30$ and in $\triangle DEF$, $\text{m}\angle E = 150$. Show that if $AB = DE$ and $BC = EF$, the areas of the two triangles are equal.

22. Aaron wants to draw $\triangle ABC$ with $AB = 15$ inches, $BC = 8$ inches, and an area of 40 square inches.

a. What must be the sine of $\angle B$?

b. Find, to the nearest tenth of a degree, the measure of $\angle B$.

c. Is it possible for Aaron to draw two triangles that are not congruent to each other that satisfy the given conditions? Explain.

23. Let $ABCD$ be a parallelogram with $AB = c$, $BC = a$, and $\text{m}\angle B = \theta$.

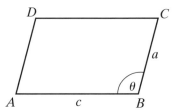

a. Write a formula for the area of parallelogram $ABCD$ in terms of c, a, and θ.

b. For what value of θ does parallelogram $ABCD$ have the greatest area?

14-5 LAW OF SINES

If we know the measures of two angles and the included side of a triangle (ASA), or if we know the measures of two angles and the side opposite one of the angles of a triangle (AAS), the size and shape of the triangle is determined. Therefore, we should be able to find the measures of the remaining sides.

In $\triangle ABC$, let $\text{m}\angle A$ and $\text{m}\angle B$ be two angles and $AC = b$ be the side opposite one of the angles. When we know these measures, is it possible to find $BC = a$?

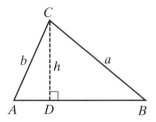

Let $\overline{CD}$ be the altitude from C to $\overline{AB}$. Let $CD = h$ and $BC = a$.

In right $\triangle ACD$,	In right $\triangle BCD$,
$\sin A = \dfrac{\text{opp}}{\text{hyp}}$	$\sin B = \dfrac{\text{opp}}{\text{hyp}}$
$\sin A = \dfrac{h}{b}$	$\sin B = \dfrac{h}{a}$
$h = b \sin A$	$h = a \sin B$

Since $a \sin B$ and $b \sin A$ are each equal to h, they are equal to each other. Therefore, $a \sin B = b \sin A$. To solve for a, divide both sides of this equation by $\sin B$.

$$a \sin B = b \sin A$$

$$\frac{a \sin B}{\sin B} = \frac{b \sin A}{\sin B}$$

$$a = \frac{b \sin A}{\sin B}$$

More generally, we can establish a proportional relationship between two angles and the sides opposite these angles in a triangle. Divide both sides of this equation by $\sin A \sin B$.

$$a \sin B = b \sin A$$

$$\frac{a \sin B}{\sin A \sin B} = \frac{b \sin A}{\sin A \sin B}$$

$$\frac{a}{\sin A} = \frac{b}{\sin B}$$

An alternative derivation of this formula begins with the formulas for the area of a triangle.

$$\text{Area } \triangle ABC = \tfrac{1}{2}bc \sin A = \tfrac{1}{2}ac \sin B = \tfrac{1}{2}ab \sin C$$

We can multiply each of the last three terms of this equality by 2.

$$2\left(\tfrac{1}{2}bc \sin A\right) = 2\left(\tfrac{1}{2}ac \sin B\right) = 2\left(\tfrac{1}{2}ab \sin C\right)$$

$$bc \sin A = ac \sin B = ab \sin C$$

Now divide each side of the equality by abc.

$$\frac{bc \sin A}{abc} = \frac{ac \sin B}{abc} = \frac{ab \sin C}{abc}$$

$$\frac{\sin A}{a} = \frac{\sin B}{b} = \frac{\sin C}{c}$$

These equal ratios are usually written in terms of their reciprocals.

$$\frac{a}{\sin A} = \frac{b}{\sin B} = \frac{c}{\sin C}$$

This equality is called the **Law of Sines**.

EXAMPLE I

In $\triangle ABC, c = 12, m\angle B = 120,$ and $m\angle C = 45.$ Find the exact value of $b.$

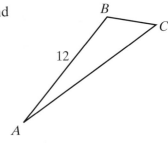

Solution   *How to Proceed*

(1) Use the ratios of the Law of Sines that use b and c:

$$\frac{b}{\sin B} = \frac{c}{\sin C}$$

(2) Substitute the given values:

$$\frac{b}{\sin 120°} = \frac{12}{\sin 45°}$$

(3) Solve for $b,$ substituting sine values:

$$b = \frac{12 \sin 120°}{\sin 45°}$$

$$b = \frac{12\left(\frac{\sqrt{3}}{2}\right)}{\frac{\sqrt{2}}{2}}$$

(4) Write the value of b in simplest form:

$$b = \frac{12\sqrt{3}}{2} \times \frac{2}{\sqrt{2}}$$

$$= \frac{12\sqrt{3}}{\sqrt{2}} \times \frac{\sqrt{2}}{\sqrt{2}}$$

$$= \frac{12\sqrt{6}}{2}$$

$$= 6\sqrt{6}$$

Answer $b = 6\sqrt{6}$

EXAMPLE 2

In $\triangle DEF, m\angle D = 50, m\angle E = 95,$ and $f = 12.6.$ Find d to the nearest tenth.

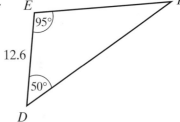

Solution Use the form of the Law of Sines in terms of the side whose measure is known, $f,$ and the side whose measure is to be found, $d.$

$$\frac{d}{\sin D} = \frac{f}{\sin F}$$

To use this formula, we need to know $m\angle F.$

$$m\angle F = 180 - (50 + 95) = 35$$

Therefore:

$$\frac{d}{\sin 50°} = \frac{12.6}{\sin 35°}$$

$$d = \frac{12.6 \sin 50°}{\sin 35°}$$

Use a calculator to evaluate d.

ENTER: (12.6 **SIN** 50))
   ÷ **SIN** 35) **ENTER**

DISPLAY: (12.6 sin(50))/si
n(35)
     16.82802739

Answer To the nearest tenth, $d = 16.8$.

Exercises

Writing About Mathematics

1. If the sine of an angle of a triangle is known, is it possible to determine the measure of the angle? Explain why or why not.

2. If the cosine of an angle of a triangle is known, is it possible to determine the measure of the angle? Explain why or why not.

Developing Skills

3. In $\triangle ABC$, if $a = 9$, m$\angle A = \frac{\pi}{3}$, and m$\angle B = \frac{\pi}{4}$, find the exact value of b in simplest form.

4. In $\triangle ABC$, if $a = 24$, m$\angle A = \frac{\pi}{6}$, and m$\angle B = \frac{\pi}{2}$, find the exact value of b in simplest form.

5. In $\triangle ABC$, if $c = 12$, m$\angle C = \frac{2\pi}{3}$, and m$\angle B = \frac{\pi}{6}$, find the exact value of b in simplest form.

6. In $\triangle ABC$, if $b = 8$, m$\angle A = \frac{\pi}{3}$, and m$\angle C = \frac{5\pi}{12}$, find the exact value of a in simplest form.

7. In $\triangle DEF$, $\sin D = 0.4$, $\sin E = 0.25$, and $d = 20$. Find the exact value of e.

8. In $\triangle PQR$, $\sin P = \frac{3}{4}$, $\sin R = \frac{2}{5}$, and $p = 40$. Find the exact value of r.

9. In $\triangle DEF$, m$\angle D = 47$, m$\angle E = 84$, and $d = 17.3$. Find e to the nearest tenth.

10. In $\triangle DEF$, m$\angle D = 56$, m$\angle E = 44$, and $d = 37.5$. Find e to the nearest tenth.

11. In $\triangle LMN$, m$\angle M = 112$, m$\angle N = 54$, and $m = 51.0$. Find n to the nearest tenth.

12. In $\triangle ABC$, m$\angle A = 102$, m$\angle B = 34$, and $a = 25.8$. Find c to the nearest tenth.

13. In $\triangle PQR$, m$\angle P = 125$, m$\angle Q = 14$, and $p = 122$. Find r to the nearest integer.

14. In $\triangle RST$, m$\angle R = 12$, m$\angle S = 75$, and $r = 3.52$. Find t to the nearest tenth.

15. In $\triangle CDE$, m$\angle D = 125$, m$\angle E = 28$, and $d = 12.5$. Find c to the nearest hundredth.

16. The base of an isosceles triangle measures 14.5 centimeters and the vertex angle measures 110 degrees.

 a. Find the measure of one of the congruent sides of the triangle to the nearest hundredth.

 b. Find the perimeter of the triangle to the nearest tenth.

17. The length of one of the equal sides of an isosceles triangle measures 25.8 inches and each base angle measures 53 degrees.

   a. Find the measure of the base of the triangle to the nearest tenth.

   b. Find the perimeter of the triangle to the nearest inch.

18. Use the Law of Sines to show that if $\angle C$ of $\triangle ABC$ is a right angle, $\sin A = \frac{a}{c}$.

Applying Skills

19. A telephone pole on a hillside makes an angle of 78 degrees with the upward slope. A wire from the top of the pole to a point up the hill is 12.0 feet long and makes an angle of 15 degrees with the pole.

   a. Find, to the nearest hundredth, the distance from the foot of the pole to the point at which the wire is fastened to the ground.

   b. Use the answer to part **a** to find, to the nearest tenth, the height of the pole.

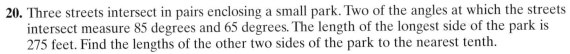

20. Three streets intersect in pairs enclosing a small park. Two of the angles at which the streets intersect measure 85 degrees and 65 degrees. The length of the longest side of the park is 275 feet. Find the lengths of the other two sides of the park to the nearest tenth.

21. On the playground, the 10-foot ladder to the top of the slide makes an angle of 48 degrees with the ground. The slide makes an angle of 32 degrees with the ground.

   a. How long is the slide to the nearest tenth?

   b. What is the distance from the foot of the ladder to the foot of the slide to the nearest tenth?

22. A distress signal from a ship, S, is received by two coast guard stations located 3.8 miles apart along a straight coastline. From station A, the signal makes an angle of 48° with the coastline and from station B the signal makes an angle of 67° with the coastline. Find, to the nearest tenth of a mile, the distance from the ship to the nearer station.

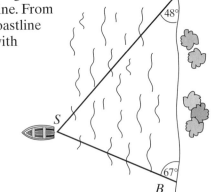

23. Two sides of a triangular lot form angles that measure 29.1° and 33.7° with the third side, which is 487 feet long. To the nearest dollar, how much will it cost to fence the lot if the fencing costs $5.59 per foot?

14-6 THE AMBIGUOUS CASE

If we know the measures of two sides of a triangle and the angle opposite one of them, the Law of Sines makes it possible for us to find the sine of the angle opposite the second side whose measure is known. However, we know that the measures of two sides and the angle opposite one of them (SSA) is not sufficient to determine the size and shape of the triangle in every case. This is often called the **ambiguous case**.

Consider the following cases in which we are given a, b, and angle A. For $0 < \sin B < 1$, there are two values of B in the interval from $0°$ to $180°$. We will call these values B and B'. Since the sum of the degree measures of the angles of a triangle is 180, the sum of the degree measures of two angles of a triangle must be less than 180.

CASE 1 *Two triangles can be drawn.*

In $\triangle ABC$, $a = 8$, $b = 12$, and $\angle A = 30°$. We can use the Law of Sines to find $\sin B$.

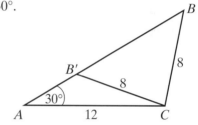

$$\frac{a}{\sin A} = \frac{b}{\sin B}$$
$$\frac{8}{\sin 30°} = \frac{12}{\sin B}$$
$$8 \sin B = 12 \sin 30°$$
$$\sin B = \frac{12\left(\frac{1}{2}\right)}{8}$$
$$\sin B = \frac{3}{4}$$

When $\sin B = \frac{3}{4}$, $m\angle B \approx 48.59$ or $m\angle B' \approx 180 - 48.59 = 131.41$. As shown in the diagram, there are two triangles, $\triangle ABC$ and $\triangle AB'C$ in which two sides measure 8 and 12 and the angle opposite the shorter of these sides measures $30°$. Two triangles can be drawn.

CASE 2 *Only one triangle can be drawn and that triangle is a right triangle.*

In $\triangle ABC$, $a = 8$, $b = 16$, and $\angle A = 30°$. We can use the Law of Sines to find $\sin B$.

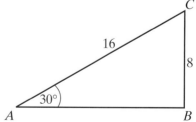

$$\frac{a}{\sin A} = \frac{b}{\sin B}$$
$$\frac{8}{\sin 30°} = \frac{16}{\sin B}$$
$$8 \sin B = 16 \sin 30°$$
$$\sin B = \frac{16\left(\frac{1}{2}\right)}{8}$$
$$\sin B = 1$$

When $\sin B = 1, \mathrm{m}\angle B = 90$. This is the only measure of $\angle B$ that can be the measure of an angle of a triangle. One triangle can be drawn and that triangle is a right triangle.

Note: If $\mathrm{m}\angle A = 150$, $\sin B = 1$ and $\mathrm{m}\angle B = 90$. There is no triangle with an obtuse angle and a right angle.

CASE 3 *Only one triangle can drawn.*

In $\triangle ABC$, $a = 16$, $b = 8$, and $\angle A = 30°$. We can use the Law of Sines to find $\sin B$.

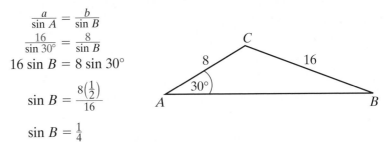

$$\frac{a}{\sin A} = \frac{b}{\sin B}$$
$$\frac{16}{\sin 30°} = \frac{8}{\sin B}$$
$$16 \sin B = 8 \sin 30°$$
$$\sin B = \frac{8\left(\frac{1}{2}\right)}{16}$$
$$\sin B = \frac{1}{4}$$

When $\sin B = \frac{1}{4}, \mathrm{m}\angle B = 14.48$ or $\mathrm{m}\angle B = 180 - 14.48 = 165.52$.

- If we let $\angle B$ be an acute angle, $\mathrm{m}\angle A + \mathrm{m}\angle B = 30 + 14.48 < 180$. There is a triangle with $\mathrm{m}\angle A = 30$ and $\mathrm{m}\angle B = 14.48$. ✔

- If we let $\angle B$ be an obtuse angle, $\mathrm{m}\angle A + \mathrm{m}\angle B = 30 + 165.52 > 180$. There is no triangle with $\mathrm{m}\angle A = 30$ and $\mathrm{m}\angle B = 165.52$. ✘

Only one triangle can be drawn.

CASE 4 *No triangle can be drawn.*

In $\triangle ABC$, $a = 8$, $b = 20$, and $\angle A = 30°$. We can use the Law of Sines to find $\sin B$.

$$\frac{a}{\sin A} = \frac{b}{\sin B}$$
$$\frac{8}{\sin 30°} = \frac{20}{\sin B}$$
$$8 \sin B = 20 \sin 30°$$
$$\sin B = \frac{20\left(\frac{1}{2}\right)}{8}$$
$$\sin B = \frac{5}{4}$$

There is no value of B for which $\sin B > 1$. No triangle can be drawn.

These four examples show that for the given lengths of two sides and the measure of an acute angle opposite one of them, two, one, or no triangles can be formed.

To determine the number of solutions given *a*, *b*, and m∠*A* in △*ABC*:

Use the Law of Sines to solve for sin *B*.

If sin *B* > 1, there is no triangle.	If sin *B* = 1, there is one right triangle if ∠*A* is acute but no triangle if ∠*A* is obtuse.
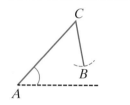	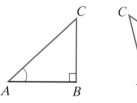

If ∠*A* is acute and sin *B* < 1, find two possible values of ∠*B*:
0 < m ∠*B* < 90 and m ∠*B*′ = 180 − m ∠*B*.

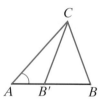	
If m∠*A* + m∠*B*′ < 180, there are two possible triangles, △*ABC* and △*AB*′*C*.	If m∠*A* + m∠*B*′ ≥ 180, △*AB*′*C* is not a triangle. There is only one possible triangle, △*ABC*.

If ∠*A* is obtuse and sin *B* < 1, ∠*B* must be acute: 0 < m∠*B* < 90.

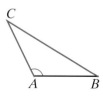	
If m∠*A* + m∠*B* < 180, there is one triangle, △*ABC*.	If m∠*A* + m∠*B* ≥ 180, there is no triangle.

Alternatively, if we let $h = b \sin A$, the height of the triangle, we can summarize the number of possible triangles given a, b, and $m\angle A$ in $\triangle ABC$:

$\angle A$ is:	Acute	Acute	Acute	Acute	Obtuse	Obtuse
	$a < h$	$h = a$	$h < a < b$	$a > b$	$a \leq b$	$a > b$
Possible triangles:	None	One, right $\triangle$	Two	One	None	One

EXAMPLE I

In $\triangle ABC$, $b = 9$, $c = 12$, and $m\angle C = 45$.

a. Find the exact value of $\sin B$.

b. For the value of $\sin B$ in **a**, find, to the nearest hundredth, the measures of two angles, $\angle B$ and $\angle B'$, that could be angles of a triangle.

c. How many triangles are possible?

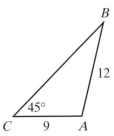

Solution **a.** (1) Use the ratios of the Law of Sines that use b and c:

$$\frac{b}{\sin B} = \frac{c}{\sin C}$$

(2) Substitute the given values:

$$\frac{9}{\sin B} = \frac{12}{\sin 45°}$$

(3) Solve for $\sin B$:

$$12 \sin B = 9 \sin 45°$$

(4) Substitute the exact value of $\sin 45$:

$$\sin B = \frac{9}{12} \sin 45°$$

$$\sin B = \frac{3}{4} \times \frac{\sqrt{2}}{2}$$

$$\sin B = \frac{3\sqrt{2}}{8}$$

b. Use a calculator to find the approximate measure of $\angle B$.

ENTER: 2nd SIN⁻¹ (3 ×   DISPLAY:
2nd √ 2))

÷ 8 ENTER

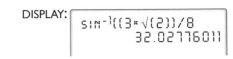

$m\angle B \approx 32.03$ and $m\angle B' \approx 180 - 32.03 = 147.97$ *Answer*

c. $m\angle A + m\angle B = 45 + 32.03 < 180$ and $\triangle ABC$ is a triangle.

$m\angle A + m\angle B' = 45 + 147.97 > 180$ and $\triangle AB'C$ is *not* a triangle.

There is one possible triangle. *Answer*

EXAMPLE 2

How many triangles can be drawn if the measures of two of the sides are 12 inches and 10 inches and the measure of the angle opposite the 10-inch side is 110 degrees?

Solution Let a possible triangle be $\triangle PQR$ with $PQ = r = 12$ and $QR = p = 10$. The angle opposite $\overline{QR}$ is $\angle P$ and m$\angle P = 110$.

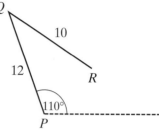

(1) We know p, r, and $\angle P$. Use the Law of Sines to solve for sin R:

$$\frac{p}{\sin P} = \frac{r}{\sin R}$$

$$\frac{10}{\sin 110°} = \frac{12}{\sin R}$$

$$10 \sin R = 12 \sin 110°$$

$$\sin R = \frac{12 \sin 110°}{10}$$

(2) Use a calculator to find sin R:

$$\sin R \approx 1.127631145$$

Since sin $R > 1$, there is no triangle.

Alternative Solution Let h represent the height of the triangle. Then

$$h = r \sin P = 12 \sin 70° \approx 11.28$$

The height of the triangle, or the altitude at Q, would be longer than side $\overline{QR}$. No such triangle exists.

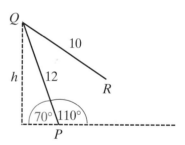

Answer 0

Exercises

Writing About Mathematics

1. Without using formulas that include the sine of an angle, is it possible to determine from the given information in Example 2 that there can be no possible triangle? Justify your answer.

2. Explain why, when the measures of two sides and an obtuse angle opposite one of them are given, it is never possible to construct two different triangles.

Developing Skills

In 3–14: **a.** Determine the number of possible triangles for each set of given measures. **b.** Find the measures of the three angles of each possible triangle. Express approximate values to the nearest degree.

3. $a = 8, b = 10, m\angle A = 20$

4. $a = 5, b = 10, m\angle A = 30$

5. $b = 12, c = 10, m\angle B = 49$

6. $a = 6, c = 10, m\angle A = 45$

7. $c = 18, b = 10, m\angle C = 120$

8. $a = 9, c = 10, m\angle C = 150$

9. $AB = 14, BC = 21, m\angle C = 75$

10. $DE = 24, EF = 18, m\angle D = 15$

11. $PQ = 12, PR = 15, m\angle R = 100$

12. $BC = 12, AC = 12\sqrt{2}, m\angle B = 135$

13. $RS = 3\sqrt{3}, ST = 3, m\angle T = 60$

14. $a = 8, b = 10, m\angle A = 45$

Applying Skills

15. A ladder that is 15 feet long is placed so that it reaches from level ground to the top of a vertical wall that is 13 feet high.

   a. Use the Law of Sines to find the angle that the ladder makes with the ground to the nearest hundredth.

   b. Is more than one position of the ladder possible? Explain your answer.

16. Max has a triangular garden. He measured two sides of the garden and the angle opposite one of these sides. He said that the two sides measured 5 feet and 8 feet and that the angle opposite the 8-foot side measured 75 degrees. Can a garden exist with these measurements? Could there be two gardens of different shapes with these measurements? Write the angle measures and lengths of the sides of the garden(s) if any.

17. Emily wants to draw a parallelogram with the measure of one side 12 centimeters, the measure of one diagonal 10 centimeters and the measure of one angle 120 degrees. Is this possible? Explain why or why not.

18. Ross said that when he jogs, his path forms a triangle. Two sides of the triangle are 2.0 kilometers and 2.5 kilometers in length and the angle opposite the shorter side measures 45 degrees. Rosa said that when she jogs, her path also forms a triangle with two sides of length 2.0 kilometers and 2.5 kilometers and an angle of 45 degrees opposite the shorter side. Rosa said that her route is longer than the route Ross follows. Is this possible? Explain your answer.

14-7 SOLVING TRIANGLES

If the known measures of any three parts of a triangle include at least one side, the measures of the remaining three parts can be determined.

The Right Triangle

When the triangle is a right triangle, the ratio of the measures of any two sides of the triangle is a trigonometric function value of one of the acute angles.

In right $\triangle ABC$ with $m\angle C = 90$:

$$\sin A = \frac{\text{opp}}{\text{hyp}} = \frac{BC}{AB}$$

$$\cos A = \frac{\text{adj}}{\text{hyp}} = \frac{AC}{AB}$$

$$\tan A = \frac{\text{opp}}{\text{adj}} = \frac{BC}{AC}$$

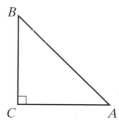

The height of a building, a tree, or any similar object is measured as the length of the perpendicular from the top of the object to the ground. The measurement of height often involves right triangles.

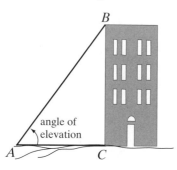

An **angle of elevation** is an angle such that one ray is part of a horizontal line and the other ray represents a line of sight raised upward from the horizontal. To visualize the angle of elevation of a building, think of some point on the same horizontal line as the base of the building. The angle of elevation is the angle through which our line of sight would rotate from the base of the building to its top. In the diagram, $\angle A$ is the angle of elevation.

An **angle of depression** is an angle such that one ray is part of a horizontal line and the other ray represents a line of sight moved downward from the horizontal. To visualize the angle of depression of a building, think of some point on the same horizontal line as the top of the building. The angle of depression is the angle through which our line of sight would rotate from the top of the building to its base. In the diagram, $\angle ABD$ is the angle of depression.

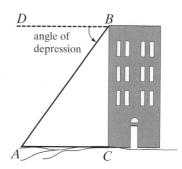

When solving right triangles, we can use the ratio of sides given above or we can use the Law of Sines or the Law of Cosines.

EXAMPLE I

From a point 12 feet from the foot of the tree, the measure of the angle of elevation to the top of the tree is 57°. Find the height of the tree to the nearest tenth of a foot.

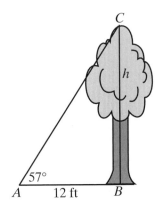

Solution   The height of the tree is the length of the perpendicular from the top of the tree to the ground. Use the ratio of sides of a right triangle.

   We know the length of the side adjacent to the angle of elevation and we want to know the height of the tree, the length of the side opposite the angle of elevation.

$$\tan \theta = \frac{\text{opp}}{\text{adj}}$$
$$\tan 57° = \frac{h}{12}$$
$$h = 12 \tan 57°$$
$$h = 18.478$$

Alternative Solution   We know the measures of two angles and the included side. Find the measure of the third angle and use the Law of Sines.

$$\text{m}\angle B = 180 - (\text{m}\angle A + \text{m}\angle C)$$
$$\text{m}\angle B = 180 - (57 + 90)$$
$$\text{m}\angle B = 33$$

$$\frac{a}{\sin A} = \frac{b}{\sin B}$$
$$\frac{h}{\sin 57°} = \frac{12}{\sin 33°}$$
$$h = \frac{12 \sin 57°}{\sin 33°}$$
$$h = 18.478$$

Answer   The tree is 18.5 feet tall.

The General Triangle

The Law of Cosines and the Law of Sines can be used to find the remaining three measures of any triangle when we know the measure of a side and the measures of any two other parts of the triangle.

CASE I   *Given*: Two sides and the included angle

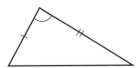

• Use the Law of Cosines to find the measure of the third side.

- Use the Law of Sines or the Law of Cosines to find the measure of another angle.
- Use the sum of the angles of a triangle to find the measure of the third angle.

CASE 2 *Given*: Three sides

- Use the Law of Cosines to find the measure of an angle.
- Use the Law of Sines or the Law of Cosines to find the measure of another angle.
- Use the sum of the angles of a triangle to find the measure of the third angle.

CASE 3 *Given*: Two angles and a side

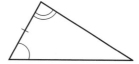

- Use the sum of the angles of a triangle to find the measure of the third angle.
- Use the Law of Sines to find the remaining sides.

CASE 4 *Given*: Two sides and an angle opposite one of them

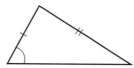

- Use the Law of Sines to find the possible measure(s) of another angle.
- Determine if there are two, one, or no possible triangles.
- If there is a triangle, use the sum of the angles of a triangle to find the measure(s) of the third angle.
- Use the Law of Sines or the Law of Cosines to find the measure(s) of the third side.

EXAMPLE 2

The Parks Department is laying out a nature trail through the woods that is to consist of three straight paths that form a triangle. The lengths of two paths measure 1.2 miles and 1.5 miles. What must be the measure of the angle between these two sections of the path in order that the total length of the nature trail will be 4.0 miles?

Solution   (1) Draw a diagram. The trail forms a triangle, $\triangle ABC$, with $AB = c = 1.2$ and $BC = a = 1.5$:

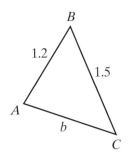

(2) The perimeter of the triangle is 4.0. Find $CA = b$:

$$P = AB + BC + CA$$
$$4.0 = 1.2 + 1.5 + b$$
$$4.0 = 2.7 + b$$
$$1.3 = b$$

(3) Use the Law of Cosines to find the measure of an angle when the measures of three sides are known. Find the measure of $\angle B$:

$$\cos B = \frac{a^2 + c^2 - b^2}{2ac}$$
$$\cos B = \frac{1.5^2 + 1.2^2 - 1.3^2}{2(1.5)(1.2)}$$
$$\cos B = 0.5555 \ldots$$
$$m\angle B = 56.25$$

Answer   56°

EXAMPLE 3

A man, standing on a cliff 85 feet high at the edge of the water, sees two ships. He positions himself so that the ships are in a straight line with a point directly below where he is standing. He estimates that the angle of depression of the closer ship is 75 degrees and the angle of depression of the farther ship is 35 degrees. How far apart are the two ships?

Solution   (1) Draw and label a diagram. Let A be the closer ship, B the farther ship, C the edge of the cliff where the man is standing, and D the point directly below C at sea level. The angle of depression is the angle between a horizontal ray and a ray downward to the ship. Determine the measure of each angle in the diagram:

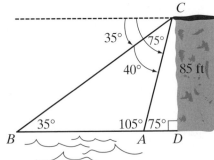

(2) Use right $\triangle ACD$ to find AC, the measure of a side of both $\triangle ABC$ and $\triangle ACD$.

$$\sin 75° = \frac{\text{opp}}{\text{hyp}}$$
$$\sin 75° = \frac{85}{AC}$$
$$AC = \frac{85}{\sin 75°}$$
$$AC \approx 87.998$$

(3) In $\triangle ABC$, we now know the measures of two angles and the side opposite one of them. Use the Law of Sines to find AB, the required distance.

$$\frac{AB}{\sin \angle ACB} = \frac{AC}{\sin \angle ABC}$$
$$\frac{AB}{\sin 40°} = \frac{87.998}{\sin 35°}$$
$$AB = \frac{87.998 \sin 40°}{\sin 35°}$$
$$AB \approx 98.62$$

Answer The ships are 99 feet apart.

Exercises

Writing About Mathematics

1. Navira said that in Example 3, it would have been possible to solve the problem by using $\triangle BDC$ to find BC first. Do you agree with Navira? Explain why or why not.

2. Explain why an angle of depression is always congruent to an angle of elevation.

Developing Skills

In 3–10: **a.** State whether each triangle should be solved by using the Law of Sines or the Law of Cosines. **b.** Solve each triangle, rounding answers to the nearest tenth. Include all possible solutions.

3.

4.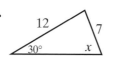

5.

6.

7.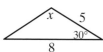

8.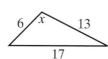

9.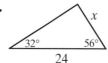

10.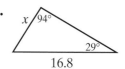

In 11–22, solve each triangle, that is, find the measures of the remaining three parts of the triangle to the nearest integer or the nearest degree.

11. In $\triangle ABC$, $a = 15$, $b = 18$, and $m\angle C = 60$.

12. In $\triangle ABC$, $a = 10$, $b = 12$, and $m\angle B = 30$.

13. In $\triangle ABC$, $b = 25$, m$\angle B = 45$, and m$\angle C = 60$.

14. In $\triangle ABC$, $a = 8$, m$\angle B = 35$, and m$\angle C = 55$.

15. In $\triangle DEF$, $d = 72$, $e = 48$, and m$\angle F = 110$.

16. In $\triangle PQR$, $p = 12$, m$\angle Q = 80$, and m$\angle R = 30$.

17. In $\triangle RST$, $r = 38$, $s = 28$, and $t = 18$.

18. In $\triangle ABC$, $a = 22$, $b = 18$, and m$\angle C = 130$.

19. In $\triangle PQR$, $p = 12$, $q = 16$, and $r = 20$.

20. In $\triangle DEF$, $d = 36$, $e = 72$, and m$\angle D = 30$.

21. In $\triangle RST$, $r = 15$, $s = 18$, and m$\angle T = 90$.

22. In $\triangle ABC$, $a = 15$, $b = 25$, and $c = 12$.

23. In the diagram, $AD = 25$, $CD = 10$, $BD = BC$, and m$\angle D = 75$. Find AB to the nearest tenth.

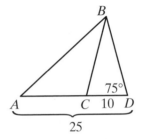

Applying Skills

24. A small park is in the shape of an isosceles trapezoid. The length of the longer of the parallel sides is 3.2 kilometers and the length of an adjacent side is 2.4 kilometers. A path from one corner of the park to an opposite corner is 3.6 kilometers long.

    a. Find, to the nearest tenth, the measure of each angle between adjacent sides of the park.

    b. Find, to the nearest tenth, the measure of each angle between the path and a side of the park.

    c. Find, to the nearest tenth, the length of the shorter of the parallel sides.

25. From a point on the ground 50 feet from the foot of a vertical monument, the measure of the angle of elevation of the top of the monument is 65 degrees. What is the height of the monument to the nearest foot?

26. A vertical telephone pole that is 15 feet high is braced by two wires from the top of the pole to two points on the ground that are 5.0 feet apart on the same side of the pole and in a straight line with the foot of the pole. The shorter wire makes an angle of 65 degrees with the ground. Find the length of each wire to the nearest tenth.

27. From point C at the top of a cliff, two points, A and B, are sited on level ground. Points A and B are on a straight line with D, a point directly below C. The angle of depression of the nearer point, A, is 72 degrees and the angle of depression of the farther point, B, is 48 degrees. If the points A and B are 20 feet apart, what is the height of the cliff to the nearest foot?

28. Mark is building a kite that is a quadrilateral with two pairs of congruent adjacent sides. One diagonal divides the kite into two unequal isosceles triangles and measures 14 inches. Each leg of one of the isosceles triangles measures 15 inches and each leg of the other measures 12 inches. Find the measures of the four angles of the quadrilateral to the nearest tenth.

CHAPTER SUMMARY

When we know the measures of three sides, two sides and the included angle, or two angles and any side of a triangle, the size and shape of the triangle are determined. We can use the Law of Cosines or the Law of Sines to find the measures of the remaining parts of the triangle.

Law of Cosines: $\quad a^2 = b^2 + c^2 - 2bc \cos A \qquad \cos A = \frac{b^2 + c^2 - a^2}{2bc}$

$\qquad\qquad\qquad\quad\ b^2 = a^2 + c^2 - 2ac \cos B \qquad \cos B = \frac{a^2 + c^2 - b^2}{2ac}$

$\qquad\qquad\qquad\quad\ c^2 = a^2 + b^2 - 2ab \cos C \qquad \cos C = \frac{a^2 + b^2 - c^2}{2ab}$

The Law of Cosines can be used to find the measure of the third side of a triangle when the measures of two sides and the included angle are known. The Law of Cosines can also be used to find the measure of any angle of a triangle when the measures of three sides are known.

Law of Sines: $\quad \frac{a}{\sin A} = \frac{b}{\sin B} = \frac{c}{\sin C}$

The Law of Sines can be used to find the measure of a side of a triangle when the measures of any side and any two angles are known. The Law of Sines can also be used to find the number of possible triangles that can be drawn when the measures of two sides and an angle opposite one of them are known and can be used to determine the measures of the remaining side and angles if one or two triangles are possible.

Area of a Triangle: $\quad$ Area of $\triangle ABC = \frac{1}{2}bc \sin A = \frac{1}{2}ac \sin B = \frac{1}{2}ab \sin C$

The area of any triangle can be found if we know the measures of two sides and the included angle.

An **angle of elevation** is an angle between a horizontal line and a line that is rotated upward from the horizontal position. An **angle of depression** is an angle between a horizontal line and a line that is rotated downward from the horizontal position.

To determine the number of solutions given a, b, and m$\angle A$ in $\triangle ABC$, use the Law of Sines to solve for $\sin B$.

- If $\sin B > 1$, there is no triangle.

- If $\sin B = 1$, there is one right triangle if $\angle A$ is acute but no triangle if $\angle A$ is obtuse.

- If $\angle A$ is acute and $\sin B < 1$, find two possible values of $\angle B$:
  $0 < $ m$\angle B < 90$ and m$\angle B' = 180 - $ m$\angle B$.

| If m$\angle A + $ m$\angle B' < 180$, there are two possible triangles, $\triangle ABC$ and $\triangle AB'C$. | If m$\angle A + $ m$\angle B' \geq 180$, $\triangle AB'C$ is not a triangle. There is only one possible triangle, $\triangle ABC$. |

• If ∠*A* is obtuse and sin *B* < 1, ∠*B* must be acute: 0 < m∠*B* < 90.

| If m∠*A* + m∠*B* < 180, there is one triangle, △*ABC*. | If m∠*A* + m∠*B* ≥ 180, there is no triangle. |

Alternatively, if we let *h* = *b* sin *A*, the height of the triangle, we can summarize the number of possible triangles given *a*, *b*, and m∠*A* in △*ABC*:

∠*A* is:	Acute	Acute	Acute	Acute	Obtuse	Obtuse
	a < *h*	*h* = *a*	*h* < *a* < *b*	*a* > *b*	*a* < *h*	*a* > *b*
Possible triangles:	None	One, right △	Two	One	None	One

REVIEW EXERCISES

1. In △*RST*, *RS* = 18, *RT* = 27, and m∠*R* = 50. Find *ST* to the nearest integer.

2. The measures of two sides of a triangle are 12.0 inches and 15.0 inches. The measure of the angle included between these two sides is 80 degrees. Find the measure of the third side of the triangle to the nearest tenth of an inch.

3. In △*DEF*, *DE* = 84, *EF* = 76, and *DF* = 94. Find, to the nearest degree, the measure of the smallest angle of the triangle.

4. The measures of three sides of a triangle are 22, 46, and 58. Find, to the nearest degree, the measure of the largest angle of the triangle.

5. Use the Law of Cosines to show that if the measures of the sides of a triangle are 10, 24, and 26, the triangle is a right triangle.

6. In $\triangle ABC$, $AB = 24$, $AC = 40$, and m$\angle A = 30$.

  a. Find the area of $\triangle ABC$.

  b. Find the length of the altitude from C to $\overline{AB}$.

7. The lengths of the sides of a triangle are 8, 11, and 15.

  a. Find the measure of the smallest angle of the triangle to the nearest tenth.

  b. Find the area of the triangle to the nearest tenth.

8. In $\triangle ABC$, $BC = 30.0$, m$\angle A = 70$, and m$\angle B = 55$. Find, to the nearest tenth, AB and AC.

9. The measures of two angles of a triangle are $100°$ and $46°$. The length of the shortest sides of the triangle is 12. Find, to the nearest integer, the lengths of the other two sides.

10. In $\triangle ABC$, $a = 14$, $b = 16$, and m$\angle A = 48$.

  a. How many different triangles are possible?

  b. Find the measures of $\angle B$ and of $\angle C$ if $\triangle ABC$ is an acute triangle.

  c. Find the measures of $\angle B$ and of $\angle C$ if $\triangle ABC$ is an obtuse triangle.

11. Show that it is not possible to draw $\triangle PQR$ with $p = 12$, $r = 15$, and m$\angle P = 66$.

12. The measure of a side of a rhombus is 28.0 inches and the measure of the longer diagonal is 50.1 inches.

  a. Find, to the nearest degree, the measure of each angle of the rhombus.

  b. Find, to the nearest tenth, the measure of the shorter diagonal.

  c. Find, to the nearest integer, the area of the rhombus.

13. Use the Law of Cosines to find two possible lengths for AB of $\triangle ABC$ if $BC = 7$, $AC = 8$, and m$\angle A = 60$.

14. Use the Law of Sines to show that there are two possible triangles if $BC = 7$, $AC = 8$, and m$\angle A = 60$.

15. A vertical pole is braced by two wires that extend from different points on the pole to the same point on level ground. One wire is fastened to the pole 5.0 feet from the top of the pole and makes an angle of 61 degrees with the ground. The second wire is fastened to the top of the pole and makes an angle of 66 degrees with the ground. Find the height of the pole to the nearest tenth.

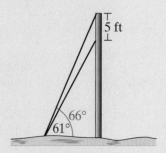

16. Coastguard station A is 12 miles west of coastguard station B along a straight coastline. A ship is sited by the crew at station A to the northeast of the station at an angle of 35 degrees with the coastline and by the crew at station B to the northwest of the station at an angle of 46 degrees with the coastline. Find, to the nearest tenth, the distance from the ship to each of the stations.

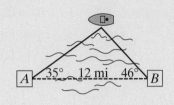

17. In the diagram, $\triangle ABC$ is a right triangle with the right angle at C, $AC = 4.0$, $BC = 3.0$, and $m\angle BAC = \theta$. Side $\overline{CB}$ is extended to D and $m\angle DAC = 2\theta$.

a. Find the exact value of $\sin 2\theta$. (Hint: Use the double-angle formulas.)

b. Find θ to the nearest degree.

c. Find DC to the nearest tenth.

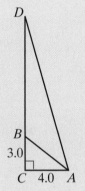

Exploration

Part A

Use software to draw any triangle, $\triangle ABC$. Trisect each angle of the triangle.
Let D be the intersection of the trisection lines from A and B that are closest to $\overline{AB}$.
Let E be the intersection of the trisection lines from B and C that are closest to $\overline{BC}$.
Let F be the intersection of the trisection lines from C and A that are closest to $\overline{CA}$.
Measure $\overline{DE}$, $\overline{EF}$, and $\overline{FD}$. What seems to be true about $\triangle DEF$?

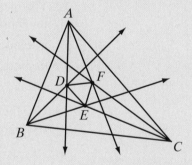

Part B

1. Draw any triangle, $\triangle ABC$. Measure the lengths of the sides and the measures of the angles.

2. Draw the trisection lines and label $\triangle DEF$ as in part A.

3. Use the Law of Sines to find AD and BD in $\triangle ABD$.

4. Use the Law of Sines to find BE and CE in $\triangle BCE$.

5. Use the Law of Sines to find CF and AF in $\triangle CAF$.

6. Use the Law of Cosines to find DF in $\triangle ADF$.

7. Use the Law of Cosines to find DE in $\triangle BDE$.

8. Use the Law of Cosines to find EF in $\triangle CEF$.

9. Is $\triangle DEF$ an equilateral triangle?

CUMULATIVE REVIEW   CHAPTERS 1–14

Part I

Answer all questions in this part. Each correct answer will receive 2 credits. No partial credit will be allowed.

1. The sum of $\sqrt{-25}$ and $2\sqrt{-16}$ is
    (1) -13 (2) $13i$ (3) -18 (4) $18i$

2. If $f(\theta) = \cot \theta$, then $f\left(\frac{\pi}{2}\right)$ is
    (1) 0 (2) 1 (3) ∞ (4) undefined

3. The expression $\displaystyle\sum_{k=1}^{3} (k + 1)^2$ is equal to

    (1) 13 (2) 14 (3) 18 (4) 29

4. If $f(x) = x - 1$ and $g(x) = x^2$, then $f(g(3))$ is equal to
    (1) 18 (2) 11 (3) 8 (4) 4

5. The sum $\sin 50° \cos 30° + \cos 50° \sin 30°$ is equal to
    (1) $\sin 80°$ (2) $\sin 20°$ (3) $\cos 80°$ (4) $\cos 20°$

6. If $\log_x \frac{1}{4} = -2$, then x equals
    (1) 16 (2) 2 (3) $\frac{1}{2}$ (4) $\frac{1}{16}$

7. The solution set of $x^2 - 2x + 5 = 0$ is
    (1) $\{-1, 3\}$ (3) $\{1 - 2i, 1 + 2i\}$
    (2) $\{-3, 1\}$ (4) $\{-1 - 2i, 1 + 2i\}$

8. In simplest form, the fraction $\dfrac{\frac{a}{b} - \frac{b}{a}}{\frac{1}{a} - \frac{1}{b}}$ is equal to

    (1) $a + b$ (2) $a - b$ (3) $-(a + b)$ (4) $b - a$

9. An angle of $\frac{7\pi}{4}$ radians is congruent to an angle of
    (1) $135°$ (2) $225°$ (3) $315°$ (4) $405°$

10. Which of the following is a geometric sequence?

(1) $1, 3, 5, 7, \ldots$ (3) $1, 0.1, 0.01, 0.001, \ldots$

(2) $1, 3, 6, 10, \ldots$ (4) $1, \frac{1}{2}, \frac{1}{3}, \frac{1}{4}, \ldots$

Part II

Answer all questions in this part. Each correct answer will receive 2 credits. Clearly indicate the necessary steps, including appropriate formula substitutions, diagrams, graphs, charts, etc. For all questions in this part, a correct numerical answer with no work shown will receive only 1 credit.

11. Use the exact values of the sin 30°, sin 45°, cos 30°, and cos 45° to find the exact value of cos 15°.

12. Write the fraction $\frac{1 - \sqrt{5}}{1 + \sqrt{5}}$ as an equivalent fraction with a rational denominator.

Part III

Answer all questions in this part. Each correct answer will receive 4 credits. Clearly indicate the necessary steps, including appropriate formula substitutions, diagrams, graphs, charts, etc. For all questions in this part, a correct numerical answer with no work shown will receive only 1 credit.

13. Solve for x and check: $3 - \sqrt{2x - 3} = x$

14. Find all values of x in the interval $0° \leq x < 360°$ that satisfy the following equation:

$$6 \sin^2 x - 5 \sin x - 4 = 0$$

Part IV

Answer all questions in this part. Each correct answer will receive 6 credits. Clearly indicate the necessary steps, including appropriate formula substitutions, diagrams, charts, etc. For all questions in this part, a correct numerical answer with no work shown will receive only 1 credit.

15. If $\log_b x = \log_b 3 + 2 \log_b 4 - \frac{1}{2} \log_b 8$, express x in simplest form.

16. a. Sketch the graph of $y = 2 \sin x$ in the interval $0 \leq x \leq 2\pi$.

    b. On the same set of axes, sketch the graph of $y = \cos 2x$.

    c. For how many values of x in the interval $0 \leq x \leq 2\pi$ does $2 \sin x = \cos 2x$?

STATISTICS

Every year the admission officers of colleges choose, from thousands of applicants, those students who will be offered a place in the incoming class for the next year. An attempt is made by the college to choose students who will be best able to succeed academically and who best fit the profile of the student body of that college. Although this choice is based not only on academic standing, the scores on standardized tests are an important part of the selection. Statistics establishes the validity of the information obtained from standardized tests and influence the interpretation of the data obtained from them.

15-1 GATHERING DATA

Important choices in our lives are often made by evaluating information, but in order to use information wisely, it is necessary to organize and condense the multitude of facts and figures that can be collected. **Statistics** is the science that deals with the collection, organization, summarization, and interpretation of related information called **data**. **Univariate statistics** consists of one number for each data value.

Collection of Data

Where do data come from? Individuals, government organizations, businesses, and other political, scientific, and social groups usually keep records of their activities. These records provide factual data. In addition to factual data, the outcome of an event such as an election, the sale of a product, or the success of a movie often depends on the opinion or choices of the public.

Common methods of collecting data include the following:

1. **Censuses:** every ten years, the government conducts a *census* to determine the U.S. population. Each year, almanacs are published that summarize and update data of general interest.

2. **Surveys:** written questionnaires, personal interviews, or telephone requests for information can be used when experience, preference, or opinions are sought.

3. **Controlled experiments:** a structured study that usually consists of two groups: one that makes use of the subject of the study (for example, a new medicine) and a control group that does not. Comparison of results for the two groups is used to indicate effectiveness.

4. **Observational studies:** similar to controlled experiments except that the researcher does not apply the treatment to the subjects. For example, to determine if a new drug causes cancer, it would be unethical to give the drug to patients. A researcher *observes* the occurrence of cancer among groups of people who *previously* took the drug.

When information is gathered, it may include data for all cases to which the result of the study is to be applied. This source of information is called the **population**. When the *entire* population can be examined, the study is a **census**. For example, a study on the age and number of accidents of every driver insured by an auto insurance company would constitute a census if all of the company's records are included. However, when it is not possible to obtain information from every case, a **sample** is used to determine data that may then be applied to every case. For example, in order to determine the quality of their product, the quality-control department of a business may study a sample of the product being produced.

In order that the sample reflect the properties of the whole group, the following conditions should exist:

1. The sample must be representative of the group being studied.

2. The sample must be large enough to be effective.

3. The selection should be random or determined in such a way as to eliminate any bias.

If a new medicine being tested is proposed for use by people of all ages, of different ethnic backgrounds, and for use by both men and women, then the sample must be made up of people who represent these differences in sufficient number to be effective. A political survey to be effective must include people of different cultural, ethnic, financial, geographic, and political backgrounds.

Potential Pitfalls of Surveys

Surveys are a very common way of collecting data. However, if not done correctly, the results of the survey can be invalid. One potential problem is with the wording of the survey questions. For example, the question, "Do you agree that teachers should make more money?" will likely lead to a person answering "Yes." A more neutral form of this question would be, "Do you believe that teachers' salaries are too high, too low, or just about right?"

Questions can also be too vague, "loaded" (that is, use words with unintended connotations), or confusing. For example, for many people, words that invoke race will likely lead to an emotional response.

Another potential problem with surveys is the way that participants are selected. For example, a magazine would like to examine the typical teenager's opinion on a pop singer in a given city. The magazine editors conduct a survey by going to a local mall. The problem with this survey is that teenagers who go to the mall are not necessarily representative of all teenagers in the city. A better survey would be done by visiting the high schools of the city.

Many surveys often rely on volunteers. However, volunteers are likely to have stronger opinions than the general population. This is why the selection of participants, if possible, should be random.

EXAMPLE 1

A new medicine intended for use by adults is being tested on five men whose ages are 22, 24, 25, 27, and 30. Does the sample provide a valid test?

Solution No:

- The sample is too small.
- The sample includes only men.
- The sample does not include adults over 30.

EXAMPLE 2 ▬▬▬▬▬▬▬▬▬▬▬▬▬▬▬▬▬▬▬

The management of a health club has received complaints about the temperature of the water in the swimming pool. They want to sample 50 of the 200 members of the club to determine if the temperature of the pool should be changed. How should this sample be chosen?

Solution One suggestion might be to poll the first 50 people who use the pool on a given day. However, this will only include people who use the pool and who can therefore tolerate the water temperature.

Another suggestion might be to place 50 questionnaires at the entrance desk and request members to respond. However, this includes only people who choose to respond and who therefore may be more interested in a change.

A third suggestion might be to contact by phone every fourth person on the membership list and ask for a response. This method will produce a random sample but will include people who have no interest in using the pool. This sample may be improved by eliminating the responses of those people. ▢

Organization of Data

In order to be more efficiently presented and more easily understood and interpreted, the data collected must be organized and summarized. Charts and graphs such as the histogram are useful tools. The **stem-and-leaf diagram** is an effective way of organizing small sets of data.

For example, the heights of the 20 children in a seventh-grade class are shown to the right. To draw a stem-and-leaf diagram, choose the tens digit as the *stem* and the units digit as the *leaf*.

Heights of Children									
61	71	58	72	60	53	74	61	68	65
72	67	64	48	70	56	65	67	59	61

(1) Draw a vertical line and list the tens digits, 4, 5, 6, and 7 (or the **stem**), from bottom to top to the left of the line:

Stem	
7	
6	
5	
4	

(2) Enter each height by writing the **leaf**, the units digit, to the right of the line, following the appropriate stem:

Stem	Leaf
7	1 2 4 2 0
6	1 0 1 8 5 7 4 5 7 1
5	8 3 6 9
4	8

(3) Arrange the leaves in numerical order after each stem:

(4) Add a key to indicate the meaning of the numbers in the diagram:

Stem	Leaf
7	0 1 2 2 4
6	0 1 1 1 4 5 5 7 7 8
5	3 6 8 9
4	8

Key: 4 | 8 = 48

For larger sets of data, a **frequency distribution table** can be drawn. Information is grouped and the *frequency*, the number of times that a particular value or group of values occurs, is stated for each group. For example, the table to the right lists the scores of 250 students on a test administered to all tenth graders in a school district.

The table tells us that 24 students scored between 91 and 100, that 82 scored between 81 and 90, that 77 scored between 71 and 80. The table also tells us that the largest number of students scored in the 80's and that ten students scored 50 or below. Note that unlike the stem-and-leaf diagram, the table does not give us the individual scores in each interval.

Score	Frequency
91–100	24
81–90	82
71–80	77
61–70	36
51–60	21
41–50	8
31–40	2

EXAMPLE 2

The prices of a gallon of milk in 15 stores are listed below.

$3.15 $3.39 $3.28 $2.98 $3.25 $3.45 $3.58 $3.24
$3.35 $3.29 $3.29 $3.30 $3.25 $3.40 $3.29

a. Organize the data in a stem-and-leaf diagram.

b. Display the data in a frequency distribution table.

c. If the 15 stores were chosen at random from the more than 100 stores that sell milk in Monroe County, does the data set represent a population or a sample?

Solution **a.** Use the first two digits of the price as the stem.

Use the last digit as the leaf.

Write the leaves in numerical order.

Stem	
3.5	
3.4	
3.3	
3.2	
3.1	
3.0	
2.9	

Stem	Leaf
3.5	8
3.4	5 0
3.3	9 5 0
3.2	8 5 4 9 9 5 9
3.1	5
3.0	
2.9	8

Stem	Leaf
3.5	8
3.4	0 5
3.3	0 5 9
3.2	4 5 5 8 9 9 9
3.1	5
3.0	
2.9	8

Key: 2.9 | 8 = 2.98

b. Divide the data into groups of length $0.10 starting with $2.90. These groups correspond with the stems of the stem-and-leaf diagram. The frequencies can be determined by the use of a tally to represent each price.

Stem	Leaf
3.5	8
3.4	0 5
3.3	0 5 9
3.2	4 5 5 8 9 9 9
3.1	5
3.0	
2.9	8

Key: 2.9 | 8 = 2.98

Price	Tally	Frequency
$3.50–$3.59	|	1
$3.40–$3.49	||	2
$3.30–$3.39	|||	3
$3.20–$3.29	⊬||	7
$3.10–$3.19	|	1
$3.00–$3.09		0
$2.90–$2.99	|	1

c. The data set is obtained from a random selection of stores from all of the stores in the study and is therefore a sample. *Answer*

Once the data have been organized, a graph can be used to visualize the intervals and their frequencies. A **histogram** is a vertical bar graph where each interval is represented by the width of the bar and the frequency of the interval is represented by the height of the bar. The bars are placed next to each other to show that, as one interval ends, the next interval begins. The histogram below shows the data of Example 2:

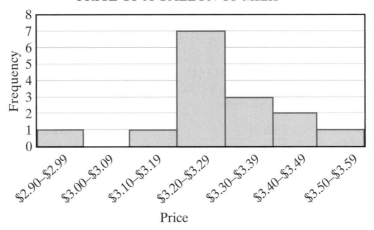

PRICE OF A GALLON OF MILK

A graphing calculator can be used to display a histogram from the data on a frequency distribution table.

(1) Clear L_1 and L_2, the lists to be used, of existing data.

ENTER: **STAT** **4** **2nd** **L1** **,** **2nd** **L2** **ENTER**

(2) Press **STAT** **1** to edit the lists. Enter the *minimum* value of each interval in L_1 and the frequencies into L_2.

(3) Clear any functions in the **Y=** menu.

(4) Turn on Plot1 from the STAT PLOT menu and select ⌷⌷ₗ for histogram. Make sure to also set *Xlist* to L_1 and *Freq* to L_2.

ENTER: **2nd** **STAT PLOT** **1** **ENTER**

▼ **▶** **▶** **ENTER** **▼** **2nd**

L1 **▼** **2nd** **L2**

(5) In the **WINDOW** menu, enter *Xmin* as 2.8, the length of one interval less than the smallest interval value, and *Xmax* as 3.7, the length of one interval more than the largest interval value. Enter *Xscl* as 0.10, the length of the interval. The *Ymin* is 0 and *Ymax* is 9 to be greater than the largest frequency.

(6) Press **GRAPH** to graph the histogram. We can view the frequency associated with each interval by pressing **TRACE**. Use the left and right arrows to move from one interval to the next.

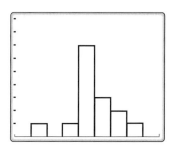

Exercises

Writing About Mathematics

1. In a controlled experiment, two groups are formed to determine the effectiveness of a new cold remedy. One group takes the medicine and one does not. Explain why the two groups are necessary.

2. In the experiment described in Exercise 1, explain why it is necessary that a participant does not know to which group he or she belongs.

Developing Skills

In 3–5, organize the data in a stem-and-leaf diagram.

3. The grades on a chemistry test:

   95 90 84 85 74 67 78 86 54 82
   75 67 92 66 90 68 88 85 76 87

4. The weights of people starting a weight-loss program:

   173 210 182 190 175 169 236 192 203 196 201
   187 205 195 224 177 195 207 188 184 196 155

5. The heights, in centimeters, of 25 ten-year-old children:

   137 134 130 144 131 141 136 140 137 129 139 137 144
   127 147 143 132 132 142 142 131 129 138 151 137

In 6–8, organize the data in a frequency distribution table.

6. The numbers of books read during the summer months by each of 25 students:

   2 2 5 1 3 0 7 2 4 3 3 1 8
   5 7 3 4 1 0 6 3 4 1 1 2

7. The sizes of 26 pairs of jeans sold during a recent sale:

   8 12 14 10 12 16 14 6 10 9 8 13 12
   8 12 10 12 14 10 12 16 10 11 15 8 14

8. The number of siblings of each of 30 students in a class:

   2 1 1 5 1 0 2 2 1 3 4 0 6 2 0
   3 1 2 2 1 1 1 0 2 1 0 1 1 2 3

In 9–11, graph the histogram of each set of data.

9.

x_i	f_i
35–39	13
30–34	19
25–29	10
20–24	13
15–19	8
10–14	19
5–9	15

10.

x_i	f_i
101–110	3
91–100	6
81–90	10
71–80	13
61–70	14
51–60	2
41–50	2

11.

x_i	f_i
$55–$59	20
$50–$54	15
$45–$49	12
$40–$44	5
$35–$39	10
$30–$34	12
$25–$29	16

Applying Skills

In 12–18, suggest a method that might be used to collect data for each study. Tell whether your method uses a population or a sample.

12. Average temperature for each month for a given city

13. Customer satisfaction at a restaurant

14. Temperature of a patient in a hospital over a period of time

15. Grades for students on a test

16. Population of each of the states of the United States

17. Heights of children entering kindergarten

18. Popularity of a new movie

19. The grades on a math test of 25 students are listed below.

> 86 92 77 84 75 95 66 88 84 53 98 87 83
> 74 61 82 93 98 87 77 86 58 72 76 89

   a. Organize the data in a stem-and-leaf diagram.

   b. Organize the data in a frequency distribution table.

   c. How many students scored 70 or above on the test?

   d. How many students scored 60 or below on the test?

20. The stem-and-leaf diagram at the right shows the ages of 30 people in an exercise class. Use the diagram to answer the following questions.

   a. How many people are 45 years old?

   b. How many people are older than 60?

   c. How many people are younger than 30?

   d. What is the age of the oldest person in the class?

   e. What is the age of the youngest person in the class?

Stem	Leaf
7	2
6	0 1 5
5	1 3 6 6 7 9
4	2 2 4 5 5 6 7
3	9
2	1 3 3 5 7 8 8
1	0 2 2 6 9

Key: 1 | 9 = 19

Hands-On Activity

In this activity, you will take a survey of 25 people. You will need a stopwatch or a clock with a second hand. Perform the following experiment with each participant to determine how each perceives the length of a minute:

1. Indicate the starting point of the minute.
2. Have the person tell you when he or she believes that a minute has passed.
3. Record the actual number of seconds that have passed.

After surveying all 25 participants, use a stem-and-leaf diagram to record your data. Keep this data. You will use it throughout your study of this chapter.

15-2 MEASURES OF CENTRAL TENDENCY

After data have been collected, it is often useful to represent the data by a single value that in some way seems to represent all of the data. This number is called the **measure of central tendency**. The most frequently used measures of central tendency are the *mean*, the *median*, and the *mode*.

The Mean

The **mean** or **arithmetic mean** is the most common measure of central tendency. The mean is the sum of all of the data values divided by the number of data values.

For example, nine members of the basketball team played during all or part of the last game. The number of points scored by each of the players was:

$$21, 15, 12, 9, 8, 7, 5, 2, 2$$

$$\text{Mean} = \frac{21 + 15 + 12 + 9 + 8 + 7 + 5 + 2 + 2}{9} = \frac{81}{9} = 9$$

Note that if each of the 9 players had scored 9 points, the total number of points scored in the game would have been the same.

The summation symbol, Σ, is often used to designate the sum of the data values. We designate a data value as x_i and the sum of n data values as

$$\sum_{i=1}^{n} x_i = x_1 + x_2 + x_3 + \cdots + x_n$$

For the set of data given above:

$$n = 9, x_1 = 21, x_2 = 15, x_3 = 12, x_4 = 9, x_5 = 8, x_6 = 7, x_7 = 5, x_8 = 2, x_9 = 2$$

$$\sum_{i=1}^{9} x_i = 81$$

The subscript for each data value indicates its position in a list of data values, not its value, although the value of i and the data value may be the same.

> **Procedure**
>
> **To find the mean of a set of data:**
>
> **1.** Add the data values.
>
> **2.** Divide the sum by n, the total number of data values.

EXAMPLE 1

An English teacher recorded the number of spelling errors in the 40 essays written by students. The table below shows the number of spelling errors and the frequency of that number of errors, that is, the number of essays that contained that number of misspellings. Find the mean number of spelling errors for these essays.

Errors	0	1	2	3	4	5	6	7	8	9	10
Frequency	1	3	2	2	6	9	7	5	2	1	2

Solution To find the total number of spelling errors, first multiply each number of errors by the frequency with which that number of errors occurred. For example, since 2 essays each contained 10 errors, there were 20 errors in these essays. Add the products in the $f_i x_i$ row to find the total number of errors in the essays. Divide this total by the total frequency.

												Total
Errors (x_i)	0	1	2	3	4	5	6	7	8	9	10	
Frequency (f_i)	1	3	2	2	6	9	7	5	2	1	2	40
Errors · Frequency ($f_i x_i$)	0	3	4	6	24	45	42	35	16	9	20	204

$$\text{Mean} = \frac{\sum_{i=0}^{10} f_i x_i}{40} = \frac{204}{40} = 5.1$$

Note that for this set of data, the data value is equal to i for each x_i.

Answer 5.1 errors

The Median

The **median** is the middle number of a data set arranged in numerical order. When the data are arranged in numerical order, the number of values less than or equal to the median is equal to the number of values greater than or equal to the median. Consider again the nine members of the basketball team who played during all or part of the last game and scored the following number of points:

$$2, 2, 5, 7, 8, 9, 12, 15, 21$$

We can write $9 = 4 + 1 + 4$. Therefore, we think of four scores below the median, the median, and four scores above the median.

$$\underbrace{2, \ 2, \ 5, \ 7,}_{\text{scores less than the median}} \quad \underset{\underset{\text{median}}{\uparrow}}{8}, \quad \underbrace{9, \ 12, \ 15, \ 21}_{\text{scores greater than the median}}$$

Note that the number of data values can be written as $2(4) + 1$. The median is the $(4 + 1)$ or 5th value from either end of the distribution. The median number of points scored is 8.

When the number of values in a set of data is even, then the median is the mean of the two middle values. For example, eight members of a basketball team played in a game and scored the following numbers of points.

$$4, 6, 6, 7, 11, 12, 18, 20$$

We can separate the eight data values into two groups of 4 values. Therefore, we average the largest score of the four lowest scores and the smallest score of the four highest scores.

$$\underbrace{4, \ 6, \ 6, \ 7,}_{\text{four lowest scores}} \quad \underset{\underset{\text{median}}{\uparrow}}{\frac{7 + 11}{2}}, \quad \underbrace{11, \ 12, \ 18, \ 20}_{\text{four highest scores}}$$

$$\text{median} = \frac{7 + 11}{2} = 9$$

Note that the number of data values can be written as $2(4)$. The median is the mean of the 4th and 5th value from either end of the distribution. The median is 9. There are four scores greater than the median and four scores lower than the median. The median is a middle mark.

Procedure

To find the median of a set of data:

1. Arrange the data in order from largest to smallest or from smallest to largest.

2. a. If the number of data values is odd, write that number as $2n + 1$. The median is the score that is the $(n + 1)$th score from either end of the distribution.

   b. If the number of data values is even, write that number as $2n$. The median is the score that is the mean of the nth score and the $(n + 1)$th score from either end of the distribution.

The Mode

The **mode** is the value or values that occur most frequently in a set of data. For example, in the set of numbers

$$3, 4, 5, 5, 6, 6, 6, 6, 7, 7, 8, 10,$$

the number that occurs most frequently is 6. Therefore, 6 is the mode for this set of data.

When a set of numbers is arranged in a frequency distribution table, the mode is the entry with the highest frequency. The table to the right shows, for a given month, the number of books read by each student in a class. The largest number of students, 12, each read four books. The mode for this distribution is 4.

A data set may have more than one mode. For example, in the set of numbers 3, 4, 5, 5, 5, 5, 6, 6, 6, 6, 7, 7, 8, 10, the numbers 5 and 6 each occur four times, more frequently than any other number. Therefore, 5 and 6 are modes for this set of data. The set of data is said to be **bimodal**.

Number of Books Read	Frequency
8	1
7	1
6	3
5	6
4	12
3	4
2	2
1	0
0	1

Quartiles

When a set of data is listed in numerical order, the median separates the data into two equal parts. The **quartiles** separate the data into four equal parts. To find the quartiles, we first separate the data into two equal parts and then separate each of these parts into two equal parts. For example, the grades of 20 students on a math test are listed below.

$$58, 60, 65, 70, 72, 75, 76, 80, 80, 81, 83, 84, 85, 87, 88, 88, 90, 93, 95, 98$$

Since there are 20 or 2(10) grades, the median grade that separates the data into two equal parts is the average of the 10th and 11th grade.

$$\underbrace{58, 60, 65, 70, 72, 75, 76, 80, 80, 81,}_{\text{Lower half}} \quad \underbrace{83, 84, 85, 87, 88, 88, 90, 93, 95, 98}_{\text{Upper half}}$$

$$\uparrow$$
$$\boxed{82}$$
Median

The 10th grade is 81 and the 11th grade is 83. Therefore, the mean of these two grades, $\frac{81 + 83}{2}$ or 82, separates the data into two equal parts. This number is the median grade.

Now separate each half into two equal parts. Find the median of the two lower quarters and the median of the two upper quarters.

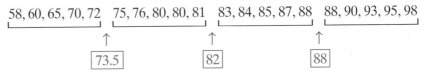

The numbers 73.5, 82, and 88 are the quartiles for this data.

- One quarter of the grades are less than or equal to 73.5. Therefore, 73.5 is the **first quartile** or the **lower quartile**.
- Two quarters of the grades are less than or equal to 82. Therefore, 82 is the **second quartile**. The second quartile is always the median.
- Three quarters of the grades are less than or equal to 88. Therefore, 88 is the **third quartile** or the **upper quartile**.

Note: The minimum, first quartile, median, third quartile, and maximum make up the **five statistical summary** of a data set.

When the data set has an odd number of values, the median or second quartile will be one of the values. This number is not included in either half of the data when finding the first and third quartiles.

For example, the heights, in inches, of 19 children are given below:

$$37, 39, 40, 42, 42, 43, 44, 44, 44, 45, 46, 47, 47, 48, 49, 49, 50, 52, 53$$

There are $19 = 2(9) + 1$ data values. Therefore, the second quartile is the 10th height or 45.

$$37, 39, 40, 42, 42, 43, 44, 44, 44, \quad \boxed{45}, \quad 46, 47, 47, 48, 49, 49, 50, 52, 53$$

There are $9 = 2(4) + 1$ heights in the lower half of the data and also in the upper half of the data. The middle height of each half is the 5th height.

$$37, 39, 40, 42, \quad \boxed{42}, \quad 43, 44, 44, 44, \quad \boxed{45}, \quad 46, 47, 47, 48, \quad \boxed{49}, \quad 49, 50, 52, 53$$

Lower quartile   Median   Upper quartile

In this example, the first or lower quartile is 42, the median or second quartile is 45, and the third or upper quartile is 49. Each of these values is one of the data values, and the remaining values are separated into four groups with the same number of heights in each group.

Box-and-Whisker Plot

A box-and-whisker plot is a diagram that is used to display the quartile values and the maximum and minimum values of a distribution. We will use the data from the set of heights given above.

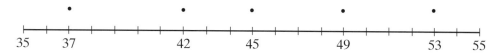

1. Choose a scale that includes the maximum and minimum values of the data. We will use a scale from 35 to 55.

2. Above the scale, place dots to represent the minimum value, the lower quartile, the mean, the upper quartile, and the maximum value.

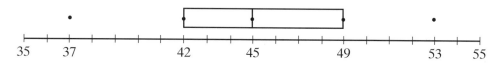

3. Draw a box with opposite sides through the lower and upper quartiles and a vertical line through the median.

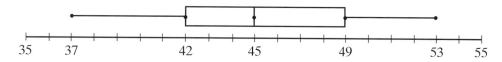

4. Draw whiskers by drawing a line to join the dot that represents the minimum value to the dot that represents the lower quartile and a line to join the dot that represents the upper quartile to the dot that represents the maximum value.

 A graphing calculator can be used to find the quartiles and to display the box-and-whisker plot.

(1) Enter the data for this set of heights into L_1.

(2) Turn off all plots and enter the required choices in Plot 1.

ENTER: 2nd | STAT PLOT | 1 | ENTER

▼ ► ► ► ► ENTER ▼

2nd | L1 | ▼ | ALPHA | 1

(3) Now display the plot by entering

[ZOOM] [9]. You can press [TRACE] to display the five statistical summary.

The five statistical summary of a set of data can also be displayed on the calculator by using 1-Var Stats.

ENTER: [STAT] [▶] [ENTER] [ENTER]

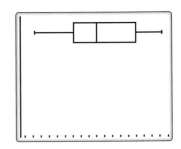

The first entry under 1-Var Stats is $\bar{x}$, the value of the mean. The next is $\sum x$, the sum of the data values. The next three entries are values that will be used in the sections that follow. The last entry is the number of data values. The arrow tells us that there is more information. Scroll down to display the minimum value, $minX = 37$, the lower quartile, $Q_1 = 42$, the median or second quartile, $Med = 45$, the upper quartile, $Q_3 = 49$, and the maximum value, $maxX = 53$.

```
1-VAR STATS
x̄=45.31578947
Σx=861
Σx²=39353
Sx=4.321170515
σX=4.205918531
↓N=19
```

```
1-VAR STATS
↑N=19
 minX=37
 Q₁=42
 MED=45
 Q₃=49
 MAXX=53
```

EXAMPLE 1

Find the mean, the median, and the mode of the following set of grades:

$$92, 90, 90, 90, 88, 87, 85, 70$$

Solution Mean $= \frac{1}{n}\sum_{i=1}^{n} x_i = \frac{1}{8}(92 + 90 + 90 + 90 + 88 + 87 + 85 + 70) = \frac{692}{8} = 86.5$

Median = the average of the 4th and 5th grades $= \frac{90 + 88}{2} = 89$

Mode = the grade that appears most frequently $= 90$

EXAMPLE 2

The following list shows the length of time, in minutes, for each of 35 employees to commute to work.

```
25  12  20  18  35  25  40  35  27  30  60  22  36  20  18
27  35  42  35  55  27  30  15  22  10  35  27  15  57  18
25  45  24  27  25
```

a. Organize the data in a stem-and-leaf diagram.

b. Find the median, lower quartile, and upper quartile.

c. Draw a box-and-whisker plot.

Solution **a.** (1) Choose the tens digit as the stem and enter the units digit as the leaf for each value.

Stem	Leaf
6	0
5	5 7
4	0 2 5
3	5 5 0 6 5 5 0 5
2	5 0 5 7 2 0 7 7 2 7 5 4 7 5
1	2 8 8 5 0 5 8

(2) Write the leaves in numerical order from smallest to largest.

Stem	Leaf
6	0
5	5 7
4	0 2 5
3	0 0 5 5 5 5 5 6
2	0 0 2 2 4 5 5 5 5 7 7 7 7 7
1	0 2 5 5 8 8 8

Key: 1 | 0 = 10

b. (1) The median is the middle value of the 35 data values when the values are arranged in order.

$$35 = 2(17) + 1$$

The median is the 18th value. Separate the data into groups of 17 from each end of the distribution. The median is 27.

Stem	Leaf
6	0
5	5 7
4	0 2 5
3	0 0 5 5 5 5 5 6
2	0 0 2 2 4 5 5 5 5 7 [7] 7 7 7
1	0 2 5 5 8 8 8

(2) There are $17 = 2(8) + 1$ data values below the median. The lower quartile is the 9th data value from the lower end. The upper quartile is the 9th data value from the upper end.

Stem	Leaf
6	0
5	5 7
4	0 2 5
3	0 0 5 5 5 [5] 5 6
2	0 [0] 2 2 4 5 5 5 5 7 [7] 7 7 7
1	0 2 5 5 8 8 8

The lower quartile is 20, the median is 27, and the upper quartile is 35.

c.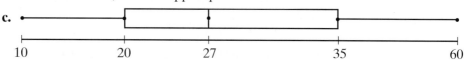

10   20   27   35   60

Note: In the stem-and-leaf diagram of Example 2, the list is read from left to right to find the lower quartile. The list is read from *right* to *left* (starting from the top) to find the upper quartile.

Exercises

Writing About Mathematics

1. Cameron said that the number of data values of any set of data that are less than the lower quartile or greater than the upper quartile is exactly 50% of the number of data values. Do you agree with Cameron? Explain why or why not.

2. Carlos said that for a set of $2n$ data values or of $2n + 1$ data values, the lower quartile is the median of the smallest n values and the upper quartile is the median of the largest n values. Do you agree with Carlos? Explain why or why not.

Developing Skills

In 3–8, find the mean, the median, and the mode of each set of data.

3. Grades: 74, 78, 78, 80, 80, 80, 82, 88, 90

4. Heights: 60, 62, 63, 63, 64, 65, 66, 68, 68, 68, 70, 75

5. Weights: 110, 112, 113, 115, 115, 116, 118, 118, 125, 134, 145, 148

6. Number of student absences: 0, 0, 0, 1, 1, 2, 2, 2, 3, 4, 5, 9

7. Hourly wages: $6.90, $7.10, $7.50, $7.50, $8.25, $9.30, $9.50, $10.00

8. Tips: $1.00, $1.50, $2.25, $3.00, $3.30, $3.50, $4.00, $4.75, $5.00, $5.00, $5.00

In 9–14, find the median and the first and third quartiles for each set of data values.

9. 2, 3, 5, 8, 9, 11, 15, 16, 17, 20, 22, 23, 25

10. 34, 35, 35, 36, 38, 40, 42, 43, 43, 43, 44, 46, 48, 50

11. 23, 27, 15, 38, 12, 17, 22, 39, 28, 20, 27, 18, 25, 28, 30, 29

12. 92, 86, 77, 85, 88, 90, 81, 83, 95, 76, 65, 88, 91, 81, 88, 87, 95

13. 75, 72, 69, 68, 66, 65, 64, 63, 63, 61, 60, 59, 59, 58, 56, 54, 52, 50

14. 32, 32, 30, 30, 29, 27, 26, 22, 20, 20, 19, 18, 17

15. A student received the following grades on six tests: 90, 92, 92, 95, 95, x.

   a. For what value(s) of x will the set of grades have no mode?

   b. For what value(s) of x will the set of grades have only one mode?

   c. For what value(s) of x will the set of grades be bimodal?

16. What are the first, second, and third quartiles for the set of integers from 1 to 100?

17. What are the first, second, and third quartiles for the set of integers from 0 to 100?

Applying Skills

18. The grades on a English test are shown in the stem-and-leaf diagram to the right.

    a. Find the mean grade.

    b. Find the median grade.

    c. Find the first and third quartiles.

    d. Draw a box-and-whisker plot for this data.

Stem	Leaf
9	0 0 1 2 5 9
8	0 2 2 5 5 5 7 8 8
7	3 5 6 6 7
6	0 5 5
5	5
4	7

Key: 4 | 7 = 47

19. The weights in pounds of the members of the football team are shown below:

    181 199 178 203 211 208 209 202 212 194
    185 208 223 206 202 213 202 186 189 203

    a. Find the mean.

    b. Find the median.

    c. Find the mode or modes.

    d. Find the first and third quartiles.

    e. Draw a box-and-whisker plot.

20. Mrs. Gillis gave a test to her two classes of algebra. The mean grade for her class of 20 students was 86 and the mean grade of her class of 15 students was 79. What is the mean grade when she combines the grades of both classes?

Hands-On Activity

Use the estimates of a minute collected in the Hands-On Activity of the previous section to determine the five statistical summary for your data. Draw a box-and-whisker plot to display the data.

15-3 MEASURES OF CENTRAL TENDENCY FOR GROUPED DATA

Most statistical studies involve much larger numbers of data values than can be conveniently displayed in a list showing each data value. Large sets of data are usually organized into a frequency distribution table.

Frequency Distribution Tables for Individual Data Values

A frequency distribution table records the individual data values and the frequency or number of times that the data value occurs in the data set. The example on page 606 illustrates this method of recording data.

Each of an English teacher's 100 students recently completed a book report. The teacher recorded the number of misspelled words in each report. The table records the number of reports for each number of misspelled words. Let x_i represent the number of misspelled words in a report and f_i represent the number of reports that contain x_i misspelled words.

												Total
x_i	0	1	2	3	4	5	6	7	8	9	10	
f_i	5	7	6	8	19	26	16	7	5	0	1	100
x_if_i	0	7	12	24	76	130	96	49	40	0	10	444

The mean of this set of data is the total number of misspelled words divided by the number of reports. To find the total number of misspelled words, we must first multiply each number of misspelled words, x_i, by the number of reports that contain that number of misspelled words, f_i. That is, we must find x_if_i for each number of misspelled words.

For this set of data, when we sum the f_i row, $\sum\limits_{i=0}^{10} f_i = 100$, and when we sum the x_if_i row $\sum\limits_{i=0}^{10} x_if_i = 444$.

$$\text{Mean} = \frac{\sum\limits_{i=0}^{10} x_if_i}{\sum\limits_{i=0}^{10} f_i} = \frac{444}{100} = 4.44$$

To find the median and the quartiles for this set of data, we will find the *cumulative frequency* for each number of misspelled words. The **cumulative frequency** is the accumulation or the sum of all frequencies less than or equal to a given frequency. For example, the cumulative frequency for 3 misspelled words on a report is the sum of the frequencies for 3 or fewer misspelled words, and the cumulative frequency for 6 misspelled words on a report is the sum of the frequencies for 6 or fewer misspelled words. The third row of the table shows that 0 misspelled words occur 5 times, 1 or fewer occur $7 + 5$ or 12 times, 2 or fewer occur $6 + 12$ or 18 times. In each case, the cumulative frequency for x_i is the frequency for x_i plus the cumulative frequency for x_{i-1}. The cumulative frequency for the largest data value is always equal to the total number of data values.

x_i	0	1	2	3	4	5	6	7	8	9	10
f_i	5	7	6	8	19	26	16	7	5	0	1
Cumulative Frequency	5	12	18	26	45	71	87	94	99	99	100

$\uparrow$ $\uparrow$ $\uparrow$

3   5   6

Lower   Median Upper
quartile   quartile

Since there are 100 data values, the median is the average of the 50th and 51st values. To find these values, look at the cumulative frequency column. There are 45 values less than or equal to 4 and 71 less than or equal to 5. Therefore, the 50th and 51st values are both 5 and the median is 5 misspelled words.

Similarly, the upper quartile is the average of the 75th and 76th values. Since there are 71 values less than or equal to 5, the 75th and 76th values are both 6 misspelled words. The lower quartile is the average of the 25th and 26th values. Since there are 18 values less than or equal to 2, the 25th and 26th values are both 3 misspelled words.

Percentiles

A **percentile** is a number that tells us what percent of the total number of data values lie at or below a given measure.

For example, let us use the data from the previous section. The table records the number of reports, f_i, that contain each number of misspelled words, x_i, on 100 essays.

x_i	0	1	2	3	4	5	6	7	8	9	10
f_i	5	7	6	8	19	26	16	7	5	0	1
Cumulative Frequency	5	12	18	26	45	71	87	94	99	99	100

To find the percentile rank of 7 misspelled words, first find the number of essays with fewer than 7 misspelled words, 87. Add to this *half* of the essays with 7 misspelled words, $\frac{7}{2}$ or 3.5. Add these two numbers and divide the sum by the number of essays, 100.

$$\frac{87 + 3.5}{100} = \frac{90.5}{100} = 90.5\%$$

Percentiles are usually not written with fractions. We say that 7 misspelled words is at the 90.5th or 91st percentile. That is, 91% of the essays had 7 or fewer misspelled words.

Frequency Distribution Tables for Grouped Data

Often the number of different data values in a set of data is too large to list each data value separately in a frequency distribution table. In this case, it is useful to list the data in terms of groups of data values rather than in terms of individual data values. The following example illustrates this method of recording data.

There are 50 members of a weight-loss program. The weights range from 181 to 285 pounds. It is convenient to arrange these weights in groups of 10 pounds starting with 180–189 and ending with 280–289. The frequency distribution table shows the frequencies of the weights for each interval.

Weights	Midpoint x_i	Frequency f_i	$x_i f_i$	Cumulative Frequency
280–289	284.5	1	284.5	50
270–279	274.5	3	823.5	49
260–269	264.5	4	1,058.0	46
250–259	254.5	8	2,036.0	42
240–249	244.5	12	2,934.0	34
230–239	234.5	10	2,345.0	22
220–229	224.5	5	1,122.5	12
210–219	214.5	3	643.5	7
200–209	204.5	2	409.0	4
190–199	194.5	1	194.5	2
180–189	184.5	1	184.5	1
		50	12,035	

In order to find the mean, assume that the weights are evenly distributed throughout the interval. The mean is found by using the midpoint of the weight intervals as representative of each value in the interval groupings.

$$\text{Mean} = \frac{\sum\limits_{i=1}^{11} x_i f_i}{\sum\limits_{i=1}^{11} f_i} = \frac{12{,}035}{50} = 240.7$$

To find the median for a set of data that is organized in intervals greater than 1, first find the interval in which the median lies by using the cumulative frequency.

There are 50 data values. Therefore, the median is the value between the 25th and the 26th values. The cumulative frequency tells us that there are 22 values less than or equal to 239 and 34 values less than or equal to 249. Therefore, the 25th and 26th values are in the interval 240–249. Can we give a better approximation for the median?

The endpoints of an interval are the lowest and highest data values to be entered in that interval. The boundary values are the values that separate intervals. The lower boundary of an interval is midway between the lower endpoint of the interval and the upper endpoint of the interval that precedes it. The lower boundary of the 240–249 is midway between 240 and 239, that is, 239.5. The upper endpoint of this interval is between 249 and 250, that is 249.5.

Since there are 34 weights less than or equal to 249 and 22 weights less than 240, the weights in the 240–249 interval are the 23rd through the 34th weights. Think of these 12 weights as being evenly spaced throughout the interval.

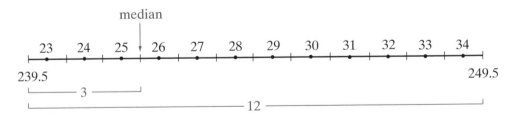

The midpoint between the 25th and 26th weights is $\frac{3}{12}$ of the distance between the boundaries of the interval, a difference of 10.

$$\text{Median} = 239.5 + \tfrac{3}{12}(10) = 239.5 + 2.5 = 242$$

The estimated median for the weights is 242 pounds. Thus, if we assume that the data values are evenly distributed within each interval, we can obtain a better approximation for the median.

EXAMPLE I

The numbers of pets owned by the children in a sixth-grade class are given in the table.

a. Find the mean.

b. Find the median.

c. Find the mode for this set of data.

d. Find the percentile rank of 4 pets.

No. of Pets	Frequency
6	2
5	1
4	3
3	5
2	8
1	14
0	7

Solution **a.** Add to the table the number of pets times the frequency, and the cumulative frequency for each row of the table.

No. of Pets x_i	Frequency f_i	$f_i x_i$	Cumulative Frequency
6	2	12	40
5	1	5	38
4	3	12	37
3	5	15	34
2	8	16	29
1	14	14	21
0	7	0	7
	40	74	

Add the numbers in the $f_i x_i$ column and the numbers in the f_i column.

$$\sum_{i=0}^{6} f_i = 40 \qquad \sum_{i=0}^{6} f_i x_i = 74$$

$$\text{Mean} = \frac{\sum_{i=0}^{6} x_i f_i}{\sum_{i=0}^{6} f_i} = \frac{74}{40} = 1.85$$

b. There are 40 data values in this set of data. The median is the average of the 20th and the 21st data values. Since there are 21 students who own 1 or fewer pets, both the 20th and the 21st data value is 1. Therefore, the median number of pets is 1.

c. The largest number of students, 14, have 1 pet. The mode is 1.

d. There are 34 students with fewer than 4 pets and 3 students with 4 pets. Add 34 and half of 3 and divide the sum by the total number of students, 40.

$$\frac{34 + \frac{3}{2}}{40} = \frac{35.5}{40} = 0.8875 = 88.75\%$$

Four pets represents the 89th percentile.

Calculator Solution for a, b, and c

Clear the lists first if necessary by keying in [STAT] [4] [2nd] [L1] [,] [2nd] [L2] [ENTER].

Enter the number of pets in L_1.

Enter the frequency for each number of pets in L_2.

Display 1-Var Stats.

ENTER: [STAT] [▶] [1] [2nd] [L1] [,] [2nd] [L2] [ENTER]

DISPLAY:

```
1-VAR STATS
x̄=1.85
Σx=74
Σx²=236
Sx=1.594059485
σx=1.574007624
↓n=40
```

```
1-VAR STATS
↑n=40
minX=0
Q₁=1
MED=1
Q₃=3
maxX=6
```

The first entry, $\bar{x}$, is the mean, 1.85. Use the down-arrow key, [▼], to display the median, *Med*. The median is 1.

Note that $\sum x = 74$ is $\sum_{i=0}^{6} f_i x_i$ and $n = 40$ is $\sum_{i=0}^{6} f_i$.

Answers   **a.** mean = 1.85 **b.** median = 1 **c.** mode = 1 **d.** 4 is the 89th percentile

Note: The other information displayed under 1-Var Stats will be used in the sections that follow.

EXAMPLE 2

A local business made the summary of the ages of 45 employees shown below. Find the mean and the median age of the employees to the nearest integer.

Age	20–24	25–29	30–34	35–39	40–44	45–49	50–54	55–59	60–64
Frequency	1	2	5	6	7	10	7	5	2

Solution   Add to the table the midpoint, the midpoint times the frequency, and the cumulative frequency for each interval.

Age	Midpoint (x_i)	Frequency (f_i)	$x_i f_i$	Cumulative Frequency
60–64	62	2	124	45
55–59	57	5	285	43
50–54	52	7	364	38
45–49	47	10	470	31
40–44	42	7	294	21
35–39	37	6	222	14
30–34	32	5	160	8
25–29	27	2	54	3
20–24	22	1	22	1
		45	1,995	

$$\text{Mean} = \frac{\sum_{i=1}^{11} x_i f_i}{\sum_{i=1}^{11} f_i} = \frac{1,995}{45} = 44.333\ldots \approx 44$$

The mean is the middle age of 45 ages or the 23rd age. Since there are 21 ages less than 45, the 23rd age is the second of 10 ages in the 45–49 interval. The boundaries of the 45–49 interval are 44.5 to 49.5

$$\text{Median} = 44.5 + \frac{1\frac{1}{2}}{10} \times 5$$

$$= 44.5 + 0.75 = 45.25 \approx 45$$

Answer mean = 44, median = 45

Exercises

Writing About Mathematics

1. Adelaide said that since, in Example 2, there are 10 employees whose ages are in the 45–49 interval, there must be two employees of age 45. Do you agree with Adelaide? Explain why or why not.

2. Gail said that since, in Example 2, there are 10 employees whose ages are in the 45–49 interval, there must at least two employees who are the same age. Do you agree with Gail? Explain why or why not.

Developing Skills

In 3–8, find the mean, the median, and the mode for each set of data.

3.

x_i	f_i
5	6
4	10
3	15
2	11
1	2
0	1

4.

x_i	f_i
50	8
40	12
30	17
20	10
10	3

5.

x_i	f_i
12	7
11	15
10	13
9	16
8	14
7	15
6	9
5	2

6.

x_i	f_i
10	1
9	1
8	3
7	7
6	6
5	2
4	2

7.

x_i	f_i
$1.10	1
$1.20	5
$1.30	8
$1.40	6
$1.50	6

8.

x_i	f_i
95	2
90	8
85	12
80	10
75	9
70	3
65	0
60	1

9. Find the percentile rank of 2 for the data in Exercise 3.

10. Find the percentile rank of 20 for the data in Exercise 4.

11. Find the percentile rank of 8 for the data in Exercise 5.

12. Find the percentile rank of 6 for the data in Exercise 6.

In 13–18, find the mean and the median for each set of data to the nearest tenth.

13.

x_i	f_i
21–25	2
16–20	3
11–15	12
6–10	6
1–5	1

14.

x_i	f_i
91–100	5
81–90	8
71–80	10
61–70	6
51–60	0
41–50	1

15.

x_i	f_i
$1.51–$1.60	2
$1.41–$1.50	5
$1.31–$1.40	14
$1.21–$1.30	4
$1.11–$1.20	2
$1.01–$1.10	3

16.

x_i	f_i
17–19	20
14–16	27
11–13	32
8–10	39
5–7	32

17.

x_i	f_i
$60–$69	16
$50–$59	5
$40–$49	16
$30–$39	2
$20–$29	5
$10–$19	37

18.

x_i	f_i
0.151–0.160	16
0.141–0.150	5
0.131–0.140	0
0.121–0.130	6
0.111–0.120	0

Applying Skills

19. The table shows the number of correct answers on a test consisting of 15 questions. Find the mean, the median, and the mode for the number of correct answers.

Correct Answers	6	7	8	9	10	11	12	13	14	15
Frequency	1	0	1	3	5	8	9	6	5	2

20. The ages of students in a calculus class at a high school are shown in the table. Find the mean and median age.

Age	Frequency
19	2
18	8
17	9
16	1
15	1

21. Each time Mrs. Taggart fills the tank of her car, she estimates, from the number of miles driven and the number of gallons of gasoline needed to fill the tank, the fuel efficiency of her car, that is, the number of miles per gallon. The table shows the result of the last 20 times that she filled the car.

a. Find the mean and the median fuel efficiency (miles per gallon) for her car.

b. Find the percentile rank of 34 miles per gallon.

Miles per Gallon	32	33	34	35	36	37	38	39	40
Frequency	1	3	2	5	3	3	2	0	1

22. The table shows the initial weights of people enrolled in a weight-loss program. Find the mean and median weight.

Weight	191–200	201–210	211–220	221–230	231–240	241–250
Frequency	1	1	5	7	10	12
Weight	251–260	261–270	271–280	281–290	291–300	
Frequency	13	16	8	5	2	

23. In order to improve customer relations, an auto-insurance company surveyed 100 people to determine the length of time needed to complete a report form following an auto accident. The result of the survey is summarized in the following table showing the number of minutes needed to complete the form. Find the mean and median amount of time needed to complete the form.

Minutes	26–30	31–35	36–40	41–45	46–50	51–55	56–60	61–65	66–70
Frequency	2	8	12	15	10	24	26	1	2

Hands-On Activity

Organize the data from the survey in the Hands-On Activity of Section 15-1 using intervals of five seconds. Use the table to find the mean number of seconds. Compare this result with the mean found using the individual data values.

15-4 MEASURES OF DISPERSION

The mean and the median of a set of data help us to describe a set of data. However, the mean and the median do not always give us enough information to draw meaningful conclusions about the data. For example, consider the following sets of data.

Ages of students on a middle-school basketball team	Ages of students in a community center tutoring program
11 12 12 13 13 13 13 13 14 14 14 14	6 8 9 9 10 13 13 15 17 18 19 19

The mean of both sets of data is 13 and the median of both sets of data is 13, but the two sets of data are quite different. We need a measure that indicates how the individual data values are scattered or spread on either side of the mean. A number that indicates the variation of the data values about the mean is called a **measure of dispersion**.

Range

The simplest of the measures of dispersion is called the *range*. The **range** is the difference between the highest value and the lowest value of a set of data. In the sets of data given above, the range of ages of students on the middle-school basketball team is 14 − 11 or 3 and the range of the ages of the students in the community center tutoring program is 19 − 6 or 13. The difference in the ranges indicates that the ages of the students on the basketball team are more closely grouped about the mean than the ages of the students in the tutoring program.

The range is dependent on only the largest data value and the smallest data value. Therefore, the range can be very misleading. For example consider the following sets of data:

Ages of members of the chess club: 11, 11, 11, 11, 15, 19, 19, 19, 19
Ages of the members of the math club: 11, 12, 13, 14, 15, 16, 17, 18, 19

For each of these sets of data, the mean is 15, the median is 15, and the range is 8. But the sets of data are very different. The range often does not tell us critical information about a set of data.

Interquartile Range

Another measure of dispersion depends on the first and third quartiles of a distribution. The difference between the first and third quartile values is the **interquartile range**. The interquartile range tells us the range of at least 50% of the data. The largest and smallest values of a set of data are often not representative of the rest of the data. The interquartile range better represents the spread of the data. It also gives us a measure for identifying extreme data values, that is, those that differ significantly from the rest of the data. For example, consider the ages of the members of a book club.

$$21, 24, 25, 27, 28, 31, 35, 35, 37, 39, 40, 41, 69$$

$$\begin{array}{ccc} \uparrow & \uparrow & \uparrow \\ 26 & 35 & 39.5 \\ Q_1 & \text{median} & Q_3 \end{array}$$

For these 13 data values, the median is the 7th value. For the six values below the median, the first quartile is the average of the 3rd and 4th values from the lower end: $\frac{25 + 27}{2} = 26$. The third quartile is the average of the 3rd and 4th values from the upper end: $\frac{39 + 40}{2} = 39.5$. The interquartile range of the ages of the members of the book club is $39.5 - 26$ or 13.5. The age of the oldest member of the club differs significantly from the ages of the others. It is more than 1.5 times the interquartile range above the upper quartile:

$$69 > 39.5 + 1.5(13.5)$$

We call this data value an *outlier*.

DEFINITION _____

An **outlier** is a data value that is greater than the upper quartile plus 1.5 times the interquartile range or less than the lower quartile minus 1.5 times the interquartile range.

When we draw the box-and-whisker plot for a set of data, the outlier is indicated by a ✷ and the whisker is drawn to the largest or smallest data value that is not an outlier. The box-and-whisker plot for the ages of the members of the book club is shown on page 616.

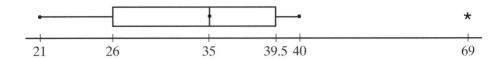

We can use the graphing calculator to graph a box-and-whisker plot with outliers. From the STAT PLOT menu, choose the □─── option. For example, with the book club data entered into L_1, the following keystrokes will graph a box-and-whisker plot with outliers.

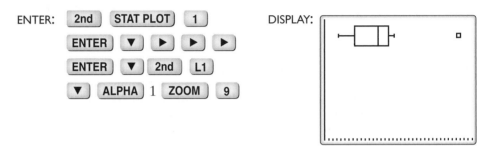

EXAMPLE I

The table shows the number of minutes, rounded to the nearest 5 minutes, needed for each of 100 people to complete a survey.

a. Find the range and the interquartile range for this set of data.

b. Does this data set include outliers?

Minutes	30	35	40	45	50	55	60	65	70	85
Frequency	3	8	12	15	10	24	17	8	2	1
Cumulative Frequency	3	11	23	38	48	72	89	97	99	100

Solution **a.** The range is the difference between the largest and smallest data value.

$$\text{Range} = 85 - 30 = 55$$

To find the interquartile range, we must first find the median and the lower and upper quartiles. Since there are 100 values, the median is the average of the 50th and the 51st values. Both of these values lie in the interval 55. Therefore, the median is 55. There are 50 values above the mean and 50 values below the mean. Of the lower 50 values, the lower quartile is the average of the 25th and 26th values. Both of these values lie in the interval 45. The lower quartile is 45. The upper quartile is the average of the 25th and the 26th from the upper end of the distribution (or the 75th and 76th from the lower end). These values lie in the interval 60. The upper quartile is 60.

$$\text{Interquartile range} = 60 - 45 = 15$$

b. An outlier is a data value that is 1.5 times the interquartile range below the first quartile or above the third quartile.

$$45 - 1.5(15) = 22.5 \qquad 60 + 1.5(15) = 82.5$$

The data value 85 is an outlier.

Answers **a.** Range = 55, interquartile range = 15 **b.** The data value 85 is an outlier.

Exercises

Writing About Mathematics

1. In any set of data, is it always true that $x_i = i$? For example, in a set of data with more than three data values, does $x_4 = 4$? Justify your answer.

2. In a set of data, $Q_1 = 12$ and $Q_3 = 18$. Is a data value equal to 2 an outlier? Explain why or why not.

Developing Skills

In 3–6, find the range and the interquartile range for each set of data.

3. 3, 5, 7, 9, 11, 13, 15, 17, 19

4. 12, 12, 14, 14, 16, 18, 20, 22, 28, 34

5. 12, 17, 23, 31, 46, 54, 67, 76, 81, 93

6. 2, 14, 33, 34, 34, 34, 35, 36, 37, 37, 38, 40, 42

In 7–9, find the mean, median, range, and interquartile range for each set of data to the nearest tenth.

7.

x_i	f_i
50	3
45	8
40	12
35	15
30	11
25	7
20	4

8.

x_i	f_i
10	2
9	4
8	6
7	9
6	3
5	3
4	2
3	0
2	1

9.

x_i	f_i
11	5
16	8
19	9
31	6
37	5
32	5
35	6

Applying Skills

10. The following data represents the yearly salaries, in thousands of dollars, of 10 basketball players.

$$533 \quad 427 \quad 800 \quad 687 \quad 264 \quad 264 \quad 125 \quad 602 \quad 249 \quad 19{,}014$$

  a. Find the mean and median salaries of the 10 players.

  b. Which measure of central tendency is more representative of the data? Explain.

  c. Find the outlier for the set of data.

  d. Remove the outlier from the set of data and recalculate the mean and median salaries.

  e. After removing the outlier from the set of data, is the mean more or less representative of the data?

11. The grades on a math test are shown in the stem-and-leaf diagram to the right.

  a. Find the mean grade.

  b. Find the median grade.

  c. Find the first and third quartiles.

  d. Find the range.

  e. Find the interquartile range.

Stem	Leaf
9	0 0 1 2 6 9
8	0 2 3 5 5 5 7
7	8 8
6	4 6 6 7 9
5	0 6 7
4	6 8

12. The ages of students in a Spanish class are shown in the table. Find the range and the interquartile range.

Age	Frequency
19	1
18	8
17	8
16	6
15	2

13. The table shows the number of hours that 40 third graders reported studying a week. Find the range and the interquartile range.

Hours	3	4	5	6	7	8	9	10	11	12
Frequency	2	1	3	3	5	8	8	5	4	1

14. The table shows the number of pounds lost during the first month by people enrolled in a weight-loss program.

  a. Find the range.

  b. Find the interquartile range.

  c. Which of the data values is an outlier?

Pounds Lost	1	2	3	4	5	6	7	8	9	11	15
Frequency	1	1	2	2	6	10	7	7	2	1	1

15. The 14 students on the track team recorded the following number of seconds as their best time for the 100-yard dash:

$$13.5 \quad 13.7 \quad 13.1 \quad 13.0 \quad 13.3 \quad 13.2 \quad 13.0$$
$$12.8 \quad 13.4 \quad 13.3 \quad 13.1 \quad 12.7 \quad 13.2 \quad 13.5$$

Find the range and the interquartile range.

16. The following data represent the waiting times, in minutes, at Post Office A and Post Office B at noon for a period of several days.

$$A: 1, 2, 2, 2, 3, 3, 3, 3, 3, 3, 3, 3, 9, 10$$
$$B: 1, 2, 2, 3, 3, 5, 5, 6, 6, 7, 7, 8, 8, 9, 10$$

a. Find the range of each set of data. Are the ranges the same?

b. Graph the box and whisker plot of each set of data.

c. Find the interquartile range of each set of data.

d. If the data values are representative of the waiting times at each post office, which post office should you go to at noon if you are in a hurry? Explain.

Hands-On Activity

Find the range and the interquartile range of the data from the survey in the Hands-On Activity of Section 15-1 estimating the length of a minute. Does your data contain an outlier?

15-5 VARIANCE AND STANDARD DEVIATION

Variance

Let us consider a more significant measure of dispersion than either the range or the interquartile range. Let x_i represent a student's grades on eight tests.

Grade (x_i)	$x_i - \bar{x}$	$(x_i - \bar{x})^2$
95	9	81
92	6	36
88	2	4
87	1	1
86	0	0
82	−4	16
80	−6	36
78	−8	64
$\displaystyle\sum_{i=1}^{8} x_i = 688$	$\displaystyle\sum_{i=1}^{8} (x_i - \bar{x})^2 = 0$	$\displaystyle\sum_{i=1}^{8} (x_i - \bar{x})^2 = 238$

$$\text{Mean} = \frac{\displaystyle\sum_{i=1}^{8} x_i}{8} = \frac{688}{8} = 86$$

The table shows the deviation, or difference, of each grade from the mean. Grades above the mean are positive and grades below the mean are negative. For any set of data, the sum of these differences is always 0.

In order to find a meaningful sum, we can use the squares of the differences so that each value will be positive. The square of the deviation of each data value from the mean is used to find another measure of dispersion called the **variance**. To find the variance, find the sum of the squares of the deviations from the mean and divide that sum by the number of data values. In symbols, the variance for a set of data that represents the entire population is given by the formula:

$$\textbf{Variance} = \frac{1}{n}\sum_{i=1}^{n}(x_i - \bar{x})^2$$

For the set of data given above, the variance is $\frac{1}{8}(238)$ or 29.75.

Note that since the *square* of the differences is involved, this method of finding a measure of dispersion gives greater weight to measures that are farther from the mean.

EXAMPLE I

A student received the following grades on five math tests: 84, 97, 92, 88, 79. Find the variance for the set of grades of the five tests.

Solution The mean of this set of grades is:

$$\frac{84 + 97 + 92 + 88 + 79}{5} = \frac{440}{5} = 88$$

x_i	$x_i - \bar{x}$	$(x_i - \bar{x})^2$
84	−4	16
97	9	81
92	4	16
88	0	0
79	−9	81
$\sum_{i=1}^{5} x_i = 440$		$\sum_{i=1}^{5}(x_i - \bar{x})^2 = 194$

$$\textbf{Variance} = \frac{1}{n}\sum_{i=1}^{5}(x_i - \bar{x})^2 = \left(\frac{1}{5}\right)194 = 38\frac{4}{5} = 38.8 \ \textit{Answer}$$

When the data representing a population is listed in a frequency distribution table, we can use the following formula to find the variance:

$$\textbf{Variance} = \frac{\displaystyle\sum_{i=1}^{n} f_i(x_i - \bar{x})^2}{\displaystyle\sum_{i=1}^{n} f_i}$$

EXAMPLE 2

In a city, there are 50 math teachers who are under the age of 30. The table below shows the number of years of experience of these teachers. Find the variance of this set of data. Let x_i represent the number of years of experience and f_i represent the frequency for that number of years.

x_i	0	1	2	3	4	5	6	7	8	
f_i	1	2	7	8	6	9	8	4	5	50
$x_i f_i$	0	2	14	24	24	45	48	28	40	225

Solution

$$\text{Mean} = \frac{\sum_{i=0}^{8} x_i f_i}{\sum_{i=0}^{8} f_i} = \frac{225}{50} = 4.5$$

For this set of data, the data value is equal to i for each x_i.

The table below shows the deviation from the mean, the square of the deviation from the mean, and the square of the deviation from the mean multiplied by the frequency.

x_i	f_i	$(x_i - \bar{x})$	$(x_i - \bar{x})^2$	$f_i(x_i - \bar{x})^2$
8	5	3.5	12.25	61.25
7	4	2.5	6.25	25.00
6	8	1.5	2.25	18.00
5	9	0.5	0.25	2.25
4	6	−0.5	0.25	1.50
3	8	−1.5	2.25	18.00
2	7	−2.5	6.25	43.75
1	2	−3.5	12.25	24.50
0	1	−4.5	20.25	20.25
	$\sum_{i=1}^{8} f_i = 50$			$\sum_{i=0}^{8} f_i(x_i - \bar{x})^2 = 214.50$

$$\text{Variance} = \frac{\sum_{i=0}^{8} f_i(x_i - \bar{x})^2}{\sum_{i=0}^{8} f_i} = \frac{214.50}{50} = 4.29 \ \text{Answer}$$

Note that the data given in this example represents a population, that is, data for all of the teachers under consideration.

Standard Deviation Based on the Population

Although the variance is a useful measure of dispersion, it is in square units. For example, if the data were a set of measures in centimeters, the variance would be in square centimeters. In order to have a measure that is in the same unit of measure as the given data, we find the square root of the variance. The square root of the variance is called the **standard deviation**. When the data represents a population, that is, all members of the group being studied:

$$\text{Standard deviation based on a population} = \sqrt{\frac{1}{n}\sum_{i=1}^{n} f_i(x_i - \bar{x})^2}$$

If the data is grouped in terms of the frequency of a given value:

$$\text{Standard deviation based on a population} = \sqrt{\frac{\sum_{i=1}^{n} f_i(x_i - \bar{x})^2}{\sum_{i=1}^{n} f_i}}$$

The symbol for the standard deviation for a set of data that represents a population is σ (lowercase Greek sigma). Many calculators use the symbol σ_x.

EXAMPLE 3

Find the standard deviation for the number of years of experience for the 50 teachers given in Example 2.

Solution   The standard deviation is the square root of the variance.

Therefore:
$$\text{Standard deviation} = \sqrt{4.29} = 2.071231518$$

Calculator   (1) Clear lists 1 and 2.
Solution
(2) Enter the number of years of experience in L_1 and the frequency in L_2.

(3) Locate the standard deviation for a population (σ_x) under 1-Var Stats.

ENTER: STAT ▶ ENTER   DISPLAY:

            2nd L1 ,

            2nd L2 ENTER

```
1-VAR STATS
x̄=4.5
Σx=225
Σx²=1227
Sx=2.092259788
σx=2.071231518
↓n=50
```

(4) $\sigma_x = 2.071231518$

Answer   The standard deviation is approximately 2.07.

Standard Deviation Based on a Sample

When the given data is information obtained from a sample of the population, the formula for standard deviation is obtained by dividing the sum of the squares of the deviation from the mean by 1 less than the number of data values.

If each data value is listed separately:

$$\textbf{Standard deviation for a sample} = \sqrt{\frac{1}{n-1}\sum_{i=1}^{n}(x_i - \bar{x})^2}$$

If the data is grouped in terms of the frequency of a given value:

$$\textbf{Standard deviation for a sample} = \sqrt{\frac{\sum_{i=1}^{n}f_i(x_i - x)^2}{\left(\sum_{i=1}^{n}f_i\right) - 1}}$$

The symbol for the standard deviation for a set of data that represents a sample is s. Many calculators use the symbol s_x.

EXAMPLE 4

From a high school, ten students are chosen at random to report their number of online friends. The data is as follows: 15, 13, 12, 10, 9, 7, 5, 4, 3, and 2.

Solution The total number of online friends for these 10 students is 80 or a mean of 8 online friends ($\bar{x} = 8$).

Online Friends (x_i)	$x_i - \bar{x}$	$(x_i - \bar{x})^2$
15	7	49
13	5	25
12	4	16
10	2	4
9	1	1
7	−1	1
5	−3	9
4	−4	16
3	−5	25
2	−6	36
		$\sum_{i=1}^{10}(x_i - \bar{x})^2 = 182$

$$\text{Standard deviation} = \sqrt{\frac{1}{10-1}\sum_{i=1}^{10}(x_i - \bar{x})^2} = \sqrt{\tfrac{1}{9}(182)} \approx 4.5 \;\; \textit{Answer}$$

EXAMPLE 5

In each of the following, tell whether the population or sample standard deviation should be used.

a. In a study of the land areas of the states of the United States, the area of each of the 50 states is used.

b. In a study of the heights of high school students in a school of 1,200 students, the heights of 100 students chosen at random were recorded.

c. In a study of the heights of high school students in the United States, the heights of 100 students from each of the 50 states were recorded.

Solution   **a.** Use the population standard deviation since every state is included. *Answer*

b. Use the sample standard deviation since only a portion of the total school population was included in the study. *Answer*

c. Use the sample standard deviation since only a portion of the total school population was included in the study. *Answer*

EXAMPLE 6

A telephone survey conducted in Monroe County obtained information about the size of the households. Telephone numbers were selected at random until a sample of 130 responses were obtained. The frequency chart below shows the result of the survey.

No. of People per Household	1	2	3	4	5	6	7	8	9
Frequency	28	37	45	8	7	3	1	0	1

Solution   The table below can be used to find the mean and the standard deviation.

x_i	f_i	$f_i x_i$	$(x_i - \bar{x})$	$(x_i - \bar{x})^2$	$f_i(x_i - \bar{x})^2$
1	28	28	−1.6	2.56	71.68
2	37	74	−0.6	0.36	13.32
3	45	135	0.4	0.16	7.20
4	8	32	1.4	1.96	15.68
5	7	35	2.4	5.76	40.32
6	3	18	3.4	11.56	34.68
7	1	7	4.4	19.36	19.36
8	0	0	5.4	29.16	0
9	1	9	6.4	40.96	40.96
	$\sum_{i=1}^{9} f_i = 130$	$\sum_{i=1}^{9} f_i x_i = 338$			$\sum_{i=1}^{9} f_i(x_i - \bar{x})^2 = 243.20$

$$\text{Mean} = \frac{\sum\limits_{i=1}^{9} f_i x_i}{\sum\limits_{i=1}^{9} f_i} = \frac{338}{130} = 2.6$$

$$\text{Standard deviation} = \sqrt{\frac{\sum\limits_{i=1}^{9} f_i(x_i - \bar{x})^2}{\left(\sum\limits_{i=1}^{9} f_i\right) - 1}} = \sqrt{\frac{243.20}{130 - 1}} = \sqrt{\frac{243.20}{129}} = 1.373051826$$

Calculator Solution Enter the number of members of the households in L_1 and the frequency for each data value in L_2.

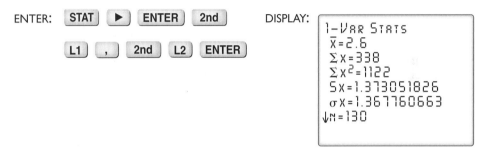

ENTER: [STAT] [▶] [ENTER] [2nd]

[L1] [,] [2nd] [L2] [ENTER]

DISPLAY:

```
1-VAR STATS
x̄=2.6
Σx=338
Σx²=1122
Sx=1.373051826
σx=1.367760663
↓n=130
```

The standard deviation based on the data from a sample is $s_x = 1.373051826$ or approximately 1.37. *Answer*

Exercises

Writing About Mathematics

1. The sets of data for two different statistical studies are identical. The first set of data represents the data for all of the cases being studied and the second represents the data for a sample of the cases being studied. Which set of data has the larger standard deviation? Explain your answer.

2. Elaine said that the variance is the square of the standard deviation. Do you agree with Elaine? Explain why or why not.

Developing Skills

In 3–9, the given values represent data for a population. Find the variance and the standard deviation for each set of data.

3. 9, 9, 10, 11, 5, 10, 12, 9, 10, 12, 6, 11, 11, 11

4. 11, 6, 7, 13, 5, 8, 7, 10, 9, 11, 13, 12, 9, 16, 10

5. 20, 19, 20, 17, 18, 19, 42, 41, 41, 39, 39, 40

6. 20, 101, 48, −5, 63, 31, 20, 50, 16, 14, −45, 9

7.

x_i	f_i
30	1
35	7
40	10
45	9
50	11
55	8
60	6

8.

x_i	f_i
2	0
4	1
6	6
8	10
10	13
12	21
14	7
16	1

9.

x_i	f_i
20	21
25	10
30	1
35	2
40	2
45	4
50	5
55	3
60	7

In 10–16, the given values represent data for a sample. Find the variance and the standard deviation based on this sample.

10. 6, 4, 9, 11, 4, 3, 22, 3, 7, 10

11. 12.1, 33.3, 45.5, 60.1, 94.2, 22.2

12. 15, 10, 16, 19, 10, 19, 14, 17

13. 1, 3, 5, 22, 30, 45, 50, 55, 60, 70

14.

x_i	f_i
55	11
50	15
45	4
40	1
35	14
30	12
25	4

15.

x_i	f_i
33	3
34	1
35	4
36	6
37	5
38	11
39	6

16.

x_i	f_i
1	3
2	3
3	3
4	3
5	3
6	3
7	3

Applying Skills

17. To commute to the high school in which Mr. Fedora teaches, he can take either the Line A or the Line B train. Both train stations are the same distance from his house and both stations report that, on average, they run 10 minutes late from the scheduled arrival time. However, the standard deviation for Line A is 1 minute and the standard deviation for Line B is 5 minutes. To arrive at approximately the same time on a regular basis, which train line should Mr. Fedora use? Explain.

18. A hospital conducts a study to determine if nurses need extra staffing at night. A random sample of 25 nights was used. The number of calls to the nurses' station each night is shown in the stem-and-leaf diagram to the right.

Stem	Leaf
9	0 2 3 4 4 6 6
8	6 8
7	0 1 2 3 4
6	4 4 6 7 8 9
5	0 3 7
4	1 9

a. Find the variance.

b. Find the standard deviation.

19. The ages of all of the students in a science class are shown in the table. Find the variance and the standard deviation.

Age	Frequency
18	1
17	2
16	9
15	9

20. The table shows the number of correct answers on a test consisting of 15 questions. The table represents correct answers for a sample of the students who took the test. Find the standard deviation based on this sample.

Correct Answers	6	7	8	9	10	11	12	13	14	15
Frequency	2	1	3	3	5	8	8	5	4	1

21. The table shows the number of robberies during a given month in 40 different towns of a state. Find the standard deviation based on this sample

Robberies	0	1	2	3	4	5	6	7	8	9	10
Frequency	1	1	1	2	2	6	10	7	7	2	1

22. Products often come with registration forms. One of the questions usually found on the registration form is household income. For a given product, the data below represents a random sample of the income (in thousands of dollars) reported on the registration form. Find the standard deviation based on this sample.

$$38 \quad 40 \quad 26 \quad 42 \quad 39 \quad 25 \quad 40 \quad 40 \quad 39 \quad 36$$
$$46 \quad 41 \quad 43 \quad 47 \quad 49 \quad 43 \quad 39 \quad 35 \quad 43 \quad 37$$

Hands-On Activity

The people in your survey from the Hands-On Activity of Section 15-1 represent a random sample of all people. Find the standard deviation based on your sample.

15-6 NORMAL DISTRIBUTION

The Normal Curve

Imagine that we were able to determine the height in centimeters of all 10-year-old children in the United States. With a scale along the horizontal axis that includes all of these heights, we will place a dot above each height for each child of that height. For example, we will place above 140 on the horizontal scale a dot for each child who is 140 centimeters tall. Do this for 139, 138, 137, and so on for each height in our data. The result would be a type of frequency histogram. If we

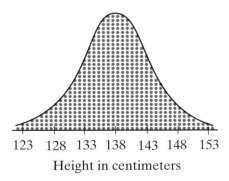

123   128   133   138   143   148   153
Height in centimeters

draw a smooth curve joining the top dot for each height, we will draw a bell-shaped curve called the **normal curve**. As the average height of 10-year-olds is approximately 138 centimeters, the data values are concentrated at 138 centimeters and the normal curve has a peak at 138 centimeters. Since for each height that is less than or greater than 138 centimeters there are fewer 10-year-olds, the normal curve progressively gets shorter as you go farther from the mean.

Scientists have found that large sets of data that occur naturally such as heights, weights, or shoe sizes have a bell-shaped or a normal curve. The highest point of the normal curve is at the mean of the data. The normal curve is symmetric with respect to a vertical line through the mean of the distribution.

Standard Deviation and the Normal Curve

A **normal distribution** is a set of data that can be represented by a normal curve. For a normal distribution, the following relationships exist.

1. The mean and the median of the data values lie on the line of symmetry of the curve.

2. Approximately 68% of the data values lie within one standard deviation from the mean.

3. Approximately 95% of the data values lie within two standard deviations from the mean.

4. Approximately 99.7% of the data values lie within three standard deviations from the mean.

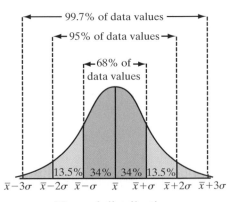

Normal distribution

EXAMPLE I

A set of data is normally distributed with a mean of 50 and a standard deviation of 2.

a. What percent of the data values are less than 50?

b. What percent of the data values are between 48 and 52?

c. What percent of the data values are between 46 and 54?

d. What percent of the data values are less than or equal to 46?

Solution   **a.** In a normal distribution, 50% of the data values are to the left and 50% to the right of the mean.

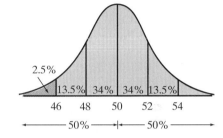

b. 48 and 52 are each 1 standard deviation away from the mean.

$$48 = 50 - 2 \qquad 52 = 50 + 2$$

Therefore, 68% of the data values are between 48 and 52.

c. 46 and 53 are each 2 standard deviations away from the mean.

$$46 = 50 - 2(2) \qquad 54 = 50 + 2(2)$$

Therefore, 95% of the data values are between 48 and 52.

d. 50% of the data values are less than 50.

47.5% of the data values are more than 46 and less than 50.

Therefore, 50% $-$ 47.5% or 2.5% of the data values are less than or equal to 46.

Answers   **a.** 50%   **b.** 68%   **c.** 95%   **d.** 2.5%

Z-Scores

The **z-score** for a data value is the deviation from the mean divided by the standard deviation. Let x be a data value of a normal distribution.

$$\text{z-score} = \frac{x - \bar{x}}{\text{standard deviation}} = \frac{x - \bar{x}}{\sigma}$$

The z-score of x, a value from a normal distribution, is positive when x is above the mean and negative when x is below the mean. The z-score tells us how many standard deviations x is above or below the mean.

1. The z-score of the mean is 0.

2. Of the data values, 34% have a z-score between 0 and 1 and 34% have a z-score between −1 and 0. Therefore, 68% have a z-score between −1 and 1.

3. Of the data values, 13.5% have a z-score between 1 and 2 and 13.5% have a z-score between −2 and −1.

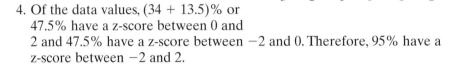

4. Of the data values, (34 + 13.5)% or 47.5% have a z-score between 0 and 2 and 47.5% have a z-score between −2 and 0. Therefore, 95% have a z-score between −2 and 2.

5. Of the data values, 99.7% have a z-score between −3 and 3.

For example, the mean height of 10-year-old children is 138 centimeters with a standard deviation of 5. Casey is 143 centimeters tall.

$$\text{z-score for Casey} = \frac{143 - 138}{5} = \frac{5}{5} = 1$$

Casey's height is 1 standard deviation above the mean. For a normal distribution, 34% of the data is between the mean and 1 standard deviation above the mean and 50% of the data is below the mean. Therefore, Casey is as tall as or taller than (34 + 50)% or 84% of 10-year-old children.

   A calculator will give us this same answer. The second entry of the DISTR menu is normalcdf(. When we use this function, we must supply a minimum value, a maximum value, the mean, and the standard deviation separated by commas:

$$\text{normalcdf}(minimum, maximum, mean, standard\ deviation)$$

For the minimum value we can use 0. To find the proportion of 10-year-old children whose height is 143 centimeters or less, use the following entries.

The calculator returns the number 0.8413447404, which can be rounded to 0.84 or 84%.

If we wanted to find the proportion of the 10-year-old children who are between 134.6 and 141.4 centimeters tall, we could make the following entry:

ENTER: [2nd] [DISTR] [2] 134.6   DISPLAY:

[,] 141.4 [,] 138 [,]

5 [ )] [ENTER]

```
NORMALCDF(134.6,
141.4,138,5)
        .5034956838
```

The calculator returns the number 0.5034956838, which can be rounded to 0.50 or 50%. Since 134.6 and 141.4 are equidistant from the mean, 25% of the data is below the mean and 25% is above the mean. Therefore, for this distribution, 134.6 centimeters is the first quartile, 138 centimeters is the median or second quartile, and 141.4 centimeters is the third quartile.

Note: 134.6 is 0.68 standard deviation below the mean and 141.4 is 0.68 standard deviation above the mean. For any normal distribution, data values with z-scores of -0.68 are approximately equal to the first quartile and data values with z-scores of 0.68 are approximately equal to the third quartile.

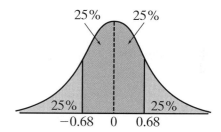

EXAMPLE I

On a standardized test, the test scores are normally distributed with a mean of 60 and a standard deviation of 6.

a. Of the data, 84% of the scores are at or below what score?

b. Of the data, 16% of the scores are at or below what score?

c. What is the z-score of a score of 48?

d. If 2,000 students took the test, how many would be expected to score at or below 48?

Solution **a.** Since 50% scored at or below the mean and 34% scored within 1 standard deviation above the mean, $(50 + 34)\%$ or 84% scored at or below 1 deviation above the mean:

$$\bar{x} + \sigma = 60 + 6 = 66$$

b. Since 50% scored at or below the mean and 34% scored within 1 standard deviation below the mean, $(50 - 34)\%$ or 16% scored at or below 1 deviation below the mean:

$$\bar{x} + \sigma = 60 - 6 = 54$$

c. z-score $= \frac{x - \bar{x}}{\sigma} = \frac{48 - 60}{6} = \frac{-12}{6} = -2$

d. A test score of 48 has a z-score of −2. Since 47.5% of the scores are between −2 and 0 and 50% of the scores are less than 0, 50% − 47.5% or 2.5% scored at or below 48.

$$2.5\% \times 2{,}000 = 0.025 \times 2{,}000 = 50 \text{ students}$$

Answers **a.** 66 **b.** 54 **c.** −2 **d.** 50

EXAMPLE 2

For a normal distribution of weights, the mean weight is 160 pounds and a weight of 186 pounds has a z-score of 2.

a. What is the standard deviation of the set of data?

b. What percent of the weights are between 155 and 165?

Solution **a.** z-score $= \dfrac{x - \bar{x}}{\sigma}$

$2 = \dfrac{186 - 160}{\sigma}$

$2\sigma = 26$

$\sigma = 13$ *Answer*

b. ENTER: [2nd] [DISTR] [2]   DISPLAY:
```
NORMALCDF(155,16
5,160,13)
            .2994775047
```
155 [,] 165 [,] 160 [,]

13 [ )] [ENTER]

About 30% of the weights are between 155 and 165. *Answer*

Exercises

Writing About Mathematics

1. A student's scores on five tests were 98, 97, 95, 93, and 67. Explain why this set of scores does not represent a normal distribution.

2. If 34% of the data for a normal distribution lies between the mean and 1 standard deviation above the mean, does 17% of the data lie between the mean and one-half standard deviation above the mean? Justify your answer.

Developing Skills

In 3–9, for a normal distribution, determine what percent of the data values are in each given range.

3. Between 1 standard deviation below the mean and 1 standard deviation above the mean

4. Between 1 standard deviation below the mean and 2 standard deviations above the mean

5. Between 2 standard deviations below the mean and 1 standard deviation above the mean

6. Above 1 standard deviation below the mean

7. Below 1 standard deviation above the mean

8. Above the mean

9. Below the mean

10. A set of data is normally distributed with a mean of 40 and a standard deviation of 5. Find a data value that is:

   a. 1 standard deviation above the mean

   b. 2.4 standard deviations above the mean

   c. 1 standard deviation below the mean

   d. 2.4 standard deviations below the mean

Applying Skills

In 11–14, select the numeral that precedes the choice that best completes the statement or answers the question.

11. The playing life of a Euclid mp3 player is normally distributed with a mean of 30,000 hours and a standard deviation of 500 hours. Matt's mp3 player lasted for 31,500 hours. His mp3 player lasted longer than what percent of other Euclid mp3 players?

   (1) 68%   (2) 95%   (3) 99.7%   (4) more than 99.8%

12. The scores of a test are normally distributed. If the mean is 50 and the standard deviation is 8, then a student who scored 38 had a z-score of

   (1) 1.5   (2) −1.5   (3) 12   (4) −12

13. The heights of 10-year-old children are normally distributed with a mean of 138 centimeters with a standard deviation of 5 centimeters. The height of a 10-year-old child who is as tall as or taller than 95.6% of all 10-year-old children is

   (1) between 138 and 140 cm.   (2) between 140 and 145 cm.
   (3) between 145 and 148 cm.   (4) taller than 148 cm.

14. The heights of 200 women are normally distributed. The mean height is 170 centimeters with a standard deviation of 10 centimeters. What is the best estimate of the number of women in this group who are between 160 and 170 centimeters tall?

   (1) 20   (2) 34   (3) 68   (4) 136

15. When coffee is packed by machine into 16-ounce cans, the amount can vary. The mean weight is 16.1 ounces and the standard deviation is 0.04 ounce. The weight of the coffee approximates a normal distribution.

   a. What percent of the cans of coffee can be expected to contain less than 16 ounces of coffee?

   b. What percent of the cans of coffee can be expected to contain between 16.0 and 16.2 ounces of coffee?

16. The length of time that it takes Ken to drive to work represents a normal distribution with a mean of 25 minutes and a standard deviation of 4.5 minutes. If Ken allows 35 minutes to get to work, what percent of the time can he expect to be late?

17. A librarian estimates that the average number of books checked out by a library patron is 4 with a standard deviation of 2 books. If the number of books checked out each day approximates a normal distribution, what percent of the library patrons checked out fewer than 7 books yesterday?

18. The heights of a group of women are normally distributed with a mean of 170 centimeters and a standard deviation of 10 centimeters. What is the z-score of a member of the group who is 165 centimeters tall?

19. The test grades for a standardized test are normally distributed with a mean of 50. A grade of 60 represents a z-score of 1.25. What is the standard deviation of the data?

20. Nora scored 88 on a math test that had a mean of 80 and a standard deviation of 5. She also scored 80 on a science test that had a mean of 70 and a standard deviation of 3. On which test did Nora perform better compared with other students who took the tests?

15-7 BIVARIATE STATISTICS

Statistics are often used to compare two sets of data. For example, a pediatrician may compare the height and weight of a child in order to monitor growth. Or the owner of a gift shop may record the number of people who enter the store with the revenue each day. Each of these sets of data is a pair of numbers and is an example of **bivariate statistics**.

Representing bivariate statistics on a two-dimensional graph or **scatter plot** can help us to observe the relationship between the variables. For example the mean value for the critical reading and for the math sections of the SAT examination for nine schools in Ontario County are listed in the table and shown on the graph on the right.

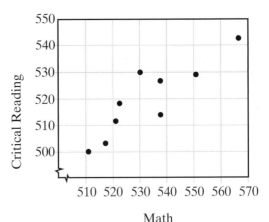

Math	530	551	521	522	537	511	516	537	566
Critical Reading	530	529	512	518	526	500	504	515	543

The graph shows that there appears to be a linear relationship between the critical reading scores and the math scores. As the math scores increase, the critical reading scores also increase. We say that there is a **correlation** between the two scores. The points of the graph approximate a line.

These data can also be shown on a calculator. Enter the math scores as L_1 and the corresponding critical reading scores as L_2. Then turn on Plot 1 and use ZoomStat from the ZOOM menu to construct a window that will include all values of x and y:

ENTER: 　2nd 　 STAT PLOT 　 1 　　　　ENTER: 　 ZOOM 　 9

　　　　 ENTER 　 ▼ 　 ENTER

　　　　 ▼ 　 2nd 　 L1 　 ▼

　　　　 2nd 　 L2

DISPLAY:　　　　　　　　　　　　　　　　DISPLAY:

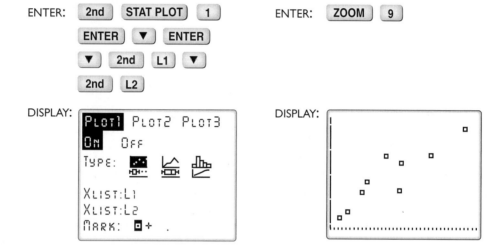

The calculator will display a graph similar to that shown above. To draw a line that approximates the data, use the **regression line** on the calculator. A regression line is a special **line of best fit** that minimizes the square of the vertical distances to each data point. In this course, you do not have to know the formula to find the regression line. The calculator can be used to determine the regression line:

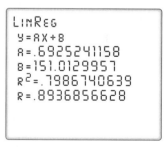

ENTER: 　STAT 　 ▶ 　 4 　 VARS 　 ▶ 　 1 　 1 　 ENTER

The calculator displays values for a and b for the linear equation $y = ax + b$ and stores the regression equation into Y_1 in the 　Y= 　 menu. Press 　ZOOM 　 9 　 to display the scatter plot and the line that best approximates the data. If we round the given values of a and b to three decimal places, the linear regression equation is:

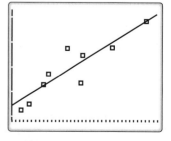

$$y = 0.693x + 151.013$$

We will study other data that can be approximated by a curve rather than a line in later sections.

The scatter plots below show possible linear correlation between elements of the pairs of bivariate data. The correlation is positive when the values of the second element of the pairs tend to increase when the values of the first elements of the pairs increase. The correlation is negative when the values of the second element of the pairs tend to decrease when the values of the first elements of the pairs increase.

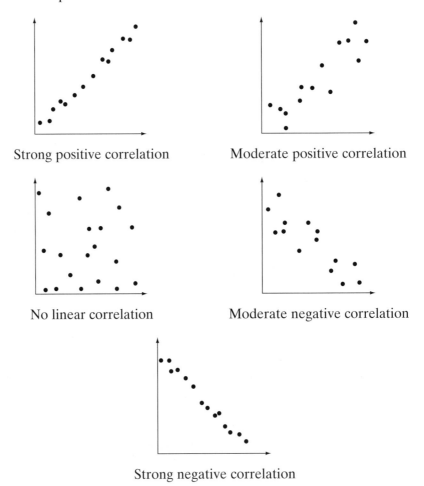

Strong positive correlation   Moderate positive correlation

No linear correlation   Moderate negative correlation

Strong negative correlation

EXAMPLE 1

Jacob joined an exercise program to try to lose weight. Each month he records the number of months in the program and his weight at the end of that month. His record for the first twelve months is shown below:

Month	1	2	3	4	5	6	7	8	9	10	11	12
Weight	248	242	237	228	222	216	213	206	197	193	185	178

a. Draw a scatter plot and describe the correlation between the data (if any).

b. Draw a line that appears to represent the data.

c. Write an equation of a line that best represents the data.

Solution **a.** Use a scale of 0 to 13 along a horizontal line for the number of months in the program and a scale from 170 to 250 along the vertical axis for his weight at the end of that month. The scatter plot is shown here. There appears to be a strong negative correlation between the number of months and Jacob's weight.

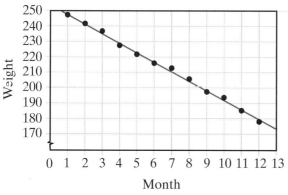

b. The line on the graph that appears to approximate the data intersects the points $(1, 248)$ and $(11, 185)$. We can use these two points to write an equation of the line.

$$y - 248 = \frac{248 - 185}{1 - 11}(x - 1)$$
$$y - 248 = -6.3x + 6.3$$
$$y = -6.3x + 254.3$$

c. On a calculator, enter the number of the month in L_1 and the weight in L_2. To use the calculator to determine a line that approximates the data, use the following sequence:

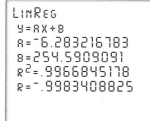

ENTER: [STAT] [▶] [4] [VARS] [▶] [1]
       [1] [ENTER]

The calculator displays values for a and b for the linear equation $y = ax + b$ and stores the equation as Y_1 in the [Y=] menu. If we round the given values of a and b to three decimal places, the linear regression equation is:

$$y = -6.283x + 254.591$$

Turn on Plot 1, then use ZoomStat to graph the data:

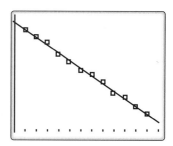

ENTER: [2nd] [STAT PLOT] [ENTER] [ENTER]
       [▼] [ENTER] [▼] [2nd] [L1]
       [ENTER] [2nd] [L2] [ENTER]
       [ZOOM] [9]

The calculator will display the scatter plot of the data along with the regression equation.

EXAMPLE 2

In order to assist travelers in planning a trip, a travel guide lists the average high temperature of most major cities. The listing for Albany, New York is given in the following table.

Month	1	2	3	4	5	6	7	8	9	10	11	12
Temperature	31	33	42	51	70	79	84	81	73	62	48	35

Can these data be represented by a regression line?

Solution Draw a scatter plot the data. The graph shows that the data can be characterized by a curve rather than a line. Finding the regression line for this data would not be appropriate.

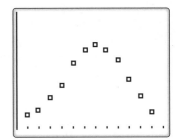

SUMMARY

- The slope of the regression line gives the direction of the correlation:
  - ○ A positive slope shows a positive correlation.
  - ○ A negative slope shows a negative correlation.
- The regression line is appropriate only for data that appears to be linearly related. Do *not* calculate a regression line for data with a scatter plot showing a non-linear relationship.
- The regression equation is sensitive to rounding. Round the coefficients to at least three decimal places.

Exercises

Writing About Mathematics

1. Explain the difference between univariate and bivariate data and give an example of each.

2. What is the relationship between slope and correlation? Can slope be used to measure the strength of a correlation? Explain.

Developing Skills

In 3–6, is the set of data to be collected univariate or bivariate?

3. The science and math grades of all students in a school

4. The weights of the 56 first-grade students in a school

5. The weights and heights of the 56 first-grade students in a school

6. The number of siblings for each student in the first grade

In 7–10, look at the scatter plots and determine if the data sets have high linear correlation, moderate linear correlation, or no linear correlation.

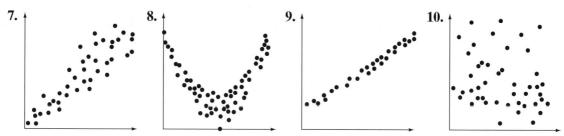

7.   **8.**   **9.**   **10.**

Applying Skills

In 11–17: **a.** Draw a scatter plot. **b.** Does the data set show strong positive linear correlation, moderate positive linear correlation, no linear correlation, moderate negative linear correlation, or strong negative linear correlation? **c.** If there is strong or moderate correlation, write the equation of the regression line that approximates the data.

11. The following table shows the number of gallons of gasoline needed to fill the tank of a car and the number of miles driven since the previous time the tank was filled.

Gallons	8.5	7.6	9.4	8.3	10.5	8.7	9.6	4.3	6.1	7.8
Miles	255	230	295	250	315	260	290	130	180	235

12. A business manager conducted a study to examine the relationship between number of ads placed for each month and sales. The results are shown below where sales are in the thousands.

Number of Ads	10	12	14	16	18	20	22	24	26	28	30
Sales	20	26.5	32	34.8	40	47.2	49.1	56.9	57.9	65.8	66.4

13. Jack Sheehan looked through some of his favorite recipes to compare the number of calories per serving to the number of grams of fat. The table below shows the results.

Calories	310	210	260	330	290	320	245	293	220	260	350
Fat	11	5	11	12	14	16	7	10	8	8	15

14. Greg did a survey to support his theory that the size of a family is related to the size of the family in which the mother of the family grew up. He asked 20 randomly selected people to list the number of their siblings and the number of their mother's siblings. Greg made the following table.

Family	2	2	3	1	0	3	5	2	1	2
Mother's Family	4	0	3	7	4	2	2	6	4	7
Family	3	6	0	2	1	4	1	0	4	1
Mother's Family	5	4	6	1	0	2	1	3	3	2

15. When Marie bakes, it takes about five and a half minutes for the temperature of the oven to reach 350°. One day, while waiting for the oven to heat, Marie recorded the temperature every 20 seconds. Her record is shown below.

Seconds	0	20	40	60	80	100	120	140	160
Temperature	100	114	126	145	160	174	193	207	222
Seconds	180	200	220	240	260	280	300	320	340
Temperature	240	255	268	287	301	318	331	342	350

16. An insurance agent is studying the records of his insurance company looking for a relationship between age of a driver and the percentage of accidents due to speeding. The table shown below summarizes the findings of the insurance agent.

Age	17	18	21	25	30	35	40	45	50	55	60	65
% of Speeding Accidents	49	49	48	39	31	33	24	25	16	10	5	6

17. A sociologist is interested in the relationship between body weight and performance on the SAT. A random sample of 10 high school students from across the country provided the following information:

Weight	197	193	194	157	159	170	149	169	157	185
Score	1,485	1,663	1,564	1,300	1,668	1,405	1,544	1,752	1,395	1,214

15-8 CORRELATION COEFFICIENT

We would like to measure the strength of the linear relationship between the variables in a set of bivariate data. The slope of the regression equation tells us the direction of the relationship but it does not tell us the strength of the relationship. The number that we use to measure both the strength and direction of the linear relationship is called the **correlation coefficient**, r. The value of the correlation coefficient does not depend on the units of measurement. In more advanced statistics courses, you will learn a formula to derive the correlation coefficient. In this course, we can use the graphing calculator to calculate the value of r.

EXAMPLE I

The coach of the basketball team made the following table of attempted and successful baskets for eight players.

Attempted Baskets (x_i)	10	12	12	13	14	15	17	19
Successful Baskets (y_i)	6	7	9	8	10	11	14	15

Find the value of the correlation coefficient.

Solution Enter the given x_i in L_1 and y_i in L_2.

Then choose LinReg(ax+b) from the CALC STAT menu:

ENTER: [STAT] [▶] [4] [ENTER]

The calculator will list both the regression equation and r, the correlation coefficient.

```
LinReg
y=ax+b
a=1.066666667
b=⁻4.933333333
r²=.9481481481
r=.9737289911
```

Answer $r = 0.97$

Note: If the correlation coefficient does not appear on your calculator, enter [2nd] [CATALOG] [D], scroll down to DiagnosticOn, press [ENTER], and press [ENTER] again.

When the absolute value of the correlation coefficient is close to 1, the data have a strong linear correlation. When the absolute value of the correlation coefficient is close to 0, there is little or no linear correlation. Values between 0 and 1 indicate various degrees of positive moderate correlation and values between 0 and −1 indicate various degrees of negative moderate correlation.

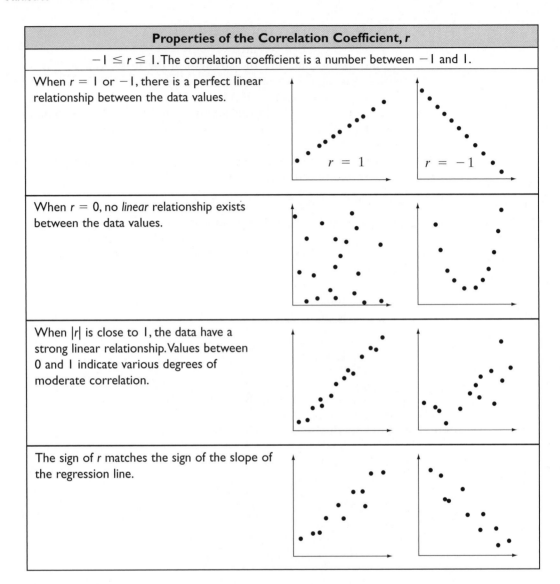

Properties of the Correlation Coefficient, r

$-1 \leq r \leq 1$. The correlation coefficient is a number between -1 and 1.

When $r = 1$ or -1, there is a perfect linear relationship between the data values.

$r = 1$ $r = -1$

When $r = 0$, no *linear* relationship exists between the data values.

When $|r|$ is close to 1, the data have a strong linear relationship. Values between 0 and 1 indicate various degrees of moderate correlation.

The sign of r matches the sign of the slope of the regression line.

EXAMPLE 2

An automotive engineer is studying the fuel efficiency of a new prototype. From a fleet of eight prototypes, he records the number of miles driven and the number of gallons of gasoline used for each trip.

Miles Driven	310	270	350	275	380	320	290	405
Gallons of Gasoline Used	10.0	9.0	11.2	8.7	12.3	10.2	9.5	12.7

a. Based on the context of the problem, do you think the correlation coefficient will be positive, negative, or close to 0?

b. Based on the scatter plot of the data, do you expect the correlation coefficient to be close to $-1, 0,$ or 1?

c. Use a calculator to find the equation of the regression line and determine the correlation coefficient.

Solution **a.** As gallons of gasoline used tend to increase with miles driven, we expect the correlation coefficient to be positive.

Enter the given data as L_1 and L_2 on a calculator.

b.

There appears to be a strong positive correlation, so r will be close to 1.

c.

$y = 0.030x + 0.748, r = 0.99$

EXAMPLE 3

The produce manager of a food store noted the relationship between the amount the store charged for a pound of fresh broccoli and the number of pounds sold in one week. His record for 11 weeks is shown in the following table.

Cost per Pound	$0.65	$0.85	$0.90	$1.00	$1.25	$1.50	$1.75	$1.99	$2.25	$2.50	$2.65
Pounds Purchased	58	43	49	23	39	16	56	32	12	35	11

What conclusion could the product manager draw from this information?

Solution   Enter the given data as L_1 and L_2 on the calculator, graph the scatter plot, and find the value of the correlation coefficient.

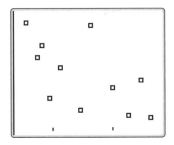

```
LinReg
y=ax+b
a=-13.82881393
b=55.73638117
R²=.331571737
R=-.5758226611
```

From the scatter plot, we can see that there is a moderate negative correlation. Since $r = -0.58$, the product manager might conclude that a lower price does explain some of the increase in sales but other factors also influence the number of sales.

A Warning About Cause-and-Effect

The correlation coefficient is a number that measures the strength of the linear relationship between two data sets. However, simply because there appears to be a strong linear correlation between two variables does *not* mean that one causes the other. There may be other variables that are the cause of the observed pattern. For example, consider a study on the population growth of a city. Although a statistician may find a linear pattern over time, this does not mean that time causes the population to grow. Other factors cause the city grow, for example, a booming economy.

SUMMARY

- $-1 \leq r \leq 1$. The correlation coefficient is a number between -1 and 1.

- When $r = 1$ or $r = -1$, there is a perfect linear relationship between the data values.

- When $r = 0$, no *linear* relationship exists between the data values.

- When $|r|$ is close to 1, the data have a strong linear relationship. Values between 0 and 1 indicate various degrees of moderate correlation.

- The sign of r matches the sign of the slope of the regression line.

- A high correlation coefficient does *not* necessarily mean that one variable causes the other.

Exercises

Writing About Mathematics

1. Does a correlation coefficient of -1 indicate a lower degree of correlation than a correlation coefficient of 0? Explain why or why not.

2. If you keep a record of the temperature in degrees Fahrenheit and in degrees Celsius for a month, what would you expect the correlation coefficient to be? Justify your answer.

Developing Skills

In 3–6, for each of the given scatter plots, determine whether the correlation coefficient would be close to $-1, 0,$ or 1.

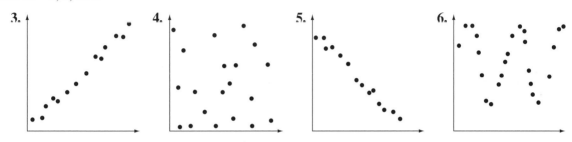

In 7–14, for each of the given correlation coefficients, describe the linear correlation as strong positive, moderate positive, none, moderate negative, or strong negative.

7. $r = 0.9$ **8.** $r = -1$ **9.** $r = -0.1$ **10.** $r = 0.3$

11. $r = 1$ **12.** $r = -0.5$ **13.** $r = 0$ **14.** $r = -0.95$

Applying Skills

In 15–19: **a.** Draw a scatter plot for each data set. **b.** Based on the scatter plot, would the correlation coefficient be close to $-1, 0,$ or 1? Explain. **c.** Use a calculator to find the correlation coefficient for each set of data.

15. The following table shows the number of gallons of gasoline needed to fill the tank of a car and the number of miles driven since the previous time the tank was filled.

Gallons	12.5	3.4	7.9	9.0	15.7	7.0	5.1	11.9	13.0	10.7
Miles	392	137	249	308	504	204	182	377	407	304

16. A man on a weight-loss program tracks the number of pounds that he lost over the course of 10 months. A negative number indicates that he actually *gained* weight for that month.

Month	1	2	3	4	5	6	7	8	9	10
No. of Pounds Lost	9.2	9.1	4.8	4.5	2.8	1.8	1.2	0	0.8	−2.6

17. An economist is studying the job market in a large city conducts of survey on the number of jobs in a given neighborhood and the number of jobs paying $100,000 or more a year. A sample of 10 randomly selected neighborhood yields the following data:

Total Number of Jobs	24	28	17	39	32	21	39	39	24	29
No. of High-Paying Jobs	3	3	4	5	7	3	4	7	7	4

18. The table below shows the same-day forecast and the actual high temperature for the day over the course of 18 days. The temperature is given in degrees Fahrenheit.

Same-Day Forecast	56	52	67	55	58	56	59	57	53
Actual Temperature	53	54	63	49	66	54	54	56	59
Same-Day Forecast	45	55	45	58	59	55	48	53	54
Actual Temperature	48	60	36	59	59	47	46	52	48

19. The table below shows the five-day forecast and the actual high temperature for the fifth day over the course of 18 days. The temperature is given in degrees Fahrenheit.

Five-Day Forecast	56	52	67	55	58	56	59	57	53
Actual Temperature	50	50	84	54	57	40	70	79	48
Five-Day Forecast	45	55	45	58	59	55	48	53	54
Actual Temperature	40	61	40	70	46	75	49	46	88

20. **a.** In Exercises 18 and 19, if the forecasts were 100% accurate, what should the value of r be?

b. Is the value of r for Exercise 18 greater than, equal to, or less than the value of r for Exercise 19? Is this what you would expect? Explain.

15-9 NON-LINEAR REGRESSION

Not all bivariate data can be represented by a linear function. Some data can be better approximated by a curve. For example, on the right is a scatter plot of the file size of a computer program called Super Type over the course of 6 different versions. The relationship does not appear to be linear. For this set of data, a linear regression would *not* be appropriate.

There are a variety of non-linear functions that can be applied to non-linear data. In a statistics course, you will learn more rigorous methods of determining the regression model. In this course, we will use the scatter plot of the data to choose the regression model:

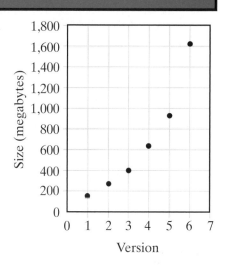

Regression to Use	Description of Scatter Plot	Examples
Exponential	• An exponential curve that does *not* pass through $(0, 0)$ • y-intercept is positive • Data constraint: $y > 0$	
Logarithmic	• A logarithmic curve that does *not* pass through $(0, 0)$ • y-intercept is positive or negative • Data constraint: $x > 0$	
Power	• Positive half of power curve passing through $(0, 0)$ • Data constraints: $x > 0, y > 0$	

(*continued on next page*)

Regression to Use	Description of Scatter Plot	Examples
Specific types of power regression:		
Quadratic	• A quadratic curve	
Cubic	• A cubic curve	

The different non-linear regression models can be found in the STAT CALC menu of the graphing calculator.

- 5:QuadReg is quadratic regression.
- 6:CubicReg is cubic regression.
- 9:LnReg is logarithmic regression.
- 0:ExpReg is exponential regression.
- A:PwrReg is power regression.

In the example of the file size of Super Type, the scatter plot appears to be exponential or power. The table below shows the data of the scatter plot:

Version	1	2	3	4	5	6
Size (megabytes)	155	240	387	630	960	1,612

To find the exponential regression model, enter the data into L_1 and L_2. Choose ExpReg from the STAT CALC menu:

ENTER: STAT ▶ 0 VARS ▶ 1 1 ENTER

The calculator will display the regression equation and store the equation into Y_1 of the Y= menu. To the nearest thousandth, the regression equation is $y = 95.699(1.596^x)$. Press ZOOM 9 to graph the scatter plot and the regression equation.

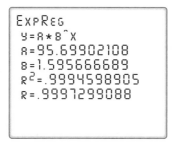

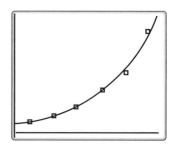

To find the power regression model, with the data in L_1 and L_2, choose PwrReg from the STAT CALC menu:

ENTER: [STAT]   [▶]   [ALPHA]   [A]   [VARS]   [▶]   [1]   [1]   [ENTER]

The calculator will display the regression equation and store the equation into Y_1 of the [Y=] menu. To the nearest thousandth, the regression equation is $y = 121.591x^{1.273}$. Press [ZOOM]   [9] to graph the scatter plot and the regression equation.

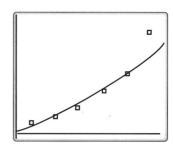

From the scatter plots, we see that the exponential regression equation is a better fit for the data.

EXAMPLE I

A stone is dropped from a height of 1,000 feet. The trajectory of the stone is recorded by a high-speed video camera in intervals of half a second. The recorded distance that the stone has fallen in the first 5 seconds in given below:

Seconds	1	1.5	2	2.5	3	3.5	4	4.5
Distance	16	23	63	105	149	191	260	321

a. Determine which regression model is most appropriate.

b. Find the regression equation. Round all values to the nearest thousandth.

Solution **a.** Draw a scatter plot of the data. The data appears to approximate an exponential function or a power function. Enter the data in L_1 and L_2. Find and graph the exponential and power models. From the displays, it appears that the power model is the better fit.

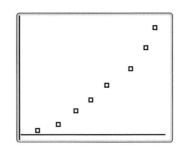

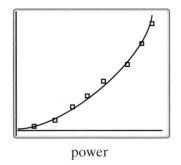

exponential   power

b. To the nearest thousandth, the calculator will display the power equation $y = ax^b$ for $a = 13.619, b = 2.122$.

Answers **a.** Power regression   **b.** $y = 13.619(x^{2.122})$

EXAMPLE 2

A pediatrician has the following table that lists the head circumferences for a group of 12 baby girls from the same extended family. The circumference is given in centimeters.

Age in Months	2	2	5	4	1	17	11	14	7	11	10	19
Circumference	36.8	37.2	38.6	38.2	35.9	40.4	39.7	39.9	39.2	41.1	39.3	40.5

a. Make a scatter plot of the data.

b. Choose what appears to be the curve that best fits the data.

c. Find the regression equation for this model.

Solution **a.**

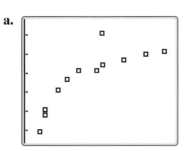

b. The data appears to approximate a log function. *Answer*

c. With the ages in L_1 and the circumferences in L_2, choose LnReg from the STAT CALC menu:

ENTER: [STAT] [▶] [9] [ENTER]

The equation of the regression equation, to the nearest thousandth, is

$$y = 35.938 + 1.627 \ln x \quad \textit{Answer}$$

Exercises

Writing About Mathematics

1. At birth, the average circumference of a child's head is 35 centimeters. If the pair $(0, 35)$ is added to the data in Example 2, the calculator returns an error message. Explain why.

2. Explain when the power function, $y = ax^b$, has only positive or only negative y-values and when it has both positive and negative y-values.

Developing Skills

In 3–8, determine the regression model that appears to be appropriate for the data.

3.

4.

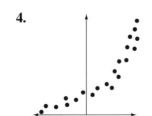

5.

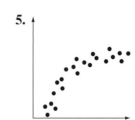

6. 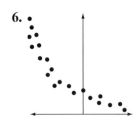 **7.** 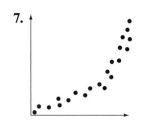 **8.**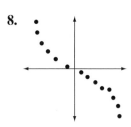

In 9–13: **a.** Create a scatter plot for the data. **b.** Determine which regression model is the most appropriate for the data. Justify your answer. **c.** Find the regression equation. Round the coefficient of the regression equation to three decimal places.

9.

x	4	7	3	8	6	5	6	3	9	4.5
y	10	7	15	9	5	6	6	14	14	8

10.

x	−2.3	1.8	−1.0	4.6	1.4	3.7	5.3	1.9	0.7	4.2
y	2.2	18.3	8.2	63.7	15.3	43.0	89.5	22.7	12.1	54.7

11.

x	3.3	−3.8	−2.1	0.4	3.5	−3.8	−1.8	−0.4	2.4	1.2
y	12.5	−17.1	−3.6	0.4	15.0	−18.9	−2.1	−0.4	4.2	3.4

12.

x	1	2	3	4	5	6	7	8	9	10
y	−7	−5.6	−4.8	−4.2	−3.8	−3.4	−3.1	−2.8	−2.6	−2.4

13.

x	1	2	3	4	5	6	7	8	9	10
y	2.7	2.3	2.0	1.7	1.5	1.2	1.1	0.9	0.8	0.7

Applying Skills

14. Mrs. Vroman bought $1,000 worth of shares in the Acme Growth Company. The table below shows the value of the investment over 10 years.

Year	1	2	3	4	5	6	7	8	9	10
Value ($)	1,045	1,092	1,141	1,192	1,246	1,302	1,361	1,422	1,486	1,553

a. Find the exponential regression equation for the data with the coefficient and base rounded to three decimal places.

b. Predict, to the nearest dollar, the value of the Vromans' investment after 11 years.

15. The growth chart below shows the average height in inches of a group of 100 children from 2 months to 36 months.

Month	2	4	6	8	10	12	14	16	18
Height in Inches	22.7	26.1	27.5	28.9	31.7	32.1	32.7	33.1	34.0
Month	20	22	24	26	28	30	32	34	36
Height in inches	34.4	34.6	34.9	35.2	35.6	36.0	36.6	37.2	37.6

a. Find the logarithmic regression equation for the data with the coefficients rounded to three decimal places.

b. Predict, to the nearest tenth of an inch, the average height of a child at 38 months.

16. The orbital speed in kilometers per second and the distance from the sun in millions of kilometers of each of six planets is given in the table.

Planet	Venus	Earth	Mars	Jupiter	Saturn	Uranus
Orbital Speed	34.8	29.6	23.9	12.9	9.6	6.6
Distance from the Sun	108.2	149.6	227.9	778.0	1,427	2,871

a. Find the regression equation that appears to be the best fit for the data with the coefficient rounded to three decimal places.

b. Neptune has an orbital speed of 5.45 km/sec and is 4,504 million kilometers from the sun. Does the equation found for the six planets given in the table fit the data for Neptune?

17. A mail order company has shipping boxes that have square bases and varying heights from 1 to 5 feet. The relationship between the height of the box and the volume is shown in the table.

Height (ft)	1	1.5	2	2.5	3	3.5	4	4.5	5
Volume (ft³)	2	7	16	31	54	86	128	182	250

a. Create a scatter plot for the data. Let the horizontal axis represent the height of the box and the vertical axis represent the volume.

b. Determine which regression model is most appropriate for the data. Justify your answer.

c. Find the regression equation. Round the coefficient of the regression equation to three decimal places.

18. In an office building the thermostats have six settings. The table below shows the average temperature in degrees Fahrenheit for a month that each setting produced.

Setting	1	2	3	4	5	6
Temperature (°F)	61	64	66	67	69	70

   a. Create a scatter plot for the data. Let the horizontal axis represent the setting and the vertical axis represent temperature.

   b. Find the equation of best fit using a power regression. Round the coefficient of the regression equation to three decimal places.

19. The following table shows the speed in megahertz of Intel computer chips over the course of 36 years. The time is given as the number of years since 1971.

Year	0	1	3	7	11	14	18	22
Speed	0.108	0.8	2	5	6	16	25	66
Year	24	26	28	29	31	34	35	36
Speed	200	300	500	1,500	1,700	3,200	2,900	3,000

   One application of *Moore's Law* is that the speed of a computer processor should double approximately every two years. Use this information to determine the regression model. Does Moore's Law hold for Intel computer chips? Explain.

Hands-On Activity: Sine Regression

If we make a scatter plot of the following set of data on a graphing calculator, we may observe that the data points appear to form a sine curve.

x	13	14	15	16	17	18	19	20	21	22	23	24
y	−11.6	−17.2	−18.0	−15.0	−8.4	6.2	12.0	20.1	18.4	11.9	3.4	−6.9

A sine function should be used to model the data. We can use a graphing calculator to find the **sinusoidal** regression equation. Enter the x-values into L_1 and the y-values into L_2. Then with the calculator in radian mode, choose SinReg from the STAT CALC menu:

ENTER: STAT ▶ ALPHA C VARS ▶

1 1 ENTER

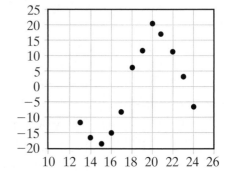

To the nearest thousandth, the sinusoidal regression equation is:

$$y = 19.251 \sin (0.551x + 2.871) - 0.029$$

Press [ZOOM] [9] to graph the scatter plot and the regression equation.

Note: The sinusoidal regression model on the graphing calculator assumes that the x-values are equally spaced and in increasing order. For arbitrary data, you need to give the calculator an estimate of the period. See your calculator manual for details.

The average high temperature of a city is recorded for 14 months. The table below shows this data.

Month	1	2	3	4	5	6	7	8	9	10	11	12	13	14
Temp (°F)	40	48	61	71	81	85	83	77	67	54	41	39	42	49

a. Create a scatter plot for the data.

b. Find the sinusoidal regression equation for the data with the coefficient and base rounded to three decimal places.

c. Predict the average temperature of the city at 15 months. Round to the nearest degree.

d. Predict the average temperature of the city at 16 months. Round to the nearest degree.

15-10 INTERPOLATION AND EXTRAPOLATION

Data are usually found for specific values of one of the variables. Often we wish to approximate values not included in the data.

Interpolation

The process of finding a function value between given values is called **interpolation**.

EXAMPLE 1

Each time Jen fills the tank of her car, she records the number of gallons of gas needed to fill the tank and the number of miles driven since the last time that she filled the tank. Her record is shown in the table.

Gallons of Gas	7.5	8.8	5.3	9.0	8.1	4.7	6.9	8.3
Miles	240	280	170	290	260	150	220	270

a. If Jen needs 8.0 gallons the next time she fills the tank, to the nearest mile, how many miles will she have driven?

b. If Jen has driven 200 miles, to the nearest tenth, how many gallons of gasoline can she expect to need?

Solution Graph the scatter plot of the data. There appears to be positive linear correlation.

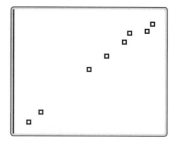

With the data in L_1 and L_2, use the calculator to find the regression equation:

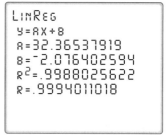

ENTER: [**STAT**] [**▶**] [**4**] [**ENTER**]

When rounded to the nearest thousandth, the linear equation returned is

$$y = 32.365x - 2.076$$

a. Substitute 8.0 for x in the equation given by the calculator.

$$y = 32.365x - 2.076$$
$$= 32.365(8.0) - 2.076$$
$$= 256.844$$

Jen will have driven approximately 257 miles. *Answer*

b. Substitute 200 for y in the equation given by the calculator.

$$y = 32.365x - 2.076$$
$$200 = 32.365x - 2.076$$
$$202.076 = 32.365x$$
$$6.244 \approx x$$

Jen will need approximately 6.2 gallons of gasoline. *Answer*

Extrapolation

Often we want to use data collected about past events to predict the future. The process of using pairs of values within a given range to approximate values outside of the given range of values is called **extrapolation**.

EXAMPLE 2

The following table shows the number of high school graduates in the U.S. in the thousands from 1992 to 2004.

Year	1992	1993	1994	1995	1996	1997	1998
No. of Graduates	2,478	2,481	2,464	2,519	2,518	2,612	2,704
Year	1999	2000	2001	2002	2003	2004	
No. of Graduates	2,759	2,833	2,848	2,906	3,016	3,081	

a. Write a linear regression equation for this data.

b. If the number of high school graduates continued to grow at this rate, how many graduates would there have been in 2006?

c. If the number of high school graduates continues to grow at this rate, when is the number of high school graduates expected to exceed 3.5 million?

Solution **a.** Enter the year using the number of years since 1990, that is, the difference between the year and 1990, in L_1. Enter the corresponding number of high school graduates in L_2. The regression equation is:

$$y = 53.984x + 2{,}277.286 \ \textit{Answer}$$

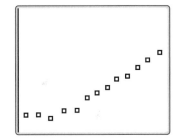

b. Use the equation $y = 53.984x + 2{,}277.286$ and let $x = 16$:

$$y = 53.984(16) + 2{,}277.286 \approx 3{,}141$$

If the increase continued at the same rate, the expected number of graduates in 2006 would have been approximately 3,141,000.

c. Use the equation $y = 53.984x + 2{,}277.286$ and let $y = 3{,}500$:

$$3{,}500 = 53.984x + 2{,}277.286$$
$$1{,}222.714 = 53.984x$$
$$22.650 \approx x$$

If the rate of increase continues, the number of high school graduates can be expected to exceed 3.5 million in the 23rd year after 1990 or in the year 2013.

Unlike interpolation, extrapolation is not usually accurate. Extrapolation is valid provided we are sure that the regression model continues to hold outside of the given range of values. Unfortunately, this is not usually the case. For instance, consider the data given in Exercise 15 of Section 15-9.

The growth chart below shows the average height in inches of a group of 100 children from 2 months to 36 months.

Month	2	4	6	8	10	12	14	16	18
Height in Inches	22.7	26.1	27.5	28.9	32.1	31.7	33.1	32.7	34.0
Month	20	22	24	26	28	30	32	34	36
Height in inches	34.4	34.6	36.0	34.6	35.2	36.6	35.6	37.2	37.6

The data appears logarithmic. When the coefficient and the exponent are rounded to three decimal places, an equation that best fits the data is $y = 19.165 + 5.026 \ln x$. If we use this equation to find the height of child who is 16 years old (192 months), the result is approximately 45.6 inches or less than 4 feet. The average 16-year-old is taller than this. The chart is intended to give average growth for very young children and extrapolation beyond the given range of ages leads to errors.

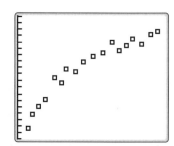

Exercises

Writing About Mathematics

1. Explain the difference between interpolation and extrapolation.

2. What are the possible sources of error when using extrapolation based on the line of best fit?

Developing Skills

In 3–5: **a.** Determine the appropriate linear regression model to use based on the scatter plot of the given data. **b.** Find an approximate value for y for the given value of x. **c.** Find an approximate value for x for the given value of y.

3. b. $x = 5.7$ **c.** $y = 1.25$

x	1	2	3	4	5	6	7	8	9	10
y	1.05	1.10	1.16	1.22	1.28	1.34	1.41	1.48	1.55	1.62

4. b. $x = 12$ **c.** $y = 140$

x	1	2	3	4	5	6	7	8	9	10
y	3.1	3.6	4.0	4.5	5.1	5.6	6.0	6.5	6.9	7.5

5. b. $x = 0.5$ **c.** $y = 0.5$

x	1	2	3	4	5	6	7	8	9	10
y	1	2.3	3.7	5.3	6.9	8.6	10.3	12.1	14.0	15.9

In 6–9: **a.** Determine the appropriate non-linear regression model to use based on the scatter plot of the given data. **b.** Find an approximate value for y for the given value of x. **c.** Find an approximate value for x for the given value of y.

6. b. $x = 1.4$ **c.** $y = 1.50$

x	1	2	3	4	5	6	7	8	9	10
y	0.80	1.09	1.25	1.37	1.46	1.54	1.60	1.66	1.71	1.75

7. a. $x = 12$ **b.** $y = 80$

x	23	26	13	14	20	17	29	18	18	17
y	11.6	33.3	52.5	43.5	4.0	18.7	84.0	8.0	12.4	11.5

8. a. $x = 10.5$ **b.** $y = 100.0$

x	-2.0	-1.0	-0.5	0.1	0.5	0.8	1.1	1.5	1.8	2.1
y	1.0	2.3	3.3	5.3	7.7	9.3	12.0	16.5	21.6	28.0

9. a. $x = 12$ **b.** $y = 215$

x	0.5	1.7	2.7	3.9	4.9	5.7	7.0	8.2	9.2	10.0
y	0.1	2.1	7.8	23.1	43.4	65.1	114.9	183.8	248.3	311.3

Applying Skills

In 10–12, determine the appropriate linear regression model to use based on the scatter plot of the given data.

10. The following table represents the percentage of the Gross Domestic Product (GDP) that a country spent on education.

Year	1960	1965	1970	1975	1980	1985	1990	1995	2000	2005
Percent	2.71	3.17	5.20	5.61	7.20	8.14	8.79	10.21	10.72	11.77

a. Estimate the percentage of the GDP spent on education in 1998.

b. Assuming this model continues to hold into the future, predict the percentage of the GDP that will be spent on education in 2015.

11. The following chart gives the average time in seconds that a group of 10 F1 racing cars went from zero to the given miles per hour.

Speed	75	100	125	150	175	200	225	250	275	300
Time	1.2	2.2	2.9	4.0	5.6	6.8	7.3	8.7	9.3	9.9

a. What was the average time it took the 10 racing cars to reach 180 miles per hour?

b. Estimate the average time it will take the 10 racing cars to reach 325 miles per hour.

12. The relationship between degrees Celsius and degrees Fahrenheit is shown in the table at intervals of 10° Fahrenheit.

Celsius	0	10	20	30	40	50	60	70	80	90	100
Fahrenheit	32	50	68	86	104	122	140	158	176	194	212

a. Find the Fahrenheit temperature when the Celsius temperature is 25°.

b. Find the Celsius temperature when the Fahrenheit temperature is −4°.

13. The following table gives the number of compact cars produced in a country over the course of several years.

Year	1981	1984	1987	1990	1993	1996	1999	2002	2005	2008
No. of Cars	100	168	471	603	124	1,780	1,768	4,195	6,680	10,910

a. Estimate the number of cars produced by the country in 2000 using an exponential model.

b. Estimate the number of cars produced by the country in 1978 using the model from part a.

14. In an office building the thermostats have six settings. The table below shows the average temperature in degrees Fahrenheit for a month that each setting produced.

Setting	1	2	3	4	5	6
Temperature (°F)	61	64	66	67	69	70

Using a power model and assuming that it is possible to choose a setting between the given settings:

a. what temperature would result from a setting halfway between 2 and 3?

b. where should the setting be placed to produce a temperature of 68 degrees?

In 15 and 16, determine the appropriate non-linear regression model to use based on the scatter plot of the given data.

15. A mail order company has shipping boxes that have square bases and varying height from 1 to 5 feet. The relationship between the height of the box and the volume in cubic feet is shown in the table.

Height	1	1.5	2	2.5	3	3.5	4	4.5	5
Volume	2	6.75	16	31.25	54	85.75	128	182.25	250

a. If the company introduces a box with a height of 1.25 feet, what would be the volume to the nearest hundredth cubic foot?

b. If the company needs a box with a volume of at least 100 cubic feet, what would be the smallest height to the nearest tenth of a foot?

c. If the company needs a box with a volume of 800 square feet, what would be the height to the nearest foot?

16. Steve kept a record of the height of a tree that he planted. The heights are shown in the table.

Age of Tree in Years	1	3	5	7	9	11	13
Height in Inches	7	12	15	16.5	17.8	19	20

a. Write an equation that best fits the data.

b. What was the height of the tree after 2 years?

c. If the height of the tree continues in this same pattern, how tall will the tree be after 20 years?

CHAPTER SUMMARY

Univariate
Statistics

Statistics is the science that deals with the collection, organization, summarization, and interpretation of related information called **data**. **Univariate statistics** consists of one number for each data value. Data can be collected by means of censuses, surveys, controlled experiments, and observational studies. The source of information in a statistical survey may be the population, all cases to which the study applies, or a sample, a representative subset of the entire group.

The *mean*, the *median*, and the *mode* are the most common measures of **central tendency**. The **mean** is the sum of all of the data values divided by the number of data values.

$$\text{Mean} = \frac{\sum\limits_{i=1}^{n} f_i x_i}{\sum\limits_{i=1}^{n} f_i}$$

When finding the mean of data grouped in intervals larger than 1, the median value of each interval is used to represent each entry in the interval.

The **median** is the middle number when the data are arranged in numerical order. In a set of $2n$ data values arranged in numerical order, the median is the average of values that are the n and $(n + 1)$ entries. In a set of $2n + 1$ data values arranged in numerical order, the median is the $n + 1$ entry.

The **mode** is the data value that occurs most frequently.

The **quartiles** separate the data into four equal parts. The **second quartile** is the median. When the data values are arranged in numerical order starting with the smallest value, the **first** or **lower quartile** is the middle value of the values that precede the median, and the **third** or **upper quartile** is the middle value of the values that follow the median.

An **outlier** is a data value that is more than 1.5 times the **interquartile range** above the third quartile or less than 1.5 times the interquartile range below the first quartile.

Measures of
Dispersion

- Range = the difference between the smallest and largest data values
- Interquartile range = $Q_3 - Q_1$

- Variance based on a population = $\dfrac{\sum\limits_{i=0}^{8} f_i (x_i - \bar{x})^2}{\sum\limits_{i=0}^{8} f_i}$

- Standard deviation based on a population = $\sigma = \sqrt{\dfrac{\sum\limits_{i=1}^{n} f_i (x_i - \bar{x})^2}{\sum\limits_{i=1}^{n} f_i}}$

- Standard deviation based on a sample $= s = \sqrt{\dfrac{\sum\limits_{i=1}^{n} f_i(x_i - \bar{x})^2}{\left(\sum\limits_{i=1}^{n} f_i\right) - 1}}$

The Normal
Distribution

A **normal curve** represents large sets of data that occur naturally, such as heights, weights, or test grades. The normal curve is bell shaped, symmetric with respect to a vertical line through its highest point, which is at the mean.

A **normal distribution** is a set of data that can be represented by a normal curve. For a normal distribution, the following relationships exist.

1. The mean and the median of the data values lie on the line of symmetry of the curve.

2. Approximately 68% of the data values lie within one standard deviation from the mean.

3. Approximately 95% of the data values lie within two standard deviations from the mean.

4. Approximately 99.7% of the data values lie within three standard deviations from the mean.

The **z-score** for a data value is the deviation from the mean divided by the standard deviation. Let x be a data value of a normal distribution.

$$\text{z-score} = \frac{x - \bar{x}}{\text{standard deviation}} = \frac{x - \bar{x}}{\sigma}$$

The z-score of x, a value from a normal distribution, is positive if x is above the mean and negative if x is below the mean. The z-score tells us how many standard deviations x is above or below the mean.

Bivariate Data

A **scatter plot** can help us to observe the relationship between elements of the pairs of **bivariate data**. The **correlation** between two variables is positive when the values of the second element of the pairs increase when the values of the first elements of the pairs increase. The correlation is negative when the values of the second element of the pairs decrease when the values of the first elements of the pairs increase.

The **correlation coefficient** is a number that indicates the strength and direction of the linear correlation between variables in a set of bivariate data. The correlation coefficient, r, is a number between -1 and 1. When the absolute value of the correlation coefficient is close to 1, the data has a strong linear correlation. When the absolute value of the correlation coefficient is close to 0, there is weak or no linear correlation.

Not all bivariate data can be represented by a linear function. Some data can be better approximated by a curve. By graphing data in a scatter plot, exponential, logarithmic, or power regressions may be identified. A calculator can also be used to find specific linear, exponential, logarithmic, or power functions that best represent the data.

The process of finding a function value between given values is called **interpolation**. The process of using pairs of values within a given range to approximate values outside of the given range of values is called **extrapolation**.

A function value can be found by substituting in the equation of the line of best fit. Extrapolation can sometimes lead to inaccurate results when the extreme values of the data do not fit the pattern of the rest of the data.

VOCABULARY

15-1 Statistics • Data • Univariate statistics • Census • Survey • Controlled experiment • Observational study • Population • Sample • Stem-and-leaf diagram • Stem • Leaf • Frequency distribution table • Histogram

15-2 Measure of central tendency • Mean • Arithmetic mean • Median • Mode • Quartiles • First quartile • Lower quartile • Second quartile • Third quartile • Upper quartile • Five statistical summary • Box-and-whisker plot

15-3 Cumulative frequency • Percentile

15-4 Measure of dispersion • Range • Interquartile range • Outlier

15-5 Variance • Standard deviation • Standard deviation for a population (σ) • Standard deviation for a sample (s)

15-6 Normal curve • Normal distribution • z-score

15-7 Bivariate statistics • Scatter plot • Correlation • Regression line • Line of best fit

15-8 Correlation coefficient (r)

15-9 Sinusoidal

15-10 Interpolation • Extrapolation

REVIEW EXERCISES

In 1–3, determine if the data to be collected is univariate or bivariate.

1. The ages of the 20 members of a book club

2. The median heights of boys for each year from age 12 to 18

3. The weight and weight-loss goals of the members of an exercise program

4. Name and describe four common ways of obtaining data for a statistical study.

5. In order to determine the average grade for all students who took a test given to all 9th-grade students in the state, a statistics student at a local college gathered the test grades for five randomly chosen students in the high school of the town where he lived.

    a. Do the data represent the population or a sample? Explain your answer.

    b. Can the person collecting the data expect that the data collected will reflect the grades of all students who took the test? Explain why or why not.

6. Sue's grades are 88, 87, 85, 82, 80, 80, 78, and 60.

    a. What is the range of her grades?

    b. What is the mean grade?

    c. What is the median grade?

    d. What are the upper and lower quartiles?

    e. What is the interquartile range?

    f. Is the grade of 60 an outlier? Justify your answer.

    g. Draw a box-and-whisker plot for Sue's grades.

7. The hours, x_i, that Peg worked for each of the last 15 weeks are shown in the table at the right. Show your work. In **a** through **h**, you are not allowed to use the $\boxed{\text{STAT}}$ menu of the calculator.

Hours x_i	Frequency f_i
42	1
40	2
39	2
38	5
37	3
36	2

    a. Find the mean.

    b. Find the median.

    c. Find the mode.

    d. What is the range?

    e. What are the first and third quartiles?

    f. What is the interquartile range?

    g. Find the variance.

    h. Find the standard deviation.

    i. Use the $\boxed{\text{STAT}}$ menu on a calculator and compare the values given with those found in **a, b, e,** and **h.**

8. Each time Kurt swims, he times a few random laps. His times, in seconds, are shown in the following table.

Time	79	80	81	82	83	84	85	86	87	88	89	90
Frequency	1	3	9	17	30	40	41	29	18	8	3	1

a. What percent of the data lie within 1 standard deviation from the mean?

b. What percent of the data lie within 2 standard deviations from the mean?

c. Does the data appear to represent a normal distribution? Justify your answer.

In 9–12, for each of the given scatter plots: **a.** Describe the correlation as strong positive linear correlation, moderate positive linear correlation, no linear correlation, moderate negative linear correlation, or strong negative linear correlation. **b.** Determine whether the correlation coefficient would be positive or negative.

9.

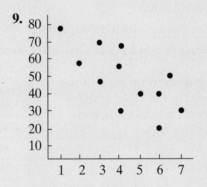

10.

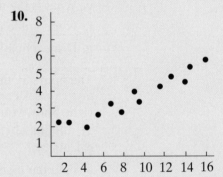

11.

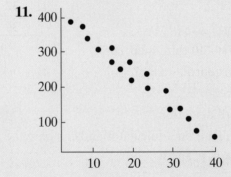

12.

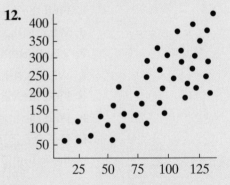

13. A school investment club chose nine stocks at the beginning of the year that it believed would represent a good investment portfolio. The chart shows the price of the stock at the beginning of the year and the gain (a positive number) or loss (a negative number) at the end of the year.

Stock Price	31.94	18.12	20.76	29.65	16.18	10.14	45.85	56.30	40.15
Gain or Loss	7.19	4.20	−3.57	−2.61	−3.32	−5.75	14.76	9.23	−3.93

a. Draw a scatter plot for the data.

b. Does there appear to be a linear correlation between the stock price and the gain or loss? If so, is the correlation strong or moderate?

In 14 and 15, find: **a.** the equation of the linear regression model that appears to be appropriate for the data. **b.** the value of the correlation coefficient.

14. The table lists the six states that had the largest percent of increase in population from 2006 to 2007. Population is given in millions, that is, 2.49 represents 2,490,000.

State	Nevada	Arizona	Utah	Idaho	Georgia	N. Carolina
Population 2006	2.49	6.17	2.58	1.46	9.34	8.87
Population 2007	2.57	6.34	2.65	1.50	9.54	9.06

15. The salaries that Aaron earned for four years are shown in the table. The salary is given to the nearest thousand dollars.

Year	1	2	3	4
Salary	52	54	61	63

In 16 and 17, determine the equation of the non-linear regression model that appears to be appropriate for the data.

16. Mrs. Brudek bakes and sells cookies. The table shows the number of dozens of cookies that she baked each week for the first seven weeks of this year.

Week	1	2	3	4	5	6	7
Dozens of Cookies	120	139	162	191	222	257	300

17. In a local park, an attempt is being made to control the deer population. The table shows the estimated number of deer in the park for the six years that the program has been in place.

Year	1	2	3	4	5	6
Number of Deer	700	525	425	350	300	250

In 18–21, use your answers to Exercises 14–17 to estimate the required values.

18. Use the information in Exercise 14 to find the approximate population of Texas in 2007 if the population in 2006 was 23.9 million.

19. Use the information in Exercise 15 to predict Aaron's salary in the 5th year.

20. If the number of dozens of cookies that Mrs. Brudek bakes continues to increase according to the pattern shown in Exercise 16, how many dozen cookies will she bake in week 8?

21. a. Use the information in Exercise 17 to find the estimated number of deer in the 8th year of the program if the number of deer continues to decrease according to the given pattern.

b. Using the information shown in Exercise 17, in what year will the number of deer be approximately 200?

Exploration

The graphing calculator fits a non-linear function to a set of data by transforming the data in such a way that the resulting data fall into a linear pattern. The calculator then fits a *linear* equation to the transformed data and applies an inverse function to find the equation of the non-linear function. In this Exploration, we will be using this procedure.

Look at the following set of data.

x	−2	−1	1	2	3	4	5	6	7	8
y	13	20	29	45	65	114	188	258	378	583
ln y	1.125	1.294	1.504	1.699	1.830	2.058	2.273	2.412	2.578	2.766

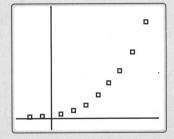 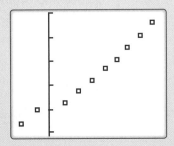

The scatter plot on the left reveals that the data is exponential. The scatter plot on the right is a graph of the ordered pairs (x, y') where $y' = \ln y$. Notice that this scatter plot indicates a clear linear pattern. The data is said to have been **linearized**. If we now find the *linear* regression equation of the ordered pairs (x, y'), the equation is

$$y' = 0.387x + 3.197$$

However, this equation is really

$$\ln y = 0.387x + 3.197$$

To find the exponential regression model, raise both sides to the power e.

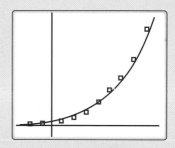

$$\ln y = 0.387x + 3.197$$
$$e^{\ln y} = e^{0.387x + 3.197}$$
$$y = e^{0.387x} \cdot e^{3.197}$$
$$y = (e^{0.387})^x (e^{3.197})$$
$$y = 24.459(1.473)^x$$

The scatter plot on the right shows the original data and the graph of the exponential regression equation.

For other types of non-linear data, we can follow a similar procedure. First we linearize the data using the appropriate substitution(s). Then we find the *linear* regression equation on the transformed data. Lastly, we undo the transformation to obtain the non-linear regression equation.

The substitutions to linearize power and logarithmic data are given below:

Power: $(x, y) \rightarrow (\ln x, \ln y)$ to obtain $\ln y = a \ln x + b$
Logarithmic: $(x, y) \rightarrow (\ln x, y)$ to obtain $y = a \ln x + b$

1. Find a power model using the techniques of this Exploration for the data given in Example 1 of Section 15-9.

2. Find a logarithmic model using the techniques of this Exploration for the data given in Example 2 of Section 15-9.

CUMULATIVE REVIEW   CHAPTERS 1–15

Part I

Answer all questions in this part. Each correct answer will receive 2 credits. No partial credit will be allowed.

1. Solve for x: $2 + \sqrt{x + 3} = 6$

  (1) 1   (2) 13   (3) 31   (4) 33

2. An angle measure of 240° is equivalent to
(1) π radians
(3) $\frac{4}{3}\pi$ radians
(2) $\frac{3}{4}\pi$ radians
(4) $\frac{3}{2}\pi$ radians

3. If f(x) = 3x + 2 and g(x) = x^2 − 2, then f(g(3)) is equal to
(1) 18
(2) 23
(3) 77
(4) 119

4. In the set of real numbers, what is the largest possible domain of the function f(x) = $\frac{2x^2 - 2}{x - 1}$?
(1) $\{x : x \neq 1\}$
(3) $\{x : x \neq 2\}$
(2) $\{x : x \neq 1 \text{ and } x \neq -1\}$
(4) $\{x : x \text{ is any real number}\}$

5. The expression $\frac{\sqrt{-36}}{-\sqrt{36}}$ is equal to
(1) 1
(2) −1
(3) i
(4) −i

6. If sin θ < 0 and tan θ = $-\frac{4}{5}$, in what quadrant does the terminal side of θ lie when θ is in standard position?
(1) I
(2) II
(3) III
(4) IV

7. For $a \neq 0, 1, -1$, the expression $\dfrac{\frac{a-1}{a}}{\frac{a^2-1}{a^2}}$ is equivalent to

(1) $\frac{a}{a+1}$
(2) $\frac{a+1}{a}$
(3) $\frac{a}{a-1}$
(4) $\frac{a-1}{a}$

8. The roots of the equation $x^2 + 7x - 8 = 0$ are
(1) real, rational, and equal
(3) real, irrational, and unequal
(2) real, rational, and unequal
(4) imaginary

9. The sum of the infinite geometric series when the first term is 8 and the common ratio is 0.2 is
(1) 0.1
(2) 1
(3) 10
(4) 40

10. The product $(-2 + 6i)(3 + 4i)$ is equal to
(1) $-6 + 24i$
(3) $18 + 10i$
(2) $-6 - 24i$
(4) $-30 + 10i$

Part II

Answer all questions in this part. Each correct answer will receive 2 credits. Clearly indicate the necessary steps, including appropriate formula substitutions, diagrams, graphs, charts, etc. For all questions in this part, a correct numerical answer with no work shown will receive only 1 credit.

11. What are the roots of $x^2 - 6x + 13 = 0$?

12. Sketch the graph of $y = 2 \cos \frac{1}{2}x$ in the interval $-2\pi \leq x \leq 2\pi$.

Part III

Answer all questions in this part. Each correct answer will receive 4 credits. Clearly indicate the necessary steps, including appropriate formula substitutions, diagrams, graphs, charts, etc. For all questions in this part, a correct numerical answer with no work shown will receive only 1 credit.

13. If $\log 2 = a$ and $\log 3 = b$, express $\log \frac{\sqrt[3]{6}}{9}$ in terms of a and b.

14. Solve for x: $27^{x+1} = 81^x$

Part IV

Answer all questions in this part. Each correct answer will receive 6 credits. Clearly indicate the necessary steps, including appropriate formula substitutions, diagrams, graphs, charts, etc. For all questions in this part, a correct numerical answer with no work shown will receive only 1 credit.

15. A tree service wants to estimate the height of a tree. At point A, the angle of elevation of the top of the tree is $50°$. From point B, 20 feet farther from the foot of the tree than point A, the angle of elevation of the top of the tree is $40°$. If points A and B lie on a line perpendicular to the tree trunk at its base, what is the height of the tree to the nearest foot?

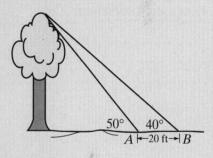

16. Solve the following system of equations graphically:

$$y = -x^2 + 2x$$
$$y = 2^x - 1$$

16

PROBABILITY AND THE BINOMIAL THEOREM

Medical facilities and pharmaceutical companies engage in research to develop new ways to cure or prevent disease and to ease pain. Often after years of experimentation a new drug or therapeutic technique will be developed. But before this new drug or technique can be accepted for general use, it is necessary to determine its effectiveness and the possible side effects by studying data from an experimental group of people who use the new discovery. The principles of statistics provide the guidelines for choosing such a group, for conducting the necessary study, and for evaluating the results.

This chapter will review some of the basic principles of probability and extend these principles to questions such as those above.

16-1 THE COUNTING PRINCIPLE

Tossing a coin is familiar to everyone. We say that when a fair coin is tossed, either side, heads or tails, is equally likely to appear face up. If a penny and a nickel are tossed, each coin can show heads, H, or tails, T. Each result is called an **outcome**. If we designate the outcomes on the penny as H and T and the outcomes on the nickel as H and T, then the set of possible outcomes for tossing the two coins contains four pairs:

$$\{(H, H), (H, T), (T, H), (T, T)\}$$

This set of all possible outcomes is the **sample space** for the activity.

If a nickel and a die are tossed, there are 2 ways in which the nickel can land and 6 ways in which the die can land. There are 2×6 or 12 possible outcomes. The sample space is shown as a graph of ordered pairs to the left and as a set of ordered pairs below:

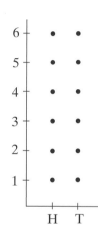

$$\{(T, 1), (T, 2), (T, 3), (T, 4), (T, 5), (T, 6), (H, 1), (H, 2),$$
$$(H, 3), (H, 4), (H, 5), (H, 6)\}$$

These examples illustrate the **Counting Principle**.

▶ *The Counting Principle*: **If one activity can occur in any of *m* ways and a second activity can occur in any of *n* ways, then both activities can occur in the order given in *m · n* ways.**

The counting principle can be applied to any number of activities. For example, consider a set of three cards lying facedown numbered 4, 6, and 8. Cards are drawn, one at a time, and not replaced. The numbers are used in the order in which they are drawn to form three-digit numbers. Note that since each draw comes from the same set of cards and the card or cards previously drawn are not replaced, the number of cards available decreases by 1 after each draw.

The table below represents the number of ways the cards may be drawn. However, it does *not* show us what those cards may be. We can draw a **tree diagram** to represent the outcomes.

Draw	Number of Cards	Number of Outcomes
1st	3	3
2nd	2	$3 \times 2 = 6$
3rd	1	$3 \times 2 \times 1 = 6$

$$8 \begin{cases} 6 \longrightarrow 4 \\ 4 \longrightarrow 6 \end{cases}$$

$$6 \begin{cases} 8 \longrightarrow 4 \\ 4 \longrightarrow 8 \end{cases}$$

$$4 \begin{cases} 8 \longrightarrow 6 \\ 6 \longrightarrow 8 \end{cases}$$

- There are 6 outcomes in the sample space: {468, 486, 684, 648, 846, 864}.
- The outcomes greater than 800 are the elements of the subset {846, 864}.
- The outcomes less than 650 are the elements of the subset {468, 486, 648}.
- The outcomes that are odd numbers are the elements of the subset { } or ∅, the empty set.
- The outcomes that are even numbers are the elements of the subset {468, 486, 684, 648, 846, 864}, that is, the sample space.

A subset of the sample space is an **event**. In the example above, cards were drawn without replacement. The outcome of each draw depended on the previous, so the draws are **dependent events**.

Consider a similar example. There are three cards lying facedown numbered 4, 6, and 8. Three cards are drawn, one at a time, and replaced after each draw. The numbers are used in the order in which they are drawn to form three-digit numbers. Note that since each card previously drawn is replaced, the number of cards available is the same for each draw.

Draw	Number of Cards	Number of Outcomes
1st	3	3
2nd	3	3 × 3 = 9
3rd	3	3 × 3 × 3 = 27

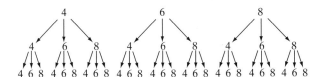

The number of possible outcomes when the drawn card is replaced and can be drawn again is significantly greater than the number of outcomes in the similar example without replacement. Since the outcome of each draw does *not* depend on the previous, we call the draws **independent events**.

EXAMPLE 1

In a debate club, there are 12 boys and 15 girls.

a. In how many ways can a boy and a girl be selected to represent the club at a meeting?

b. Are the selection of a boy and the selection of a girl dependent or independent events?

Solution **a.** Use the counting principle. There are 12 possible boys and 15 possible girls who can be selected. There are 12 × 15 or 180 possible pairs of a boy and a girl that can be selected.

b. The selection of a boy does not affect the selection of a girl, and vice versa. Therefore, these are independent events.

EXAMPLE 2

The skating club is holding its annual competition at which a gold, a silver, and a bronze medal are awarded.

a. In how many different ways can the medals be awarded if 12 skaters are participating in the competition?

b. Are the awarding of medals dependent or independent events?

Solution **a.** • The gold medal can be awarded to one of the 12 skaters.

• After the gold medal is awarded, the silver medal can be awarded to one of 11 skaters.

• After the silver medal is awarded, the bronze medal can be awarded to one of 10 skaters.

There are $12 \times 11 \times 10$ or 1,320 possible ways the medals can be awarded.

b. Since the silver medalist cannot have won the gold and the bronze medalist cannot have won either the gold or silver, the awarding of medals are dependent events.

Exercises

Writing About Mathematics

1. How is choosing a boy and a girl from 12 boys and 12 girls to represent a club different from choosing two girls from 12 girls to be president and treasurer of the club?

2. Compare the number of ordered pairs of cards that can be drawn from a deck of 52 cards without replacement to the number of ordered pairs of cards that can be drawn from a deck of 52 cards with replacement.

Developing Skills

In 3–6, for the given values of r and n, find the number of ordered selections of r objects from a collection of n objects without replacement.

3. $r = 4, n = 4$ **4.** $r = 4, n = 5$ **5.** $r = 3, n = 8$ **6.** $r = 2, n = 12$

In 7–10, for the given values of r and n, find the number of ordered selections of r objects from a collection of n objects with replacement.

7. $r = 4, n = 4$ **8.** $r = 4, n = 5$ **9.** $r = 2, n = 8$ **10.** $r = 3, n = 5$

11. For the sample space {A, B, C, D}, determine how many events are possible.

In 12–17, state whether the events are independent or dependent.

12. Tossing a coin and rolling a die

13. Selecting a winner and a runner-up from 7 contestants

14. Picking 2 cards from a standard deck without replacing the first card

15. Buying a magazine and a snack for a train trip

16. Choosing a size and a color for a sweater

17. Writing a 5-letter word using the alphabet if letters cannot be repeated

In 18–27, determine the number of possible outcomes.

18. Tossing a die 3 times

19. Tossing a coin 5 times

20. Choosing breakfast of juice, cereal, and fruit from 3 juices, 5 cereals, and 4 fruits

21. Assembling an outfit from 6 shirts, 4 pairs of pants, and 2 pairs of shoes

22. Choosing a 4-digit entry code using 0–9 if digits cannot be repeated

23. Choosing from 5 flavors of iced tea in 3 different sizes with or without sugar

24. Ordering an ice cream cone from a choice of 31 flavors, 3 types of cone, with or without one of 4 toppings

25. Making a 7-character license plate using the letters of the alphabet and the digits 1–4 if the first three characters must be non-repeating letters and the remaining four are digits that may repeat

26. Seating Andy, Brenda, Carlos, Dabeed, and Eileen in a row of 5 seats

27. Seating Andy, Brenda, Carlos, Dabeed, and Eileen in a row of 5 seats with Andy and Eileen occupying end seats

Applying Skills

28. In how many ways can 5 new bus passengers choose seats if there are 8 empty seats?

29. In how many ways can 3 different job openings be filled if there are 8 applicants?

30. In how many ways can 5 roles in the school play be filled if there are 9 possible people trying out? (Each role is to be played by a different person.)

31. From the set {2, 3, 4, 5, 6}, a number is drawn and replaced. Then a second number is drawn. How many two-digit numbers can be formed?

32. In a class of 22 students, the teacher calls on a student to give the answer to the first homework problem and then calls on a student to give the answer to the second homework problem.

    a. How many possible choices could the teacher have made if the same student was not called on twice?

    b. How many possible choices could the teacher have made if the same student may have been called on twice?

33. A manufacturer produces jeans in 9 sizes, 7 different shades of blue, and 6 different leg widths. If a branch store manager orders two pairs of each possible type, how many pairs of jeans will be in stock?

34. A restaurant offers a special in which a diner can have a choice of appetizer, entree, and dessert for $11.99.

Appetizer	Entree	Dessert
Mozzarella sticks Garden salad Tomato soup	Roast chicken Steamed fish Roast beef Vegetable curry	Chocolate pie Apple pie Cherry pie Blueberry pie Strawberry-rhubarb pie Lemon meringue pie

a. How many different meals can a diner order?

b. How many of these choices contain vegetable curry?

c. If a diner does not want to eat the soup or the fish, how many different meals can she order?

35. On a certain ski run, there are 8 places where a skier can choose to go to the left or to the right. In how many different ways can the skier cover the run?

36. Stan has letter tiles $\boxed{A}$ $\boxed{M}$ $\boxed{T}$ $\boxed{E}$.

a. How many different ways can Stan arrange all four tiles?

b. How many different arrangements begin with a vowel?

37. Six students will be seated in a row in the classroom.

a. How many different ways can they be seated?

b. If one student forgot his eyeglasses and must occupy the front seat, how many different seatings are possible?

38. To duplicate a key, a locksmith begins with a dummy key that has several sections. The locksmith grinds a specific pattern into each section.

a. A particular brand of house key includes 6 sections, and there are 4 possible patterns for each section. How many different house keys are possible?

b. A desk key has 3 sections, and 64 different keys are possible. How many patterns are available for each section if each section has the same number of possible patterns?

39. A physicist, a chemist, a biologist, an astronomer, a geologist, and a mathematician are guest speakers at a government-sponsored forum on scientific research in the twenty-first century. The speakers will be seated in a row on a raised platform at the front of the meeting room.

   a. How many different ways can the speakers be seated?

   b. The astronomer and the mathematician have co-written a paper that they will present. How many ways can the speakers be seated if the astronomer and the mathematician wish to sit side-by-side?

40. A given region has the telephone prefix 472.

   a. How many 7-digit telephone numbers are possible with this prefix?

   b. How many of these possibilities end with 4 unique digits?

   c. How many 7-digit telephone numbers with this prefix form an even number?

41. How many 2-letter patterns can be formed from the alphabet if the first letter is a vowel (A, E, I, O, U) and the second letter is a consonant that occurs later in the alphabet than the vowel in the first position?

42. How many 2-letter patterns may be formed from the alphabet if the first letter is a consonant and the second letter is a vowel that occurs earlier than the first letter?

16-2 PERMUTATIONS AND COMBINATIONS

Permutations

A **permutation** is an arrangement of objects in a specific order. For example, if four students are scheduled to give a report in class, then each possible order in which the students give their reports is a permutation. The number of possible permutations is shown in the following table.

	Number of Students	Number of Outcomes
1st	4	4
2nd	3	$4 \cdot 3 = 12$
3rd	2	$4 \cdot 3 \cdot 2 = 24$
4th	1	$4 \cdot 3 \cdot 2 \cdot 1 = 24$

We say that this is *the number of permutations of 4 things taken 4 at a time*, which can be symbolized as $_4P_4 = 4 \cdot 3 \cdot 2 \cdot 1 = 24$. There are 24 permutations; that is, there are 24 different orders in which the 4 students can give their reports.

We know that for any natural number n, n factorial is the product of the natural number from n to 1. Recall that the symbol for n factorial is $n!$:

$$n! = n(n - 1)(n - 2) \cdot \cdots \cdot 3 \cdot 2 \cdot 1$$

Therefore, the number of ways 4 students can give their reports is:

$$_4P_4 = 4 \cdot 3 \cdot 2 \cdot 1 = 4!$$

▶ **For any positive integer n, the number of arrangements of n objects taken n at a time can be represented as:**

$$_nP_n = n!$$

If there are 20 students in the class and any 4 of them may be asked to give a report today, then this permutation includes both a selection of 4 of the 20 students and the arrangement of those 4 in some order. In this case, there are 20 possible choices for the first report, 19 for the second, 18 for the third and 17 for the fourth.

	Number of Students	Number of Outcomes
1st	20	20
2nd	19	20 × 19
3rd	18	20 × 19 × 18
4th	17	20 × 19 × 18 × 17

The number of orders in which 4 of the 20 students can give a report is the number of permutations of 20 things taken 4 at a time. This permutation is symbolized as:

$$_{20}P_4 = 20 \times 19 \times 18 \times 17$$

Note that the last factor is $(20 - 4 + 1)$.

We can think of $20 \cdot 19 \cdot 18 \cdot 17$ as $20!$ with the last 16 factors canceled.

$$\frac{20!}{16!} = \frac{20 \times 19 \times 18 \times 17 \times 16 \times 15 \times \cdots \times 3 \times 2 \times 1}{16 \times 15 \times \cdots \times 3 \times 2 \times 1}$$

$$= 20 \times 19 \times 18 \times 17 = {_{20}P_4}$$

▶ **In general, for $r \le n$, the number of permutations of n things taken r at a time is:**

$$_nP_r = n(n - 1)(n - 2) \cdots (n - r + 1) = \frac{n!}{(n - r)!}$$

EXAMPLE 1

In how many different ways can the letters of the word PENCIL be arranged?

Solution   This is the number of permutations of 6 things taken 6 at a time.

$$_6P_6 = 6!$$

Use a calculator to evaluate 6!. Factorial is entry 4 of the PRB menu located under **MATH** .

ENTER: 6 **MATH** **◄** **4** **ENTER**

DISPLAY:

```
6!
          720
```

Alternative Solution   The value of $_6P_6$ can also be found directly by using the *nPr* function of the calculator. The *nPr* function is entry 2 of the PRB menu located under **MATH** . The value of *n* is entered before accessing this symbol.

ENTER: 6 **MATH** **◄** **2** 6 **ENTER**

DISPLAY:

```
6 nPr 6
          720
```

Answer   6! = 720 ways

EXAMPLE 2

In how many different ways can the letters of the word PENCIL be arranged if the first letter must be a consonant?

Solution   Since the first letter must be a consonant and there are 4 consonants, there are 4 possible choices for the first letter. Then, after the first letter has been chosen, there are 5 letters to be arranged.

$$4 \times {}_5P_5 = 4 \times 5! = 4 \times 5 \times 4 \times 3 \times 2 \times 1$$

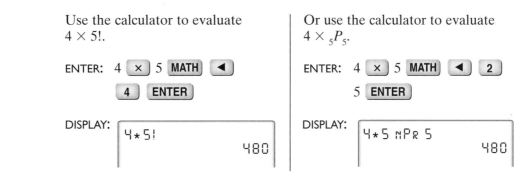

Use the calculator to evaluate $4 \times 5!$.

ENTER: 4 ⊠ 5 **MATH** ◄
4 **ENTER**

DISPLAY:

```
4*5!
              480
```

Or use the calculator to evaluate $4 \times {}_5P_5$.

ENTER: 4 ⊠ 5 **MATH** ◄ 2
5 **ENTER**

DISPLAY:

```
4*5 nPr 5
              480
```

Answer   480 ways

Permutations with Repetition

In the two examples given above, there are 6 different letters to be arranged. Would the problem have been different if the word had been ELEVEN? This is again a 6-letter word, but 3 of the letters are the same. Begin by thinking of the E's as different letters: E_1, E_2, and E_3. There are 480 arrangements of the letters. Consider all the arrangements in which the consonants are in given places. For example:

$$E_1LE_2VE_3N \qquad E_2LE_1VE_3N \qquad E_3LE_2VE_1N$$
$$E_1LE_3VE_2N \qquad E_2LE_3VE_1N \qquad E_3LE_1VE_2N$$

For every arrangement in which L, V, and N are fixed, there are 3! or 6 different arrangements of the E's. Divide the number of arrangements of all of the letters by the number of arrangements of the E's.

The number of arrangements of the 6 letters in which 3 of them are alike is:

$$\frac{6!}{3!} = \frac{6 \times 5 \times 4 \times 3 \times 2 \times 1}{3 \times 2 \times 1} = \frac{720}{6} = 120$$

We can think of this as placing just the 3 consonants in the 6 available positions. There are 6 possible positions for the first consonant, 5 possible positions for the second consonant, and 4 possible positions for the third consonant.

The number of arrangements of the 6 letters in which 3 of them are alike is:

$$6 \times 5 \times 4 = 120$$

▶ **In general, the number of permutations of *n* things taken *n* at a time when *a* are identical is:**

$$\frac{n!}{a!}$$

EXAMPLE 3

Helene is lining up beads to plan a necklace. She has a total of 36 beads and 32 of them are identical. How many different arrangements of the 36 beads can she make?

Solution   The number of arrangements of 36 beads with 32 identical beads is $\frac{36!}{32!}$.

Use the calculator to evaluate $\frac{36!}{32!}$.

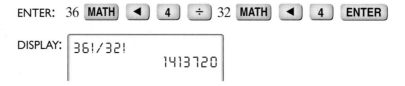

ENTER: 36 [MATH] [◄] [4] [÷] 32 [MATH] [◄] [4] [ENTER]

DISPLAY:
```
36!/32!
          1413720
```

Answer   1,413,720 arrangements

For some permutations of n things taken n at a time, more than one thing repeats. Let the number of times each of r things repeats be $a_1, a_2, \ldots, a_r$. Then we can say that the number of permutations of n things taken n at a time with multiple repetitions is:

$$\frac{n!}{a_1! \cdot a_2! \cdot \cdots \cdot a_r!}$$

EXAMPLE 4

A music teacher is arranging a recital for her students. There are 7 students, each with his or her own instrument: 3 play the piano, 2 play the violin, 1 plays the flute, and 1 plays the cello. In how many ways can the order of the instruments be arranged?

Solution   This is an arrangement of 7 instruments with 3 alike and 2 alike. Divide the number of arrangements of 7 instruments by the number arrangements of the 3 pianos and by the number arrangements of the 2 violins.

$$\frac{7!}{3! \times 2!} = \frac{7 \times 6 \times 5 \times 4 \times 3 \times 2 \times 1}{(3 \times 2 \times 1) \times (2 \times 1)} = 7 \times 6 \times 5 \times 2 = 420 \; \textit{Answer}$$

EXAMPLE 5

The number of arrangements of 5 letters from a given set of letters is 12 times the number of arrangements of 3 letters from the set. How many letters are in the set?

Solution Let x be the number of letters in the set.

$$_xP_5 = 12(_xP_3)$$

$$\frac{x!}{(x-5)!} = 12\left(\frac{x!}{(x-3!)}\right)$$

$$\frac{x!}{(x-5)!}\left[\frac{(x-3)!}{x!}\right] = 12$$

$$\frac{x!}{x!}\left[\frac{(x-3)!}{(x-5)!}\right] = 12$$

$$1\left[\frac{(x-3)(x-4)\cancel{(x-5)}\cancel{(x-6)}\cdots\cancel{(3)(2)(1)}}{\cancel{(x-5)}\cancel{(x-6)}\cdots\cancel{(3)(2)(1)}}\right] = 12$$

$$(x-3)(x-4) = 12$$

$$x^2 - 7x + 12 = 12$$

$$x^2 - 7x = 0$$

$$x(x-7) = 0$$

$$x = 0 \ \text{ or } \ x = 7$$

There cannot be 0 letters. Therefore, there are 7 letters in the set. *Answer*

Combinations

Five members of a club, Adam, Brian, Celia, Donna, and Elaine, have volunteered to represent the club at a competition, but the club can send only two members. The names of the volunteers are put in a box and two are drawn at random.

The first column of the table to the right shows all possible orders in which two names can be drawn. Each pair of names occurs in two different orders. The second column lists all possible pairs of names when order is not important.

Selection	Teams
(A, B), (B, A)	{A, B}
(A, C), (C, A)	{A, C}
(A, D), (D, A)	{A, D}
(A, E), (E, A)	{A, E}
(B, C), (C, B)	{B, C}
(B, D), (D, B)	{B, D}
(B, E), (E, B)	{B, E}
(C, D), (D, C)	{C, D}
(C, E), (E, C)	{C, E}
(D, E), (E, D)	{D, E}

Each ordered pair is a permutation. Each element of a set in which order is *not* important is a **combination**. The number of combinations of n things taken r at a time is symbolized $_nC_r$ or as $\binom{n}{r}$.

In order to find the number of combinations, we must first find the number of permutations and then divide that number by the number of rearrangements of the selected objects among themselves.

▶ In general, for $r \le n$, the number of combinations of n things taken r at a time is:

$$_nC_r = \frac{_nP_r}{_rP_r} \ \text{ or } \ \binom{n}{r} = \frac{_nP_r}{r!}$$

In the example above, we can say that the number of teams of two club members chosen from a group of five volunteers is:

$$_5C_2 = \frac{_5P_2}{_2P_2} = \frac{5 \times 4}{2 \times 1} = 10$$

 On a calculator, we can find the number of combinations by using keys similar to those we used for permutations. The symbol nCr is entry 3 of the PRB menu located under **MATH** . To evaluate $_5C_2$:

ENTER: 5 **MATH** **◄** **3** 2 **ENTER**

DISPLAY:

    5 nCr 2
                        10

The calculator displays the answer, 10.

EXAMPLE 6

How many different combinations of five letters can be selected from the alphabet?

Solution   This is a combination of 26 things taken 5 at a time.

$$_{26}C_5 = \frac{26 \times 25 \times 24 \times 23 \times 22}{5 \times 4 \times 3 \times 2 \times 1}$$

Use a calculator to evaluate.

ENTER: 26 **MATH** **◄** **3** 5 **ENTER**

DISPLAY:

    26 nCr 5
                        65780

Answer 65,780

EXAMPLE 7

How many different combinations of 5 letters can be drawn from the alphabet if 3 are consonants and 2 are vowels?

Solution   There are 21 consonants and 5 vowels in the alphabet.

First find the number of combinations of 3 out of the 21 consonants.

$$_{21}C_3 = \frac{21 \times 20 \times 19}{3 \times 2 \times 1} = 1,330$$

Then find the number of combinations of 2 out of 5 vowels.

$$_5C_2 = \frac{5 \times 4}{2 \times 1} = 10$$

Now use the counting principle: $1,330 \times 10 = 13,300$ *Answer*

Exercises

Writing About Mathematics

1. Show that $_nC_r = \frac{n!}{(n-r)! \times r!}$.

2. Show that $n! = n(n-1)!$.

Developing Skills

In 3–22, evaluate each expression.

3. $5!$

4. $12!$

5. $8! \div 3!$

6. $9! \div 8!$

7. $\frac{10!}{3!}$

8. $_6P_6$

9. $_8P_4$

10. $_6P_5$

11. $_5P_2$

12. $_4C_3$

13. $_{12}C_7$

14. $_{12}C_5$

15. $_{10}P_4 \div 4!$

16. $\binom{15}{5}$

17. $\binom{15}{10}$

18. $\binom{4}{4}$

19. $\binom{12}{0}$

20. $_5P_2 \times _3P_2$

21. $_8C_3 \div _8C_5$

22. $_{20}C_5 \div _{20}C_{15}$

In 23–30, find the number of different arrangements that are possible for the letters of each of the following words.

23. FACTOR

24. DIVIDE

25. EXCEED

26. ABSCISSA

27. RANDOMLY

28. REMEMBERED

29. STATISTICS

30. MATHEMATICS

31. A box contains 9 red, 4 blue, and 6 yellow chips. In how many ways can 6 chips be chosen if:

a. all 6 chips are red?

b. all 6 chips are yellow?

c. 2 chips are blue?

d. 3 chips are red?

e. 4 chips are yellow?

f. there are 2 chips of each color?

g. 3 chips are red and 3 chips are blue?

h. 5 chips are red and 1 chip is yellow?

In 32–37, determine the number of different arrangements.

32. The finishing order of 7 runners in a race

33. Seating of 5 students in a row of 9 chairs

34. Stacking 6 red, 4 yellow, and 2 blue t-shirts

35. The order in which a student answers 8 out of 10 test questions

36. Making a row of coins using 4 pennies, 3 nickels, and 3 dimes

37. Rating 6 employees in order of their friendliness

In 38–43, solve for x.

38. $_6P_2 = x(_5P_3)$

39. $_xP_6 = 30(_xP_4)$

40. $_{13}P_5 = 1{,}287(_xP_x)$

41. $_xC_6 = _xC_4$

42. $\binom{x}{8} = \binom{x}{7}$

43. $_{12}C_4 = _xC_8$

Applying Skills

44. Show that $_nC_r = _nC_{n-r}$.

45. At the library, Jordan selects 8 books that he would like to read but decides to check out just 5 of them. How many different selections can he make?

46. Eli has homework assignments for 5 subjects but decides to complete 4 of them today and complete the fifth before class tomorrow. In how many different orders can he choose 4 of the 5 assignments to complete today?

47. There are 12 boys and 13 girls assigned to a class. In how many ways can 2 boys and 3 girls be selected to transfer to a different class?

48. In how many ways can the letters of CIRCLE be arranged if the first and last must be consonants?

49. In how many ways can 6 transfer students be assigned to seats in a classroom that has 10 empty seats?

50. From a standard deck of 52 cards, how many hands of 5 cards can be dealt?

51. A local convenience store hires three students to work after school. Next month, there are 20 days on which they will work. Alex will work 8 days, Rosa will work 6 days, and Carla will work 6 days. In how many ways can their schedule for the month be arranged?

52. Marie has an English assignment and a history assignment that require research in the library. She makes a schedule for the 5 days of this week to work on the English assignment for 2 days and the history assignment for 3 days. How many different schedules are possible?

53. Roy and Valerie are playing a game of tic-tac-toe. In how many different ways can the first 4 moves in the game be made?

54. Rafael is running for mayor of his town. He sent out a survey asking his constituents to rank the following issues in order of importance to them: crime, unemployment, air quality, schools, public transportation, parking, and taxes.

   a. How many different rankings are possible?

   b. What fraction of the possible rankings put crime first?

   c. How many of the possible rankings are in alphabetical order?

55. The Electronic Depot received 108 mp3 players, 12 of which are defective. In how many different ways can 6 players be selected for display such that:

  a. none are defective?

  b. half are defective?

  c. all are defective?

56. The program director at the local sports center wants to schedule the following classes on Wednesdays: aerobics, tai chi, yoga, modern dance, fitness, and weight training. Starting at 9:00 A.M., each class begins on the hour and ends at 10 minutes before the hour.

  a. How many different ways can the classes be scheduled?

  b. If the yoga instructor is available only for a 9:00 A.M. class, how many different ways can the classes be scheduled?

57. A medical researcher has received approval to test a new combination drug therapy. A total of 6 doses of drug X, 3 doses of drug Y, and 4 doses of drug Z are to be given on successive days. The researcher wants to determine if some orders are more effective than others. How many different orders are possible?

16-3 PROBABILITY

Probability is an estimate of the frequency of an outcome in repeated trials. **Theoretical probability** is the ratio of the number of elements in an event to the number of equally likely elements in the sample space. For example, the sample space for tossing two coins, $\{(H, H), (H, T), (T, H), (T, T)\}$, has four equally likely elements and the event of landing two heads, $\{(H, H)\}$, has one element. The theoretical probability that two heads will show when two coins are tossed is $\frac{1}{4}$.

Let $P(E)$ = the probability of an event E

$n(E)$ = the number of element in event E

$n(S)$ = the number of elements in the sample space

$$P(E) = \frac{n(E)}{n(S)}$$

An event can be any subset of the sample space, including the empty set and the sample space itself: $0 \leq n(E) \leq n(S)$. Therefore, since $n(S)$ is a positive number,

$$\frac{0}{n(S)} \leq \frac{n(E)}{n(S)} \leq \frac{n(S)}{n(S)} \quad \text{or} \quad 0 \leq P(E) \leq 1$$

- $P(E) = 0$ when the event is the empty set, that is, there are no outcomes.
- $P(E) = 1$ when the event is the sample space, that is, certainty.
- $P(E) + P(\text{not } E) = 1$

For example, when a die is tossed, the probability of the die showing a number greater than 7 is 0 and the probability of the die showing a number less than 7 is 1. The probability of the die showing 2 is $\frac{1}{6}$ and the probability of a die showing a number that is not 2 is $\frac{5}{6}$.

$$P(2) + P(\text{not } 2) = \frac{1}{6} + \frac{5}{6} = \frac{6}{6} = 1$$

Some probabilities are arrived at by considering observed data or the results of an experiment or simulation. **Empirical probability** or **experimental probability** is the best estimate, based on a large number of trials, of the ratio of the number of successful results to the number trials. If we toss two coins four times, we cannot expect that two heads will occur exactly once. However, if this experiment, tossing two coins, is repeated thousands of times, the empirical probability, that is the ratio of the number of times two heads appear to the number of times the coins are tossed, will be approximately $\frac{1}{4}$.

Permutations, Combinations, and Probability

There are 8 boys and 12 girls in a chess club. Two boys and two girls are to be selected at random to represent the club at a tournament. What is the probability that Jacob and Emily will both be chosen? We can answer this question by using combinations.

There are $\frac{8 \times 7}{2 \times 1}$ or 28 possible combinations of 2 boys and $\frac{12 \times 11}{2 \times 1}$ or 66 possible combinations of 2 girls. Therefore there are 28×66 or 1,848 possible choices: $n(S) = 1,848$.

If Jacob is chosen, the pair of boys can be Jacob and one of the 7 other boys. Therefore, there are 7 pairs that include Jacob.

If Emily is chosen, the pair of girls can be Emily and one of the 11 other girls. Therefore, there are 11 pairs that include Emily.

- There are 7×11 or 77 choices that include Jacob and Emily: $n(E) = 77$.
- The probability that Jacob and Emily are chosen is $\frac{77}{1,848} = \frac{1}{24}$.

We can also determine this probability in another way.

- The probability that Jacob is chosen is $\frac{7}{28} = \frac{1}{4}$.
- The probability that Emily is chosen is $\frac{11}{66} = \frac{1}{6}$.
- The probability that both Jacob and Emily are chosen is $\frac{1}{4} \times \frac{1}{6} = \frac{1}{24}$.

There are 4 possible results of this choice: both Jacob and Emily are chosen, Jacob is chosen but not Emily, Emily is chosen but not Jacob, and neither Jacob nor Emily is chosen.

$$P(\text{Jacob is chosen}) = \tfrac{1}{4} \qquad P(\text{Jacob is not chosen}) = 1 - \tfrac{1}{4} = \tfrac{3}{4}$$

$$P(\text{Emily is chosen}) = \tfrac{1}{6} \qquad P(\text{Emily is not chosen}) = 1 - \tfrac{1}{6} = \tfrac{5}{6}$$

- $P(\text{Jacob and Emily are chosen}) = \tfrac{1}{4} \times \tfrac{1}{6} = \tfrac{1}{24}$
- $P(\text{Jacob is chosen but not Emily}) = \tfrac{1}{4} \times \tfrac{5}{6} = \tfrac{5}{24}$
- $P(\text{Emily is chosen but not Jacob}) = \tfrac{1}{6} \times \tfrac{3}{4} = \tfrac{3}{24}$
- $P(\text{neither Jacob nor Emily is chosen}) = \tfrac{3}{4} \times \tfrac{5}{6} = \tfrac{15}{24}$

The sum of all possible probabilities must be 1. To check:

$$\tfrac{1}{24} + \tfrac{5}{24} + \tfrac{3}{24} + \tfrac{15}{24} = \tfrac{24}{24} = 1 \ \checkmark$$

In the solution of this problem we have used the following relationship:

▶ **If $P(E) = p$, then, $P(\text{not } E) = 1 - p$.**

We have also used the probability form of the Counting Principle:

▶ **If E_1 and E_2 are independent events, and if $P(E_1) = p$ and $P(E_2) = q$, then the probability of both E_1 and E_2 occurring is $p \times q$.**

EXAMPLE I

The letters of the word CABIN are rearranged at random. What is the theoretical probability that one arrangement chosen at random will begin and end with a vowel?

Solution The sample space has 5! elements: $n(S) = 120$.

Since the first and last letters must be vowels, there are 2 possible letters for the first place and 1 possible letter for the last. There are 3! permutations of the remaining three letters: $n(E) = 2 \times 3 \times 2 \times 1 \times 1 = 12$.

$$P(E) = \frac{n(E)}{n(S)} = \frac{12}{120} = \frac{1}{10} \ \textit{Answer}$$

EXAMPLE 2

The student attendance record for one semester is shown on page 690. If two students are chosen at random, what is the probability that both students have fewer than two absences? Express the answer to the nearest hundredth.

											Total
Absences	10	9	8	7	6	5	4	3	2	1	0	
Frequency	2	1	4	8	5	15	12	18	26	37	12	$\sum\limits_{i=0}^{10} f_i = 140$

Solution   The frequencies tell us the number of students.

There are 140 students and 49 students have fewer than 2 absences.

$$n(S) = {}_{140}C_2 = \frac{140 \times 139}{2 \times 1} = 9{,}730$$

$$n(E) = {}_{49}C_2 = \frac{49 \times 48}{2 \times 1} = 1{,}176$$

The probability that both students have fewer than 2 absences $= \frac{1{,}176}{9{,}730} \approx .12$.

Answer   .12

Geometric Probabilities

There are geometric applications to probability.

For example, the graph shows the region on the coordinate axis that is enclosed by the rectangle whose vertices have the coordinates $(-a, 0)$, $(a, 0)$, (a, a), and $(-a, a)$, and the graph shows the inequality $y \geq |x|$. If a point within the rectangle is chosen at random, what is the probability that the coordinates of the point make the inequality $y \geq |x|$ true? To answer this question we must assume that regions of equal area contain the same number of points.

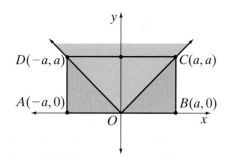

Let S, the sample space, be the set of points in the rectangle.

Let E, the event, be the set of points in the rectangle that are a point of $y \geq |x|$.

$$P(E) = \frac{n(E)}{n(S)} = \frac{\text{Area of } \triangle CDO}{\text{Area of rectangle } ABCD} = \frac{\frac{1}{2}a(2a)}{a(2a)} = \frac{1}{2}$$

EXAMPLE 3

A student wrote a computer program that will generate random pairs of numbers. The first element of each pair is between 0 and 4 and the second element is between 0 and 3. What is the theoretical probability that a pair of numbers generated by the program is in the interior of a circle whose equation is $(x - 1)^2 + (y - 1)^2 = 1$?

Solution
- Every pair of numbers generated by the computer program is in the interior of a rectangle whose length is 4 and whose width is 3. The area of this rectangle is 12.

- The center of the circle is the point $(1, 1)$ and the radius is 1.

- The area of this circle is $\pi(1)^2 = \pi$.

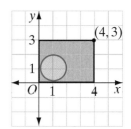

The interior of the circle is a subset of the interior of the rectangle as shown in the graph at the right.

$$P(\text{point is in the interior of the circle}) = \frac{\text{Area of the circle}}{\text{Area of the rectangle}} = \frac{\pi}{12} \quad \textit{Answer}$$

Exercises

Writing About Mathematics

1. A seed company tests 2,000 tomato seeds and obtains 1,954 plants. The seed company advertises that their seed is 97% productive.

a. Is the seed company's claim justified?

b. Is the seed company's claim based on theoretical or empirical probability?

2. Three students, Alex, Beth, and Casey, auditioned for the lead in the spring play. The director felt that the probability that Alex would be given the part was .35 and that the probability that Beth would be given the part was .25. Is it possible to determine the probability that Casey would be given the part? Justify your answer.

Developing Skills

3. What is the probability of getting a 2 on a single throw of a fair die?

4. What is the probability of getting a number greater than 2 on a single throw of a fair die?

5. What is the probability of getting a sum of 8 when a pair of dice are thrown?

6. The buyer of a lottery ticket chooses four numbers from the numbers 1 to 32. Repetition is not allowed.

a. How many combinations of four numbers are possible?

b. What is the probability of choosing all four of the winning numbers?

7. A standard deck of cards contains 52 cards divided into 4 suits. There are two red suits: hearts and diamonds, and two black suits: clubs and spades. Each suit contains 13 cards; ace, king, queen, jack, and cards numbered 2 through 10. A card is drawn from a standard deck without replacement. What is the probability that the card is a king?

8. Two cards are drawn from a standard deck of 52 cards without replacement. What is the probability that both cards are kings?

9. Three cards are drawn from a standard deck of 52 cards without replacement. What is the probability that all three cards are kings?

10. The letters of the word SEED are arranged at random. What is the probability that the arrangement begins and ends with E?

11. The letters of the word TOMATO are arranged at random. What is the probability that the arrangement begins and ends with T?

12. The letters of the word MATHEMATICS are arranged at random. What is the probability that the arrangement begins and ends with M?

13. The letters of the word STRAWBERRIES are written on cards and the cards are then shuffled. A card is picked at random. Find the probability that the card contains:

a. the letter S   **b.** the letter R

c. a consonant   **d.** a letter in the word CREAM

14. A piggy bank contains 60 pennies, 25 nickels, 10 dimes, and 5 quarters. If it is equally likely that any one of the coins will fall out when the bank is turned upside down, what is the probability that the coin is:

a. a penny?   **b.** a dime?

c. not a quarter?   **d.** a nickel or a dime?

Applying Skills

15. Three construction companies bid on a renovation project for a supermarket. The owner of Well-Built estimates that the competing companies have probabilities of .2 and .4, respectively, of getting the job. What is the probability that Well-Built will get the job?

16. A coin that was tossed 1,000 times came up heads 526 times. Determine the experimental probabilities for heads and tails.

17. A die was rolled 1,200 times and a 5 came up 429 times.

a. Find the experimental probability for rolling a 5.

b. Based on a comparison of the experimental and theoretical probabilities, do you think the die is fair? Explain your answer.

18. The weather report gives the probability of rain on Saturday as 40% and the probability of rain on Sunday as 60%. What is the probability that it will rain on Saturday but not rain on Sunday?

19. A variety box of instant oatmeal contains 10 plain, 6 maple, and 4 apple-cinnamon flavored packets. Ernestine reaches in and takes 3 packets without looking. Find each probability:

    a. P(2 plain)   **b.** P(1 maple, 1 apple-cinnamon)

    c. P(2 plain, 1 maple)   **d.** P(1 of each flavor)

20. The quality control department of a company that produces flashbulbs finds that 1 out of 1,000 bulbs tested fails to function properly. The flashbulbs are sold in packages of four. What is the probability that all the bulbs in a package will function properly?

21. There are 3 seniors and 15 juniors in Mrs. Gillis's math class. Three students are chosen at random from the class.

    a. What is the probability that the group consists of a senior and two juniors?

    b. If the group consists of a senior and two juniors, what is the probability that Stephanie, a senior, and Jan, a junior, are chosen?

22. Of the 18 students in Mrs. Shusda's math class, 12 take chemistry. If three students are absent from the class today, what is the probability that none of them take chemistry?

23. A play area for children has activities that are designed for taller and others for shorter children. The heights in inches of children at this play area last Saturday are given in the table.

Height	40	41	42	43	44	45	46	47	48	49	50
Frequency	1	1	2	5	9	14	8	6	3	0	1

    a. For one of the activities, a child must be at least 42 inches tall. What is the probability that a group of four children who want participate in this activity are all tall enough?

    b. For another activity, the child's height must be less than 45 inches. What is the probability that a group of five children who want to participate in this activity are all short enough?

24. At a carnival, Chrystal is managing a game in which a dart is thrown at a square board with a bull's-eye in the center. The board measures 3 feet by 3 feet and the bull's-eye has a 1-inch radius. Players who hit the bull's-eye receive a prize.

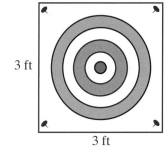

3 ft

3 ft

    a. Assume that each player is unskilled and each throw is random but always lands within the square. What is the theoretical probability that a player will hit the bull's-eye?

    b. At the end of the day, Chrystal finds that she gave out 72 prizes for 1,270 throws. What is the experimental probability that a player hit the bull's-eye?

    c. Are the theoretical probability and the empirical probability the same? If not, explain why they are different.

25. An irregular figure is drawn on a graph within a square that measures 6 inches on each side. The theoretical probability that the coordinates of a point that lies within the square also lies within the irregular figure is $\frac{4}{9}$. What is the area of the irregular figure?

26. A group of friends are playing a game in which each player chooses a number from 1 to 6 and takes a turn tossing a die until the chosen number appears on the die. The required number of tosses is the person's score for that round of play.

a. What is the probability that a player gets a score of 3 on the first round?

b. What is the probability that a player gets a score of 5 on the first round?

c. What is the probability of a score of n?

d. Show that the probabilities of scores that are consecutive integers from 1 to 5 form a geometric sequence.

27. A carton of 24 eggs contains 4 eggs with double yolks. If 3 eggs are selected at random, determine each probability to the nearest hundredth:

a. All 3 eggs will have single yolks.

b. 2 eggs will have single yolks and 1 egg will have a double yolk.

28. Inheritance of different genes can produce 6 different types of coats in cattle: straight red (SSRR or SsRR), curly red (ssRR), straight white (SSrr or Ssrr), curly white (ssrr), straight roan (SSRr or SsRr), or curly roan (ssRr).

Straight Roan Parent

		SR	Sr	sR	sr
Curly Roan Parent	**sR**	SsRR	SsRr	ssRR	ssRr
	sr	SsRr	Ssrr	ssRr	ssrr
	sR	SsRR	SsRr	ssRR	ssRr
	sr	SsRr	Ssrr	ssRr	ssrr

Use the Punnett square given above to determine the probability that one parent with a straight roan coat (SsRr) and one with a curly roan coat (ssRr) will produce an offspring with:

a. a straight roan coat   **b.** a curly roan coat

16-4 PROBABILITY WITH TWO OUTCOMES

In previous sections, we found the probability of a particular result on all trials. For example, if the probability that a team will win a game is $\frac{3}{4}$, then the probability of losing a game is $1 - \frac{3}{4}$ or $\frac{1}{4}$.

- $P(\text{win all of the first five games}) = \frac{3}{4} \times \frac{3}{4} \times \frac{3}{4} \times \frac{3}{4} \times \frac{3}{4} = \frac{243}{1,024}$

- $P(\text{lose all of the first five games}) = \frac{1}{4} \times \frac{1}{4} \times \frac{1}{4} \times \frac{1}{4} \times \frac{1}{4} = \frac{1}{1,024}$

What is the probability that the team will win some games and lose other games? We will examine all possibilities.

CASE 1 *Win four out of five games*

There are five ways this can happen:

$\{W, W, W, W, L\}$ $\quad P(\text{win all but the fifth}) = \frac{3}{4} \times \frac{3}{4} \times \frac{3}{4} \times \frac{3}{4} \times \frac{1}{4} = \frac{81}{1,024}$

$\{W, W, W, L, W\}$ $\quad P(\text{win all but the fourth}) = \frac{3}{4} \times \frac{3}{4} \times \frac{3}{4} \times \frac{1}{4} \times \frac{3}{4} = \frac{81}{1,024}$

$\{W, W, L, W, W\}$ $\quad P(\text{win all but the third}) = \frac{3}{4} \times \frac{3}{4} \times \frac{1}{4} \times \frac{3}{4} \times \frac{3}{4} = \frac{81}{1,024}$

$\{W, L, W, W, W\}$ $\quad P(\text{win all but the second}) = \frac{3}{4} \times \frac{1}{4} \times \frac{3}{4} \times \frac{3}{4} \times \frac{3}{4} = \frac{81}{1,024}$

$\{L, W, W, W, W\}$ $\quad P(\text{win all but the first}) = \frac{1}{4} \times \frac{3}{4} \times \frac{3}{4} \times \frac{3}{4} \times \frac{3}{4} = \frac{81}{1,024}$

The number of ways that the team can win four out of five games is $_5C_4 = \frac{5!}{4!} = 5$.

$$P(\text{win 4 out of 5 games}) = {_5C_4}\left(\frac{3}{4}\right)^4\left(\frac{1}{4}\right)^1 = 5\left(\frac{81}{256}\right)\left(\frac{1}{4}\right) = \frac{405}{1,024}$$

We can use the same reasoning for each of the other possible combinations of wins and losses.

CASE 2 *Win three out of five games*

The team can win the first three games and lose the fourth and fifth:

$$P(\{W, W, W, L, L\}) = \frac{3}{4} \times \frac{3}{4} \times \frac{3}{4} \times \frac{1}{4} \times \frac{1}{4} = \left(\frac{3}{4}\right)^3\left(\frac{1}{4}\right)^2 = \frac{27}{1,024}$$

There are other orders in which the team can win three and lose two. The number of combinations of three wins out of five games is $_5C_3 = \frac{5 \times 4 \times 3}{3 \times 2 \times 1} = 10$. For each of these 10 possibilities, the probability is $\frac{27}{1,024}$.

$$P(\text{3 wins out of 5 games}) = {_5C_3}\left(\frac{3}{4}\right)^3\left(\frac{1}{4}\right)^2 = 10\left(\frac{27}{64}\right)\left(\frac{1}{16}\right) = \frac{270}{1,024}$$

CASE 3 *Win two out of five games*

The team can win the first two games and lose the third, fourth and fifth:

$$P(\{W, W, L, L, L\}) = \tfrac{3}{4} \times \tfrac{3}{4} \times \tfrac{1}{4} \times \tfrac{1}{4} \times \tfrac{1}{4} = \left(\tfrac{3}{4}\right)^2\left(\tfrac{1}{4}\right)^3 = \tfrac{9}{1,024}$$

There are other orders in which the team can win two and lose three. The number of combinations of two wins out of five games is $_5C_2 = \tfrac{5 \times 4}{2 \times 1} = 10$. For each of these 10 possibilities, the probability is $\tfrac{9}{1,024}$.

$$P(2 \text{ wins out of 5 games}) = {_5}C_2\left(\tfrac{3}{4}\right)^2\left(\tfrac{1}{4}\right)^3 = 10\left(\tfrac{9}{16}\right)\left(\tfrac{1}{64}\right) = \tfrac{90}{1,024}$$

CASE 4 *Win one out of five games*

The team can win the first game and lose the last four:

$$P(\{W, L, L, L, L\}) = \tfrac{3}{4} \times \tfrac{1}{4} \times \tfrac{1}{4} \times \tfrac{1}{4} \times \tfrac{1}{4} = \left(\tfrac{3}{4}\right)^1\left(\tfrac{1}{4}\right)^4 = \tfrac{3}{1,024}$$

There are other orders in which the team can win one and lose four. The number of combinations of one win out of five games is $_5C_1 = \tfrac{5}{1} = 5$. For each of these 5 possibilities, the probability is $\tfrac{3}{1,024}$.

$$P(1 \text{ win out of 5 games}) = {_5}C_1\left(\tfrac{3}{4}\right)^1\left(\tfrac{1}{4}\right)^4 = 5\left(\tfrac{3}{4}\right)\left(\tfrac{1}{256}\right) = \tfrac{15}{1,024}$$

We have found the probability that the team will win 5, 4, 3, 2, 1, or 0 games. The sum of the probabilities must be 1. *Check:*

$$P(5 \text{ wins}) + P(4 \text{ wins}) + P(3 \text{ wins}) + P(2 \text{ wins}) + P(1 \text{ win}) + P(0 \text{ wins}) = 1$$

$$\tfrac{243}{1,024} + \tfrac{405}{1,024} + \tfrac{270}{1,024} + \tfrac{90}{1,024} + \tfrac{15}{1,024} + \tfrac{1}{1,024} = \tfrac{1,024}{1,024} = 1 \; ✔$$

The example given above illustrates the solution to finding exactly r successes in n independent trials in a **Bernoulli experiment** or **binomial experiment**. In a Bernoulli experiment, the n trials are independent and each trial has only two outcomes: success and failure. For instance, if a die is tossed, and success is showing 1, then failure is not showing 1. If a card is drawn from a deck of 52 cards, the trials are independent only if the card drawn is replaced after each draw. If success is drawing a king, then failure is not drawing a king.

In general, for a Bernoulli experiment, if the probability of success is p and the probability of failure is $1 - p = q$, then the probability of r successes in n independent trials is:

$$_nC_r p^r q^{n-r}$$

We refer to this formula as the **binomial probability formula**.

EXAMPLE 1

A waiter knows from experience that 7 out of 10 people who dine alone will leave a tip. Tuesday evening, the waiter served 12 lone diners.

a. Is this a Bernoulli experiment?

b. What is the probability, to the nearest thousandth, that the waiter received a tip from 9 of these diners?

Solution **a.** There are two outcomes to this experiment: the diner may leave a tip or the diner may not leave a tip. For each diner, leaving a tip or not leaving a tip is an independent outcome. Therefore, this is a Bernoulli experiment. *Answer*

b. If 7 out of 10 lone diners leave a tip, the probability that a lone diner will leave a tip is $\frac{7}{10}$ or .7 and the probability that a lone diner will not leave a tip is $1 - .7 = .3$.

$$P(9 \text{ tips from 12 diners}) = {}_{12}C_9 \, (.7)^9 \, (.3)^3 \approx .240 \; Answer$$

EXAMPLE 2

What is the probability that 2 shows on three of the dice when four dice are tossed?

Solution These are independent trials with $n = 4$ and $r = 3$. The number shown on any die is independent of that shown on any other.

$$P(2) = p = \tfrac{1}{6}$$
$$P(\text{not } 2) = q = 1 - \tfrac{1}{6} = \tfrac{5}{6}$$
$$P(2 \text{ exactly three out of four times}) = {}_4C_3 \, p^3 q^1$$
$$= \tfrac{4 \times 3 \times 2}{3 \times 2 \times 1}\left(\tfrac{1}{6}\right)^3\left(\tfrac{5}{6}\right)^1$$
$$= 4\left(\tfrac{1}{216}\right)\left(\tfrac{5}{6}\right) = \tfrac{20}{1,296} \; Answer$$

EXAMPLE 3

A company that makes breakfast cereal puts a coupon for a free box of cereal in 3 out of every 20 boxes. What is the probability that Mrs. Sullivan will find 2 coupons in the next 5 boxes of cereal that she buys?

Solution $P(\text{box contains a coupon}) = \tfrac{3}{20}$

$$P(\text{box does not contain a coupon}) = 1 - \tfrac{3}{20} = \tfrac{17}{20}$$
$$P(2 \text{ coupons in 5 boxes}) = {}_5C_2 \, p^2 q^3$$
$$= \tfrac{5 \times 4}{2 \times 1}\left(\tfrac{3}{20}\right)^2\left(\tfrac{17}{20}\right)^3$$
$$= 10\left(\tfrac{9}{400}\right)\left(\tfrac{4,913}{8,000}\right) = \tfrac{442,170}{3,200,000} \; Answer$$

Binomial Probabilities and the Graphing Calculator

We can use the calculator to find a binomial probability with the binompdf(function of the DISTR menu. To use the binompdf(function, we must supply the number of trials n, the probability of success p, and the number of successes r:

$$\text{binompdf}(n, p, r)$$

For example, to find the probability of winning 2 out of 3 games $\left(\text{where the probability of winning is } \frac{3}{4}\right)$, enter:

ENTER: **2nd**   **DISTR**

Scroll to binompdf(and press ENTER.

ENTER: 3 **,** 3 **÷** 4 **,** 2 **)** **ENTER**

DISPLAY:

```
binompdf(3,3/4,2
)
              .421875
```

The probability of winning 2 out of 3 games is approximately .4219.

To calculate the probabilities of winning 0 to 3 games, use the cumulative probability function, binomcdf(. To use the binomcdf(function, we must supply the number of trials n, the probability of success p, and the number of successes r:

$$\text{binomcdf}(n, p, r)$$

If r is not supplied, a list of probabilities for 0 to n trials is returned.

ENTER: **2nd**   **DISTR**

Scroll to binomcdf(and press ENTER.

ENTER: 3 **,** 3 **÷** 4 **)** **ENTER**

DISPLAY:

```
binomcdf(3,3/4)
 {.015625  .15625...
```

The calculator returns the set {.015625, .15625, .578125, 1}.

$$P(0 \text{ games out of } 3) \approx .015625 \qquad P(0 \text{ or } 1 \text{ games out of } 3) \approx .15625$$
$$P(0, 1, \text{ or } 2 \text{ games out of } 3) \approx .578125 \qquad P(0, 1, 2, \text{ or } 3 \text{ games out of } 3) \approx 1$$

We can use these cumulative probabilities to find the individual probabilities.

$$P(0 \text{ wins}) \approx .015625$$
$$P(1 \text{ win}) \approx .15625 - .015625 = .140625$$
$$P(2 \text{ wins}) \approx .578125 - .15625 = .421875$$
$$P(3 \text{ wins}) \approx 1 - .578125 = .421875$$

Exercises

Writing About Mathematics

1. There are 20 students in a club, 12 boys and 8 girls. If five members of the club are chosen at random to represent the club at a competition, what is the probability that in the group chosen there are exactly 2 boys? Explain why this is not a Bernoulli experiment.

2. Hunter said that the number of combinations of n things taken r at a time is equal to the number of permutations of n things taken n at a time when r are identical; that is, $_nC_r = \frac{n!}{r!}$. Do you agree with Hunter? Explain why or why not.

Developing Skills

In 3–6, find exact probabilities showing all required computation.

3. A fair coin is tossed five times. What is the probability that the coin lands heads:

  a. exactly once?   **b.** exactly twice?   **c.** exactly three times?

  d. exactly four times?   **e.** exactly five times?   **f.** zero times?

  g. Which is the most likely event(s) when tossing a coin five times?

4. A fair die is tossed five times. What is the probability of tossing a 6:

  a. exactly once?   **b.** exactly twice?   **c.** exactly three times?

  d. exactly four times?   **e.** exactly five times?   **f.** zero times?

  g. Which is the most likely event(s) when tossing a die five times?

5. A card is drawn and replaced four times from a standard deck of 52 cards. What is the probability of drawing a king: **a.** exactly once?   **b.** exactly twice?

  c. exactly three times?   **d.** exactly four times?   **e.** zero times?

  f. Which is the most likely event(s) when drawing a card four times with replacement?

6. A multiple-choice test of 10 questions has 4 choices for each question. Only one choice is correct. A student who did not study for the test guesses at each answer.

  a. What is the probability that that student will have exactly 5 correct answers?

  b. What is the probability that that student will have only 1 correct answer?

Applying Skills

In 7–14, answers can be rounded to four decimal places.

7. The probability that our team will win a basketball game is $\frac{2}{3}$. What is the probability that they will win exactly 5 of the next 7 games?

8. Jack's batting average is .2. What is the probability that he will get one hit in the next three times at bat?

9. Marie plays solitaire on her computer and keeps a record of her wins and losses. Her record shows that she won 950 of the last 1,000 games that she played. What is the probability that she will win three of the next four games that she plays?

10. A fast-food restaurant gives coupons for 10% off of the next purchase with 1 out of every 5 purchases.

  a. What is the probability that Zoe will receive a coupon with her next purchase?

  b. What is the probability that Zoe will receive just one coupon with her next three purchases?

11. A store estimates that 1 out of every 25 customers is returning or exchanging merchandise. In the last hour, a cashier had 5 customers. What is the probability that exactly 2 of those were making returns or exchanges?

12. Assume that there are an equal number of births in each month so that the probability is $\frac{1}{12}$ that a person chosen at random was born in May. A group of 20 friends meet monthly and celebrate the birthdays for that month. What is the probability that at the May celebration, exactly two members of the group have May birthdays?

13. Using a special type of fishing hook, researchers discovered that the hooking mortality rate was approximately 8%. That is, of the fish caught and released with this hook, about 8% eventually died. A fisherman caught and released 5 fish using this hook. What is the probability that all 5 fish will live?

14. Statisticians investigating the Internet-surfing habits of students at a large high school discovered that only 35% are interested in protecting their privacy online. What is the probability that none of four randomly chosen students are interested in protecting their privacy?

16-5 BINOMIAL PROBABILITY AND THE NORMAL CURVE

At Least and At Most

When we are anticipating success or failure in future events, we are often interested in success in at least r of the next n trials or failure in at most r of the next n trials.

What is the probability of getting at least 8 heads when a coin is tossed 10 times? We will get at least 8 heads if we land 8, 9, or 10 heads.

$$P(\text{at least 8 heads}) = P(\text{8 heads}) + P(\text{9 heads}) + P(\text{10 heads})$$
$$= {}_{10}C_8\left(\tfrac{1}{2}\right)^8\left(\tfrac{1}{2}\right)^2 + {}_{10}C_9\left(\tfrac{1}{2}\right)^9\left(\tfrac{1}{2}\right)^1 + {}_{10}C_{10}\left(\tfrac{1}{2}\right)^{10}\left(\tfrac{1}{2}\right)^0$$
$$= 45\left(\tfrac{1}{1,024}\right) + 10\left(\tfrac{1}{1,024}\right) + 1\left(\tfrac{1}{1,024}\right)$$
$$= \tfrac{56}{1,024} = .0546875$$

In this example, we are asked to compute the probability of landing at least 8 heads or at least 8 successes. The table below shows common English expressions that are equivalent to "at least 8 successes."

Expression	Binomial Probability
At least 8 successes 8 or more successes No fewer than 8 successes	$r \geq 8,\ P\left(\genfrac{}{}{0pt}{}{\text{at least}}{\text{8 successes}}\right) = P(\text{8 successes}) + P(\text{9 successes}) + \cdots + P(n \text{ successes})$
At most 8 successes 8 or fewer successes No more than 8 successes	$r \leq 8,\ P\left(\genfrac{}{}{0pt}{}{\text{at most}}{\text{8 successes}}\right) = P(\text{0 successes}) + P(\text{1 success}) + \cdots + P(\text{8 successes})$
More than 8 successes Exceeding 8 successes	$r > 8,\ P\left(\genfrac{}{}{0pt}{}{\text{more than}}{\text{8 successes}}\right) = P(\text{9 successes}) + P(\text{10 successes}) + \cdots + P(n \text{ successes})$
Fewer than 8 successes Under 8 successes	$r < 8,\ P\left(\genfrac{}{}{0pt}{}{\text{fewer than}}{\text{8 successes}}\right) = P(\text{0 successes}) + P(\text{1 success}) + \cdots + P(\text{7 successes})$

EXAMPLE 1

Find the probability that when nine dice are cast, 5 will show on at most seven dice.

Solution The probability of success on one die is $\tfrac{1}{6}$.

The probability of failure on one die is $\tfrac{5}{6}$.

Success means that 5 shows on 0, 1, 2, 3, 4, 5, 6, or 7 dice.

Failure means that 5 shows on 8 or 9 dice.

Probability of success and of failure can be written in sigma notation.

$$P(\text{success}) = \sum_{r=0}^{7} {}_9C_r\left(\tfrac{1}{6}\right)^r\left(\tfrac{5}{6}\right)^{9-r} \qquad P(\text{failure}) = \sum_{r=8}^{9} {}_9C_r\left(\tfrac{1}{6}\right)^r\left(\tfrac{5}{6}\right)^{9-r}$$

Since there are fewer cases for the probability of failure, we will evaluate $P(\text{failure})$ and use the fact that $P(\text{success}) + P(\text{failure}) = 1$.

$$P(\text{failure}) = \sum_{r=8}^{9} {}_9C_r\left(\tfrac{1}{6}\right)^r\left(\tfrac{5}{6}\right)^{9-r} = {}_9C_8\left(\tfrac{1}{6}\right)^8\left(\tfrac{5}{6}\right)^1 + {}_9C_9\left(\tfrac{1}{6}\right)^9\left(\tfrac{5}{6}\right)^0$$

$$= 9\left(\tfrac{1}{6}\right)^8\left(\tfrac{5}{6}\right)^1 + 1\left(\tfrac{1}{6}\right)^9\left(\tfrac{5}{6}\right)^0$$

$$= \frac{45}{10{,}077{,}696} + \frac{1}{10{,}077{,}696}$$

$$= \frac{46}{10{,}077{,}696} \approx 0.0000045645$$

$$P(\text{success}) = 1 - P(\text{failure}) = 1 - \frac{46}{10{,}077{,}696}$$

$$= \frac{10{,}077{,}650}{10{,}077{,}696}$$

$$\approx 0.9999954355 \ \textit{Answer}$$

Area Under a Normal Curve

In Section 15-6, we learned about the normal distribution, a special type of distribution that frequently occurs in nature. Recall that in a normal distribution:

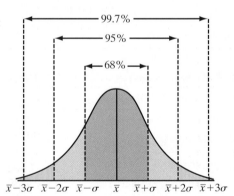

- Approximately 68% of the data falls within one standard deviation of the mean.

- Approximately 95% of the data falls within two standard deviations of the mean.

- Approximately 99.7% of the data falls within three standard deviations of the mean.

The area under a normal curve can be interpreted as a probability. The area can be found by using the normalcdf(function of the graphing calculator. The normalcdf(function requires the following inputs:

$$\text{normalcdf}(low, high, mean, standard\ devation)$$

This will find the area or probability under the normal curve from *low* to *high* with the given *mean* and *standard deviation*.

EXAMPLE 2

The height of a certain variety of plant is normally distributed with a mean of 50 centimeters and a standard deviation of 15 centimeters. What is the probability that the height of a randomly selected plant is between 40 and 55 centimeters?

Solution

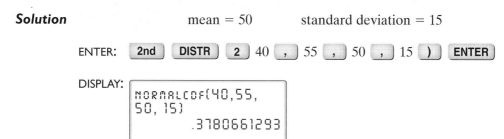

mean = 50   standard deviation = 15

ENTER: **2nd**   **DISTR**   **2**   40 **,**   55 **,**   50 **,**   15 **)**   **ENTER**

DISPLAY:

NORMALCDF(40,55,
50, 15)
          .3780661293

The probability that the height of a randomly selected plant is between 40 and 55 centimeters is approximately .378. *Answer*

The Normal Approximation to the Binomial Distribution

When 10 coins are tossed, the outcome can show 0, 1, 2, 3, 4, 5, 6, 7, 8, 9, or 10 heads. The table gives the number of possible combination of coins on which heads occur for each number of heads and the probability of that number of heads.

No. of Heads	No. of Combinations	Probability
0	1	.00098
1	10	.00977
2	45	.04395
3	120	.11719
4	210	.20508
5	252	.24609
6	210	.20508
7	120	.11719
8	45	.04395
9	10	.00977
10	1	.00098

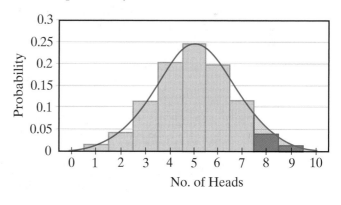

The graph above shows the probability of landing each number of heads. When we draw a smooth curve through these points, the figure is a bell-shaped curve that approximates a normal curve. Geometrically, each bar has width 1 and each height is equal to the corresponding probability at that value. Thus, the probability of getting at least 8 heads is equal to the sum of the *areas* of the bars labeled 8, 9, and 10. Alternatively, since the distribution can be approximated by a normal curve, the probability of getting at least 8 heads is approximately equal to the area under the normal curve as shown in the graph at the bottom of page 703. However, before we can use the normal distribution in this way, we first need to find the mean and standard deviation of this distribution.

It can be shown that for a **binomial distribution** of n trials with a probability of success of p, the mean is

$$np$$

and the standard deviation is

$$\sqrt{np(1 - p)}$$

For this distribution

$$np = 10\left(\tfrac{1}{2}\right) = 5 \qquad \sqrt{np(1 - p)} = \sqrt{10\left(\tfrac{1}{2}\right)\left(1 - \tfrac{1}{2}\right)} \approx 1.58$$

The normal distribution with mean 5 and standard deviation 1.58 is said to make a good **normal approximation to the binomial distribution**. The probability of getting at least 8 heads when a coin is tossed 10 times is approximately equal to the area under this normal curve from 7.5 (the boundary value between 7 and 8) to 10.5 (the boundary value for the largest value). The area under the curve can be found by using the normalcdf(function of the graphing calculator:

ENTER: 2nd   DISTR   2   7.5 ,   10.5 ,   5 ,   1.58)   ENTER

DISPLAY:

```
NORMALCDF(7.5, 10
.5, 5, 1.58)
           .0565432019
```

The calculator gives the probability as .0565432019. Compare this value with the exact value of .0546875 found at the beginning of the section. As the number of trials increases, the normal approximation more closely approximates the exact value.

Why should we use the normal approximation? Suppose, for example, that we want to find the probability of getting at least 60 heads in 100 tosses of a fair coin. Here any number of heads greater than or equal to 60 is a success. There are 41 possible results:

$$P(\text{at least 60 heads}) = \sum_{r=60}^{100} {}_{100}C_r \left(\tfrac{1}{2}\right)^r \left(\tfrac{1}{2}\right)^{100-r}$$

We can obtain an approximate answer by considering the area under the normal curve. The mean is $np = 100(.5)$ or 50 and the standard deviation is $\sqrt{np(1 - p)} = \sqrt{100(.5)(1 - .5)}$ or 5. A calculator can be used to evaluate the answer.

ENTER: [**2nd**] [**DISTR**] [**2**] 59.5 [**,**] 100.5 [**,**] 50 [**,**] 5 [**)**] [**ENTER**]

DISPLAY:
```
NORMALCDF(59.5, 1
00.5, 50, 5)
          .0287164928
```

The probability of getting at least 60 heads in 100 tosses of a fair coin is approximately equal to .0287164928.

EXAMPLE 3

The probability that a team will win a game is $\frac{3}{5}$. Use the normal approximation to estimate the probability that the team will win at least 10 of its next 25 games.

Solution

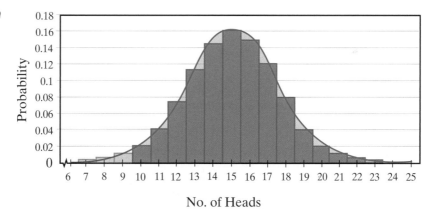

No. of Heads

$$p = \text{probability of success} = \tfrac{3}{5} \qquad q = \text{probability of failure} = \tfrac{2}{5}$$
$$n = 25 \qquad\qquad\qquad r = 10$$

The mean and standard deviation of this binomial distribution are

$$np = (25)\left(\tfrac{3}{5}\right) = 15 \qquad \sqrt{npq} = \sqrt{25\left(\tfrac{3}{5}\right)\left(\tfrac{2}{5}\right)} \approx 2.4495$$

The probability of winning at least 10 of the next 25 games is approximately equal to the area under the normal curve from 9.5 to 25.5.

ENTER: [2nd] [DISTR] [2] 9.5 [,] 25.5 [,] 15 [,] 2.4495 [)] [ENTER]

DISPLAY:

```
NORMALCDF(9.5, 25
.5, 15, 2.4495)
          .9876183243
```

Answer .988

Note: When using the normal approximation, the standard deviation needs to be rounded to at least one more decimal place than the desired probability.

Exercises

Writing About Mathematics

1. Emma said that if the probability of success in at least r out of n trials is z, then the probability of success in at most r out of n trials is $1 - z$. Do you agree with Emma? Explain why or why not.

2. If the probability of success for one trial is .1, does the area under the normal curve give a good approximation of the probability of success in at least 8 out of 10 trials? Explain why or why not. (*Hint:* Consider the shape of the curve.)

Developing Skills

In 3–5, find exact probabilities showing all required computation.

3. Five fair coins are tossed. Find the probability of the coins showing:

    a. at least four heads   **b.** at least three heads

    c. at least two heads   **d.** at least one head

4. Three fair dice are tossed. Find the probability of the dice showing:

    a. at most one 6   **b.** at most two 6's

    c. at least two 6's   **d.** at least one 6

5. A spinner is divided into five equal sections numbered 1 through 5. The arrow is equally likely to land on any section. Find the probability of:

    a. an odd number on any one spin

    b. at least three odd numbers on four spins

    c. at least two odd numbers on four spins

    d. at least one odd number on four spins

In 6–9, write an expression using sigma notation that can be used to find each probability.

6. At least 10 heads when 15 coins are tossed

7. At most 7 tails when 10 coins are tossed

8. At least 5 wins in the next 20 games when $P(\text{win}) = \frac{2}{3}$

9. At most 3 losses in the next 10 games when $P(\text{win}) = \frac{2}{3}$

In 10–13, the mean and standard deviation of a normal distribution are given. Find each probability to the nearest hundredth.

10. mean $= 80$, standard deviation $= 10$, $P(50 \le x \le 95)$

11. mean $= 50$, standard deviation $= 2.5$, $P(44 \le x \le 50)$

12. mean $= 6$, standard deviation $= 3$, $P(2 \le x \le 7)$

13. mean $= 8$, standard deviation $= 1$, $P(7 \le x)$

In 14–17, use the normal approximation to estimate each probability. Round your answers to three decimal places.

14. $P(\text{at least 20 successes})$, $p = \frac{1}{6}$, $n = 100$ **15.** $P(\text{more than 20 successes})$, $p = \frac{1}{10}$, $n = 200$

16. $P(\text{fewer than 60 successes})$, $p = \frac{5}{6}$, $n = 75$ **17.** $P(\text{at most 100 successes})$, $p = \frac{6}{7}$, $n = 125$

Applying Skills

18. A family with 5 children is selected at random. What is the probability that the family has at least 3 boys?

19. A quality control department tests a product and finds that the probability of a defective part is .001. What is the probability that in a run of 100 items, no more than 2 are defective?

20. A landscape company will replace any shrub that they plant if it fails to grow. They estimate that the probability of failure to grow is .02. What is the probability that, of the 200 shrubs planted this week, at most 3 must be replaced?

21. When adjusted for inflation, the monthly amount that a historic restaurant spent on cleaning for a 30-year period was normally distributed with a mean of $2,100 and a standard deviation of $240.

   a. What is the probability that the restaurant will spend between $2,500 and $2,800 on cleaning for the month of January?

   b. If the restaurant has only $2,400 allotted for cleaning for the month of January, what is the probability that it will exceed its budget for cleaning?

In 22–24, use the normal approximation to estimate each probability. Round your answers to three decimal places.

22. A fair coin is tossed 100 times. What is the probability of at most 48 heads?

23. A fair die is tossed 70 times. What is the probability of at least 40 even numbers?

24. The probability that a basketball team will win any game is .5. What is the probability that the team will win at least 12 out of 25 games?

16-6 THE BINOMIAL THEOREM

The algebraic expressions $5a + 3$, $a^2 - 2a$, $3p + 2q$, and $b + c$ are all binomials. Any binomial can be expressed as $x + y$ with x equal to the first term and y equal to the second term. When n is a positive integer, $(x + y)^n$ can be expressed as a polynomial called the **binomial expansion**.

$$(x + y)^1 = 1x + 1y$$
$$(x + y)^2 = 1x^2 + 2xy + 1y^2$$
$$(x + y)^3 = 1x^3 + 3x^2y + 3xy^2 + 1y^3$$
$$(x + y)^4 = 1x^4 + 4x^3y + 6x^2y^2 + 4xy^3 + 1y^4$$
$$(x + y)^5 = 1x^5 + 5x^4y + 10x^3y^2 + 10x^2y^3 + 5xy^4 + 1y^5$$

A pattern of powers of x and y develop as we write the binomial expansions. Consider the last expansion shown above: $(x + y)^5$.

Write $(x + y)^5$ as a product of five factors:

$$(x + y)(x + y)(x + y)(x + y)(x + y)$$

This product is the sum of all possible combinations of an x or a y from each term.

- We can choose the y from zero factors and x from all five factors to obtain the term in x^5. There is $_5C_0$ or 1 way to do this, so the term is $1x^5$.
- We can choose y from one factor and x from four factors to obtain the term in x^4y. There are $_5C_1$ or 5 ways to do this, so the term is $5x^4y$.
- We can choose y from two factors and x from three factors to obtain the term in x^3y^2. There are $_5C_2$ or 10 ways to do this, so the term is $10x^3y^2$.
- We can choose y from three factors and x from two factors to obtain the term in x^2y^3. There are $_5C_3$ or 10 ways to do this, so the term is $10x^2y^3$.
- We can choose y from four factors and x from one factor to obtain the term in x^1y^4. There are $_5C_4$ or 5 ways to do this, so the term is $5xy^4$.
- We can choose y from all five factors in one way to obtain the term in y^5. There is $_5C_5$ or 1 way to do this, so the term is $1y^5$.

Compare the coefficients of the terms of a binomial expansion with the combinations $_nC_r$. Include $(x + y)^0 = 1$ in this comparison.

$$
\begin{array}{lll}
(x + y)^0 = & 1 & 1 \\
(x + y)^1 = & _1C_0\,x + {}_1C_1\,y & 1\ 1 \\
(x + y)^2 = & _2C_0\,x^2 + {}_2C_1\,xy + {}_2C_2\,y^2 & 1\ 2\ 1 \\
(x + y)^3 = & _3C_0\,x^3 + {}_3C_1\,x^2y + {}_3C_2\,x^1y^2 + {}_3C_3\,y^3 & 1\ 3\ 3\ 1 \\
(x + y)^4 = & _4C_0\,x^4 + {}_4C_1\,x^3y + {}_4C_2\,x^2y^2 + {}_4C_3\,x^1y^3 + {}_4C_4\,y^4 & 1\ 4\ 6\ 4\ 1 \\
(x + y)^5 = & _5C_0\,x^5 + {}_5C_1\,x^4y + {}_5C_2\,x^3y^2 + {}_5C_3\,x^2y^3 + {}_5C_4\,x^1y^4 + {}_5C_5\,y^5 & 1\ 5\ 10\ 10\ 5\ 1
\end{array}
$$

The pattern shown at the right above is called **Pascal's Triangle**. Each row of Pascal's Triangle has one more entry than the row above. Each row begins and ends with 1. Each entry between the first and last entry are the sum of the two entries from the row above. For example, in the fourth row:

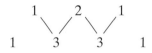

The next row of Pascal's Triangle would be:

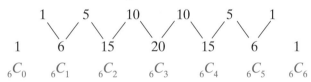

We can write the general expression for the expansion of a binomial:

$$(x + y)^n = \sum_{i=0}^{n} {}_nC_i\,x^{n-i}y^i$$

Note that in the expansion of $(x + y)^n$, there are $n + 1$ terms and the rth term is:

$$_nC_{r-1}\,x^{n-r+1}\,y^{r-1}$$

EXAMPLE I

Write the expansion of $(x + y)^8$.

Solution
$$
\begin{aligned}
(x + y)^8 &= \sum_{i=0}^{8} {}_8C_i\,x^{8-i}y^i \\
&= {}_8C_0\,x^8 + {}_8C_1\,x^7y + {}_8C_2\,x^6y^2 + {}_8C_3\,x^5y^3 + {}_8C_4\,x^4y^4 + {}_8C_5\,x^3y^5 \\
&\quad + {}_8C_6\,x^2y^6 + {}_8C_7\,xy^7 + {}_8C_8\,y^8 \\
&= 1x^8 + 8x^7y + 28x^6y^2 + 56x^5y^3 + 70x^4y^4 + 56x^3y^5 + 28x^2y^6 \\
&\quad + 8xy^7 + 1y^8 \ \textit{Answer}
\end{aligned}
$$

EXAMPLE 2

Write the expansion of $(3a - 2)^4$

Solution
$$(x + y)^4 = {}_4C_0\, x^4 + {}_4C_1\, x^3y + {}_4C_2\, x^2y^2 + {}_4C_3\, x^1y^3 + {}_4C_4\, y^4$$
$$= 1x^4 + 4x^3y + 6x^2y^2 + 4x^1y^3 + 1y^4$$

Let $x = 3a$ and $y = -2$.
$$(3a - 2)^4 = 1(3a)^4 + 4(3a)^3(-2) + 6(3a)^2(-2)^2 + 4(3a)(-2)^3 + (-2)^4$$
$$= 81a^4 - 216a^3 + 216a^2 - 96a + 16 \ \ \textit{Answer}$$

EXAMPLE 3

Write the 4th term of the expansion of $(b + 5)^{10}$.

Solution Let $n = 10$, $r = 4$, $x = b$, and $y = 5$.

METHOD I	**METHOD 2**
The 4th term of $(x + y)^{10}$ is the term that chooses $(10 - 4 + 1)$ or 7 factors of x and $(4 - 1)$ or 3 factors of y:	Use the formula:

METHOD I

The 4th term of $(x + y)^{10}$ is the term that chooses $(10 - 4 + 1)$ or 7 factors of x and $(4 - 1)$ or 3 factors of y:

$${}_{10}C_3 x^7y^3$$
$$= {}_{10}C_3(b)^7(5)^3$$
$$= 120b^7(125)$$
$$= 15{,}000b^7$$

METHOD 2

Use the formula:

The 4th term $= {}_nC_{r-1}\, x^{n-r+1}y^{r-1}$
$$= {}_{10}C_{4-1}\,(b)^{10-4+1}(5)^{4-1}$$
$$= {}_{10}C_3\, b^7 5^3$$
$$= 120b^7(125)$$
$$= 15{,}000b^7$$

Answer $15{,}000b^7$

Exercises

Writing About Mathematics

1. Explain why the expansion of $(x + y)^n = \displaystyle\sum_{i=0}^{n} {}_nC_i x^{n-i}y^i$ can also be written as

$$(x + y)^n = \sum_{i=0}^{n} {}_nC_{n-i}x^{n-i}y^i.$$

2. Ariel said that $\left(x + \frac{1}{x}\right)^n$ is equal to $\displaystyle\sum_{i=0}^{n} {}_nC_i x^{n-2i}$. Do you agree with Ariel? Explain why or why not.

Developing Skills

In 3–10, write the expansion of each binomial.

3. $(x + y)^6$ **4.** $(x + y)^7$ **5.** $(1 + y)^5$ **6.** $(x + 2)^5$

7. $(a + 3)^4$ **8.** $(2 + a)^4$ **9.** $(2b - 1)^3$ **10.** $(1 - i)^5; i = \sqrt{-1}$

11. Write 10 lines of the Pascal Triangle, starting with 1.

In 12–17, write the nth term of each binomial expansion.

12. $(x + y)^{15}, n = 3$ **13.** $(x + y)^{10}, n = 7$ **14.** $(2x + y)^6, n = 4$

15. $(x - y)^9, n = 5$ **16.** $(3a + 2b)^7, n = 6$ **17.** $\left(y - \frac{1}{y}\right)^8, n = 4$

In 18–20, for the given expansion, identify which term is shown and write the next term.

18. $(a - 2b)^8; 1{,}792a^3b^5$ **19.** $(5c - 2d)^4; -160cd^3$ **20.** $\left(\frac{x}{2} - 2y\right)^{11}; \frac{165}{4}x^7y^4$

Applying Skills

21. The length of a side of a cube is represented by $(3x - 1)$. Use the binomial theorem to write a polynomial that represents the volume of the cube.

22. When \$100 are invested at 4% interest compounded quarterly, the value of the investment after 3 years is $100(1 + 0.01)^{12}$. Use the binomial theorem to express the value in sigma notation.

23. A machine purchased for \$75,000 is expected to decrease in value by 20% each year. The value of the machine after n years is $75{,}000(1 - 0.20)^n$. Use the binomial theorem to express the value of the machine after 5 years in sigma notation.

CHAPTER SUMMARY

The set of all possible **outcomes** for an activity is called the **sample space**. An **event** is a subset of the sample space.

The **Counting Principle**: If one activity can occur in any of m ways and a second activity can occur in any of n ways, then both activities can occur in the order given in $m \cdot n$ ways.

A **permutation** is an arrangement of objects in a specific order. The number of permutations of n things taken r at a time, $r \leq n$, is written $_nP_r$.

$$_nP_r = n(n - 1)(n - 2) \ldots (n - r + 1) = \frac{n!}{(n - r)!}$$

The number of permutations of n things taken n at a time when a are identical is $\frac{n!}{a!}$.

A **combination** is a selection of objects in which order is not important. The number of combinations of n things taken r at a time is symbolized $_nC_r$ or as $\binom{n}{r}$.

$$_nC_r = \frac{_nP_r}{_rP_r} \text{ or } \binom{n}{r} = \frac{_nP_r}{r!}.$$

Probability is an estimate of the frequency of an outcome in repeated trials. **Theoretical probability** is the ratio of the number of elements in an event to the number of equally likely elements in the sample space. **Empirical probability** or **experimental probability** is the best estimate, based on a large number of trials, of the ratio of the number of successful results to the number trials.

Let $P(E)$ represent the probability of an event E.

$n(E)$ be the number of element in event E.

$n(S)$ be the number of elements in the sample space.

$$P(E) = \frac{n(E)}{n(S)}$$

The probability form of the **Counting Principle**: If E_1 and E_2 are independent events, and if $P(E_1) = p$ and $P(E_2) = q$, then the probability of both E_1 and E_2 occurring is $p \times q$.

An event can be any subset of the sample space, including the empty set and the sample space itself: $0 \le n(E) \le n(S)$. Therefore since $n(S)$ is a positive number, $\frac{0}{n(S)} \le \frac{n(E)}{n(S)} \le \frac{n(S)}{n(S)}$ or $0 \le P(E) \le 1$. $P(E) = 0$ when the event is the empty set, that is, there are no outcomes. $P(E) = 1$ when the event is the sample space, that is, certainty. $P(E) + P(\text{not } E) = 1$.

A **Bernoulli experiment** or **binomial experiment** is an experiment with two independent outcomes: success or failure. In general, for a Bernoulli experiment, if the probability of success is p and the probability of failure is $1 - p = q$, then the probability of r successes in n independent trials is:

$$_nC_r p^r q^{n-r}$$

- The probability of at most r successes in n trials is $\sum_{i=0}^{r} {_nC_i} p^i q^{n-i}$.

- The probability of at least r successes in n trials is $\sum_{i=r}^{n} {_nC_i} p^i q^{n-i}$.

The area under a normal curve between two x-values is equal to the probability that a randomly selected x-value is between the two given values.

For a binomial distribution of n trials with a probability of success of p, the mean is np and the standard deviation is $\sqrt{np(1 - p)}$. A binomial distribution can be approximated by a normal distribution with this given mean and standard deviation.

When n is a positive integer, $(x + y)^n$ can be expressed as a polynomial called the **binomial expansion**:

$$(x + y)^n = \sum_{i=0}^{n} {}_nC_i x^{n-i} y^i$$

In the expansion of $(x + y)^n$, there are $n + 1$ terms and the rth term is:

$${}_nC_{r-1} x^{n-r+1} y^{r-1}$$

VOCABULARY

16-1 Outcome • Sample space • Counting Principle • Tree diagram • Event • Dependent events • Independent events

16-2 Permutation • Combination

16-3 Probability • Theoretical probability • Empirical probability • Experimental probability

16-4 Bernoulli experiment • Binomial experiment • Binomial probability formula

16-5 Binomial distribution • Normal approximation to the binomial distribution

16-6 Binomial expansion • Pascal's Triangle

REVIEW EXERCISES

In 1–12, evaluate each expression.

1. $6!$

2. ${}_5P_5$

3. ${}_5C_5$

4. ${}_8P_6$

5. ${}_8C_6$

6. ${}_8C_0$

7. ${}_8C_8$

8. $\binom{10}{6}$

9. $\binom{50}{49}$

10. $\frac{12!}{10!}$

11. $\frac{12!}{5!\,7!}$

12. ${}_{40}C_{39}$

13. In how many different ways can 8 students choose a seat if there are 8 seats in the classroom?

14. In how many different ways can 8 students choose a seat if there are 12 seats in the classroom?

15. Randy is trying to arrange the letters T, R, O, E, E, M to form distinct arrangements. How many possible arrangements are there?

16. A buffet offers six flavors of ice cream, four different syrups, and five toppings. How many different sundaes consisting of ice cream, syrup, and topping are possible?

17. A dish contains eight different candies. How many different choices of three can be made?

18. At a carnival, the probability that a person will win a prize at the ring-toss game has been found to be $\frac{1}{20}$. What is the probability that a prize will be won by exactly

    a. 1 of the next 5 players?   **b.** 3 of the next 5 players?

    c. 0 of the next 5 players?   **d.** 2 of the next 20 players?

19. A fair coin is tossed 12 times. What is the probability of the coin showing

    a. exactly 10 heads?   **b.** at least 10 heads?   **c.** at most 10 heads?

20. A convenience store makes over \$200 approximately 75% of the days it is open. Use the normal approximation to determine the probability that the store makes over \$200 at least 265 out of 365 days. Round your answer to the nearest thousandth.

21. The math portion of the SAT is designed so that the scores are normally distributed with a mean of 500 and a standard deviation of 100. The maximum possible score is 800. To the nearest thousandth, what is the probability that a randomly selected student's score on the math section is

    a. more than 700?   **b.** less than 300?   **c.** between 400 and 600?

In 22–25, write the expansion of each binomial.

22. $(x + y)^4$ **23.** $(2a + 1)^7$ **24.** $(3 - x)^6$ **25.** $\left(b - \frac{1}{b}\right)^8$

26. When $i = \sqrt{-1}$, express $(1 + i)^5$ in $a + bi$ form.

27. Write in simplest form the fourth term of $(x - y)^9$.

28. Write in simplest form the seventh term of $(a + 3)^{10}$.

29. Write in simplest form the middle term of $(2x - 1)^{12}$.

Exploration

The winner of the World Series is determined by a *seven-game series*, where two teams play against each other until one team wins four games. Let p represent the probability that the American League team will win a game in the World Series.

a. Explain why the probability that a team will win the World Series is *not*

$$_7C_4\, p^4(1-p)^3.$$

b. Write an expression for the probability that the American League team will win the World Series in exactly 4 games.

c. Write an expression for the probability that the American League team will win the World Series in exactly 5 games. (*Hint:* In order for the series to run 5 games, the winning team must win 3 out of the first 4 games and then win the last game.)

d. Write an expression for the probability that the American League team will win the World series in exactly 6 games. (*Hint:* In order for the series to run 6 games, the winning team must win 3 out of the first 5 games and then win the last game.)

e. Write an expression for the probability that the American League team will win the World Series in exactly 7 games.

f. Write an expression for the probability that the American League team will win the World Series.

CUMULATIVE REVIEW   CHAPTERS 1–16

Part I

Answer all questions in this part. Each correct answer will receive 2 credits. No partial credit will be allowed.

1. In simplest form, $(3x - 9) \div \left[12\left(x - \frac{9}{x}\right)\right]$ is equal to

(1) $\frac{5}{4x}$

(3) $\frac{1}{4(x + 3)}$

(2) $\frac{x}{4x + 3}$

(4) $\frac{x}{4(x + 3)}$

2. Which of the following sets of ordered pairs is *not* a function?

(1) $\{(1, 2), (2, 4), (3, 6), (4, 8), (5, 10)\}$

(2) $\{(1, 2), (2, 2), (3, 2), (4, 2), (5, 2)\}$

(3) $\{(1, 2), (1, 4), (1, 6), (1, 8), (1, 10)\}$

(4) $\{(1, 1), (2, 2), (3, 3), (4, 4), (5, 5)\}$

3. The sum of the roots of a quadratic equation is -2 and the product is -6. Which of the following could be the equation?

(1) $x^2 - 2x + 6 = 0$

(3) $x^2 + 2x + 6 = 0$

(2) $x^2 - 2x - 6 = 0$

(4) $x^2 + 2x - 6 = 0$

4. If $f(x) = x^2$ and $g(x) = 2x - 1$, then $f(g(-2))$ is
(1) 25 (2) 15 (3) 9 (4) 7

5. The product $(\sqrt{-4})(\sqrt{-12})$ is equal to
(1) $4\sqrt{3}$ (2) $\sqrt{48}$ (3) $4i\sqrt{3}$ (4) $-4\sqrt{3}$

6. Which of the following is an arithmetic sequence?
(1) $3, 1, -1, -3, \ldots$
(2) $3, -1, 1, -3, \ldots$
(3) $3, -1, \frac{1}{3}, -\frac{1}{3}, \ldots$
(4) $1, 1, 2, 3, 5, 8, \ldots$

7. Which of the following is equivalent to $\log a + \log b = 2 \log c$?
(1) $\log (a + b) = \log 2c$
(2) $\log (a + b) = \log c^2$
(3) $\log ab = \log 2c$
(4) $\log ab = \log c^2$

8. Which of the following is an identity?
(1) $\sec^2 \theta - 1 = \tan^2 \theta$
(2) $\tan^2 \theta - 1 = \sec^2 \theta$
(3) $\tan^2 \theta - \cot^2 \theta = 1$
(4) $\cos \theta = \sqrt{1 + \sin^2 \theta}$

9. Which is not an equation of the graph shown at the right?
(1) $y = \sin \left(x - \frac{\pi}{2}\right)$
(2) $y = -\cos x$
(3) $y = \sin \left(x + \frac{\pi}{2}\right)$
(4) $y = \cos (x + \pi)$

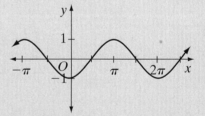

10. The scores of a standardized test have a normal distribution with a mean of 75 and a standard deviation of 6. Which interval contains 95% of the scores?
(1) 57–93 (2) 63–87 (3) 69–81 (4) 57–87

Part II

Answer all questions in this part. Each correct answer will receive 2 credits. Clearly indicate the necessary steps, including appropriate formula substitutions, diagrams, graphs, charts, etc. For all questions in this part, a correct numerical answer with no work shown will receive only 1 credit.

11. If $\log_2 \frac{1}{4} = \log_8 x$, what is the value of x?

12. Find the solution set of the equation $\sqrt{3x + 1} = 2x$.

Part III

Answer all questions in this part. Each correct answer will receive 4 credits. Clearly indicate the necessary steps, including appropriate formula substitutions, diagrams, graphs, charts, etc. For all questions in this part, a correct numerical answer with no work shown will receive only 1 credit.

13. a. Draw the graph of $f(x) = (1.6)^x$ from $x = -1$ to $x = 4$.

b. Write an expression for $f^{-1}(x)$ and draw the graph of f^{-1} for $0 < x \leq 5$.

14. The length of a rectangle is 3 feet less than twice the width. The area of the rectangle is less than 20 square feet. Find the possible widths of the rectangle.

Part IV

Answer all questions in this part. Each correct answer will receive 6 credits. Clearly indicate the necessary steps, including appropriate formula substitutions, diagrams, graphs, charts, etc. For all questions in this part, a correct numerical answer with no work shown will receive only 1 credit.

15. An environmental group organized a campaign to clean up debris in the city. Each Saturday, volunteer crews worked at the project. The number of volunteers increased each week as shown in the following record.

Week	1	2	3	4	5	6	7	8	9
Volunteers	30	50	67	78	85	87	91	94	97

a. Draw a scatter plot to display the data.

b. Write the non-linear equation that best represents this data.

c. What would be the expected number of volunteers on the 10th Saturday?

16. Find, to the nearest degree, all of the values of θ in the interval $0° \leq \theta \leq 360°$ that satisfy the equation $2 \cos^2 \theta + 2 \sin \theta - 1 = 0$.

INDEX